Jacob Bear

Modeling Phenomena of Flow and Transport in Porous Media

Springer

Jacob Bear
Department of Civil and Environmental Engineering
Technion—Israel Institute of Technology
Haifa
Israel

ISSN 0924-6118 ISSN 2213-6940 (electronic)
Theory and Applications of Transport in Porous Media
ISBN 978-3-319-72825-4 ISBN 978-3-319-72826-1 (eBook)
https://doi.org/10.1007/978-3-319-72826-1

Library of Congress Control Number: 2017961494

Printed on acid-free paper

This Springer imprint is published by Springer Nature
The registered company is Springer International Publishing AG
The registered company address is: Gewerbestrasse 11, 6330 Cham, Switzerland

To Siona,
Eitan and Jennifer,
Alon, Lior, Yoav, and Ido,
Iris and Moshe,
Sharon, Efrat, and Gilad,
with love.

Preface

The objective of this book is to present and discuss the construction of mathematical models that describe *phenomena of flow and transport in porous media* as encountered in many disciplines, e.g., civil and environmental engineering, soil mechanics, groundwater hydrology, petroleum engineering, drainage and irrigation in agricultural engineering, reactors in chemical engineering, and geothermal engineering. The book can also serve as a text for courses on modeling in these disciplines.

Soil, sand, fissured and fractured rock, cemented sandstone, Karstic limestone, ceramics, filters, foam rubber, bread, wood, concrete, and kidneys, are just a few examples of the large variety of natural and man-made porous materials encountered in these disciplines.

Phenomena of transport of extensive quantities, like mass of fluid phases, mass of chemical species dissolved in fluid phases, momentum and energy of the solid matrix and of fluid phases, as well as electric charge, in porous medium domains, are encountered in the disciplines mentioned above and in others not mentioned.

For example, civil engineers deal with the movement of moisture through concrete walls and with water flow through and under hydraulic structures. They also deal with the movement of heat in structures, and with stresses under building foundations. Hydrogeologists deal with the flow of water in aquifers. Environmental engineers deal with the transport of contaminants in the subsurface. Agricultural engineers deal with the movement of water and solutes in the root zone during irrigation and drainage. Heat and mass transport in packed bed reactors are encountered in chemical engineering. Petroleum engineers deal with the flow of (liquid and gaseous) hydrocarbons and water in petroleum reservoirs. The movement of fluids and chemical species in lungs and kidneys is studied in biomedical engineering.

Part of the void space in the subsurface may be occupied by a nonaqueous toxic liquid phase. Components of such a liquid may dissolve in and move with percolating water, thus constituting a source for groundwater contamination. Often, dissolved chemical species interact with each other and with the soil, especially with the clay and organic fractions of the latter. Phenomena, such as adsorption, ion exchange, dissolution, volatilization, and biological decay, continuously affect the concentration of chemical constituents in the water. The solid matrix itself may be porous, with tiny pores. Obviously, modeling all these phenomena is essential in any effort to understand and combat subsurface contamination, or to produce hydrocarbons from petroleum reservoirs.

Land subsidence, as a consequence of pumping, or upheaval as a consequence of fluid injection, are examples involving solid matrix deformation as a result of changes in pressure distribution in the subsurface. Non-isothermal flow occurs and the solid matrix may undergo deformation when hot or cold water is injected into the subsurface for energy storage purposes, when CO_2, captured from power plants' emissions, is injected for disposal in depleted petroleum reservoirs, or in deep brine-containing formations, and when compressed air or gas is injected into anticlinal geological formations for short or long term storage purposes.

In all the above examples, *extensive quantities* (mass, momentum, energy, or electric charge) are transported through porous material domains. The term *transport* is used here to describe the *movement, storage, and transformation* of a considered extensive quantity.

To investigate all these phenomena, as encountered when solving problems in practice, mathematical models have to be constructed. By solving the latter, forecasts are obtained of the *response* of the considered *system* to *excitations* in the form of changes in controllable source terms, or in boundary conditions. Pressure, stress, strain, velocity, solute concentration, temperature, etc., for each phase in the system, and for the porous medium as a whole, serve as examples of state variables. In this way, models serve as an essential step in the solution of real-life problems.

In this book, we present and discuss various phenomena of flow, transport, and transformation that take place in porous medium domains, and the construction of conceptual and well-posed *mathematical models* that describe them. The objective is to develop models of phenomena of transport in porous medium domains that are based on visualizing such domains as *continua*.

Analytical solutions of the mathematical models considered here are seldom possible for problems of practical interest. The usual way of solving models of such cases is to transform the mathematical model into a *numerical* one, and then use *computer programs* to solve the latter. Thus, the mathematical models developed in this book should serve as the background to computer codes such as TOUGH (developed at LBNL), NUFT (developed at LLNL), PFLOTRAN (open source, multi-institution code), and many other freely available or commercial codes. Numerical models and computer codes are not discussed in this book.

The book is divided into nine chapters:

Chapter 1 starts with the question "What is a porous medium?". We then present the *continuum approach*, define *microscopic*, *macroscopic*, and *megascopic* levels of description of porous media and phenomena of transport that take place in them. The continuum description is obtained by employing the *phenomenological approach*, or by implementing an *averaging* or *homogenization* approach. We also mention the molecular (or nano-), microscopic (or pore-), macroscopic (or laboratory, field-), and megascopic (or formation-) scales of description and the issue of upscaling. Finally the role, content and construction of transport models are presented.

Chapter 2 introduces some fundamentals of thermodynamics of phases and chemical species that are required for the presentation in this book. Among the concepts defined and discussed are the pressure, density, chemical and other potentials, Gibbs function, internal energy, enthalpy and capillary pressure. Equations of state and their role in modeling are discussed.

In Chap. 3, we start by presenting the *general balance equation* for any *extensive quantity*, such as mass momentum and energy, of phases and chemical species encountered in porous medium domains. The equation is stated at the *microscopic level*, i.e., at a point within a fluid phase present in the void space, and at a point in the solid matrix. We make use of the phenomenological approach to present this balance equation at the *macroscopic level*. The macroscopic balance equation can also be obtained by volume averaging, or by homogenization. These approaches are briefly described. We then derive equations for mass, momentum and energy.

Chapter 4 is devoted to the macroscopic fluid motion equation, emphasizing its origin as a momentum balance equation. We start with the momentum balance equation of a phase. Then, under certain simplifying assumptions, this equation is reduced to the well known (linear) Darcy's law. Other forms of the motion equation, e.g., which take into account also inertial effects, are also presented.

Chapter 5 is devoted to modeling mass transport of a single fluid phase, liquid or gas, that completely occupies the void space. The core of such model is the mass balance equation of the phase. The specific storativity is introduced to account for fluid and solid matrix compressibility. Boundary and initial conditions are presented, leading to a well-posed flow (= mass transport) model.

In Chap. 6, we construct mass transport models for the case of multiple (i.e., two or three) fluid phases that occupy the void space simultaneously. Again, the objective is to lead to complete, well-posed mathematical models, taking into account the coupling that takes place between the phases.

Chapter 7 is devoted to the transport of *chemical species* dissolved in fluid phases that occupy the void space, with or without chemical reactions. The phenomenon of dispersion is introduced and solute fluxes due to diffusion, dispersion, and advection, are discussed. Like any other model of transport, the core of the solute transport model is the balance of mass of the considered chemical species. Various sources and sinks, as well as chemical reactions and interphase transfers are presented. The discussion leads to a well-posed model of the solute transport problem.

Chapter 8 deals with non-isothermal situations. This requires the construction of a model that also describes the transport of energy, or heat, together with appropriate initial and boundary conditions.

In Chap. 9, we deal with cases of transport of mass and energy in which the solid matrix, as the entire porous medium domain, undergoes deformation. Models of consolidation as a consequence of construction, and models of land subsidence due to heavy pumping, are developed and presented. Also, the propagation of waves in a porous medium domain, in which soil deformability plays an essential role, is briefly discussed.

Three appendices, written by appropriate experts, are added at the end of the book. Appendix A, written mainly by Prof. Raphael Semiat (Chemical Engineering, Technion) demonstrates how the material presented in the book is applied in Chemical Engineering, especially in the design of reactors. Appendix B, by Dr. Jonathan Ajo-Franklin and Dr. Marco Voltolini (LBNL) deals with modeling at the microscopic level, while Appendix C, by Dr. David Trebotich (LBNL), discusses the advent of supercomputing.

The book is written as a text suitable for graduate and upper level undergraduate students, and for practitioners. Although one of the main objectives of this book is to construct *mathematical models*, the amount of mathematical knowledge required is kept minimal. The emphasis is on understanding the physical and chemical phenomena that take place in porous medium domains. Mathematics is used only as a compact language for expressing these phenomena.

Much of the material included in this book has been developed in the period 2009–2017, during which I have been an active researcher in three projects: MUSTANG. PANACEA and TRUST, conducted and funded by the European Commission within the 7th program of the European Union Framework Programme for Research and Innovation. I wish to acknowledge the support I received from these three projects which dealt with CO_2 disposal in deep saline water containing geological formations. Obviously, an important part of these research programs, conducted by a team of experts and researchers from more than 20 universities and industries, has been the development and extensive use of models that describe the transport of mass–energy and momentum, as well as chemical reactions, encountered in such projects. These models can be used for predicting the spreading and trapping of CO_2 injected into geological formations. Many of these developments are incorporated in this book.

Finally, I wish to thank Prof. Dr. S. Majid Hassanizadeh (Utrecht University, NL), Prof. George J. Moridis (LBNL), Prof. Brian Berkowitz (Weizmann Institute of Science, Israel), and Prof. Henry Power (Nottingham University, UK) for their important suggestions and contributions. Special thanks are due to Dr. Peter Lichtner of OFM Research for his expert advice and for devoting many days to review parts of the book. Thanks are also due to Dr-Ing. Patrick Kurzeja (TU Dortmund), Dr. Hui-Hai Liu (ARAMCO), Dr. Ehsan Nikooee (Shiraz University), Yin Xiaoguang (Utrecht University), Dr. Vahid Joekar-Niasar (University of Manchester), Dr. Chaozong Qin (Eindhoven Technical University), Prof. Alex Cheng (University of Mississippi), Prof. Alex Furman (IIT, Israel), and Prof. Dr. Ing. Jennifer Niessner (Heilbronn University) for their important comments. Thanks are also due to Diana Swantek (LBNL) for her devoted and meticulous work on some of the figures presented in this book.

Haifa, Israel Jacob Bear
2017

Contents

Symbols[1]

a	Dispersivity of a porous medium (fourth rank tensor); (dims. L; u: m)
a_{ijkl}	Component of **a;** (dims: L; u: m).
a_L, a_T	Longitudinal and transversal dispersivities (isotropic medium), respectively (dims: L; u: m).
$\mathcal{B}^E_\alpha$	Balance operator for an extensive quantity E in an α-phase, ($\equiv \partial\theta_\alpha e_\alpha/\partial t + \nabla\cdot\theta_\alpha(e_\alpha \mathrm{V}_\alpha + \mathrm{J}^E_{dif} + \mathrm{J}^E_{dis})$)..
c^γ_α	Mass concentration of a γ-species in an α-phase (dims: ML^{-3}; u: $\mathrm{kg/m}^3$).
c_p, c_v	Specific heat capacity at constant pressure and at constant volume, resp. (dims: $\mathrm{L}^2\mathrm{T}^{-2}\Theta^{-1}$; u: J/kg K)
C_p, C_v	Heat capacity at constant pressure and constant volume, resp. (dims: $\mathrm{ML}^2\mathrm{T}^{-2}\Theta^{-1}$; u: J/K).
d	Size of void space. Grain diameter (dims: L; u: m).
Da	Darcy number (Dimensionless).
Dm	Damköhler number (Dimensionless).
$\mathbf{D}^\gamma_\alpha$	Coefficient of mass dispersion of a γ-species in an α-phase (dims: $\mathrm{L}^2\mathrm{T}^{-1}$; u: m^2/s).
$\mathcal{D}^\gamma_\alpha$	Coefficient of molecular diffusion of a γ-species in an α-phase (dims: $\mathrm{L}^2\mathrm{T}^{-1}$; u: m^2/s).
$\boldsymbol{\mathcal{D}}^{*\gamma}_\alpha$	(= $\mathcal{D}^\gamma_\alpha \mathbf{T}^*$) Coefficient of molecular diffusion of a γ-species in an α-phase in a porous medium (dims: $\mathrm{L}^2\mathrm{T}^{-1}$; u: m^2/s).
e	Specific value of E (= E per unit mass).
e'	Density of E (= E per unit volume).

[1]1 Dimensions (indicated by dims.) are indicated by M, L, T, Θ, for mass, length, time and temperature, resp. (2) All units (indicated by u:) are in the SI System. Basic units: m (for length in meters), s (for time in seconds), kg (for mass in kilograms), mol (for quantity of substance in moles), N (for force in Newtons), Pa (for pressure, in Pascals) J (for energy, work, in Joules), K (for temperature, Θ, in Kelvin degrees), (3) Note: $1\mathrm{N} = 1\mathrm{kg\ ms}^{-2}$; $1\mathrm{Pa} = 1\mathrm{N\ m}^{-2}$; $1\mathrm{J} = 1\mathrm{N\ m}$.

$\breve{e}$	Molar density of E (= E per mole).
e'^{γ}_{α}	Density of E^{γ}_{α} (= E of γ in α, per unit volume of α-phase).
e^{γ}_{α}	Specific value of E^{γ}_{α} (= E of γ in α, per unit mass of α-phase).
$\breve{e}$	Void ratio (= $\mathbb{V}_{ov}/\mathbb{V}_{os}$).
E	An extensive quantity. Young's modulus (dims: $ML^{-1}T^{-2}$; u: Pa).
E^{γ}_{α}	An extensive quantity, E, of a γ-species in an α-phase.
$\mathbb{E}$	Energy (dims: ML^2T^{-2}; u: J).
$f^{E}_{\alpha\rightarrow\beta}$	Rate of transfer of E from an α-phase to a β-phase, across their common microscopic interface, per unit volume of porous medium.
f	Fugacity (dims: $ML^{-1}T^{-2}$; u: Pa).
F	Concentration of a species adsorbed on a solid (= m^{γ}_{s}/m_s) (dims: dimensionless; u: kg/kg); Force per unit mass of porous medium (dims: LT^{-2}; u: N/kg).
Fo	Fourier number (Dimensionless).
Fr	Froude number (Dimensionless).
$\mathbb{F}$	Force per unit volume (dims: $ML^{-2}T^{-2}$; u: N/m^3).
g	Gravity acceleration (dims: L^2T^{-1}; u: m^2/s). Specific Gibbs free energy (dims: L^2T^{-2}; u: J/kg).
$g^{\cdot}$	Molar Gibbs free energy (dims: L^2T^{-2}; u: J/mole.
G	Shear modulus of solid (dims: ML^{-1} T^{-2}; u: Pa).
$\mathbb{G}$	Gibbs free energy (dims: ML^2T^{-2}; u: J).
h	Piezometric head (dims: L; u: m). Specific enthalpy (dims: L^2T^{-2}; u: J/kg).
h'	Enthalpy density (dims: $ML^{-1}T^{-2}$; u: J/m^3).
$h^{\cdot}$	Molar enthalpy (dims: L^2T^{-2}; u: J/mole).
h_r	Relative humidity.
$\mathcal{H}$	Henry's coefficient (dims: $L^{-1}T^{-2}$; u: Pa/mole).
$\mathbb{H}$	Enthalpy (dims: ML^2T^{-2}; u: J).
$\mathrm{j}^{\mathrm{E}}_{\alpha,\mathrm{y}}$	Microscopic flux of E in an α-phase (E per unit area of α-phase per unit time), $y = adv$ for advection, $y = dif$ for diffusion.
$\mathbf{J}^{E}_{\alpha,y}$	Macroscopic flux of E in an α-phase (E per unit area of α-phase in the porous medium cross-section, per unit time), $y = adv$ for advection, $y = dif$ for diffusion, $y = dis$ for dispersion.
$\mathcal{J}$	Hydraulic gradient. Jacobian (dimensionless).
$\mathbf{k}$	Permeability (second rank tensor) (dims: L^2; u: m^2, or darcy).
$\mathbf{k}_{\alpha}$	Effective permeability of an α-phase (dims: L^2; u: m^2, or darcy).
k^{γ}_{α}	Degradation rate constant of a γ-species in an α-phase.
$\mathbf{K}$	Hydraulic conductivity (second rank tensor)(dims: LT^{-1}; u: m/s).
K_d	Partitioning coefficient (dims: ML^{-3}; u: kg m^{-3}).
K_{eq}	Equilibrium coefficient.
K_{sp}	Solubility product.

Kn	Knudsen number (Dimensionless).
L^*	Characteristic size of domain (dims: L; u: m).
L	Latent heat (dims: L^2T-2; u: J/kg).
m	Mass (dims: M; u: kg).
m_α^γ	Mass of a γ-species in an α-phase (dims: M; u: kg).
$\hat{m}_\alpha^\gamma$	Molality of a γ-species in an α-phase (u: moles/kg).
M^γ	Molecular mass of a γ-species (dimensionless atomic mass units).
M_α	Momentum of an α-phase (dims: MLT^{-1}; u: kg m/s).
n_α^γ	Number of moles of γ-species in an α-phase (Dimensionless); u: Pa).
p	Pressure (dims: $ML^{-1}T^{-2}$;u: Pa).
p_c	Capillary pressure, macroscopic (dims: $ML^{-1}T^{-2}$; u: Pa).
p_α	Pressure in an α-phase (dims: $ML^{-1}T^{-2}$; u: Pa).
p_α^γ	Partial pressure of a γ-species in an α-phase (dims: $ML^{-1}T^{-2}$; u: Pa).
p^v	Average void space pressure. Vapor pressure (dims: $ML^{-1}T^{-2}$; u: Pa).
Pe	Peclet number (Dimensionless).
q_α	Specific discharge (of volume) of an α-phase (= $\theta_\alpha \mathrm{V}_\alpha$) (= discharge of α-phase per unit area of porous medium) (dims: LT^{-1}; u: m/s).
$\mathrm{q}_{\alpha r}$	Specific discharge of an α-phase, relative to the solid matrix $(= \theta_\alpha(\mathrm{V} - \mathrm{V}_s))$, (dims: LT^{-1}; u: m/s).
q_α^E	Specific discharge of E in an α-phase (i.e., E per unit area of p.m).
q_α^γ	Specific mass discharge, i.e., mass of γ per unit area of porous medium, per unit time in α-phase (dims: $ML^{-2}T^{-1}$; u: kg/m^2s).
Q	Fluid discharge (dims: L^3T^{-1}; u: m^3/s). Heat, or thermal energy (dims: ML^2T^{-2}; u: J).
r	Radial distance (dims: L; u: m).
R	Universal gas constant (= 8.3145 J/mol K).
R_α^γ	Solubility of a γ-species in an α-phase.
R_d, R_v	Partitioning coefficient.
R_r	Reaction rate.
Re	Reynolds number (Dimensionless).
s'	Entropy density (dims: $ML^{-1}T^{-2}\ \Theta^{-1}$; u: J/m^3K).
$s^{\cdot}$	Molar entropy (dims: $ML^{-1}T^{-2}\ \Theta^{-1}$; u: J/mole K).
s	Specific entropy (dims: $L^2T^{-2}\ \Theta^{-1}$; u: J/kg K).
S	Aquifer storativity (dims: dimensionless).
S_o	Specific storativity of a porous medium (dims: L^{-1}: u: m^{-1}).
S_y	Specific yield (dimensionless).
S_r	Specific retention (dimensionless).
S_α	Saturation of an α-phase (dimensionless).
$S_{\alpha r}$	Irreducible, or residual saturation of an α-phase (dimensionless).
S_{we}	Effective water saturation.
$\mathcal{S}_o$, $\mathcal{S}_{o\alpha}$	Surfaces, of areas, S_o and $S_{o\alpha}$, respectively.
St	Strouhal number (dimensionless).

$\mathbb{S}$	Entropy (dims: $ML^2T^{-2}\ \Theta^{-1}$: u: J/K).
t	Time (dims: T; u: s).
T	Temperature (dims: Θ; u: K).
$\mathbf{T}$	Aquifer transmissivity (second rank tensor) (dims: L^2T^{-1}; u: m^2s^{-1}).
$\mathbf{T}^*$	Tortuosity of void space or of phase in it (dimensionless).
$\mathbf{u}_\alpha$	Velocity of an interphase surface (dims: LT^{-1},; u: m/s).
u'	Internal energy density (dims: $ML^{-1}T^{-2}$; u: J/m^3).
u	Specific internal energy (dims: L^2T^{-2}; u: J/kg).
$u^{\cdot}$	Molar internal energy (dims: $L^{-1}T^{-2}$; u: J/mole).
$\mathbb{U}$	Internal energy (dims: ML^2T^{-2}; u: J).
v	Specific volume (= $1/\rho$).
V	Mass weighted velocity of a fluid phase (dims: LT^{-1}).
V_α	Velocity of an α-phase, (dims: LT^{-1}; u: m/s).
$V^{E_\alpha^\gamma}$	Velocity of an E_α^γ-continuum.
V_α^γ	Velocity of a γ-species in an α-phase.
$\mathbb{V}$	Volume (dims: L^3; u: m^3).
w_s, w	Solid matrix displacement vector (dims: L; u: m).
W^γ	Molar mass of a γ-species (u: kg/mole).
x	Horizontal coordinate (dims: L; u: m).
x, x′	Position vectors (dims: L; u: m).
$\mathring{x}$	Deviation ($= x - x_o$) (dims: L; u: m).
x_o	Position vector of the centroid of an REV (dims: L; u: m).
X_α^γ	Mole fraction of a γ-species in an α-phase (dimensionless).
y	Horizontal coordinate (dims: L; u: m).
z	Vertical coordinate (positive upward) (dims: L; u: m).
z^γ	Electrical charge of ion of γ species (dims: L; u: m).
Z	Compressibility factor (dimensionless).

Greek Letters

α	Symbol/subscript for an α-phase.
α_T	Coefficient of linear thermal expansion (dims: Θ^{-1}; u: K^{-1}).
α_B	Biot Coefficient (dimensionless).
α_{pm}	Coefficient of porous medium compressibility (dims: $M^{-1}LT^2$; u: Pa^{-1}).
$\alpha^{\star E}$	Transfer coefficient of E.
β_p, β_T	Fluid compressibility (= bulk modulus) at constant pressure. and constant T, respectively (dims: $M^{-1}L^1T^2$; u: Pa^{-1}).
β_T	Coefficient of thermal expansion (dims: Θ^{-1}, u: K^{-1}).
γ	Symbol/superscript denoting a γ-species.
$\gamma(x)$	Characteristic function of the void space.
$\gamma_{\alpha\beta}$	Surface tension between α- and β-phases (dims: M T^{-2}; u: Pa m).

$\Gamma^{E_\alpha^\gamma}$	Rate of production of E_α^γ, per unit mass of an α-phase.
δ	Unit tensor; Kronecker delta. Components δ_{ij}.
Δ	Characteristic distance from solid to fluid in REV (dims: L; u: m).
Δ_α	Hydraulic radius of an α-phase (dims: L. u: m).
ε	Strain tensor (Components ε_{ij}) (dimensionless).
η_α^γ	Molar γ-species concentration in an α-phase,; i.e., per unit volume (dims: moles L^{-3}; u: moles/m^3).
η_α	Molar α-phase concentration (dims: moles L^{-3}; u: moles/m^3).
θ_{LG}	Contact angle (between liquid and gas).
θ_α	Volumetric fraction of an α-phase ($\equiv V_\alpha/V = \phi S_\alpha$).
$\theta_{\alpha r}$	Irreducible, or residual volumetric fraction of an α-phase.
λ	Coefficient of radioactive decay (dims: T^{-1}; u: 1/s). Coefficient of thermal conductivity (dims: $MLT^{-3}\ \Theta^{-1}$; u: J m^{-1} s^{-1} K^{-1}).
λ_s''	Lamé constant of an elastic solid matrix (dims: $ML^{-1}T^{-2}$; u: Pa).
Λ^H	Cofficient of thermal conductivity of porous medium; (dims: ML T^{-3} Θ^{-1}u: J $m^{-1}s^{-1}$ K^{-1}).
μ_s''	Lamè constant of an elastic solid matrix ((dims: $ML^{-1}T^{-2}$; u: Pa)).
μ_α	Dynamic viscosity of an α-fluid phase (dims: $ML^{-1}T^{-1}$; u: Pa s).
μ_α^γ	Chemical potential of a γ-species of an α-phase.
ν_s''	Poisson ratio.
ν_α	Kinematic viscosity of an α-phase (dims: L^2T^{-1}; u: m^2s^{-1}).
ν^γ	Stoichiometric coefficients of a γ-species.
$\boldsymbol{\nu}$	Unit outward normal vector on a boundary surface.
ξ	Position vector of a point at the microscopic level.
ρ_b	Bulk mass density of the solid matrix ($= (1-\phi)\rho_s$); (dims: ML^{-3}; u: kg/m^3).
ρ_α	Mass density of an α-phase (dims: ML^{-3}; u: kg/m^3).
ρ_α^γ	Mass density of a γ-species in an α-phase (dims: ML^{-3}; u: kg/m^3).
$\boldsymbol{\sigma}$	Stress tensor (dims: $ML^{-1}T^{-2}$; u: Pa).
$\boldsymbol{\sigma}_s'$	Effective stress ($\equiv \overline{\sigma_s'}$) (dims: $ML^{-1}T^{-2}$; u: Pa).
$\Sigma_{\alpha\beta}$	Specific area of $\mathcal{S}_{\alpha\beta}$ (dims: L^{-1}; u: m^{-1}).
$\boldsymbol{\tau}$	Shear stress. Deviatoric stress (dims: $ML^{-1}T^{-2}$; u: Pa).
ϕ	Porosity (dimensionless).
ϕ'	Angle of internal friction.
χ^γ	Stoichiometric coefficients of γ species.
ψ	Suction, or matric suction (dims: L; u: m).
Ψ_g, Ψ_m	Gravity and matric potentials, respectively.
Ψ_p, Ψ_{sol}	Pressure and solute (osmotic) potentials, respectively.
Ψ_{sw}, Ψ_T	Soil-water and thermal potentials, respectively.
Ψ_{total}	Total potential.

ω_α	Mass fraction of an α-phase (dimensionless).
ω_α^γ	Mass fraction of a γ-species in an α-phase (dimensionless).
Ω	Porous medium domain (bounded by surface. $\partial\Omega$).

Subscripts

a	Air.
adv	Advective.
c	Characteristic value.
dif	Diffusive.
dis	Dispersive.
g	Gas.
f	Fluid.
i	Component of a vector. Intermediate wetting fluid.
im	Immobile phase.
l	Liquid phase.
n	Nonwetting fluid phase.
o	Organic liquid. Oil.
pm	Porous medium.
s	Solid phase, or solid matrix. Component of a vector in 1s direction.
v	Void space.
w	Water. Wetting phase.
α	α -phase.
β	β -phase. Also, a symbol for all non- α phases.

Superscripts

a	Air, as a species.
H	Heat.
T	Transpose of a matrix.
v	Vapor.
w	Water (H_2O), as a component.
γ	γ -species.

Special Symbols

$\overline{(..)}$	Average, volume average, or phase average of (..) ($= \frac{1}{\mathbb{V}_o}\int_{\mathbb{V}_o}(..)\,d\mathbb{V}$).
$\overline{(..)}^\alpha$	Intrinsic phase average of (..) $= \frac{1}{\mathbb{V}_{o\alpha}}\int_{\mathbb{V}_o}(..)\,d\mathbb{V}$.
$\overset{\circ}{(..)}$	Deviation of (..) from its intrinsic volume average over an REV.
$\breve{(..)}$	Deviation of (..) from its intrinsic mass average over an REV.
$\dot{(..)}$	Total (= advective) derivative of (..) ($\equiv$ D../Dt).
$\widehat{(..)}$	Average over an aquifer thickness.
$\widehat{(..)}^{\alpha\beta}$	Average of (..) over the $(S)_{\alpha\beta}$-surface, or over an aquifer thickness.

$\widetilde{(..)}^{\alpha}$	Intrinsic mass average of (..).
$\nabla \cdot \boldsymbol{A}$	Divergence of a vector $\boldsymbol{A}$ ($\equiv$ div $\boldsymbol{A}$).
∇A	Gradient of a scalar A ($\equiv$ grad A).
$\frac{D_E(..)}{Dt}$	Material derivative of (..), as observed by the E-continuum.
$[\![A]\!]_{\alpha,\beta}$	Jump in A across an α-β-interface ($\equiv A\|_{\alpha} - A\|_{\beta}$). Also: $[A]_{\alpha,\beta}$.
[A]	Molar concentration of species A (= $c^{\mathrm{A}}/M^{\mathrm{A}}$) (dims: moles L^{-3}; u: moles/m^3).
$\{A\}$	Activity of a species A.

Chapter 1
Porous Media

The objective of this book is to present and discuss the construction of conceptual and well-posed mathematical models that describe phenomena of flow and transport in porous medium domains. In Chemical Engineering, such domain is called "packed bed". Until we define a porous medium more precisely in Sect. 1.1.2, we shall understand a *porous medium domain* as a spatial domain occupied partly by a solid matrix, with the remaining part, referred to as pore or void space, occupied by one or more fluid phases. Porous medium domains are encountered in a large number of industrial and engineering disciplines, e.g., paper and diaper industries, petroleum reservoir engineering, hydrogeology, chemical engineering, and biomedical engineering. In all these disciplines, *thermodynamic extensive quantities*, e.g., mass, momentum and energy, are transported through porous medium domains, and decisions, associated with such quantities, have to be made concerning activities and operations in such domains. The transport of such *extensive quantities*, will be considered. We shall use the symbol E to denote the amount of a considered extensive quantity.

It is important to emphasize from the outset that in many cases of practical interest, we have to deal with *coupled processes*, i.e., the simultaneous transport of more that one extensive quantity, taking into account the interaction among such quantities. An example is fluid mass transport under non-isothermal conditions, where mass and energy are transported simultaneously. Fluid momentum and mass transport are always coupled, but, as we shall see throughout this book, fluid's momentum transport is usually expressed as a flux equation. Other examples are coupling of flow and deformation, solute transport under non-isothermal conditions, etc. We note that The term 'processes' is often used to describe phenomena that occur within the framework of transport of an extensive quantity. Examples are interphase mass,

J. Bear, *Modeling Phenomena of Flow and Transport in Porous Media*,
Theory and Applications of Transport in Porous Media 31,
https://doi.org/10.1007/978-3-319-72826-1_1

energy or momentum transfer, and phase change. In fact, as we shall see throughout this book, every term that appears in the balance equation of an extensive quantity expresses a process.

Obviously, we do not intend to present any new ideas and concepts in *Thermodynamics*, nor in *Continuum Mechanics*. Our intention is to show and discuss how these theories and their mathematical representation can be applied to modelling phenomena of transport of extensive quantities in the special domain referred to as 'porous medium domain'.

The term *mathematical model* is usually (and here) used for the tool that uses mathematics to describe phenomena of flow and transport in a specified real domain. Its usefulness to decision makers stems from its ability to effectively and economically *predict* the future behavior in a considered domain, in response to planned activities. Thus, a decision maker can use a model of a considered porous medium system to predict the consequences of implementing contemplated activities, and the relative importance of the factors, parameters and conditions affecting them. By comparing the consequences of implementing various decision alternatives, the decision maker can select the best one, according to some preferred criteria.

The main objective of this first chapter is to introduce some basic concepts, primarily:

- The idea of treating a spatial heterogeneous domain as a *continuum*.
- The idea of treating an *intensive quantity* within a phase that occupies a specified spatial domain also as a continuum.
- Microscopic versus macroscopic level continua and approaches for obtaining macroscopic level models.
- The idea of treating multiple extensive quantities within a phase as *overlapping continua*.
- The mathematical model as a tool for describing phenomena of flow and transport of various extensive quantities through such domains.

In addition, we shall, briefly, also introduce a specific type of porous medium–the fractured rock. Many oil and gas bearing geological formations are made up of fractured rocks.

1.1 The Continuum Approach to Porous Media

A *thermodynamical extensive quantity*, E, is a physical quantity, whose value at every instant is proportional to the mass or volume of the spatial domain it occupies. Such a quantity is *additive*, i.e., we can determine its value in a given spatial domain by summing up the values over its constituent subdomains. In this book, the main considered extensive quantities are the mass, linear momentum, enthalpy and energy of a fluid (liquid or gas) phase, and a solid phase, as well as of a chemical species within such phases. The terms 'phase' and 'chemical species' will be defined in Sect. 1.1.1.

At every point within a spatial domain regarded as continuum for a considered extensive quantity, we may define a corresponding *intensive quantity*, e, as the extensive quantity per unit mass, or e' as the extensive quantity per unit volume, of the considered phase. Obviously, intensive quantities are *non-additive*. We shall also use the symbol $\check{e}$ for molar density of E, i.e., E per mole.

By *phenomena of flow and transport*, we mean the movement, accumulation and transformation of extensive quantities within a specified domain. In this book, we discuss the flow and transport of such extensive quantities as mass, linear momentum, enthalpy and energy, all *of a phase within a porous medium domain*.

Actually, the word 'transport' includes also the notion of flow. Nevertheless, as is common among most of those dealing with this subject, we shall often use the more commonly used expression 'flow and transport', whenever we consider the movement, transformation and accumulation of an extensive quantity transported by a moving fluid. However, we shall continue to use 'flow', whenever we focus only on the movement of fluid mass, whether density and temperature are constant or not.

1.1.1 Phases, Chemical Species and Components

A *phase*, from the Greek word $\varphi\alpha\sigma\iota\varsigma$, meaning 'appearance', is the *spatial domain* occupied by a material such that a *single* set of *equations of state*, e.g., the relationship between density, composition, pressure, and temperature, describes the behavior *at all points* within that domain. In other words, the physical properties are continuous over the domain occupied by a phase.

Another, perhaps less rigorous, definition of a phase is a portion of space that is separated from other such portions by a definite physical boundary (= *interface*, or *interphase boundary*). According to the second definition, many globules of oil, each surrounded by water, are considered separate phases, unless we declare them as a *single multiply connected phase domain*. There can be *only one gaseous phase in a system*, as all gaseous phases are completely *miscible* and do not maintain a distinct interface between them. We may, however, have more than one fluid phase in the void space, e.g., oil and water, or water and air. Such fluid phases are often referred to as *immiscible phases*, or *fluids*. Actually, *all* phases are miscible to some extent, but as long as a *visible* interface separates adjacent fluids, during the time frame of the observation, they are, usually, referred to as 'immiscible fluids'.

With the above in mind, we may now introduce the term, or concept, of a *continuum*. A spatial domain is regarded as a *continuum* with respect to a considered property, say mass density of a certain substance, if a value of that property can be assigned to *every* point within that domain. In what follows, we shall elaborate on this term, or concept.

A *chemical species* is an *identifiable* chemical compound (atom, molecule, or ion) that participates as an entity, whether as a reactant, or as a product, in a chemical reaction that takes place *within a phase*. A chemical species is distinguishable by its chemical composition, and by the phase in which it is present. Thus, the term 'chemical species', or just 'species' refers to the actual form in which a molecule, ion, or a group of molecules is present in a phase. For example, *iodine* in an aqueous solution may exist in the form of one or more species, e.g., I_2, I^-, HIO, IO^-, IO_3^-. *Oxygen* may serve as another example. It may be present in the air as $O_{2(g)}$, and in water as dissolved $O_{2(aq)}$, or as the ion $O^{-2}_{(aq)}$. Similarly, in the strict chemical sense, the same compound present in different fluid phases is regarded as different species. However, the same compound in solution, say in an aqueous liquid and as an *adsorbate* on a solid are considered the same species, but distributed over the two phases with different concentrations, say, (in molar fractions) $X_{aq}^{O_2}$ and $X_s^{O_2}$.

A phase may be composed of a large number of different chemical species, or just of a single species. For example, water, as a phase, is made up of hydrogen ions, H^+, and hydroxyl ones, OH^-. An organic phase (e.g., oil in a petroleum reservoir) may involve over 100 different chemical species. An oil phase may also contain large amounts of dissolved species, e.g., in the form of gaseous hydrocarbons, such as CH_4. The gaseous phase 'air' contains the species O_2, N_2, CO_2, and some additional species.

An important concept is that of a *component*. Under conditions of *chemical equilibrium*, the *minimum* number of independent chemical species required to completely describe the composition of a given phase may be much smaller than the number of species present in a solution. We use the term *component*, or *chemical component*, to denote a set of chemical species that belongs to the smallest set of species that is required to completely define the chemical composition of a phase *under equilibrium conditions*. While interacting, a component contains the same group of chemical species *throughout a considered process*. Thus, CO_2, H_2O, CH_4 are components, but an *aqueous phase* is not a component because it may contain H_2O, and various dissolved gases and other dissolved species. *Oil* is generally not a component; however, quite often reservoir engineers do consider a mixture of several hydrocarbons with similar properties as a component. *Air* may be considered as a component if we assume that the O_2 to N_2 ratio remains change. However, if wish to track O_2 and N_2 (e.g., because O_2 solubility in H_2O is much higher than that of N_2), we cannot use *air* as a component; but have to track these two species separately.

As another example, consider liquid water in contact with its vapour. Water is composed of oxygen and hydrogen ions, but these two species are *always* present in fixed and definite proportions. Therefore, the system is described by a *single component* only-'water', or H_2O. Once we know the concentration of this component, we also know the concentrations of the species comprising it. When chemical equilibrium *is not* assumed, all species are defined to be components. In an unsaturated (air-water) flow problem, 'water' may be regarded as a component that always contains the species H^+ and OH^-, while air, containing N_2 and O_2, can also be considered as a component. Various chemical species may still be present in the water and in the air (and even as adsorbed on the solid). More on phases, components and species can be found in Sect. 3.8.

If, within a closed container, we have liquid water, and by changing temperature and/or pressure, the liquid water evaporates, or freezes, we have two or *three phases* of H_2O: liquid, vapour and ice. However, we still have *one component*–water, but that component is present in 1, 2, or 3 phases (see details in Sect. 2.3.1 A).

The term component is often used also in numerical models and computer codes of reactive transport, to indicate a set of *one or more species that always appear as a group* in a considered problem with equilibrium reactions, maintaining fixed concentration proportions among them.

Note that we are using the term 'component' also for the components of a vector, e.g., V_i, or of a tensor, σ_{ij}, with i, j indicating coordinates.

As background leading to the continuum approach, let us introduce here one more concept, namely *mixture theory*. We start by considering a spatial domain, Ω, occupied by a fluid, say, a gaseous phase. Let this gas be a *mixture* of a number of gases, say gases A, B and C. In a liquid phase, we would have considered three chemical species A, B, C. Every A molecule has its properties: its mass, energy, etc. However, for obvious reasons, we cannot model and predict the behavior of these individual molecules. Instead, we consider a point inside Ω and a small spatial (say, spherical) domain, $\Delta\Omega$ around it. We make the latter sufficiently large such that it will always contain a sufficiently large number of molecules, to the extent that although molecules continuously move in and out of the considered small volume $\Delta\Omega$ around the point, their average properties, e.g., mass per unit volume, will always remain *practically* unchanged. These averages are then assigned as the description of the *averaged behavior of gas* A *at the point*. Repeating this process for all points within Ω, we can declare Ω as a continuum with respect to the mass density of A, ρ_A: to every point within Ω we can assign a value of $\rho_A(\mathbf{x}, t)$.

For the same Ω-domain, we can now repeat the above procedure for B- and for C-molecules. The same domain will then be regarded as continua also for $\rho_B(\mathbf{x}, t)$ and for $\rho_C(\mathbf{x}, t)$. *All three continua are overlapping*. In fact, the same domain is also a continuum for $\rho_\Omega(= \rho_A + \rho_B + \rho_C)$. This is the essence of mixture theory. We have used the density as an example, but the concept is applicable also to other intensive quantities of matter.

Next, consider a domain Ω of volume $\mathbb{V}$ composed of a *very large* number of *irregularly shaped and randomly distributed* sub-domains, Ω_Δ, with $\mathbb{V}_\Delta \ll \mathbb{V}$. Some of these sub-domains are occupied by an A-phase, e.g., a solid ($\Delta = s$), while the others by a B-phase, e.g., a liquid ($\Delta = \ell$). Each Ω_Δ is a continuum of the phase occupying it, say with respect to the mass density. Again, since (or if) *we do not know the boundaries between the subdomains*, and we still want to investigate and predict the behavior of the phases within Ω, we follow the mixture theory ideas presented above. At any point within Ω, we select a representative (say, spherical) volume $\mathbb{V}_o$, such that $\mathbb{V}_\Delta \ll \mathbb{V}_o \ll \mathbb{V}$. Within each such representative volume, we can determine the mass density of the phases, $\bar{\rho}_A$ and $\bar{\rho}_B$, as well as $\bar{\rho}$ 'for all phases at the point'. In this way, for the same Ω-domain, we obtain *three overlapping continua*, one for the A-phase, with a density $\bar{\rho}_A(\mathbf{x}, t)$, the second for $\bar{\rho}_B(\mathbf{x}, t)$, and the third one for $\bar{\rho}(\mathbf{x}, t)$.

Of course, we could have presented the concept of mixture theory also through other densities, e.g., momentum density (= momentum per unit mass of substance $\equiv$ phase velocity).

"There is only one reality" is a very trivial statement. Reality is usually too complicated, and often involves parts and details that are unknown to us. Therefore, we are forced to describe reality "approximately". Since there are many ways to approximate the same reality, phenomena that occur in it can also be described in many ways. Usually, the selection depends on the objective of that description. Is the color of the solid matrix in a porous medium domain of any importance? is the shape of any individual sand grain of any significance? Obviously, the amount of details included in the description depends on the objective of the latter.

The concept of scale (as in "'Consider a phenomenon described at a certain scale..."') will be defined and discussed in Sect. 1.4. For the time being, we suggest that a spatial domain, occupied *only* by a fluid, or *only* by a solid, will be regarded as a *continuum* at the *microscopic level*, or at the *microscopic (or micro-) scale* of description, if values of kinematical and dynamical state variables (e.g., velocity, density, pressure, temperature), as well as values of phase properties (e.g., viscosity, or thermal conductivity) can be assigned to *every* point within that domain. These values are obtained, by averaging the behavior at the *molecular level*, or *molecular scale*, over a *Microscopic Representative Elementary Volume*, denoted as μREV (= microREV), around *every* point within the domain occupied by the phase. In practice, however, they are determined experimentally. The term 'pore-scale model' is often used for this level of modeling.

The size of the μREV of a phase must be such that it includes a sufficiently large number of molecules. This means that the μREV must be much larger than the *mean free path* of the molecules. This will ensure that meaningful phase thermodynamic properties are obtained at every point. Altogether, except for extreme cases, e.g., *at* the front of a shock wave, or in a rarefied gas in a relatively narrow domain (such as nano-pores in shales), a μREV can be defined for any considered phase domain. For a multi-species fluid, we require that a μREV exists for every chemical species, and that a common μREV can be found for all of them. An upper limit for the size of an μREV is that it must be much smaller than the domain of interest occupied by the phase. Thus, a 'fluid phase' and a 'solid phase' are *continuum concepts*, obtained from the molecular level by volume averaging, at least conceptually, over a μREV at every point within the phase domain. Bear (1972, p. 18) mentions also a Representative Elementary Time interval (RET) that is needed (conceptually) to overcome the random movement of molecules, leading to a fluid as a continuum.

1.1.2 The Porous Medium

We distinguish between three types of porous materials (Fig. 1.1):

- **Naturally occurring porous media** are those encountered as *geological formations* below ground surface. They are of interest, for example, to hydrogeologists and (petroleum) reservoir engineers. Thus, soil is a porous medium and so are geological formations composed, for example, of sand, fissured rock, fractured

porous rock, cemented sandstone, and Karstic limestone. Such formations serve as aquifers and as gas and petroleum reservoirs.
When considering a large domain within such a (natural) formation, we usually find it to be *highly heterogeneous* and often also anisotropic (see Sect. 1.1.7 C, D).
- **Manufactured porous media**, e.g., paper, ceramics, foam rubber, bread, and filters. Usually, they are homogeneous, and they may be anisotropic, with heterogeneity and anisotropy being controlled at desired degrees.
- **Organic porous media**, e.g., bones, lungs and kidneys.

In all three types, but in particular in manufactured porous media, although we may also think of a thin slice of a consolidated sandstone, we may encounter a situation to which we refer as *a thin porous medium* (see discussion following (1.1.28)).

What is common to all these examples? In fact, we may now ask: "what is the definition of a *porous medium*"?

Intuitively, considering these examples, a porous medium domain is a spatial domain that always contains two parts: one part is occupied by a (possibly deformable) solid phase. As already introduced in Sect. 1.1.2, this part is called the *solid matrix*. The remaining part is a *void space* (i.e., void of solids) that is occupied by one or more fluid phases–gas, or liquid.

The term *solid* here should be construed to represent a wider spectrum than what its strict definition describes. Thus, in addition, say, to a crystalline substance, materials like manufactured polymers, rubber, and organic materials, such as the tissue matrix of lungs and kidneys, are also regarded here as 'solid'. In all cases considered in this book, the solid portion of the porous medium domain is assumed to be *interconnected*. It is then also referred to as 'solid matrix'. It may be deformable. Actually, a *fluidized bed* (see Appendix A) may also be regarded as a *porous medium continuum*, although the solid particles do not constitute an *interconnected* domain.

The void space, occupied by one or more fluids, need not be *interconnected*. However, when considering fluid flow, an implicit assumption (actually, a necessity) is that at least part of the void space must be interconnected, with portions of the interconnected void space on the domain's boundary surface. However, heat can be transported through both the solid matrix and the fluid phase(s) occupying the void space, even if the latter is not continuous. In Sect. 1.1.7 A, we shall define porosity and introduce the concept of 'dead-end pores'.

Unless otherwise stated, in this book, we shall assume that both the solid matrix and the void space are continuous within the considered porous medium domain, with portions on the external bounding surface of the latter.

Another essential feature of a porous medium is that *both the solid matrix and the void space are distributed all over the considered (porous medium) domain*, albeit, not necessarily uniformly. Figure 1.2 shows two cross-sections through a spatial domain. In Fig. 1.2b, the cross-section is not of a porous medium domain, as the solid matrix in it is not distributed, more or less, over the entire domain, nor is the void space. Obviously, the above statements should be understood as 'for the purpose of the discussion presented here'. So far, we have required (1) that a porous medium should contain a solid matrix and a void space sub-domain, and (2) that both

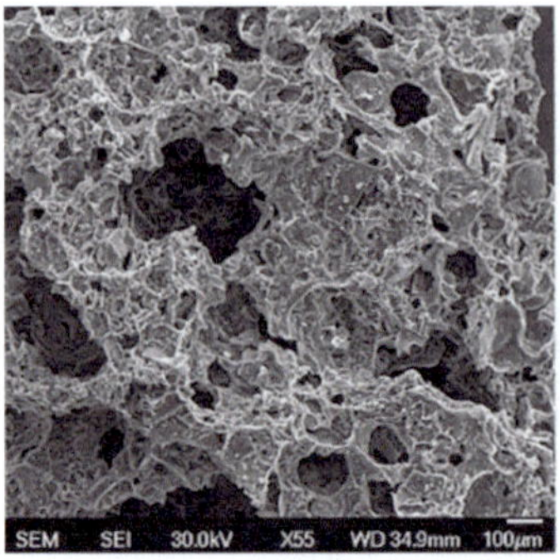

(a) Ceramics

(b) Bioclastic limestone

(c) High porosity sandstone

(d) Franciscan fractured mudstone

(e) Domogene sandstone

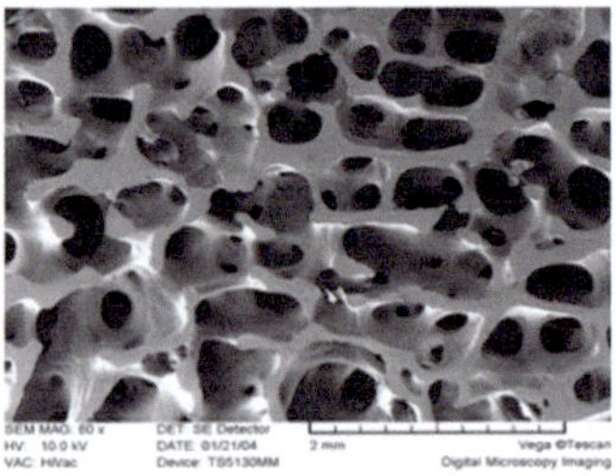

(f) Bone

Fig. 1.1 Typical geological, industrial and biological materials (Courtesy of Dr. George J. Moridis, LBNL (**a**, **d**–**f**), and Dr. Michel Séranne, Univ. of Montpellier (**b**, **c**))

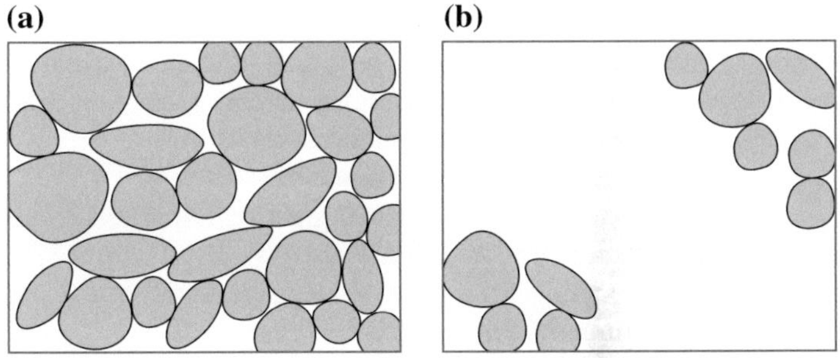

Fig. 1.2 Definition of a domain: **a** a porous medium, and **b** not a porous medium

sub-domains should be distributed throughout the considered domain. To examine whether condition (2) is satisfied for a given domain, we use the concept of a 'sample'. Given a spatial solid matrix plus void space domain, by taking samples of some constant volume at many arbitrarily selected points within this domain, we require that both a solid matrix and a void space be present in *all* samples. Obviously, we note that when a sample is sufficiently small, it contains only a solid, or only a void space.

Altogether, at this point, we can summarize that a porous medium is a spatial domain that (1) always contains a persistent solid portion and a void space (meaning 'void of solid'), and (2) that it should be possible to find for that domain a sample, referred to as a *Representative Elementary Volume* (abbreviated REV), of a size such that wherever we place it within the domain, it will *always contain both a solid matrix and a void space*.

The issue of the appropriate size of the REV will be discussed in Sect. 1.1.6.

In what follows, we shall explain why we need to introduce the concept of an REV for a given porous medium domain, and how do we make use of the REV in developing continuum models of phenomena of transport.

The main objective of this book is (a) to develop models of phenomena of transport in porous medium domains that are based on envisioning the latter as continua, and (b) to describe the phenomena in such domains by a modified form of Continuum (fluid and solid) Mechanics. Here and elsewhere in this book. "phenomena of transport" include also chemical reactions and phase transformation.

1.1.3 The Porous Medium Domain as a Continuum

Now that we have a definition for a porous medium, consider the flow and transport of an extensive quantity of a fluid phase that occupies the entire void space of a porous medium domain, or part of it. In Sect. 1.1.1, we have defined the spatial domain occupied by a fluid phase as a *continuum*. To describe the flow and transport of an extensive quantity through such a continuum, we construct and solve the mathematical model that describes this transport. Throughout this book, following the ideas underlying *continuum mechanics*, we shall stipulate that the core of any (mathematical) model that describes a transport problem is the balance equation of that extensive quantity. This balance equation takes the form of either a partial differential equation, or an integro-differential equation, that describes the local (i.e., at a point) balance of the considered extensive quantity. Some researchers use the term 'conservation law' instead of 'balance equation'. Here we consider a phase continuum at the microscopic level. Later we shall show that the above statement is valid also for a phase continuum at the macroscopic level.

We refer to a model that describes what happens at a *point within a fluid (or solid) phase continuum*, as a model at the *microscopic level*, or *microscopic scale*. At this level, a considered fluid phase domain is bounded by solid-fluid interfaces and by surfaces on the external boundary of the considered domain. When the void space is occupied by more than one fluid phase, then the domain occupied by a specific fluid will also be partly bounded by fluid–fluid interfaces.

For example, for mass transport of a fluid phase in the void space, i.e., $E = m$, with $\mathbf{V}$, p and ρ denoting the velocity (of the mass) pressure and mass density, respectively, the model consists of a momentum balance equation and a mass balance one, both for the fluid phase:

$$\frac{\partial \rho \mathbf{V}}{\partial t} = -\nabla \cdot (\rho \mathbf{V}\mathbf{V} - \boldsymbol{\tau}) + \nabla p - \mathbf{F}, \quad \mathbf{F} = -\rho g \nabla z, \tag{1.1.1}$$

$$\frac{\partial \rho}{\partial t} = -\nabla \cdot \rho \mathbf{V}, \tag{1.1.2}$$

$$\text{with :} \quad \rho = \rho(p), \quad \boldsymbol{\tau} = \mu[\nabla \mathbf{V} + (\nabla \mathbf{V})^T], \tag{1.1.3}$$

to be discussed later in the book. Note that the above equations are written in *vector notation*. In this book, we shall use this kind of notation as well as *indicial notation*, interchangeably. Both equations have to be satisfied at all points in the fluid phase. Furthermore, we require information (1) on the fluid's constitutive relationship $\rho = \rho(p)$, (2) on the relationship between *shear stress* and velocity, $\boldsymbol{\tau} = \boldsymbol{\tau}(\mathbf{V})$, which depends on the fluid's nature, and (3) on the body force, $\mathbf{F}$, which, usually, is assumed to be due only to gravity.

In addition, to obtain a complete model for a specific case, we have to specify initial and boundary conditions, for example, no flow normal to the solid grain surfaces, i.e., $\mathbf{V} \cdot \boldsymbol{\nu} = 0$ on the fluid–solid interface, and specified conditions on the external boundaries of the considered flow domain. Obviously, it is not possible to specify these conditions unless the shape of the fluid–solid boundary is known. Do we know the shape of this boundary for a given porous medium domain? In two or three phase flow, each phase is surrounded partly by an inter-phase boundary surface and conditions on it are also required. Again, even when conditions on such interfaces are known, the shape and location of these fluid–fluid interfaces are not known. The conclusion is that *we cannot write and solve well posed 'pore-scale models'*.

Until recently, the above conclusion was obvious, and this led to the introduction of the 'macroscopic' level of description of phenomena of transport in porous medium domains, as at that level, information on microscopic solid-fluid and fluid–fluid surfaces is not required. In fact, as will be shown later, this information is replaced by various porous medium 'coefficients'. This approach will be introduced below; it underlies the presentation throughout this book.

However, in recent years, with the advancement in both *imaging* and *computing power*, modeling of phenomena of transport at the microscopic level has been made possible, albeit, for domains of a rather limited size. Nowadays, it is actually possible to see and measure the microscopic geometry of interphase boundaries, so that problems can be set-up at the microscopic level, actually observing what happens inside individual pores and grains, and solved, i.e., determining values of state variables, when the system is excited by known boundary conditions. Some comments on this world of imaging, describing porous material at the microscopic level, and solving transport problems at that level, using the power of computing, is presented in the Sect. 1.2. More on this subject is presented in Appendix B.

Altogether, with the exception of fundamental studies of micro-flow analysis discussed above (which seek answers to fundamental physics), in almost all cases of practical interest, *microscopic interphase boundaries cannot be specified and problems cannot be solved at the microscopic level.* This is particularly true for geological formations for which we do not have information on the geometry of the void-solid boundaries. In fact, we do not have such information even for a small core, or sand column. Under such conditions, writing *complete* mathematical flow and transport models at the microscopic level is not possible. We need another approach—the macroscopic one—to be introduced below. This book is devoted to modeling at the *macroscopic level.*

At the macroscopic level, sometimes referred to as *macro-scale*, a given porous medium domain as a whole, involving a solid matrix and a void space, occupied by one or more fluids, is regarded as a *single continuum*. This means that to *every* point in such domain, values can be assigned to variables of state and to properties that correspond to *any* phase present in the domain. Furthermore, each phase may also be regarded as a continuum that occupies the *entire* domain, albeit at various volumetric fractions (to be defined below).

The REV, introduced in Sect. 1.1.2 (and to be further discussed in Sect. 1.1.4), can now be used to transform the microscopic model of a given domain into a macroscopic one. In the description of phenomena at the latter level, every point within the considered porous medium domain is regarded as associated with, and represented by, the centroid of an REV the size of which has been selected as appropriate for that considered domain. We can then average the behavior of the phases (and of chemical species) within every REV, e.g., the pressure, solute concentration, temperature, velocity, or their time and space derivatives, to obtain an averaged, or *macroscopic* description of phenomena within the porous medium domain. The average values are assigned to the center of the REV. By the process of averaging, information on the local variations, i.e., at the microscopic scale, is lost. We recall that, actually, the microscopic level information is not available. The advantage of the macroscopic

(= averaged) level of description is that *it does not require information on the geometry of the microscopic interphase surfaces*. Another important feature is that this is also the level at which we *measure*, or *monitor*, relevant variables of state, like through an instrument window. In fact, this is the only practical approach to obtain measured values that are required in such studies. The 'price' for achieving this goal is that various *coefficients* are created in the process of averaging. They reflect, at the macroscopic level, the effects of the microscopic interphase (solid-fluid and fluid–fluid) boundaries, recalling that, anyway, we do not have the detailed information about the geometry of such boundaries. Thus, by this process of averaging, we obtain a model that circumvents the need to know the microscopic geometry of the fluid–solid boundary, and, in two-phase flow, also that of the fluid–fluid interfaces. Depending on how the averaging is performed, it is possible to obtain information on the structure and composition of these coefficients, i.e., on their tensorial nature and on their dependence on various geometrical features of the void space, or phase configuration. Unless the void-space has a known well-defined geometry, a situation which may exist in the case of a manufactured porous medium, but is rare, and usually does not exist in natural geological media, the numerical values of these coefficients cannot be determined as part of the averaging process; they must be determined *experimentally*.

To summarize, by averaging the microscopic description of phenomena over an REV, we obtain a macroscopic description in which the entire multi-phase domain behaves as a *continuum*. This means that macroscopic values of state variables of any of the extensive quantities of any of the phases present in the domain, as well as coefficients associated with each of the phases and their spatial distribution, can be assigned to *every* point within this continuum. In fact, we represent a considered multiphase domain, in which each phase occupies only part of the domain in some random spatial distribution, as *multiple overlapping continua*. To every point within such continuum we assign properties, variable of state, etc., for each of the phases present at that point in the domain. Phases may exchange extensive quantities *at a point*, although they belong to different continua. As we shall see, this approach of describing the behavior of multiple phases occupying a spatial domain as occurring in multiple interacting overlapping continua, requires that certain conditions be satisfied. We shall present and discuss these conditions. As we shall show, the continuum approach will be shown to be valid not only to granular-type porous medium domains, but also to fractured rock domains.

Although we have referred to 'volume (or mass) averaging over an REV' as the technique for passing from the microscopic level to the macroscopic one, in Sect. 1.4.3 we shall mention *homogenization* as an additional technique for achieving this goal. Some authors (e.g., Blunt 2001; Raoof et al. 2010; Raoof and Hassanizadeh 2012; and Joekar-Niasar et al. 2008, 2010) use pore-network models as the starting point for developing continuum models for transport in porous media.

The *phenomenological approach* employed throughout this book, is another option that does not involve formal REV-averaging (see Sect. 1.4.4).

So far, we have shown that in order to overcome the difficulties stemming from the lack of information on solid-void space boundaries within a porous medium

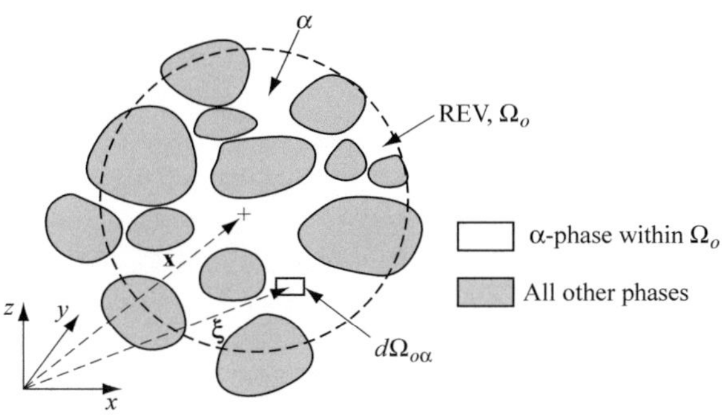

Fig. 1.3 Nomenclature for averaging over an REV

domain, we have used the concepts of an REV and 'averaging over an REV' in order to pass from the microscopic level of problem description to the macroscopic one. The latter may still be heterogeneous with respect to macro-level properties such as porosity and permeability. We may refer to the process as *upscaling*, in this case, from the microscopic scale of description to the macroscopic one. In fact, we have also mentioned the molecular level. By averaging at that level, we get the microscopic level at which we have, for example, fluid and solid phases as continua. In Sect. 1.4, we shall discuss the subject of upscaling in more detail.

The definition of volume averaging is presented in the next subsection.

1.1.4 Volume and Mass Averages

Figure 1.3 shows an REV, occupied by a solid phase and a void space (occupied by a number of fluid phases). We use subscript α (= 1, 2, 3), to denote each of the fluid phases. Each α-phase occupies at time t a domain $\Omega_{o\alpha}$ of volume $\mathbb{V}_{o\alpha}$ within the REV domain Ω_o, of volume $\mathbb{V}_o$. Note that we are using subscripts s and $\boldsymbol{v}$ to denote, respectively, the solid phase and the void space, occupied by fluid phases. The *porosity*, ϕ, and the *volumetric fraction* of the α-phase, θ_α, within the REV, both assigned to the latter's centroid, $\mathbf{x}$, are defined by:

$$\theta_\alpha(\mathbf{x}, t) = \frac{\mathbb{V}_{o\alpha}(\mathbf{x}, t)}{\mathbb{V}_o(\mathbf{x}, t)}, \quad \phi = \frac{\mathbb{V}_{ov}}{\mathbb{V}_o}, \quad \mathbb{V}_o = \mathbb{V}_{os} + \mathbb{V}_{ov}, \tag{1.1.4}$$

with

$$\sum_{(\alpha)} \mathbb{V}_{o\alpha} = \mathbb{V}_{ov}, \qquad \sum_{(\alpha)} \theta_\alpha = \phi. \tag{1.1.5}$$

In Chemical Engineering, the term "voidage" is often used for porosity.

Let $e'_\alpha(\boldsymbol{\xi}, t)(= dE_\alpha/d\mathbb{V}_\alpha)$ denote the (volumetric) *density* of some extensive quantity E of an α-phase (= amount of E per unit phase volume) at the microscopic point, $\boldsymbol{\xi}$. We use here the position vector $\boldsymbol{\xi}$ to emphasize that the field, e.g., $e'_\alpha(\boldsymbol{\xi}, t)$, is at the microscopic level. For certain extensive quantities, the monitored value is not the *density*, e', of E, but the *specific value*, e, of E, i.e., E per unit mass of the phase, with

$$e'_\alpha = \rho\, e_\alpha. \tag{1.1.6}$$

For example, for mass of a phase, the specific value is $e_\alpha = 1$, and e'_α is the mass density, ρ. Another example, is $\mathbf{V}$, which is *specific momentum*, i.e., momentum per unit mass It is assumed that both e_α and e' are finite, continuous and differentiable everywhere within the sub-domain occupied by the α-phase.

Three kinds of volumetric averages of e'_α can be defined:

A. Intrinsic Phase Average

The (volumetric) intrinsic phase average of e'_α, taken over the α-phase in an Ω_o-domain:

$$\overline{e'_\alpha}^\alpha(\mathbf{x}, t) = \frac{1}{\mathbb{V}_{o\alpha}(\mathbf{x}, t)} \int_{\Omega_{o\alpha}(\mathbf{x},t)} e'_\alpha(\boldsymbol{\xi}, t; \mathbf{x})\, d\mathbb{V}(\boldsymbol{\xi}), \tag{1.1.7}$$

where the $\mathbf{x}$ in $(\boldsymbol{\xi}, t; \mathbf{x})$ indicates that we consider points $\boldsymbol{\xi}$ belonging to an REV centered at $\mathbf{x}$; $d\mathbb{V}$ denotes a volume element of $\mathbb{V}_{o\alpha}(\mathbf{x}, t)$.

It is possible to replace the above equation by:

$$\overline{e'_\alpha}^\alpha(\mathbf{x}, t) = \frac{1}{\mathbb{V}_{o\alpha}(\mathbf{x}, t)} \int_{\Omega_{o}(\mathbf{x},t)} e'_\alpha(\boldsymbol{\xi}, t; \mathbf{x})\, \gamma_\alpha(\boldsymbol{\xi}) d\mathbb{V}(\boldsymbol{\xi}), \tag{1.1.8}$$

where $\gamma_\alpha(\boldsymbol{\xi})$ is the *indicator function* for an α-phase that occupies part of the REV, with s denoting the solid phase, and

$$\gamma_\alpha(\boldsymbol{\xi}) = \begin{cases} 1, & \text{for } \boldsymbol{\xi} \text{ within } \Omega_\alpha, \\ 0, & \text{for } \boldsymbol{\xi} \text{ within } \Omega - \Omega_\alpha. \end{cases} \tag{1.1.9}$$

Equations (1.1.7) and (1.1.8) express the same intrinsic phase average.

The average $\overline{e'_\alpha}^\alpha$ is a function of the macroscopic coordinates, $\mathbf{x}$. The symbol α next to the bar in $\overline{(..)}^\alpha$ indicates that this average is over $\mathbb{V}_{o\alpha}$.

B. Volumetric Phase Average

The volumetric phase average of e'_α,

$$\overline{e'_\alpha}(\mathbf{x}, t) = \frac{1}{\mathbb{V}_{o}(\mathbf{x}, t)} \int_{\Omega_{o\alpha}(\mathbf{x},t)} e'_\alpha(\boldsymbol{\xi}, t; \mathbf{x})\, d\mathbb{V}(\boldsymbol{\xi}). \tag{1.1.10}$$

Here, the total amount of the extensive quantity of the α-phase is averaged over the *entire* sub-domain, Ω_o, of the REV. Since we deal with a property of the α-phase only, the integrations in both (1.1.7) and (1.1.10) are over the sub-domain $\Omega_{o\alpha}$ only.

From (1.1.7) and (1.1.10), with θ_α defined by (1.1.4), it follows that the two kinds of averages are related to each other by:

$$\overline{e'_\alpha} = \theta_\alpha \overline{e'_\alpha}^\alpha. \tag{1.1.11}$$

C. Volume Average

When E is an extensive quantity that is defined for all α-phases present in $\mathbb{V}_o$, e.g., mass, we define the *volume average* of E by:

$$\begin{aligned}\overline{e'} &= \frac{1}{\mathbb{V}_o}\int_{\Omega_o} e'\, d\mathbb{V} = \frac{1}{\mathbb{V}_o}\sum_{(\alpha)}\int_{\Omega_{o\alpha}} e'_\alpha\, d\mathbb{V}\\ &= \frac{1}{\mathbb{V}_o}\sum_{(\alpha)} \mathbb{V}_{o\alpha}\overline{e'_\alpha}^\alpha = \sum_{(\alpha)} \theta_\alpha \overline{e'_\alpha}^\alpha = \sum_{(\alpha)} \overline{e'_\alpha}.\end{aligned} \tag{1.1.12}$$

where $\int_{\Omega_{o\alpha}} e'_\alpha d\mathbb{V}_\alpha$ represents the total amount of E within $\Omega_{o\alpha}$.

The quantity:

$$\mathring{e}'_\alpha(\boldsymbol{\xi}, t; \mathbf{x}) \equiv e'_\alpha(\boldsymbol{\xi}, t; \mathbf{x}) - \overline{e'_\alpha}^\alpha(\mathbf{x}, t) \tag{1.1.13}$$

defines the *deviation* of e'_α at a point $\boldsymbol{\xi}$ within the REV centered at $\mathbf{x}$ from its intrinsic phase average over that REV.

By taking the volumetric intrinsic phase average of (1.1.13), we obtain:

$$\overline{\mathring{e}'_\alpha}^\alpha = 0. \tag{1.1.14}$$

D. Intrinsic Mass Average

Let $e_\alpha(\boldsymbol{,}, t)$ denote the *specific value* of E_α (i.e., the quantity of E of an α-phase per unit *mass* of that phase), with $m_\alpha = \overline{\rho_\alpha}^\alpha \mathbb{V}_{o\alpha}$, and $dm_\alpha = \rho_\alpha d\mathbb{V}_\alpha$. The *intrinsic mass average*, $\widetilde{e_\alpha}^\alpha$ of e_α may then be defined by:

$$\begin{aligned}\widetilde{e_\alpha}^\alpha(\mathbf{x}, t) &= \frac{1}{m_{o\alpha}}\int_{m_{o\alpha}} e_\alpha\, dm = \frac{1}{\overline{\rho_\alpha}^\alpha \mathbb{V}_{o\alpha}}\int_{\Omega_{o\alpha}} e_\alpha(\boldsymbol{\xi}, t)\rho_\alpha(\boldsymbol{\xi}, t)\, d\mathbb{V}(\boldsymbol{\xi}; \mathbf{x})\\ &= \frac{\overline{\rho_\alpha e_\alpha}^\alpha}{\overline{\rho_\alpha}^\alpha} = \frac{\overline{e'_\alpha}^\alpha(\mathbf{x}, t)}{\overline{\rho_\alpha}^\alpha(\mathbf{x}, t)}, \quad \Rightarrow \quad \overline{\rho_\alpha}^\alpha \widetilde{e_\alpha}^\alpha = \overline{e'_\alpha}^\alpha = \overline{\rho_\alpha e_\alpha}^\alpha,\end{aligned} \tag{1.1.15}$$

where $\overline{\rho_\alpha}^\alpha$ is the *intrinsic phase average mass density* of the α-phase.

Similar to (1.1.13), the quantity:

$$\check{e}_\alpha(\boldsymbol{\xi}, t; \mathbf{x}) \equiv e_\alpha(\boldsymbol{\xi}, t; \mathbf{x}) - \widetilde{e_\alpha}^\alpha(\mathbf{x}, t), \quad \widetilde{\check{e}_\alpha}^\alpha = 0, \tag{1.1.16}$$

defines the *deviation* of e_α at a point $\boldsymbol{\xi}$ within the REV centered at $\mathbf{x}$ from its intrinsic mass average over that REV. As an example, consider the case of momentum, $E = \mathbf{M}_\alpha = m_\alpha \mathbf{V}_\alpha \Rightarrow \mathbf{V}_\alpha = \mathbf{M}_\alpha / m_\alpha = e_\alpha$, and $\widetilde{\mathbf{V}_\alpha}^\alpha$ is the *intrinsic mass averaged velocity* of the α-phase. In this book, when considering phenomena of transport at the macroscopic level, and it is obvious that a considered equation is at the macroscopic level, we shall use the symbol $\mathbf{V}$ to denote the *mass averaged velocity*.

It is important to reiterate that when performing either of the two kinds of averaging, leading to $\overline{(e_\alpha)}^\alpha$, or $\widetilde{e_\alpha}^\alpha$, the integrand $e'_\alpha d\mathbb{V}$, or $e_\alpha dm$, has to be *physically meaningful* and the respective average has to be *measurable*. Thus, for $E = m_\alpha$ and fluid velocity, $\mathbf{V}_\alpha$, the integral $\int_\Omega \rho V d\mathbb{V}$ is the total momentum in Ω, while $\int_\Omega \rho d\mathbb{V}$ is the total mass in the same volume. The, the ratio, which is the momentum per unit mass is also the *mass averaged velocity*.

E. A Comment on Monitoring Averaged Values

The kind of average to be used in each case, i.e., intrinsic phase volume or mass average, or volume phase average, depends on the way the averaged, or up-scaled, quantity is actually measured, or monitored. It is of interest to note that the two averages are equal when the mass density is constant. For example, if at a point in a porous medium domain we take a liquid sample out of a porous medium domain, in order to determine its solute concentration, the latter is an intrinsic phase average, as it is taken over a volume of liquid phase. It is also possible to take a certain mass of the exiting fluid, and determine the average mass flux. We can then consider the volume and determine the density of the exiting liquid. The sample serves as an 'instrument window'. In fact the above statement on monitoring applies to any upscaling technique. If values at the upscaled level cannot be monitored, there is no way to verify the model!

The term 'flux' has been used above to express fluid volume passing through a unit area normal to the flow direction, per unit time. Throughout the book, we shall refer also to 'mass flux', 'momentum flux' and 'energy flux', as the quantity of the considered extensive quantity per unit area per unit time.

In the discussion above, we have been referring to an intensive quantity, e_α, of an α-phase and to its volume or mass average. In practice, as presented in Chaps. 5–9, we shall refer to the specific cases of phase density (ρ_α), concentration (e.g., c_α), fluid velocity ($\mathbf{V}_\alpha$), and temperature (T_α). A comment on the *physical* meaning of the averages defined above, especially *velocity* and *pressure*, is appropriate. Concerning fluid (and solid) velocity, we actually never measure velocity directly, as 'distance per unit time', certainly not at the macroscopic level. In fact, in view of fluid compressibility, even the definition of velocity as 'fluid volume per unit porous medium area, per unit time' is meaningless, unless associated with the fluid's density. What we often measure at an outlet (from a porous medium domain), is the mass of fluid leaving the domain per unit area per unit time, namely 'fluid mass flux', meaning 'mass per unit area per unit time'. If we also know the fluid's (instantaneous) density, then what we really *infer to* is the *fluid's mass averaged velocity*.

A fundamental thermodynamic variable like fluid density and fluid temperature is the pressure. All three are attributes of a given mass of fluid. A comment on the terms/definitions of 'mechanical pressure' and 'thermodynamic pressure' is presented in Sect. 2.2.2

It is obvious that pressure propagates very quickly, so that there cannot be large differences in pressure at points within an REV, and the notion of an average is reasonable. However, when we consider solute concentration, or temperature, the smoothing out is by the diffusive fluxes of solute mass and of heat; this is a much slower process. For the case of solute concentration, the Peclet number, defined by (7.2.39) as the ratio of advective to diffusive transfers will dictate the relationship between these two fluxes. For large Pe, the effect of advection dominates and we may encounter big differences in concentration within the REV. We shall return to this subject when considering higher level averaging in Sect. 7.2.4 A.

Finally, a comment on the relationship between the size of the measuring device and the monitored averaged, or upscaled value, especially in very heterogeneous (e.g., geological) formations is appropriate. We have already emphasized that when upscaling is performed by averaging over an REV, e.g., from microscopic to macroscopic levels, the monitoring device should have the size of an REV at the macroscopic level. Then, we can compare observed and predicted values of variables of state. The observed, or monitored, result depends also on the size of the measuring device. We usually monitor, say concentration and temperature, in an observation well, but the actual device may be much smaller and we have to make sure that what we monitor is the averaged value that appears in the model, or sufficiently close to it. Concentration and temperature, for example, vary continuously, and, in principle, also within the well, and we assume that the device exhibits the average over the observation well and the latter represents the value at the point in the domain. This value is to be compared with the (averaged) value appearing in the model.

Upscaling, especially in connection with solute transport in large geological formations, is performed by other techniques, e.g., CTRW (Sect. 7.6.2 B), we usually use the same monitoring device (i.e., observation well). Can such observation be compared with what is predicted (e.g., pressure, temperature or concentration) by the model?

F. Measuring Mass in Moles

So far, we have referred to mass of a fluid or of a solid as measured, say, in kg. However, when dealing with the transport of possibly reacting chemical species dissolved in a fluid phase, or constituting part of a solid, the mass of the these species is usually measured in *moles*. The averaged fluid velocity can then be based on the molar fractions of the various species.

1.1.5 Areal Average

So far, in this section, we have discussed volume and mass averages of intensive quantities that are additive over volume. However, certain quantities, like stress and flux, are additive only over area. This means that for such quantities, averages per unit area have to be defined.

Let A_o and $A_{o\alpha}$ denote the area of an arbitrarily oriented planar Representative Elementary Area (Abbrev. REA), and the α-area, within it, respectively, centered at some point, $\mathbf{x}$, within a porous medium domain. We can define two kinds of areal averages for any component of a tensorial α-phase quantity, $\boldsymbol{\pi}_\alpha$ associated with area (e.g., flux, stress), such that $\boldsymbol{\pi} \cdot d\mathbf{A}_\alpha$ is physically meaningful and additive over $A_{o\alpha}$.

The first is:

$$\widehat{\pi_\alpha}^\alpha(\mathbf{x}, t) = \frac{1}{A_{o\alpha}(\mathbf{x}, t)} \int_{\mathcal{A}_{o\alpha}(\mathbf{x})} \pi_\alpha(\boldsymbol{\xi}, t; \mathbf{x}) \cdot d\mathbf{A}(\boldsymbol{\xi}; \mathbf{x}), \tag{1.1.17}$$

called the *intrinsic phase areal average* of π_α, taken over the area $\mathbf{A}_{o\alpha}$. The second is:

$$\widehat{\pi_\alpha}(\mathbf{x}, t) = \frac{1}{A_o(\mathbf{x}, t)} \int_{\mathcal{A}_{o\alpha}(\mathbf{x})} \pi_\alpha(\boldsymbol{\xi}, t; \mathbf{x}) \cdot d\mathbf{A}(\boldsymbol{\xi}; \mathbf{x}), \tag{1.1.18}$$

called the *areal average* of π_α, taken over the area $\mathbf{A}_o$.

The two averages are related to each other through the *areal fraction* $\theta_\alpha^A (= A_{o\alpha}/A_o)$, i.e.,

$$\widehat{\pi_\alpha} = \theta_\alpha^A \widehat{\pi_\alpha}^\alpha, \tag{1.1.19}$$

where the areas $\mathbf{A}_o$ and $\mathbf{A}_{o\alpha}$ are facing some direction denoted by the unit vector $\boldsymbol{\nu}(= \mathbf{A}_o/|\mathbf{A}_o|)$.

In Continuum Mechanics, the product $\rho\mathbf{V}$ is the *linear momentum density of mass*. By analogy, the product $e'_\alpha \mathbf{V}^{E_\alpha}$ may be regarded as the *linear momentum density of an extensive quantity E of an α-phase*. Hence, $e'_\alpha \mathbf{V}^{E_\alpha} d\mathbb{V}_\alpha$ is additive over volume, and taking a volume average of it is permissible.

However, $e'_\alpha \mathbf{V}^{E_\alpha}$ also represents an *E-flux* (= amount of E passing through a unit area of the α-phase, normal to $\mathbf{V}^{E_\alpha}$, per unit time). This means that $e'_\alpha \mathbf{V}^{E_\alpha} \cdot d\mathbf{A}_\alpha$ is additive over area. Hence, taking an average of $e'_\alpha \mathbf{V}^{E_\alpha}$ over the area $\mathbf{A}_{o\alpha}$ of an REA (which is normal to the direction of $\mathbf{V}^{E_\alpha}$) *is also permissible*.

In the spirit of assigning the REV average to a point, taking an areal average at a point does not really represent what happens in cross sections of the REV centered at the point. As an example, we consider the flux of E_α at a point, denoted as $\mathbf{j}^{E_\alpha} \equiv e_\alpha \mathbf{V}^{E_\alpha}$.

Let us examine the conditions under which, at a given point $\mathbf{x}$, the areal average of the flux, $\widehat{(e_\alpha \mathbf{V}^{E_\alpha})}\big|_{\mathbf{x}} \equiv \widehat{\mathbf{j}^{E_\alpha}}\big|_{\mathbf{x}}$, and the volume average, $\overline{e_\alpha \mathbf{V}^{E_\alpha}}\big|_{\mathbf{x}}$, of the momentum, are identical.

For a point $\mathbf{x}$, which is the centroid of $\mathbb{V}_o(\mathbf{x}, t)$, the volume average of $e_\alpha V_1^{E_\alpha}$ is given by:

$$\overline{e_\alpha V_1^{E_\alpha}}(\mathbf{x}, t) = \frac{1}{\mathbb{V}_o(\mathbf{x}, t)} \int_{\Omega_{o\alpha}(\mathbf{x}, t)} (e_\alpha V_1^{E_\alpha})\big|_{\mathbf{x}', t} \, d\mathbb{V}_\alpha. \tag{1.1.20}$$

Averaging of a flux over a single cross-section at a point does not really represent what happens at the point. Instead we have to average over a number of parallel cross sections around the point. Accordingly, let us choose an REV in the form of a cylinder of constant cross-sectional area, A_o, equal to the REA normal to the unit vector $\mathbf{1}x_1$ in the direction of the x_1-axis, and length $s_o = (\mathbb{V}_o/A_0)$ in the direction of $\mathbf{1}x_1$. We may rewrite (1.1.20) in the form:

$$\overline{j_1^{E_\alpha}}(\mathbf{x},t) = \frac{1}{s_o}\int_{x_1-\frac{s_o}{2},x_2,x_3}^{x_1+\frac{s_o}{2},x_2,x_3} dx_1' \frac{1}{A_o}\int_{A_{o\alpha}(x_1',x_2,x_3)} j_1^{E_\alpha}\Big|_{x_1',x_2',x_3'} dx_2'\,dx_3'. \tag{1.1.21}$$

With:

$$\widehat{j_1^{E_\alpha}}\Big|_{x_1',x_2,x_3} = \frac{1}{A_o}\int_{A_{o\alpha}(x_1',x_2,x_3)} j_1^{E_\alpha}\Big|_{x_1',x_2',x_3'} dx_2'\,dx_3', \tag{1.1.22}$$

we obtain from (1.1.21):

$$\overline{j_1^{E_\alpha}}\Big|_{\mathbf{x}} = \frac{1}{s_o}\int_{x_1-\frac{s_o}{2},x_2,x_3}^{x_1+\frac{s_o}{2},x_2,x_3} \widehat{j_1^{tE_\alpha}}(x_1',x_2,x_3)\,dx_1' = \left\lfloor \widehat{j_1^{E_\alpha}} \right\rfloor^{s_o}\Big|_{\mathbf{x}}, \tag{1.1.23}$$

where $\lfloor\ \rfloor^{s_o}$ indicates an average over the length s_o. In words, (1.1.23) states that the volume average at $\mathbf{x}$ is equal to the average over the length s_o of the areal averages, each taken over a cross-section, A_o.

By developing the areal average $\widehat{j_1^{E_\alpha}}\Big|_{x_1',x_2,x_3}$ into a power series about the point $\mathbf{x}$, Bear and Bachmat (1991, p. 35) show that:

$$\left\lfloor \widehat{j_1^{E_\alpha}} \right\rfloor^{s_o}\Big|_{\mathbf{x}} = \widehat{j_1^{E_\alpha}}\Big|_{\mathbf{x}}, \tag{1.1.24}$$

up to an accuracy of $O((\ell/L^*)^2)$. They proceed to show that for $s_o \ll L^*$:

$$\overline{j_1^{E_\alpha}}^{\alpha}\Big|_{\mathbf{x}} = \widehat{j_1^{E_\alpha}}^{\alpha}\Big|_{\mathbf{x}}. \tag{1.1.25}$$

This means that subject to all the constraints imposed on the sizes of the REV and REA, *the volumetric intrinsic phase average and the areal intrinsic phase average of* $\mathbf{j}^{tE_\alpha}(\equiv e_\alpha'\mathbf{V}^{E_\alpha})$ *at a point are identical.*

In principle, we should also make a distinction between the *porosity*, ϕ (= volume of void space per unit volume of porous medium), and the *areal porosity*, ϕ^A (= area of void in a planar cross-section, per unit area of cross section), and in the case of multiple phases between θ_α and θ_α^A, with $\phi^A = \phi^A(\nu_1,\nu_2,\nu_3)$, in which ν_1,ν_2,ν_3 denote the components of the unit vector, $\boldsymbol{\nu}$, normal to the considered cross-section. However, in view of the above developments, it is usually assumed that $\phi \approx \phi^A$. This is the approach undertaken in this book.

1.1.6 Size of REV

The REV is an essential concept in the definition of a porous medium (Sect. 1.1.2). For a considered porous medium domain, the size of the REV is also needed in order to determine the size of the 'instrument window' required for observing, or monitoring, the (macroscopic) values of state variables that appear in our predictive macroscopic models. Such measurements should provide the values of these variables at any monitored point, i.e., the average value taken over the REV assigned to its center. These are the values of state variables that appear in the mathematical models. For example, in order to determine the concentration of a chemical species at a point in a porous medium domain, we should extract a volume of fluid which is equal to the size of the void space within the REV and determine the concentration of the considered species in that volume. The resulting value is assigned to the point. However, often, the measuring device (e.g., a well for observing pressure, a thermistor for determining temperature, or an electric charge for determining the salinity of water in a geological formation) is much smaller or much larger than the size of the REV. The use of such measuring devices is justified only when we may assume that pressure, temperature and concentration vary (practically) linearly across the REV, so that *practically*, the device, albeit small, provides the sought average value. Otherwise, we have to be careful in the interpretation of what we monitor.

For a considered porous medium, the size of the REV is selected such that:

- The average value of *any relevant geometrical characteristic of the microstructure of the void space*, at any point in a porous medium, will be *a unique function of the location of that point only*.
- The measured averaged value should be *independent of small variations in the size of the* REV.

This means that the average value *at a point* should remain, more or less, constant over a range of REV volumes that corresponds to the range of variation in the sample size, or in the instrument that monitors that average.

Note that we have emphasized above that the selected REV should be common to all *relevant geometrical characteristics of the microscopic structure of the void space*. The reason is that in the passage from the microscopic to the macroscopic levels, the (unknown) detailed geometry of the solid-void interface is replaced by various (geometrical) coefficients (e.g., porosity, hydraulic radius of void space, specific solid-void area, permeability, tortuosity, fluid saturation, etc.) We have to ensure that the value of each of these properties at a point in the porous medium domain is independent of the selected REV size.

In the case of *thin porous media*, e.g., a sheet of paper, one dimension of the REV is the thickness of the porous medium domain, while the aforementioned considerations apply to the other two dimensions of the REV.

Denoting the characteristic size of an REV by ℓ, e.g., diameter of a spherical REV, and the length characterizing the microscopic structure of the void space by d (say, the typical size of grain or pore, or the *hydraulic radius of the void space*), a necessary condition for obtaining non-random estimates of the geometrical characteristics of the void space at any macroscopic point within a porous medium domain is:

$$\ell \gg d, \tag{1.1.26}$$

say, $\ell > 10d$. Dagan (1989) estimated the radius of a spherical domain Ω_{min} as about 50 times the pore radius.

Another condition that sets an upper limit to the size of the REV is:

$$\ell \ll \ell_{\max}, \tag{1.1.27}$$

where $\ell_{\max}$ is the distance beyond which the spatial distribution of the relevant macroscopic coefficients that characterize the configuration of the void space (e.g., porosity, permeability) deviates from the linear one by not more than some acceptable value (Bear and Bachmat 1991, p. 22). The selection of the size of the REV is also constrained by the requirement that:

$$\ell \ll L, \tag{1.1.28}$$

where L is a characteristic length of the porous medium domain. For example, $\ell < L/100$.

To understand the above requirement, we recall that within half an REV next to a boundary surface, the domain cannot be regarded as a continuum. The selection of 100 is, of course, arbitrary, just to say 'many times the size of an REV'. The objective is to minimize the effect of the fact that very close to the domain's boundary, the behavior in the domain is far from the macroscopic averaged description.

A comment on the above constraint is appropriate here. Suppose, for a considered granular porous domain (as an example), condition (1.1.28) is satisfied in only two orthogonal space directions, say x_1 and x_2, but not in the third, x_3-direction, which is mutually orthogonal to x_1 and x_2. Can an REV be determined? Can such domain be regarded and treated as a continuum? Indeed, intuitively, if in the x_3-direction the domain's thickness is only *a few* grains' thick, the domain cannot be treated as a 3-d continuum. However, if the thickness is 10 or more 'grain sizes', but still less that the '100' figure mentioned above, it can be treated as a two-dimensional continuum. We shall refer to the latter case as a *thin porous medium domain*, which can be treated as a *two-dimensional continuum*. One can write transport models for such continuum. However, we shall not discuss thin porous media in this book (see Qin and Hassanizadeh 2013, 2015).

To gain a better understanding of the above discussion of an REV, let us follow Bear and Bachmat (1991, pp. 16–27) and consider a porous medium domain, focusing first on *porosity*, ϕ, as a typical porous medium property. Let $\boldsymbol{\xi}$ denote a point within a small porous medium domain, Ω, of volume $\mathbb{V}$ centered at $\mathbf{x}$. The Ω-domain is composed of two portions (see Fig. 1.3): the subdomain of void space, Ω_v, of volume $\mathbb{V}_v$, and that of the solid matrix, Ω_s, of volume $\mathbb{V}_s$ $(\equiv \mathbb{V} - \mathbb{V}_v)$. We make use of the *indicator function*, γ, defined in (1.1.9). This γ-function includes *all* the information on the void-solid distribution within a porous medium domain. Here,

$$\gamma(\boldsymbol{\xi}) = \begin{cases} 1, & \text{for } \boldsymbol{\xi} \text{ within } \Omega_{ov}, \\ 0, & \text{for } \boldsymbol{\xi} \text{ within } \Omega - \Omega_{ov} \end{cases} \tag{1.1.29}$$

is the *indicator function of the void-space*.

An average, $\overline{\gamma}(\mathbf{x})$, may be defined by:

$$\overline{\gamma}(\mathbf{x}) = \frac{1}{\mathbb{V}_o}\int_{\Omega_o(\mathbf{x})} \gamma(\boldsymbol{\xi};\mathbf{x})\, d\,\mathbb{V} = \frac{1}{\mathbb{V}_o}\int_{\Omega_{ov}(\mathbf{x})} d\,\mathbb{V} = \left.\frac{\mathbb{V}_v}{\mathbb{V}}\right|_{\mathbf{x}}, \tag{1.1.30}$$

with a *deviation* from the average, $\mathring{\gamma}(\boldsymbol{\xi};\mathbf{x})$, defined as:

$$\mathring{\gamma}(\boldsymbol{\xi};\mathbf{x}) = \gamma(\boldsymbol{\xi};\mathbf{x}) - \overline{\gamma}(\mathbf{x}), \tag{1.1.31}$$

where the $\boldsymbol{\xi}$ in $\mathring{\gamma}(\boldsymbol{\xi};\mathbf{x})$ and in $\gamma(\boldsymbol{\xi};\mathbf{x})$ indicates that these values correspond to a point $\boldsymbol{\xi}$ located within the domain Ω centered at $\mathbf{x}$. From (1.1.30) and (1.1.31), and for an Ω_o that denotes an REV domain, it follows that:

$$\overline{\gamma}(\mathbf{x}) = \left.\frac{\mathbb{V}_{ov}}{\mathbb{V}_o}\right|_{\mathbf{x}} \equiv \phi(\mathbf{x}) \quad \text{and} \quad \overline{\mathring{\gamma}}(\mathbf{x}) = 0. \tag{1.1.32}$$

In this way, the porosity, ϕ, has been defined as the *volume average of the indicator function* $\gamma(\boldsymbol{\xi},\mathbf{x})$. Recall that while the property $\gamma(\boldsymbol{\xi},\mathbf{x})$ is a microscopic level property, the porosity, $\phi(\mathbf{x})$, is a macroscopic one.

Let us now introduce the concept of *ergodicity of a stationary random function* (e.g., Yaglom 1965). A stationary random function is said to be *ergodic*, if any statistical characteristic of the function, taken over a sufficiently large domain of its argument in a *single realization*, is an *unbiased and consistent estimate* of the same characteristic over the entire set of possible realizations of the function. An estimate of a population parameter is said to be *unbiased* if its *expected value* is equal to the value of the parameter. An estimate of a parameter is said to be *consistent* if it approaches, probabilistically, the value of the parameter as the sample size increases.

If the characteristic function $\gamma(\boldsymbol{\xi},\mathbf{x})$ defined above possesses the *ergodic property* within the domain Ω_o, of volume $\mathbb{V}_o$, centered at $\mathbf{x}_o$, then:

$$\overline{\gamma}(\mathbf{x}_o) \equiv \frac{1}{\mathbb{V}_o(\mathbf{x}_o)}\int_{\Omega_o(\mathbf{x}_o)} \gamma(\boldsymbol{\xi};\mathbf{x}_o) d\mathbb{V} = \phi(\mathbf{x}_o). \tag{1.1.33}$$

The spatial distribution of the void space within Ω_o can be described by various geometrical characteristics. One family of such characteristics is composed of spatial averages of products of γ-values taken at different points within the REV. For example:

$$\begin{aligned}
&\left.\overline{\mathring{\gamma}(\mathbf{x})\mathring{\gamma}(\mathbf{x}+\mathbf{h})}\right|_{\mathbf{x}_o,\mathbb{V}_o,\mathbf{h}} \\
&\equiv \frac{1}{\mathbb{V}_o}\int_{\Omega_o(\mathbf{x}_o)} [\gamma(\mathbf{x}) - \overline{\gamma}(\mathbf{x}_o)][\gamma(\mathbf{x}+\mathbf{h}) - \overline{\gamma}(\mathbf{x}_o+\mathbf{h})]\, d\mathbb{V} \\
&= \frac{1}{\mathbb{V}_o}\int_{\Omega_o(\mathbf{x}_o)} \gamma(\mathbf{x})\gamma(\mathbf{x}+\mathbf{h})\, d\mathbb{V} - \phi(\mathbf{x}_o,\mathbb{V}_o)\phi(\mathbf{x}_o+\mathbf{h},\mathbb{V}_o),
\end{aligned} \tag{1.1.34}$$

where $\mathbf{h}$ is an oriented distance between any two points within the REV. The average $\overline{\mathring{\gamma}(\mathbf{x})\mathring{\gamma}(\mathbf{x}+\mathbf{h})}$ characterizes the configuration of the void space within Ω_o. A particular case of this average is obtained for $h(\equiv |\mathbf{h}|) = 0$. Then, (1.1.34) reduces to:

$$\overline{\mathring{\gamma}^2}\Big|_{\mathbf{x}_o} = \frac{1}{\Omega_o}\int_{\Omega_o(\mathbf{x}_o)} [\gamma^2(\mathbf{x}) - (\overline{\gamma})^2(\mathbf{x}_o)]\, d\mathbb{V} = \phi(1-\phi)\Big|_{\mathbf{x}_o,\mathbb{V}_o}. \tag{1.1.35}$$

This parameter represents the spread of the void space about its average density expressed by ϕ.

We have still to define the size of the 'porous medium sample' that should be selected in order to represent the porosity *at a point* in a porous medium domain. In other words, 'how large should the $\mathbb{V}_o$-volume be so that ϕ will represent the porosity at a point that serves as the centroid of an REV'?

As emphasized earlier, the selection of the REV size, $\mathbb{V}_o$, should be such that not only the porosity, but *the values of all relevant averaged geometrical characteristics of the microstructure of the porous medium at any point in the porous medium domain should be single valued functions of the location of that point and of time only*, independent of the size of the REV. This requirement can be expressed by:

$$\frac{\partial\phi(\mathbf{x}_o,\mathbb{V})}{\partial\mathbb{V}}\bigg|_{\mathbb{V}=\mathbb{V}_o} \equiv \frac{\partial\overline{\gamma}(\mathbf{x}_o,\mathbb{V})}{\partial\mathbb{V}}\bigg|_{\mathbb{V}=\mathbb{V}_0} = 0, \tag{1.1.36}$$

and:

$$\frac{\partial\overline{\mathring{\gamma}(\mathbf{x})\mathring{\gamma}(\mathbf{x}+\mathbf{h})}\big|_{\mathbf{x}_o,\mathbb{V},\mathbf{h}}}{\partial\mathbb{V}}\Bigg|_{\mathbb{V}=\mathbb{V}_o} = 0. \tag{1.1.37}$$

In principle, for every point $\mathbf{x}_o$ within a given domain, Ω, one can visualize an experiment consisting of a succession of gradually increasing volumes $\mathbb{V}_1 < \mathbb{V}_2 < \mathbb{V}_3, \ldots,$ all centered at $\mathbf{x}_o$, and a concurrent determination of $\overline{\gamma}(\equiv\phi)$ and $\overline{\mathring{\gamma}(\mathbf{x})\mathring{\gamma}(\mathbf{x}+\mathbf{h})}$ for each such volume, hoping that a volume $\mathbb{V} = \mathbb{V}_o$, which satisfies both (1.1.36) and (1.1.37) will be found. After repeating this procedure for determining $\mathbb{V}_o$ at all points $\mathbf{x} \in \Omega$, one can replace the actual porous medium within Ω, by a *model* of a fictitious continuum, provided $\mathbb{V}_o$ is uniform throughout Ω. Obviously, this is an impossible task, since it is impractical to observe *all* points within Ω.

Instead, let us try to arrive at the size of an REV from its relationships with *measurable* macroscopic parameters of the microscopic configuration of the void space (Bachmat and Bear, 1986). To this end, we make use of the indicator function, $\gamma(\mathbf{x})$ which is a *random function of position*, i.e., at any point $\mathbf{x}_p$ in a porous medium domain, Ω, the characteristic function, $\gamma(\mathbf{x}_p)$, is a *random variable* which may attain the values zero or one. We define the probabilities θ and $(1-\theta)$ as:

$$\mathrm{P}(\gamma|_{\mathbf{x}_p} = 1) = \theta|_{\mathbf{x}_p}, \qquad \mathrm{P}(\gamma|_{\mathbf{x}_p} = 0) = 1-\theta|_{\mathbf{x}_p}.$$

Traversing once the domain Ω, we obtain a *non-random* function $\gamma^{(1)}(\mathbf{x})$. We call $\gamma^{(1)}(\mathbf{x})$ a *realization* of $\gamma(\mathbf{x})$. Repeating this process N times, we obtain additional realizations, $\gamma^{(2)}(\mathbf{x}), \gamma^{(3)}(\mathbf{x}), \ldots, \gamma^{(N)}(\mathbf{x})$.

We shall refer to $\gamma(\mathbf{x})$ as a *stationary random function* in Ω if:

(a) The expected value of γ, given $\mathbf{x}$, is such that:

$$E[\gamma(\mathbf{x})] = \theta = \text{const.}$$

(b) The covariance of γ-values at any two points, $\mathbf{x}_p$ and $\mathbf{x}_q$, is such that

$$\text{Cov}[\gamma(\mathbf{x}_p), \gamma(\mathbf{x}_q)] \equiv E\{[\gamma(\mathbf{x}_p) - \theta][\gamma(\mathbf{x}_q) - \theta]\} = f(\mathbf{h}_{pq}),$$

for all $\mathbf{x} \in \Omega$, where $\mathbf{h}_{pq} = \mathbf{x}_p - \mathbf{x}_q$ is the oriented distance between points $\mathbf{x}_p$ and $\mathbf{x}_q$.

(c) $\text{Var}[\gamma] = E\{[\gamma(\mathbf{x}) - \theta]^2\} = f(0) = \text{const.}$ This is a consequence of (b).

A domain, Ω, for which (a) and (b) hold, is referred to as *macroscopically homogeneous* with respect to $\gamma(\mathbf{x})$.

We shall refer to Ω as *isotropic* with respect to $\gamma(\mathbf{x})$, if:

$$\text{Cov}_\gamma[\mathbf{h}_{pq}] = \text{Cov}_\gamma[h], \quad h = |\mathbf{h}_{pq}|,$$

i.e., the *correlation function* between the values of γ at different points within Ω depends only on the distance, h, between them, and not on their relative orientation.

Our next objective is to establish a relationship between moments of the indicator function $\gamma(\mathbf{x})$ and the spatial averages of γ over an REV. To this end, we employ the notion of *ergodicity of a stationary random function* (Yaglom 1965) defined earlier in this subsection. If the function $\gamma(\mathbf{x})$ possesses the ergodic property within $\mathbb{V}_o$ centered at $\mathbf{x}_o$, then from (1.1.33) we have:

$$\overline{\gamma}(\mathbf{x}_o) \equiv \frac{1}{\mathbb{V}_o} \int_{\Omega_o} \gamma(\mathbf{x}) d\mathbb{V} = \phi(\mathbf{x}_o) \simeq E(\gamma\big|_{\mathbf{x}_o}) = \theta\big|_{\mathbf{x}_o} \tag{1.1.38}$$

$$\begin{aligned} \overline{\mathring{\gamma}(\mathbf{x})\mathring{\gamma}(\mathbf{x}+\mathbf{h})}\Big|_{\mathbb{V}_o,\mathbf{h}} &= \frac{1}{\mathbb{V}_o} \int_{\Omega_o} \mathring{\gamma}(\mathbf{x})\mathring{\gamma}(\mathbf{x}+\mathbf{h})\, d\mathbb{V} \\ &\simeq \text{Cov}_\gamma(\mathbf{h})\big|_{\mathbf{x}_o} = \text{Var}_\gamma(\mathbf{x}_o)\tau_\gamma(\mathbf{h}) \\ &= \phi(1-\phi)\tau_\gamma(\mathbf{h}), \end{aligned} \tag{1.1.39}$$

where $\text{Cov}_\gamma(\mathbf{h})\big|_{\mathbf{x}_o}$ is the covariance of γ in Ω_o for points spaced an oriented distance $\mathbf{h}$ apart, $\text{Var}_\gamma(\mathbf{x}_o) = \phi(1-\phi)$, the symbol $\tau_\gamma(\mathbf{h})$ denotes the *correlation coefficient* at $\mathbf{x}_o$, between values of γ at points spaced an oriented distance $\mathbf{h}$ apart and $\phi(= \mathbb{V}_{ov}/\mathbb{V}_o)$ is the porosity at $\mathbf{x}_o$, where $\mathbb{V}_{ov}$ denotes the volume of void space within Ω_o.

In fact, the volume $\mathbb{V}_o$ of an REV should be sufficiently large, so that the volumetric averages, e.g., those appearing in (1.1.38) and (1.1.39), can be considered as satisfactory estimates of the relevant population parameters of the void space configuration at $\mathbf{x}_o$, i.e., estimates which are free of errors caused by the size of the sample and its random choice.

In the present case, the sufficient condition for (1.1.38) and (1.1.39) to hold is that $|\int_o^\infty \tau_\gamma(h)\,dh| < \infty$. By definition,

$$\tau_\gamma(0) = 1.$$

As shown by Debye et al. (1957), *for an isotropic porous medium* and for any function $\tau_\gamma(h)$, the relation:

$$\left.\frac{\partial \tau_\gamma}{\partial h}\right|_{h=0} = -\frac{1}{4\Delta_v(1-\phi)} \tag{1.1.40}$$

always holds, where $\Delta_v(= \mathbb{V}_{ov}/S_{vs})$ is the *hydraulic radius* of the void space (of volume $\mathbb{V}_{ov}$ and area of contact, S_{vs}, with the solid). An example of an approximate expression for $\tau_\gamma(h)$ for an isotropic porous medium with a random distribution of void and solid spaces, is given by Debye et al. (1957), in the form:

$$\tau_\gamma(h) \simeq \exp\left\{-\frac{h}{4\Delta_v(1-\phi)}\right\}, \qquad h = |\mathbf{h}|. \tag{1.1.41}$$

From (1.1.41) it follows that $\tau_\gamma(h) \to 0$ as $h \to \infty$, ensuring that the sufficiency condition given above holds.

From the above discussion it follows that a necessary condition for obtaining *non-random estimates* (i.e., ones that are not subject to sampling errors) of the geometrical characteristics of the void space at any point $\mathbf{x}_o$ which serves as a centroid of a *sphere* of volume $\mathbb{V}_o$ and diameter ℓ, is:

$$h_{max} = \ell_{min} \gg \Delta_v. \tag{1.1.42}$$

The magnitude of ℓ_{min} is determined by the chosen accuracy and reliability levels of the parameter estimates. Thus, as a conceptual experiment for estimating the porosity, ϕ, of a porous medium at a point $\mathbf{x}_o$, let the volume $\mathbb{V}_o$ of a cubical REV centered at that point be split into N disjoint elementary subdomains, each of volume $\delta\mathbb{V} = \mathbb{V}_o/N$, such that in each of them one may encounter (more or less) *either solid or void*. The average of γ over the N samples is taken as an estimate, $\hat{\phi}$, of the porosity, ϕ, at $\mathbf{x}_o$, i.e.,

$$\hat{\phi} = \sum_{i=1}^{N} \frac{\gamma_i}{N}\left(= \frac{1}{N\delta\mathbb{V}}\sum_{i=1}^{N}\gamma_i\delta\mathbb{V}\right). \tag{1.1.43}$$

By definition, and by (1.1.38) and (1.1.39), we have:

$$\sigma_{\hat{\phi}}^2 = \frac{1}{N^2}\sum_{p=1}^{N}\sum_{q=1}^{N}\mathrm{Cov}(\gamma_p, \gamma_q) = \frac{\phi(1-\phi)}{N^2}\sum_{p=1}^{N}\sum_{q=1}^{N}\tau_\gamma(h_{pq}), \tag{1.1.44}$$

where $\sigma_{\hat{\phi}}^2$ is the *variance* of the estimate of ϕ, and $h_{pq} = |\mathbf{x}_p - \mathbf{x}_q|$ is the distance between points $\mathbf{x}_p$ and $\mathbf{x}_q$.

Employing (1.1.41) we obtain:

$$\sigma_{\hat{\phi}}^2 = \frac{\phi(1-\phi)}{N^2}\left[N + \sum_{p=1}^{N}\sum_{q=1, p\neq q}^{N}\exp\left\{-\frac{h_{pq}}{4\Delta_v(1-\phi)}\right\}\right]. \tag{1.1.45}$$

From (1.1.45) it follows that if h_{pq} is expressed in units of Δ_v, we have $N = N(\phi, \sigma_{\hat{\phi}}^2)$.

Bear and Bachmat (1991, p. 21) present a figure that shows an example of the relationship $\sigma_{\hat{\phi}}^2 = \sigma_{\hat{\phi}}^2(\phi, N)$, leading, in the case of a cubical REV, to:

$$\ell_{min}^{(\phi)} = \left\{N(\phi, \sigma_{\hat{\phi}})^2\right\}^{1/3} C_\Delta \Delta_v, \tag{1.1.46}$$

where $C_\Delta \equiv \left(\ell_{min}^{(\phi)}/\Delta_v\right)\big|_{N=1}$ is a numerical coefficient. In the above example, this means $\ell_{min}^{\phi} = 20 C_\Delta \Delta_v$, where we have added the superscript (ϕ) to emphasize that we have been considering the porosity, as the macroscopic geometrical characteristic in determining ℓ_{min}.

Following *Chebyschev's inequality* (e.g., Feller 1957), Bear and Bachmat (1991, p. 20) note that the probability that the magnitude of the estimation error exceeds a prescribed level, say ϵ, is bounded from above by:

$$\mathrm{P}(|\hat{\phi} - \phi| \geq \epsilon) < \sigma_{\hat{\phi}}^2/\epsilon^2. \tag{1.1.47}$$

We use the symbol P^* to denote the probability such that $\sigma_{\hat{\phi}}^2/\epsilon^2 = \mathrm{P}^*$. Then, N would represent the smallest number of subdomains of Ω_o which is sufficient to ensure, with a reliability $1 - \mathrm{P}^*$, that the estimation error $|\hat{\phi} - \phi|$ will not exceed ϵ. In the above case, this means, for example, that for $\epsilon = 0.1$, $\mathrm{P}^* = 0.32$. Obviously, any reduction in ϵ and in P^*, in this example, would require a much larger value of N.

In order to determine $\ell_{min}^{(\phi)}$ for a given porous medium by (1.1.46), for a selected value of $\sigma_{\hat{\phi}}^2 = \epsilon^2 \mathrm{P}^*$, one has to make use of a preliminary estimate of ϕ and $C_\Delta \Delta_v$. However, the effect of differences in ϕ on the value of N and, hence on $\ell_{min}^{(\phi)}$ decreases as $\sigma_{\hat{\phi}}^2$ decreases.

The requirement of ergodicity also sets an upper bound on the size of the REV. We require $\ell < \ell_{max}$, where ℓ_{max} is the distance between points in the porous medium domain beyond which the domain of averaging ceases to be *statistically homogeneous* with respect to the moments of $\gamma(\mathbf{x})$.

In reality, the requirement of homogeneity is seldom satisfied, as the macroscopic parameters of the void space geometry usually vary from point to point. This is especially true in geological formations. However, even for a domain that is heterogeneous with respect to these parameters, it is possible to define around every point, a sufficiently small subdomain, within which these parameters may still be considered uniform, *up to a prescribed error level*. The size of such a subdomain around a given point serves as the upper bound for the size of the REV at that point.

In order to determine this upper bound, and following the definitions presented above, for a stationary random function, a domain Ω centered at a point $\mathbf{x}_o$ is called *homogeneous* with respect to the statistical parameters of $\gamma(\mathbf{x})$, if

$$\mathrm{E}_\gamma(\mathbf{x}) \equiv \phi(\mathbf{x}) = \text{const.} = \phi_o,$$

$$\mathrm{Cov}[\gamma(\mathbf{x}+\mathbf{h}), \gamma(\mathbf{x})] = f(\mathbf{h}),$$

i.e., a function of $\mathbf{h}$ only for all $\mathbf{x} \in \Omega$. Then

$$\mathrm{Var}\,\gamma(\mathbf{x}) = f(0) = \text{const.} = \phi_o(1-\phi_o).$$

For a heterogeneous domain, we have $\phi = \phi(\mathbf{x})$. We shall refer to the domain Ω as *approximately homogeneous* (here, with respect to porosity), if within it:

$$\frac{\phi_{max} - \phi_{min}}{\overline{\phi}} \equiv \delta \ll 1, \tag{1.1.48}$$

where ϕ_{max}, ϕ_{min} and $\overline{\phi}$ are the largest, the smallest and the average values of ϕ, respectively, within Ω, and δ (with $0 < \delta \leq 1$) is an arbitrarily selected small number.

For a sufficiently small domain around $\mathbf{x}_o$, any differentiable function, $\phi(\mathbf{x})$, can be approximated by its linear part (Fig. 1.4), i.e.,

$$\hat{\phi}(\mathbf{x}) = \phi_o + (\nabla\,\phi)\big|_{\mathbf{x}_o}\cdot(\mathbf{x}-\mathbf{x}_o), \tag{1.1.49}$$

where $\phi_o = \phi\big|_{\mathbf{x}_o}$.

Introducing the definition $\ell^{(\phi)}_{max} = 2Max|\mathbf{x}-\mathbf{x}_o|$, and employing (1.1.49), Eq. (1.1.48) yields:

$$\ell^{(\phi)}_{max} = \frac{\phi_o}{|\nabla\,\phi|_{\mathbf{x}_o}}\,\hat{\delta}, \tag{1.1.50}$$

where $\hat{\delta} = (\hat{\phi}_{max} - \hat{\phi}_{min})/\overline{\phi}$. The distance $\ell^{(\phi)}_{max}$ (based on porosity) is, thus, the upper limit for the size of the REV at a point $\mathbf{x}_o$, within a porous medium domain at the selected error level.

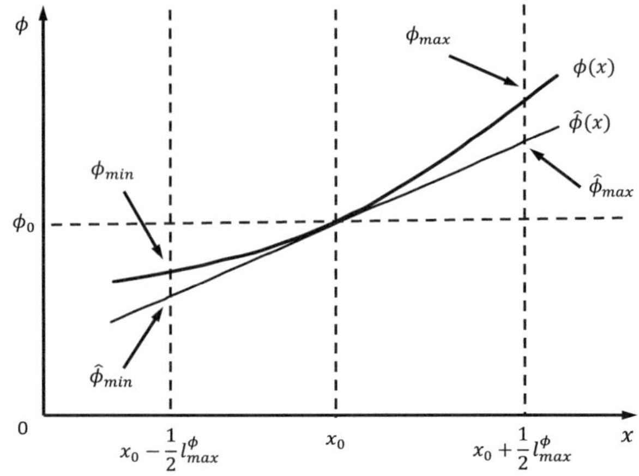

Fig. 1.4 Conceptual determination of $\ell_{max}^{(\phi)}$ by (1.1.50)

Altogether, $\ell^{(\phi)}$ has to satisfy the condition

$$\ell_{min}^{(\phi)} \ll \ell^{(\phi)} \ll \ell_{max}^{(\phi)} \tag{1.1.51}$$

at all points, $\mathbf{x}_o$, of the given domain.

If a non-zero range of $\ell^{(\phi)}$ *can be found, which is common to all points within a given spatial domain, one can adopt the continuum model for the porous medium within that domain.*

Finally, we have to relate $\ell^{(\phi)}$ to the size of the considered domain. If L^* is a characteristic length of the domain, we require that

$$\ell^{(\phi)} \ll L^*, \tag{1.1.52}$$

in order to ensure that the boundary region of the domain, which has a width $\ell^{(\phi)}$, and in which the continuum approach is not applicable (see Sect. 2.7.1), be small compared to the size of the domain itself. The size of the REV in a domain Ω is thus determined by the porosity and the specific surface of the void space in Ω, by prescribed acceptable reliability and error levels in estimating ϕ, by the size of the domain, by the spatial variation of ϕ within the REV and by a prescribed tolerable deviation of ϕ from uniformity within it. If a range satisfying (1.1.51) cannot be found, the domain Ω cannot be represented as a continuum.

So far, the concept and size of the REV have been related to porosity as a geometrical porous medium property. We have indicated this fact by using the superscript (ϕ). Whenever additional geometrical characteristics of the porous medium appear in the macroscopic model in the form of coefficients that are associated with a transport problem, e.g., permeability, a range for the REV has to be determined for each of them. *If a common REV range can be found, a continuum model of the porous medium can be employed.*

Our interest, however, is in the description of transport of extensive quantities in single or multiphase flow, at the continuum level. At such level, the state of each phase is specified by a set of relevant *state variables*, (e.g., density, pressure, temperature). This means that to describe phenomena of transport in porous media, an analogous approach should be undertaken with respect to these variables. A range for REV should be selected for each state variable, following considerations similar to those associated with the geometrical characteristics of the void space. The size, ℓ, of the REV corresponding to each state variable will be bounded by $\ell_{\rm min}$ that depends on the spatial distribution of the microscopic values of that variable within the phase, and on $\ell_{\rm max}$ that depends on the spatial variation of its macroscopic counterpart.

In most cases, we have a number of relevant state variables. Hence, the continuum description of the process involving them can be employed only if a common range of REV can be found for all of them. The same range should also be common with that associated with the configuration of the void space.

An inherent difficulty is that the spatial and temporal variations of state variables vary from one problem to the next. Furthermore, the values of state variables are both time and space dependent, within each case.

Two examples of interest may be mentioned. One is pressure wave propagation. In view of the above considerations, if the length of a pressure wave is smaller than $\ell_{\rm min}$ of the REV of the porous medium (say, it is of the order of magnitude of the pore size, or less), that process of wave propagation in a porous medium cannot be described by means of the continuum approach.

A second example is the spreading of a chemical species within a phase (e.g., a solute) from a point source. In that part of the domain where the size of the REV associated with the species' spatial concentration distribution is smaller than the lower bound of the size of the REV of the porous medium, the continuum approach to the spreading of the species is not applicable.

An interesting case is that of two phase flow (Chap. 6), where the effective-permeability-saturation and the capillary-pressure-saturation relationships constitute *porous medium geometrical properties*. However, both are functions of fluid saturation, say, of the wetting fluid, which is a *variable of state*. Selecting an appropriate REV must take both into account. Joekar-Niasar et al. (2008) investigated such a case. They studied capillary pressure curves, using pore-and-throat network models composed of tubes (representing pore-throats) and spheres (representing pore bodies). They showed that 40 nodes in each direction were needed in order to have a representative network, i.e., an REV.

To summarize, if an REV of volume $\mathbb{V}_o$ can be found which is common to *all* points within a given spatial domain, both for all relevant geometrical characteristics of the void space and for all phases occupying it, and for all relevant state variables, we can define fields of these state variables throughout the domain and treat the latter as a *continuum* for all of them.

The requirement that across the REV, any *macroscopic property* (whether one of the void space, or of a state variable) *should vary linearly, or approximately so, justifies the assignment of the averaged values taken over the REV to the latter's centroid.*

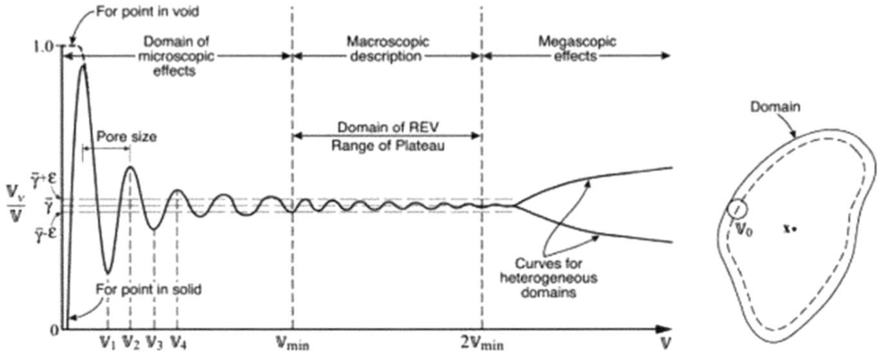

Fig. 1.5 Variation of $\mathbb{V}_v/\mathbb{V}$ in the vicinity of a point, as a function of the averaging volume

To get some feeling of how the size of an REV can be determined in the case of porosity, we envisage a *gedankenexperiment* in which, at an arbitrary point **x** within a given porous medium domain, we consider sequence of gradually increasing volumes, $\mathbb{V}_1 < \mathbb{V}_2 < \mathbb{V}_3 < \ldots$, all centered at **x**, up to a volume which is many times the volume of a (typical) grain or pore. The point **x** may fall in the void space, or in the solid matrix. For each such volume, we calculate the ratio $\mathbb{V}_v/\mathbb{V}$. The results are drawn as Fig. 1.5.

We note that as the volume $\mathbb{V}$ increases, the ratio $\mathbb{V}_v/\mathbb{V}$ fluctuates with a decreasing amplitude. Then, within a certain range of volumes, say, $\mathbb{V}_{min} \leq \mathbb{V} \leq 2\mathbb{V}_{min}$, we may observe a *plateau* in the ratio $\mathbb{V}_v/\mathbb{V}$, where '*plateau*' means a range of $\mathbb{V}$ within which $\mathbb{V}_v/\mathbb{V}$ fluctuates in within a specified small range, $\pm\epsilon$.

We then repeat this experiment for all points **x** within the considered domain. If (1) such a plateau exists *for all points* within the considered domain, and (2) a common range of volumes can be found for all points in the domain for which the $\bar{\gamma} - \epsilon \leq \mathbb{V}_v/\mathbb{V} \leq \bar{\gamma} + \epsilon$, we may choose a volume, $\mathbb{V}_o$, within that common range as the volume of the REV of the considered porous medium domain. In Fig. 1.5, the plateau exists between $\mathbb{V}_{min}$ and $2\mathbb{V}_{min}$. The figure also shows what happens when the considered porous medium is heterogenous. As we increase the volume of the examined domain, we may reach another (higher or lower) plateau. Recall that in selecting the size of the REV, we have also to obey condition (1.1.28).

Obviously, the value of $\mathbb{V}_o$ depends also on the selected ϵ. Once we have selected the size of an REV (i.e., $\mathbb{V}_o$), we have $\bar{\gamma} = \phi$.

The above procedure for determining the size of the REV should be repeated (in our *gedankenexperiment*) for all points within the considered domain.

If a common REV can be found for the entire considered domain, then the latter may be regarded as a porous medium, in which the behavior of each phase and chemical species can be described as that of a continuum at the macroscopic level. However, *it is possible that no plateau will be reached for a considered spatial domain occupied by solid and void spaces, i.e., an REV cannot be defined. Then,*

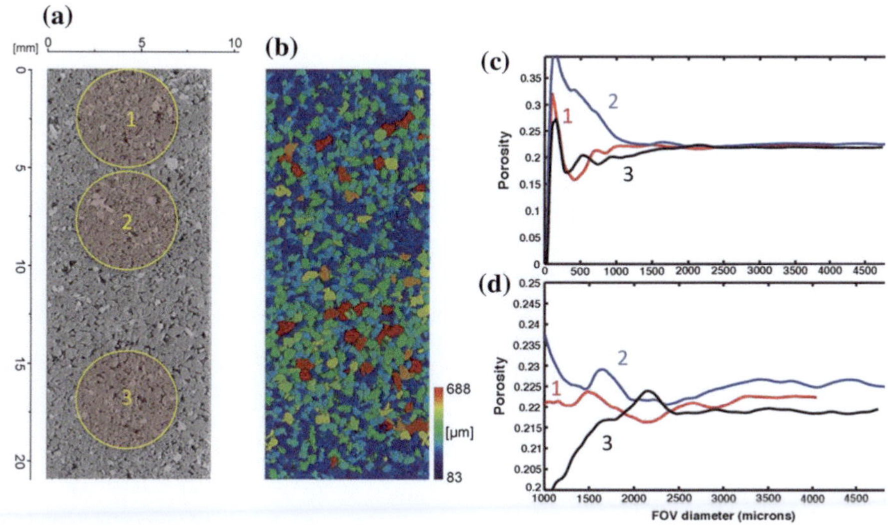

Fig. 1.6 Example of direct porosity REV calculation for a sandstone using synchrotron micro-tomography (Courtesy of Dr. Jonathan Ajo-Franklin, LBNL)

that domain cannot be treated as a porous medium; the continuum approach at the macroscopic level is not applicable to such a domain.

We have determined the size of the REV by making use of the ratio $\mathbb{V}_v/\mathbb{V}_{pm}$ as a characteristic property. However, as we shall see throughout the book, there exist additional fundamental geometrical features such as characteristic pore size, specific interphase area, and void space tortuosity. Transport and storage properties of the phases comprising the porous medium depend on these fundamental features. Hence, the procedure described above should be repeated also for these features, or, at least, for the relevant features for each case, in an effort to determine a common REV for all such features that are relevant to a considered problem. Kjetil and Ringrose (2008) present an interesting article about determining the REV for permeability in Heterolithic deposits.

Figure 1.6 presents an example of determining the size of an REV for a sandstone using synchrotron micro-tomography. Panel A shows a vertical slice of the sandstone sample (Domengine Formation, Antioch, CA) with highlighted regions showing the zones of porosity calculation. Panel B depicts individual grain diameters (color coded) after segmentation and grain shape analysis for the same slice shown in panel A. Panel C shows porosity estimates for different prospective REVs via the expanding sphere approach for three sub-volumes, labeled with reference to the zones shown in panel A. Panel D depicts a zoom of panel C for larger calculation volumes. As can be seen in panels C and D. Porosity estimates reach a plateau (± 0.02) at approximately 2.5 mm, an appropriate REV for porosity in this particular facies. In this case, the mean grain diameter is approximately 300 microns, hence an REV is achieved after averaging over a volume of 8.3 grains. All calculations are performed on 3D volumes.

The example illustrates the application of *microCT* to test continuum rules-of-thumb directly from micro-structure.

Although we have presented above a technique for determining the size of an REV for a given porous medium domain, we wish to emphasize that this actual size is not important. We do not use it anywhere, except that, in principle, we should use it for determining the size of the monitoring device.

Some researchers (e.g., Murdoch and Hassanizadeh 2002) avoided the idea of an REV and averaging over an REV, by defining the average by using an integrable non-negative *weighting function*, $m(\mathbf{x}')$, such that when integrated over the entire space (= convolution) yields 1. Then, they define the averaged (= macroscopic) value $\langle \mathbf{a}\rangle(\mathbf{x})$ at the macroscopic point $\mathbf{x}$, by

$$\langle a\rangle(\mathbf{x}) = \int a(\mathbf{x}+\mathbf{x}')m(\mathbf{x}')d\mathbf{x}'. \qquad (1.1.53)$$

The REV average is then a special case, obtained by an appropriate choice of m. To obtain the porosity, we use $a = \gamma$, with γ defined by (1.1.30). Marle (1981, p. 16) doubted the use of averaging as the basis for developing a theory for flow through porous media.

Geological formations are usually heterogeneous. For example, we often encounter geological formations that are made up of continuous (or discontinuous) layers of different materials, e.g., clay, sand, and gravel. Randomly distributed lenses of sand present in a clay formation, or lenses of clay present in a sandy formation, may serve as additional examples.

In the case of such, 'slightly heterogeneous' domains, an REV can still be identified, possibly with a larger $\mathbb{V}_o$, following the procedure described above. As an example, Fig. 1.7 shows a porous medium domain with three sub-domains. It is possible to seek a plateau, $\mathbb{V} > \mathbb{V}_{min1}$ for Ω_1, $\mathbb{V} > \mathbb{V}_{min2}$ for $\Omega_2, \ldots$, and $\mathbb{V} > \mathbb{V}_{min3}$ for $\Omega_3, \ldots$, following the procedure described above. If a *common plateau* can be found, it is then the common REV for the domain, and the latter can be treated as a continuum, with the selected common REV. Such inhomogeneous porous medium domain may be regarded as a continuum, with a porosity and other relevant coefficients that vary within the considered domain, but always with a common REV of volume $\mathbb{V}_o$. If we use a spherical sampling domain, its diameter must be much larger, say at least 10 times, than the *scale of heterogeneity* at the macroscopic level, say the length of correlation of permeability, and much smaller than the size of the considered domain.

For a *highly heterogeneous domain*, as geological formations often are, it is possible that a common REV cannot be found for the entire domain. One possible conclusion then is that *we cannot treat that domain as a single continuum* to which we apply flow and transport models obtained by averaging, say, over an RMV (= Representative Macroscopic Volume). Other approaches are also possible. An introduction to such approaches is presented in Sect. 7.6. The reason for presenting this material in Chap. 7 is that these approaches have been introduced primarily in connection with modeling solute transport in geological formations.

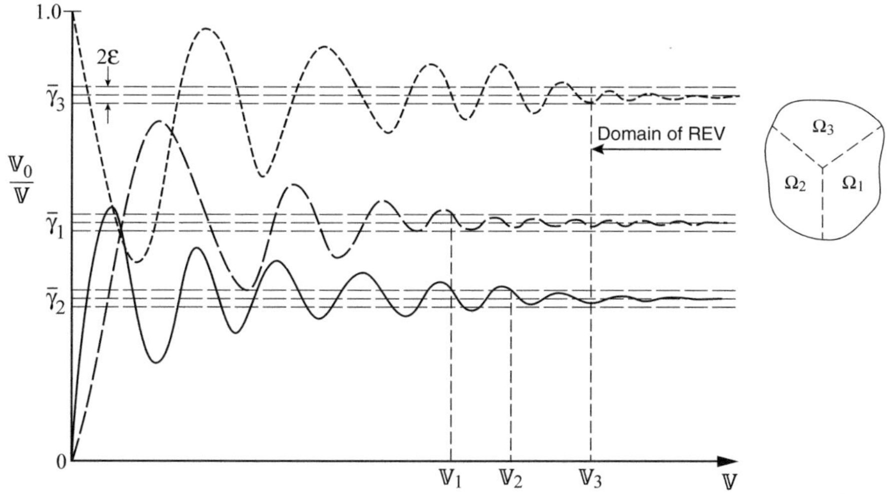

Fig. 1.7 Selecting an REV for a slightly heterogeneous domain

Zhang et al. (2000) presented results of REV based on porosity, permeability, and specific surface area, obtained by using Boltzmann simulations on pore geometries reconstructed from micro-tomographic images. They studied both homogeneous and heterogeneous domains.

So far, we have determined whether a domain may be regarded as a porous medium, and treated as a continuum, on the basis of the void space, or solid matrix configuration. It is *assumed* that the same REV, of volume $\mathbb{V}_o$, can then be used for averaging values of state variables of the considered problem. In principle, we have to make sure that the selected REV is the same also for all other relevant void-space geometrical properties. e.g., specific surface area of void space, hydraulic radius and tortuosity, as these are the building blocks for the macroscopic coefficients, like permeability (see Sect. 4.2.5). In the case of multi-phase flow, we have to refer to the portion of the void space occupied by a wetting and non-wetting fluids, so that we have to consider hydraulic radius, solid-fluid specific surface of each fluid, etc. We then have to require a common REV also at all saturations. It is usually assumed (intuitively!) that the (conceptual) REV based on the porosity as described above is valid for all these coefficients. However, some authors (justifiably!) suggest that 'porosity REV' is considerably different from 'permeability REV', as permeability is much more sensitive to the details of pore structures.

We can summarize the above discussion as follows:

- A domain composed of a solid matrix and a void space can be regarded as a *porous medium domain* for which the *continuum approach* is applicable (i.e., averaged geometrical properties can be associated with *every* point in that domain) if an REV can be found which is common to all points within the domain.
- The size of the REV, should satisfy

$$d \ll l \ll L, \tag{1.1.54}$$

where d is a characteristic size of the void space (e.g., its hydraulic radius), or of the solid matrix (e.g., a characteristic grain size), and L is a characteristic length of the considered domain.

- It is possible that no common REV can be found for a given domain occupied by a solid matrix and a void-space. Then, the considered domain cannot be regarded as a single continuum, and the single continuum approach of modeling is not applicable. Furthermore, in such cases, no information can be obtained on flow coefficients. Another approach, like CTRW (Sect. 7.6.2 B) may be required.

Actually, in practice, we do not really follow the process described above for determining the existence and the size of an REV for a considered investigated domain. In most cases of practical interest, by studying the considered domain, e.g., by studying cores of geological formations, we reach the conclusion about the applicability of the continuum approach. By modeling phenomena of transport at the continuum level, we assume, *implicitly*, that an REV does exist for the investigated domain, as otherwise, a continuum approach cannot be justified. Nevertheless, estimating the size of *the* REV *is important as it dictates the size of the monitoring devices*. We should always make sure that when we measure, or determine the magnitude of a variable, it is really the (averaged) variable that appears in the mathematical model that we solve for predicting the behavior within a given domain. Thus, the REV determines the size of the 'window' of observation or the monitoring instrument.

Once an REV has been identified for a considered porous medium domain, macroscopic (= continuum) models of transport can be written for it in terms of REV-averaged values of the various state variables. Verification of model behavior must be based on monitoring of such values of state variables.

Furthermore, we recall that when performing a formal averaging, the effects of the geometry of the solid-fluid and fluid–fluid interfaces appear in the form of coefficients. Except for cases where such geometry can explicitly be defined, all such coefficients have to be determined *experimentally*, by solving an 'inverse problem' (see, for example, Sun 1999; Sun and Sun 2015), or by 'model calibration'. This means that we obtain model coefficients by requiring that what we measure in the field should be the same values that appear in the (REV-averaged) model.

The various inverse techniques for identifying formation coefficients are based on a comparison between field measured variables of state (e.g., pressure or solute concentration) and their model predicted values, and determining model coefficients such that this difference will be minimal.

It is obvious that with the little available field-measured information, and in view of the strong heterogeneity usually encountered in geological formations, just using interpolation to obtain the missing information is unacceptable. Instead, it is possible, on the basis of the available, albeit meager, information to obtain the *statistical distributions* of the considered coefficients within the domain. For example, for a given geological formation, we do not know the spatial distribution of $k = k(x, y, z)$, but, based on the available information, we can estimate the probability of k, and

Table 1.1 Typical porosities of natural materials

Material	Porosity	Material	Porosity
Soils	0.5–0.6	Gravel	0.3–0.4
Clay	0.45–0.55	Uniform sand	0.3–0.4
Silt	0.4–0.5	Fine to medium mixed sand	0.3–0.35
Shale	0.01–0.1	Gravel and sand	0.3–0.35
Basalt	0.01-0.25	Limestone	0.01–0.1
Sandstone	0.1–0.2	Dolomite	0.001–0.15
Chalk	0.15–0.45	Fractured igneous rock	0.01–0.1
Karst limestone	0.05–0.5	Diatomite earth	0.25–0.65
Bone (cortical)	0.05–0.10	Bone (trabecular)	0.50–0.90

of the other relevant coefficients within the considered domain. This information is then used in *stochastic models*. This kind of modeling is beyond the scope of this book.

1.1.7 Phase Saturation and Solid Matrix Properties

Following are some definitions of commonly used porous medium properties:

A. Porosity, Void Ratio and Sphericity

Porosity, ϕ, *at a point* in a porous medium domain, is defined as the volume of void space per unit volume of porous medium *at that point*,

$$\phi(\mathbf{x}, t) = \frac{\mathbb{V}_{ov}(\mathbf{x}, t)}{\mathbb{V}_o(\mathbf{x})}, \tag{1.1.55}$$

where $\mathbb{V}_o$ and $\mathbb{V}_{ov}$ are the volumes of the REV centered at point $\mathbf{x}$ and of the void space within that REV, respectively. Obviously, this is a macroscopic porous medium property. The porosity depends on the texture and structure of the porous medium. Soil porosity varies over a wide range of values. Table 1.1 gives typical porosity values for a number of natural (geological) materials.

Sometimes, the void space is made up of two portions: an *interconnected* portion, through which a fluid can move, and a *non-interconnected* (or *occluded*) portion. The latter portion is occupied by an immobile fluid. Because, usually, we are interested in the transport of mass of fluid phases, often carrying chemical species, within the void space, unless otherwise specified, we shall use the term 'porosity' as an indication of 'interconnected porosity', i.e., indicating only the interconnected portion of the void space.

Sometimes, the solid matrix is also porous. Grains comprising the solid matrix may sometimes be porous. In such cases, in addition to the 'macro porosity' due to the void space between grains, there exists a 'microporosity' due to void space *within* the grains themselves. *Diamotite earth* is an example of such a medium; it is usually characterized by a very large overall porosity (often >60%). Although the pores comprising the micro-porosity can be tiny, to the extent that they have very low permeability and, consequently, fluid exchange between the former and the latter is limited, they may play an important role because of their very large specific surface area.

In some porous media, the configuration of the interconnected portion of the void space is such that most of the fluid flow takes place through only part of the interconnected void space, with a small fraction of the flow taking place through the remaining *dead-end* (*cul-de-sac*) portion. We often approximate the situation by assuming that in the latter portion of the void space, the fluid is *practically* immobile. This may happen, for example, when pores have the shape of a *dead-end*, or when very small pores, say between very small grains, are mixed with very large ones. The term 'effective porosity' is often used to describe that part of the total void space through which (most of the) flow takes place. However, the volumetric fraction of the dead-end pores may play an important role in solute transport problems.

In certain porous media, the void space appears in the form of 'pores' of two (or more) sizes. A simple example is when the solid matrix is granular, but the grains are porous, with pores that are much smaller than those between the grains. We refer to such case as a 'double porosity' porous medium. In a fractured porous rock (Sect. 1.3), the void space of the fractures and the void space in the rock blocks surrounded by the fractures are also regarded as two porosity and the fractured porous rock is said to exhibit double or dual porosity

In soil mechanics, the term *void ratio*, $\breve{e}$, defined as

$$\breve{e}(\mathbf{x},t)=\frac{\mathbb{V}_{ov}(\mathbf{x},t)}{\mathbb{V}_{os}(\mathbf{x},t)}=\frac{\phi}{1-\phi}, \tag{1.1.56}$$

is used, with $\mathbb{V}_{os}$ denoting the volume of the solid matrix within the REV.

The term sphericity $\mathfrak{s}$ is sometimes used for unconsolidated porous media (i.e., made up of individual particles) indicate how close is the shape of the individual solid particles to that of a sphere:

$$\mathfrak{s}=\frac{\text{Surface area of sphere having the same volume as particle}}{\text{Surface area of particle}}.$$

Obviously, as everywhere in the continuum approach, sphericity at a point means the average value of $\mathfrak{s}$ for all particles in the REV centered at that point.

The *bulk density* of the solid matrix in a porous medium, ρ_b (= mass of the solid per unit volume of porous medium), is defined as

$$\rho_b = \frac{m_{os}}{\mathbb{V}_o} = \rho_s \frac{\mathbb{V}_{os}}{\mathbb{V}_o} = \rho_s(1 - \phi). \tag{1.1.57}$$

The porosity as defined in (1.1.55) may be referred to as 'Eulerian porosity', where the 'Eulerian approach' means that we are focusing on a fixed point, $\mathbf{x}$, and determine the porosity as it changes with time, *at that point*: $\mathbb{V}_v(\mathbf{x}, t) \rightarrow \mathbb{V}_v(\mathbf{x}, t + \Delta t)$. However, according to Coussy (2004, 2007), it is possible to define another kind of porosity – the *Lagrangian porosity*, φ. In the latter case, we consider a point $\boldsymbol{\xi}$ in a porous medium domain. At that point, we have, *initially*, a certain volume of solid matrix, $\mathbb{V}_{os}$ and of void space, $\mathbb{V}_{ov}$. As flow occurs, accompanied by solid matrix deformation, we follow the mass of solid initially within $\mathbb{V}_o$, centered at $\boldsymbol{\xi}|_{t=0}$, and determine the porosity associated with this solid mass *as flow (and deformation) occur*. In this case, as in the usual (Eulerian) definition, porosity at a point is still defined as the pore volume per unit volume of porous medium; however, the Eulerian porosity is defined in the reference state, while the Lagrangian porosity refers to the actual (deformed) state. The Lagrangian definition of porosity is not used in this book.

B. Saturation and Fluid Content

Let a number of fluid phases occupy the entire void space. The quantity of an α-phase at time t at a point $\mathbf{x}$ (i.e., 'within the REV centered at $\mathbf{x}$') can be described by one of the following two definitions:

- **Volumetric fraction**

$$\theta_\alpha(\mathbf{x}, t) = \frac{\text{Volume of } \alpha\text{-fluid in REV}}{\text{Volume of REV}}, \qquad 0 \leq \theta_\alpha \leq \phi, \quad \sum_{(\alpha)} \theta_\alpha = \phi, \tag{1.1.58}$$

In the particular case of water, it is also called *moisture content*,

- **Fluid saturation**

$$S_\alpha(\mathbf{x}, t) = \frac{\text{Volume of} \alpha\text{-fluid in REV}}{\text{Volume of void space in REV}}, \qquad 0 \leq S_\alpha \leq 1, \quad \sum_{(\alpha)} S_\alpha = 1. \tag{1.1.59}$$

In both definitions, the summation is over all fluid phases, present in the void space.

The above definitions are related to each other by

$$\theta_\alpha = \phi S_\alpha, \tag{1.1.60}$$

When the porous medium is inhomogeneous with respect to porosity, or when it undergoes deformation, which alters the porosity, the fluid saturation should be used to describe the quantity of a fluid within the void space at a point. This will enable a separate treatment of changes in porosity.

C. Specific Surface

The specific surface area, $\Sigma_{sv} = S_{sv}/m_s$, where S_{sv} is the total (internal and external) surface area of the solid matrix (= solid-void interface) and m_s is the mass of the solid matrix, is defined as the surface area of the solid matrix per unit mass of soil (e.g., in m^2/g). It is a very important soil characteristic, especially in connection with surface phenomena such as adsorption and ion-exchange. Fine soils, e.g., clay, are characterized by a huge specific surface area.

To estimate Σ_{sv}, consider a soil made up of spherical particles of diameter d. For such spheres, the area per unit mass is given by $6/\rho_s d$, where ρ_s is the mass density of the solid. For a soil composed of a number of fractions of particle sizes, with m_i denoting the mass of solid in the ith fraction, we have

$$\Sigma_{sv} = \frac{6}{\rho_s} \sum_{(i)} \frac{m_i}{m_s} \frac{1}{d_i},$$

For soil particles in the form of platelets $\ell \times \ell \times b$, the specific area is

$$\Sigma_{sv} = \frac{2(\ell + 2b)}{\rho_s \ell b}.$$

For very thin platelets, $\Sigma_{sv} \approx 2/\rho_s b$.

D. Heterogeneity

A porous medium domain is said to be *homogeneous* with respect to some property defined for that domain if the value of that property remains unchanged *at all points* in the domain. Otherwise, the domain is said to be *inhomogeneous*, or *heterogeneous*. When considering phenomena of flow and transport in porous media, permeability and porosity are the two main parameters of interest. Later, we shall introduce the *dispersivity* as another fundamental soil parameter. In most cases, the homogeneity or heterogeneity of a domain is related to these parameters.

Whether a given porous medium domain is characterized as homogeneous or not depends on the *scale of heterogeneity*, in comparison to the scale of the problem domain defined as the length of interest within the context of the problem. The distance between a pumping well and a point of observation, or the length of a plume of contaminants spreading out from a given source serves as examples of problem scale. In the latter case, the length of the plume varies with time and, typically, so does the length of interest. Note that we have not identified the 'length of interest' of a problem with some dimension of the entire domain, say, the thickness of an aquifer, because the length of interest may be much smaller.

In Sect. 1.4.1 B, we shall introduce heterogeneities at the microscopic scale, caused by the presence of voids and solids. There, the grain, or pore size, serves as a typical scale of heterogeneity. The REV is a tool for passing from the microscopic level of description to the macroscopic one. By requiring that the size of the REV be much larger than the scale of microscopic heterogeneity, and much smaller than the length scale of interest, we provide a method for homogenizing these heterogeneities.

Heterogeneities may take different forms. Consider permeability. One form is a gradual change in permeability from point to point within the considered domain. Another form is when the domain may be subdivided by surfaces of discontinuity into well defined subdomains, each having a different, constant or gradually varying, permeability. A layered aquifer may serve as an example. Another example is the case of low permeability lenses (say, silt) embedded in a domain of higher permeability, say sand.

Most subsurface domains are highly heterogeneous. This is the result of the geological processes that have been active over long periods of time to produce the domains that we encounter now.

To solve problems of flow and contaminant transport in an inhomogeneous domain, the detailed variations in the permeability and other relevant parameters within the domain of interest must be known. Unfortunately, such information is rarely available, and its acquisition is very costly. Various approximate approaches are used instead. One approach is to homogenize again, over a new, larger, scale of heterogeneity, thus smoothing out variations at the lower scale. Other approaches involve various forms of stochastic modeling. We shall not discuss this approach in this book, although there is no doubt that heterogeneity may strongly affect the conclusions drawn from modeling efforts. Stochastic approaches to modeling are reviewed and discussed, for example, by Gelhar (1993), Gelhar and Axnes (1983), Dagan (1989), Rubin (2003), and by a number of authors in a volume edited by Dagan and Neuman (1997).

E. Anisotropy

When the value of a porous medium property *at a point* within a domain varies with direction, the domain is said to be *anisotropic* with respect to that property. If the value of the property at a point is independent of direction, the porous medium is said to be *isotropic* at that point, with respect to the considered property. Permeability is an important example of a property that may be anisotropic. Whether a porous medium domain is isotropic or not depends on the geometry of the void space, or of the solid matrix at the microscopic level. Actually, as will be shown in Chaps. 3 and 4, the permeability depends on a fundamental porous medium property, which is the tortuosity of the void space. We shall also encounter anisotropy in the thermal conductivity of a porous medium, and the various coefficients of porous medium elasticity in the case of a deformable porous medium. In the case of natural porous media, anisotropy reflects processes of movement and accumulation of the materials that constitute natural depositional formations.

1.2 Microscopic Level Imaging and Modeling

In the previous section, we have justified the passage from the microscopic level of modeling phenomena of transport in porous media to the macroscopic one noting that we cannot describe the void-solid geometry in a porous medium domain, nor can we predict (by solving a mathematical model) and monitor variables of state, like pressure, at that level. In recent years, this statement is no more true. The objective of this section[1] is to provide a glimpse into the world that makes such monitoring and predictions possible, albeit, still for rather small domains. As such, and interesting and useful as it is, the practice of modeling phenomena of transport in large porous medium domains still requires modeling at the macroscopic level, as discussed in this book.

1.2.1 Objectives of Imaging

The goal of pore-scale imaging is to provide an accurate digital 3D model of the porous media at the smallest scale relevant to the modeled process, e.g., diffusion, advection, etc. A secondary goal, which only recently has become feasible, is to provide a time sequence of images that actually capture the modification in either the solid or fluid geometry, as required for testing a descriptive model. With a static or time variable pore-scale model, an investigator can either extract microscopic geometrical information (e.g., porosity, pore-throat statistics, two-phase saturation), or directly calculate flow properties, including permeability, from pore-scale images by solving the appropriate microscopic governing equations. While a range of techniques has recently become available for achieving these goals, hard x-ray micro-computed tomography (mCT), using either synchrotron or conventional tube sources, is currently the dominant technology for both static and dynamic imaging of opaque porous samples at resolutions from 0.5 microns and sample dimensions in the centimeter range, sufficient for imaging the larger pores in sandstones and soils (Wildenschild and Sheppard 2013; Cnudde and Boone 2013). In mCT, a large number of 2D x-ray projections are acquired and then reconstructed into a 3D image volume, depicting the object's internal x-ray attenuation structure, as a proxy for density and z-number (Stock 2008). Recently, these techniques have been extended to higher time resolutions (>1s), allowing dynamic monitoring of fast hydrologic processes such as *Haines jumps* (Berg et al. 2013; Armstrong et al. 2014). Other available 3D pore scale imaging modalities include *Neutron Tomography* (Strobl et al. 2009), ablative imaging techniques, such as Focused Ion Beam/Scanning Electron Microscopy (FIB-SEM; Holzer and Cantoni 2011), and 3D optical methods for thin samples, e.g., Confocal Microscopy (Fredrich 1999). In all these techniques, beyond data acquisition, the challenges are efficient processing and utilization of the

[1]Contributed by Dr. Jonathan Ajo-Franklin, LBNL.

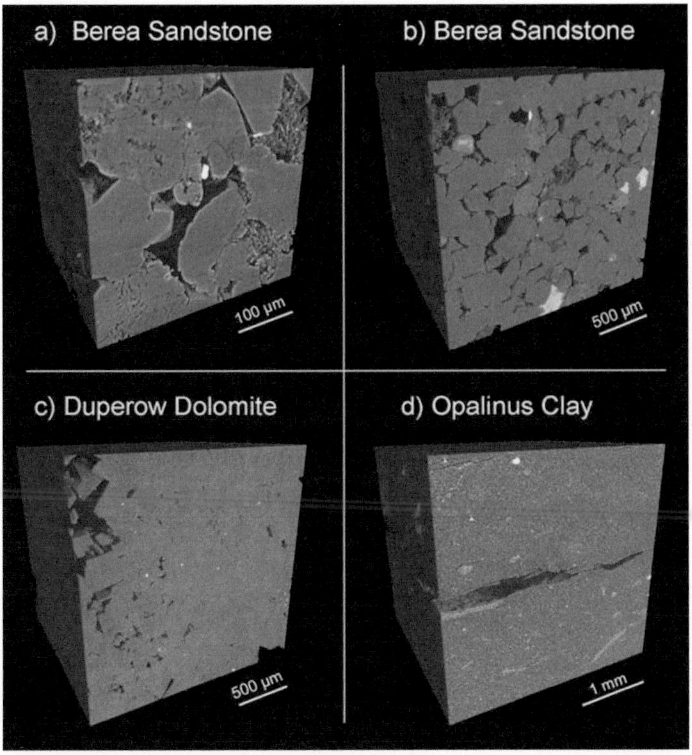

Fig. 1.8 Rock samples scanned at various resolutions: two images of *Berea Sandstone* (**a**, **b**), a sample of the *Duperow Dolomite* (**c**), a fractured *Opalinus Clay* (**d**) (Courtesy of Dr. Jonathan Ajo-Franklin, LBNL)

resulting large high-resolution data sets, particularly conversion into either reduced forms, or modeling of flow processes (e.g. Schluter et al. 2014).

1.2.2 Examples of 3D Imaging of Porous Materials

The volumes of data that result from modern pore-scale imaging approaches capture substantial details of the structure of a geological sample, including pore morphology and topology, grains, mineralogical differences, and fracture geometry. Typical examples are shown in Figure micro-x. All samples were scanned at beamline 8.3.2 (MacDowell et al. 2012) at the Advanced Light Source, a synchrotron facility at Lawrence Berkeley National Laboratory.

Figure 1.8a, b, show two scans at different resolutions of the *Berea sandstone* sample: 325 nm/voxel (a), and 1.3 microns/voxel (b). As can be seen, the resolution in such measurements is generally inversely proportional to the Field of View (FOV);

hence, imaging of finer scale features, such as clay aggregates in panel (a), will sacrifice the extent required for capturing the more extensive pore network. The volume shown in panel (b) is more than sufficient to obtain an estimate of permeability using an appropriate flow modeling code. Texture and particle shape can also be retrieved from 3D image volumes, thus allowing quantitative retrieval of angular pore shapes such as the inter-crystalline porosity shown in the dolomite sample in panel (c).

Finally, the geometry of fractures and fracture networks can be obtained from such image volumes, including fracture aperture maps and near-fracture weathering zones. Panel (d) shows a fracture in a hard Opalinus clay sample of obtained from the *Mont Teri Rock Laboratory*. While the resolution of current mCT beamlines is insufficient to image *single clay crystallites*, or *tactoids*, fractures, *bioclasts*, and mineralogical features such as *pyrite nodules* are often visible in *shales*.

1.2.3 Microscopic Level Modeling

Beyond the 3D characterization of porous media discussed above, the use of microscopic level models to obtain *quantitative* estimates of flow and mechanical properties has rapidly expanded in recent years, driven and supported by the increase in computational performance and in the availability of high-quality data sets. Early models, based on network analysis of pore systems (e.g., Celia et al. 1995; Blunt 2001; Blunt et al. 2013), have gradually given way to direct numerical simulation of flow utilizing a variety of techniques (e.g. Raeini et al. 2014).

At present, reliable modeling of porous medium permeability is routine and modeling multiphase flow processes and mechanics is becoming more common. Recent synthesis of pore-scale numerical flow models with reactive chemistry is also providing a route to examine the non-linear coupling between pore geometry and reaction rate (Molins et al. 2012) at scales approaching 109 grid cells for state-of-the-art high performance computing facilities (Molins et al. 2014). In all of these cases, pore-scale numerical modeling can serve as a testing ground for improving the underlying physics and chemistry included in macroscopic level continuum scale models as well as providing a virtual laboratory to explore the impact of pore-scale alterations on flow processes.

More on microscopic level imaging is presented in Appendix B.

1.3 Soil and Fractured Rock Domains

In Sect. 1.1.2, we have mentioned fresh and saline water aquifers and hydrocarbon reservoirs as major kinds of naturally occurring porous medium domains. These formations, made up of clay, sand, sandstone, cemented sandstone, gravel, etc., occur as geological formations below ground surface, down to depths of thousands of meters. A special kind of such formations includes those that are made of fissured rock, fractured rock and fractured porous rock. This kind of porous medium will be presented in Sect. 1.3.3–1.3.6. Another kind of natural porous medium is *soil*. We use this term to denote the layer that is encountered just below ground surface, down to a depth of a few meters. This is the layer where vegetation roots occur. Thus, this also the zone where irrigation and drainage take place. It is of interest primarily to agronomists, soil physicists, and agriculture engineers. Because of the very wide range of topics involved in modeling phenomena of transport that occur in this soil domain, we shall not deal with it in this book, beyond a short discussion on soil structure presented in the next subsection.

1.3.1 Soil Structure

The physical properties of soil depend to a large extent on the size of soil particles, on the manner in which individual soil particles are arranged and on the strength of particle interactions. Structural features may occur at various scales. It is, especially useful to distinguish between soil microstructure and macrostructure; the former occurs at the level of individual particles, while the latter relates to aggregates of many particles.

Soil microstructure depends on the size and shape of soil particles, as these impose certain constraints on the physical orientation and packing geometry, and on interparticle forces acting between particles. These forces, in turn, control the stability of the microstructure. The total force acting between particles increases with contact area, with coarse-grained soils exhibiting little interparticle attraction, or *cohesion*, unless secondary minerals precipitate at particle contacts, bridging and cementing the grain contacts. Otherwise, such soils are cohesionless and exhibit a microstructure that is largely controlled by particle geometry. It is also strongly influenced by the amount of moisture and historical stresses imposed on the grains due to gravitational forces (overburden weight) and transient external loads (e.g., compaction).

The microstructure of soils containing small fractions of clay minerals is controlled, to a large degree, by the composition of the soil solution. The major cations included in the soil solution are: calcium, magnesium, sodium, and potassium. The major anions are: chloride, sulfate, carbonate, and bicarbonate. Surfaces of amorphous oxides and edges of *phyllosilicate minerals* have the capacity to exhibit a positive or negative electrical charge, depending on the soil pH and ionic strength. Since most soils contain at least some phyllosilicate minerals, with a permanent negative charge on planar surfaces, lowering the pH to a value below the *zero point of charge* (ZPC) for mineral edges, or for *amorphous oxides*, will produce electrostatic binding between negatively charged planar surfaces and positively charged ones. Thus, clay minerals may be bound together in clumps, sometimes referred to as *floccules*. The process by which particles group into floccules is referred to as *flocculation*. The opposite condition, in which particles tend to repel each other and maintain their separate identity, is referred to as *soil dispersion* (not to be

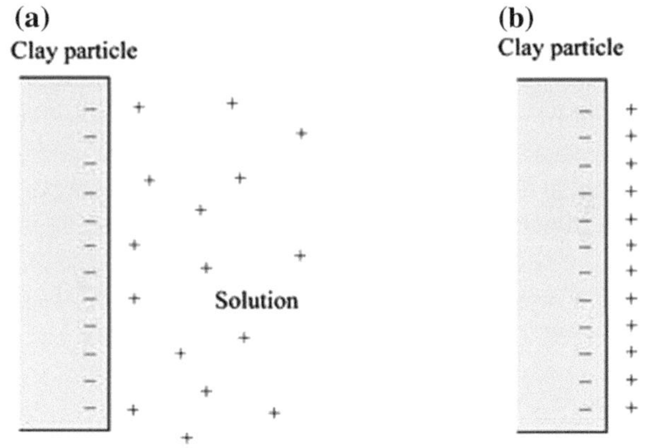

Fig. 1.9 Formation of a diffuse double layer in soil: **a** hydrated state, and **b** dry state

confused with the mechanism of transport by the same name, discussed in Chap. 7). The magnitude of electrostatic binding increases with the strength of the induced positive charge, which, in turn, increases with decreasing pH and increasing ionic strength. Thus, flocculation is usually enhanced by decreasing pH and increasing ionic strength, while soil dispersion has the opposite trends.

The *soil solution* composition also affects flocculation and soil dispersion phenomena through its control of the *electrical double layer*. This term is used to denote a charged surface and the ions associated with it. The surface here is the interface between the solid matrix and liquid occupied void space.

At a dry colloidal surface (say, of clay or humus), the counterions are attached to the surface, thus rendering it neutral. Upon wetting, some of the ions dissociate from the surface and enter the solution. The hydrated colloidal particle thus forms a *micelle* in which the adsorbed ions are spatially separated from the negatively charged particles. The adsorbed ions occur, at least partially, as a diffuse ionic cloud, the concentration of which decreases gradually with increasing distance from the surface. Together, the particle surface, acting as a multiple anion, and the cloud of cations hovering about it, form an *electrostatic double layer* (Fig. 1.9). The above phenomena occur also near clay particles.

The adsorbed ions, in the immediate proximity of the surface are known as the *Stern layer*. The cations in solution are distributed over some distance from the surface, with a concentration decreasing with distance. This distribution is a result of the attraction to the negatively charged ions on the surface, and the spreading due to Brownian motion of the liquid molecules, inducing diffusion of the adsorbed cations away from the surface. Figure 1.10 shows how positive and negative ions are distributed in the vicinity of a clay surface.

The hydrogen atoms in a water molecule are not arranged symmetrically around the oxygen atom. Figure 1.11 shows the structure of water molecule behaves like a rod or *dipole*, with positive and negative charges at its opposite ends. Hence, dipolar water molecules can also be electrically-attracted to the surface of clay particles. In general, water molecules can be electrically attracted to the charged surfaces of

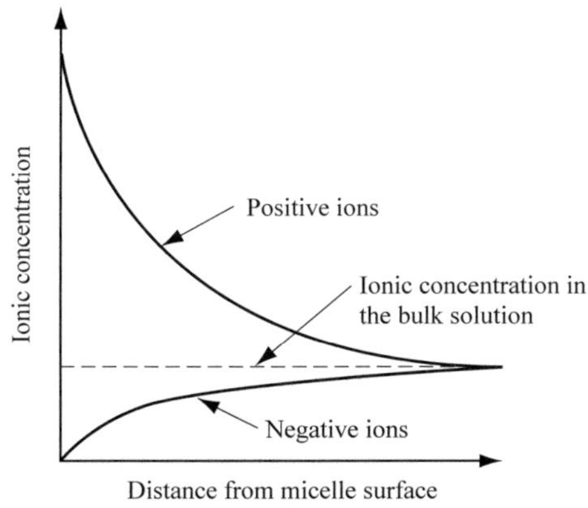

Fig. 1.10 Distribution of ions in the vicinity of a charged clay surface

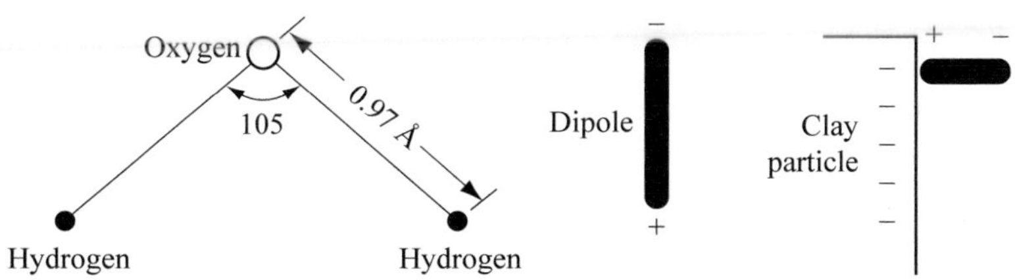

Fig. 1.11 Dipolar structure of water

clay particles by the above mentioned attraction between the negatively charged clay surface. They can also be attracted by attraction forces between the cations in the double layer and the negatively ends of the dipoles, and by the sharing of the hydrogen atom in the water molecules. This sharing produces a hydrogen bonding between the oxygen atoms in the clay and those in the water molecules. This electrically attracted water surrounding clay particles is known as *double-layer water*. The innermost layer of double-layer water, which is very strongly held by the clay particle, is called *adsorbed water*.

If the ion valence or the bulk ionic strength is increased, the thickness of an ionic cloud or a double layer is compressed. Fluids with low dielectric constants can also induce double layer compression (most organic liquids have dielectric constants that are 5–10 time lower than that of water). If the compression is sufficiently large, particles may be able to get close enough to each other, so that short-range attractive *van der Waals* forces can enable flocculation. Thus, high ionic strength and the presence of higher valence cations can promote flocculation due to double layer compression.

These changes in particle interactions at the microscopic scale can have significant effects on the macroscopic behavior of the soil. For example, if the solution composition induces soil dispersion, the pore space may become clogged with loose

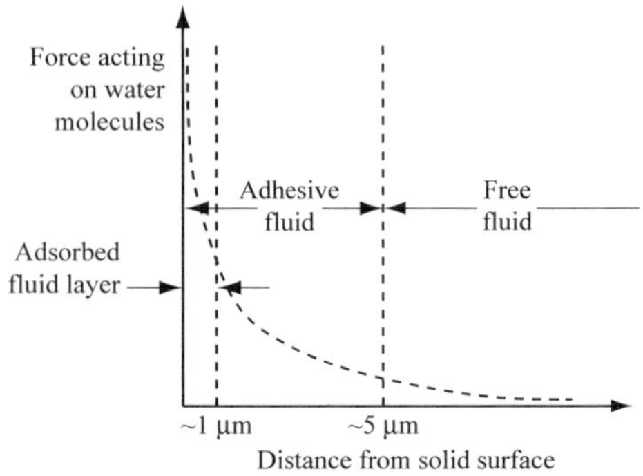

Fig. 1.12 Schematic diagram of adhesive fluid near a solid surface

particles, resulting in a significant reduction in the permeability to fluids. For this reason, permeability may exhibit marked dependence on solution composition, especially for fine-grained soils (and particularly for soils with expansive 2:1 minerals which exhibit a very high surface area). In such soils, increasing the ionic strength or *counterion valence* for aqueous solutions, or replacing the pore fluids by organic liquids, may result in increases in permeability.

Figure 1.12 shows, schematically, the magnitude of the forces of molecular attraction between a solid and a wetting fluid that is adjacent to it. These forces decrease rapidly with the distance from the solid wall. Various explanations have been given for the resulting 'adsorbed water'. One hypothesis is that positively charged ions, which are surrounded by water molecules, are attracted to the mineral surfaces, which, as is well known, are usually negatively charged. Another explanation is that water molecules form hydrogen bonds with clay surfaces, thus facilitating hydrogen bonding between the water molecules and the solid (Low 1961). Other forces that attract water molecules to the solid surface are the *van der Waals forces*, and interactions between the electric field produced by the solid and water dipoles. There is evidence that adsorbed water can have different transport and thermodynamic properties than those of bulk water (Parker 1986). For example, it has been hypothesized that water molecules next to clay particles may have a preferential orientation because of their polar nature and that this orientation can propagate for some significant distance into the fluid (Low 1961). The result is a layer of water, perhaps only a few molecules thick, with thermodynamic properties, such as density and viscosity, which are different from those of the bulk water at the same pressure and temperature (e.g., Low 1976).

The term *adhesive fluid* is used to denote the fluid layer in which the above forces are significant. This fluid layer is made up of two sub-layers. The first, next to the solid, is referred to as an *adsorbed fluid layer*. Its thickness may reach a few tens of molecules. When the fluid is water, its *bipolar* (= *dipolar*) structure (Fig. 1.11) causes water molecules to be oriented perpendicular to the solid surface. In this layer, the properties of the water differ significantly from those of ordinary water.

For example, density and viscosity are much larger. The stress created by the forces of attraction is very strong, but decreases rapidly with distance.

The second sub-layer of adhesive fluid (say, from 0.1 to 0.5 μm for water), forms a transition zone in which the forces of attraction still play a role in making the fluid in this layer relatively *immobile*. Beyond this layer, the fluid is said to be *free*, or *mobile*.

In two-phase flow, molecular forces prevent the complete drainage of the wetting fluid from the void space. A thin *film* of adsorbed wetting fluid will always remain on the solid. As we shall see below, some wetting fluid may also remain in the void space for other reasons.

Soil macrostructure involves features which are visually identifiable. Some macrostructure features may be inherited from the parent material. For example, fractures in the parent rock or fine stratification in sediments may lead to similar relic features in the soil. Additional features may develop by pedogenetic (meaning soil-forming) processes. Repeated wetting and drying, and associated swelling and shrinking, may produce aggregates. These can be stabilized by organic matter which acts as a binding agent, or by thin films of clay, washed from upper layers.

Biotic activities can produce structural features that greatly influence the movement of fluids in soil. Root channels, which remain after roots decay, and holes made by earthworms, ants, and even moles (or other subterranean inhabitants), may produce a complex network of *macropores*.

In addition to the void space between aggregates, a void space may be present *within* (porous) soil particles. The former are, in general, much larger than the latter. The total void space may be represented as a bi-modal pore size distribution. Moisture tends to be retained to a greater extent *within* the aggregates, whereas water moves more rapidly through the larger inter-aggregate voids. This may provide favorable conditions for many plants which need water as well as air for root respiration.

The ability of the soil to absorb and conduct fluids applied at ground surface is greatly affected by soil macrostructure, since the main contribution to the soil's permeability is due to the large inter-aggregate pores. When the latter form a continuous pathway that happens also to be connected to ground surface, liquids applied at the surface can infiltrate very rapidly through them. Stable aggregates at the soil surface that can sustain the impact of raindrops, will prevent the surface from turning into a puddle of mud that could limit infiltration and increase surface runoff. The transport of dissolved chemicals or suspended solids through the soil, is, thus, affected by the soil structure's control of the amount infiltrating water, and by the subsequent rate of movement through the subsurface.

1.3.2 Clay Minerals and Soil Colloids

Clay-size particles exert a disproportionately great influence on the physical and chemical properties of soils, because of their very large *specific surface*. For typical *kaolinite* particles of thickness ≈50 nm, the specific surface is about 15 m^2/g.

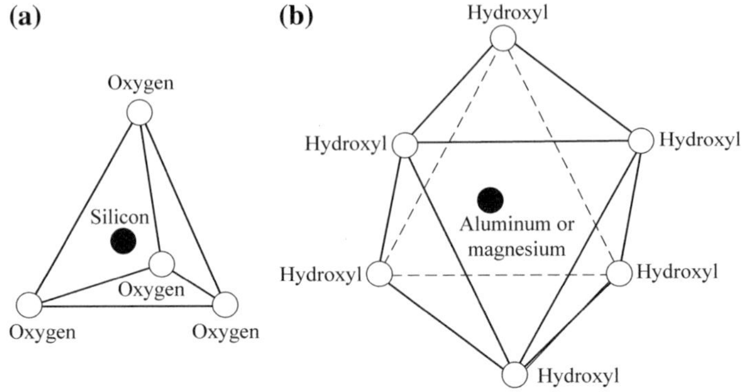

Fig. 1.13 **a** A silicon-oxygen tetrahedral unit, **b** an aluminum or magnesium octahedral unit

The clay-size fraction is, generally, predominantly composed of secondary minerals, formed by the alteration and weathering of primary minerals in the soil. Clay minerals, may be divided into two broad types:

- *Aluminosilicate* or *phyllosilicate* (*phyllo* means sheetlike) *clay* minerals, which are the most prevalent minerals in the clay fraction of temperate region soils.
- *Hydrous oxide* minerals, or hydrated oxides of iron and aluminum, which are prevalent in soils of tropical regions.

The typical *aluminosilicate clay* mineral has a laminated, or sheetlike crystal structure composed of layers of *aluminum hydroxide* octahedral units, and *silica oxide* tetrahedral units in either 2:1 (tetrahedral—octahedral—-tetrahedral), or 1:1 (tetrahedral-octahedral) stacking sequences. The tetrahedral units are composed of six oxygen atoms or hydroxyls, surrounding a larger cation, usually aluminum (Al^{3+}), or magnesium (Mg^{2+}). The octahedral units are composed of four oxygen atoms surrounding a central cation, usually silicon (Si^{4+}). Figure 1.13 shows a silicon-oxygen tetrahedral unit and an aluminum or magnesium octahedral unit.

Expandable 2:1 minerals, such as *montmorillonite*, have very weak bonds between crystal layers. This results in a very high accessible surface area. Water is free to enter between the crystallites, thus causing expansion and contraction as the water content changes. Soils with a large amount of *montmorillonite clay* will shrink and crack when dried, and will swell when wetted. This is a troublesome characteristic for the geotechnical engineer. Since the permeability of such soils depends on the degree of cracking or swelling, it is also a troublesome feature to the hydrologist and soil physicist interested in predicting water movement through such soils,

Other 2:1 minerals include *vermiculite*, *chlorite*, and *hydrous micas*, which have sufficiently strong bonds between crystallites to preclude intercalation of water. As a result, crystals with a large number of individual crystallites will form. These have a lower surface area, and are of a less expansive nature than *montmorillonite*. The most common 1:1 mineral is *kaolinite*. It is built of quite strongly bonded stacks of 1:1 crystallites. As a result, *kaolinite* crystals are relatively large and nonswelling.

The class of *hydrous oxide* clay minerals includes a number of species. Some exhibit very little crystalline structure; they are referred to as *amorphous* (meaning formless) oxides. *Limonite* and *goethite* are common crystalline hydrous iron oxides, while *gibbsite* is a common crystalline aluminum hydroxide. *Iron oxides* may occur as discrete minerals, or as a coating on the surface of other minerals. They produce the red and yellow colors that are typical of many well-oxidized mineral soils.

Clay minerals typically exhibit defects in their crystal structure, involving substitution of Mg^{2+}, Fe^{2+}, or other ions, for Al^{3+} and Fe^{3+}, and substitution of Al^{3+}, or other ions, for Si^{4+}. The latter result in a net negative electric charge on the surface. This surface charge is largely associated with the planar surfaces of clay minerals. On edges of the sheet silicates, and over the entire surface of oxides, a surface charge may also develop. This is due to ionization of polyprotic acidic functional groups, resulting in a charge that varies with the soil's pH. At high pH levels, H^+ ionizes from the surface, leaving an increasingly negative surface charge, while at low pH, OH^- ionizes and may lead to a net positive surface charge. The crossover pH value is called *zero point of charge*. This degree of surface ionization increases with increasing ionic strength, so that at a given pH, the magnitude of surface charge (positive or negative) is greater if the ionic strength is higher.

Surface charge sites arising from either crystal substitution or surface ionization are balanced by ions of opposite charge which are attracted to the surface from the soil solution. Since the ions are held on the surface by relatively weak electrostatic attraction, they may be displaced from the surface by other ions present in the solution. These are called *exchangeable ions*. Some are more strongly attracted than others. The cations can be arranged in a series, in terms of their affinity for attraction to the negatively charged clay surface:

$$Al^{3+} > Ca^{2+} > Mg^{2+} > NH_4^+ > K^+ > H^+ > Na^+ > Li^+.$$

This means, for example, that Al^{3+} ions can replace Ca^{2+} ions, and that Ca^{2+} ions can replace Na^+ ions. For example,

$$2Na^+_{clay} + CaCl_2 \rightarrow Ca^{2+}_{clay} + 2NaCl.$$

The capacity of the solid phase to adsorb cations, on negatively charged surface sites, is referred to as *cation exchange capacity* (CEC). The capacity to retain anions on positively charged sites is referred to as *anion exchange capacity* (AEC).

Due to the large accessible surface area of expansive 2:1 minerals, *montmorillonite* has a rather high CEC, of about 80–100 milliequivalent (meq) per 100 g of soil. *Hydrous mica*, *kaolinite*, and *hydrous oxides* may have CECs of about 15–40, 3–15 and 0–5 meq/100 g. Depending on the amount of ionic substitution in the crystal, the surface area and the pH, substantial variability may occur. For the 1:1 and *hydrous oxide* minerals, the CEC will be more markedly pH-dependent. They may also exhibit a net positive charge resulting in an anion exchange capacity (AEC).

In addition to clay minerals, *colloidal* organic matter may occur in soils in significant quantities, particularly in horizons near ground surface. Colloidal material is composed of very fine (solid) granular particles, ranging in size from a few *nanometers* to a few micrometers (say, less than 10μm). Because of their small size, their gravitational settling velocity is less than 0.01 cm/s, and, usually, under suitable chemical conditions, they remain suspended in water. Under such conditions, thermal or Brownian motion provides enough kinetic energy to overcome gravitational forces which would otherwise cause larger particles to settle. Solid phase colloidal organic matter is often referred to as *humus*. It is the result of biological decomposition of plant and animal remains in the soil. Such material has a very high specific surface, and its CEC, which is strongly pH-dependent, may be as high as 200 meq/100 g. Although the organic matter content of mineral soils is usually only a few percent in the soil close to ground surface, and much less in deeper horizons, it may have a considerable effect on the chemical properties of the soil.

1.3.3 Fractured Domains

In this subsection[2] we shall focus on fractured domains and fractured porous medium domains. The emphasis is on natural geological domains. In recent years, the subject of fractures has been receiving significant attention because of its economic importance in connection with hydrocarbon resources. Man-made fractured domains are also considered here.

As will be emphasized below, in principle, fractured domains are a special kind of porous medium, and the flow and transport through them can be modeled by the same kind of models presented throughout this book. However, because of their importance as petroleum reservoirs, we have decided to devote this special section to their description, with additional sections throughout the book on modeling phenomena of transport through them. More information can be found in extensive literature available on the subject (e.g., Adler and Thovert 1999; Bear et al. 1993).

In this section, we introduce fractured porous medium domains, whether natural or man-made, which contain (1) a network of fractures, and (2) a solid matrix, whether porous or not. We shall focus our attention on *naturally occurring fractured porous media*. They comprise the geological formations which are of interest to a wide spectrum of applications, including groundwater hydrology, petroleum engineering, and geothermal engineering.

A *fracture* is a part of the void space in a porous medium domain that has a special geometry: one of its dimensions is much smaller than the other two. In a consolidated, lithified granular porous medium, in which the grains comprising the solid matrix are usually cemented to each other, a fracture interrupts the grain-to-grain contact in the solid matrix. Fractures do not exist in unconsolidated media. The fracture geometry is significantly different from that of the void space in a granular material; it is also different from that of long continuous pores that look like *wormholes*. Fractures can be caused by a variety of reasons, e.g., accumulation of locally high mechanical or

---

[2]Dr. George J. Moridis of LBNL co-authored Sects. 1.3.3–1.3.6.

thermal stresses that exceed the mechanical strength of the rock, or by the presence of zones of mechanical weakness in the rock. The process of fracture creation also involves (often significant) displacement in which the two faces of a fracture move relative to each other, thus preventing the complete closure of the fracture once the fracture-creating stresses are relaxed. However, when this happens at a later stage, fine materials may be deposited in the fracture, preventing its closure when stresses are relaxed. The distance between the opposite faces of a fracture is referred to as *aperture*.

Fractures are a commonly occurring phenomenon in geological formations. It is not an exaggeration to state that it is difficult to encounter large portions of geological formations that are free of fractures. Fractures are characterized by several geometrical attributes: (1) their extent (reach), (2) their *aperture*, (3) their surface area, (4) their frequency of occurrence in the solid matrix comprising the formation (i.e., the number of fractures per unit length), and (5) their overall porosity (i.e., their volume as a fraction of the bulk volume of the rock). Their hydraulic properties as fluid carrying conduits, whether for water, brine, oil or gas, are distinctly different from those of the porous matrix in which they are imbedded. They often exhibit a high permeability, relative to that of the porous rock surrounding them, thus making them the main fluid pathway. In very tight media, such as shales, they constitute the fluid's dominant pathway.

A rock domain, or a porous medium domain in which part of the void space is in the form of fractures is referred to as a 'fractured rock', or a 'fractured porous medium'. These are more precisely defined in Sect. 1.3.5. We recall that in order to refer to such domains: the void space subdomain, the solid matrix subdomain, and the fractured porous medium domain, as continua, we have to make use of the REV-concept introduced earlier in this chapter. Once these domains are envisioned as a continua, the continuum modeling approach can be applied to them. In whet follows, we shall emphasize the difficulties encountered in taking this route.

As defined above, a naturally occurring *fractured rock domain*, or a man-made one, is a porous domain in which part of the void space has a special geometry–it takes the form of a network of (often-interconnected) *fractures*, or fissures. We use the term *fractured rock* when the rock blocks, surrounded by fractures, contain no void space (disregarding the presence of micro- or nano-pores). We use the term *fractured porous domain*, or fractured porous rock domain, when the domain intersected by the fractures is porous. In reality, almost all fractured media are fractured porous media.

Actually, there is no need for presenting a special discussion on the case of a solid domain, whether natural or artificially produced, which is intersected by a network of fractures. The discussion presented in the previous section will dictate the conditions, mainly the existence of an REV, under which such a domain may be *visualized as a porous medium and treated as a continuum*. The same ideas underlying the passage to a (macroscopic) continuum remain valid also when the pore space takes the form of fractures. In fact we shall apply the methodology of representing a porous medium as multiple overlapping continua also to fractured media and to fractured porous media. Nevertheless, because of certain special features and because of its importance in connection with hydrocarbon reservoirs, we shall present this separate discussion here. Thus, in this chapter, we shall focus on cases in which a porous medium domain is intersected by fractures. We shall discuss model that describe phenomena of transport in such rock domains as occurring in multiple overlapping interacting continua.

Karstic domains constitute another type of geological porous rock formation. A *Karst* is a continuous channel-cave system produced in a soluble rock, e.g., dolomite, limestone, or gypsum, whenever rainwater, which is usually slightly acidic, infiltrates through the soil layer below ground surface and continues to percolate through the rock formation towards a drainage base (e.g., Ten Dam and Erentz 2011). Cracks and bedding planes serve as conveyers for the percolating water. As the rock (e.g., limestone) dissolves by the acidic water, the cracks and fractures tend to get bigger and wider by dissolution, until, eventually, a drainage system is created. As this underground drainage system develops, the development of additional karst channels is enhanced. The interconnected karsts constitute a network of pipe-like channels, dolines and caves that provide a passage for water from the region of infiltration from precipitation to natural water outlets. Although bearing some similarity to fractured rock domains, karst formations will not be discussed in the current chapter.

1.3.4 Natural and Induced Fractures

Fractured domains are anything but rare in geologic systems. Naturally fractured rock reservoirs constitute more than 20% of the world's hydrocarbon reserves (Aguilera 1995). However, since the beginning of the 21 century, following the development of efficient artificial fracturing techniques, production of most of the world's vast *unconventional petroleum resources*, e.g., tight-gas, shale gas, and shale oil reservoirs, has been made economically and technically feasible. This led to an explosion of hydrocarbon production and to a dramatic increase in the estimates of hydrocarbon reserves. In fact, in the U.S.A., tight-sand and shale gas reservoirs are currently the main unconventional resources upon which the bulk of production activity is focusing (Warlick 2006). The production from such resources in the U.S. has skyrocketed, from virtually nil at the beginning of 2000, to 6% of the gas produced in 2005, to 23% in 2010; it is expected to reach 49% by 2035 (The Annual Energy Review (AER) issued by the U.S. Energy Information Administration (EIA), 2011).

Low- and Ultra-Low Permeability Reservoirs (LULPR) are considered unconventional reservoirs. They are best described as oil and/or gas accumulations that are difficult to characterize and commercially produce by conventional exploration and production technologies. Typically, these resources are located in very 'tight', heterogeneous, extremely complex, and often poorly understood geological systems; they are easy to find, but difficult to produce from. The main feature of all such tight and ultra-tight reservoirs is the unavoidable need for well and reservoir *stimulation*. Their matrix permeability is extremely low (often at the nano-Darcy level) such that even in the presence of a system of natural fractures, it cannot support flow at commercially viable rates without permeability enhancement. Such enhancement/stimulation is provided by a number of methods, all of which are designed to develop a new system of artificial fractures. These increase the system's permeability, in addition to increasing the surface area over which reservoir fluids flow from the porous matrix to the induced fractures, thus providing access to larger fluid volumes. Altogether, stimulation techniques are the only means for rendering such resource-rich, but unproductive, natural reservoirs commercially viable entities. Conventional stimulation techniques involve the creation of a system of individual fractures emanating from particular points along a (usually horizontal, except in the case of very thick shale reservoirs) wellbore.

The most common stimulation methods are variants of *hydro-fracturing*, or fracking in which the (almost negligible) incompressibility of water is exploited to deliver a pressure shock that induces rock fracturing stemming from the target point (Sutton et al. 2010; Cipolla et al. 2010). Often, *proppants*, in the form of sand grains, or grains of a hard material (e.g., ceramics), are added to the injected fluid in order to keep the fractures open after their creation.

In recent years, *fracturing* (or "*Fracking*") techniques have been developed that use injected liquid, primarily water ('fracking fluid'), injected at high pressure through wellbores and horizontal wells to produce fractures in gas and petroleum-containing deep rock and shale formations. When the hydraulic pressure is removed from the well, small grains called *proppants*, i.e. small particle materials (e.g., sand or aluminium oxide) that hold the fractures open, prevent hydraulic fractures from closing once the fracturing pressure is removed, The increased permeability enables natural gas and petroleum (and the saline water) saturating the formation to move towards extraction wells.

Acid-fracturing treatments are an alternative to the standard practice of hydraulic fracturing and to the use of proppants. Williams et al. (1979) present a thorough explanation of the fundamentals of acid fracturing. The latter, in the form of '*Fracture-acidizing*' or '*acid-fracturing*', in which the injection results in pressures that exceed the permitted rock fracturing pressure, need to be clearly differentiated from 'fracture acidizing', which occurs at pressures below the fracturing pressure. The main difference between standard proppant-assisted hydraulic fracturing and acid fracturing is the mechanism that creates and maintains enhanced fracture permeability. In acid fracturing, fractures are induced by high-pressure acid injection, and the acid etched channels in the rock. These channels are not kept open by the use of obstacles (proppants) that prevent closure, but through the removal of rock material through chemical dissolution. For obvious reasons, this method is applicable only to rocks that are soluble in the used acids. This method is widely used in low-permeability carbonate reservoirs. However, it offers no advantages in reservoirs of different mineralogy (e.g., sandstone, shale, or coal bed methane reservoirs).

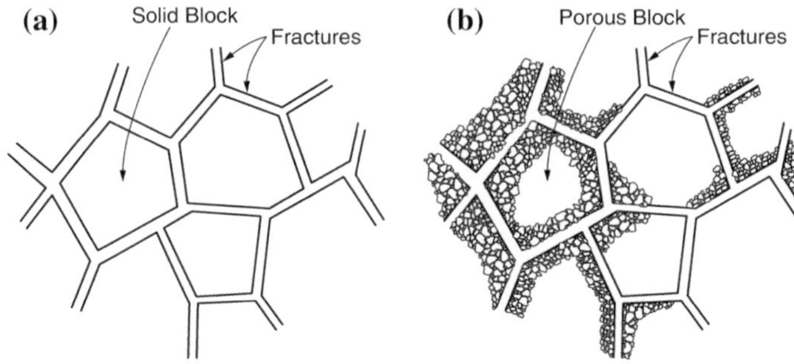

Fig. 1.14 A schematic section of **a** A fractured rock domain, **b** a fractured porous rock (or porous medium) domain

1.3.5 Fractures–Porous Blocks Interactions

When considering a fractured porous rock domain, composed of a subdomain of fractures and another which includes the porous blocks, we encounter cases in which:

- Flow takes place overwhelmingly in the fractures, because of their much higher permeability compared to that of the porous blocks (even when the matrix permeability and porosity are not insignificant). Fluid exchange between the fractures and the porous blocks is negligible. This type of system will be hereafter referred to as *fractured rock*.
- Flow occurs in both the fractures and the porous matrix, and fluid can be exchanged between the two systems. Flow in fractured shale gas and oil reservoirs, and in fractured permeable sandstones, may serve as examples of such cases. This type of system will be hereafter referred to as *fractured porous medium*. Note that in reality practically all geological fractured systems belong in this category, and fractured rock systems are a limiting case that involves but very low rock (matrix) porosity and/or permeability compared to that of the fracture that prevent meaningful fluid exchange between the two.

In both cases, heat exchange by conduction will always occur between the fractures and the porous matrix as long as there is a temperature difference between the two systems, and there is no thermal insulator between them unless the fracture is entirely occupied by gases (which have a very low thermal conductivity). However, advective heat exchange will occur as long as fluid advection is not negligible. These cases will be discussed in Chap. 8.

Figure 1.14a and b show, schematically, segments of a fractured rock domain, and of a fractured porous rock domain, respectively.

Figure 1.15 shows a variety of examples of fractured rocks and fractured porous rocks encountered at or near ground surface or extracted as cores from deep geological formations.

(a) Fractured marl

(b) Fractured granite

(c) Fractured Harzburgite

(d) Franciscan fractured mudstone

(e) Potsdam sandstone

(f) Fractured sandstone

Fig. 1.15 Fractured rocks and fractured porous rock (Courtesy of Dr. Phillipe Gouze, Univ. of Montpellier (**a**–**c**), Dr. George J. Moridis, LBNL (**d**), Dr. Timothy Kneafsey, LBNL (**e**), and the Potsdam Public Museum (**f**))

A verbal description of these figures is presented later in this section. Note the complexity of these real rock systems. Obviously, the only hope for modeling flow and transport in such domains is if we may regard them as continua, which means identifying an REV for each considered rock or porous rock domain.

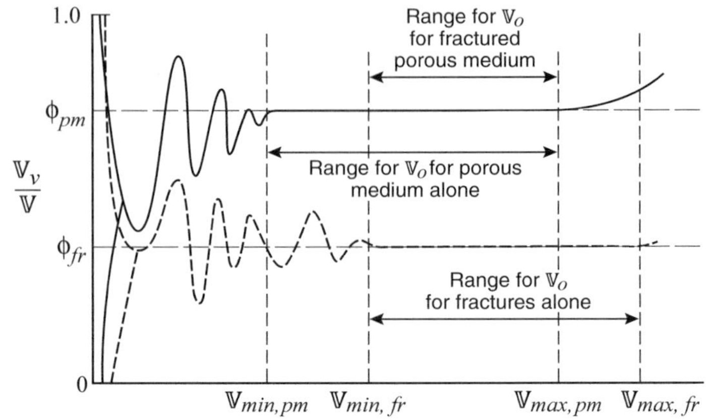

Fig. 1.16 Selecting an REV for a fractured porous medium domain

There is no need to repeat the explanation why we cannot deal with phenomena of transport in fractured domains, and in fractured porous domains, at the microscopic level. Thus, we need to examine the option of envisioning such domains as continua. Following the discussion in Sect. 1.1, the conceptual procedure for examining whether or not a given fractured domain can be regarded as a continuum, involves the possibility of identifying an appropriate REV for such domain. Figure 1.16 explains the selection of an REV for the porous fractured block shown in Fig. 1.14b.

Once an REV has been identified for a fractured rock domain, or for a fractured porous rock domain, we may regard it as a *continuum*. This means that a common REV exists for both the network of fractures and for the porous rock domain. In fact, we have here *four* overlapping continua: (1) *the void space in the domain as a whole* is regarded as a continuum that occupies the entire domain at the domain's porosity, (2) *the solid matrix* is regarded as a continuum that occupies the entire domain, at the solid's porosity (= volume of solid per unit volume of porous medium), (3) *the fractures* constitute a domain that covers the entire domain, and (4) *the void space in the porous blocks* also constitute a continuum that covers the entire considered domain. All these continua *overlap* each other.

Since, in this section, we are considering fluid flow only, we shall assume that the void spaces in both the fractures and in the porous blocks are interconnected, such that we may have (1) fluid mass storage in both domains, (2) fluid flow in both domains, and (3) fluid mass exchange between them. However, in most cases of fractured porous media,

- Most of the fluid storage is in the porous blocks,
- Most of the flow is in the fractures.

In fractured porous rocks, the storage in the matrix may be large, but, because of very small pore size, permeability is very small, so that flow into or out of storage in the matrix is low, and, unless enhanced, may be insignificant for practical purposes. For shale gas reservoir, matrix flow is the limiting factor, hence its importance.

1.3.6 Approaches to the Description of Fractured Media

In this subsection, we shall briefly discuss the more common concepts and approaches to the description of transport in fractured porous media. Note that we consider here not only the transport of one or two fluids, but also of heat.

A. Equivalent Continuum Model (ECM)

The Equivalent Continuum Model (ECM) does not differentiate between the void space in fractures and that in the porous blocks. Instead, it represents the void space in both the fractures and the porous blocks as a *single void space continuum*. This single continuum is characterized by a single set of flow and transport properties (e.g., porosity, permeability, dispersivity) that take into account the contributions of both the fractures and the porous blocks. Furthermore, at any point in this domain and at any time, fluid flow, heat flow and solute transport are represented by a single set of variables of state, such as pressure, phase saturation, solute concentration and temperature. It is an appropriate approximation when studying flow in large domains. However, in small domains, it is incapable of capturing the difference between the fluid behavior in the fractures and in the porous blocks.

Altogether, the ECM method leads to good results when the time scale for flow between fracture and matrix is much smaller than the time scale for flow in the whole domain.

B. Double-Porosity Model

This model, introduced by Barenblatt et al. (1960), and expanded by Warren and Root (1963), represents the first conceptual and corresponding mathematical model of flow through a porous fractured system; it is still widely used in studies and analyses of flow through porous fractured petroleum reservoirs. In this approach, the total void space of the fractured rock system, and hence the associated porosity, is partitioned into (1) a primary porosity, which consists of the interconnected void space within the porous blocks (referred to as 'matrix porosity'), and (2) a secondary void space, consisting of interconnected fractures and joints, referred to as 'fracture porosity'. Each of the two void spaces, and the associated porosities, is treated as a continuum which is characterized by the usual porous medium properties, e.g., permeability, porosity, and compressibility. The two continua overlap each other.

The fractured rock is treated as a porous medium continuum. It has its own properties and characteristics. These are distinctly different from those of the (overlapping) matrix continuum domain.

Figure 1.14 shows a porous fractured rock and its representation as a double porosity domain. In it, the fractured porous rock contains the blocks of porous rock, with every block surrounded on all sides by fractures. Fluid flow and storage take place *within* each of the two interacting domains–fractures and matrix blocks (although limited storage may occur in the fractures).

An identifying characteristic of the double porosity model is that it involves flow both from matrix blocks to fractures and within the system of interconnected fractures. However, there is no direct flow from one matrix block to an adjacent one as these are separated by fractures.

C. Dual Permeability Model

This model is an extension of the double porosity model. It maintains the partitioning of the porous rock domain into two interacting (overlapping) continua: the matrix and the fractures. However, unlike the double porosity model, which allows flow only (1) between the matrix and the fractures, and (2) in the network of interconnected fractures, the*dual-permeability model* permits also inter-block (i.e., matrix-to-matrix) flow. The direct matrix-to-matrix flow indicates that the dual-permeability model is applicable to fractured media in which fracturing is not extensive and/or is limited to one or two general directions.

Doughty (1999) presents a thorough discussion on the double-porosity versus the dual permeability models, as well as a comparison to other conceptual and numerical approaches for evaluating mass and heat transport in fractured media.

D. Triple Porosity Models

The *double-porosity* and *dual-permeability* models described above are limited to cases in which, within each block, the fractures and the matrix have more or less homogeneous porosity and permeability. However, block and fracture properties may still vary from one block to the other.

This model fails when the heterogeneity within each matrix block is so pronounced that it cannot be represented by a single (medium) REV, and a second porous medium with distinct properties has to be defined for the same block. A domain contains two kinds of fractures–e.g., narrow and wide (e.g., large cavities or vugs, or low permeability occlusions) is an example of such a case. Such internal spatial heterogeneity may occur also when the domain contains a single type of fractures, but the matrix is constituted of *two* types of rock with distinctly different sets of properties (e.g., porosity and permeability). An example is a rock matrix with high porosity (say, due to micro-fractures) and a permeability that is in contact with the fractures, on one hand and with another rock matrix type deeper within the rock block (i.e., away from the fractures) that is characterized by lower porosity and permeability. Another case is a rock matrix with regularly-distributed occlusions of very different properties (e.g., pyrite inclusions in shales). Under such conditions of internal heterogeneity, the domain can be represented as composed of *three* overlapping continua: the porous blocks, the major network of fractured, and either a system of narrow fractures in the block or as a fraction of the matrix block with different properties. Fluid (under a pressure difference) and heat (under a temperature difference) can be exchanged between the various combinations of such continua. We refer to a model of this kind as a *triple-porosity model*. It has found numerous applications in petroleum

(a) Large vugs (b) Small vugs

Fig. 1.17 A fractured rock domain with large and small vugs (Courtesy of Dr. George J. Moridis, LBNL)

engineering (e.g., Abdassah and Ershaghis 1986; Al-Ghamdi and Ershaghi 1996; Wu et al. 2004),

E. Fractures and Vugs

Vugs are large voids (cavities) occluded in the porous matrix of geological formations. They are very different (in terms of their geometry, physical appearance and effect on flow behavior) from fractures. They are three-dimensional void spaces that may be small but may reach several meters across and with volumes that can be as large as several tens of cubic meters.

Vugs are created by a variety of causes, e.g., rock dissolution in carbonate formations, intrusion of acidic water, or vesicles created by accumulating and/or escaping gases during the cooling of volcanic magma. Vugs act as local reservoirs of occluded fluid domains, which, in turn, significantly affect the flow behavior in the geological system. They can be isolated within the porous matrix, or can be connected through fractures with neighboring vugs and fractures.

Figure 1.17 shows small and large vugs. When the distance between them is small relative to the considered domain, and there are many of them, such that we may regard the vugs as a continuum of a certain porosity and permeability that can carry fluid and exchange fluid with the surrounding continuum of fractures and/or porous blocks, we can regard them as another layer in the multi-continuum model.

However, when vugs are large, they are usually also widely and irregularly spaced to the extent that we cannot describe them as a continuum. Large vugs, connected to a fracture network, create very complex flow regimes and require different approaches. Flow through vuggy reservoirs has been described by triple- and multiple-porosity models (e.g., Wu et al. 2006), but this remains a challenging problem that tests the limits of the ability to mathematically describe flow in such complex natural domains.

1.4 Scales and Upscaling

We have already introduced the concept of upscaling in Sect. 1.1.3, where we have discussed how we overcome the lack of information on how the solid and fluid phases are distributed throughout a considered porous medium domain. In particular, we have considered space or mass averaging as a tool for overcoming the lack of detailed information about the geometry of interphase boundaries. In fact, we have also mentioned the passage from the molecular distribution of matter to smooth phases by averaging of the molecular behavior. In what follows, we shall extend the concept of upscaling to higher levels.

1.4.1 Scales of Heterogeneity

So far in this book, we have mentioned four levels of describing phenomena:

- **Molecular level** (also referred to as 'nano-level' or 'nano-scale'). At this level we refer to consider molecules and their movement.
- **Nanometric level**. This level contains porous media in which the pores are of nanometer size. Certain shales belong to this group and so are rocks with nanometer size apertures.
- **Microscopic level**. At this level, also referred to as 'pore scale', each considered phase within a domain of interest is regarded as a continuum. We consider what happens at points within a considered phase. Typical distances of interest are μm. The phase may be a solid, a liquid, or a gas. Variables of state are defined for every point within a considered phase. In Sect. 1.1.1 we have introduced this as a definition of a continuum.
- **Macroscopic level**. The entire porous medium domain of interest is regarded as a continuum. Each of the phases present in that domain is also considered as a continuum. Variables of state (= average over an REV) are assigned to *every* point within the considered porous medium domain. Some authors refer to this scale as 'laboratory scale', or 'Darcy scale', although this is the level at which most porous medium domains, including geological formations, are usually modelled. Typical distance at this scale is cm, or m.
- **Megascopic level**. This level, often referred to as 'field scale' or 'formation scale', is used to describe phenomena of transport in geological formations which, typically, are highly heterogeneous. Typical distances of interest are from meters to kilometers. This kind of porous medium domains requires special attention.

The passage from one level of description of phenomena of transport to the next is referred to as 'upscaling'. The reason for upscaling is mainly lack of information about the domain properties, or about the behavior of the phases within a considered domain. At the higher level, the missing information is represented by certain properties, referred to as *coefficients*.

It may be interesting to note that the term 'scale' is often used to mean something different from that described above. Sometimes the 'scale of a problem' is used to express the 'size of the considered geological domain'. In the case of solute transport in the form of an advancing plume, the 'scale of the plume' is an expression often used for the length of the solute plume. Finally, 'scale of heterogeneity' may be used to indicate the distance over which a certain property, e.g., permeability, is still correlated.

Our main interest in this book is modeling at the macroscopic level. Nevertheless, we have added some material also on modeling at the megascopic scale.

Accordingly, we shall start from modeling phenomena of flow and transport at the microscopic level, i.e., the level at which a fluid and a solid are already defined, overlooking their molecular structure, and briefly introduce two techniques, averaging and homogenization, for passing from the microscopic level to the macroscopic one. There is no need to present these approaches in detail as they are described in numerous books and publications (e.g., Bear and Bachmat 1991; Cushman 1997; Whitaker 1999; Selyakov and Kadet 1996). We shall then focus on the phenomenological approach to modeling (directly at the macroscopic level). This approach will be introduced in the present section and employed throughout the book (see Sect. 1.4.4).

Let us add some comments on these four scales, or levels of description:

A. Molecular Level

At this level, say in a liquid or a gas that occupies part of the void space, we note the presence of molecules that are continuously in motion. The *mean travel path* (between collisions) of the moving molecules may serve as a measure of the spatial heterogeneity in the fluid domain.

B. Nanometer Level

Gas or liquid flow, as well as molecular diffusion of dissolved chemical species through shale, tight sandstone formations, or through powder and engineered porous media, may serve as examples of nano-scale transport in porous medium domains. Not all the material presented in this book is valid without modification also for such porous media. The *Klinkenberg effect* (Sect. 4.3.3) my play an important role.

C. Microscopic Level

Consider a fluid moving through the void space, or part of it, in a porous medium domain. We are interested in describing what happens *at points within the fluid*. This is the microscopic level of description. Another example is heat flow and we are interested in the temperature at points inside the solid matrix. In such cases, the heterogeneity of the domain through which the transport occurs is due to the very presence of void space and solid matrix. This heterogeneity is characterized by the average grain size, or the average pore size, or, for non-granular material, by the hydraulic radius of the void space. Typical values of this scale of heterogeneity may be a few *mm*'s (in the case of sand) and a few *cm* in the case of gravel. The material on imaging presented in Sect. 1.2 is applicable.

D. Macroscopic Level

Here, we are interested in the averaged behavior of fluid flow, solute transport, heat transport and stresses in the porous medium domain, visualized as a continuum that has various transport properties. Typically, natural porous medium domains (e.g., aquifers, petroleum reservoirs) are heterogeneous, due to spatial variations in porous medium properties, such as porosity, and permeability. A length characterizing this heterogeneity is the length of correlation of the spatial distribution of a porous medium property, e.g., permeability. When the porous medium domain is homogeneous (say, with respect to permeability), its smallest dimension serves as its characteristic length (e.g., the thickness in an extensive oil reservoir).

E. Megascopic Level

Here, the heterogeneity is of porosity, permeability, and other relevant porous medium transport properties. Most geological formations, especially when our interest is in modeling large domains within such formations, belong to this level. An example of such heterogeneity is that introduced by the existence of layers and lenses of different permeability. In the latter case, the scale (of heterogeneity) is the length of correlation between lenses and layers of different permeability.

In principle, whenever we cannot describe a transport phenomenon (by writing its complete well-posed model) at one level, say because we cannot describe the way properties like permeability and porosity vary, we describe the considered phenomenon at a higher level, obviously, if this is possible. We have already referred to such process as *up-scaling*.

The objective of this book is to model phenomena of flow and transport at the macroscopic level. Because the main interest in upscaling due to porous medium heterogeneity is in connection with solute (or reactive) transport, we shall present more about up-scaling in Sect. 7.2.4 A.

1.4.2 REV Averaging

Volume (or REV) averaging facilitates the passage from the microscopic level to the macroscopic one. In this method we start by writing a transport model of interest at the microscopic level and then, making use of certain *averaging rules*, transform that model to one at the macroscopic level.

A. The Bear–Bachmat Approach

Because the basic balance equation for any extensive quantity includes a sum of terms, products of terms and partial space and time derivatives, the Bear–Bachmat approach (Bachmat and Bear 1964; Bear and Bachmat 1991), like all volume averaging approaches, starts by presenting a number of *averaging rules*. By integrating (or averaging) the microscopic balance equation for any extensive quantity, over an REV, and using these averaging rules, the corresponding macroscopic equation are obtained.

1. Average of a Sum

Let $g_1(\boldsymbol{\xi}, t; \mathbf{x})$ and $g_2(\boldsymbol{\xi}, t; \mathbf{x})$ be two quantities pertinent to a phase, and $\overline{g_1}^\alpha(\mathbf{x}, t)$ and $\overline{g_2}^\alpha(\mathbf{x}, t)$ be their corresponding intrinsic phase averages, respectively. We use $\boldsymbol{\xi}$ to denote a point within a phase, while $\mathbf{x}$ denotes a point at the macroscopic level; it is the centroid of the REV.

$$\frac{1}{\mathbb{V}_{o\alpha}} \int_{\mathbb{V}_o(\mathbf{x})} \{g_1(\boldsymbol{\xi}, t; \mathbf{x}) + g_2(\boldsymbol{\xi}, t; \mathbf{x})\} d\mathbb{V}(\boldsymbol{\xi})$$
$$= \frac{1}{\mathbb{V}_{o\alpha}} \int_{\mathbb{V}_{o\alpha}(\mathbf{x})} g_1(\boldsymbol{\xi}, t; \mathbf{x})\, d\mathbb{V}(\boldsymbol{\xi}) + \frac{1}{\mathbb{V}_{o\alpha}} \int_{\mathbb{V}_{o\alpha}(\mathbf{x})} g_2(\boldsymbol{\xi}, t; \mathbf{x})\, d\mathbb{V}(\boldsymbol{\xi}),$$

from which it follows that

$$\overline{g_1(\boldsymbol{\xi}, t; \mathbf{x}) + g_2(\boldsymbol{\xi}, t; \mathbf{x})}^\alpha = \overline{g_1}^\alpha(\mathbf{x}, t) + \overline{g_2}^\alpha(\mathbf{x}, t). \tag{1.4.1}$$

Recall the comment following (1.1.15) that $g(\boldsymbol{\xi}, t)\, d\mathbb{V}(\boldsymbol{\xi})$ must be physically meaningful.

2. Average of a Product

With deviation from the average defined by (1.1.13), we have:

$$g_1(\boldsymbol{\xi}, t; \mathbf{x}) = \overline{g_1}^\alpha(\mathbf{x}, t) + \mathring{g}_1(\boldsymbol{\xi}, t; \mathbf{x}), \qquad g_2(\boldsymbol{\xi}, t; \mathbf{x}) = \overline{g_2}^\alpha(\mathbf{x}, t) + \mathring{g}_2(\boldsymbol{\xi}, t; \mathbf{x}),$$

and, employing (1.1.14), we obtain (Bear and Bachmat 1991, p. 116),

$$\overline{g_1 g_2}^\alpha(\mathbf{x}) = \overline{g_1}^\alpha(\mathbf{x})\overline{g_2}^\alpha(\mathbf{x}) + \overline{\mathring{g}_1\, \mathring{g}_2}^\alpha(\mathbf{x}), \tag{1.4.2}$$

i.e., *the intrinsic average of a product is equal to the sum of the product of the averages and the average of the product of the deviations.*

3. Average of a Time Derivative

Figure (1.18) shows an REV, Ω_o, of volume $\mathbb{V}_o$, centered at $\mathbf{x}$. It contains a subdomain $\mathbb{V}_{o\alpha}$ of an α-phase; the symbol β denotes the union of all other phases present in $\mathbb{V}_o$. Let $e'(\equiv e'_\alpha)$ denote the density of an extensive quantity E of the α-phase. We assume that e' is differentiable with respect to time, and has no discontinuity within $\mathbb{V}_{o\alpha}$. We

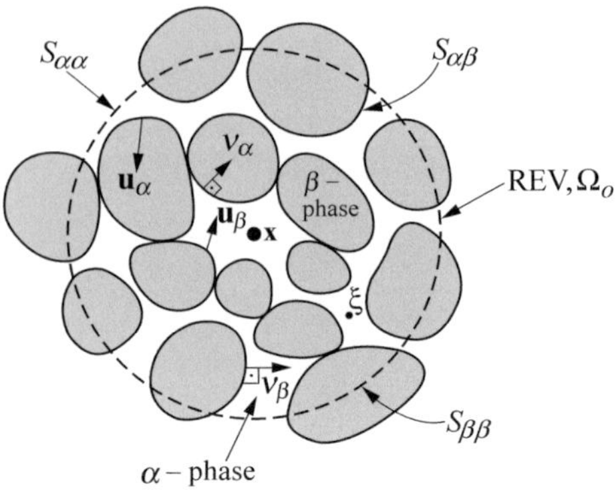

Fig. 1.18 Definition sketch for REV

regard $\mathbb{V}_{o\alpha}$ as a *material volume* with respect to E, so that the *Reynolds Transport Theorem*, (see an appropriate book on mathematics, e.g., Leal 2007; Marsden and Tromba 2003) is applicable to E within the domain $\Omega_{o\alpha}$ at time t. We obtain

$$\frac{D_E}{Dt}\int_{\mathbb{V}_{o\alpha}(t)} e'\,d\mathbb{V} = \int_{\mathbb{V}_{o\alpha}(t)} \frac{\partial e'}{\partial t}\,d\mathbb{V} + \int_{\mathcal{S}_{\alpha\beta}(t)} e'\mathbf{u}\cdot\boldsymbol{\nu}_\alpha\,d\mathcal{S} + \int_{\mathcal{S}_{\alpha\alpha}(t)} e'\mathbf{V}^E\cdot\boldsymbol{\nu}_\alpha\,d\mathcal{S}, \qquad (1.4.3)$$

where $\mathcal{S}_{o\alpha} = \mathcal{S}_{\alpha\alpha} + \mathcal{S}_{\alpha\beta}$ is the total surface bounding $\Omega_{o\alpha}$ (see Fig. 1.18), $\mathbf{u}$ is the velocity at which the $\mathcal{S}_{\alpha\beta}$-surface is being displaced and $\boldsymbol{\nu}_\alpha$ denotes the outward normal unit vector to the α-phase on $\mathcal{S}_{\alpha\beta}$. We recall that $e'd\mathbb{V}$ must be physically meaningful and that because $\mathcal{S}_{\alpha\beta}$ is regarded here as a material surface, we replace on it $\mathbf{V}^E\cdot\boldsymbol{\nu}$ by $\mathbf{u}\cdot\boldsymbol{\nu}$.

On the other hand, we may focus our attention on the material rate of change of E in the α-phase within Ω_o as a whole. We express this rate by:

$$\frac{D_E}{Dt}\int_{\mathbb{V}_o} \gamma_\alpha e'\,d\mathbb{V} = \frac{\partial}{\partial t}\int_{\mathbb{V}_o} \gamma_\alpha e'\,d\mathbb{V} + \int_{\mathcal{S}_o} \gamma_\alpha e'\mathbf{V}^E\cdot\boldsymbol{\nu}\,d\mathcal{S}, \qquad (1.4.4)$$

where γ_α is the *characteristic function* defined in (1.1.9). Here,

$$\gamma(\boldsymbol{\xi}) = \begin{cases} 1 & \text{for } \boldsymbol{\xi} \text{ within } \Omega_{o\alpha}, \\ 0 & \text{for } \boldsymbol{\xi} \text{ outside } \Omega_{o\alpha}. \end{cases} \qquad (1.4.5)$$

In words, (1.4.4) states that the material rate of change of the total quantity of E in the α-phase within Ω_o is equal to the rate of change of E *instantaneously* within Ω_o, plus the net efflux of E leaving Ω_o through its boundary surface $\mathcal{S}_o (= \mathcal{S}_{\alpha\alpha} + \mathcal{S}_{\beta\beta})$; $\boldsymbol{\nu}_\alpha$ is the outward normal unit vector on $\mathcal{S}_o$.

Since e' vanishes outside the phase, (1.4.4) may be rewritten in the form:

$$\frac{D_E}{Dt}\int_{\mathbb{V}_{o\alpha}} e'\,d\mathbb{V} = \frac{\partial}{\partial t}\int_{\mathbb{V}_{o\alpha}} e'\,d\mathbb{V} + \int_{\mathcal{S}_{\alpha\alpha}} e'\mathbf{V}^E\cdot\boldsymbol{\nu}_\alpha\,d\mathcal{S}. \tag{1.4.6}$$

By comparing (1.4.3) with (1.4.6), we obtain:

$$\frac{\partial}{\partial t}\int_{\mathbb{V}_{o\alpha}(t)} e'\,d\mathbb{V} = \int_{\mathbb{V}_{o\alpha}(t)} \frac{\partial e'}{\partial t}\,d\mathbb{V} + \int_{\mathcal{S}_{\alpha\beta}(t)} e'\mathbf{u}\cdot\boldsymbol{\nu}_\alpha\,d\mathcal{S}. \tag{1.4.7}$$

Actually, (1.4.7), known as the *Reynolds Transport Theorem*, may be considered as an extension of *Leibnitz' rule* (of taking the derivative of an integral with respect to a parameter), with the surface integral representing the effect of the rate of displacement of the boundary of integration, $\mathcal{S}_{\alpha\beta}$.

Using the definition (1.1.7) for an intrinsic phase average, we obtain from (1.4.7):

$$\frac{\partial}{\partial t}(\mathbb{V}_{o\alpha}\overline{e'}^\alpha) = \mathbb{V}_{o\alpha}\left(\overline{\frac{\partial e'}{\partial t}}^\alpha\right) + \int_{\mathcal{S}_{\alpha\beta}} e'\mathbf{u}\cdot\boldsymbol{\nu}_\alpha\,d\mathcal{S}. \tag{1.4.8}$$

After dividing by $\mathbb{V}_o$, we obtain:

$$\frac{\partial\theta_\alpha\overline{e'}^\alpha}{\partial t} = \theta_\alpha\left(\overline{\frac{\partial e'}{\partial t}}^\alpha\right) + \frac{1}{\mathbb{V}_o}\int_{\mathcal{S}_{\alpha\beta}} e'\mathbf{u}\cdot\boldsymbol{\nu}_\alpha\,d\mathcal{S}. \tag{1.4.9}$$

which relates the time derivative of an average value to the average of the time derivative, with $\theta_\alpha(\equiv \mathbb{V}_{o\alpha}/\mathbb{V}_o)$ denoting the volumetric fraction of the α-phase.

4. Average of a Spatial Derivative

The averaging rule for a spatial derivative can now be derived by making use of Gauss' theorem (3.2.7), rewritten here in the form:

$$\int_{\mathbb{V}_{o\alpha}} \frac{\partial g_{jkl\ldots}}{\partial x_i}\,d\mathbb{V} = \int_{\mathcal{S}_{\alpha\alpha}} g_{jkl\ldots}\nu_{\alpha i}\,d\mathcal{S} + \int_{\mathcal{S}_{\alpha\beta}} g_{jkl\ldots}\nu_{\alpha i}\;d\mathcal{S}, \tag{1.4.10}$$

where β denotes all non-α phases. We note that:

$$\begin{aligned}\int_{\mathcal{S}_{\alpha\alpha}} g_{jkl\ldots}\nu_{\alpha i}\,d\mathcal{S} &= \int_{\mathcal{S}_o} g_{jkl\ldots}\gamma_\alpha\nu_{\alpha i}\,d\mathcal{S} = \frac{\partial}{\partial x_i}\int_{\mathbb{V}_o} g_{jkl\ldots}\gamma_\alpha\,d\mathbb{V}\\ &= \frac{\partial(\overline{g_{jkl\ldots}}^\alpha\mathbb{V}_{o\alpha})}{\partial x_i} = \mathbb{V}_o\frac{\partial(\theta\overline{g_{jkl\ldots}}^\alpha)}{\partial x_i},\end{aligned} \tag{1.4.11}$$

where γ_α is defined by (1.4.5). By combining (1.4.10) and (1.4.11), we obtain the averaging rule:

$$\theta\overline{\frac{\partial g_{jkl\ldots}}{\partial x_i}}^\alpha = \frac{\partial}{\partial x_i}\theta\overline{g_{jkl\ldots}}^\alpha + \frac{1}{\mathbb{V}_o}\int_{\mathcal{S}_{\alpha\beta}} g_{jkl\ldots}\nu_i\,d\mathcal{S}, \tag{1.4.12}$$

where $g_{jkl...}$ stands for a tensorial quantity of any rank (of the α-phase). Some introductory remarks on tensors are presented in Sect. 9.1 A.

The evaluation of the surface integral, in (1.4.12), requires information on the configuration of the $\mathcal{S}_{\alpha\beta}$-surface and on the distribution of g (or g_α) on it. This information is nothing but the boundary condition of the first kind, i.e., the value of the variable g_α on the boundary, that is on the solid surface bounding the void space. Usually, this information is not available.

Bear and Bachmat (1991, p. 125) consider the special case in which this information is available in the form:

$$\nabla^2 g_\alpha = 0, \quad \text{in } \mathbb{V}_{o\alpha}, \quad \text{and} \quad \frac{1}{\mathcal{S}_{\alpha\alpha}} \int_{\mathcal{S}_{\alpha\alpha}} \frac{\partial g_\alpha}{\partial x_i} d\mathcal{S}_{\alpha\alpha} \approx \frac{\partial \overline{g_\alpha}^\alpha}{\partial x_i}. \tag{1.4.13}$$

They show that under such conditions, the averaging rule (1.4.12) reduces to:

$$\overline{\frac{\partial g_\alpha}{\partial x_j}}^\alpha = \frac{\partial \overline{g_\alpha}^\alpha}{\partial x_i} T^*_{\alpha ij} + \frac{1}{\mathbb{V}_{o\alpha}} \int_{\mathcal{S}_{\alpha\beta}} \mathring{x}_j \frac{\partial g_\alpha}{\partial x_i} \nu_i \, d\mathcal{S}, \tag{1.4.14}$$

where $\mathring{\mathbf{x}} = \boldsymbol{\xi} - \mathbf{x}$, and:

$$T^*_{\alpha ij} = \frac{1}{\mathbb{V}_{o\alpha}} \int_{\mathcal{S}_{\alpha\alpha}} \nu_i \mathring{x}_j \, d\mathcal{S}, \tag{1.4.15}$$

in which $\mathcal{S}_{\alpha\alpha}$ denotes the $\alpha - \alpha$ surface on the external surface of the REV. The subscript α is used in $T^*_{\alpha ij}$, to emphasize that this coefficient depends only on the configuration of the α-phase within $\mathbb{V}_o$.

Bear and Bachmat (1991, p. 126) refer to $T^*_{\alpha ij}$ as the *tortuosity* of the α-phase (thus referring also to multiphase flow within the void-space). The definition in (1.4.15) defines the tortuosity as *the static moment of the oriented elementary surfaces comprising the $\mathcal{S}_{\alpha\alpha}$-surface, with respect to planes passing through the centroid of the REV, per unit volume of the α-phase within $\mathbb{V}_o$.*

Note that the requirement that $\nabla^2 g_\alpha = 0$ limits the discussion to *diffusive* type of flow (e.g., of solute or of heat, but also the case of *creeping fluid flow* governed by $\nabla^2 p_\alpha = 0$, thus also limiting the use of the term tortuosity suggested above. We may refer to $T^*_{\alpha ij}$ as *diffusive tortuosity.*

They consider three cases of boundary conditions on the $\mathcal{S}_{\alpha\beta}$-surface that partly surrounds $\mathbb{V}_{o\alpha}$. The REV contains a fluid α-phase and an s-solid phase.

CASE A: The condition on $\mathcal{S}_{\alpha s}$ is

$$\nabla g_\alpha \cdot \boldsymbol{\nu}_\alpha = 0 \quad \text{on} \quad \mathcal{S}_{\alpha\beta}. \tag{1.4.16}$$

By inserting (1.4.16) into (1.4.14), we obtain

$$\overline{\frac{\partial g_\alpha}{\partial x_j}}^\alpha = \frac{\partial \overline{g_\alpha}^\alpha}{\partial x_i} T^*_{\alpha ij} \tag{1.4.17}$$

as the relationship between the average of a gradient and the gradient of an average for this particular case.

CASE B: The conditions on $\mathcal{S}_{\alpha s}$ are:

$$-(\lambda_\alpha \nabla g_\alpha \cdot \boldsymbol{\nu}_\alpha)\Big|_{\substack{\alpha \text{ side}\\ \text{of } \mathcal{S}_{\alpha s}}} = (\lambda_s \nabla g_s \cdot \boldsymbol{\nu}_s)\Big|_{\substack{s \text{ side}\\ \text{of } \mathcal{S}_{\alpha s}}}, \tag{1.4.18}$$

$$g_\alpha\Big|_{\substack{\alpha \text{ side}\\ \text{of } \mathcal{S}_{\alpha s}}} = g_s\Big|_{\substack{s \text{ side}\\ \text{of } \mathcal{S}_{\alpha s}}}, \tag{1.4.19}$$

where λ_α and λ_β are constant coefficients that depend on the nature of G and on the nature of the α- and s-phases, respectively, and g_s denotes the value of g in $\mathbb{V}_{os}$.

By applying (1.4.14), first to the α-phase, and multiplying the equation by λ_α, then to the s-phase, and multiplying the equation by λ_s, and then adding the two resulting equations, making use of condition (1.4.18), we obtain

$$\lambda_\alpha \theta_\alpha \overline{\frac{\partial g_\alpha}{\partial x_j}}^\alpha + \lambda_s \theta_s \overline{\frac{\partial g_s}{\partial x_j}}^s = \lambda_\alpha \theta_\alpha \frac{\partial \overline{g_\alpha}^\alpha}{\partial x_i} T^*_{\alpha ij} + \lambda_s \theta_s \frac{\partial \overline{g_s}^s}{\partial x_i} T^*_{sij}. \tag{1.4.20}$$

From (1.4.15), we obtain

$$\begin{aligned}\int_{\mathcal{S}_o} \mathring{x}_j \nu_{\alpha i}\, d\mathcal{S} &= \int_{\mathcal{S}_{\alpha\alpha}} \mathring{x}_j \nu_{\alpha i}\, d\mathcal{S} + \int_{\mathcal{S}_{ss}} \mathring{x}_j \nu_{si}\, d\mathcal{S}\\ &= \mathbb{V}_{o\alpha} T^*_{\alpha ji} + \mathbb{V}_{o\beta} T^*_{sji} = \mathbb{V}_o \delta_{ij},\end{aligned} \tag{1.4.21}$$

whence we have a relationship between $T^*_{\alpha ij}$ and T^*_{sij} in the form:

$$\theta_\alpha T^*_{\alpha ji} + \theta_s T^*_{sji} = \delta_{ji}. \tag{1.4.22}$$

Next, we write (1.4.12) twice, once for the α-phase and then for the s-phase, add the two equations and employ condition (1.4.19). We obtain

$$\theta_\alpha \overline{\frac{\partial g_\alpha}{\partial x_j}}^\alpha + \theta_\beta \overline{\frac{\partial g_s}{\partial x_j}}^\beta = \frac{\partial(\theta_\alpha \overline{g_\alpha}^\alpha)}{\partial x_j} + \frac{\partial(\theta_s \overline{g_s}^s)}{\partial x_j}. \tag{1.4.23}$$

Finally, multiplying (1.4.23) by λ_β and subtracting the result from (1.4.20), yields

$$\begin{aligned}\overline{\frac{\partial g_\alpha}{\partial x_j}}^\alpha = \frac{1}{\lambda_\alpha - \lambda_s}\Bigg\{ &\left(\lambda_\alpha \frac{\partial \overline{g_\alpha}^\alpha}{\partial x_i} - \lambda_s \frac{\partial \overline{g_s}^s}{\partial x_i}\right) T^*_{\alpha ij}\\ &- \frac{\lambda_s}{\theta_\alpha} \frac{\partial}{\partial x_j} \theta_\alpha \left(\overline{g_\alpha}^\alpha - \overline{g_s}^\beta\right)\Bigg\},\end{aligned} \tag{1.4.24}$$

where we have employed (1.4.22), together with the relationship $\theta_\alpha + \theta_s = 1$.

In the particular case of $\lambda_s = 0$, Eq. (1.4.24) reduces to (1.4.17), corresponding to *CASE A*. The same holds when $\overline{g_\alpha}^\alpha = \overline{g_s}^s$. We shall later see that the fact that the *averaged* g's are equal, means that *on the average* (but not locally!) there is no exchange between the phases across $\mathcal{S}_{\alpha s}$.

CASE C: Again, a single fluid occupies the entire void space, but the condition on the $\alpha - s$-interface is

$$\nabla g_\alpha \cdot \boldsymbol{\nu}_\alpha = \frac{g_\alpha - \overline{g_\alpha}^\alpha}{\Delta}, \tag{1.4.25}$$

where Δ is a microscopic elementary distance between the $\alpha - s$ surface and the interior of the α-phase occupying $\mathbb{V}_{o\alpha}$. For example, we may view it as proportional to the *hydraulic radius* $\Delta_\alpha = \mathbb{V}_{o\alpha}/\mathcal{S}_{\alpha s}$, such that $\Delta = \Delta_\alpha / C_\alpha$, where C_α is a coefficient that varies with the orientation of elements of the $\mathcal{S}_{\alpha s}$-surface.

By inserting (1.4.25) into (1.4.14), we obtain

$$\overline{\frac{\partial g_\alpha}{\partial x_j}}^\alpha \simeq \frac{\partial \overline{g_\alpha}^\alpha}{\partial x_i} T^*_{\alpha ij} + M_j \frac{\widetilde{g_\alpha}^{\alpha s} - \overline{g_\alpha}^\alpha}{\Delta_\alpha^2}, \tag{1.4.26}$$

where $\widetilde{g_\alpha}^{\alpha s}$ denotes the average of g_α on the $\mathcal{S}_{\alpha s}$-surface. We note that the r.h.s. of the above equation involves two coefficients. The first, $T^*_{\alpha ij}$, is the tortuosity as considered so far, associated with the tortuous pathways, as compared to the macroscopic distance between points. The second,

$$M_j = \frac{1}{\mathcal{S}_{\alpha\beta}} \int_{\mathcal{S}_{\alpha\beta}} C_\alpha \mathring{x}_j \, d\mathcal{S}, \tag{1.4.27}$$

with $\mathring{\mathbf{x}} = \mathbf{x} - \mathbf{x}_o$, is a macroscopic coefficient associated with the configuration of the $\mathcal{S}_{\alpha s}$-surface within the REV. We note that in (1.4.26), we have introduced another macroscopic state variable, viz., $\widetilde{g_\alpha}^{\alpha s}$. Thus, the average of the gradient of g_α depends also on the difference $\widetilde{g_\alpha}^{\alpha s} - \overline{g_\alpha}^\alpha$.

5. The Macroscopic E-Balance Equation

Although the fundamental microscopic balance equation for any extensive quantity, E (density e'), will be discussed only in Chap. 3, we shall present it here for the purpose of demonstrating the Bear–Bachmat REV averaging approach. This equation takes the form of (3.2.12), repeated here as:

$$\frac{\partial e'}{\partial t} = -\nabla \cdot (e' \mathbf{V} + \mathbf{j}^E) + \rho \Gamma^E. \tag{1.4.28}$$

In this equation, $\mathbf{j}^E$ denotes the diffusive flux of E, defined in (3.1.10), and Γ denotes the rate of production of E per unit mass of the considered phase.

According to the Bear–Bachmat approach, to obtain the macroscopic level equation, we start by 'averaging' the microscopic balance equation (1.4.28), i.e., integrate this equation over the volume of the α-phase within the REV, and divide the result by the latter's volume. We obtain

$$\frac{1}{\mathbb{V}_o}\int_{\mathbb{V}_{o\alpha}}\frac{\partial e'}{\partial t}\,d\mathbb{V} = -\frac{1}{\mathbb{V}_o}\int_{\mathbb{V}_{o\alpha}}\nabla\cdot(e'\mathbf{V}+\mathbf{j}^E)\,d\mathbb{V} + \frac{1}{\mathbb{V}_o}\int_{\mathbb{V}_{o\alpha}}\rho\Gamma^E\,d\mathbb{V}. \qquad (1.4.29)$$

By using the averaging rules, the above equation is rewritten in the form:

$$\begin{array}{ccccc} \dfrac{\partial\theta\overline{e'}^{\alpha}}{\partial t} & = -\nabla\cdot\theta\overline{(e'\mathbf{V}+\mathbf{j}^E)}^{\alpha} & - & \frac{1}{\mathbb{V}_o}\int_{\mathcal{S}_{\alpha\beta}} e'(\mathbf{V}-\mathbf{u})\cdot\boldsymbol{\nu}\,dS & \\ \text{(a)} & \text{(b)} & & \text{(c)} & \qquad (1.4.30)\\ & -\frac{1}{\mathbb{V}_o}\int_{\mathcal{S}_{\alpha\beta}}\mathbf{j}^E\cdot\boldsymbol{\nu}\,dS & + & \theta\overline{\rho\Gamma^E}^{\alpha}, & \\ & \text{(d)} & & \text{(e)} & \end{array}$$

in which

(a) Rate of increase of E (in the phase), per unit volume of porous medium.
(b) Net influx of E by averaged advection and diffusion, per unit volume of porous medium.
(c) Amount of E entering the phase, through the interface surface, $\mathcal{S}_{\alpha\beta}$, of the phase within $\mathbb{V}_o$, per unit volume of porous medium and per unit time, by advection with respect to the (possibly moving) $\mathcal{S}_{\alpha\beta}$-surface.
(d) Same as (c), but by diffusion through $\mathcal{S}_{\alpha\beta}$.
(e) Amount of E generated by sources of E *within* $\mathbb{V}_{o\alpha}$, per unit volume of porous medium and per unit time.

By (1.4.2), the (intrinsic-phase-) averaged advective flux, $\overline{e'\mathbf{V}}^{\alpha}$, may be decomposed into two fluxes: a macroscopic advective flux, $\overline{e'}^{\alpha}\overline{\mathbf{V}}^{\alpha}$, and a flux $\overline{\mathring{e}'\mathring{\mathbf{V}}}^{\alpha}$. We shall refer to this second flux as the *dispersive flux* of E (= the amount of E per unit area of phase in the cross section, per unit time).

With these fluxes, (1.4.30) is rewritten in the form:

$$\begin{aligned}\frac{\partial\theta\overline{e'}^{\alpha}}{\partial t} &= -\ \nabla\cdot\theta\left(\overline{e'}^{\alpha}\overline{\mathbf{V}}^{\alpha} + \overline{\mathring{e}'\mathring{\mathbf{V}}}^{\alpha} + \overline{\mathbf{j}^E}^{\alpha}\right)\\ &\quad -\frac{1}{\mathbb{V}_o}\int_{\mathcal{S}_{\alpha\beta}}\left\{e'(\mathbf{V}-\mathbf{u})+\mathbf{j}^E\right\}\cdot\boldsymbol{\nu} + \theta\overline{\rho\Gamma^E}^{\alpha}. \qquad (1.4.31)\end{aligned}$$

Equation (1.4.31) is the *general (macroscopic) differential balance equation* of an extensive quantity, E, of a phase, written in terms of e'.

It is interesting to emphasize two terms that appear in the above macroscopic level E-balance equation, but do not appear in the microscopic level one:

- The term $\overline{\mathring{e}'\mathring{\mathbf{V}}}^{\alpha}$ that expresses the dispersive flux of E,

- The surface integral on the second line of the equation which expresses interphase E-transfer.

The *dispersive flux of* E, $\overline{\mathring{e}'\mathring{\mathbf{V}}}^{\alpha}$, will be discussed in detail in Sect. 3.4.3 and especially in Sect. 7.2.3, in connection with the dispersive flux of a chemical species in an α-fluid phase.

Bear and Bachmat (1991, p. 135) apply (1.4.31) to the particular cases: $E = m, m^{\gamma}, \mathbf{M}$ and $\mathbb{E}$. Because of their interest primarily in ground water hydrology, where ρ is essentially constant, their models are based on the intrinsic phase average, $\overline{e'}^{\alpha}$, and not on the specific value e of E and the mass averaged value, $\widetilde{e}^{\alpha}$, of the specific value e of E, as defined by (1.1.7).

An advantage of the Bear–Bachmat REV averaging approach, as of other averaging approaches, say over the phenomenological approach presented in Sect. 1.4.4, and employed throughout this book, is that it provides a better understanding of the structure of the various porous medium coefficients, albeit on the basis of quite a large number of simplifying assumptions, and their relationships to geometrical features of the solid-void space interface.

B. Whitaker's Approach

Whitaker (e.g., Whitaker 1999) suggested another, slightly different method for volume averaging, or 'spatial smoothing', of transport models. Let us demonstrate Whitaker's approach through the example of solute transport presented by Whitaker (1999, p. 1).

Whitaker demonstrates his approach by considering the case diffusion of a chemical γ-species, of concentration c_{α}^{γ}, within a fluid α-phase that occupies the entire void space of a porous medium domain. The internal surface of that domain is a *catalytic surface*.

The microscopic mass balance equation for a diffusing γ-species in an α-phase, is

$$\frac{\partial c_{\alpha}^{\gamma}}{\partial t} = -\nabla \cdot \mathbf{j}_{dif,\alpha}^{\gamma} + \Gamma_{\alpha}^{\gamma}, \qquad \gamma = 1, 2 \ldots N, \tag{1.4.32}$$

in which, the diffusive γ-flux is expressed by:

$$\mathbf{j}_{dif,\alpha}^{\gamma} (\equiv c_{\alpha}^{\gamma} \mathbf{V}_{\alpha}^{\gamma}) = -\eta_{\alpha} \mathcal{D}_{\alpha}^{\gamma} \nabla X_{\alpha}^{\gamma}, \qquad \frac{1}{\mathcal{D}_{mixture}^{\gamma}} = \sum_{\delta=1}^{\delta=N, \delta \neq \gamma} \frac{X_{\alpha}^{\delta}}{\mathcal{D}^{\gamma\delta}}, \tag{1.4.33}$$

where η_{α} is the molar concentration (in moles per unit volume) of the α-phase, c_{α}^{γ} is the molar concentration of the γ-species in the α-phase, and X_{α}^{γ} (dimensionless) denotes mole fraction, of γ in α, $\mathbf{j}_{dif,\alpha}^{\gamma} \equiv c_{\alpha}^{\gamma} \mathbf{V}_{\alpha}^{\gamma}$ is the diffusive flux of γ in α, with $\mathbf{V}_{\alpha}^{\gamma}$ denoting the velocity of the γ-species in the α-phase; Γ_{α}^{γ} is the molar rate of production of γ by *heterogenous chemical reactions* (in moles per unit volume per unit time). At the $\alpha - s$ - interface, with s denoting the solid, we have the *interfacial flux constitutive relationship*:

$$\mathbf{j}_\alpha^\gamma|_{(\alpha,s)}\cdot\boldsymbol{\nu}_{\alpha,s} = k_1 c_\alpha^\gamma - k_{1'} c_s^\gamma, \tag{1.4.34}$$

in which k_1 and $k_{1'}$ denote *coefficients of adsorption and desorption*, respectively. This condition expresses a linear adsorption isotherm.

Altogether, the mathematical model of a diffusing γ chemical in an α fluid occupying the void space of a porous medium domain consists of the mass balance of the γ-species in the void-space, boundary condition (B.C.) on the $\alpha - s$ surface area and on the fluid portion of the boundary domain, $\mathcal{S}_{\alpha e}$, and initial conditions (I.C.) in the fluid occupying the void space within the domain:

$$\frac{\partial c_\alpha^\gamma}{\partial t} = \nabla\cdot\left(\mathcal{D}^\gamma \nabla c_\alpha^\gamma\right), \qquad \text{in the } \alpha\text{-phase}, \tag{1.4.35}$$

$$\text{I.C.} \qquad c_\alpha^\gamma = \mathcal{F}_1(\mathbf{x}), \quad \text{at } t=0 \text{ in the } \alpha \text{ phase}, \tag{1.4.36}$$

$$\text{B.C.1} \qquad -\mathcal{D}^\gamma \nabla c_\alpha^\gamma \cdot \boldsymbol{\nu}|_{\alpha,s} = k c_\alpha^\gamma, \quad \text{on } \mathcal{S}_{\alpha,s}, \tag{1.4.37}$$

$$\text{B.C.2} \qquad c_\alpha^\gamma = \mathcal{F}_2(\mathbf{x}, t), \text{ on } \mathcal{S}_{\alpha,e}, \tag{1.4.38}$$

where $\mathcal{S}_{\alpha,s}$ denotes the α-solid surface, and $\mathcal{S}_{\alpha,e}$ denotes the fluid–fluid portion on the surface that bounds the REV.

The problem as stated above is presented at the *microscopic level*. Obviously, we cannot solve it as we do not have the information on the detailed configuration of the $\alpha - s$ surfaces.

Like Bear and Bachmat (1991), Whitaker also considers a point $\mathbf{x}$ within the considered porous medium domain, which serves as the center of an REV of volume $\mathbb{V}_o$; $\mathbb{V}_{o\alpha}$ denotes the volume of the α-phase within $\mathbb{V}_o$. Whitaker defines two kinds of averages for c_α^γ (as, obviously, for any variable):

- **Intrinsic phase average concentration of** c_α^γ:

$$\overline{c_\alpha^\gamma}^\alpha(\mathbf{x}, t) = \frac{1}{\mathbb{V}_{o\alpha}(\mathbf{x}, t)} \int_{\mathbb{V}_{o\alpha}(\mathbf{x},t)} c_\alpha^\gamma(\boldsymbol{\xi}, t; \mathbf{x})\, d\mathbb{V}_\alpha(\boldsymbol{\xi}), \tag{1.4.39}$$

i.e., the average $\overline{c_\alpha^\gamma}^\alpha$ is a function of the macroscopic space coordinates, $\mathbf{x}$. Note that $\boldsymbol{\xi}$ is used to denote the location of a point (at the microscopic level) *within* the $\mathbb{V}_{o\alpha}$-domain, while $\mathbf{x}$ is used to denote the centroid of the REV.

- **Volumetric phase average** of c_α^γ:

$$\overline{c_\alpha^\gamma}(\mathbf{x}, t) = \frac{1}{\mathbb{V}_o(\mathbf{x}, t)} \int_{\mathbb{V}_{o\alpha}(\mathbf{x},t)} c_\alpha^\gamma(\boldsymbol{\xi}, t; \mathbf{x})\, d\mathbb{V}_\alpha(\boldsymbol{\xi}). \tag{1.4.40}$$

Here, the total amount of the extensive quantity of the α-phase is averaged over the *entire* volume $\mathbb{V}_o$ of the REV. Since we deal with a property of the α-phase only, the integrations in both (1.4.39) and (1.4.40) are over the sub-domain $\mathbb{V}_{o\alpha}$ only.

From the above two equations, it follows that the two kinds of averages are related to each other by

$$\overline{c_\alpha^\gamma}(\mathbf{x}, t) = \theta_\alpha \overline{c_\alpha^\gamma}^\alpha(\mathbf{x}, t), \tag{1.4.41}$$

where θ_α denotes the volumetric fraction of the α-phase within the REV.

With the above definition, we now average (1.4.35), i.e.,

$$\frac{1}{\mathbb{V}_o}\int_{\mathbb{V}_{o\alpha}} \frac{\partial c_\alpha^\gamma}{\partial t} d\mathbb{V}_\alpha = \frac{1}{\mathbb{V}_o}\int_{\mathbb{V}_{o\alpha}} \nabla\cdot\left(\mathcal{D}^\gamma \nabla c_\alpha^\gamma\right) d\mathbb{V}_\alpha, \tag{1.4.42}$$

In single phase flow in a nondeformable porous medium, $\mathbb{V}_{o\alpha}$ =const. Then, using the Reynolds transport theorem (1.4.7), Whitaker replaces (1.4.42) by

$$\frac{\partial \overline{c_\alpha^\gamma}}{\partial t} = \overline{\nabla\cdot\left(\mathcal{D}_\alpha \nabla c_\alpha^\gamma\right)}, \tag{1.4.43}$$

or, since the solid matrix is assumed to be non-deformable, i.e., $\partial \phi_\alpha \partial t = 0$,

$$\phi_\alpha \frac{\partial \overline{c_\alpha^\gamma}^\alpha}{\partial t} = \overline{\nabla\cdot\left(\mathcal{D}_\alpha \nabla c_\alpha^\gamma\right)}. \tag{1.4.44}$$

where, for single phase flow, $\phi_\alpha \equiv \phi$. We recall that the l.h.s. of the above equation expresses the rate of increase of the γ-species, in moles per unit volume of porous medium. We use this opportunity to mention that a situation for which $\partial(..)/\partial t = 0$ is referred to as a *steady state* situation with respect to (..).

Similar to the steps taken by Bear and Bachmat (1991, see Sect. 1.4.2 A), Whitaker's next step is to introduce 'averaging theorems', i.e., rules that relate the average of (time and space) derivatives to the derivatives of an average. For e_α,

$$\overline{\nabla e_\alpha}(\mathbf{x}, t) = \nabla \overline{e_\alpha}(\mathbf{x}, t) + \frac{1}{\mathbb{V}_o}\int_{\mathcal{S}_{\alpha\beta}} e_\alpha(\mathbf{x}, t; \boldsymbol{\xi})\, \boldsymbol{\nu}\, d\mathcal{S}(\mathbf{x}, t; \boldsymbol{\xi}), \tag{1.4.45}$$

in which $\mathcal{S}$ denotes the area of the interface between the α-phase and all other β-phases within $\mathbb{V}$ and $\boldsymbol{\nu}$ is the unit vector normal to $\mathcal{S}$. For a single fluid phase that occupies the entire void space, $\mathcal{S}_{\alpha\beta} \equiv \mathcal{S}_{\alpha s}$. The above rule is an extension to three dimensions of Leibnitz rule for interchanging differentiation and integration (see discussion leading to (1.4.9) and (1.4.12)).

Rewriting the last equation for a vector variable, say $\mathbf{V}_\alpha$, we obtain:

$$\overline{\nabla\cdot\mathbf{V}_\alpha}(\mathbf{x}, t) = \nabla\cdot\overline{\mathbf{V}_\alpha}(\mathbf{x}, t) + \frac{1}{\mathbb{V}_o}\int_{\mathcal{S}_{\alpha\beta}} \mathbf{V}_\alpha(\mathbf{x}, t; \boldsymbol{\xi})\cdot\boldsymbol{\nu}\, d\mathcal{S}(\mathbf{x}, t; \boldsymbol{\xi}), \tag{1.4.46}$$

With the above rules, the averaged equation (1.4.44) leads to the macroscopic equation:

$$\phi_\alpha \frac{\partial \overline{c_\alpha^\gamma}^\alpha}{\partial t} = \nabla \cdot \overline{\mathcal{D}_\alpha \nabla c_\alpha^\gamma} + \frac{1}{\mathbb{V}_o} \int_{\mathcal{S}_{\alpha\beta}} \mathcal{D}_\alpha \nabla c_\alpha^\gamma \cdot \boldsymbol{\nu}_S \, d\mathcal{S}. \tag{1.4.47}$$

By making use of the boundary condition (1.4.37), we can rewrite the above equation in the form:

$$\phi_\alpha \frac{\partial \overline{c_\alpha^\gamma}^\alpha}{\partial t} = \nabla \cdot \overline{\mathcal{D}_\alpha \nabla c_\alpha^\gamma} - \frac{1}{\mathbb{V}_o} \int_{\mathcal{S}_{\alpha\beta}} k \, c_\alpha^\gamma \, d\mathcal{S}. \tag{1.4.48}$$

We recall that the second term on the r.h.s. represents heterogenous reactions.

Whitaker approximates (1.4.48) by assuming that k is a constant within the REV and that $\mathcal{D}_\alpha$ also remains approximately constant within the REV. These lead to:

$$\phi_\alpha \frac{\partial \overline{c_\alpha^\gamma}^\alpha}{\partial t} = \nabla \cdot \left(\mathcal{D}_\alpha \overline{\nabla c_\alpha^\gamma} \right) - \frac{k}{\mathbb{V}_o} \int_{\mathcal{S}_{\alpha\beta}} c_\alpha^\gamma \, d\mathcal{S}, \tag{1.4.49}$$

which is, eventually approximated as:

$$\phi_\alpha \frac{\partial \overline{c_\alpha^\gamma}^\alpha}{\partial t} = \nabla \cdot \left(\mathcal{D}_\alpha \overline{\nabla c_\alpha^\gamma} \right) - \frac{k}{\mathbb{V}_o} \int_{\mathcal{S}_{\alpha\beta}} c_\alpha^\gamma \, d\mathcal{S}, \tag{1.4.50}$$

or:

$$\phi_\alpha \frac{\partial \overline{c_\alpha^\gamma}^\alpha}{\partial t} = \nabla \cdot \left(\mathcal{D}_\alpha \overline{\nabla c_\alpha^\gamma} \right) - \Sigma_{\alpha s} \, k \, \widetilde{c_\alpha^\gamma}, \tag{1.4.51}$$

$$\widetilde{c_\alpha^\gamma} = \frac{1}{\mathcal{S}_{\alpha,s}} \int_{\mathcal{S}_{\alpha,s}} c_\alpha^\gamma \, d\mathcal{S}_{\alpha,s}. \tag{1.4.52}$$

where $\Sigma_{\alpha s}$ is the specific area of the $\alpha - s$ surface ($= \mathcal{S}_{\alpha,s}/\mathbb{V}$), and $\widetilde{c_\alpha^\gamma}$ denotes the average concentration on $\mathcal{S}_{\alpha,s}$.

Finally, Whitaker rewrites the mass balance of the γ-species in the α-phase in the form:

$$\phi_\alpha \frac{\partial \overline{c_\alpha^\gamma}^\alpha}{\partial t} = \nabla \cdot \left[\mathcal{D}_\alpha \left(\phi_\alpha \nabla \overline{c_\alpha^\gamma}^\alpha + \overline{c_\alpha^\gamma}^\alpha \nabla \phi_\alpha + \frac{1}{\mathbb{V}} \int_{\mathcal{S}_{\alpha\beta}} c_\alpha^\gamma \boldsymbol{\nu}_{\alpha,s} \, d\mathcal{S} \right) \right]$$
$$- \Sigma_{\alpha s} \, k \, \widetilde{c_\alpha^\gamma}. \tag{1.4.53}$$

In this equation, we note that the value of c_α^γ on the interface $\mathcal{S}_{\alpha\beta}$ is not known, as this requires the solution of the problem as stated microscopic level in Eqs. (1.4.35)–(1.4.38). Obviously, a passage for the macroscopic description of the problem is a called for.

Following Gray (1975), Whitaker introduces a separation of length scales in the form of a relationship between the average value of a state variable, say c_α^γ at a *point* *within the α-phase in the REV* and its average value assigned to the REV's center. The former is a *local* (= microscopic value), while the latter is associated with the large length-scale. Written, for example for the concentration c_α^γ, we have:

$$c_\alpha^\gamma = \overline{c_\alpha^\gamma}^\alpha + \overset{\circ}{c}{}_\alpha^\gamma, \tag{1.4.54}$$

in which $\overset{\circ}{c}{}_\alpha^\gamma$ denotes the *spatial deviation concentration*. With (1.4.54), Eq. (1.4.53) can be rewritten as:

$$\phi_\alpha \frac{\partial \overline{c_\alpha^\gamma}^\alpha}{\partial t} = \nabla \cdot \left[\mathcal{D}_\alpha \left(\phi_\alpha \nabla \overline{c_\alpha^\gamma}^\alpha + \overline{c_\alpha^\gamma}^\alpha \nabla \phi_\alpha + \frac{1}{\mathbb{V}} \int_{\mathcal{S}_{\alpha\beta}} \overline{c_\alpha^\gamma}^\alpha \boldsymbol{\nu}_{\alpha,s}\, d\mathcal{S} \right.\right.$$
$$\left.\left. + \frac{1}{\mathbb{V}} \int_{\mathcal{S}_{\alpha\beta}} \overset{\circ}{c}{}_\alpha^\gamma \boldsymbol{\nu}_{\alpha,s}\, d\mathcal{S} \right) \right] - \Sigma_{\alpha s}\, k\, \widetilde{c_\alpha^\gamma}. \tag{1.4.55}$$

Whitaker indicates three difficulties in the above equation: (1) the presence of the volume averaged (= macroscopic) value inside the surface integral, (2) the presence of the surface averaged concentration in the heterogenous reaction rate term, and (3) the presence of the spatial deviation concentration in the above (macroscopic) equation for $\overline{\mathcal{D}_\alpha c_\alpha^\gamma}^\alpha$. Concerning item (1), he develops $\overline{c_\alpha^\gamma}$, which, unlike the average in the Bear and Bachmat approach is not a constant over the REV, in a Taylor power series about the centroid of the REV, and, following an order of magnitude analysis, leads to the conclusion that this term may be eliminated from (1.4.55).

Whitaker develops also a macroscopic level model for convective transport. With $\mathbf{V}_\alpha$ denoting the velocity of the fluid α-phase, the leads to the macroscopic balance equation for the concentration of a γ-species in a fluid α-phase in the form:

$$\phi_\alpha \frac{\partial \overline{c_\alpha^\gamma}^\alpha}{\partial t} = \nabla \cdot \left[\mathcal{D}_\alpha \left(\phi_\alpha \nabla \overline{c_\alpha^\gamma}^\alpha + \overline{c_\alpha^\gamma}^\alpha \nabla \phi_\alpha + \frac{1}{\mathbb{V}} \int_{\mathcal{S}_{\alpha\beta}} \overline{c_\alpha^\gamma}^\alpha \boldsymbol{\nu}_{\alpha,s}\, d\mathcal{S} \right.\right.$$
$$\left.\left. + \frac{1}{\mathbb{V}} \int_{\mathcal{S}_{\alpha\beta}} \overset{\circ}{c}{}_\alpha^\gamma \boldsymbol{\nu}_{\alpha,s}\, d\mathcal{S} \right) \right] - \Sigma_{\alpha s}\, k\, \widetilde{c_\alpha^\gamma}. \tag{1.4.56}$$

Once Whitaker has reached (1.4.55), his next step is to solve the *closure problem* that will lead to a closed form of this equation. This means deriving a closed form for the spatial deviation concentration $\overset{\circ}{c}{}_\alpha^\gamma$. He derives this form by solving the boundary value problem for the spatial deviation concentration. This is typical of the approach suggested by him.

The analysis referred to here is based on a comparison between two characteristic length scales: that of the averaged concentration as a whole and that related to the size, say radius, of the REV.

In his book (1999) and in many papers, Whitaker deals also with the construction of averaged models for two-phase mass, energy and momentum balance equations, deriving Darcy's law from the momentum balance equation.

C. Hassanizadeh-Gray's Approach

In a series of papers, Hassanizadeh and Gray (e.g., 1979a, b, 1980, 1990) present an approach, which they characterize as 'thermodynamically correct', to the passage from the microscopic level to the macroscopic one.

Hassanizadeh (1979a), henceforth abbreviated by H&G, consider a multiphase transport in porous medium systems. Their approach consists of three main parts. In the first part, similar to Bear and Bachmat's and Whittaker's approaches described above, microscopic balance laws for extensive quantities are stated and averaged, yielding macroscopic balance laws that are valid for any phase present in a porous medium domain. In addition to phase domains, they also develop microscopic and averaged balance equations for extensive properties that are associated with *interphase boundaries*, i.e., *interfaces* such as fluid–fluid and solid-fluid interfaces (Gray and Hassanizadeh 1989). At the microscopic level, such interfaces are between phases. However, at the macroscopic level, they are added to the phases within the REV, as a *continuum of interphase domains*. They have their own E-balance equations, except that the intensive properties express E per unit area of interface. Actually, in their models, they also take into account the curves along which interphase surfaces intersect each other and suggest balance equation along such curves and averages over all such intersection curves within an REV.

In the third part of their work, they employ *rational thermodynamics* (Truesdell 1977; Eringen 1980) in order to develop macro-scale constitutive equations that describe the behavior of phases in porous medium domains.

The averaging procedure in the H&G approach follows, basically, that of Bear and Bachmat (1991) described above. They also start by stating the microscopic balance equations at a point *within a phase*, and use REV-averaging laws to derive averaged, macroscopic balance equations at points within a *porous medium domain, regarded as a continuum*, or as multiple overlapping continua. However, in their models, they use the *specific value* of E (i.e., E per unit mass of a phase) as variable to be solved for.

The microscopic balance equation for an extensive thermodynamic property E, of a phase, with ρ and $\mathbf{V}$ denoting the phase mass density and velocity, takes the general form:

$$\frac{\partial \rho e}{\partial t} = -\nabla\cdot(\rho e\mathbf{V} + \mathbf{j}^{E}_{dif}) + \rho(\Phi^{E} + \Gamma^{E}), \tag{1.4.57}$$

in which e denotes E per unit mass of a phase occupying the void space, or part of it, ρe denotes E per unit volume, $\mathbf{j}_{dif}$ denotes the diffusive flux of E (as defined by (3.1.10)), Γ^{E} denotes external sources of E, and Φ^{E} denotes internal sources, both per unit mass. However, in general, $\Phi^{E} = 0$, except for entropy. Each term in the above equation expresses the rate of added E per unit phase volume.

Examples of E and e are: for $E = m$, $e = 1$; for $E = m^\gamma$, $e^\gamma = m^\gamma/m = \omega^\gamma$; for $E = \mathbf{M} \equiv m\mathbf{V}$, $e = \mathbf{V}$; for $E = \mathbb{E}$, $e = u + \frac{1}{2}\mathbf{V}^2$, with u denoting internal energy per unit mass.

H&G consider also the case of an interface, or a *thin interphase zone* in two-phase flow (e.g., Hassanizadeh and Gray 1990), e.g., between an α and a β fluids, or between an α-fluid and the solid (s). Such surfaces may play an important role in the E balance at a point within a porous medium domain. We shall use subscripts $\alpha\beta$ and αs to indicate values on an $\alpha\beta$- and on an αs-surface, respectively.

Rather unique in H&G's work is the way they regard all interphase surfaces within a porous medium domain, with the void space occupied by one or two fluid phases (α, β). They introduce *specific interfacial areas* (= area of interface between adjacent phases per unit porous medium volume: $a_{\alpha\beta}$, $a_{\alpha s}$, $a_{\beta s}$) as new variables. Extensive quantities can (1) move within these surfaces, (2) be transported from or to adjacent phases, and (3) accumulate on the surfaces. In addition, they also consider the line along which phases intersect and consider the balance of E on such line.

For a point on such an interface, with $\rho^{\alpha\beta}$ denoting the areal mass density (= mass per unit area) of the interface, and $e_{\alpha\beta}$ denoting the density (actually, the specific value) of E, i.e., E per unit interface mass, they write the microscopic E-balance equation in the form:

$$\underset{(a)}{\frac{\partial \rho_{\alpha\beta} e_{\alpha\beta}}{\partial t}} = \underset{(b)}{-\nabla_{\alpha\beta}\cdot(\rho_{\alpha\beta} e_{\alpha\beta}\mathbf{V}_{\alpha\beta} + \mathbf{j}^{\alpha\beta}_{dif})} + \underset{(c)}{2\rho_{\alpha\beta} e_{\alpha\beta}\mathcal{K}_{\alpha\beta}\mathbf{V}_{\alpha\beta}\cdot\mathbf{n}}$$
$$\underset{(d)}{-\sum_{\alpha=1}^{2}\left[\rho_\alpha e_\alpha(\mathbf{V}_\alpha - \mathbf{V}_{\alpha\beta}) + \mathbf{j}_{\alpha,dif}\right]\cdot\boldsymbol{\nu}_\alpha} + \underset{(e)}{\rho_{\alpha\beta}(\Phi_{\alpha\beta} + \Gamma_{\alpha\beta})}, \qquad (1.4.58)$$

in which the divergence ($\nabla_{\alpha\beta}\cdot$) is only in the two-dimensional interface domain, $\mathbf{V}_{\alpha\beta}$ denotes the velocity of the material comprising the interface domain, $\mathbf{u}$ denotes the velocity of the (possibly moving) interface, $\mathcal{K}_{\alpha\beta}$ denotes the mean curvature of the interface, with

$$\mathcal{K}_{\alpha\beta} = \frac{\mathcal{K}_1 + \mathcal{K}_2}{2}, \quad \text{or} \quad \frac{2}{\mathcal{R}_{\alpha\beta}} = \frac{1}{\mathcal{R}_1} + \frac{1}{\mathcal{R}_2},$$

$\mathcal{R}_1, \mathcal{R}_2$ denoting the principal radii of curvature, and $\mathcal{R}_{\alpha\beta}$ denoting the mean one, $\boldsymbol{\nu}$ denotes the unit vector normal to the interface, and $\rho_{\alpha\beta}\Gamma_{\alpha\beta}$ denoting external sources of E per unit mass of the interface domain. Note that all quantities with subscript $\alpha\beta$ are surface material quantities; they are defined within the two-dimensional domain of the interface. Only the interface velocity $\mathbf{V}_{\alpha\beta}$ has a component normal to the interface.

The various terms in the balance equation (1.4.58) can be interpreted as follows:

(a) Accumulation of E per unit area of interface, per unit time.

(b) Added E per unit area of interface, per unit time, due to advection and diffusion of E *within* the thin domain.
(c) Added E per unit area of interface, per unit time, due to the displacement of the interface, i.e., $\mathbf{V}_{\alpha\beta}\cdot\mathbf{n}$.
(d) Exchange of E with the two neighboring phases that are present on both sides of the thin interface domain.
(e) Added E by sources within the thin domain. H&G make a distinction between external and internal source. The $\Phi_{\alpha\beta}$-source is added to facilitate the discussion on entropy, regarded as an internal source.

H&G emphasize that (1.4.57) does not hold on interphase boundaries. Note that (1.4.58) replaces the 'no jump condition' discussed in Sect. 5.2.3. Indeed, when the interface zone has no thermodynamic extensive quantities, (1.4.58) reduces to the no-jump condition discussed in Sect. 5.2.3.

Similar to Bear and Bachmat (1991), they define two kinds of volume averages of an intensive quantity, e, of an α-phase that occupies a volumetric fraction θ_α of an REV of volume $\mathbb{V}(\mathbf{x})$ satisfying (1.1.54). These averages, similar to (1.1.7) and (1.1.8), are:

Volume average of e:

$$\overline{e_\alpha}(\mathbf{x},t)=\frac{1}{\mathbb{V}_o}\int_{\mathbb{V}_o} e(\boldsymbol{\xi},t;\mathbf{x})\gamma_\alpha(\boldsymbol{\xi},t)d\mathbb{V},\quad \int_{\mathbb{V}_{o\alpha}} ed\mathbb{V}\equiv\int_{\mathbb{V}_o}\psi\gamma_\alpha d\mathbb{V}, \tag{1.4.59}$$

with:

$$\overline{\rho_\alpha}(\mathbf{x},t)=\frac{1}{\mathbb{V}_o}\int_{\mathbb{V}_o}\rho(\boldsymbol{\xi},t;\mathbf{x})\gamma_\alpha(\boldsymbol{\xi},t)d\mathbb{V}. \tag{1.4.60}$$

This kind of average is the same as that defined by Bear and Bachmat.

Intrinsic volume average of ψ:

$$\overline{e_\alpha}^\alpha(\mathbf{x},t)=\frac{1}{\mathbb{V}_{o\alpha}}\int_{\mathbb{V}_o} e(\boldsymbol{\xi},t;\mathbf{x})\gamma_\alpha(\boldsymbol{\xi},t)d\mathbb{V} \tag{1.4.61}$$

Obviously,

$$\overline{e}=\theta_\alpha\overline{e}^\alpha,\quad \text{e.g.,}\quad \overline{\rho_\alpha}=\theta_\alpha\overline{\rho_\alpha}^\alpha, \tag{1.4.62}$$

where θ_α denotes the volumetric fraction of the α-phase in the void space ($\equiv \mathbb{V}_{o\alpha}/\mathbb{V}_o$). This kind of average is equivalent to Bear and Bachmat's intrinsic phase average, $\overline{(..)}^\alpha$.

H&G also define a mass average for a variable density fluid:

Intrinsic mass-average of e:

$$\widetilde{e}^\alpha(\mathbf{x},t)=\frac{\int_{\mathbb{V}_o(\mathbf{x},t)}\rho(\boldsymbol{\xi},t;\mathbf{x})e(\boldsymbol{\xi},t;\mathbf{x})\gamma_\alpha(\boldsymbol{\xi},t;\mathbf{x})d\mathbb{V}}{\int_{\mathbb{V}_o(\mathbf{x},t)}\rho(\boldsymbol{\xi},t;\mathbf{x})\gamma_\alpha(\boldsymbol{\xi},t;\mathbf{x})d\mathbb{V}}. \tag{1.4.63}$$

Recalling that $\gamma_\alpha d\mathbb{V} = d\mathbb{V}_\alpha$, we have:

$$\widetilde{e}^\alpha(\mathbf{x}, t) = \frac{1}{\overline{\rho_\alpha}(\mathbf{x}, t)\mathbb{V}_o} \int_{\mathbb{V}_o} \rho(\boldsymbol{\xi}, t; \mathbf{x}) e(\boldsymbol{\xi}, t; \mathbf{x}) \gamma_\alpha(\boldsymbol{\xi}, t; \mathbf{x}) d\mathbb{V}. \tag{1.4.64}$$

Note that the last equation leads to:

$$\overline{\rho_\alpha}\, \widetilde{e}^\alpha = \overline{(\rho e)_\alpha}^\alpha. \tag{1.4.65}$$

The quantity:

$$\check{e}_\alpha(\boldsymbol{\xi}, t; \mathbf{x}) \equiv e_\alpha(\boldsymbol{\xi}, t; \mathbf{x}) - \widetilde{e_\alpha}^\alpha(\mathbf{x}, t) \tag{1.4.66}$$

defines the *deviation* of e_α at a point ¸ within the REV centered at **x** from its intrinsic phase average, over that REV, centered at **x**. The above definition of deviation, $\check{e}$, leads to the relationship between the product of an average and the average of a product:

$$\widetilde{e_\alpha g_\alpha}^\alpha = \widetilde{e_\alpha}^\alpha \widetilde{g_\alpha}^\alpha + \widetilde{\check{e}_\alpha \check{g}_\alpha}^\alpha, \tag{1.4.67}$$

to be compared with B&B's (1.4.2). However, we note that in the balance equation (1.4.57), the E-flux is expressed by a product of three variables: $\rho e\mathbf{V}$. Hence, considering three variables ρ_α, ψ_α, and g_α, and making use of (1.1.15), we obtain:

$$\overline{\rho_\alpha e_\alpha g_\alpha}^\alpha = \overline{\rho_\alpha}^\alpha \widetilde{e_\alpha g_\alpha}^\alpha = \overline{\rho_\alpha}^\alpha \left(\widetilde{e_\alpha}^\alpha \widetilde{g_\alpha}^\alpha + \widetilde{\check{e}_\alpha \check{g}_\alpha}^\alpha \right). \tag{1.4.68}$$

H&G also define an **area-average operator** of **e** over an area $\mathbb{A}$:

$$\ddot{\mathbf{e}}^\alpha(\mathbf{x}, t) = \frac{1}{\mathbb{A}} \int_{\mathbb{A}} \mathbf{e}(\boldsymbol{\xi}, t; \mathbf{x}) \gamma_\alpha(\boldsymbol{\xi}, t; \mathbf{x}) \cdot \boldsymbol{\nu} d\mathbb{A}. \tag{1.4.69}$$

Hassanizadeh and Gray develop rules for averages of space and time derivatives which are similar to those developed by B&B (henceforth abbreviation for Bear and Bachmat) in the form of (1.4.11) and (1.4.9):

$$\langle \nabla \cdot \varsigma_\alpha \rangle = \nabla \cdot \langle \varsigma_\alpha \rangle + \frac{1}{\mathbb{V}_\alpha} \int_{\mathcal{S}_{\alpha\beta}} \varsigma_\alpha \cdot \mathbf{u} \cdot \boldsymbol{\nu}_\alpha d\mathcal{S}, \tag{1.4.70}$$

where β denotes all non-α phases, and Note that the above rules are presented for ς_α that is a vector.

They apply these rules to the fundamental e-balance equation (1.4.57).

The macroscopic balance equation for any extensive quantity, E of an α phase that occupies part of the void space at the volumetric fraction θ_α, takes the form:

$$\frac{\partial(\theta_\alpha \overline{\rho_\alpha}^\alpha \widetilde{e_\alpha}^\alpha)}{\partial t} = -\nabla\cdot\left[\theta_\alpha(\overline{\rho_\alpha}^\alpha \widetilde{e_\alpha}^\alpha \widetilde{\mathbf{V}_\alpha}^\alpha + \widetilde{\hat{\mathbf{J}}_\alpha}^\alpha)\right] + \sum_{(\beta\neq\alpha)}\left[e_\alpha^{\alpha\beta}(\rho_\alpha e_\alpha) + \hat{I}_\alpha^{\alpha\beta}\right] + \theta_\alpha \overline{\rho_\alpha}^\alpha(\widetilde{\Phi_\alpha}^\alpha + \widetilde{\Gamma_\alpha}^\alpha), \quad (1.4.71)$$

in which $e_\alpha = E_\alpha/m_\alpha$, $\rho_\alpha = m_\alpha/\mathbb{V}_\alpha$, $(\rho e)_\alpha = E_\alpha/\mathbb{V}_\alpha$, $\widetilde{e_\alpha}^\alpha(\mathbf{x}, t)$ is the intrinsic mass averaged value of e_α, defined by (1.4.64), $\widetilde{\mathbf{V}_\alpha}^\alpha$ is the intrinsic mass averaged α-fluid velocity, $\widetilde{\hat{\mathbf{J}}_\alpha}^\alpha$ denotes the sum of all macroscopic non-advective E_α-fluxes (by diffusion and by dispersion, see Chap. 3),

$$e_\alpha^{\alpha\beta}(\rho_\alpha e_\alpha) = \frac{1}{d\mathbb{V}}\int_{\mathcal{S}_{\alpha\beta}} \rho_\alpha e_\alpha(\mathbf{V}_\alpha - \mathbf{V}_{\alpha\beta})\cdot\boldsymbol{\nu}_{\alpha\beta} d\mathcal{S}, \quad \beta\neq\alpha, \quad (1.4.72)$$

denotes the transport of E to the α-phase from all non-α phases present in the REV, through the surface bounding the α phase, and

$$\hat{I}_\alpha^{\alpha\beta} = \frac{1}{d\mathbb{V}}\int_{d\mathcal{S}_{\alpha\beta}} \mathbf{j}_{dif}^E\cdot\boldsymbol{\nu}_{\alpha\beta} d\mathcal{S}, \quad \beta\neq\alpha, \quad (1.4.73)$$

is the same as $e^{\alpha\beta}(\rho_\alpha e_\alpha)$, except that here the symbol expresses transport by E-diffusive fluxes.

The last term on the r.h.s. of (1.4.71) denotes internal and external sources, noting that $\Phi \equiv 0$, except for the case in which E represents entropy.

Each term in (1.4.71) represents an added quantity of E of α per unit volume of the α phase that occupies the entire void space, or part of it. This equation should be compared with the microscopic balance equation (1.4.57).

H&G use this equation to write balance equations for the mass of an α fluid or solid phase, for the momentum of a phase, for energy and for entropy.

Mass Balance of a Fluid, or Solid α-phase

With $\widetilde{e_\alpha}^\alpha = 1$, $\widetilde{\hat{\mathbf{J}}_\alpha}^\alpha = 0$, and $\widetilde{\Phi_\alpha}^\alpha = \widetilde{\Gamma_\alpha}^\alpha = 0$, H&G obtain the macroscopic mass balance equation:

$$\frac{\partial(\theta_\alpha \overline{\rho_\alpha}^\alpha)}{\partial t} = -\nabla\cdot(\theta_\alpha \overline{\rho_\alpha}^\alpha \widetilde{\mathbf{V}_\alpha}^\alpha) + \sum_{\beta\neq\alpha} \theta_\alpha \overline{\rho_\alpha}^\alpha e_\alpha^{\alpha\beta}(\rho_\alpha), \quad (1.4.74)$$

where $e_\alpha^{\alpha\beta}(\rho_\alpha)$ expresses mass exchange between an α-phase and its interfaces with other (β-)phases.

They also suggest the no-jump condition on an external domain's boundary segment ($\mathcal{S}_\Sigma$) moving at the macroscopic velocity $\mathbf{u}_\Sigma$:

$$[\![\ \theta_\alpha \rho_\alpha(\mathbf{V}_\alpha - \mathbf{u}_\Sigma \ ]\!]\cdot\boldsymbol{\nu} = 0, \quad on \quad \mathcal{S}_\Sigma. \quad (1.4.75)$$

Mass of a Dissolved γ-Species in an α-Phase

With $e_\alpha = \omega^\gamma_\alpha$, Hassanizadeh (1986a) obtains:

$$\frac{\partial \theta_\alpha \overline{\rho_\alpha}^\alpha \widetilde{\omega^\gamma_\alpha}^\alpha}{\partial t} = -\nabla \cdot \left[\theta_\alpha (\overline{\rho_\alpha}^\alpha \widetilde{\omega^\gamma_\alpha}^\alpha \widetilde{\mathbf{V}_\alpha}^\alpha) + \widetilde{\hat{\mathbf{J}}^\gamma_\alpha}^\alpha\right] + \sum_{(\beta \neq \alpha)} \left[e^{\alpha\beta}_\alpha (\rho_\alpha \omega^\gamma_\alpha) + \hat{I}^{\alpha\beta}_\alpha\right] + \theta_\alpha \overline{\rho_\alpha}^\alpha (\widetilde{\Phi_\alpha}^\alpha + \widetilde{\Gamma_\alpha}^\alpha), \quad (1.4.76)$$

in which $\widetilde{\hat{\mathbf{J}}^\gamma_\alpha}^\alpha$ denotes the sum of diffusive and dispersive fluxes of γ in α.

For **momentum balance** of an α-phase, with $\widetilde{e_\alpha}^\alpha = \widetilde{\mathbf{V}_\alpha}^\alpha$ (i.e., momentum per unit mass), H&G write the macroscopic equation:

$$\frac{\partial \theta_\alpha \overline{\rho_\alpha}^\alpha \widetilde{\mathbf{V}_\alpha}^\alpha}{\partial t} = -\nabla \cdot \left[\theta_\alpha (\overline{\rho_\alpha}^\alpha \widetilde{\mathbf{V}_\alpha}^\alpha \widetilde{\mathbf{V}_\alpha}^\alpha + \widetilde{\boldsymbol{\sigma}_\alpha}^\alpha)\right] + \theta_\alpha \overline{\rho_\alpha}^\alpha \mathbf{g} + \sum_{(\beta \neq \alpha)} \left[e^{\alpha\beta}_\alpha (\rho_\alpha \mathbf{V}_\alpha) + \hat{\boldsymbol{\sigma}}^{\alpha\beta}_\alpha\right], \quad (1.4.77)$$

in which the diffusive momentum flux is expressed by the fluid's stress, $\boldsymbol{\sigma}_\alpha$, and the sole external source is gravity (**g**). The last two terms account for the exchange of momentum between the α-phase and its interphase boundaries.

For **energy balance** of an α-phase, with $e_\alpha = u_\alpha + \frac{1}{2}V^2_\alpha$, H&G write:

$$\frac{\partial}{\partial t}\left[\theta_\alpha \overline{\rho_\alpha}^\alpha (\widetilde{u_\alpha}^\alpha + \tfrac{1}{2}\widetilde{V^2_\alpha}^\alpha)\right] = -\nabla \cdot \left[\theta_\alpha [\overline{\rho_\alpha}^\alpha (\widetilde{u_\alpha}^\alpha + \tfrac{1}{2}\widetilde{V^2_\alpha}^\alpha) \widetilde{\mathbf{V}_\alpha}^\alpha + \hat{\boldsymbol{\sigma}}_\alpha \cdot \widetilde{\mathbf{V}_\alpha}^\alpha + \mathbf{J}^H_\alpha]\right] + \theta_\alpha \overline{\rho_\alpha}^\alpha (\mathbf{g}\widetilde{\mathbf{V}_\alpha}^\alpha + \widetilde{\hat{h}_\alpha}^\alpha) + \sum_{\beta \neq \alpha} \left[e^{\alpha\beta}_\alpha (\rho_\alpha u_\alpha) + \hat{\boldsymbol{\sigma}}^{\alpha\beta}_\alpha \cdot \mathbf{V}_\alpha + \tfrac{1}{2} e^{\alpha\beta}_\alpha (\rho_\alpha) V^2_\alpha\right], \quad (1.4.78)$$

in which **g** denotes the gravity vector and the diffusive flux of energy consists of the work of internal stress, due to the motion, plus the internal heat flux. The external source of energy is due to the work of gravity as well as that due to radiation ($\hat{h}$). The last three terms account for the exchange of energy between α-phase and its interphase boundaries.

H&G also introduce an **entropy balance** equation for each phase, with s_α denoting the internal entropy per unit mass:

$$\frac{\partial}{\partial t}(\theta_\alpha \overline{\rho_\alpha}^\alpha \overline{s_\alpha}^\alpha) = -\nabla \cdot \left(\theta_\alpha (\overline{\rho_\alpha}^\alpha \widetilde{s_\alpha}^\alpha \widetilde{\mathbf{V}_\alpha}^\alpha + \frac{1}{T}\mathbf{J}^H)\right) + \frac{1}{T}\widetilde{\hat{h}_\alpha}^\alpha + e^{\alpha\beta}_\alpha (\rho_\alpha s_\alpha) + \Lambda_\alpha, \quad (1.4.79)$$

where it is assumed that the diffusive flux of entropy is proportional to the heat flux, and the external entropy source is proportional to the external energy source, with the proportionality ratio being the inverse of temperature, T. The last term in (1.4.79) is the net production of entropy of the α-phase, which is nonzero, because entropy is not conserved. The second law of thermodynamics will put restrictions on this term, as explained below.

As mentioned earlier, an essential element in the H&G approach is that interfaces, say between fluid phases and between the solid and a fluid, are also considered as domains within which transport occurs and on which extensive quantities may accumulate (Gray and Hassanizadeh 1989). Furthermore, lines along which such surfaces intersect each other are considered as domains on which E can accumulate as a result of transfers from the intersecting surfaces. Equation (1.4.58) expresses the balance of an extensive quantity at a point on such an interface, say between an α and a β-phase domains. H&G average this equation to obtain the balance equation at the macroscopic level, i.e., an average of the considered interface within an REV. This equation takes the form:

$$\underset{(a)}{\frac{\partial a^{\alpha\beta}\rho^{\alpha\beta}e^{\alpha\beta}}{\partial t}} = \underset{(b)}{-\nabla\cdot(a^{\alpha\beta}\rho^{\alpha\beta}e^{\alpha\beta}\mathbf{V}^{\alpha\beta} + \mathbf{J}^{\alpha\beta}_{dif+dis})} + \underset{(c)}{a^{\alpha\beta}\rho^{\alpha\beta}\left(\Phi^{\alpha\beta} + \Gamma^{\alpha\beta}\right)}$$

$$-\Big[\underset{(d)}{(e^{\alpha\beta}_{\alpha}(\rho_\alpha e_\alpha) + \hat{I}^{\alpha\beta}_{\alpha})} - \underset{(e)}{(e^{\alpha\beta}_{\beta}(\rho_\beta e_\beta) + \hat{I}^{\alpha\beta}_{\beta})}\Big] + \underset{(f)}{e^{\alpha\beta\gamma}_{\alpha\beta}(\rho^{\alpha\beta}e^{\alpha\beta})} + \underset{(g)}{\hat{I}^{\alpha\beta\gamma}_{\alpha\beta}}, \qquad (1.4.80)$$

in which, as is common in this book, we have not used any averaging symbols, as it is obvious that we are considering here an averaged balance equation. In this equation,

$$a^{\alpha\beta} = \frac{1}{\mathbb{V}_o}\int_{\mathbb{V}_o} dS_{\alpha\beta},$$

denotes the specific surface area of the α-β interface, $e^{\alpha\beta}$ denotes E per unit mass of the surface, $\rho^{\alpha\beta}$ denotes the interface mass per unit area, $\mathbf{V}^{\alpha\beta}$ denotes the velocity of matter comprising the interface domain, and $\mathbf{J}^{\alpha\beta}_{dif+dis}$ denotes the non-advective flux of E in the considered surface.

Terms (d) and (e) in (1.4.80) denote the transfer of E from the surface to the two neighboring phases, while terms (f) and (g) express the transfer of E from the surfaces to the lines (usually, curves) along which these surfaces intersect each other. Assuming that these lines do not possess thermodynamic properties, these terms are subject to the following constraints:

$$\sum_{(\beta,\alpha)}\left[e^{\alpha\beta\gamma}_{\alpha\beta}(\rho^{\alpha\beta}e^{\alpha\beta}) + \hat{I}^{\alpha\beta\gamma}_{\alpha\beta}\right] = 0. \qquad (1.4.81)$$

Specific balance laws for mass, momentum, energy, and entropy of interfaces are obtained from (1.4.80). For example, for **mass balance**, we obtain:

$$\frac{\partial a^{\alpha\beta}\rho^{\alpha\beta}}{\partial t} = -\nabla\cdot(a^{\alpha\beta}\rho^{\alpha\beta}\mathbf{V}^{\alpha\beta}) \\ - \left[e_{\alpha}^{\alpha\beta}(\rho_\alpha e_\alpha) - e_{\beta}^{\alpha\beta}(\rho_\beta e_\beta)\right] + e_{\alpha\beta}^{\alpha\beta\gamma}(\rho^{\alpha\beta}), \tag{1.4.82}$$

where the last term accounts for exchange of mass between the $\alpha\beta$-interface and other interfaces via the common line. They are subject to the following restriction:

$$\sum_{(\alpha,\beta)} e_{\alpha\beta}^{\alpha\beta\gamma}(\rho^{\alpha\beta}) = 0. \tag{1.4.83}$$

The collection of balance laws for phases and interfaces form the basis for an extended theory of multi-phase flow and transport in porous media that explicitly models interfaces at the macro-scale. In these theories, specific interfacial area is a macro-scale state variable similar to porosity and saturation. These equations must be supplemented by constitutive equations that describe the dependency on state variables of the stress tensor, the heat flux vector, and exchange fluxes. The derivation of constitutive equations for single-phase flow, two-phase flow, and multi-species solute transport can be found in Hassanizadeh and Gray (1980, 1990), and Hassanizadeh (1986b), respectively. A major element in these derivations is the second law of thermodynamics, which prescribes that the total rate of production of entropy of all phases and interfaces must be nonnegative. Their results include a definition of capillary pressure in terms of the change of free energy of the system, due to a change in saturation. This replaces hysteretic capillary pressure-saturation curves by a three-dimensional capillary pressure-saturation-interfacial surface area.

1.4.3 Homogenization

Although the volume averaging technique discussed above has been widely used for passing from the microscopic level to the macroscopic one, another technique is often employed for handling multiple-scale heterogeneity. This technique, known as the *mathematical theory of homogenization*, has been applied since the 1970s to a wide range of physical problems that involve composite materials, heterogeneous geological media, and porous media (Bensoussan et al. 1978; Sanchez-Palencia 1974; Sanchez-Palencia 1980; Lions 1981; Bakhvalov and Panasenko 1989; Jikov et al. 1994; Mikelic 2000).

Briefly, homogenization is a mathematical technique applied to differential equations that describe physical phenomena associated with a domain exhibiting heterogeneities at two scales. By homogenization, we obtain a domain which is more homogeneous, at least locally. The coefficients, which characterize this 'homogenized' medium, are referred to as 'effective' ones. In the process of homogenization, each of the equations that constitute the model of the transport problem is replaced

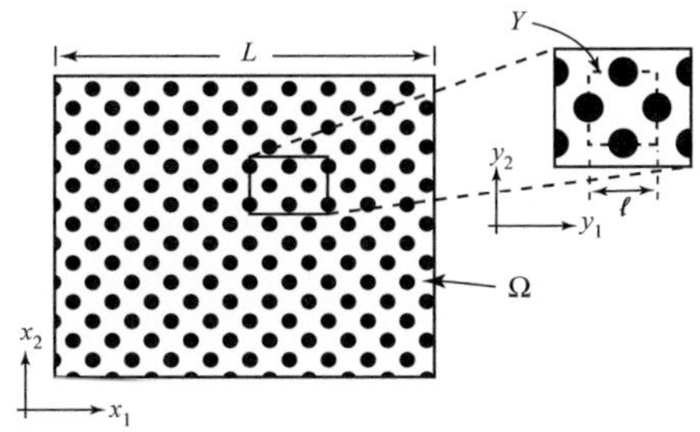

Fig. 1.19 A domain (Ω) with periodic cells (Y) with period ℓ; $\ell/L \ll 1$

by a number of equations, each addressing the dominant physical phenomena at one of the two considered scales.

A necessary condition for the application of the homogenization technique is the existence of a *periodic* structure. This condition is needed for a rigorous mathematical proof of the existence and uniqueness of the solution. Physically, this means a structure based on a *repeated pattern*, e.g., a certain form of stacking of spheres. This means that the application of the homogenization approach to porous media is based on the approximation of the geometry of the solid matrix as *periodic*. Figure 1.19 illustrates a domain of size L containing a periodic pattern, with the period denoted by ℓ. For the application of the homogenization technique, we require that the ratio $\varepsilon(=\ell/L)$ be a *small parameter*. In fact, the homogenization seeks the *asymptotic solution* in the limit $\varepsilon \to 0$. It is possible to apply the homogenization technique to a domain with multiple scales.

The homogenization approach, which is based on periodicity of the porous medium structure, facilitates the upscaling from the microscopic to the macroscopic scale. It is interesting to comment that porous media comprising geological formation are very far from a periodic structure.

It is interesting to compare the REV approach presented in earlier subsections, where homogenization is achieved by averaging over an REV, and the homogenization achieved here, where the REV is the unit cell.

Following Bear and Cheng (2010, p. 57), let us demonstrate how the homogenization technique is applied to mathematical model composed of a steady state ordinary differential equation in one spatial dimension, with appropriate boundary conditions. The considered ODE is

$$\frac{d}{dx}\left[a^{\varepsilon}(x)\frac{du^{\varepsilon}(x)}{dx}\right]=0, \quad 0 \leq x \leq L, \tag{1.4.84}$$

subject to the boundary conditions

$$u^{\varepsilon}(0)=0, \quad \text{and} \quad u^{\varepsilon}(L)=1. \tag{1.4.85}$$

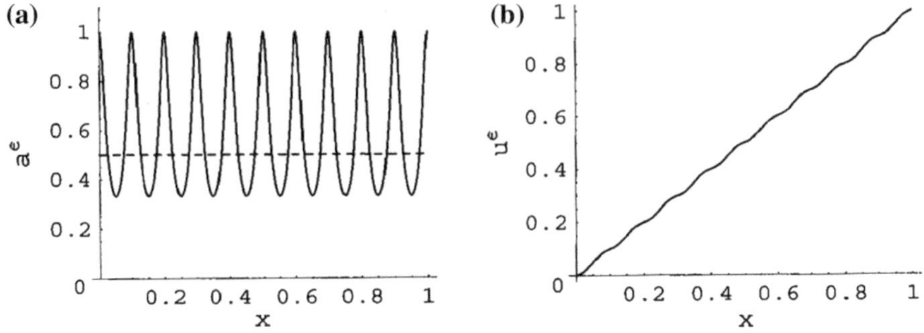

Fig. 1.20 Solution of (1.4.84) with a rapidly fluctuating coefficient: **a** A plot of a^ε, with $\varepsilon = 0.1$, in which the dashed line indicates the 'effective coefficient' $a_o = \frac{1}{2}$, and **b** The solution (1.4.88) for u^ε

Here, L is the length of the domain of interest, considered as the large scale. Within the considered domain, there exist repeated small scale features of size ℓ, introduced through the functional relation of the coefficient a^ε, and $\varepsilon = \frac{\ell}{L} \ll 1$.

As a consequence, the solution u^ε depends also on ε; the superscript is used to emphasize this fact. As will be demonstrate below, even if the variations in the a^ε are large, expressed in terms of large amplitude fluctuations, their effect on the solution of the considered equation is small; it is of the order $\mathcal{O}(\varepsilon)$ only! In fact, the purpose of homogenization is to seek the asymptotic solution as $\varepsilon \to 0$,

$$u(x) = \lim_{\varepsilon \to 0} u^\varepsilon(x). \tag{1.4.86}$$

In the present example, without loss of generality, we assume $L = 1$, and hence $\ell = \varepsilon$. Then, assuming that a^ϵ is a periodic function, with period ε, we have:

$$a^\varepsilon(x) = \frac{1}{1 + 2\sin^2 \frac{\pi x}{\varepsilon}}. \tag{1.4.87}$$

This coefficient is plotted as Fig. 1.20a for a small ε, say, $\varepsilon = 0.1$. We observe 'rapid' fluctuations (small period) with a large amplitude.

Before applying the homogenization technique, it would be instructional to examine the exact solution of the problem represented by (1.4.84) and (1.4.85). For $1/\varepsilon$ an integer, this solution is

$$u^\varepsilon(x) = \frac{\int_0^x \frac{dx}{a^\varepsilon(x)}}{\int_0^1 \frac{dx}{a^\varepsilon(x)}} = \frac{4\pi x - \varepsilon \sin \frac{2\pi x}{\varepsilon}}{4\pi - \varepsilon \sin \frac{2\pi}{\varepsilon}} = x - \frac{\varepsilon}{4\pi} \sin \frac{2\pi x}{\varepsilon}, \tag{1.4.88}$$

plotted as Fig. 1.20b for $\varepsilon = 0.1$. As observed in the figure, and also in (1.4.88), the solution consists of two parts: a slowly varying part (linear), with rapidly fluctuating, small amplitude 'ripples' superposed on top. Indeed, the magnitude of the 'disturbance' in the solution, caused by the fluctuating coefficient a^{ε}, is controlled not by the coefficient's amplitude, but by its period, which is small.

Homogenization requires the existence of *two scales*. At the larger scale, we denote the domain of size L ($\equiv 1$ in the current case) as ω, and use the coordinate system $0 \leq x \leq 1$. At the small scale, characterized by the periodic cells of size ℓ ($\equiv \varepsilon$) (see Fig. 1.19 for a two-dimensional conceptualization), we denote the repeated domain as Y, and use the scaled coordinate $y = x/\varepsilon$, such that $0 \leq y \leq 1$ in a Y-cell. With the above definition, we now express the coefficient a^{ε}, defined in (1.4.87), as

$$a^{\varepsilon}(x) = a(y) = \frac{1}{1 + 2\sin^2 \pi y}. \tag{1.4.89}$$

We can now express $u^{\varepsilon}(x)$ as a two-scale function, $u = u(x, y)$, and expand it into a power series in terms of the small parameter, ε,

$$u^{\varepsilon}(x) = u(x, y) = u^{(o)}(x, y) + \varepsilon u^{(1)}(x, y) + \varepsilon^2 u^{(2)}(x, y) + \ldots \tag{1.4.90}$$

This is known as the *perturbation technique* (Nayfeh 2000). Substituting the above expression into (1.4.84), and applying the chain rule

$$\frac{d}{dx} = \frac{\partial}{\partial x} + \frac{1}{\varepsilon}\frac{\partial}{\partial y},$$

to the two-scale functions, we can expand and separate the resulting equation into several equations, each corresponding to the same power of ε:

$$\mathcal{O}(\varepsilon^{-2}): \quad \frac{\partial}{\partial y}\left[a(y)\frac{\partial u^{(o)}(x, y)}{\partial y}\right] = 0, \tag{1.4.91}$$

$$\mathcal{O}(\varepsilon^{-1}): \quad \frac{\partial}{\partial x}\left[a(y)\frac{\partial u^{(o)}(x, y)}{\partial y}\right] + \frac{\partial}{\partial y}\left[a(y)\frac{\partial u^{(o)}(x, y)}{\partial x}\right] + \frac{\partial}{\partial y}\left[a(y)\frac{\partial u^{(1)}(x, y)}{\partial y}\right] = 0, \tag{1.4.92}$$

$$\mathcal{O}(\varepsilon^{0}): \quad \frac{\partial}{\partial x}\left[a(y)\frac{\partial u^{(o)}(x, y)}{\partial x}\right] + \frac{\partial}{\partial x}\left[a(y)\frac{\partial u^{(1)}(x, y)}{\partial y}\right] + \frac{\partial}{\partial y}\left[a(y)\frac{\partial u^{(1)}(x, y)}{\partial x}\right] + \frac{\partial}{\partial y}\left[a(y)\frac{\partial u^{(2)}(x, y)}{\partial y}\right] = 0 \tag{1.4.93}$$

and higher order equations. The boundary conditions (1.4.85), are assigned to the leading terms, such that the higher order terms take the null boundary conditions:

$$u^{(o)}(0, y) = 0, \quad u^{(o)}(1, y) = 1; \quad u^{(1)}(0, y) = u^{(1)}(1, y) = 0; \quad u^{(2)}(0, y) = u^{(2)}(1, y) = 0; \quad \ldots \tag{1.4.94}$$

Also, the periodicity of the Y-cells requires that

$$\begin{aligned} &u^{(1)}(x,0)=u^{(1)}(x,1); \quad u^{(2)}(x,0)=u^{(2)}(x,1); \quad \ldots \\ &\left.\frac{\partial u^{(1)}(x,y)}{\partial y}\right|_{y=0}=\left.\frac{\partial u^{(1)}(x,y)}{\partial y}\right|_{y=1}; \\ &\left.\frac{\partial u^{(2)}(x,y)}{\partial y}\right|_{y=0}=\left.\frac{\partial u^{(2)}(x,y)}{\partial y}\right|_{y=1}; \quad \ldots \end{aligned} \tag{1.4.95}$$

A quick inspection of the $\mathcal{O}(\varepsilon^{-2})$-equation, (1.4.91), shows that

$$u^{(o)}(x,y)=u^{(o)}(x) \tag{1.4.96}$$

is an admissible solution. In fact, this is, the *only* admissible (unique) solution of (1.4.91).

Bear and Cheng (2010, p. 60) solve also the $\mathcal{O}(\varepsilon^{-1})$-equation, (1.4.92), and the $\mathcal{O}(\varepsilon^{0})$-equation, (1.4.93). They lead to

$$u^{(o)}(x)=x. \tag{1.4.97}$$

As observed in Fig. 1.20b, the above linear term is exactly the anticipated large scale behavior. It is now possible to determine the 'flux',

$$q^{\varepsilon}=-a^{\varepsilon}\frac{du^{\varepsilon}}{dx}, \quad \rightarrow q^{\varepsilon}=-\frac{1}{2}. \tag{1.4.98}$$

This behavior is not obvious from the coefficient a^{ε} (see also Fig. 1.20a). The homogenization process, however, correctly captures this behavior by providing the effective coefficient a_o. Then,

$$q_o=-a_o\frac{du^{(o)}}{dx}=-\frac{1}{2}. \tag{1.4.99}$$

Just as expected, the solutions (1.4.97) and (1.4.99) are independent of ε, because these are the asymptotic solutions of u^{ε} and q^{ε} as $\varepsilon \rightarrow 0$. It is now possible to find the solution of higher order terms.

To summarize,

- Homogenization requires the assumption of periodicity at the smaller scale (Fig. 1.20a). Although this requirement may be viewed as a restriction, as natural materials are not periodic, this assumption provides the necessary boundary condition for a rigorous mathematical analysis, ensuring the existence and uniqueness of the solution.
- The example demonstrates that in a problem that involves two (or multiple) scales, the asymptotic expansion procedure, which is the basic tool of the homogenization technique, allows the separation of the governing equation into a number of scales, each governing the process at a specified scale. The periodic assumption allows the

smaller scale effects to be averaged (= integrated) to produce lumped (or effective) coefficients.

1.4.4 The Phenomenological Approach

Phenomenology is a term used in science to establish relationships among empirical observations of phenomena, in a way that is consistent with fundamental theory, but is not directly derived from such first principles. According to the *Concise Dictionary of Physics* (Thewlis 1973):

> Phenomenological Theory is a theory that expresses mathematically the results of observed phenomena, without paying detailed attention to their fundamental significance. It is an approach that is applicable to any science, including the physical sciences.

In the above definitions, we wish to emphasize that

- The approach is based on observations.
- The derived theory or model must obey scientific fundamentals.
- The theory is verifiable by experiments.

In the context of this book, we have been using the term 'model' for the relations among observed phenomena, and, as we shall see below, the 'scientific fundamentals' are those of continuum mechanics (e.g., conservation laws) and thermodynamics. Fourier's law of heat conduction, $j^H = -\lambda\nabla T$, may serve as an example of a phenomenological law, suggested by Fourier on the basis of experiments.

Thus, unlike the various approaches described above (e.g., volume averaging, where we begin from the microscopic level and lead to the macroscopic one, which is our goal), in the phenomenological approach we observe the phenomena of interest at the macroscopic level (and we have discussed above the meaning of such observations), and we construct the models that express how such observations are interrelated directly at that level. In this way, we obtain models that are simple and at the same time, physically, chemically and thermodynamically correct.

The phenomenological approach in flow through porous media is not new. In fact, *Darcy's law*, obtained on the basis of sand column experiments, is a phenomenological law. In the area of groundwater flow, Dupuit (1848, 1863), Forchheimer (1886, 1901) and Boussinesq (1904), and many others, followed the same path. Richards (1931) extended Darcy's law to unsaturated flow. In petroleum engineering, Muskat (1949) wrote the classical book on *Physical Principles of Oil Production*, in which Darcy's law was extended to two phase flow (water-oil, or water-gas). Bear et al. (1968, Sect. 1.6) presented a brief historical review on the development of the theory and applications of phenomena of transport in porous media. Most of the developments were based on the phenomenological approach. Other approaches, e.g., based on mixture theory, on REV averaging and on homogenization techniques started mainly only around the 1950s.

In this book, we shall make use of the phenomenological approach for any extensive quantity by undertaking the following steps, all *directly at the macroscopic level:*

- Write the balance equation for the considered extensive quantity.
- Identify the driving forces of the considered extensive quantity and express the fluxes of that quantity in terms of these forces.
- Identify all interphase exchanges of the considered extensive quantity and the phenomena that produce them.
- Express all sources of the considered extensive quantity in terms of the relevant state variables, and
- Write the initial and boundary conditions of the considered transport problem.

Accordingly, in Chap. 3, we shall present the fundamental balance equations, both at the microscopic and macroscopic levels of the mass of a single phase

1.5 Modeling Procedure

The primary objective of this book is to present and discuss the construction of mathematical models that describe phenomena of flow and transport in porous medium domains which are regarded as continua at the macroscopic level. Such models are constructed for two purposes: (1) to summarize and organize our understanding of phenomena that take place within considered domains, and (2) to enable the prediction of the behavior of fluids and of the solid matrix in response to excitations (forward modeling), and (3) to determine significant properties and parameters through an optimization process involving minimization of the deviations between observations and model-based predictions (backward/inverse modeling or history matching). For example, we may wish to predict the deformation of a porous medium domain, saturated by one or more fluids to applied stresses, or we may wish to predict the spreading of a contaminant within an aquifer in response to a specified pumping regime.

In this section, we shall start with a definition of a model, discuss the important step of establishing the conceptual model of a problem, present the standard content of a mathematical model, and outline the major steps included in the modeling process.

For the purpose of this chapter, a model is defined as as a *selected simplified (abstract, or physical) version of a real system and phenomena that take place within it, which approximately simulates the system's excitation-response (or input–output) relationships that are of interest*. For example, a petroleum reservoir may be 'excited' by pumping, or by injection, and its 'response' takes the form of spatial and temporal changes in its state and conditions, e.g., pressure, temperature, and fluid saturations. As we shall emphasize below, the model for a particular problem is established once we have selected its conceptual model. In order to enable a prediction of the system's behavior, we translate the conceptual model into a mathematical one. Very often, the mathematical model is not amenable to analytical solution. In such cases, the mathematical model is transformed into a *numerical model*, which, in turn, is solved by means of a computer and a *computer code*. The development of numerical models is not included in this book.

1.5.1 The Conceptual Model

The considered system and its behavior may be very complicated, depending on the amount of details we wish, or need, to include to adequately describe it. Certain features may be of no practical interest, while others may be significant in governing those aspects which are important from the point of view of the modeling objectives. We should also bear in mind that gathering information concerning the modeled system, for the purpose of model validation and determination of model coefficients, is always costly, so that a balance should be sought between additional information and the benefits to be derived from it. The *art* of modeling is to simplify the model's description of the considered system and its behavior to a degree that will still provide useful predictions.

The 'simplification' is introduced in the form of a *set of assumptions* that express our understanding of the nature of the system and its behavior. Because the model is a *simplified* version of the actual system, no unique model exists for describing it. Different sets of simplifying assumptions will result in different models, each approximating the considered domain and the phenomena that occur in it in a different way. This set of assumptions is referred to as the *conceptual model* of the problem and problem domain. Examples of such assumptions are:

- The surface that bounds the domain of interest and the problem's dimensionality.
- The temporal behavior of the system: steady state or time-dependent.
- The kind of porous material comprising the domain, as well as inhomogeneity, anisotropy, and deformability of such material.
- The number and kinds and properties of the fluid phases and of the relevant chemical species.
- The extensive quantities of interest (mass, mass of a contaminant, energy) transported within the domain.
- The relevant flow and transport mechanisms within the domain.
- The possibility of phase change and exchange of chemical species between adjacent phases.
- The relevant chemical, physical, and biological processes that take place in the domain.
- The flow regimes of the fluids involved (e.g., laminar or non-laminar).
- The existence of isothermal or non-isothermal conditions (and their influence on fluid and solid properties and on chemical–biological processes).
- The relevant state variables, and the areas or volumes over which averages of such variables should be taken.

- The presence of sources and sinks of fluids and solutes within the domain, and their nature (spatial distribution and temporal variation).
- The initial conditions within the domain, and conditions on its boundaries.

Obviously, more items may be included in the conceptual model of specific cases.

Selecting the appropriate conceptual model for a given problem is the most important step in the modeling process. If we oversimplify, we may not adequately describe its defining characteristics and, thus, may not produce the required information. If we under-simplify, we may have neither the information required for model calibration (see below), nor the resources to solve it. If we select inappropriate or wrong assumptions, our model may not represent the relevant features of the system's behavior.

1.5.2 The Modeling Process

Once we have (1) identified the information we expect the model to provide, and (2) established the conceptual model, we can start the process of modeling that involves the following steps:

A. Development of a Mathematical Model

In this step, the conceptual model is expressed in the form of a *mathematical model*. The *continuum type of mathematical model*, introduced earlier in the current section is employed. In principle, the mathematical model at the macroscopic level, can be obtained by starting from the microscopic level model and using some averaging technique (as described in Sect. 1.4.2), or by making use of the phenomenological approach, discussed in Sect. 1.4.4. The resulting mathematical model consists of:

- Identifying the *geometry* of the surface that bounds the considered domain.
- *Equations* that express the *balances* of the considered extensive quantities (e.g., mass of fluids, mass of chemical species, energy).
- *Equations* that express the fluxes of the considered extensive quantities in terms of the relevant state variables of the problem (e.g., advective mass flux and Fick's law for the diffusive mass flux of a chemical species in a fluid phase).
- *Constitutive equations* that define the behavior of the particular phases and chemical species involved (e.g., dependence of density on pressure, temperature, and solute concentration, and the relationship between stress and strain of a porous medium).
- *Sources and sinks* of the relevant extensive quantities.
- *Initial conditions* that describe the *known* state of the considered system at some initial time.
- *Boundary conditions* that describe the interaction of the considered domain with its environment (i.e., outside the delineated domain) across their common boundaries.

The above content of a mathematical model is straightforward as long as we are dealing with a single specific extensive quantity and a single PDE that expresses the balance of that quantity in terms a single variable, say pressure or temperature. However, an analytical solution is seldom possible because of the complexity resulting from the dependence of phase properties on the considered variable.

When considering multiple interacting extensive quantities, like the mass of two fluids that together occupy the void space, and temperature within the multiphase domain, we need to take into account the continuous interaction between the hydraulic and the thermal system, e.g., temperature affects density and viscosity, which, in turn, affects flow, which affects temperature, etc. In practice, solutions of multiple E's models are solved numerically. In such solution, at every time step the individual equations are solved sequentially, and solutions for properties that depend on pressure and temperature are updated. The process is repeated until convergence is achieved (at a desired level of accuracy) and the system moves to solving for the next time step.

In passing from a model at the microscopic scale to the macroscopic model, using some averaging process, or by employing the phenomenological approach, various *coefficients of flow and transport*, of interphase transfer and of *storage* of the considered extensive quantities are introduced. The permeability of a porous medium (Sect. 4.2.3) and dispersivity (Sect. 7.2.3) are examples of such coefficients. The numerical values of these coefficients can be obtained only *experimentally*, in a process referred to as 'inverse modeling', *model calibration*, or *coefficient estimation*, through the 'history-matching process' that involves inverse modeling, i.e., the use of a model in which the sought coefficients appear, not to predict the simulated system behavior, but to determine the values of the coefficients that appear in the model by minimizing the deviations between model predictions and field measurements (e.g., Bear and Cheng 2010, p. 36).

B. Development of a Numerical Model and Code

Having constructed a mathematical model, in terms of the relevant state variables, it has to be solved for cases of interest. The preferred method of solution is the *analytical* one, as it provides a general solution that can be applied (for the same domain geometry) to various sets of parameters and coefficients. However, because of the complexity of most problems of practical interest (shape of the domain and its boundaries, heterogeneity, non-linearity, irregular source functions, etc.), generally, it is not possible to derive analytical solutions except for relatively simple cases. Numerical methods are usually employed for solving the mathematical model. This means that various methods are used in order to transform the mathematical model into a numerical one, in which the partial differential equations are represented by their numerical counterparts. A computer program, or a *code* is then required in order to solve the problem numerically.

C. Code Verification

When a new numerical model and a code are developed for solving a mathematical model, the code is not considered ready for use unless it undergoes a proper verification procedure. Here, *verification* means checking that the code does what it proclaims to do, namely, to provide a solution which is identical, or sufficiently close to the solution of the mathematical model. Verification involves comparing solutions obtained by using the code with those obtained by analytical methods, whenever such solutions are possible. This is usually done for some simplified domain geometry, homogeneous materials, etc. In many cases, analytical solutions cannot be derived. The only procedure, then, is to compare code solutions with solutions obtained by other codes.

D. Model Validation

Once a model has been selected *for a particular problem*, the model must be *validated*. Model validation is the process of ensuring that the model correctly describes all the relevant processes that affect the excitation-response relations of interest to an acceptable degree of accuracy. The only way to validate a model is an *experiment* conducted on the considered porous medium domain. However, we often validate the model *in principle*, i.e., ensuring that it represents the considered phenomena, by conducting controlled field or laboratory experiments. Unfortunately, unlike laboratory experiments, many features encountered in field experiments, such as field heterogeneity and anisotropy, cannot be controlled or identified, although, in many cases they dominate the system's behavior.

Another problem that arises during model validation may stem from phenomena that occur during the experiment, but are not represented in the model. Fingering due to density and viscosity difference in two phase flow (Sect. 6.4.4) may serve as an example. One should be careful to distinguish between fingering caused by these physical phenomena and those produced by numerical roundoff errors. Finally, problems may arise from the application of models to space and time domains, which are much larger than the ones used for model validation.

E. Model Calibration and Parameter Estimation

Obviously, no model can be employed in any *particular case of interest*, unless *numerical values* are assigned to all the coefficients and parameters that appear in it. We refer to the activity of identifying the values of such model coefficients as the *identification problem*, *inverse problem*, or *parameter estimation problem*.

We use the term *model calibration* for the combines model validation and parameter estimation for a specific problem of interest. Both activities are actually executed simultaneously. Thus, in the procedure of calibration, the values of model coefficients for a considered porous medium domain are determined by solving an inverse problem, using measured data from that investigated domain.

As emphasized above, the only way to obtain the values of these coefficients for a considered porous medium domain is to investigate the real system. Data can be obtained from planned experiments, but, for aquifers and petroleum reservoirs, historical data can also be envisioned as experimental results. Basically, the technique involves finding a set of coefficients that will minimize the difference between observed and predicted data.

It should be emphasized that because the model is only an approximation of the real system, we should never expect these two sets of values—predicted by the model and measured in the actual domain, to be *exactly* identical. Instead, we search for a 'best fit' between them, according to some criteria, and plausible reasonable explanations for possible deviations.

Let us add a few words about what we mean by stating (Sect. 1.1.3) that geometrical porous medium coefficients have to be determined 'experimentally'. As stated above, porous medium coefficients, like permeability, dispersivity, specific area, etc., cannot be determined by direct measurements (see discussion in Sect. 1.2.1), they are obtained by the process of *model calibration* outlined above. This means that we monitor what happens in the domain in terms of the relevant variables of state. Then we solve the model 'backward', seeking the values of the coefficients that will transform the system from its initial to its final state. In this way, we determine the values of the relevant coefficients.

Because field measurements are essential, the question "what do we really measure?" is essential. When we state: 'we measured a certain pressure at a point', say in an observation well, what do we really measure? Is that the pressure p or $\bar{p}$ that appears in the PDE that is part of the model? Or, when we take a one liter water sample from a well to determine its solute concentration, is that concentration represented by the symbol c^γ or $\overline{c^\gamma}$ that appears in the solute balance equation? In order to make the predicted value comparable with the monitored one, both have to be an average over an REV. Is this kind of information provided by field monitoring? When considering pressure, the difference may be small, as pressure propagates fast. However, in the case of *diffusive phenomena*, like solute or heat, spreading is slow, and the difference between predicted and observed values may be significant.

F. Model Applications

Once we have a calibrated model for a considered problem (and this includes all the required site-specific coefficients), the model is ready for use.

G. Analysis of Model Uncertainty and Sensitivity Analysis

This is an important feature of modeling, closely associated with the problem of parameter identification. The term *sensitivity analysis* is used here to describe tools that help the modeler evaluate the impact of uncertainty, say, in the values of model coefficients, on the results predicted by the model.

A sensitivity analysis, and an analysis of the effects of uncertainty must accompany every modeling effort. We are uncertain about many elements associated with the model, e.g., the assumptions included in the conceptual model, and values of model coefficients, including their spatial variations.

The uncertainty in the spatial variability of coefficients' values, leads to various stochastic modeling techniques. The consequence of uncertainty in model parameters and coefficients, is uncertainty in model predictions. In this book we focus on deterministic modeling, i.e., assuming that we know the values of the various coefficients appearing in the model.

1.5.3 Existence, Uniqueness and Stability of Solution

The content of a mathematical model has already been presented in Sect. 1.5.2 A.

The set of equations (balance equations, flux equations, and constitutive relations) must be a *closed* one, i.e., it should contain a sufficient number of independent equations to permit the simultaneous solution for all the state (= dependent) variables associated with the problem.

However, after writing the closed set of equations, we should use the discussion in Sect. 3.9.2 on *degrees of freedom* to determine the number of *primary variables* of the problem and to select the most convenient ones. We then identify an equal number of (partial differential) balance equations (e.g., of mass of fluid phases, mass of chemical species, and energy) that have to be solved in order to determine the values of these variables (whether they appear explicitly in the selected equations or not). All the remaining equations and relationships, including partial differential equations, are then employed in order to determine the remaining variables. Initial and boundary conditions have to be specified only for the partial differential equations that have to be solved.

The solution of a mathematical model of a problem takes the form of spatial and temporal distributions of the state variables of interest within the prescribed space and time domains of the problem.

Not every set of conditions imposed on the boundaries of a problem domain is satisfactory (for a given set of partial differential equations) from the mathematical point of view. This is even more so because, often, we have to resort to estimates of coefficients and simplifications of the mathematical models in specifying the boundary conditions for a considered problem. A mathematical model that represents a *physical reality* (and only such cases are considered here) is said to be *well-posed* if it satisfies the following requirements (e.g., Courant and Hilbert 1962):

- A solution to the problem exists. (*existence*).
- The solution is unique (*uniqueness*).
- The solution is stable (*stability*).

The first requirement simply states that at least one solution does exist. The second one stipulates completeness of the problem statement, with no ambiguity. The third requirement means that small variations in data (e.g., initial and boundary conditions, and/or values of coefficients) should lead to small changes in the resulting solution. If small errors in the data do not lead to correspondingly small errors in the solution, then the mathematical model is *ill-posed*. This last requirement is of particular interest, as all our observations have always some measurement error. A model will be meaningless if these small errors will significantly affect the solution and, hence, the prediction obtained by the model.

Thus, once a complete mathematical model has been stated, the next step is to ensure that it is well-posed. Only then should a solution be sought.

We shall not go into the mathematical analysis of whether a model developed here is well-posed, or not, although, as stated above, this analysis is an essential step in the modeling process. The mathematical models developed and presented in this book, since they are based on a thorough analysis of a physical reality and on the description of this reality, albeit with certain simplifying assumptions, are implicitly *assumed* to be always well-posed. Therefore, they should provide unique, stable solutions. The techniques used in this analysis can be found in appropriate mathematical texts on partial differential equations.

Finally, we should mentioned that in most cases of practical interest, especially when the considered domain is geological formation, the actual solution of the model is implemented by numerical methods of solution. These are not discussed in this book.

References

Abdassah D, Ershaghis I (1986) Triple-porosity system for representing naturally fractured reservoirs. SPE Form Eval 1:113–127

Adler PM, Thovert J-M (1999) Fractures and networks. Kluwer Academic Publishers, Dordrecht, 429 p

Aguilera R (1995) Naturally fractured reservoirs, 2nd edn. Pennwell Books, Tulsa, 520 p

Al-Ghamdi A, Ershaghi I (1996) Pressure transient analysis of dually fractured reservoirs. SPE J 1(1):93–100

Armstrong RT, Ott H, Georgiadis A, Rucker M, Schwing A, Berg S (2014) Subsecond pore-scale displacement processes and relaxation dynamics in multiphase flow. Water Resour Res 50:9162–9176

Bachmat Y, Bear J (1964) The general equation of hydrodynamic dispersion in homogeneous isotropic porous mediums. J Geophys Res 69:2561–2567

Bakhvalov N, Panasenko G (1989) Homogenisation: averaging processes in periodic media. Kluwer Publishing Company, Dordrecht

Barenblatt GI, Zheltov IP, Kochina IN (1960) Basic concepts in the theory of seepage of homogeneous liquids in fissured rocks. J Appl Math Mech (PMM) 24:852–864

Bear J (1972) Dynamics of fluids in porous media. American Elsevier, 764 p (also published by Dover Publications, 1988; translated into Chinese)

Bear J, Bachmat Y (1991) Introduction to modeling phenomena of transport in porous media. Kluwer Publishing Company, Dordrecht, 553 p

Bear J, Cheng AH-D (2010) Modeling groundwater flow and contaminant transport. Springer, Berlin, 834 p

Bear J, Zaslavsky D, Irmay S (1968) Physical principles of water percolation and seepage. UNESCO, 465 p

Bear J, Tsang CF, de Marsily G (eds) (1993) Flow and contaminant transport in fractured rock. Academic Press Inc, San Diego, 560 p

Bensoussan A, Lions JL, Papanicolaou G (1978) Asymptotic analysis of periodic structures. North-Holland, Amsterdam

Berg S, Ott H, Klapp SA, Schwing A, Neiteler R, Brusse N, Makurat A, Leu L, Enzmann F, Schwarz J-O, Kersten M, Irvine S, Stanpanoni M (2013) Real-time 3D imaging of Haines jumps in porous media flow. Proc Nat Acad Sci 110(10):3755–3759

Blunt MJ (2001) Flow in porous media - pore-network models and multiphase flow, current opinion. Colloid Interface Sci 6:197–207

Blunt MJ, Bijeljic B, Dong H, Gharbi O, Iglauer S, Mostaghimi P, Paluszny A, Pentland C (2013) Pore-scale imaging and modelling. Adv Water Resour 51:197–216

Boussinesq J (1904) Recherches théoriques sur l'ecoulement des nappes d'eau infiltrées dans les sol, et sur le débit des sources. *CRH Acad Sci,* Paris, J Math Pure et Appl 10:5–78 (1903), 1(0):363–394
Celia M, Reeves PC, Ferrand LA (1995) Recent advances in pore scale models for multiphase flow in porous media. Rev Geophys 33:1049–1057
Cipolla C, Mack M, Maxwell S (2010) Reducing exploration and appraisal risk in low-permeability reservoirs using microseismic fracture mapping. In: SPE 137437, presented at the Canadian unconventional resources and international petroleum conference, Calgary, Canada, October, 10–21, 2010
Cnudde V, Boone MN (2013) High-resolution X-ray computed tomography in geosciences: a review of the current technology and applications. Earth Sci Revs 123:1–17
Courant R, Hilbert D (1962) Methods of mathematical physics. Wiley Interscience, N.Y., 560 p
Coussy O (2004) Poromechanics. Wiley, New York, 312 p
Coussy O (2007) Revisiting the constitutive equations of unsaturated porous solids using a Lagrangian saturation concept. Int J Numer Anal Meth Geomech 31(15):1675–1694
Cushman JH (1997) The physics of fluids in hierarchial porous media. Kluwer Academic Publishers., Dordrecht, 467 p
Dagan G (1989) Flow and transport in porous formations. Springer, New York, 465 p
Dagan G, Neuman SP (eds) (1997) Subsurface flow and transport: a stochastic approach. Cambridge University Press, Cambridge, 241 p
Debye P, Anderson HR Jr, Brumberger H (1957) Scattering by an inhomogeneous solid *II*. The correlation function and its application. Appl Phys 28(6):679–683
Doughty C (1999) Investigation of conceptual and numerical approaches for evaluating moisture, gas, chemical, and heat transport in fractured unsaturated rock. J Contam Hydrol 38(1–3):69–106
Dupuit J (1863) Études Théoriques et Pratiques sur les Mouvementdes des Eaux dans les Cannaux Découverts et à Travers les Terrains Perméables, 1st edn 1948, 2nd edn. Dunod, Paris, 304 p
Eringen AC (1980) Mechanics of continua, 2nd edn. Krieger Publishing Company, Malabar, 592 p
Feller W (1957) An introduction to probability theory and its applications, 2nd edn. Wiley, New York
Forchheimer P (1886) Uber die Ergiebigkeit von Brunnenanlagen und Sickerschlitzen. Hannover Zeitz. d. Archit u Ing Ver 8(2):538–564
Forchheimer P (1901) Wasserbewegung durch boden. Z Ver Deutsch Ing 45:1782–1788
Fredrich JT (1999) 3D imaging of porous media using laser scanning confocal microscopy with application to microscale transport processes. Phys Chem Earth (A) 24(7):551–561
Gelhar LW (1993) Stochastic subsurface hydrology. Prentice-Hall, Englewood Cliffs
Gelhar LW, Axnes CL (1983) Three dimensional stochastic analysis of macrodispersion in aquifers. Water Resour Res 19:161–180
Gray WG (1975) A derivation of the equations for multi-phase transport. Chem Eng Sci 30:229
Gray WG, Hassanizadeh SM (1989) Averaging theorems and averaged equations for transport of interface properties in multiphase systems. Int J Multiphase Flow 15:81–95
Hassanizadeh SM (1986a) Derivation of basic equations of mass transport in porous media; Part I. Macroscopic balance laws. Adv Water Resour 9:196–206
Hassanizadeh SM (1986b) Derivation of basic equations of mass transport in porous media; Part II. Generalized Darcy's law and Fick's law. Adv Water Resour 9:207–222
Hassanizadeh M, Gray W (1979a) General conservation equations for multiphase systems: 1. Averaging procedure. Adv Water Resour 2:131–144
Hassanizadeh M, Gray W (1979b) General conservation equations for multiphase systems: 2. Mass, momentum, energy and entropy equations. Adv Water Resour 2:191–203
Hassanizadeh SM, Gray WG (1980) General conservation equations for multi-phase systems, 3. Constitutive theory for porous media flow. Adv Water Resour 3:25–40
Hassanizadeh SM, Gray WG (1990) Mechanics and thermodynamics of multiphase flow in porous media including interphase boundaries. Adv Water Res 13:169–186

Holzer L, Cantoni M (2011) Review of FIB-tomography. Nano-fabrication using focused ion and electron beams: principles and applications. Oxford University Press, Oxford
Jikov VV, Kozlov SM, Oleinik OA (1994) Homogenization of differential operators and integral functionals. Springer, Berlin
Joekar-Niasar V, Hassanizadeh SM, Leijnse A (2008) Insights into the relationships among capillary pressure, saturation, interfacial area and relative permeability using pore-scale network modelling. Trans Porous Media 74:201–219
Joekar-Niasar V, Hassanizadeh SM, Dahle HK (2010) Dynamic pore-network modeling of drainage in two-phase flow. J Fluid Mech 655:38–71
Kjetil N, Ringrose SR (2008) Identifying the representative elementary volume for permeability of heterolithic deposits using numerical rock models. Math Geosci 40:753–771
Leal LG (2007) Advanced transport phenomena: fluid mechanics and convective transport processes, Cambridge University Press, Cambridge
Lions J-L (1981) Some methods in the mathematical analysis of systems and their control. Kexue Chubanshe Science Press, Beijing
Low PF (1961) Physical chemistry of clay-water interactions. Adv Agron 13:269–327
Low PF (1976) Viscosity of interlayer water in montmorillonite. Soil Sci Soc Am J 40:500–505
MacDowell AA, Parkinsona DY, Habouba A, Schaiblea E, Nasiatkaa JR, Yeea CA, Jamesona JR, Ajo-Franklina JB, Brodersenb CA, McElronec AJ (2012) X-ray micro-tomography at the advanced light source, developments in X-ray tomography VIII. In: Stock SR (ed) Proceedings of SPIE, p 850618
Marle CM (1981) Multiphase flow in porous media. Editions Technip, Paris, 267 p
Marsden JE, Tromba A (2003) Vector calculus, 5th edn. W. H. Freeman, New York, 790 p
Mikelic A (2000) Homogenization theory and applications to filtration through porous media. In: Fasano A (ed) Filtration in porous media and industrial application, vol 1734. Lecture notes in mathematics. Springer, Berlin, pp 127–214
Molins S, Trebotich D, Steefel CI, Shen C (2012) An investigation of the effect of pore-scale flow on average geochemical reaction rates using direct numerical simulation. Water Resour Res 48(3)
Molins S, Trebotich D, Yang L, Ajo-Franklin J, Ligocki TJ, Shen C, Steefel CI (2014) Pore-scale controls on calcite dissolution rates from flow-through laboratory and numerical experiments. Environ Sci Tech 48(13):7453–7460
Murdoch AI, Hassanizadeh SM (2002) Macroscale balance relations for bulk, interfacial, and common line systems in multiphase flow through porous media on the basis of molecular considerations. Int J Multiph Flow 28:1091–1123
Muskat M (1949) Physical principles of oil production. McGraw-Hill, New York, 922 p
Nayfeh AH (2000) Perturbation methods. Wiley, New York, 437 p
Parker JG (1986) Hydrostatics of water in porous media. In: Sparks DL (ed) Soil physical chemistry. CRC Press, Boca Raton
Qin C, Hassanizadeh SM (2013) Multiphase flow through multilayers of thin porous media: General balance equations and constitutive relationships for a solid-gas-liquid three-phase system, Int J Heat Mass Tran, 70: 693–708
Qin C, Hassanizadeh SM (2015) A new approach to modelling water flooding in a polymer electrolyte fuel cell. Int J Hydrog Energy 40(8):3348–3358
Raeini AQ, Blunt MJ, Bijeljic B (2014) Direct simulations of two-phase flow on micro-CT images of porous media and upscaling of pore-scale forces. Adv Water Resour 74:116–126
Raoof A, Hassanizadeh SM (2012) A new formulation for pore-network modeling of two-phase flow. Water Resour Res 48:13
Raoof A, Hassanizadeh SM, Leijnse A (2010) Upscaling transport of adsorbing solutes in porous media: pore network modeling. Vadose Zone J 9:624–636
Richards LA (1931) Capillary conduction of liquids through porous medium. Physics 1:318–333
Rubin Y (2003) Applied Stochastic Hydrogeology. Oxford University Press, Oxford, 391 p
Sanchez-Palencia E (1974) Comportement local et macroscopique d'un type de milieu physiques héterogénes. Int J Eng Sci 12:331–352

Sanchez-Palencia E (1980) Non-homogeneous media and vibration theory, vol 127. Lecture notes in physics. Springer, N.Y
Schluter S, Sheppard A, Brown K, Wildenschild D (2014) Image processing of multiphase images obtained via X-ray microtomography: a review. Water Resour Res 50:3615–3639
Selyakov VI, Kadet VV (1996) Percolation models for transport in porous media. Kluwer Academic Publishers, Dordrecht, 241 p
Stock SR (2008) Recent advances in X-ray micro-tomography applied to materials. Intern Mater Rev 53(3):129–181
Strobl M, Manke I, Kardjilov N, Hilger A, Dawson M, Banhart J (2009) Advances in neutron radiography and tomography. J Phys D Appl Phys 42(24):3001–3022
Sun N-Z (1999) Inverse problems in groundwater modeling, 2nd edn. Springer, Berlin, 338 p
Sun N-Z, Sun A (2015) Model calibration and parameter estimation: for environmental and water resource systems. Springer, Berlin, 621 p
Sutton RP, Cox SA, Barre, RD (2010) SPE paper 138447 presented at the SPE tight gas completions conference held in San Antonio, Texas, USA, 2–3 November 2010
Ten Dam A, Erentz C (1970) Kizildere geothermal field, Western Anatolia. Geothermics 2(1):124–129 U.S. EIA, 2011
Thewlis J (ed) (1973) Concise dictionary of physics. Pergamon Press, Oxford, 248 p
Truesdell C (1977) Rational thermodynamics. Springer, Berlin, 578 p
Warlick DN (2006) Gas shale and CBM development in North America. Oil Gas J 3(11):1–8
Warren JE, Root PJ (1963) The behavior of naturally fractured reservoirs. J Soc Pet Eng 3:245–255
Whitaker S (1999) The method of volume averaging. Kluwer Acad. Publishers, Dordrecht, 219 p
Wildenschild D, Sheppard AP (2013) X-ray imaging and analysis techniques for quantifying pore-scale structure and processes in subsurface porous medium systems. Adv Water Resour 51:217–246
Williams BB, Gidley JL, Schechter RS (1979) Acidizing fundamentals. Soc Petrol Eng Monograph Ser. 6(2):124 p
Wu YS, Liu HH, Bodvarsson GS (2004) A triple-continuum approach for modeling flow and transport processes in fractured rock. J Contam Hydrol 73(1):145–179
Wu YS, Efendiev Y, Kang Z, Ren Y (2006) A multiple-continuum approach for modeling multiphase flow in naturally fractured vuggy petroleum reservoirs. In: SPE. Beijing Conference
Yaglom AM (1965) Theory of stationary random functions. Prentice Hall, Englewood Cliff
Zhang D, Zhang R, Chen S, Soll WE (2000) Pore scale study of flow in porous media: scale dependency, REV, and statistical REV. Geophys Res Lett 27(8):1195–1198

Chapter 2
Some Elements of Thermodynamics

The objective of this chapter is to introduce a selection of thermodynamics' topics, which are required for the understanding and constructing the flow and transport models considered in this book. Emphasis will be on equilibrium conditions, but not exclusively. No effort will be made to present a complete review of the considered subjects, nor their proofs. These can easily be found in texts on Thermodynamics (e.g., Smith et al. 2005).

Phases and chemical species were defined in Sect. 1.1.1. When a considered fluid phase is composed of N γ-chemical species, with n^γ denoting the number of γ-species moles in a fluid phase, we shall use the following symbols:

$$n = \sum_{\gamma=1}^{N} n^\gamma, \quad m = \sum_{\gamma=1}^{N} m^\gamma, \quad n^\gamma = \frac{m^\gamma}{M^\gamma}, \quad \omega^\gamma = \frac{m^\gamma}{m}, \quad X^\gamma = \frac{n^\gamma}{n},$$

where ω^γ and X^γ denote the *mass fraction* and the *molar fraction* of the γ-species, respectively, and the symbols m^γ, and M^γ denote the mass and molecular mass of the γ-species, respectively (see definitions in Sect. 7.1.1). We add the subscript α (e.g., in n^γ_α) to denote an α-phase. In the case of multi-phase flow (Chap. 6), we often use subscripts w, n and i for wetting, non-wetting and intermediate wetting fluids, respectively.

Throughout the book, we shall introduce many variables of state, coefficients, and material properties. Each of these is meaningless unless it is accompanied by an appropriate unit. In this book, we shall use the *International System* of units (abbreviated *SI*). In the List of Main Symbols, preceding Chap. 1, we present the main symbols used in this book, adding the dimensions and the units employed for that symbol. Occasionally, the unit will be mentioned also in the text itself.

J. Bear, *Modeling Phenomena of Flow and Transport in Porous Media*,
Theory and Applications of Transport in Porous Media 31,
https://doi.org/10.1007/978-3-319-72826-1_2

2.1 Equilibrium

The continuum approach is presented in Sect. 1.1. There, we have introduced the μREV as the volume over which the average of molecular behavior yields the behavior of the microscopic level continuum, while the REV was introduced as the volume over which averages of microscopic values yield a macroscopic continuum description.

As a first step in the discussion of any thermodynamic concept, it is important to understand the meaning of *thermodynamic equilibrium* (TA), and *approximate thermodynamic equilibrium* (ATE).

The assumption of 'local thermodynamic equilibrium' often underlies the discussion on the values of *thermodynamic state variables* at a point within a fluid phase. *Local thermodynamic equilibrium* means that when a μREV centered at a point within a phase is isolated from its surroundings, equilibrium prevails among all chemical species within that μREV. This also means that *mechanical, chemical and thermal* equilibria prevail and properties and conditions within that μREV remain time-invariant. Actually, equilibrium requires uniformity of temperature and of all chemical potentials. When chemical reactions take place, local thermodynamic equilibrium also includes chemical equilibrium among the participating species. Thus, 'local thermodynamic equilibrium' is equivalent to 'local equilibrium at the microscopic level'. Such equilibrium guarantees that the standard *thermodynamic variables*, e.g., temperature, pressure, and chemical potential (as well as density, internal energy, and entropy) can be uniquely defined at every point in the phase continuum. Altogether, the term 'thermodynamic equilibrium' describes a situation in which *thermal*, *chemical*, and *mechanical equilibria* prevail simultaneously. At equilibrium, in the absence of external forces, the entire considered system is at the same pressure. Note that the pressure referred to here should be referred to as *thermodynamic pressure* (see comment in Sect. 2.2.2).

The driving force for local thermodynamic equilibrium is molecular motion and collisions. Thus, the time interval (= *relaxation time*) required for a sufficient number of molecular collisions per unit time to occur determines how fast equilibrium can be reached. Local equilibrium is violated only under severe non-equilibrium conditions, when phenomena take place over a time span that is much shorter than the relaxation time. Shock waves may serve as an example. By definition, the size of the μREV of a phase is selected such that it includes a sufficiently large number of molecules. This means that the μREV must be much larger than the *mean free path* of the molecules so as to enable a sufficiently large number of collisions to occur. This will ensure that meaningful thermodynamic properties are obtained at every point and at every instant of time (in the sense of an average over some Representative Time Interval) (RET).

This is the lower limit of the μREV size. For example, if we have a sufficiently narrow passage, or a rarefied gas, a μREV may not exist, and the gas may not be treated as a continuum. We then have *Knudsen gas flow* (Knudsen 1934), which requires special treatment, using the theory of non-equilibrium statistical mechanics (see Sect. 4.3.3).

An upper limit for the size of a μREV is that it must be much smaller than the domain of interest occupied by the phase. Once we have ascertained that a μREV exists for a considered fluid in a considered microscopic domain, the values of the thermodynamical variables for every point within that domain are obtained by averaging the molecular behavior over the μREV centered at the point.

When considering a multi-species fluid phase, we require (1) that a μREV exists for every chemical species, and (2) that a common μREV can be found for all of them. Then, except for special cases like those mentioned above, 'equilibrium at a (microscopic) point', or 'local thermodynamic equilibrium at a point' means that all chemical species *at a microscopic point*, and their partitioning in the various phases, are in equilibrium.

Henceforth, in this book, we shall assume that local thermodynamic equilibrium always prevails at every *microscopic point* within every phase present in the void space, and that meaningful thermodynamic variables of the phase (e.g., temperature, pressure, phase velocity and species concentrations and partitioning) can be defined at every such point.

Microscopic thermodynamic equilibrium within a phase occupying a domain within an REV means that the temperature and the chemical potentials (to be discussed in Sect. 2.5) of all chemical species comprising the phase are uniform within that domain and that velocities of all phases are zero. Bear and Nitao (1995) and Nitao and Bear (1996) discuss conditions for thermal, chemical, and mechanical equilibria among phases and chemical species within an REV. They show that under conditions of microscopic thermodynamic equilibrium, we have a single value of chemical potential (discussed in Sect. 2.2.6) for every chemical species present in the REV, and a single temperature within the REV.

From entropy considerations, Bear and Nitao (1995) show that in the absence of gravity and surface forces, the pressure within any phase must be uniform, and that the Laplace formula, (2.4.12), discussed below, describes the jump in pressure across an interphase boundary. It may be of interest to note that in the absence of gravity the mean radius of curvature of this interface will be uniform everywhere within an REV, as long as both fluid phases are continuous.

Under *microscopic thermodynamic equilibrium* within a phase domain inside an REV, there exists no net transport of mass of chemical species (chemical equilibrium), of energy (thermal equilibrium), and of linear momentum (mechanical equilibrium) within that phase domain. To accommodate real situations, with the possibility of transport of extensive quantities within a porous medium domain, Bear and Nitao (1995) introduce the concept of *approximate thermodynamic equilibrium*, which allows small gradients in these state variables. They suggest conditions for the existence of such thermodynamic equilibrium. They define *approximate thermodynamic equilibrium* to mean that every REV is sufficiently close, thermodynamically, to an identical, but sealed, system in complete thermodynamic equilibrium. Here 'identical' means that the void space geometrical configurations in the two systems are the same, that they both have the same mass of every species and phase, and that both contain the same amount of internal energy. They also require that the boundary of the sealed system be rigid, so that no work be done on or by the system with respect to its surroundings, and that any force field, such as gravity or surface forces, be the same as for the equivalent system. Since the sealed system is at complete equilibrium, all phase velocities in the sealed system are zero.

Since the macroscopic level of description is obtained by averaging the behavior at the microscopic one, under conditions of microscopic thermodynamic equilibrium, macroscopic values of state variables are equal to their microscopic counterparts. However, we recall that thermodynamic quantities such as pressure and concentration will be uniformly distributed over an REV only if we assume the absence of gravitational and surface forces. The latter always exist between a solid and the thin film of fluid that covers it.

When it is obvious from the text that the discussion is at the macroscopic level, the adjective 'local' can be omitted, and we refer to the situation as 'thermodynamic equilibrium at a point'.

Under the assumed microscopic equilibrium conditions, and in the absence of gravity and surface forces, because pressure, temperature, and chemical potentials are uniform within a phase in the REV, the respective microscopic values are also identical to their macroscopic counterparts.

2.2 Energy, Work, Entropy and Enthalpy

Work, entropy and enthalpy are three fundamental concepts associated with energy and its transport in any phase continuum. In this section, we introduce these concepts only to the extent that is required in order to model the transport of energy in porous medium domains. The discussion follows the concepts and definitions as applicable to a (microscopic) point in a phase continuum; however, we shall follow the phenomenological approach and extend the same concepts to the phase continuum at the macroscopic level.

2.2.1 *Entropy*

A. Work, Energy and the First Law of Thermodynamics

In mechanics, *work*, W, is defined as the scalar product of the two vectors: a *force*, $\mathbf{F}$, and the resulting *displacement*, $\mathbf{w}$, of the point of application, with $dW = \mathbf{F} \cdot d\mathbf{w}$. In thermodynamics, when a *system* applies a force on its surroundings, causing a displacement at the boundary, the above scalar product expresses the work *of the system*. We note that we have always *two* systems that interact. The work of a considered system is considered positive when the latter is doing work on its surrounding; it is negative when the surrounding is doing work on the considered system.

Consider a system and its surroundings. When a force, $\mathbf{F}$, is applied by the surroundings on the boundary of that system, resulting in motion, the work performed by the system is $W = -\int_A^B \mathbf{F} \cdot d\mathbf{w}$.

An *adiabatic process* is one that involves no heat exchange between a considered system and its surroundings. *The first law of thermodynamics states that the work of a system connecting two end states in an adiabatic process depends only on the end states;* it does not depend on the detailed path of the process. Hence, we can define *work* as an *extensive quantity*. Another way to express the *first law is: the work done on a body in an adiabatic process, not involving changes in kinetic or potential energy, is equal to an increase in the internal energy (see below), which is a function of the state of the body.*

Accordingly, *energy*, $\mathbb{E}$, is an extensive quantity that measures the adiabatic work of a system between two end states, $\Delta\mathbb{E} = -W_{adiabatic}$. However if *non-adiabatic heat interactions*, Q, are involved in the passage from one state to the other, then $\Delta\mathbb{E} = Q - W$, i.e., the change in the energy of any process, is equal to the work on the system and the heat input into it. Another form of writing the above balance is $d\mathbb{U} = dQ + dW$, i.e., the change in internal energy is equal to the heat added to the system plus the work done *on* it. The above is another, equivalent, form of the *first law of thermodynamics*.

B. Entropy and the Second Law of Thermodynamics

Clausius inequality states that for any *closed system* (i.e., a system that is not interacting with its surroundings by exchanging mass) undergoing a cyclic process, we have:

$$\oint \left(\frac{\delta Q}{T} \right) \leq 0, \tag{2.2.1}$$

where δQ denotes the heat absorbed by the body (= system) at a thermodynamic temperature, T. Here, and elsewhere in this chapter, the symbol δ is used to denote a small quantity which is *not an exact differential*. This means that integration requires detailed information on the pathway/evolution of the process. The $<$ sign is applicable to an*irreversible* process, while the equal sign applies to a *reversible* one. A consequence of the above statement is that the integral of $(\delta Q/T)_{rev}$ between two points (= states) in a reversible process is independent of the path selected for the transition. Hence, the quantity $(\delta Q/T)_{rev}$ qualifies as an *extensive quantity*, with:

$$d\mathbb{S} \equiv \left. \frac{\delta Q}{T} \right|_{rev}, \tag{2.2.2}$$

where $\mathbb{S}$ is called *entropy*. Entropy represents a measure of the availability (or the lack thereof) of a system's thermal energy for conversion into mechanical work. It can also provide a measure of the degree of randomness in a system. It expresses the added heat per degree K and has the dimensions of $ML^2T^{-2}\Theta^{-1}$, where Θ represents temperature. In the *International System of Units*, it is measured in Joules per degree Kelvin ($\equiv$ *Joule/Kelvin*). The change in entropy of a closed system, or body, between two states, A and B, is given by:

$$d\mathbb{S} = \mathbb{S}_B - \mathbb{S}_A = \int_A^B \frac{\delta Q}{T}\bigg|_{rev}, \tag{2.2.3}$$

which is an extensive quantity; its value is assigned relative to an arbitrarily selected state. The corresponding intensive quantity is the specific entropy (= entropy per unit mass, here per mole), denoted by s.

It can be shown that for any process in which a body is transformed from state A to state B:

$$\Delta\mathbb{S} \equiv \mathbb{S}_B - \mathbb{S}_A \geq \int_A^B \left(\frac{\delta Q}{T}\right). \tag{2.2.4}$$

For an infinitesimal change:

$$d\mathbb{S} = \frac{\delta Q}{T}\bigg|_{rev}, \quad d\mathbb{S} > \frac{\delta Q}{T}\bigg|_{irrev}. \tag{2.2.5}$$

A consequence of the above statements is that for an isolated system, or for a system undergoing an adiabatic process, i.e., one in which the net heat transfer to or from the considered system is zero, $\delta Q = 0$, and therefore:

$$d\mathbb{S}_{adiabatic} \geq 0, \tag{2.2.6}$$

where the equal sign holds for a reversible process.

We may now introduce the *second law of thermodynamics*, which states that *the entropy of a system in an adiabatic enclosure can never decrease; it increases in an irreversible process and remains constant in a reversible one,* i.e., $d\mathbb{S} \geq 0$, or $d\mathbb{S}/dt \geq 0$. Often, the second law is presented in the form:

$$d\mathbb{S} \geq \frac{dQ}{T}\bigg|_{closed\ system}. \tag{2.2.7}$$

2.2.2 Enthalpy and Internal Energy

The *Enthalpy*, $\mathbb{H}$, is an extensive property; it is also a *function of state*. It is a useful expression for energy in many chemical and physical systems, because it simplifies the description of energy transfer. Its usefulness stems from its ability to take into account energy changes that are due to the change of volume of a considered system. It is defined as:

$$\mathbb{H} = \mathbb{U} + p\mathbb{V}, \quad \Rightarrow \quad h = u + p\,v, \tag{2.2.8}$$

in which $\mathbb{U}$ and u denote *internal energy* and *specific internal energy*, respectively, $\mathbb{H}$ and h denote *enthalpy* and *specific enthalpy*, respectively, $\mathbb{V}$ and v denote volume and *specific volume*, respectively, and p denotes *thermodynamic pressure*, defined as energy per unit volume. The specific values (e.g., *specific enthalpy*) in (2.2.8) are per unit mass. However, often it is more convenient to express specific values per mole, e.g., *molar enthalpy* means enthalpy per mole. We use the symbol $\check{e}$ for molar density of E, i.e., E per mole.

A comment on mechanical pressure vs. thermodynamic pressure is appropriate here. The former is always associated with an area, as it is defined as *force per unit area*. The latter, in the *kinetic theory of gases*, is associated with the random motion of molecules in a fluid occupying a closed container, e.g., $p = (N/\mathbb{V})\mathrm{m}\overline{V^2}$, where m is the mass of a molecule, N is the number of molecules in the container of volume $\mathbb{V}$, and V denotes their velocity. This pressure acts (as force per unit area) on the walls of the container. Thus, although these two pressures are defined differently, they are actually the same. In practice, as in this book, we shall overlook the difference between them. It is interesting to mention that pressure may be interpreted also as energy per unit volume.

A most useful property of enthalpy and of internal energy is that they are completely additive and path-insensitive, depending only on the initial and final states of the considered system.

The internal energy of a system is another extensive quantity. The corresponding intensive quantity is the specific internal energy (= internal energy per unit mass), $u = u(T, v)$. Hence, the change in u can be expressed as:

$$du = \left.\frac{\partial u}{\partial T}\right|_v dT + \left.\frac{\partial u}{\partial v}\right|_T dv, \quad v = \frac{1}{\rho}. \tag{2.2.9}$$

With *heat capacity at constant volume*:

$$C_v = \left.\frac{\partial \mathbb{U}}{\partial T}\right|_v, \tag{2.2.10}$$

and *specific heat capacity at constant volume*:

$$c_v = \left.\frac{\partial u}{\partial T}\right|_v, \tag{2.2.11}$$

the change in *specific internal energy* is given by:

$$du = c_v dT + \left.\frac{\partial u}{\partial v}\right|_T dv. \tag{2.2.12}$$

The *specific enthalpy*: h, obeys:

$$h = h(T, p) = u + \frac{p}{\rho} = u + p\,v, \quad v = \frac{1}{\rho}, \tag{2.2.13}$$

in which u denotes the specific internal energy. In fact, the relationship $h = h(T, p)$ is an *equation of state* that varies from one substance to the other. Thus, in the general case, we have:

$$dh = \left.\frac{\partial h}{\partial T}\right|_p dT + \left.\frac{\partial h}{\partial p}\right|_T dp. \tag{2.2.14}$$

The variation of enthalpy with temperature is also an extensive quantity:

$$C_p = \left.\frac{\partial \mathbb{H}}{\partial T}\right|_p, \tag{2.2.15}$$

called *heat capacity at constant pressure*. The corresponding intensive quantity is the *specific heat capacity*, c_p: defined by:

$$c_p = \left.\frac{\partial h}{\partial T}\right|_p. \tag{2.2.16}$$

From (2.2.2), we have for any process;

$$\delta Q_{rev} = T\,d\mathbb{S}. \tag{2.2.17}$$

With $\delta W_{rev} = -p\,d\mathbb{V}$, the first law for a reversible process takes the form:

$$d\mathbb{U} = T d\mathbb{S} - p\,d\mathbb{V}. \tag{2.2.18}$$

From the above relationships it follows that:

$$du = T ds - p\,dv, \quad \text{and} \quad dh = T ds + v\,dp. \tag{2.2.19}$$

Equations (2.2.18) and (2.2.19) are valid for a single species phase, or when the composition of the phase (say, in terms of species concentrations) is unchanged.

When we wish to take into account the composition of a liquid phase, in terms of its chemical species, each with its mass fraction ω^γ, $\gamma = 1, 2, \ldots, NC$, and chemical potential μ^γ, discussed in Sect. 2.5, then $u = u(s, v, \omega^\gamma; \gamma = 1, 2, \ldots, NC)$, and:

$$du = T ds - p\,dv + \sum_{(\gamma)} \mu^\gamma d\omega^\gamma. \tag{2.2.20}$$

Since:

$$\left.\frac{\partial u}{\partial s}\right|_{v,\omega^\gamma} = T, \quad \left.\frac{\partial u}{\partial v}\right|_{s,\omega^\gamma} = -p, \quad \left.\frac{\partial u}{\partial \omega^\gamma}\right|_{s,v,\omega^\delta,\delta\neq\gamma} = \mu^\gamma. \tag{2.2.21}$$

Without (or when neglecting) changes in γ-concentration, we have:

$$ds = \left.\frac{\partial s}{\partial T}\right|_{v,\omega^\gamma} dT + \left.\frac{\partial s}{\partial v}\right|_{T,\omega^\gamma} dv. \tag{2.2.22}$$

Since:

$$\left.\frac{\partial s}{\partial v}\right|_{T,\omega^\gamma} = \left.\frac{\partial p}{\partial T}\right|_{v,\omega^\gamma}, \quad c_v = \left.\frac{\partial \mathbb{U}}{\partial T}\right|_{v,\omega^\gamma} = T\left.\frac{\partial s}{\partial T}\right|_{v,\omega^\gamma}, \tag{2.2.23}$$

where c_v is the *specific heat* of the considered phase per unit mass at constant volume. Hence,

$$du = \left(T\left.\frac{\partial p}{\partial T}\right|_{v,\omega^\gamma} - p\right)dv + v_v dT + \sum_{(\gamma)} \mu^\gamma d\omega^\gamma. \tag{2.2.24}$$

More about enthalpy changes associated with chemical reactions will be presented in the following subsection.

2.2.3 Gibbs Free Energy

In this subsection, we present the definition and a brief discussion on the concept of *Gibbs free energy*, introduced by Gibbs (1873; e.g., in Denbigh 1981, p. 231). We shall demonstrate the usefulness of this tool for determining the mass action constant of any chemical reaction when we know its value for minerals, gases and dissolved chemical species. The Gibbs free energy can also be used to express the energy released or added during a chemical reaction.

The thermodynamic quantity, $\mathbb{G} = \mathbb{G}(p, T, n^\gamma, \gamma = 1, \ldots, NC)$, is called *Gibbs free energy*. It is an extensive quantity that depends on the temperature, the pressure and the phase composition of a given material body. It is defined as:

$$\mathbb{G} = \mathbb{H} - T\mathbb{S}, \tag{2.2.25}$$

where $\mathbb{H}$, T, and $\mathbb{S}$ denote the enthalpy (or heat content), the temperature, and the entropy, respectively.

The enthalpy is related to the internal energy, $\mathbb{U}$, and to the pressure, p, by (2.2.8). Thus, for a given system, we have:

$$\mathbb{G} = \mathbb{U} + p\mathbb{V} - T\mathbb{S}. \tag{2.2.26}$$

In the above two equations, using the SI system, $\mathbb{G}$, $\mathbb{U}$ and $\mathbb{H}$ are in joule, p is in pascal, $\mathbb{V}$ is in m^3, T is in kelvin, and $\mathbb{S}$ is in joule per kelvin.

For a differential change in $\mathbb{G}$, we have:

$$d\mathbb{G} = d\mathbb{U} + pd\mathbb{V} + \mathbb{V}dp - Td\mathbb{S} - \mathbb{S}dT,$$

in which $p\,d\mathbb{V}$ describes the work done *by* the system.

From the *first law of thermodynamics*, it follows that for a system that undergoes a reversible transformation upon the application of heat, δQ, we have

$$\delta Q = d\mathbb{U} + p\,d\mathbb{V} + \delta W'_{rev},$$

where $\delta W'_{rev}$ denotes all the reversible work other than pressure-volume work (i.e., work due to expansion) performed by the system.

From the *second law*, we have $\delta Q = T\delta\mathbb{S}$, thus leading to

$$d\mathbb{U} = Td\mathbb{S} - p\,d\mathbb{V} - \delta W'_{rev}.$$

Hence, we may express the *differential change* in $\mathbb{G}$ in the form:

$$d\mathbb{G} = -\mathbb{S}dT + \mathbb{V}dp - \delta W'_{rev}. \tag{2.2.27}$$

In the absence of work other than that due to expansion, the last term on the right-hand side vanishes, and we have:

$$d\mathbb{G} = -\mathbb{S}dT + \mathbb{V}dp. \tag{2.2.28}$$

If we consider a finite isothermal change of state of a system, say from state A to state B, then, since $dT = 0$, we obtain from (2.2.27):

$$\Delta\mathbb{G} = \int_A^B d\mathbb{G} = \mathbb{G}\big|_B - \mathbb{G}\big|_A = \int_A^B \mathbb{V}dp - \int_A^B dW'_{rev}. \tag{2.2.29}$$

If the pressure also remains unchanged during the isothermal change, then:

$$\Delta\mathbb{G} = \int_A^B d\mathbb{G} = \mathbb{G}\big|_B - \mathbb{G}\big|_A = -\int_A^B dW'_{rev} = -\Delta W'_{rev}. \tag{2.2.30}$$

At constant pressure and temperature, the only change is due to chemical reactions, $d\mathbb{G} = -\delta W'_{rev}$. This may serve as a definition for $\mathbb{G}$. The negative of $d\mathbb{G}$ gives the reversible energy available to perform work (e.g., chemical work) other than that associated with pressure. In other words, if a reversible change is taking place in a system at constant pressure and temperature, the work done by the system, $\delta W'_{rev}$, excluding the work of expansion against constant pressure, equals the decrease in the free energy of the system. Actually, at constant p and T, the only change in the system would come from chemical reactions, thus making it almost impossible to maintain constant p, T.

Starting from (2.2.28) and moving along a constant temperature line, we have $d\mathbb{G}|_T = \mathbb{V}dp$, which can also be expressed in the form:

$$d\mathbb{G}|_{T=const.} = ZRTd\ln p. \quad \text{For an ideal gas:} \quad Z = 1. \tag{2.2.31}$$

Recall that an ideal gas is one in which the individual molecules are assumed not to interact with each other, an assumption that significantly simplifies the equation of state.

2.2.4 Chemical Potential and Fugacity

Another approach for handling the difference between the behavior of an ideal gas and that of a real one is the use of the notion of *fugacity*. This concept plays an important role in thermodynamics, especially in connection with a gas, when we consider transport with chemical reactions. It may be regarded as a fictitious pressure equal to the pressure of an ideal gas which has the same chemical potential as the real gas (Denbigh 1981, p. 122). It serves as an effective pressure which replaces the true (mechanical) pressure in accurate chemical equilibrium calculations. Thus, the fugacity of a gas, $f = f(p, T)$, is a pseudo-pressure, such that:

$$d\mathbb{G}|_{T=const.} = RTd(\ln f). \tag{2.2.32}$$

By requiring also that as pressure approaches zero, and the gas' behavior approaches that of an ideal gas, the fugacity will approach the value of the pressure itself. This leads to:

$$\lim_{p\to 0}\left(\frac{f}{p}\right) = 1. \tag{2.2.33}$$

Consider a species A in an ideal gas mixture. Its chemical potential is given by (e.g., Denbigh 1981):

$$\mu^A = \mu^{*A} + RT\ln\frac{p^A}{p^*}, \tag{2.2.34}$$

where p^A is the *partial pressure* of A, and p^* is a reference pressure. Here, μ^{*A} is the chemical potential of an ideal gas at a reference pressure p^* that consists entirely of the single species A. This state is sometimes called the '*ideal gas state*'. It can be shown that μ^{*A} depends only on the temperature, that is, $\mu^{*A} = \mu^{*A}(T)$. We shall follow the convention of setting $p^* = 1$ in the units that are being used for pressure, so that (2.2.34) becomes:

$$\mu^A = \mu^{*A} + RT\ln p^A. \tag{2.2.35}$$

The generalization of the above expression to a non-ideal gas involves introducing the *fugacity*, f^A, of the species A. It is defined by

$$f^A \equiv f^{*A} \exp\left\{(\mu^A - \mu^{*A})/RT\right\},$$

where μ^{*A} is the same term as in (2.2.35), and f^{*A} is the *reference fugacity*. The unit of fugacity is the same as that of pressure.

2.2.5 Partial Pressure in a Gas-Liquid System

The partial pressure of a chemical species, γ, in a mixture of species (= phase), is the pressure, p^γ, that would prevail in that species, at the same temperature, T, when it is the sole species in the considered phase.

The material presented in this subsection will be useful when considering inter-phase mass transfers in porous medium domains (Sect. 7.4).

Consider two adjacent domains–a *solution* (= liquid, *sol*) and a *vapour* (= gas, *vap*)–separated by an interface, and assume that the two domains are in *equilibrium*. A point on that (microscopic) interface may be assumed to belong to both phases. Then, the chemical potentials at such point obey:

$$\mu^\gamma_{sol} = \mu^\gamma_{vap}, \tag{2.2.36}$$

where we have regarded each gaseous species in a vapour (= gas) phase as a 'chemical species'. Under conditions at which the vapour may be assumed to behave as an *ideal gas*, e.g., sufficiently low pressure, we may use (2.2.35), to determine the chemical potential of any chemical species in solution in terms of its partial pressure, p^γ:

$$\mu^\gamma_{sol} = \mu^\gamma_o + RT \ln p^\gamma, \quad \Rightarrow \quad \mu^\gamma_{vap} = \mu^\gamma_o + RT \ln p^\gamma, \quad \mu^\gamma_o = \mu^\gamma_o(T). \tag{2.2.37}$$

Following are three laws that deal with partial pressure of multi-species gas liquid phases.

- *Dalton's law*

This law deals with a mixture of gases. It states that the (total) pressure of a gas phase is equal to the sum of the partial pressures of the individual γ-species comprising that gas phase:

$$p_g = \sum_{(\gamma)} p^\gamma_g. \tag{2.2.38}$$

The above statement is equivalent to saying that there is no interaction among the species in the gas; each gas species behaves independently within the domain occupied by the gaseous phase. We may also write: $p_g = \sum_{(\gamma)} p^\gamma_g$. With $m_g = \sum_{(\gamma)} m^\gamma_g$, or $m_g = \sum_\gamma M^\gamma_g$, we have:

$$p_g^\gamma = X_g^\gamma p_g, \quad p_g = \sum_{(\gamma)} p_g^\gamma, \tag{2.2.39}$$

where X_g^γ denotes the molar fraction of the γ-species in the gas.

- *Raoult's law*

Consider a domain occupied by a mixture of two gases, γ and δ. Raoult's law states that the partial pressures of these gases are:

$$p_g^\gamma = X_g^\gamma p_g^{o\gamma}, \quad p_g^\delta = X_g^\delta p_g^{o\delta}, \tag{2.2.40}$$

where $p_g^{o\gamma}$ and $p_g^{o\delta}$ denote the vapour pressures of pure γ and δ gases, respectively, all the same temperature. Since $X_g^\gamma + X_g^\delta = 1$, it follows from the above two equations that:

$$p_g \equiv p_g^\gamma + p_g^\delta = p_g^{o\gamma} X_g^\gamma + (1 - X_g^\gamma) p_g^{o\delta} \tag{2.2.41}$$

Denbigh (1981, p. 223) comments that "mixtures that obey the law over the whole range of composition are the exception rather than the rule, but an approximation to the ideal behaviour is usually found whenever the components are closely similar in molecular structure".

- *Henry's law*

We are considering liquid-gas exchange across a common interface. A volatile chemical species dissolved in a liquid is present in the adjacent gaseous phase. At a liquid gas interface, *volatilization*, and dissolution phenomena occur, driven by the difference between the partial pressure of the considered substance in the liquid and in the gas bordering that liquid.

For the partial pressure of a γ-species in a liquid phase (ℓ), e.g., a γ-solute in a ℓ-solvent, Henry's law states that:

$$p_\ell^\gamma = \mathcal{H}_\ell^\gamma X_\ell^\gamma. \tag{2.2.42}$$

This law is valid for dilute solutions. It is interesting to note that Denbigh (1981, p. 225) writes $p_\ell^\gamma \rightarrow \mathcal{H}_\ell^\gamma X_\ell^\gamma$ as $X_\ell^\gamma \rightarrow O$, or $\partial p_\ell^\gamma / \partial X_\ell^\gamma \rightarrow \mathcal{H}_\ell^\gamma$ as $X_\ell^\gamma \rightarrow 0$.

In Sect. 7.4.3, in the discussion on Henry's law, we present the case of the volatile species γ (e.g., benzene, $C_6H_{6(g)}$) partitioned between water (w) and air (a). For that case, Appelo and Postma (2005, p. 491) present Henry's law for water (w) and air (a) also in the form:

$$\mathcal{H}'^{\gamma}_{w,a} = \frac{\mathcal{H}^\gamma_{w,a}}{RT} = \frac{X_a^\gamma}{X_w^\gamma}, \tag{2.2.43}$$

where $\mathcal{H}'^{\gamma}_{w,a}$ is another Henry coefficient. In fact, many forms of Henry's law can be found in the literature. In many of them, Henry's coefficient is denoted by K_H.

2.2.6 Gibbs Free Energy and Chemical Reactions

At one atmosphere, the enthalpy, say of one mole of ice at 0 °C is different from that of one mole of liquid water. The difference is the enthalpy change during fusion, or *heat of fusion*, $\Delta\check{h}_{fus}$, where $\check{h}$ indicates the *molar enthalpy density*, i.e., $\mathbb{H}$ (in joules) per mole. When liquid water is transformed into water vapour under constant pressure, there is an enthalpy (or heat) of vaporization, Δh_{vap}. For a solid-to-vapour transition under constant pressure, there is an enthalpy of sublimation, Δh_{sub}. According to convention, the sign of these $\Delta\check{h}$-values are positive when heat is absorbed in the transition, i.e., from solid to liquid, solid to vapour and liquid to vapour. For example, for H_2O at 1 atm,

$$\begin{aligned} H_2O_{(l)} &\rightarrow H_2O_{(g)}, \quad \Delta\check{h}_{373,vap} = 40656\,\text{J/mol}; \\ H_2O_{(s)} &\rightarrow H_2O_{(l)}, \quad \Delta\check{h}_{273,fus} = 6008\,\text{J/mol}; \\ H_2O_{(s)} &\rightarrow H_2O_{(g)}, \quad \Delta\check{h}_{273,sub} = 44919\,\text{J/mol}; \end{aligned}$$

Another example is the burning of coal (graphite) at 25 °C:

$$C_{(s,gr)} + O_{2(g)} \rightarrow CO_{2(g)}, \quad \Delta\check{h}_{298} = -393.5\,\text{kJ/mol}$$

i.e., when 1 mol of carbon (graphite) at 25 °C reacts with 1 mol of oxygen gas, also at 25 °C, to form one mole of carbon dioxide, and when the product, one mole of $CO_{2(g)}$, is cooled back to 25 °C, the system, in its final state, contains less enthalpy than it did initially in the form of the elements carbon and oxygen. The system before and after the reaction is assumed to be at a constant pressure of 1 atmosphere. This means that to achieve the cooling, we need to remove heat at the rate of 393.5 kJ/mole from the formed $CO_{2(g)}$.

When a chemical reaction is accompanied by a decrease in enthalpy, a quantity of heat equal to the enthalpy change is transferred to the surrounding and the reaction is said to be *exothermic*. When the chemical reaction results in an increase in enthalpy of the system, a quantity of heat equal to the enthalpy change must be supplied from the surrounding, and the reaction is then *endothermic*. It is also possible to express the energy released or added during a chemical reaction in terms of the Gibbs free energy, $\mathbb{G}$, rather than in terms of the enthalpy, $\mathbb{H}$.

Chemical reactions will be introduced and discussed in Chap. 7. However, for the purpose of the discussion here, let us introduce the chemical reaction as described by the generalized stoichiometric equation:

$$a\text{A} + b\text{B} + \cdots \rightleftharpoons p\text{P} + q\text{Q} + \cdots, \tag{2.2.44}$$

in which $A, B, \ldots, Q, ..$ are chemical compounds. For this reaction, the *law of mass action* (see any book on Chemistry) takes the form:

$$K_{\mathrm{eq}} = \frac{\{\mathrm{P}\}^p\{\mathrm{Q}\}^q \ldots}{\{\mathrm{A}\}^a\{\mathrm{B}\}^b \ldots}, \tag{2.2.45}$$

in which a bracketed quantity $\{\cdot\}$ denotes *activity* (which serves as a measure of the effective concentration of a species in a mixture) and K_{eq} denotes the *equilibrium constant* (Sect. 7.3.3 B).

In a multi-species phase, the Gibbs free energy, $\mathbb{G}$, depends on the amount of each chemical species, expressed in terms of molar concentration defined in Sect. 7.1, n^γ, $\gamma = 1, 2, \ldots, N$, the pressure, and the temperature, i.e.,

$$\mathbb{G} = \mathbb{G}(p, T, n^1, n^2, \ldots, n^N). \tag{2.2.46}$$

With $\check{g}$ denoting the *molar specific Gibbs free energy*, $\Delta\check{g}_r$, of the r'th reaction is expressed in the form:

$$\Delta\check{g}_r = \Delta\check{g}_r^o + RT \ln \frac{\{\mathrm{P}\}^p\{\mathrm{Q}\}^q \ldots}{\{\mathrm{A}\}^a\{\mathrm{B}\}^b \ldots}, \quad \Delta\check{g}_r^o = -RT \ln K_{eq}, \tag{2.2.47}$$

in which $\Delta\check{g}_r^o$ is the *molar standard specific Gibbs free energy* of the reaction, which is equal to $\Delta\check{g}_r^o$ when each product and reactant is present in the solution at unit activity (defined in (7.3.55)) and at standard conditions (25 °C, 10^5 Pa), R is the gas constant (= 8.314 J/(K mol)), and T is the absolute temperature in Kelvin degrees (= ^{o}C + 273.15). At equilibrium,

$$\Delta\check{g}_r^o = -RT \ln K_{eq}, \quad \Longrightarrow \quad \ln K_{eq} = -\frac{\Delta\check{g}_r^o}{RT}. \tag{2.2.48}$$

According to Appelo and Postma (2005, p. 133), Eq. (2.2.48) facilitates the determination of the mass action constant for any reaction from tabulated data on the molar (i.e., per mole) free energy of formation, $\Delta\check{g}_f^o$, i.e., the energy required to produce one mole of a substance from pure elements in their most stable form. By definition, the latter, and the H^+ ion, have zero value. Thus, the $\Delta\check{g}_r^o$'s are calculated from;

$$\Delta\check{g}_r^o = \Delta\check{g}_{f,products}^o - \Delta\check{g}_{f,reactants}^o.$$

For example, to determine the solubility product of *calcite* from the Gibbs free energy of formation at 25 °C,

$$CaCO_3 \rightleftharpoons Ca^{2+} + CO_3^{2-},$$

we have (Wagman et al. 1982):

$$\Delta\check{g}_{f,CaCO_3}^o = -1128.8\,\text{kJ/mol},$$

$$\Delta\check{g}_{f,Ca^{2+}}^o = -553.6\,\text{kJ/mol},$$

$$\Delta \check{g}^o_{f,CO_3^{2-}} = -527.8\,\text{kJ/mol},$$

and, therefore, the relationship:

$$\Delta \check{g}^o_r = \Delta \check{g}^o_{f,Ca^{2+}} + \Delta \check{g}^o_{f,CO_3^2} - \Delta \check{g}^o_{f,CaCO_3},$$

together with (2.2.48), leads to $\log K_{eq} = -8.30$.

Referring to (2.2.47), in view of the above relationships, we have:

$$\begin{aligned} \Delta \mathbb{G}_r &> 0, && \text{the reaction proceeds from left to right,}\\ \Delta \mathbb{G}_r &= 0, && \text{the reaction is at equilibrium,}\\ \Delta \mathbb{G}_r &< 0, && \text{the reaction proceeds from right to left.} \end{aligned}$$

From (2.2.46) it follows that the total differential, $d\mathbb{G}$, is expressed by:

$$d\mathbb{G} = \left(\frac{\partial \mathbb{G}}{\partial T}\right)\bigg|_{p,n^\gamma} dT + \left(\frac{\partial \mathbb{G}}{\partial p}\right)\bigg|_{T,n^\gamma} dp + \sum_{(\gamma)} \left(\frac{\partial \mathbb{G}}{\partial n^\gamma}\right)\bigg|_{p,T,n^\delta;\delta\neq\gamma} dn^\gamma. \quad (2.2.49)$$

The first term on the right-hand side expresses the change in $\mathbb{G}$ under constant pressure and composition. The second term expresses the change in $\mathbb{G}$ under constant temperature and composition. The third term expresses the change in $\mathbb{G}$ for a γ-species under constant pressure, temperature, and concentration of all other species except γ. By comparison with (2.2.28) for constant composition, we obtain:

$$\left(\frac{\partial \mathbb{G}}{\partial T}\right)\bigg|_{p,n^\gamma} = -\mathbb{S}, \qquad \left(\frac{\partial \mathbb{G}}{\partial p}\right)\bigg|_{T,n^\gamma} = \mathbb{V}. \quad (2.2.50)$$

Hence, we may write (2.2.49) in the form:

$$d\mathbb{G} = -\mathbb{S}dT + \mathbb{V}dp + \sum_{(\gamma)} \mu^\gamma dn^\gamma, \quad (2.2.51)$$

where μ^γ, defined by

$$\mu^\gamma \equiv \left(\frac{\partial \mathbb{G}}{\partial n^\gamma}\right)\bigg|_{p,T,n^\delta;\delta\neq\gamma}, \quad (2.2.52)$$

is the *chemical potential*, or *molar free energy* (as energy per mole) of the γ-species (Sect. 2.2.6).

The *chemical potential*, introduced by Gibbs (1873), is analogous to temperature and pressure. While a temperature difference determines the rate and direction of heat movement from one body to another, and a pressure difference determines the motion of a body, the difference in chemical potential will produce movement of a chemical species within a phase, or from one phase to another. It will also determine the direction of chemical reactions.

From (2.2.51), it follows that the chemical potential of a chemical species expresses the change (increase or decrease) in the capacity of a species to do work (other than work of expansion) of added species, at constant temperature and pressure.

Finally, two phases, α and β, with a common microscopic interphase boundary, will be in *chemical equilibrium* with respect to a γ-species, when:

$$\mu_\alpha^\gamma = \mu_\beta^\gamma \quad \text{for all} \quad \gamma. \tag{2.2.53}$$

In general, a system is said to be in *thermodynamic equilibrium* if the thermodynamic potentials (= chemical potentials and temperature) are uniform within each homogeneous region of the system and do not change with time. We often distinguish between thermodynamic equilibrium, thermal equilibrium (when the temperature is uniform), chemical equilibrium (when the chemical potential is uniform), and mechanical equilibrium (when the pressure is uniform). More detailed definitions are given by Bear and Nitao (1995).

It is interesting to note the relationship between μ of an ideal gas and that of a real gas, using the concept of fugacity, f, introduced in above. With μ denoting the chemical potential of a pure real gas at a temperature T, we have:

$$\mu_g = \mu_g^o + RT \ln f, \quad \text{with} \quad \frac{f}{p_g} \rightarrow 1, \quad \text{as} \quad p_g \rightarrow 0. \tag{2.2.54}$$

* * *

Altogether, we have introduced the temperature, T, which can be used to determine thermal equilibrium, and four extensive quantities: the internal energy, $\mathbb{U}$, the entropy, $\mathbb{S}$, the enthalpy, $\mathbb{H}$, and the Gibbs free energy, $\mathbb{G}$, and their corresponding intensive quantities (= functions of state), u, s, h, and g (i.e., per unit mass). We recall that u, s and h, and their changes, are independent of the path between their initial and the final states. This means that we can write balance equations not only for mass, momentum and energy, but also for internal energy, entropy and enthalpy.

2.3 Phase Behavior

As already emphasized in Sect. 1.1.2 A, the core of any transport model contains the partial differential equations that describe the balances of the transported extensive quantities of interest of the relevant phases. However, these balance equations are general. They contain no information on the nature and behavior of the *particular* fluid, solid and gaseous phases involved in any specific investigated case. This observation is important in view of the fact that different fluids and solids behave differently as pressure, temperature, stress, and solute concentrations vary. Moreover, their *state of aggregation*– solid, liquid, gas – may also vary during a considered case. For example, the density of a liquid and that of a solid vary in different ways when conditions (e.g., pressure and temperature) vary. Furthermore, within each of these states, different materials behave differently. In what follows, we shall focus on specific equations of state.

2.3.1 Phase Change Under Equilibrium

We consider a spatial domain, referred to as a *system*, containing one or more phases. The vicinity of a point in a porous medium domain (e.g., an REV) may serve as an example of such a system. The system may contain a number of substances in the form of different phases. It may also contain a number of species within each of the phases. Phase changes may occur. In the course of time, each substance may change its *thermodynamic state*– gas, liquid, solid. Nevertheless, let us assume, at least as a first approximation, that the processes of transport and transformation within a considered system, from one state to another, are sufficiently slow, so as to allow for spatial variations of the relevant state variables within the considered domain to smooth out and to bring the phases at a point (meaning in the close vicinity of the point) to a state of equilibrium with each other, or close to it. Under such conditions, the composition of the system is subjected to the relationship, known as *Gibbs phase rule*,

$$\mathrm{NF} = \mathrm{NC} + 2 - \mathrm{NP}, \tag{2.3.1}$$

where NF is usually referred to as the number of *degrees of freedom* (see Sect. 3.9). It is also the number of independent state variables, i.e., variables that characterize the state of the system. NP denotes the number of phases comprising the system, and NC denotes the number of different chemical constituents (= thermodynamic species) within the considered system. More on the number of degrees of freedom is presented in Sect. 3.9.

It is important to emphasize that here we are considering the notion of a *thermodynamic species*, which means that *the same chemical substance found in different phases represents only a* single species. Since NF ≥ 0, the phase rule imposes a restriction on the coexistence of phases and, hence, also on the possibility of phase change.

As an example, consider a domain containing only a single chemical compound, say CO_2, as a single phase, say liquid, *or* vapour, or frozen solid. In this case NC = NP = 1, and the state of the system at equilibrium is fully determined by $2 + 1 - 1 = 2$ independent variables, say p and T, or p and ρ. Since $\rho = \rho(p, T)$, two variables will determine the third. Every point on a *pressure-temperature diagram*, e.g., the one shown as Fig. A-10a, represents a *possible state* of the considered substance. This is a *phase diagram*. It is a chart that shows the thermodynamic state of a substance under various $p - T$ conditions. The diagram also indicates the regions of stability of the various distinct thermodynamically phases, of the considered substance. These

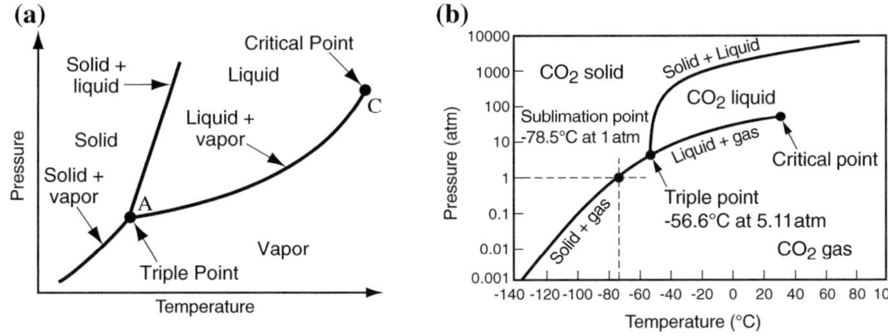

Fig. 2.1 Pressure-temperature zones for a single chemical species substance: **a** schematic, and **b** for CO_2

regions are delineated by curves of phase coexistence, at which the adjacent phases are in thermodynamic equilibrium. For example, a phase diagram shows the phases (or states of aggregation) of a considered substance under different $p - T$ conditions. Several typical phase diagrams are shown and discussed below.

Let the equation of state of a substance be $\rho = \rho(p, T)$, and the substance be present, simultaneously, as *two* phases, say, a liquid and its vapour. Then, the number of independent state variables reduces to ($NF = 2 + 1$ component $-$ 2 phases $= 1$) one, say, T. However, since we actually have 2 phases (liquid and gas), i.e., NP $= 2$ (and not 1 as above), because p and T depend on each other (vaporization line), the second variable is one of the saturations, say S_ℓ.

The process of phase change and its mathematical representation in a model can be understood in terms of two entities: (a) the physical-chemical-thermodynamic process that causes a phase to change, and (b) the primary variables (defined in Sect. 3.9), in terms of which the equation of mass balance of a given component is written. Knowledge of the thermo-physical processes that controls the phase change (e.g., the phase diagram of H_2O, discussed below) allows the selection of the appropriate primary variables for all possible states (i.e., under phase co-existence), making possible the change of primary variables when called for.

Following is a discussion of examples of phase changes under a variety of conditions, as well as the mathematical/physical criteria characterizing such phase changes, the primary variables and their changes, and how, in a considered case, all remaining thermo-physical properties can be determined from the primary variables.

A. Single Mass Component in 2 or 3 Phases

Figure 2.1a shows schematically, a typical phase diagram. On it, we note the possible states of a two-phase system and the curve $p = p(T)$. Figure 2.1b shows a phase diagram for CO_2.

Another example, for the component *water*, is shown on Fig. 2.2. In what follows, we shall use this example to elaborate on the processes that are described by this figure.

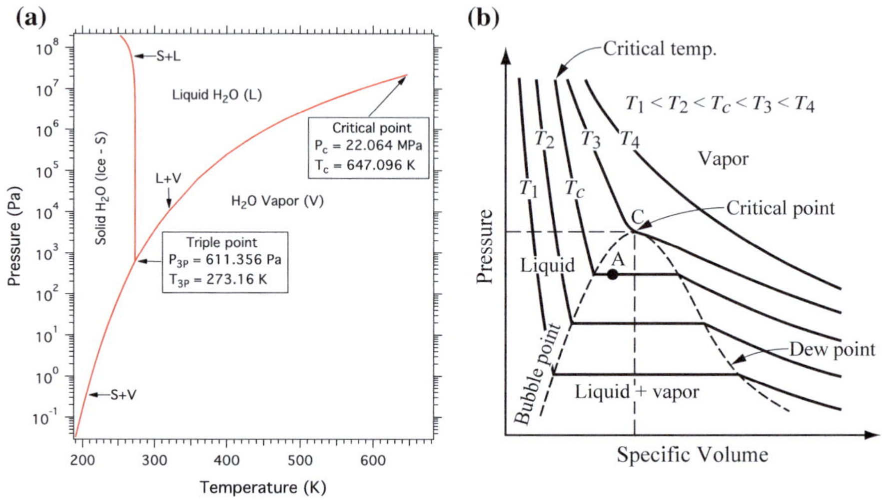

Fig. 2.2 **a** Pressure-temperature zones for H_2O, **b** schematic pressure-specific volume-temperature diagram for a single species system

The phase changes of H_2O (a single component) is an appropriate example. Inspection of the phase diagram of H_2O, shown in Fig. 2.2a, clearly identifies seven possible states: three single-phase states (V: Vapor, L: Liquid and S representing solid $\equiv$ ice), three two-phase states Vapor + Liquid, (V + L), Liquid + solid, (L + S), Solid + Vapor (S + V), and the *triple point* at which vapour, liquid and solid coexist. Knowledge of both pressure and temperature allows the complete definition of the thermo-physical properties of water in its L, V or S states. Thus, p and T are appropriate primary variables to define the aqueous and the vapour phase. However, these are inappropriate primary variables during any of the two-phase states of the liquid-vapour phase coexistence, because they are no longer independent, as there is a well-defined and unique relationship between p and T along the equilibrium line. Thus, in addition to one of the two, an alternative primary variable is required to uniquely describe the system. In the case of the triple point, both the pressure and the temperature at which it occurs are well known, so neither p nor T can serve as primary variables. Altogether, in the same problem domain, albeit in different parts of the latter, we may encounter all three states of water: Liquid, vapour and ice.

The curves L + V, S + V and S + L also indicate phase changes that almost invariably are associated with a thermal process, e.g., heating of the porous system. This means that, in addition to the mass balance equation for the H_2O, we also need to model heat transport, e.g., using the energy balance equation.

The criterion for the change of phase from L to L + V is the relationship between the pressure p in the modeled system and the *vapour pressure*, p_{vap}, at the system's temperature. In the absence of solutes in the water, or under very strong capillary pressures, p_{vap} is a function of T only, as shown on Fig. 2.2a. Thus, vapour evolves and the state of the system changes from L to L + V when $p \leq p_{vap}(T)$. Wherever this situation occurs, vapour evolves and the primary variables describing the mass balance equation change from p and T in the L state to p or T and S_ℓ (or $S_v = 1 - S_\ell$), denoting the saturation of either the liquid or the vapour phase) in the L + V state.

Continuation of the vaporization process requires the supply of additional heat to provide the latent heat needed for the phase change. As long as $S_\ell > 0$, the L + V state persists and the $p - T$ trace of the vaporization process remains confined to the L + V equilibrium line). In the L + V state, knowing p also means knowing T (and vice versa), because of the well-established $p - T$ relationship on the L + V equilibrium line; with the additional knowledge of S, all thermo-physical properties of the system of phases (density, viscosity, enthalpy, thermal conductivity, etc.) can be computed. Then the mass balance equation of H_2O and the heat balance of the system (solid and fluids) include the contributions of both the L and the V phases.

Transition from the L + V state to the V state (i.e., the single vapour phase) in any subdomain of the system occurs when the liquid is exhausted; it is triggered when $S_\ell \leq 1$. When this happens, the primary variables have to be changed to p and T, as both of them are required to determine the thermo-physical properties and to fully define the mass and energy balance of the system.

Note that the transition from the V to the L + V and then to the L state proceeds in the opposite direction, involving cooling (heat removal), but making use of the same criteria for the phase appearance and disappearance and the same primary variables to describe any of the states. All other phase transitions (S to S + L to L and L to S + V to V, recalling that solid here means ice) are described by a similar process, the same phase change criteria, and the same primary variables. In the case of H_2O, either p or T can be used interchangeably as one of the primary variables to describe the L + V and S + V two-phase states (the other being an appropriate phase saturation), but only p may be used in the S + L state, because of the near insensitivity of T to p over a very large range of p (as indicated by the long vertical component of the $p - T$ curve on the I-L equilibrium line, which makes it an unacceptable primary variable).

Finally, cooling an L + V two-phase system (removing heat, as demonstrated by a temperature decline) occurs exclusively on the L + V equilibrium line, and the triple point may eventually be reached. The transition depends on whether $p \leq p_{triple}$ or $T \leq T_{triple}$, whichever is the primary variable. When this occurs, the primary variables change from p or T and S_ℓ (or $S_v = 1 - S_\ell$ to any two of the following: S_ℓ, S_v or S_{ice}.

The dashed curve on Fig. 2.2b encloses the zone in which both liquid and vapour coexist. The solid lines are isotherms. The *bubble point* curve on this figure defines (for a single species phase) the state at which the phase is liquid, and any reduction in pressure (or increase in specific volume), at the fixed temperature, produces a vapour. Similarly, at a fixed pressure and volume, a slight increase in temperature produces a vapour.

The figure also shows the *dew point* defined for a single species substance, as that set of conditions under which a substance which is entirely in the vapour phase, any slight decrease in pressure (or reduction in specific volume), produces a liquid phase

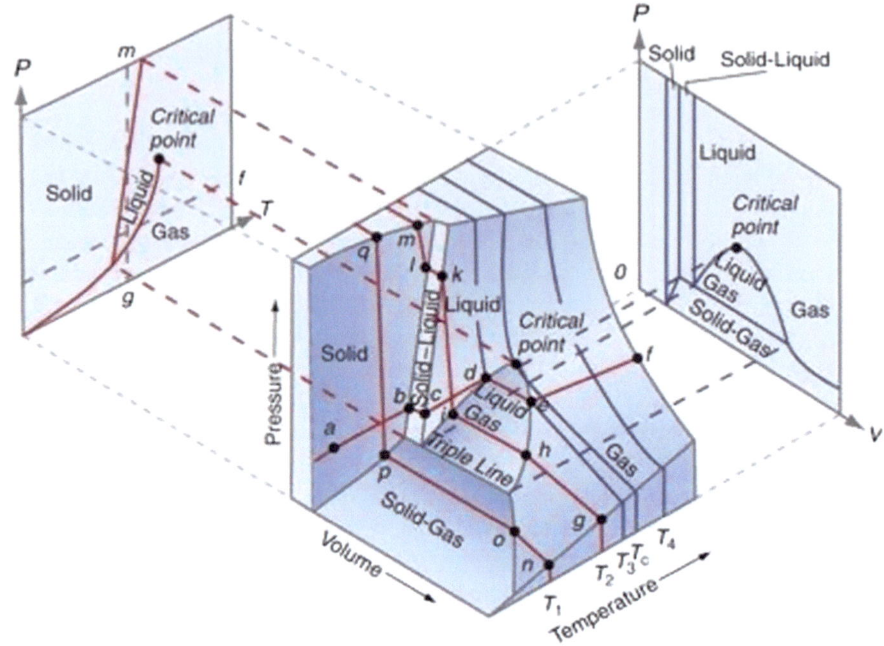

Fig. 2.3 Phase diagram for H_2O

(e.g., in the form of droplets) at the prevailing temperature. Similarly, in a vapour phase at a fixed pressure and volume, a slight reduction in temperature produces a liquid phase. On the other side of the dew point curve, liquid and vapour coexist in equilibrium. As the *bubble point curve* is crossed, all the vapour condenses and the entire system is in the liquid state.

It is interesting to see the phase diagram in its three-dimensional format. Figure 2.3 shows such diagram for water.

B. Two Mass Components in 2 or 3 Phases

As a representative example, we shall consider a system involving a liquid (oil), treated as a single component, and a gas, *methane*, CH_4 (superscript m), which is a non-condensable gas that dissolves in the liquid–oil. Assuming petroleum reservoir conditions (i.e., precluding the possibility of ice formation), the possible states are: two single-phase states (G: Gas and L: Liquid) and a single two-phase state (G + L: Gas-Liquid). In addition to the known thermodynamical properties of water mentioned above, this case involves an additional process: that of the dissolution of CH_4 in oil. This process is governed by *Henry's Law*:

$$p_G^m = \mathcal{H}^m(T)X_L^m, \quad \text{m represents here } CH_4, \tag{2.3.2}$$

where X_L^m denotes the mole fraction of methane gas in the liquid (oil), p_G^m denotes the partial pressure of CH_4 in the gas phase, and $\mathcal{H}$ is Henry's temperature-dependent gas solubility factor.

For the single gaseous (G) phase, an appropriate set of primary variables includes p, T and the CH_4 mole fraction in the gas phase, X_G^m. Then, the partial pressure of the CH_4 in the gas phase is p_G^m, defined (2.3.2), and the pressure, p_o, of the liquid oil component in the gas phase is $p_o = p - p_G^m$. Knowing these primary variable allows the computation of all oil and CH_4 properties.

Given the assumption of non-condensable CH_4 under the condition of the example considered here, the transition from the G to the G + L state occurs when the vapour pressure of the oil $p_{o,vap} \leq p_o$. The pressure $p_{o,vap}$ is a function of T only. It is well known for practically all pure substances. In the G + L state, an appropriate set of primary variables includes p, T and the gas saturation S_G. Because of the co-existence of the liquid and the gas phases, the partial pressure of the oil is by definition $p_o = p_{o,vap}$, and that of the CH_4 is $p_m = p - p_{o,vap}$. The known p_m is used to compute the amount of dissolved CH_4 in the oil equation (2.3.2). Thus, all the thermo-physical properties of the liquid and gas phases in the G-L state can be determined using these primary variables.

Further transition to the liquid state (L) is attained when S_G is reduced. In such case the primary variables have to be changed to p, T and X_L^m. The transition from the L to the G + L state indicates gas evolution and calls for monitoring the value of X_L^m, comparing the result to the maximum possible dissolved of CH_4 (i.e., the solubility limit) at the considered temperature T. At the solubility limit, $p^m = p - p_{o,vap}$, and $\max\{X_L^m\}$ is computed from (2.3.2). In this case, gas evolves and a transition to the G + L state from the L one is observed when $\{X_L^m\} \geq \max\{X_L^m\}$.

The same process is used in any system involving two components and two possible phases (i.e., 3 states, e.g., liquid H_2O and CH_4, liquid H_2O and air), etc. In such cases, when H_2O remains liquid during the study, the heat balance equation may not be required because of the very low solubility of CH_4, or air, in the H_2O.

The procedure for determining phase transition in a 2-component, 3-phase system (e.g., H_2O and air in shallow permafrost, in which H_2O can exist in any of its 3 possible phases) is analogous; its description requires the use of appropriate primary variables and primary variable change when phase changes occur. We wish to emphasize that it is possible to use more than one set of primary variables to describe a given state, but their numerical behavior may vary significantly during the simulations.

C. Multiple Substances

We wish now to consider a multi-substance system. As an example, we consider water in equilibrium with its vapour and we add CO_2. The mass of the added CO_2 may go partially into solution in the liquid water and partially remain in the gaseous phase as water vapour. As we shall show below, the distribution of CO_2 between the two phases, at equilibrium, will depend only on the pressure and the temperature.

To generalize the discussion, let us consider the definition of *solubility*. We consider a substance A that can dissolve in two mutually immiscible substances, B and C. We define the *solubility*, of A in B as:

$$R_{\mathrm{A\,in\,B}} = \frac{n_{\mathrm{A\,in\,B}}}{n_{\mathrm{A\,in\,B}} + n_{\mathrm{B}}}, \tag{2.3.3}$$

where $n_{\mathrm{A\,in\,B}}$ denotes the number of moles of A dissolved in n_{B} moles of B. Similarly,

$$R_{\mathrm{A\,in\,C}} = \frac{n_{\mathrm{A\,in\,C}}}{n_{\mathrm{A\,in\,C}} + n_{\mathrm{C}}}, \tag{2.3.4}$$

where $n_{\mathrm{A\,in\,C}}$ denotes the number of moles of A dissolved in n_{C} moles of C. These solubilities are functions of pressure and temperature. The ratio

$$\mathcal{K}_{\mathrm{A}\,in\,\mathrm{B\&C}} = \frac{R_{\mathrm{A}\,in\,\mathrm{B}}}{R_{\mathrm{A\,in\,C}}} \tag{2.3.5}$$

is called *partitioning factor* for A between B and C.

When substance C is a liquid and B is its vapour, we use the term *equilibrium ratio* to denote the ratio:

$$\mathcal{K}^{\mathrm{A}} = \frac{X_{\mathrm{V}}^{\mathrm{A}}}{X_{\mathrm{L}}^{\mathrm{A}}}, \tag{2.3.6}$$

where $X_{\mathrm{V}}^{\mathrm{A}}$ and $X_{\mathrm{L}}^{\mathrm{A}}$ are the *mole fractions* of A in the vapour and in the liquid, respectively. A similar equilibrium ratio can be defined for any γ-species in a two phase N-species system:

$$\mathcal{K}^{\gamma} = \frac{X_{\mathrm{V}}^{\gamma}}{X_{\mathrm{L}}^{\gamma}}, \tag{2.3.7}$$

where

$$X_{\alpha}^{\gamma} = \frac{n_{\alpha}^{\gamma}}{\sum_{j=1}^{N} n_{\alpha}^{j}}, \qquad \alpha = \mathrm{V}, \mathrm{L}.$$

All these equilibrium ratios are functions of pressure and temperature, as well as of the entire composition of the system. The overall composition can be specified by the mole fractions of the system as a whole. Thus,

$$X^{\gamma} = \frac{n^{\gamma}}{\sum_{j=1}^{N} n_{j}^{\gamma}}, \qquad \gamma, j = 1, 2, \ldots, N, \qquad n^{\gamma} = n_{\mathrm{V}}^{\gamma} + n_{\mathrm{L}}^{\gamma},$$

for each species, where n^{γ} is the number of moles of the γ-species in the entire system.

Under certain conditions, e.g., an *ideal gas* and dilute solutions, the above equilibrium ratios can be computed theoretically. In practice, however, they must be determined experimentally.

With $X_{\rm L}$ and $X_{\rm V}$ denoting the mole fractions of a composite system in the liquid state and in the vapour one, respectively, with:

$$X_{\rm L} + X_{\rm V} = 1, \qquad \text{and} \qquad X^\gamma = X^\gamma_{\rm L} X_{\rm L} + X^\gamma_{\rm V} X_{\rm V},$$

and since

$$\sum_{(\gamma)} X^\gamma_{\rm V} = 1,$$

we have:

$$\sum_{(\gamma)} \frac{X^\gamma}{1 + X_{\rm V}(\mathcal{K}^\gamma - 1)} = 1, \qquad \sum_{(\gamma)} \frac{X^\gamma \mathcal{K}^\gamma}{1 + X_{\rm V}(\mathcal{K}^\gamma - 1)} = 1, \tag{2.3.8}$$

The *dew point* introduced above is a term often used in petroleum engineering. It is defined for a given pressure as the temperature at which the least volatile species of a gaseous mixture begins to condense into a liquid at the same rate at which it evaporates. At this point, $X_{\rm V} = 1$, and the first part in (2.3.8) gives:

$$\sum_{(\gamma)} \frac{X^\gamma}{\mathcal{K}^\gamma} = 1. \tag{2.3.9}$$

For a body of water in contact with air, the *boiling point* is the temperature at which the vapour pressure is equal to the prevailing air (e.g., atmospheric) one. In the petroleum industry, the *bubble point*, introduced earlier, is the temperature, at a given pressure, where the first bubble of vapour is formed when heating a liquid, consisting of at least two species (Smith et al. 2005). *Bubble point pressure* is another term used in the petroleum industry for the pressure at which a natural gas begins to come out of solution and form bubbles. For a single species, we use the term *boiling point* defined as the state at which the substance is entirely in the liquid phase, and any slight reduction in pressure (or increase in specific volume), at the substance's fixed temperature, produces a vapour phase. Similarly, at a fixed pressure and volume, a slight increase in temperature produces a vapour phase.

At the bubble point pressure, $X_{\rm V} = 0$, and the second part of (2.3.8) gives:

$$\sum_{(\gamma)} X^\gamma \mathcal{K}^\gamma = 1. \tag{2.3.10}$$

In both cases, $\sum_{(\gamma)} X^\gamma = 1$.

Figure 2.4a presents a pressure-temperature diagram that shows two states, a liquid and a vapour. In reservoir engineering, for example, the liquid phase may be a mixture

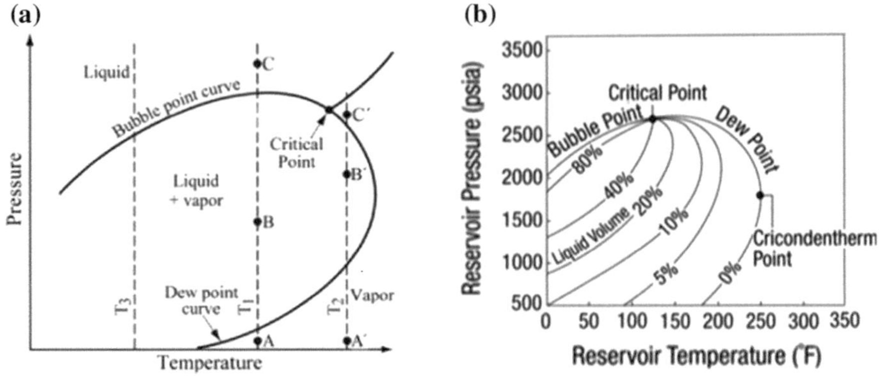

Fig. 2.4 **a** Schematic pressure–temperature diagram for a single species system, **b** pressure-temperature diagram for H_2O

of several liquid hydrocarbons, while the gas phase is a mixture of several light ones. Examining the T_1-isotherm, we note that as the pressure is reduced, we have only liquid at the point, say at C. A further reduction in pressure, leads to point B, where liquid and vapour coexist. A further reduction in pressure, leads to point A where we have only vapour. Along the T_2-isotherm, we start at point C′, at which the system is only in the vapour phase. As the pressure is reduced, the system moves to point B′, at which we have both liquid and vapour. A further reduction in pressure will lead to point A′, where only vapour exists. This change from vapour to liquid and back to vapour, is called *retrograde condensation*.

From Fig. 2.4b it follows that when a fluid, say, a hydrocarbon, is brought from the high temperature and high pressure that prevail in a deep oil reservoir, to the surface, where it is exposed to a different pressure and temperature (atmospheric pressure and some standard atmospheric temperature), a certain quantity of vapour would evolve. When the vapour is continuously removed from contact with the remaining liquid, as it is formed, the process is called *differential vaporization*. If the evolving vapour is not removed, the process is referred to as *flash evaporation*. In either case, when *standard atmospheric conditions* of pressure and temperature are reached, a certain volume of vapour would result, leaving a certain quantity of residual liquid. We regard the vapour, or gas, that has been produced, as having been dissolved in the volume of liquid at the original (reservoir) pressure and temperature.

Denoting the residual liquid, say, oil o, by $\mathbb{V}_o$, and the gas volume by $\mathbb{V}$, both measured at atmospheric conditions, we define the *gas solubility* as

$$R_o^g(p, T) = \frac{\mathbb{V}_{g,\text{SC}}}{\mathbb{V}_{o,\text{SC}}}. \tag{2.3.11}$$

i.e., the amount of gas dissolved in oil per unit volume of oil (sometimes denoted by the symbol R_{so}). This ratio is also called *solution gas-oil ratio*, and denoted by R_s.

In a petroleum reservoir, an oil *formation volume factor* is defined as:

$$B_o(p, T) = \frac{\mathbb{V}_o(p, T)}{\mathbb{V}_{o,\mathrm{SC}}}, \tag{2.3.12}$$

where $\mathbb{V}_o(p, T)$ denotes the volume of oil at the p and T conditions prevailing in the reservoir, and $\mathbb{V}_{o,\mathrm{SC}}$ denotes the oil volume under SC conditions. We recall that $\mathbb{V}_o(p, T)$ includes in it a certain mass of dissolved gas which will come out of solution when an oil sample pressure is reduced to the atmospheric one. Similarly, in an air water system, $\mathbb{V}_w(p, T)$ may include dissolved air. Both the gas solubility, $R_o^g(p, T)$, and the volume formation factor, B_o, are different for the two vaporization processes mentioned above, and so are the equilibrium ratios, defined above. Generally, $R_o^g(p, T)$ decreases as the oil's density increases (i.e., pressure increases). Actually, it more complicated, because the amount of dissolved gas changes, which changes the density. For a system of fixed composition, a pressure is reached at which no more gas can go into solution.

Other often used definitions are the *formation volume factors* for gas and for water:

$$B_g(p, T) = \frac{\mathbb{V}_g(p, T)}{\mathbb{V}_{g,\mathrm{SC}}} \quad \text{and} \quad B_w(p, T) = \frac{\mathbb{V}_w(p, T)}{\mathbb{V}_{w,\mathrm{SC}}}. \tag{2.3.13}$$

We note that the actual physical composition of a gas in a petroleum reservoir is different under different p, T conditions. In the definitions of B_o, B_g and B_w, the volumes $\mathbb{V}_o$, $\mathbb{V}_g$ and $\mathbb{V}_w$ at p, T, refer to a fixed mass of the involved substances, while the corresponding volumes at SC are those occupied by the same mass at *stock tank*, or *standard conditions*.

2.3.2 *Equations of State for Liquids*

The discussion in this subsection is applicable to both liquids and gases, except where liquids or gases are specifically referred to. However, because of the high compressibility of gases, (2.3.14) is, generally, not applicable, and the *ideal gas law*, or the *real gas law* are used as equations of state. This subject is discussed in detail in Sect. 2.3.3.

A. Fluid Density

In general, for a fluid phase composed of N γ-species, with $\sum_{(\gamma)} \rho^\gamma = \rho$, the *equation of state*, often abbreviated as EOS, can be written, symbolically, in the form:

$$\rho = \rho(p, T, \rho^\gamma; \gamma = 1, 2, \ldots, N), \tag{2.3.14}$$

where $\rho^\gamma(\equiv c^\gamma)$ denotes the density (or concentration) of the γ-species. Equation (2.3.14) states that the density, ρ, is a specific *known* function of the fluid's pressure, p, concentrations, ρ^γ, of the various γ-species (say, dissolved salts in an aqueous phase), and the temperature, T. Implicit in (2.3.14) is the assumption that for a specified T, there is a well-defined relationship between ρ and the various ρ^γ's in the fluid.

For a single-species fluid phase, the equation of state reduces to $\rho = \rho(p, T)$. Under isothermal conditions, this expression is reduced to $\rho = \rho(p)$. Sometimes, the above relationships are written in terms of the specific volume $v(\equiv 1/\rho)$, rather than in terms of the density, ρ.

From (2.3.14), it follows that:

$$d\rho = \sum_{\gamma=1}^{N}\left(\left.\frac{\partial\rho}{\partial\rho^\gamma}\right|_{p,T} d\rho^\gamma + \left.\frac{\partial\rho}{\partial p}\right|_{T,\rho^\gamma} dp + \left.\frac{\partial\rho}{\partial T}\right|_{p,\rho^\gamma} dT\right)$$

$$= \rho\left(\sum_{\gamma=1}^{N}\beta_{\rho^\gamma} d\rho^\gamma + \beta_p dp - \beta_T d\,T\right), \quad \gamma = 1, \ldots, N, \qquad (2.3.15)$$

where

$$\beta_p \equiv \frac{1}{\rho}\left.\frac{\partial\rho}{\partial p}\right|_{T,\rho^\gamma} \left(\equiv -\frac{1}{v}\left.\frac{\partial v}{\partial p}\right|_{T,\rho^\gamma}\right) \qquad (2.3.16)$$

is the *coefficient of compressibility* (often referred to as *compressibility*) of the fluid at constant temperature and γ-concentrations. The coefficient:

$$\beta_{\rho^\gamma} \equiv \frac{1}{\rho}\left.\frac{\partial\rho}{\partial\rho^\gamma}\right|_{p,T} \left(\equiv -\frac{1}{v}\left.\frac{\partial v}{\partial\rho^\gamma}\right|_{p,T}\right) \qquad (2.3.17)$$

introduces the effect of a change in ρ, or v, as a result of a change in concentration of a γ-species at constant temperature and pressure, and:

$$\beta_T \equiv -\frac{1}{\rho}\left.\frac{\partial\rho}{\partial T}\right|_{p,\rho^\gamma} \left(\equiv -\frac{1}{v}\left.\frac{\partial v}{\partial T}\right|_{p,\rho^\gamma}\right) \qquad (2.3.18)$$

is called the *coefficient of thermal expansion* at constant pressure and concentration. It is a negative number because the thermal expansion of the fluid reduces its density. Note that, in general, β_p, $\beta_{c\rho^\gamma}$ and β_T vary with p, ρ^γ, T, recalling that $\rho = \sum_{\gamma=1}^{N}\rho^\gamma$.

The use of (2.3.16)–(2.3.18) presupposes (1) a dominant fluid (e.g., water in an aqueous solution), or a fluid mixture, that can be adequately described as pseudo-homogeneous (e.g., an 'oil', i.e., a mixture of organic reservoir liquids considered collectively as the dominant species, with dissolved gases representing the non-dominant species), (2) a weak dependence of ρ on the various non-dominant γ-species, and (3) low ρ^γ-concentrations.

Returning now to Eqs. (2.3.16)–(2.3.18), if, within certain ranges of p, ρ^γ and T, the coefficients β_p, β_{ρ^γ} and β_T of a given fluid are constant (or can be approximated as such, usually for small changes in p, ρ^γ or T), the equation of state (2.3.15) assumes the specific form:

$$\rho = \rho_o \exp \left\{ \beta_p (p - p_o) - \beta_T (T - T_o) + \sum_{\gamma=1}^{N} \beta_{\rho^\gamma} (\rho^\gamma - \rho_o^\gamma) \right\}, \tag{2.3.19}$$

where $\rho_o = \rho(p_o, \rho_o^\gamma, T_o)$, and the subscript o denotes a reference state.

For small values of β_p, β_{ρ^γ} and β_T, and/or for small changes Δp, $\Delta \rho^\gamma$ and ΔT, Eq. (2.3.19) is further simplified by approximating the exponential by its linearized form:

$$\rho = \rho_o \left\{ 1 + \beta_p (p - p_o) - \beta_T (T - T_o) + \sum_{\gamma=1}^{N} \beta_{\rho^\gamma} (\rho^\gamma - \rho_o^\gamma) \right\}. \tag{2.3.20}$$

Equations (2.3.19) and (2.3.20) are widely used to describe fluid density in oil reservoirs and in aquifers, because of the usually low compressibility and thermal expansivity of such fluids (making fluid densities weak functions of p and T). Without dissolved chemical species, we can express (2.3.20) by the linear approximation: (2.3.20)

$$\rho = \rho_o (1 + \beta_p (p - p_o) - \beta_T (T - T_o)). \tag{2.3.21}$$

Special cases of (2.3.19) and (2.3.20), and further simplification of the equations, can be made for isothermal processes ($T = T_o =$ const.), homogeneous fluids ($\rho^\gamma = \rho = \rho_o =$ const., and $\beta_{\rho^\gamma} = 0$).

For larger ΔT, the expressions (2.3.19) and (2.3.20), exponential, or their linear approximation, may no longer be applicable. In that case, there are several empirical and semi-empirical equations that relate liquid density to temperature. An extensively used general relationship that covers a wide range of organic and inorganic liquids is (Yaws 1999):

$$\rho = A_\rho \, B_\rho^{(1 - T/T_c)^n}, \tag{2.3.22}$$

in which A_ρ is a term incorporating both a reference density, ρ_o, and the dependence of ρ on pressure; T_c is the critical temperature, and B_ρ and n are substance-specific constants.

Figure 2.5 presents the density of water (H_2O) as it varies with pressure and temperature.

B. Fluid Viscosity

Fluid's dynamic viscosity, μ (measured in Pa.s) expresses the ease at which a fluid undergoes deformation under shear stress. An *inviscid* fluid is one that has no resistance to shear stress. For a *Newtonian fluid*, μ is the coefficient that appears

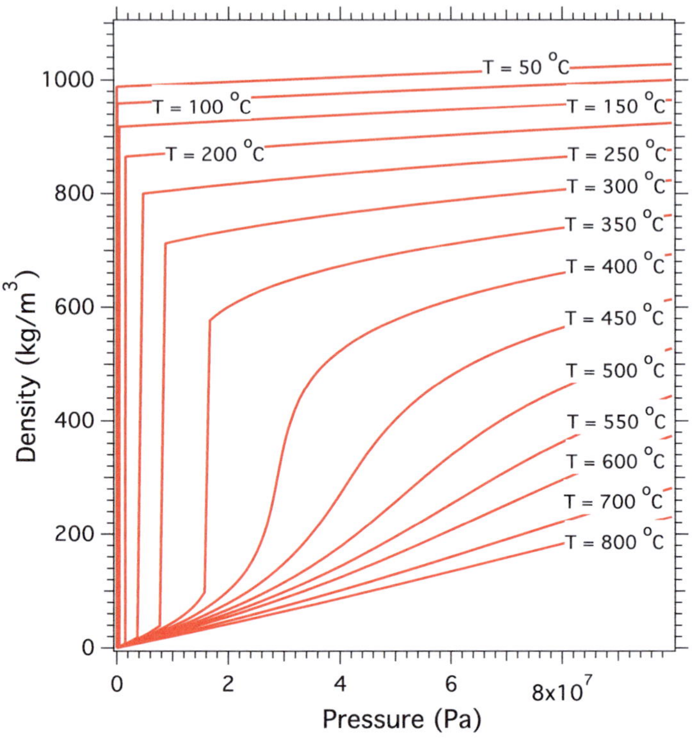

Fig. 2.5 Density-pressure-temperature diagram for water

in the relationship (3.2.26). The kinematic viscosity is defined by $\nu = \mu/\rho$. The dynamic viscosity varies with temperature, pressure and solute concentration, $\mu = \mu(p.T.c^{\gamma})$. For example, for supercritical CO_2 (denoted by subscript g). Altunin and Sakhabetdinov (1972) suggested:

$$\mu_g = \mu_g^o \exp\left(\sum_{i=1}^{4}\sum_{j=0}^{1} a_{ij}\frac{\rho_r^i}{T_r^j}\right), \qquad (2.3.23)$$

where $\rho_r = \rho_g/\rho_c$, $T_r = T_g/T_c$, are the reduced density and temperature, respectively, and μ_g^o (in $mi\ Ps.s$) is expressed by:

$$\mu_g^o = \sqrt{T_r}\left(27.2246461 - 16.6346068/T_r + 4.66920556/T_r^2\right), \qquad (2.3.24)$$

where $a_{10} = 0.248566120$, $a_{11} = 0.004894942$, $a_{20} = -0.373300660$, $a_{21} = 1.22753488, a_{30} = 0.363854523, a_{31} = -0.774229021, a_{40} = -0.0639070755$, and $a_{41} = 0.142507049$ were evaluated for the temperature range of $220 < T < 1300$ K and pressures up to 1200 bar.

C. Fluid Enthalpy

In most engineering applications, when considering non-isothermal conditions, we make use of the concepts of *enthalpy* $\mathbb{H}$ and *internal energy* $\mathbb{U}$, and their corresponding specific (i.e., per unit mass) quantities, h and u (and we used $\check{h}$ for enthalpy per mole). These were fully described in Sect. 2.2.2.

Both h and u are path-independent properties. They are always used in a differential form, i.e., as the difference between h at two states. Under isobaric condition ($dp = 0$), or when $(\partial h/\partial p)_T$ is sufficiently small, (as is the case in low-compressibility liquids), or when pressure changes are small, the specific enthalpy can be computed from:

$$h = \int_{T_o}^{T} c_p(T)\, dT = h(T) - h(T_o), \tag{2.3.25}$$

where T_o is a reference temperature. Further simplifications are possible when $\rho(T)$, or $C_p(T)$, or both, are temperature-insensitive within the considered temperature range.

For an *ideal gas* $u = u(T)$, and, $p\,v = RT$, so that

$$h = h(T) = u(T) + RT\,. \tag{2.3.26}$$

For liquids, C_p is often provided as the polynomial function,

$$C_p = \sum_{k=0}^{m} a_{\theta k} T^k, \tag{2.3.27}$$

where T is in Kelvin degrees, the polynomial order, m, is usually 3 (Yaws 1999), or 4 (Poling et al. 2000), and the $a_{\theta k}$'s are material- and equation-specific constants.

Figure 2.6 shows how enthalpy and the entropy of water, are related to pressure and temperature.

Figure 2.7 show the relationship $h = h(p, T)$ for CO_2.

The internal energy can be determined from the definition

$$\mathbb{U} = \mathbb{H} - \frac{p}{\rho}. \tag{2.3.28}$$

Both $\mathbb{H}$ and $\mathbb{U}$ are additive, path-independent properties, and they are used always in a differential form (i.e., as the difference between H at two states 1 and 2), as opposed to an absolute sense, i.e.,

$$\Delta\mathbb{H} = [\mathbb{H}(p, T)]_2 - [\mathbb{H}(p, T)]_1\,. \tag{2.3.29}$$

When equilibrium exists between pure liquid water and water vapour within the void space of a porous medium domain, the *chemical potential*, μ, for the water and

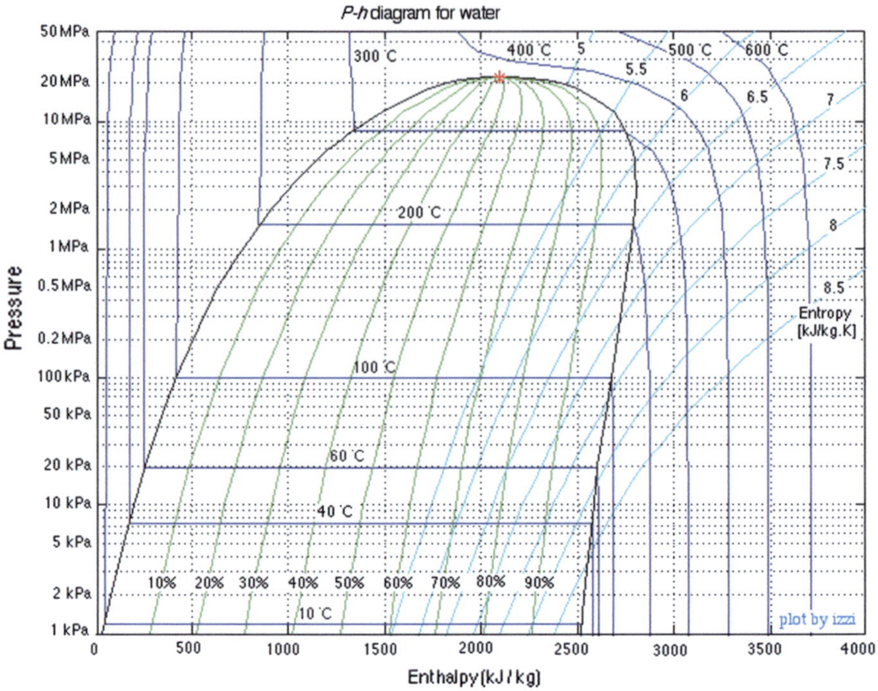

Fig. 2.6 Enthalpy diagram, $h = h(p, T)$, and entropy, $s = s(p, T)$, for H_2O

for the vapour must be the same. Obviously, water ad water vapour is used here as an example; the same phenomenon will occur for other substances.

The change in Gibbs molar free energy, $\Delta\breve{g}$, caused by raising the pressure by Δp, at constant temperature and composition, is expressed by

$$\Delta\breve{g} = \int_{\Delta p} \breve{\mathbb{V}} dp, \tag{2.3.30}$$

where $\breve{\mathbb{V}}$ is the *molar volume* (= volume per mole) of the water component in the liquid.

The *capillary pressure*, p_c', is introduced in Sect. 2.4.3 as the pressure difference across a liquid-gas interface. Hence, considering a process in which we move a small volume of liquid water from a reservoir at zero capillary pressure to a porous medium domain such that v^{mol} remains constant, we obtain the free energy of the water in the porous medium, relative to water at zero capillary pressure, in the form:

$$\Delta\mu = -\breve{V} p_c' = -\frac{m^w}{\rho_w} p_c', \tag{2.3.31}$$

where m_w is the water mass in moles.

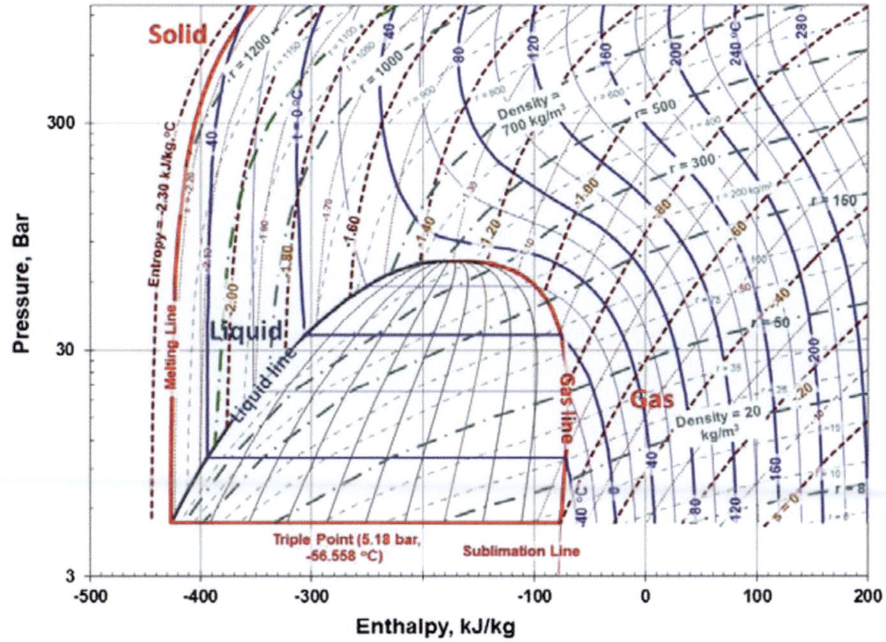

Fig. 2.7 Molier Diagram (Enthalpy-Entropy-density-pressure-temperature of CO_2, created from data provided by ChemicalLogic (1999) Courtesy of Prof. Jesus Carrera, CSIC, Spain

It is convenient to relate the chemical potential of a substance to its vapour pressure. Let us move vapour, considered as an *ideal gas*, from a reservoir, where we have a flat water surface, to one where the surface is curved. For an ideal gas,

$$\mu^{\gamma} = \mu^{\gamma o} + RT \ln \frac{p_g^{\gamma}}{p_g}. \tag{2.3.32}$$

For the vapour in the first reservoir, we have

$$\mu = \mu^{o} + RT \ln \frac{p^{v_o}}{p_g}, \tag{2.3.33}$$

where R is the universal gas constant, and T is the absolute temperature. Hence, a change in chemical potential can be related to vapour pressure by

$$\Delta\mu = RT \ln \frac{p^{v}}{p^{v_o}}, \tag{2.3.34}$$

where p^v is the partial pressure of water in the gas phase at a curved meniscus, p^{v_o} is the partial pressure of water in the absence of interfacial effects, and the ratio p^v/p^{v_o}

is the *relative vapour pressure*, or *relative humidity*, h_r. The above equation is also known as *Dalton's law*.

Thus, when (pure) water as a liquid and as a vapour are in equilibrium, we obtain from (2.3.31) and (2.3.34):

$$p'_c = -\frac{\rho_w RT}{M^w} \ln \frac{p^v}{p^{v_o}}, \tag{2.3.35}$$

known as *Kelvin's equation*. It relates the capillary pressure to the vapour pressure under equilibrium conditions. Note that $p'_c \geq 0$, and, therefore, $p^v/p^{v_o} \leq 1$ under all circumstances. Edelfsen and Anderson (1943) wrote (2.3.35) for an air–water system, at the *macroscopic level*, in the form:

$$\psi = \frac{RT}{gM^w} \ln h_r, \tag{2.3.36}$$

where ψ is the *suction*, defined by (6.1.3), and h_r denotes the *relative humidity* within the void space. In Sect. 2.5, we shall rewrite Kelvin's equation for a water phase that contains dissolved solids.

Making use of *Laplace's formula*, (2.4.12), Eq. (2.3.35) leads to:

$$\frac{2\gamma_{wn}}{r^*} = -\frac{\rho_w RT}{M^w} \ln \frac{p^v}{p^{v_o}}, \tag{2.3.37}$$

which relates the vapour pressure at a curved water surface to the latter's radius of curvature, r^*. The Laplace formula is presented and discussed in Sect. 2.4.3.

In Sect. 2.4.3, we shall consider the behavior of a wetting phase, say water, in the void space as if the latter was a capillary tube. The behavior will be related only to the liquid-gas-solid interactions. This is a good approximation for large pores, such as in sandy soil. However, the presence of a thin film of *adsorbed water* will be overlooked. In fine grained soils, especially in clay, this film may significantly affect the behavior of water in the soil. In such a soil, the effect of the film, depends on the presence of the *double layer*, introduced in Sect. 1.1, and on the presence of exchangeable cations on the solid surface. The pressures of the aqueous and gaseous phases no longer characterize the system at the macroscopic level, but the chemical potential per unit volume, known as *matric potential*, Ψ_m^w, defined in Sect. 2.5.1 B, is required (Nitao and Bear 1996).

When written in terms of the matric potential, (2.3.35) takes the form:

$$\Psi_m^w = -\rho_w \frac{RT}{M^w} \ln \frac{p^v}{p^{v_o}}. \tag{2.3.38}$$

For a pure water phase at a sufficiently high saturation, such that no water exists as films exposed to the gas phase, we have $\Psi_m^w = p_c$. Note that all potentials denoted by the symbol Ψ are expressed in terms of energy per unit volume, i.e., the same as pressure, and *not* in terms of head.

From (2.3.35), it may be readily concluded that p^v/p^{v_o} is very close to 1 in a porous medium under all but the most extreme conditions of very low saturation. For example, at a matric potential of 15 atm., which corresponds to soil conditions at which most plants cannot effectively remove water from the soil (the so called *wilting point*), the relative vapour pressure computed from (2.3.35) is 0.997. Extremely large water potentials will prevail if the relative vapour pressure is substantially less than 1. Under such conditions, the assumption that water is held only by capillary forces (viz., forces due to surface tension) may no longer be valid.

Short range interactions between water molecules, ions and solid surfaces will increasingly control the water potential as water films become less than 10 to 20 molecular diameters in thickness. Also, it is apparent that (2.3.37) can only be applied at a scale much greater than molecular dimensions, since surface tension has a meaning only for an ensemble of molecules. If the limit of the Laplace formula's applicability is taken at 10^3 diameters of water molecules (assuming 0.28 nm/molecule), the cutoff for applying (2.4.12) or (2.3.37) would be at a capillary pressure of about 10 atm. Thus, although the term 'capillary pressure' is usually used beyond this range, it should be recognized that the potential for water may actually be dominated by non-capillary phenomena, and the term 'matric potential' should be used instead.

2.3.3 *Equations of State for Gases*

For an *ideal gas*, the EOS takes the form:

$$\rho = \rho(p, T, X^\gamma; \gamma = 1, \ldots, N) = \frac{pM}{RT} \quad M = \sum_{\gamma=1}^{N} X^\gamma M^\gamma, \tag{2.3.39}$$

called the *ideal gas law*, in which R is the *universal gas constant*, M is the molecular mass of the gas, M^γ is the molecular mass of the γ-species, and X^γ is the mole fraction of a γ-species in the gas phase.

Equation (2.3.39) is valid for most noble and low-M gases over a wide range of p and T. Practically, it is applicable to every gas at low p, say, atmospheric, and a limited (but usually elevated) T-range (a few degrees). However, it may introduce significant errors at higher pressures and over extended p and T-ranges. The problem is alleviated by making use of the *real gas law*, which has the form:

$$\rho = \rho(p, T, X^\gamma; \gamma = 1, 2, \ldots, N) = \frac{p\bar{M}Z}{RT}, \quad \bar{M} = \sum_{\gamma=1}^{N} X^\gamma M^\gamma, \tag{2.3.40}$$

where $\bar{M}$ is the value of M for the multi-species gas, and $Z = Z(p, T)$ is an empirical correction factor called the *compressibility factor*. It introduces the effect of the deviation of a real gas behavior from that of an ideal gas by accounting for molecular interactions at higher pressures. For an ideal gas, $Z = 1$. Equation (2.3.40) is usually used in the petroleum industry.

There are several mathematical representations of the EOS that take into account the compressibility of real gases, i.e., gases for which $Z \neq 1$. Because of their simplicity and usefulness in both analytical and numerical solutions, the family of cubic EOS's (Walas 1985; Poling et al. 2000) has found widespread use in the study of real gases. An additional reason for their wide acceptance is the suitability of the cubic equations of state to predict the properties of a substance not only in its gaseous state, but also in its liquid state. The general cubic equation for the compressibility, Z, of a gas mixture, is given by:

$$Z^3 + A_{Z2} Z^2 + A_{Z1} Z - A_{Z0} = 0, \tag{2.3.41}$$

where A_{Z0}, A_{Z1} and A_{Z2} are parameters which are specific to the type of cubic EOS used. Some of the best known and widely used cubic EOS's are those of Redlich and Kwong (1949), Soave (1972), and Peng and Robinson (1976). Denoting these references as R-K, So and P-R, respectively, the coefficients of (2.3.41) are described as:

$$A_{Z0} = \begin{cases} A\,B & \text{in R-K and So} \\ A\,B - B^2 - B^3 & \text{in P-R,} \end{cases} \tag{2.3.42}$$

$$A_{Z1} = \begin{cases} A - B - B^2 & \text{in R-K and So} \\ A - 2B - 3B^2 & \text{in P-R,} \end{cases} \tag{2.3.43}$$

$$A_{Z2} = \begin{cases} -1 & \text{in R-K and So} \\ B - 1 & \text{in P-R,} \end{cases} \tag{2.3.44}$$

where:

$$A = \frac{(a\alpha)^*\,p}{R^2\,T^\nu}, \quad B = \frac{b^*\,p}{R\,T}, \quad \nu = \begin{cases} 2.5 & \text{in R-K} \\ 2 & \text{in So and P-R,} \end{cases} \tag{2.3.45}$$

$$(a\alpha)^* = \sum_\gamma \sum_\delta X^\gamma\,X^\delta (a\alpha)_{\gamma\delta}, \qquad b^* = \sum_\gamma X^\gamma\,b_\gamma, \tag{2.3.46}$$

$$(a\alpha)_{\gamma\delta} = \begin{cases} a_{\gamma\delta} = \sqrt{a_\gamma a_\gamma}, \quad (\alpha \equiv 1) & \text{in R-K} \\ (1 - \kappa_{\gamma\delta})\sqrt{(a\alpha)_\gamma (a\alpha)_\delta}, \quad (a\alpha)_\gamma = a_\gamma \alpha_\gamma & \text{in So} \\ (1 - \kappa^*_{\gamma\delta})\sqrt{(a\alpha)_\gamma (a\alpha)_\delta}, \quad (a\alpha)_\gamma = a_\gamma \alpha_\gamma & \text{in P-R,} \end{cases} \tag{2.3.47}$$

$$\alpha_\gamma = \begin{cases} 1 & \text{in R-K} \\ \left[1 + e_\gamma (1 - T_{r\gamma}^{0.5})\right]^2, \quad T_{r\gamma} = T/T_{c\gamma} & \text{in So and P-R} \end{cases} \tag{2.3.48}$$

$$e_\gamma = \begin{cases} 0.48508 + 1.55171\hat{a}_\gamma - 0.15613\hat{a}_\gamma^2 & \text{in So} \\ 0.37464 + 1.54226\hat{a}_\gamma - 0.26992\hat{a}_\gamma^2 & \text{in P-R,} \end{cases} \tag{2.3.49}$$

$$a_\gamma = \begin{cases} 0.42748\, R^2\, (T_{cr}^\nu/p_{cr})_\gamma & \text{in R-K} \\ 0.42747\, R^2\, (T_{cr}^\nu/p_{cr})_\gamma & \text{in So} \\ 0.45724\, R^2\, (T_{cr}^\nu/p_{cr})_\gamma & \text{in P-R,} \end{cases} \tag{2.3.50}$$

$$b_\gamma = \begin{cases} 0.08664\, R\, (T_{cr}/p_{cr})_\gamma & \text{in R-K and So} \\ 0.07780\, R\, (T_{cr}/p_{cr})_\gamma & \text{in P-R,} \end{cases} \tag{2.3.51}$$

in which X^γ and X^δ denote the mole fractions of any two species γ and δ of the mixture, $\kappa_{\gamma\delta}$ and $\kappa^*_{\gamma\delta}$ are the *binary interaction parameters* (Walas 1985) for substances γ and δ in the Soave and Peng–Robinson equations (different for each equation, and with $\kappa_{\gamma\gamma} = \kappa^*_{\gamma\gamma} = 0$), respectively, $\hat{a}$ is the *eccentric factor*, and subscript cr indicates critical conditions. The *acentric factor* is a gas property, unique for each gas, that expresses the deviation of the molecular shape from a sphere.

Note that, in addition to other cubic EOS's, higher-order and parametric EOS's have been proposed. Delving further into these EOS is beyond the scope of this book. Additional information, can be found in Walas (1985) and Poling et al. (2000).

After computing A_{Z0}, A_{Z1} and A_{Z2}, the closed forms of the roots of the cubic EOS in (2.3.41) are given by:

$$\begin{aligned} Z_1 &= (s_1 + s_2) - \frac{A_{Z2}}{3} \\ Z_2 &= -\frac{1}{2}(s_1 + s_2) - \frac{A_{Z2}}{3} + \frac{i\sqrt{3}}{2}(s_1 - s_2), \end{aligned} \tag{2.3.52}$$

where $i = \sqrt{-1}$ and:

$$s_1 = \left(r_Z + \sqrt{q_Z^3 + r_Z^2}\right)^{1/3}, \qquad s_2 = \left(r_Z - \sqrt{q_Z^3 + r_Z^2}\right)^{1/3}, \tag{2.3.53}$$

$$q_Z = \frac{A_{Z1}}{3} - \frac{A_{Z2}^2}{9} \quad \text{and} \quad r_Z = \frac{1}{6}\left(A_{Z1}A_{Z2} - 3A_{Z0}\right) - \frac{A_{Z2}^3}{27}. \tag{2.3.54}$$

Thus, (2.3.41) can have either a single real root and two complex ones, or three real roots. For a gas, the compressibility factor, Z, to be used in the computations (e.g., (2.3.39) for a single-species gas, or (2.3.40) for a mixture) is the single real root of (2.3.41) or largest real root from among the three real roots Z_1, Z_2, Z_3. In the latter case, the existence of three real roots indicates coexistence of the liquid and gas states, and the liquid compressibility is described by the smallest of the three roots.

The usefulness of the cubic EOS extends well beyond the computation of Z, as it forms the basis for the computation of a wide range of thermodynamic and thermo-physical properties, e.g., fugacity, density, enthalpy, entropy, viscosity, heat capacity, and the binary diffusion coefficients. Using the cubic EOS, the specific internal energy (u) and specific enthalpy (h) of a gas mixture, when no phase change occurs, is determined from:

$$u = u_{ideal} - u_{dep} \quad \text{and} \quad h = h_{ideal} - h_{dep}, \tag{2.3.55}$$

where subscripts $ideal$ and dep denote ideal and departure-from-the-ideal gas properties. Note that all the terms in (2.3.55) indicate differences between the (p, T) conditions under consideration and a reference state of (p_o, T_o). In their general form, for a gas mixture, the ideal parts of u and h are computed from the general equations:

$$u_{ideal} = \sum_{(\gamma)} X^\gamma \int_{T_o}^{T} C_v^\gamma \, dT, \quad \text{and} \quad h_{ideal} = \sum_{(\gamma)} X^\gamma \int_{T_o}^{T} C_p^\gamma \, dT, \tag{2.3.56}$$

where T_o is a reference temperature, $C_v^\gamma = C_v^\gamma(T) = (\partial h/\partial T)|_v^\gamma$ and $C_p^\gamma = C_p^\gamma(T) = (\partial h/\partial T)|_p^\gamma$ are the specific heat capacities of the gaseous γ-species under constant volume and constant pressure, respectively. The coefficients C_v^γ and C_p^γ for an ideal gas (and for the computation of ideal species u_{ideal} and h_{iedal} of a real gas) are related to each other through the well-known equation:

$$C_p^\gamma - C_v^\gamma = R. \tag{2.3.57}$$

in which C_p^γ is usually computed from a 4-th order polynomial (Poling et al. 2000) as:

$$\frac{C_p^\gamma}{R} = \sum_{k=0}^{4} (b_{\theta k})^\gamma \, T^k, \tag{2.3.58}$$

where $(b_{\theta k})^\gamma$ are substance-specific constants of a γ-gas.

Equation (2.3.58) is entirely analogous to (2.3.22); the only difference is the use of different parameters ($b_{\theta j}$ instead of $a_{\theta i}$) introduced in order to maintain consistency with (2.3.57). Note that (2.3.58) applies under certain circumstances to liquids, as discussed in Sect. 2.3.2 (see (2.3.27)).

The departure internal energy, u_{dep}, and the departure specific enthalpy, h_{dep}, both from the ideal behavior, are determined from the cubic EOS, using parameters computed in the course of the Z-estimation. Thus,

$$u_{dep} = \begin{cases} RT\left[1.5\dfrac{A}{B}\ln\left(1+\dfrac{B}{Z}\right)\right] & \text{for R-K} \\ RT\left[\dfrac{A}{B}\left(1+\dfrac{D}{(a\alpha)^*}\right)\ln\left(1+\dfrac{B}{Z}\right)\right] & \text{for So} \\ RT\left[\dfrac{A}{2.828B}\left(1+\dfrac{D}{(a\alpha)^*}\right)\ln\left(\dfrac{Z+2.414B}{Z-0.414B}\right)\right] & \text{for P-R,} \end{cases} \tag{2.3.59}$$

and:

$$h_{dep} = u_{dep} + RT\,(1-Z)\,, \tag{2.3.60}$$

where:

$$D = \sum_{(i)}\sum_{(j)} X_i X_j e_j (1-\kappa_{ij})\sqrt{a_i\alpha_i}\sqrt{a_j T_{rj}}. \tag{2.3.61}$$

All the terms in (2.3.59)–(2.3.61) are as obtained and computed from (2.3.45) to (2.3.51).

2.3.4 Introduction to Stress, Strain, and Tensors

Consider a bounded domain regarded as a continuum. The domain is acted upon by two kinds of forces:

- **Body force** that acts on every point *within* this domain, where we may think of a point as representing a small volume of matter centered at the point. Gravity may serve as an example of such force.
- **Traction** that *acts on every point on the surface* that surrounds the considered body. The force resulting from the weight of a soil column (overburden) above a unit horizontal area at some depth within the soil layer may serve as an example. Forces per unit area produce stresses which, in turn produce strains, manifested as displacements.

The objective of this section is to discuss these topics.

A. Stress

A *stress*, $\boldsymbol{\sigma}$, also referred to as *Cauchy's stress*, at a point within a material domain, *regarded as a continuum*, is defined as a 'force per unit area'. However, force is a *vector* (with a magnitude and a direction) and so is the *area*, which also has a magnitude, and may face different directions at the same point. How do we divide a vector by a vector? Let us use this opportunity to introduce an *informal explanation* of the notion of a *tensor* through the example of a second rank tensor.

Consider two vectors at a point in a given (fluid or solid) domain: a force, $\mathbf{f}$, and an area, $\mathbf{A}$. To obtain the stress, we would have liked to divide the force by the area,

$\mathbf{f}/\mathbf{A}$. However, division of a vector by a vector is undefined. Instead, we seek a way to express $\mathbf{f}$ through the value of $\mathbf{A}$. In other words, we introduce the stress, denoted by $\boldsymbol{\sigma}$ such that when multiplying it by the area $\mathbf{A}$, we obtain the force, i.e., $\mathbf{f} = \boldsymbol{\sigma}\cdot\mathbf{A}$. Thus, the entity $\boldsymbol{\sigma}$ is like the ratio $\mathbf{f}/\mathbf{A}$, but only in the sense that $\mathbf{f} = \boldsymbol{\sigma}\cdot\mathbf{A}$. However, we must emphasize again that the above *quotient is not defined*, and we have brought it here only to demonstrate the meaning of $\boldsymbol{\sigma}$. This mathematical entity–the stress at the considered point—is a *stress tensor*. It is a tensor of the second rank. Recalling that a vector is associated with one direction in space, a second rank tensor is associated with *two* directions: one of the force and one of the area. Nine numbers/values are requires to specify it. These numbers correspond to the nine possible combinations of components of the two directed vectors.

We can now be more formal. Tensors are classified by *rank* (or *order*) which also determines the number of their *components*. Thus, in the physical three-dimensional space considered here, a tensor of rank n has 3^n components, independent of the selected coordinate system. Accordingly, a tensor of order zero has only one component. We refer to it as a *scalar*. A tensor of order one is a *vector*; it has 3 components, whereas a second rank tensor has $3^2 = 9$ components.

In a three-dimensional space, a *Cartesian tensor* of order n is a quantity represented in any *rectangular Cartesian coordinate system*, x_i, $i = 1, 2, 3$, by an ordered set of 3^n numbers, $A_{ij...\ell m}$, called components of the tensor, which, upon transition to another rectangular Cartesian coordinate system x'_k, obtained from the first by rotation of the coordinate system, transforms (i.e., the numerical values of the components transform) to a new set of 3^n components, $A'_{rs...uv}$, according to the rule:

$$\underbrace{A'_{rs...uv}}_{n \text{ subscripts}} = \underbrace{\frac{\partial x'_r}{\partial x_i}\frac{\partial x'_s}{\partial x_j}\cdots\frac{\partial x'_u}{\partial x_\ell}\frac{\partial x'_v}{\partial x_m}}_{n \text{ derivatives}}\underbrace{A_{ij...\ell m}}_{n \text{ subscripts}}, \tag{2.3.62}$$

where:

$$\frac{\partial x'_r}{\partial x_i} \equiv \cos(\mathbf{1x}'_r, \mathbf{1x}_i)$$

with $\mathbf{1x}_r$ denoting a unit vector in the $\mathbf{x}_r$-direction, is the cosine of the angle between the positive directions of the x'_r-axis in the new coordinate system and the x_i-axis in the old one.

Thus, tensors are characterized by a linear transformation of their components upon transition from one coordinate system to the next. In fact, (2.3.62) can be used as a definition for an n-rank tensor.

We wish to emphasize again: given a tensor at a point in space, as we rotate the coordinate system, *the tensor is not changed*, but the magnitude of its *components varies with the changes in the coordinate system*.

With the above introductory remarks on tensors, we shall now focus on two *second rank symmetric tensors*: the stress, $\boldsymbol{\sigma}$ (force per unit area), and the strain, ε (dimensionless). Figure 2.8 shows the components of $\boldsymbol{\sigma}$ at a point P of a domain, i.e., acting on a parallelepiped body for which P serves as a center. This body is part of

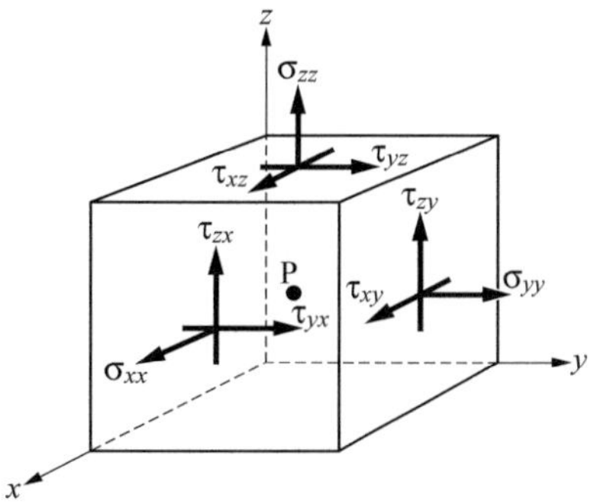

Fig. 2.8 Normal (σ_{ii}) and shear (τ_{ij}) components of the stress tensor at a point (P)

a domain regarded as a continuum. Note that the figure displays the σ_{ij}-components only on the sides that face the direction $+x$, $+y$, and $+z$. The other sides are not shown.

At a point in a continuum, a stress component, σ_{ij}, of $\boldsymbol{\sigma}$, is considered positive if it is associated with a surface that faces the positive i-direction and is acting in the positive j-direction. For example, we note on the figure that the stress component σ_{xx} is positive as it acts in the $+x$-direction on a surface that faces the $+x$-direction. Stress components for which $i = j$, are referred to as *normal stress* components. Stress components for which $i \neq j$ are referred to as *shear stress* components; they are usually denoted by τ_{ij}. For shear stresses, we also have $\tau_{ij} = \tau_{ji}$ (see any text on Fluid Mechanics). Thus, the full description of stress at a point in a 3-d domain requires information on nine scalar components:

$$\sigma_{ij} = \begin{bmatrix} \sigma_{xx} & \sigma_{xy} & \sigma_{xz} \\ \sigma_{yx} & \sigma_{yy} & \sigma_{yz} \\ \sigma_{zx} & \sigma_{zy} & \sigma_{zz} \end{bmatrix} \equiv \begin{bmatrix} \sigma_{xx} & \tau_{xy} & \tau_{xz} \\ \tau_{yx} & \sigma_{yy} & \tau_{yz} \\ \tau_{zx} & \tau_{zy} & \sigma_{zz} \end{bmatrix}, \quad \begin{aligned} \sigma_{ij} &= \sigma_{ji}, \\ \tau_{ij} &= \tau_{ji}. \end{aligned} \tag{2.3.63}$$

Sometimes, we use subscripts 1, 2, 3 instead of x, y, z. Note the use of the shear stress symbol, τ_{ij}, $i \neq j$.

A second rank tensor has two *invariants*, i.e., combinations of tensor components that are independent of the coordinate system employed. One is the sum of the diagonal components, $\sum_{(i)} \sigma_{ii} = \sigma_{11} + \sigma_{22} + \sigma_{33}$ (referred to as the *first invariant*). Recall that with *Einstein's summation convention*, employed in this book, $\sigma_{ii} \equiv \sigma_{11} + \sigma_{22} + \sigma_{33}$, i.e., without the sum symbol. One third of this quantity is a scalar quantity, often called *mean stress*

$$\sigma_m = \frac{1}{3}(\sigma_{11} + \sigma_{22} + \sigma_{33}) \equiv \frac{1}{3} tr(\boldsymbol{\sigma}). \tag{2.3.64}$$

Read: trace of sigma. It is a scalar that has no directional properties; it does not vary as the coordinate system is rotated.

A second invariant is the *deviatoric stress*, $\boldsymbol{\sigma}_d$, which is the portion of the stress tensor that is complementary to the mean stress, i.e.,

$$\boldsymbol{\sigma}_d = \boldsymbol{\sigma} - \boldsymbol{\sigma}_m \mathsf{I},$$

where I is the *identity matrix*, or the *unit tensor*, denoted also by $\boldsymbol{\delta}$ (components δ_{ij}).

$$\mathbf{I} = \begin{bmatrix} 1 & 0 & 0 \\ 0 & 1 & 0 \\ 0 & 0 & 1 \end{bmatrix} \equiv \boldsymbol{\delta}. \tag{2.3.65}$$

Consider a saturated porous medium, visualized as three overlapping continua: the solid matrix, the fluid occupying the void space, and, *at the macroscopic level*, the porous medium as a whole. Let us focus on the third case. How is the stress defined above related to the pressure at a point in a fluid phase that occupies the void space, with the fluid visualized as a continuum? Since for a non-viscous fluid, the normal stresses at a point are equal to each other, each of them must be equal to σ_m defined in (2.3.64). For such a fluid, as for a Newtonian fluid, when we invoke the *Stokes assumption*, $\lambda + (2/3)\mu = 0$, where λ is the second coefficient of viscosity, we have:

$$-p = \sigma_m, \quad \text{i.e.,} \quad p = -\frac{1}{3}\left(\sigma_{11} + \sigma_{22} + \sigma_{33}\right), \tag{2.3.66}$$

where the minus sign is a consequence of the fact that negative normal stresses occur in a fluid. Hence, for a fluid at rest, or in the absence of shear stress, the stress at a point in a fluid is described by the matrix

$$\begin{pmatrix} -p & 0 & 0 \\ 0 & -p & 0 \\ 0 & 0 & -p \end{pmatrix},$$

in which p is called the *hydrostatic pressure*. Thus, for the kind of fluids considered here, the *pressure is defined as minus the mean of the normal stresses*. With the above definition of pressure, we may express a typical stress component in a fluid, σ_{ij}, in the form:

$$\sigma_{ij} = \tau_{ij} - p\delta_{ij}, \tag{2.3.67}$$

where δ_{ij} is the *Kronecker delta* ($\equiv$ the *unit tensor*) defined in Sect. 3.2.2 C, and τ_{ij} is a component of the viscous stress, or shear tensor. Note that here, and elsewhere in this book, we consider fluid pressure, as is common in fluid mechanics, as *positive for compression*.

As will be shown below, knowledge of the (macroscopic) stress and the resulting (macroscopic) strain is required when considering flow through a deformable porous medium domain, porous medium deformation, and failure in a porous medium as a result of excessive stress.

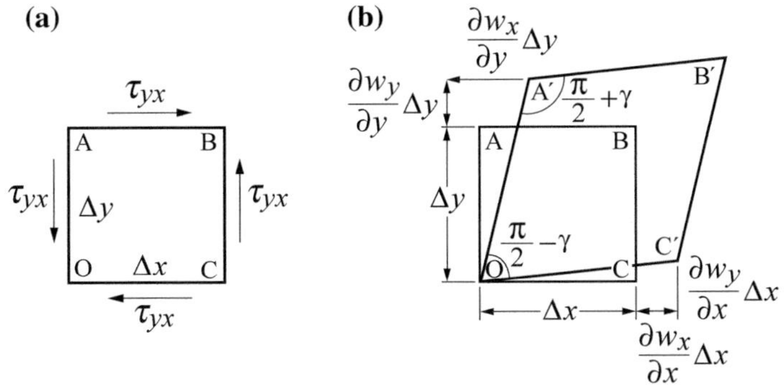

Fig. 2.9 **a** Shear components in an xy domain, and **b** deformed domain (stress components are not shown)

Before leaving the above introductory remarks on stress, it may be interesting to recall that in (3.2.22) we have defined the stress (in fact, $-\boldsymbol{\sigma}$) as the *diffusive flux of linear momentum*.

B. Strain

In (2.3.67), we have related the stress at a point in a 3-d stationary fluid continuum, to the shear stress and to the fluid pressure. This relationship is valid when the considered point belongs to a domain that is occupied by two overlapping continua in a porous medium domain: one of the solid matrix and the other of the fluid phase.

When, in a porous medium domain, the total stress, or the fluid pressure, or both, are perturbed (e.g., by producing fluid motion, or by loading the porous medium domain), the stress in the solid matrix is changed, producing *displacement* and *distortion*. In other words, the solid matrix is "strained". Let us discuss these phenomena.

Figure 2.9 shows what happens when a stress is acting on a rectangular planar domain, Δx, Δy, centered at a point P in a continuum. For the sake of simplicity, the demonstration is in a 2-d, x, y, domain. In this figure, a stress, $\boldsymbol{\sigma}$ acts on the rectangle OABC, keeping the point O clamped. The result is that points A, B, C, will be displaced to A', B', C', respectively. The original rectangle OABC will be deformed to the shape O A'B'C'. The figure shows the distortion of the original rectangle and the change of angles. The four right angles will be changed:

$$\begin{aligned}
&\sphericalangle \mathrm{O} : \pi/2 \Rightarrow \sphericalangle \mathrm{O}' : \pi/2 - \gamma,\\
&\sphericalangle \mathrm{A} : \pi/2 \Rightarrow \sphericalangle \mathrm{A}' : \pi/2 + \gamma,\\
&\sphericalangle \mathrm{B} : \pi/2 \Rightarrow \sphericalangle \mathrm{B}' : \pi/2 - \gamma,\\
&\sphericalangle \mathrm{C} : \pi/2 \Rightarrow \sphericalangle \mathrm{C}' : \pi/2 + \gamma,
\end{aligned}$$

where, denoting the *displacement vector* by $\mathbf{w}$, with components: w_x, w_y, w_z, and the *strain tensor* (which is a second rank symmetric tensor) by ε, we have (in indicial notation):

$$\varepsilon_{xx} = \frac{\partial w_x}{\partial x}, \quad \varepsilon_{yy} = \frac{\partial w_y}{\partial y}, \quad \varepsilon_{xy} = \varepsilon_{yx} \left(\equiv \frac{\gamma}{2}\right) = \frac{1}{2}\left(\frac{\partial w_x}{\partial y} + \frac{\partial w_y}{\partial x}\right). \quad (2.3.68)$$

Thus, the *shear strain*, ε_{xy}, represents the change in the angle between the two initially perpendicular lines (actually, half the angle), while the *normal strain* represents the relative change in length of the solid box dimensions. Altogether, *strain components are dimensionless.*

The above relationships can be extended to a 3-d domain. It can easily be shown that the strain tensor, $\boldsymbol{\varepsilon}$, is related to the skeleton's displacement vector, $\mathbf{w}$, by:

$$\boldsymbol{\varepsilon} = \tfrac{1}{2}\left[\nabla \mathbf{w} + (\nabla \mathbf{w})^T\right], \quad (2.3.69)$$

or, in indicial notation:

$$\varepsilon_{ij} = \frac{1}{2}\left(\frac{\partial w_i}{\partial x_j} + \frac{\partial w_j}{\partial x_i}\right). \quad (2.3.70)$$

Volume change is also related to the strain tensor, noting that shear strain does not produce volume change. Thus, for a considered continuum, the relative volume change, $\varepsilon_v \equiv \Delta\mathbb{V}/\mathbb{V}$, is expressed by:

$$\varepsilon_v \equiv \frac{\Delta\mathbb{V}}{\mathbb{V}} = \varepsilon_{xx} + \varepsilon_{yy} + \varepsilon_{zz} \equiv 3\varepsilon_m, \quad \varepsilon_v = \nabla\cdot\mathbf{w}, \quad (2.3.71)$$

where ε_m denotes the *mean strain.*

2.3.5 Stress-Strain Relationship for a Solid

The constitutive equation for a solid expresses its stress-strain relationship at a point in a solid domain. We shall use tensor concepts to describe these terms. In Sect. 9.1 we shall discuss these concepts for a solid matrix of a porous medium domain. A brief introduction to the concepts of tensor, stress and strain are presented at the beginning of this section.

While for fluids, the deviatoric stress-tensor is regarded as a force (per unit area) which arises in a moving fluid and depends on the rate of strain, a solid, even at rest, can be deformed by an applied stress. Thus, for an *anisotropic, linearly elastic solid* under isothermal conditions, the stress-strain relation is given by the *generalized Hooke's law*:

$$\sigma_{ij} = C''_{ijk\ell}\,\varepsilon_{k\ell}, \quad (2.3.72)$$

where $C''_{ijk\ell}$ is the *elasticity tensor*, or *stiffness tensor*, and $\varepsilon_{k\ell}$ denotes the $k\ell$-component of the *(Eulerian infinitesimal) strain tensor*. Here, and everywhere in this book, we are making use of Einstein's summation convention, i.e., when a subscript is regarded twice and only twice, it is regarded as a sum over the range of values of that subscript or superscript (e.g., 1, 2, 3 in three dimensions).

More on the $C''_{ijk\ell}$ for an anisotropic porous medium is presented in Sect. 9.1.4.

Common ways to express Hooke's law for an isotropic elastic solid in indicial notation are:

$$\sigma_{ij} = \lambda''_s \varepsilon_{kk}\delta_{ij} + 2\mu''_s \varepsilon_{ij} = C''_{ijk\ell}\varepsilon_{k\ell},$$
$$C''_{ijk\ell} = \lambda''_s \delta_{ij}\delta_{k\ell} + \mu''_s \left(\delta_{ik}\delta_{j\ell} + \delta_{i\ell}\delta_{jk}\right), \tag{2.3.73}$$

in which δ_{ij} is the *Kronecker delta*, and:

$$\sigma_{ij} = \frac{E}{1+\nu''_s}\left(\varepsilon_{ij} + \frac{\nu''_s}{1-2\nu''_s}\varepsilon_{kk}\delta_{ij}\right), \tag{2.3.74}$$

in which superscript $(.)''_s$ denotes solid matrix properties (here at the microscopic level), or:

$$\varepsilon_{ij} = \frac{1}{E}\left[\sigma_{ij} - \nu''_s(\sigma_{kk}\delta_{ij} - \sigma_{ij})\right], \tag{2.3.75}$$

in which E is *Young's elasticity modulus, Young's!elasticity modulus* $\lambda''_s(= K - \frac{2}{3}G)$ and $\mu''_s(\equiv G)$ are the *Lamé constants*, ν''_s is *Poisson's ratio*, K is the *bulk modulus*, and G is the *shear modulus*.

In the particular case of a uniaxial state of stress in the x_1-direction (i.e., $\sigma_{11} \neq 0$, $\sigma_{22} = \sigma_{33} = 0$), we have:

$$E = \sigma_{11}/\varepsilon_{11} \quad \text{and} \quad \nu'' = -\varepsilon_{22}/\varepsilon_{11} = -\varepsilon_{33}/\varepsilon_{11}. \tag{2.3.76}$$

In this case:

$$E = \frac{\mu''_s(3\lambda''_s + 2\mu'')}{\lambda''_s + \mu''_s}, \qquad \nu''_s = \frac{\lambda''_s}{2(\lambda''_s + \mu''_s)}. \tag{2.3.77}$$

Examples of stress-strain relations of solids are given in Fig. 2.10.

Note that $C''_{ijk\ell}$ in (2.3.72) is a fourth rank tensor. In principle, such tensor has 81 components. However, because of various symmetries, e.g., because $\sigma_{ij} = \sigma_{ji}$ and $\varepsilon_{ij} = \varepsilon_{ji}$, we have actually only *21 different independent components*. For an isotropic solid, these reduce to only *two*. These are the Lamè coefficients introduced below.

The solid's velocity at a point, $\mathbf{V}_s$, is related to the displacement at that point, $\mathbf{w}$, by:

$$\mathbf{V}_s = \frac{D_s\mathbf{w}}{Dt} \equiv \dot{\mathbf{w}}. \tag{2.3.78}$$

The *dilatation*, ε, is given by

$$\varepsilon = \varepsilon_{ii} = \frac{\partial w_i}{\partial x_i}, \tag{2.3.79}$$

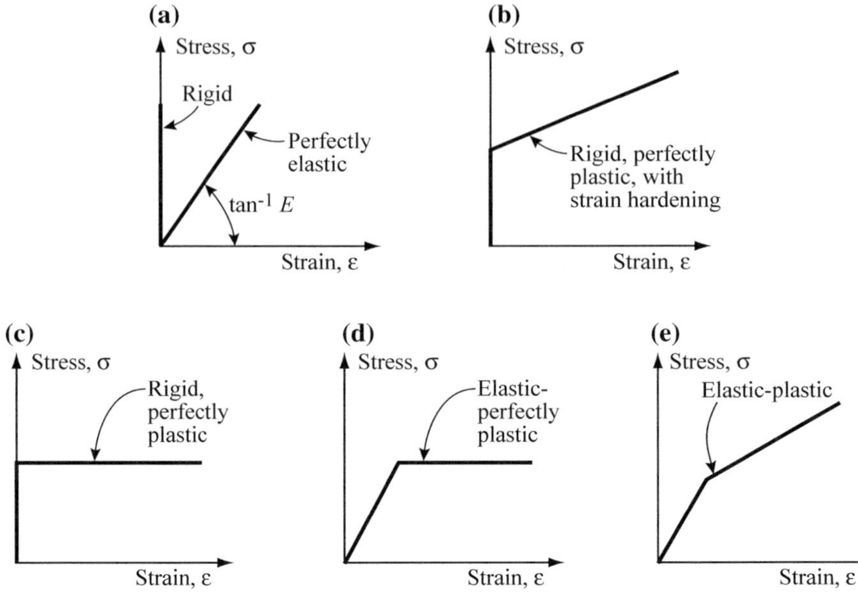

Fig. 2.10 Stress-strain relations for various solids

recalling that, as everywhere in this book, the summation convention is implied.

Thus, for an *isotropic linearly elastic solid*, (2.3.72) takes the form:

$$\sigma_{ij} = \mu_s''\left(\frac{\partial w_i}{\partial x_j} + \frac{\partial w_j}{\partial x_i}\right) + \lambda_s''\frac{\partial w_k}{\partial x_k}\delta_{ij}. \tag{2.3.80}$$

It is interesting to note that altogether, we have introduced here five material coefficients: E, ν_s'', G, λ_s'' and μ_s''. However, only two of them are independent. We can choose any two to be independent; the other three are computable from these two.

For a *linearly thermoelastic solid*, i.e., when changes in temperature are taken into account, the stress-strain relationship (2.3.72) is extended to the form:

$$\sigma_{ij} = C''_{ijk\ell}\left[\varepsilon_{k\ell} - \eta_{k\ell}(T - T_o)\right], \tag{2.3.81}$$

where $\eta_{k\ell}(T - T_o)$ is the strain contributed by the temperature field and T_o is a reference temperature. The coefficient C''_{ijkl} is a *fourth rank tensor*. Because of various symmetries, only 21 of its components are non-zero. For an isotropic porous medium, the number of components is reduced to 2. The above equation can then be written in the form:

$$C''_{ijkl} = \frac{E}{2(1+\nu_s'')}\left(\delta_{il}\delta_{jk} + \delta_{ik}\delta_{jl}\right) + \frac{E\nu_s''}{(1+\nu_s'')(1-2\nu_s'')}\delta_{ij}\delta_{kl}. \tag{2.3.82}$$

For an *isotropic solid*:

$$\eta_{k\ell} = \alpha_T \delta_{k\ell}, \tag{2.3.83}$$

where α_T is the *coefficient of linear thermal expansion*. By analogy with (2.3.80) the stress-strain relationship then becomes

$$\begin{aligned}\sigma_{ij} &= 2\mu_s''\big[\varepsilon_{ij} - \alpha_T(T - T_o)\delta_{ij}\big] + \lambda_s''\big[\varepsilon - 3\alpha_T(T - T_o)\big]\delta_{ij} \\ &= 2\mu_s''\varepsilon_{ij} + \lambda_s''\varepsilon\delta_{ij} - (3\lambda_s'' + 2\mu_s'')\alpha_T(T - T_o)\delta_{ij},\end{aligned} \tag{2.3.84}$$

where $\varepsilon(\equiv \varepsilon_{ii} \equiv \sum_i \varepsilon_{ii})$ is the volumetric strain ($\equiv$ dilatation). Equation (2.3.84) is the *constitutive equation of a linearly thermoelastic isotropic solid*. It may be inverted to yield the strain as a function of the stress:

$$\varepsilon_{ij} = \frac{1}{2\mu_s''}\left(\sigma_{ij} - \frac{\lambda_s''}{3\lambda_s'' + 2\mu_s''}\sigma_{kk}\delta_{ij}\right) + \alpha_T(T - T_o)\delta_{ij}. \tag{2.3.85}$$

Other forms are:

$$\varepsilon_{ij} = \frac{1 + \nu_s''}{E}\sigma_{ij} - \frac{\nu_s''}{E}\sigma_{kk}\delta_{ij} + \alpha_T \Delta T\,\delta_{ij}, \tag{2.3.86}$$

or

$$\sigma_{ij} = \frac{E}{1 + \nu_s''}\left(\varepsilon_{ij} + \frac{\nu_s''}{1 - 2\nu_s''}\varepsilon_{kk}\delta_{ij}\right) - \alpha_T \frac{E}{1 - 2\nu_s''}\Delta T. \tag{2.3.87}$$

2.3.6 *Enthalpy of a Solid*

The state variable enthalpy of a liquid was introduced in Sect. 2.3.2 C.

For a solid, the specific enthalpy, h, and the specific internal energy, u, under isobaric conditions, or for small pressure variations, $\Delta p = p - p_o$, with respect to a reference pressure p_0, can be described by (4.36) and (2.3.25). Note that for a solid, the pressure, p is defined by (2.3.66).

For solids, the heat capacity $C_p(= C_p(T))$ is often described by the general quadratic function in $C_p = \sum_{m=0}^{m=2} A_{\theta m} T^m$ (Yaws 1999), in which $A_{\theta m}$ $(m = 0, 1, 2)$ denotes material-specific constants. For large Δp, the effect on h of solids can be described by the general equation:

$$h = h_{po} + \beta_{po}\left[v_{po} - v + v \ln\left(\frac{v}{v_{po}}\right)\right]. \tag{2.3.88}$$

There are several general EOS's for a single-species solid under non-isothermal conditions, e.g., the *universal* equation of solids proposed by Vinet et al. (1987). However, these are cumbersome to use, and were developed for solids exposed to very large pressures and undergoing significant compression, i.e., beyond the range of conditions of most processes in flow and transport through porous media.

2.4 Interphase Surfaces and Transfers

In this section we consider the microscopic interfaces between phases within a porous medium domain and the transfer of extensive quantities across these surfaces. In single phase flow, we have only fluid-solid interfaces. In multi-phase flow, we also have fluid-fluid interfaces. The macroscopic description of these phenomena is presented in Chaps. 5–9.

2.4.1 Fluid-Fluid Interface

We are consider a fluid-fluid interface, or inter-phase boundary, at the microscopic level. The shape of the microscopic interface, $F(x, y, z, t) = 0$, is not discussed here, but the discussion in Sect. 5.2.1 D is applicable also here, with x, y, z denoting coordinates at the microscopic level.

A. The Inter-phase Zone

For a fluid-solid interface, the presence of an interface is obvious, although the fluid next to the solid may behave differently from the rest of the fluid. The situation is different when we consider the interface between two immiscible fluids—two liquids, or a liquid and a gas—within the void space, i.e., at the microscopic level. In principle, there are no 'immiscible fluids', as any two fluids are always miscible with each other, at least to some extent. However, when the degree of miscibility is very small, the concept of 'immiscible fluids' may be applied. This 'interface' is actually a *very thin zone* of transition between the two fluid phases. Close to this surface, say, within a distance of a few molecules on either side of the latter (i.e., within the transition zone), the properties of the fluid differ significantly from those within the body of either fluid. In fact, the idea of a thin film as an inter-phase zone is introduced in Sect. 7.4.3, where the concept of a thin film is approximates the 'sharp interface' between two immiscible fluids.

It is interesting to note that Hassanizadeh and Gray (1989) envision the interface between phases as a 'domain' for which balance equations of thermodynamic extensive quantities are considered in a way which is analogous to balance equations written for phase domains. Their approach to modeling transport in porous media is summarized in Sect. 1.4.2 C.

To understand this phenomenon, we have to recall what happens in the fluids at the *molecular level*.

B. Surface Tension

Molecules of a fluid are attracted to each other by an attractive force. Consider the interface between two domains: one occupied by a liquid, the other occupied by the latter's vapour. Because a molecule in the interior of the liquid body is surrounded by liquid molecules of the same kind, having a similar mean spacing, it is attracted, on the average, equally in all directions, and the resultant attractive force acting on it vanishes. The same is true for a molecule in the interior of the vapour. However, the situation changes as we approach the interface from either side. A molecule belonging to the interface is subjected to a stronger resultant attractive force towards the interior of the liquid body. As a consequence of the pull towards the liquid's interior, work must be performed in order to increase the surface of the interface by bringing liquid molecules from the interior of the liquid body to the interface. Left alone, the surface will tend to assume, *spontaneously*, the shape that corresponds to a state of minimum energy under the prevailing conditions. Thus, the surface of the liquid always tends to contract to the smallest area possible under the prevailing circumstances. The same phenomenon takes place at the interface between a liquid and a gas, and between any two immiscible liquids.

In reality, because of the continuous motion of the molecules, no sharp surface of separation exists. Instead, a transition takes place across a relatively thin zone, from the domain occupied primarily by one kind of molecules to that occupied primarily by molecules of the other kind. The properties of the transition zone vary across its width. As explained above, because molecules in this transition zone behave differently from those in the interior of the respective fluid bodies, this zone is *conceptually* replaced, *as an approximation*, by a sharp *interface* that is assumed to separate the two domains (Gibbs 1906). At the macroscopic level, the exact position of this dividing surface within the transition zone is arbitrary. Although molecules are continuously joining and leaving this interface, we regard it as a distinct surface that separates the two fluid phases. In fact, this is an observable surface.

Transport of extensive quantities (e.g., heat, or mass of a phase) may take place through this interface. In order to increase the area of the interface between two immiscible fluids, molecules from the interior of the two fluid bodies must be brought into the surface. This requires that work be done against the net *cohesive force* among the molecules in the two fluids. On the other hand, energy is gained when the area of an interface is reduced. The work required in order to increase the surface area of an interface by one unit is called *surface* (or *interfacial*) *free energy*.

The tendency of a surface to contract may be regarded as a manifestation of the *surface free energy*. The molecules at the surface behave *as if* they belong to a thin, skin-like elastic layer, or *membrane*, under tension, that adjusts its geometry to attain the smallest possible surface area under the prevailing conditions. Obviously, the 'membrane' is only a *model* of the behavior of the interfacial boundary surface, and no such membrane actually exists. This property of interfaces causes a liquid droplet to assume a spherical shape (which has the smallest surface area for a given volume), in the absence of any other forces.

We have to be careful with the analogy to a 'stretched membrane'. The tension in the latter, generally, increases with increased surface area, whereas the surface tension is independent of area. Furthermore, contrary to the case of the interface between two fluids, molecules are not added to a real membrane as it is being stretched.

The interfacial free energy manifests itself as an *interfacial tension* (inside the fictitious 'membrane'), measured as energy per unit area. For a pair of substances, α and β, the interfacial tension, $\gamma_{\alpha\beta}$, is defined as *the amount of work that must be performed in order to separate a unit area of substance α from substance β, or, equivalently, to increase their interface by a unit area.* For air (a) and water (w) at 20 °C, $\gamma_{wa} = 72.8\,\text{erg/cm}^2$ ($\equiv 72.8 \times 10^{-3}$ J/m^{-2}).

Surface tension, $\Gamma_{\alpha\beta}$ between two fluid phases, α and β, separated by an interface of area A, can also be expresses as a change in Gibbs free energy (Sect. 2.2.3) G of a system composed of the two phases and the interface separating them, by:

$$\gamma_{\alpha\beta} = \left.\frac{\partial G}{\partial A_{\alpha\beta}}\right|_{p,T}. \tag{2.4.1}$$

Equivalently, the interfacial tension can also be expressed, as force per unit distance along the membrane's surface, i.e., $\gamma_{\alpha\beta} = 72.8\,\text{dyne/cm}$ ($\equiv 72.8 \times 10^{-3}$ N/m). The interfacial tension between an α-substance and its own vapour is called *surface tension*, γ_α. For example, a general equation describing the temperature dependence of surface tension of organic substances is the relationship proposed by Escobedo and Ali Mansoori (1996):

$$\gamma_\alpha = [\gamma_{R\alpha}(\rho_{L\alpha} - \rho_{G\alpha})]^4 , \tag{2.4.2}$$

where $\rho_{L\alpha} = \rho_{L\alpha}(p, T)$ and $\rho_{G\alpha} = \rho_{G\alpha}(p, T)$ are the densities of the liquid and gas α-phases, respectively, and:

$$\gamma_{R\alpha} = \gamma_{o\alpha}(1 - T_r)^{0.37} T_r \exp\left(\frac{0.30066}{T_r} + 0.86442\, T_r^9\right), \quad T_r = T/T_c, \tag{2.4.3}$$

$$\gamma_{o\alpha} = \left(\frac{39.643}{P_c^{5/6}}\right)\left[0.22217 - 2.91042 \times 10^{-3}\left(\frac{R^*}{T_{br}^2}\right)\right] T_c^{13/12}, \tag{2.4.4}$$

where T_c denotes critical temperature, $R^* = R_{m\alpha}/R_{m,ref}$, $R_{m\alpha}$ is the molar refraction of the α-substance, $R_{m,ref}$ is the molar refraction of a reference fluid (usually methane), $T_{br} = T_{b\alpha}/T_c$, and $T_{b\alpha}$ is the boiling point of the α-substance.

The term *surface tension* is often used to indicate the *interfacial tension* associated with the interface between two immiscible liquids, or between any liquid and a gas. Henceforth, we shall also use the term 'surface tension' to indicate 'interfacial tension'.

The interface tension $\gamma_{\alpha\beta}$ of two immiscible fluids α and β can be obtained as a function of the surface tensions of the individual fluids (phases) using the equation of Girifalco and Good (1957):

$$\gamma_{\alpha\beta} = \gamma_\alpha + \gamma_\beta - 2\Phi\sqrt{\gamma_\alpha\gamma_\beta}, \qquad (2.4.5)$$

where γ_α and γ_β are the surface tensions of the immiscible fluids α and β, respectively, and Φ is a constant which is equal to the ratio of energies of adhesion and cohesion for the two phases. The values of Φ for a number of different liquid/liquid systems can be found in Girifalco and Good (1957).

As is evident from Eqs. (2.4.2) to (2.4.5), the interfacial tension depends on the temperature; it decreases by approximately 5.5×10^{-5} N/m/°C for a crude oil-water interface. It is strongly affected by surface active agents (called *surfactants*), by gas in solution, and by the pH (Schowalter 1979). A survey of methods for estimating surface tension values between various fluids is given by Poling et al. (2000).

As an illustration, consider a soap (s) bubble of radius r, with gas (= air) on both sides. Actually, in a soap bubble, we have two interfaces: a soap-(internal) air interface, and a soap-(external) air one. With $\gamma(= 2\gamma_{as})$ denoting the surface tension in the film, measured as energy per unit area, the total energy in the film surface is $4\pi r^2\gamma$. If the radius will be increased by dr, the added energy will amount to $8\pi r\gamma dr$. This increase in film area is produced by increasing the pressure difference, Δp, say, by increasing the inner pressure, p_{in}, more than the outer one, p_{out}. The added energy is due to the work of Δp. Thus,

$$(\Delta p) \times (4\pi r^2 dr) = 8\pi r\gamma dr, \quad \text{or} \quad \Delta p = 2\frac{\gamma}{r}. \qquad (2.4.6)$$

Since $\gamma > 0$, and $r > 0$, we must have $\Delta p > 0$, or $p_{in} > p_{out}$.

2.4.2 Wettability and Spreading

When any two fluids are in contact with a solid, one of them will tend to adhere to the solid. Figure 2.11a shows two immiscible fluids in contact with a planar solid surface (S). The point M in the figure is the trace of the line (perpendicular to the figure) along which the three phases are in contact with each other. Due to interfacial tension, three forces act at this line, each being directed along the tangent to the interface between adjacent phases. The magnitude of each force, per unit length of the contact line, is equal to the corresponding interfacial tension: γ_{SG}, γ_{SL}, and γ_{LG}.

The angle, θ_{LG}, called *contact angle*, or *wetting angle*, denotes the angle between the solid surface and the fluid-fluid interface, measured through the denser fluid. It depends on the properties of the two fluids, and expresses the affinity of the fluids for the solid. For a perfectly wetting fluid, $\theta_{LG} = 0$. Which means that the fluid tends

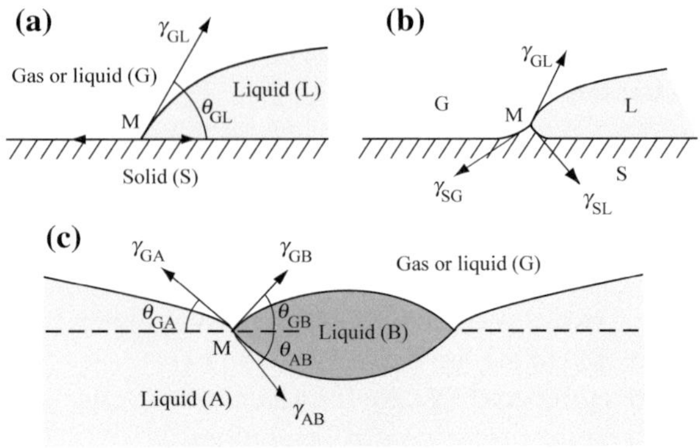

Fig. 2.11 Interfacial tension

to spread over the solid surface. For a perfectly non-wetting fluid, $\theta_{LG} = 180^o$; this fluid will form droplets on the solid surface.

At equilibrium, the equation

$$\gamma_{LG} \cos\theta + \gamma_{SL} = \gamma_{SG}, \quad \text{or} \quad \cos\theta = \frac{\gamma_{SG} - \gamma_{SL}}{\gamma_{LG}}, \tag{2.4.7}$$

called *Young's*, or *Dupré equation*, states that $\cos\theta$ is the ratio of the work required to change a unit area of S–G-interface into a unit area of S–L-interface to the work required to form a unit area between the L- and the G-phases. From (2.4.7), it follows that no equilibrium is possible if:

$$\frac{\gamma_{SG} - \gamma_{SL}}{\gamma_{LG}} > 1, \quad \text{or} \quad S_{LSG} \equiv \gamma_{SG} - \gamma_{SL} - \gamma_{LG} \geq 0, \tag{2.4.8}$$

where S_{LSG} is the *spreading coefficient*. In such a case, the L-liquid in Fig. 2.11a will spontaneously spread indefinitely over the solid surface (Adamson 1982)

If $\theta = 0$, (2.4.7) is no longer valid, and if (2.4.8) is satisfied, the imbalance in surface free energy will cause spreading. For the case of two liquids, A and B, the spreading coefficient, $S_{A/B}$, is expressed by:

$$S_{A/B} = \gamma_A - \gamma_B - \gamma_{AB}. \tag{2.4.9}$$

Spontaneous spreading occurs when this coefficient is positive. The spreading coefficient is positive if spreading is accompanied by a decrease in free energy. The surface tensions in the above equations are those of pure fluids. In reality, as fluids come into contact with each other, and with time, they gradually become mutually saturated, so that γ_A and γ_B will change, and, as a consequence, the spreading coefficient will also change. The extent of the change may be such that the sign of this coefficient will also change.

Actually, Young's equation considers only equilibrium of the force components along the tangent to the solid surface. Requiring equilibrium of force components also along the normal to the surface, would mean that, in principle, we must have the situation shown in Fig. 2.11b. We note that the surface is not planar, although the actual deviation from a plane may be very small.

The spreading coefficient may have a significant effect on whether a nonwetting fluid, when reaching residual saturation in a porous medium domain, will take the form of isolated droplets, or extended ganglia. In very fine-grained materials, e.g., clays, most of the wetting fluid–water–may take the form of films. The trend may be influenced by certain chemicals, called *surfactants* (to be discussed below), which affect surface tension (e.g., Schramm 2000).

The product $\gamma_{LG}\cos\theta$, appearing in (2.4.7), is called *adhesion tension*. It determines which of the two fluids, L or G, will preferentially wet the solid, i.e., adhere to it and spread over it. When the wetting face is replace by a nonwetting one, It is assume that the portion of the solid surface which is contact with the nonwetting fluid (see Fig. 2.11) is always devoid of wetting film. In reality, due to adhesive forces, such film does exist (see Fig. 1.12).

The fluid for which $\theta < 90°$ (e.g., L in Fig. 2.11a), is said to *wet* the solid and is called the *wetting fluid*. Gases are almost invariably nonwetting phases. When $\theta > 90°$, the fluid (G in Fig. 2.11a) is called *nonwetting fluid*. In any system similar to that shown in Fig. 2.11a, it is possible to have either a L-fluid-wet, or a G-fluid-wet solid surface, depending on the chemical composition of the two fluids and of the solid.

In the unsaturated (air–water) zone in the soil, water is, usually, the wetting phase, while air is the nonwetting one. Gas is always the non-wetting phase. It may be interesting to note that petroleum reservoir rocks are usually water-wet, but they can be oil wet, or have mixed wettability.

Sometimes, due to the heterogeneity of the mineral composition of a solid matrix, we encounter *fractional wettability*, defined as the fraction of the total surface area that is preferentially wet by one of the phases (e.g., Anderson 1987; Demond et al. 1994; Dekker and Ritsema 1994). This phenomenon may strongly affect the transport of fluid phases, and of dissolved chemical species.

Additives, called *surfactants*, or *surface active agents*, mentioned earlier, tend to accumulate in the liquid close to and at the interface. We say that they 'adsorb' on the interface. They reduce the interfacial tension, sometimes significantly, and may alter the contact angle, mainly due to modifications of solid surface properties. The presence of surfactants, even in minute quantities, may significantly change the capillary behavior of water in soil.

Figure 2.11c shows the balance of forces between three fluid phases, or between two liquid phases and a gas, with one of the liquid phases taking the form of a lens that rests on the other liquid. Equilibrium of force components tangent to the solid surface requires that:

$$\gamma_{AG} \cos\theta_{AG} = \gamma_{AB} \cos\theta_{AB} + \gamma_{GB} \cos\theta_{GB}. \tag{2.4.10}$$

It follows that when $\gamma_{AG} < (\gamma_{AB} + \gamma_{GB})$, a *lens* of the intermediate wettability B-liquid will be formed between the wetting A-phase and the nonwetting G-phase. The B-liquid is referred to as a 'nonspreading liquid'. If $\gamma_{AG} > (\gamma_{AB} + \gamma_{GB})$, the B-liquid will spread out between the A-liquid and the gas, or the G-liquid. In this case, the B-liquid is called a 'spreading liquid'.

The magnitude of the surface tension, γ_{AB}, depends on the temperature, composition, and pressure of the fluids. It is very sensitive to impurities. In general, the effect of pressure is very small and can be neglected, so that surface tension can be assumed to depend primarily on the temperature and the composition of the fluids.

It is generally assumed that all fluid phases within an REV are in *thermodynamic equilibrium, or* approximate thermodynamic equilibrium. According to our nomenclature, this is equilibrium at the *microscopic level*, obtained by averaging the molecular one. However, when modeling transport phenomena in porous media, we average the microscopic behavior (over an REV) to obtain the macroscopic one. This means that we have to consider the concept of equilibrium also at the macroscopic level. A discussion on equilibrium is presented in Sect. 2.1.

Water has a very high surface tension, relative to most other liquids. For a mixture of miscible liquids (e.g., a mixture of hydrophobic organic liquids), the surface tension is approximately equal to the weighted average of the liquids' fractional volumes. However, surfactants, because they accumulate in the interfacial region, may reduce the surface tension disproportionate to their volumetric fraction. Many hydrophobic compounds, which have a low solubility in water, prefer to accumulate in the interfacial region, rather than mingle freely with molecules in the interior of a water domain. Some examples of interfacial tension values at 20° are:

Water–water vapour	72.88 dyne/cm $= 7.288\times10^{-2}$ N/m.
Benzene–benzene vapour	28.88 dyne/cm $= 2.888\times10^{-2}$ N/m.
Water–benzaldehyde	15.50 dyne/cm $= 1.550\times10^{-2}$ N/m.

Various theoretical and empirical methods for estimating surface tension of mixtures, spreading tension and interfacial tension have been summarized by Lyman et al. (1982).

2.4.3 Capillary Pressure

In this subsection, we are dealing only with capillary pressure *at the microscopic level*, i.e., at a point on the (assumed sharp) interface between two fluids. Capillary pressure at the macroscopic level will be presented and discussed in Chap. 6, which deals with multiphase flow.

As discussed above, a discontinuity in fluid stress exists across a *curved* interface that separates any two immiscible fluids (say, air and water). The jump in the normal stress, or pressure, is a consequence of the interfacial tension which exists at every point of such an interface. The difference between the pressure $p_{concave}$, in the fluid

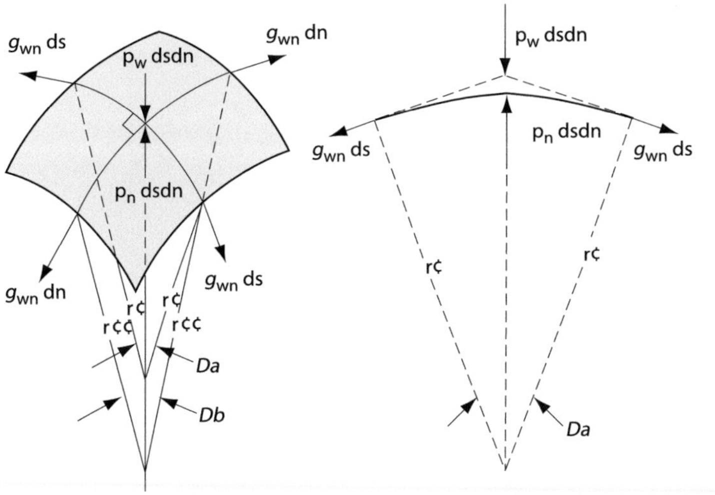

Fig. 2.12 Force balance at a curved interface

which occupies the *concave* side of an interface, and the pressure p_{convex}, which occupies the fluid on the *convex* side of the latter, is called *capillary pressure*, p_c:

$$p_c = p_{concave} - p_{convex}. \tag{2.4.11}$$

In this equation, the pressures are taken as the interface is approached within the appropriate phase. Note that we have used the prime symbol to indicate the capillary pressure at the microscopic level, i.e., at a point on the interface. In Sect. 6.1 we shall consider the capillary pressure at the macroscopic level.

The magnitude of the pressure difference at a point on an interface depends on the radius of curvature of the latter and on the surface tension at that point. The special case of a spherical surface was given by (2.4.6). Let us now consider the relationship for a general surface. There are several ways for developing this relationship. For example, it is possible to consider the work required in order to increase the area of the interface. In all approaches, the interface is visualized as a (two-dimensional) material body (actually, surface) which has rheological properties of its own. Its behavior is similar to a that of a 'stretched membrane' under tension. In fact, this assumption alone already leads to the conclusion that under conditions of force equilibrium, the normal components of the fluid's stress, or pressure, must be discontinuous as a (curved) interface is crossed.

As an example, let us use Fig. 2.12, which shows an infinitesimal element of a curved interface between a wetting-fluid (w), which occupies the convex side of the interface, and a nonwetting-fluid (n), which occupies the concave side of the latter. The figure shows the various forces acting on this element. Assuming the interfacial tension between these two fluids, γ_{wn}, to be constant, a balance of force components

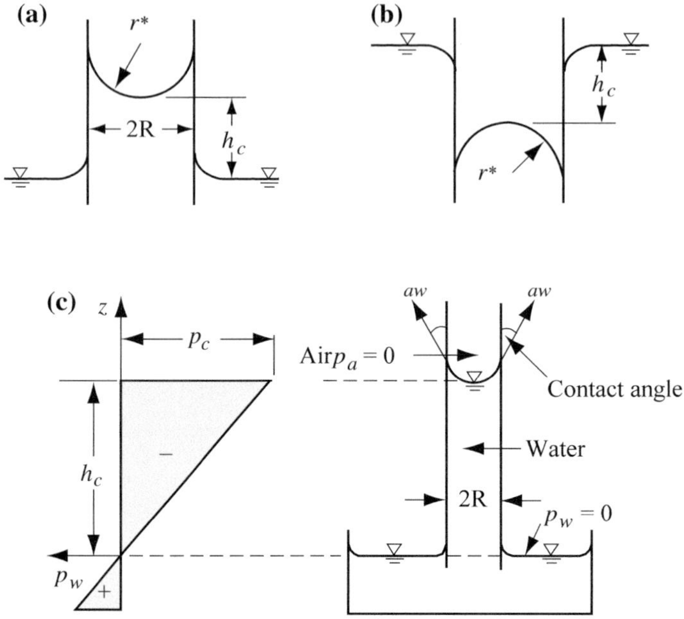

Fig. 2.13 Water (w) - air (n) interface in a capillary tube: **a** Capillary rise, $\theta = 0°$, **b** capillary depression, $\theta = 180°$, and **c** capillary rise, $\theta \neq 0°$

normal to this element requires (Adamson 1982) that at equilibrium, under steady state conditions:

$$p_c' = p_n - p_w = \gamma_{wn}\left(\frac{1}{r'} + \frac{1}{r''}\right) = \frac{2}{r^*}\gamma_{wn}, \tag{2.4.12}$$

where subscripts w and n denote the wetting and the non-wetting fluids. In the above expression, r' and r'' are the two *principal radii* of curvature of the surface, with a radius considered positive when it lies within the n-fluid. The radius r^* denotes the *mean radius of curvature*, defined by $2/r^* = 1/r' + 1/r''$. Equation (2.4.12) is known as the *Laplace*, or *Young–Laplace formula* for capillary pressure. Because γ_{nw} is positive when both radii are positive, the pressure is greater in the n-fluid, for which the surface is convex.

An example of two fluids in a capillary tube is shown in Fig. 2.13. The definition of capillary pressure presented above still holds, with subscripts w and n denoting 'wetting' and 'non-wetting' fluids, respectively.

Let us consider the effect of the interaction between two fluids and a solid surface. For this purpose, let us consider what happens to the interface between a pair of two immiscible fluids, say, a liquid and a gas, in a capillary tube. The two fluids occupy the convex and the concave sides of the interface, respectively, as shown in Fig. 2.13a. If the liquid perfectly wets the solid surface of the tube, i.e., the *contact angle* θ

$= 0°$, and the gas-liquid surface is tangent to the tube's wall at their point of contact. When the liquid is nonwetting with respect to the solid, i.e., $\theta = 180°$, a situation of *capillary depression* (Fig. 2.13b) is observed, with a convex meniscus.

In the general case of a wetting liquid, $\theta \neq 0°$, we have the situation shown in Fig. 2.13c. If the capillary tube has a circular cross-section, with a radius R that is not too large, the curved interface (= meniscus) will be approximately in the shape of a hemisphere. In this case, $r' = r''$, and $r^* = R/\cos\theta$. Then, in a small diameter circular tube, with an approximately spherical meniscus, the pressure difference given by (2.4.12) can be written in terms of the tube's radius, R, in the form:

$$p'_c = \frac{2\gamma_{wn}\cos\theta}{R}. \tag{2.4.13}$$

Figure 2.13c, shows the rise of water in a vertical capillary tube with an air–water meniscus. With h'_c (= *capillary rise*) denoting the height of the water column above a horizontal, flat water surface in a large container (i.e., $r^* = \infty$), p_c must equal the hydrostatic pressure drop along the column of length h_c. Hence:

$$\pi R^2 h'_c \rho_w g = 2\pi R\gamma_{aw}\cos\theta, \qquad h'_c = \frac{2\gamma_{aw}\cos\theta}{\rho_w g R}, \tag{2.4.14}$$

where γ_{aw} denotes the air–water surface tension. In terms of the radius of curvature of the meniscus, $r^* = R/\cos\theta$, we obtain:

$$h'_c = \frac{2\gamma_{aw}}{\rho_w g r^*} = \frac{p_c}{\rho_w g}. \tag{2.4.15}$$

2.4.4 Interphase Mass Transfer

In this subsection we consider mass transfer across a microscopic interphase surface. In Chap. 7 we shall consider interphase mass transfer at the macroscopic level.

A. Gas-Liquid Mass Exchange and Vapor Pressure

Understanding phase changes (e.g., evaporation and condensation) requires a firm grasp of the physics and thermodynamics of *vapour pressure*. Some relevant elements on this subject are presented below.

We consider a liquid and a gas separated by an interface. Both contain a substance γ, although both the liquid and the gas may contain also other species. A continuous transfer of the γ-species takes place between the gaseous and liquid phases across their common interface. For example, consider a free water (primarily H_2O, but may contain other dissolved species) surface exposed to air (which contains mostly oxygen and nitrogen, but also water vapour). As long as the air, at the system's temperature, is not saturated by water vapour, which means that the system is not at equilibrium, a continuous transfer of H_2O molecules takes place from the aqueous phase to the gaseous one, and vice versa, leading to either evaporation of liquid water, or condensation of water vapour.

Vapor pressure, p_{vap}, is the pressure exerted by the gaseous phase of a single substance on the liquid (or solid) phase of the same substance with which it is in contact and in equilibrium in a closed system. It is the pressure of the vapour resulting from the evaporation of the liquid phase present above the latter's surface. This pressure is a strong nonlinear function of the temperature. The vapour pressure is also affected by other factors, e.g., salinity and capillarity. Thus, rocks with very high capillary pressures lead to vapour-pressure lowering, a phenomenon that has significant implications in the behavior and management of geothermal reservoirs.

In a closed system, the equilibrium attained between the pressure of the liquid phase and that of its vapour indicates that the pressure in the latter is constant at a given temperature, with a balance established between molecules of the substance exchanged between the two phases across their common interface.

Denoting the molar *Gibbs free energy*, $\mathbb{G}$, defined by (2.2.28), of the liquid and the gas by subscripts l and g, respectively, we have across the interface for this case, at a specified p and T,

$$\mathbb{G}_l = \mathbb{G}_g, \qquad d\mathbb{G}_\alpha = -\mathbb{S}_\alpha dT_\alpha + \mathbb{V}_\alpha dp_\alpha. \tag{2.4.16}$$

Henceforth, to emphasize that the gas considered here is the vapour of the associated liquid phase, we shall use the subscript v instead of g.

From (2.4.16), it follows that:

$$\frac{dp}{dT} = \frac{S_l - \mathbb{S}_v}{\mathbb{V}_l - \mathbb{V}_v} \equiv \frac{\Delta\mathbb{H}_v}{T(\mathbb{V}_l - \mathbb{V}_v)}. \tag{2.4.17}$$

We note that the difference between the molar entropy of the vapour and that of the liquid at a given temperature is the *entropy of vaporization*, $\Delta\mathbb{S}_v$. Equation (2.4.17) is known as the *Clapeyron equation* (e.g., Smith et al. 2005; Poling et al. 2000).

If, in the Clapeyron equation (2.4.17), as an approximation, we neglect $\mathbb{V}_l$, (relative to $\mathbb{V}_v$, we obtain for the two co-existing phases:

$$\frac{dp}{dT} = \frac{\Delta \dot{h}_v}{T\mathbb{V}_v}, \tag{2.4.18}$$

where $\check{h}_v$ denotes molar enthalpy.

For a liquid and its vapour, if, in addition to the assumption $\mathbb{V}_l \ll \mathbb{V}_v$, we assume also that the vapour behaves as an *ideal gas* (Sect. 2.3.3), we obtain:

$$\frac{dp}{dT} = \frac{p\Delta \dot{h}_v}{RT^2\mathbb{V}_v}, \tag{2.4.19}$$

known as the *Clapeyron–Clausius equation*.

By rewriting (2.4.19) in the form:

$$\frac{dp}{p} = \frac{\Delta h^{\cdot}_{v}}{RT^2} dT, \tag{2.4.20}$$

and integrating from (p_1, T_1) to (p_2, T_2), we obtain the dependence of pressure: on temperature:

$$\ln \frac{p_2}{p_1} = \frac{\Delta h^{\cdot}_{v}}{R} \left(\frac{1}{T_1} - \frac{1}{T_2} \right). \tag{2.4.21}$$

Gas phase evolves in a system at a pressure p and temperature T initially containing only a liquid phase of a chemical species γ, when $p_v(T) \geq p$; coexistence of the vapour and liquid phases requires $p_v(T) = p$, with both phases occurring on the vapour-liquid saturation line in the phase diagram (e.g., Fig. 2.3). Conversely, a gas phase disappears if $p_{vap}(T) < p$. These are the criteria used in thermal processes involving phase changes in porous media, e.g., steam injection for heavy oil recovery, in-situ combustion.

There are several expressions describing the dependence of v_{vap} on T. The simplest one is the *Clausius–Clapeyron equation*:

$$\ln(p_v) = A - \frac{B}{T}, \tag{2.4.22}$$

where $B = \Delta \mathbb{H}_v / R \Delta Z_v$. Because of the weak dependence of $\Delta \mathbb{H}_v$ and ΔZ_v on temperature (with the exception of the vicinity of the critical point), this equation is used extensively, but it may be inaccurate when applied over a wide temperature range, especially when extrapolated below the normal boiling point (Poling et al. 2000). For improved results, Antoine (1888) proposed a slight modification of (2.4.22):

$$\ln p_{vap} = A + \frac{B}{T + C}, \tag{2.4.23}$$

where A, B and C are substance-specific parameters that can be found in several reference books, e.g., Poling et al. (2000) and Yaws (1999). This equation is applicable over a rather narrow range of temperatures. For a wider temperature range, other parametric equations are routinely used. These include the Wagner equation (Wagner 1973), the method of Ambrose and Walton (1989), the parametric equation of Yaws (1999). The general form of some of these equations are shown below:

Wagner (1973)	$\ln p_{vap} = (AT_v + BT_v^{1.5} + CT_v^3 + DT_v^6)/T_r,$
Riedel (1954)	$\ln p_{vap} = A + B/T + C\ln(T) + DT^6,$
Yaws (1999)	$\log_{10} p_{vap} = A + B/T + C\log_{10}(T) + DT + ET^2,$

where $T_v = 1 - T_r$, T_c denotes critical temperature, with $T_r = T/T_c$. The substance-specific constants (A to E) corresponding to each equation can be found in reference books such as Poling et al. (2000) and Yaws (1999).

B. Gas Solubility in a Liquid

We consider the concentration of a gaseous γ-species, e.g., methane (CH_4), dissolved in a liquid phase of a different composition. The concentration of a dissolved gaseous γ-species in a gas phase, g, that is in contact with a liquid α-phase (such as water), is related to the partial pressure, p_g^γ, of the γ-species in the gas, through *Henry's law* (see Sect. 2.2.5):

$$p_g^\gamma = \mathcal{H}_{aq}^\gamma(T, c^\delta, \delta \neq \gamma) X_{aq}^\gamma, \tag{2.4.24}$$

where p_g^γ is the partial pressure of the γ-species in the gas phase, X_{aq}^γ is the *mole fraction* of the γ-species dissolved in the aq-liquid phase, and $\mathcal{H}_{aq}^\gamma$ is *Henry's coefficient*. While $\mathcal{H}^\gamma$ may be approximated by a constant for small changes in p and T, it would be more appropriate to refer to it as *Henry's factor*, since, in general, $\mathcal{H}^\gamma (= \mathcal{H}_\alpha^\gamma(T, c_\alpha^\delta, \delta \neq \gamma))$, where c_α^δ denotes the concentration of non-γ species in the α-solution. Values of, and equations for, $\mathcal{H}_\alpha^\gamma$ can be found in the literature (e.g., Yaws 2003; Truesdell et al. 1960).

It is interesting to note that actually, Henry's law applies within a thin surface liquid layer between a gaseous domain and a liquid one, or to a gaseous phase in contact with a "well mixed" liquid domain, or only to a thin layer in contact with the gas. In a gas liquid system in a porous medium domain, Henry's law is applicable at a point in the porous medium continuum.

In general, currently available expressions for $\mathcal{H}_\alpha^\gamma(T, c_\alpha^\delta, \delta \neq \gamma)$ are quite accurate over a rather wide range of T. More accurate estimates of gas solubility may be obtained from the equality of fugacities (Sect. 2.2.4) in the aqueous and the gaseous phases. Although the use of fugacity provides a more accurate estimate, the complexity of the calculations increases significantly and the difference does not exceed a few percents, in a wide range of practical problems.

Gas solubility controls the dissolution and emergence from solution of a γ-species in a gas phase in contact with a liquid. Thus, a γ-gas species will emerge from solution when

$$X_g^\gamma > \omega_{g,max}^\gamma = \frac{p_g^\gamma}{\mathcal{H}^\gamma(T, c^\delta, \delta \neq \gamma)}, \tag{2.4.25}$$

i.e., when the dissolved mass fraction of a γ-species exceeds its maximum solubility in the liquid phase at specified temperature and a partial pressure, p_g^γ.

C. Liquid-Solid Mass Exchange

Two kinds of liquid-solid mass exchange may be considered here:

- Dissolution of the solid comprising the solid matrix in the liquid occupying the void space.
- Adsorption/desorption phenomena which describe the movement of a chemical species dissolved in the solution to become adsorbed on the solid surface, or desorb back into the solution. Surface complexation and ion exchange phenomena also belong to this group.

These phenomena are described in Sect. 7.4.

2.5 Soil Potentials and Osmotic Pressure

The *potential* is a concept that expresses the *ability to do work*. The potential energy of a mass particle, say of a fluid, is associated with its position in a force field (e.g., gravity field). The value of the potential is a measure of the work that would be done in a particular force field on the fluid particle that moves in that field from one point to another, provided this work could be done *reversibly* (Corey 1977, p. 77). A force field $\mathbf{f}$ for which a potential exists is said to be *conservative*. A necessary and sufficient condition for a force field to be conservative is that the work integral over any closed path will vanish, i.e.:

$$\oint \mathbf{f}\cdot d\mathbf{s} = 0, \tag{2.5.1}$$

where $d\mathbf{s}$ is a differential displacement vector in the field. If the above condition is satisfied, then it is possible to define a *scalar*:

$$\Psi(\mathbf{s}) = \int_{\mathbf{s}_o}^{\mathbf{s}} (-\mathbf{f})\cdot d\mathbf{s}, \tag{2.5.2}$$

where $\mathbf{s}_o$ represents a datum point for the potential point, and the minus sign is associated with the way the potential is commonly defined such that the negative gradient of the potential represents the force acting at any point in the considered domain. It is interesting to note that clay layers act equivalent to membranes, so osmotic pressure does develop across them.

The presentation in this section follows the presentation by Nitao and Bear (1996). Although in this chapter we are focussing on microscopic level phenomena, occasionally we shall also refer to macroscopic level potentials.

2.5.1 *Soil Potentials*

The term *'potential'—total potential, matric potential, osmotic potential*, etc.—is often encountered in the soil science literature. It is of special interest when conditions of equilibrium are assumed to exist between phases and species in the soil's void space. Our interest here is in the behavior of fluids and chemical species in a porous medium domain (i.e., at the macroscopic level). An extensive discussion on potentials in the soil is presented by Parker (1986). *Chemical potential*, which is an essential concept in thermodynamics is introduced and discussed in Sect. 2.2.6.

The concept of a *potential* is well established in thermodynamics, where, in this book's terminology, it is discussed 'at the microscopic level'. The basic ideas have been extended also to the discussion of flow and other phenomena of transport in porous media, i.e., at the macroscopic level of description, often without a rigorous proof. Nitao and Bear (1996) have presented a rigorous discussion on potentials at the macroscopic level. As in the theory of thermodynamics, they also start from definitions at the *microscopic* level, and then average the potentials defined at the microscopic level to obtain their macroscopic counterparts.

A chemical species in the fluid phase within the void space is acted upon by a number of forces. These forces arise from gravity, from the interaction of fluid molecules with those of the solid matrix, and from the presence of dissolved matter. Although this statement is valid for any fluid, we shall use here water as an example, because it is often the main fluid of interest. Work has to be expended in order to change the system in a direction which is opposite to any or all of the forces mentioned above.

The *potential* of the water is a concept which facilitates the discussion of such changes. It expresses *the reversible work that has to be expended in order to transform a given system from some reference state to its current one*. This amount of work is, thus, equivalent to the increase in the energy of the system. Because of the different nature of the various forces, some being non-mechanical, it is more convenient to define a number of potentials, each corresponding to a specific force, or a combination of forces.

Each potential is expressed as an *intensive quantity*, either per unit volume, per unit mass, or per unit weight of the considered fluid phase, e.g., water in the void space.

A. Total Potential, Ψ^w_{total}

This potential of the water, H_2O, as a species at a point in the soil is defined by Commission I of the International Soil Science Society (Aslyng 1963) as

> the amount of work that must be done per unit quantity of pure water in order to transport reversibly and isothermally to the soil water at a considered point, an infinitesimal quantity of pure water from a pool that contains pure water. The pool is at a specified elevation, and with the same temperature and external gas pressure as at the considered point.

Here 'pure water' refers to the water (H_2O) as a *substance*, or a *species*, while the 'soil water' is the liquid phase (or soil solution) in the unsaturated zone, which contains both water and dissolved matter.

The above definition refers to 'soil water at a considered point', where the 'point' is within a phase, i.e., at the *microscopic level*. A macroscopic potential is defined as the average of the microscopic one over all (microscopic) points within a representative elementary volume (REV) centered at a (macroscopic) point (= centroid of an REV) within a porous medium domain. The same extension from microscopic to macroscopic levels, may be applied to other types of potentials.

The total potential of a fluid phase is composed of a number of potentials:

B. Matric Potential, Ψ^w_m

The matric potential at a point within a fluid phase that occupies part of the void space in the soil, is defined as *the amount of work that must be done, per unit quantity of pure water (as a species, w), in order to transport reversibly and isothermally to* *the soil water at a considered point, an infinitesimal quantity of pure water from a reference pool. The latter is at the elevation, the temperature, and the external gas pressure of the considered point, and contains water (= soil solution) identical in composition to that present in the soil at the considered point* (Commission I, ISSS. See Aslyng 1963).

However, the matric potential as defined above refers to a *chemical species* at the microscopic level, while our interest is really (a) in solutions, or phases, and (b) in phase behavior within the void space, i.e., at the macroscopic level.

The matric potential, Ψ_m, can also be defined for a phase as a whole. At a point, it is defined as the sum of the (reversible) work that has to be expended in order to move each of the individual species comprising the phase, from a reservoir at the same elevation, temperature, external gas pressure, and composition, to the considered point.

The macroscopic matric potential of a phase is the average over the REV of the microscopic one (defined above). This matric potential, often used in soil science, when dealing with the unsaturated (air–water) zone of the soil, is a consequence of two phenomena:

- Unbalanced forces across water–air interfaces, manifested as *surface tension*.
- Attraction of molecules in the phase to the solid surface, manifested as thin films that coat the latter.

The presence of air–water interfaces gives rise to the phenomenon of capillary pressure, viz., the jump in pressure across the microscopic water-air interfaces inside the void space (Sect. 2.4.3). The microscopic capillary pressure is then averaged to obtain its macroscopic counterpart. In the simplified model of a curved meniscus, with a sufficiently thick fluid layer on each side, the concept of a surface tension is valid, and so is the resulting Laplace formula, (2.4.12), for the relationship between the radius of curvature of the meniscus and the (microscopic) capillary pressure. However, in developing an expression for the capillary pressure, say, the Laplace formula, the presence of a film of adsorbed water on the solid surface, and its effect on the relationship between water and gas pressures, was overlooked. This approach is not justified, as the portion of the void space from which water has been drained always contains some water in *pendular rings* and in *thin films* that coat the solid, and, therefore, the effect of the solid surface in the unsaturated zone cannot be ignored.

The definition of matric potential incorporates also the effect of the attractive forces acting upon these films. Nitao and Bear (1996) showed that

$$\Psi_m = -\frac{\widetilde{2\gamma_{\ell g}}^{\ell g}}{r^*} + \widetilde{\varphi}^{\ell g},$$

where $\widetilde{A}^{\ell g}$ denotes an average of A over the ℓ-g-interface, and φ denotes the *surface potential* to be discussed below. The matric potential is, thus, not identical to the *capillary potential*, which is associated with capillary forces only. The surface potential at the $\ell - g$-interface becomes negligible at high saturations as the distance from the interface to the solid surface increases. Consequently, the effect of the adsorbed water films becomes negligible. The matric potential is, then, essentially equal to the capillary potential. If we define the matric potential *per unit volume of water*, the matric potential in this range of saturations is identical to the capillary pressure.

The effect of the films becomes more significant as a soil is drained, and the water saturation approaches the irreducible one. As water is further removed by evaporation to below the irreducible saturation, the effect of adsorbed water films, and its contribution to the matric potential, becomes even more significant, as the only remaining water occurs as films. In fact, these films become discontinuous as drying continues. At some low water saturation, water in the void space can be present only as films that coat the soil surface (and pendular rings at points of contact between grains). They become thinner as water is removed by evaporation, or transpiration. In this range, the behavior is dominated not by the air–water surface tension, but by the interaction between water and solid molecules.

When expressed per unit volume of water, the matric potential at sufficiently high saturations is equal to the pressure in the water (as a liquid phase), $p_w - p_0$, with respect to a datum pressure, p_0, of the reservoir. It is positive (i.e., above the reference atmospheric pressure) in the saturated zone, and negative (i.e., below the reference atmospheric pressure) in the unsaturated zone above it. When the gaseous phase in the void space (primarily air) is at the pressure $p_g = p_0$, the matric potential per unit volume of water is expressed by the difference $(p_w - p_g)$, or by $-p_c$. The matric potential per unit weight of water is, thus, expressed by $(p_w - p_g)/\rho_w g$.

For a given soil, the relationship between the matric potential and saturation has to be derived experimentally, because it depends on the pore size distribution and on the complex geometry of the pore space. For sufficiently coarse soils, at the irreducible water saturation, $S_w \gg S_{wr}$, the $\Psi_m(S_w)$-curve (with Ψ_m expressed per unit volume of water) is essentially that for the capillary pressure presented and discussed in Sect. 2.4.3. At full saturation, the matric potential becomes zero.

C. Solute Potential, Ψ_s^w

This potential is also referred to as *osmotic potential*. The liquid in the soil, referred to by soil scientists as 'soil solution', usually contains dissolved matter. The concentration of dissolved matter affects both the surface tension and the forces that attract water molecules to solid surfaces. It also affects the energy relationships that determine the equilibrium among phases and chemical species. Thus, the solute potential at a point in the soil (i.e., microscopic level) obeys a definition similar to that of the matric potential, except that the reference pool contains pure water at the same pressure, temperature and elevation, as the considered point, while the void space at the point contains a soil solution. We wish to emphasize that the potential is defined for the water *as a chemical species* and not as a *liquid phase*. For the latter, we have to sum over all species, and the pool has to contain a dilute solution.

D. Soil Water Potential, Ψ_{sw}^w

This potential combines the work required to overcome the forces due to both surface forces, pressure, and concentration differences between the reference reservoir and the considered point. The soil-water potential is, thus, the sum of the matric and the solute potentials.

Taking into account the effect of dissolved matter, and following a derivation similar to that for Kelvin's equation (2.3.35), we may write:

$$\Psi^w_{sw} = -\frac{\rho^w_w RT}{M^w} \ln \frac{p^v}{p^{v_o}}, \tag{2.5.3}$$

in which Ψ^w_{sw} is per unit volume of water in the void space. Note that superscript w refers to water as a chemical species, while subscript w refers to water as a liquid phase, which consists primarily of water, but may contain other species. For the sake of clarity, we could have used subscript ℓ instead of w to denote the liquid phase. The above equation is valid at both the microscopic and the macroscopic levels.

E. Gravity Potential, Ψ^w_g

This potential expresses the change in the potential energy associated with the elevation of the considered point above the reference reservoir. Thus, we can use the same definition as that of the matric potential, except that the considered point and the reference reservoir are at different elevations, and both reservoirs have flat interfaces. When expressed per unit weight, the gravity potential is equal to the elevation, $z - z_o$, of the point. When expressed per unit volume, the gravity potential for the w-species is given by $\rho^w_w g(z - z_o)$, assuming that ρ^w_w does not change appreciably from z_o to z.

F. Thermal Potential, Ψ^w_T

This potential expresses the change in the free energy associated with the temperature of the considered point above that prevailing in the reference reservoir. Thus, we can use the same definition as that of the matric potential, except that the considered point and the reference pure water reservoir are at different temperatures.

G. Total Potential, Ψ^w_{total}

The total potential for a w-species, may now be defined as:

$$\Psi^w_{total} = \Psi^w_m + \Psi^w_s + \Psi^w_g + \Psi^w_T = \Psi^w_{sw} + \Psi^w_g + \Psi^w_T. \tag{2.5.4}$$

Many of these potentials depend on the liquid's saturation. The total potential, Ψ^w_{total}, is nothing but the chemical potential of water, as a chemical species in the soil, per unit volume of the water phase.

2.5.2 *Osmotic Pressure and Chemical Potential*

Osmosis is the movement of a solvent through a membrane (= a relatively thin porous medium domain of very low permeability) which is impervious to the solute.

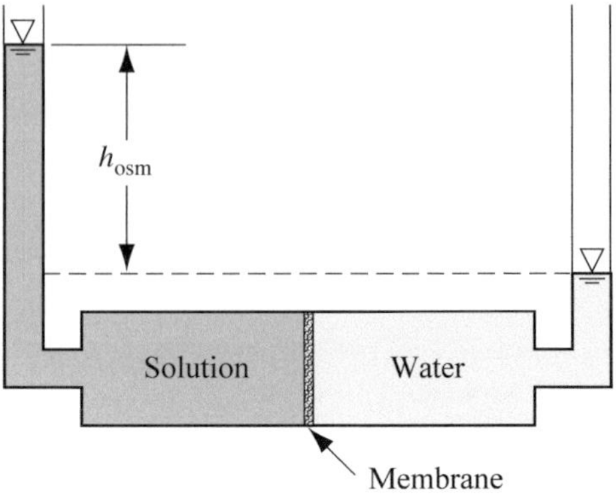

Fig. 2.14 Definition sketch for osmotic pressure

Such a membrane is often referred to as *semipervious*, or *semipermeable*. The direction of flow *of the solvent* is from the more dilute solute concentration to the more concentrated one.

For example, if a salt (= solute) in a salt solution is separated from pure water (= solvent) by a membrane that is impervious to the salt, water will pass through it (by diffusion from high to low *water* concentration), with a net flux in the direction of the salt solution. As a consequence, a difference in hydrostatic pressure develops (Fig. 2.14), with a higher pressure on the solution side. The pressure difference will grow until the pressure in the solution will be sufficient to prevent the flux of water (by diffusion) through the membrane. Eventually, equilibrium is reached, with a pressure difference that is sufficient to counteract the effect of the solute concentration difference. Under the equilibrium conditions, the solution is under a greater hydrostatic pressure than that in the pure solvent. This pressure difference is called *osmotic pressure difference*, Δp_{osm}. The corresponding osmotic potential (in units of head) is $h_{osm} = \Delta p_{osm}/\rho_w g$. It is important to note that osmotic pressure exists only in the presence of a semipervious or partially semi-pervious membrane, or relatively thin semi-pervious domain.

By applying a pressure that exceeds the osmotic pressure, the reverse effect occurs. Fluids are pressed back through the membrane, while dissolved solids stay behind. This is referred to as *reverse osmosis*.

To remove dissolved chemicals from water by using a reverse osmosis membrane, the natural osmosis effect must be *reversed*. In order to force the water with the high salt concentration to flow towards the low salt concentration reservoir, it must be pressurized to an operating pressure greater than the osmotic pressure. As a result, the high salt concentration side will get more concentrated. This is the process underlying the desalinization of sea water, brackish or polluted water.

Let us find an expression for the osmotic pressure in terms of temperature and concentration. At equilibrium, the chemical potential (= *molar free energy*) of a solvent in a solution, μ, is the same as that of the pure solvent. The chemical potential of an *ideal solvent*, $\mu^{solvent}$, of mole fraction $X^{solvent}$, is (e.g., Denbigh 1981)

$$\mu^{solvent} = \mu^{o,solvent} + RT \ln X^{solvent}, \tag{2.5.5}$$

where $\mu^{o,solvent}$ denotes the chemical potential of the pure solvent, R (= 0.082 liter-atm/mole-deg) is the universal gas constant, and T is the absolute temperature in Kelvin degrees. Since $X^{solvent} < 1$, we have $\mu^{solvent} < \mu^{o,solvent}$.

As pressure is applied to the solution (Fig. 2.14), the free energy of the solvent increases. It may be raised to the point where it is equal to that of the pure solvent. From (2.2.53), it follows that in a process at constant temperature, the effect of an increase in pressure on the chemical potential is expressed by $\Delta\mu = v^{mol} p_{osm}$. Since the osmotic pressure, p_{osm}, is the pressure in the solution, in excess of that exerted on the pure solvent, the increase in the chemical potential (Sect. 2.2.6) due to this pressure is given by:

$$\Delta\mu^{solvent} = v^{mol} \int_{p_{atm}}^{p_{osm}} \mathrm{d}p = p_{osm} v^{mol}, \tag{2.5.6}$$

where we have assumed that the solvent is incompressible, or may approximately be considered as such. The combined effect of dilution and external pressure is expressed by:

$$\mu^{solvent} = \mu^{o,solvent} + RT \ln X^{solvent} + \Delta p_{osm} v^{mol}. \tag{2.5.7}$$

When the solvent is in equilibrium with the pure solution, $\mu^{solvent} = \mu^{o,\,solvent}$. For $X^{solvent} \approx 1$, we have $\ln X^{solvent} \equiv \ln(1 - X^{solute}) \approx -X^{solute}$. For a dilute solution, $X^{solute} \approx N^{solute}/N^{solvent}$ and $v^{mol} = \mathbb{V}/N^{solvent}$, where $\mathbb{V}$ is the volume of the solution, and $N^{solvent}$ and N^{solute} are the number of moles of solvent and of solute in $\mathbb{V}$, respectively. We obtain:

$$p_{osm} = \frac{RT}{\mathbb{V}} N^{solute} = n^{solute} RT, \tag{2.5.8}$$

where n^{solute} is the molar concentration of the solute (in moles per liter), and p_{osm} is measured in atmospheres. Thus, (2.5.8) relates the osmotic pressure to the concentration and temperature of a dilute solution.

Osmotic pressure occurs in a porous medium domain whenever a layer of soil behaves as a semipervious membrane with respect to certain chemical species that are present in the water solution. In addition, the roots of vegetation behave as such a membrane. The osmotic potential was discussed in the previous subsection.

In natural solutions, like seawater and brackish water, the different ions do not have the same concentration. For example, in seawater the molar concentration of the Na^+-ions is usually lower than the molar concentration of the Cl^--ions, so when we calculate the actual osmotic pressure of the solution we have to use:

$$p_{osm} = RT \sum_{i=1}^{N} \ln n_i \approx RT \sum_{i=1}^{N} \ln \alpha_i, \tag{2.5.9}$$

where N denotes the number of the different ions in solution, n_i denotes the molar concentration of each ion, α_i is the activity of the ith ion.

2.6 Onsager's Theory of Coupled Processes

In this subsection, we present Onsager's theory (Onsager 1931) as described by de Groot (1963). The presentation is limited to the microscopic level, i.e., a point within a phase, regarded as a continuum. However, using the phenomenological approach, it is then possible to extend the theory to the macroscopic level.

A consequence of the ideas underlying the concept of equilibrium is that all (microscopic) *fluxes of extensive quantities* (defined in Sect. 1.1) vanish, simultaneously, when the thermodynamic forces vanish. And vice versa, thermodynamic forces produce fluxes. Phenomenologically, it has been known that under a wide range of experimental conditions, irreversible (i.e., diffusive, or molecular driven) fluxes are linear functions of the thermodynamic forces. Examples are Fick's law, where the components of the flux of a γ-species are linear functions of the components of the γ-concentration gradient, and Fourier's laws, where the components of the heat flux are linear functions of the components of the temperature gradient. Later (Sect. 3.4.1), we shall see that the above observation can also be applied to the flux of linear momentum as driven (for a Newtonian fluid) by components of the velocity gradient (Newton's law).

Each of the three diffusive flux laws mentioned above, also referred to as *phenomenological laws*, is a particular case of the general linear law:

$$j_i^n = -\sum_{j=1}^{3} L_{ij}^{nn} \frac{\partial \Phi^n}{\partial x_j}, \qquad i, j = 1, 2, 3, \tag{2.6.1}$$

where $j_i^n (\equiv j_i^{E^n})$ denotes the ith component of the flux of an extensive quantity E^n of a phase, Φ^n is a state variable associated with E^n, and L_{ij}^{nn} is a coefficient of proportionality, which is a second rank symmetric tensor for any given n. Since only one phase is being considered, no special symbol is used to indicate it. Some introductory remarks about tensors are presented in Sect. 9.1.

The three phenomenological laws mentioned above state that a nonuniform distribution of a state variable, Φ^n (e.g., temperature), produces a flux of only the corresponding extensive quantity (e.g., heat). However, *experimental evidence* suggests that, since *all diffusive fluxes are associated with the same molecular motion*, the gradient of any state variable, associated with one of the extensive quantities, should produce a flux also of all other extensive quantities. Phenomena of this kind are referred to as *coupled phenomena* (or *cross-effects*). Common examples of such phenomena are the *Soret* (or *thermodiffusion*) effect, in which mass flux of a solute in a liquid phase is produced by a temperature gradient, *in addition* to the flux produced by the gradient of the solute's concentration according to Fick's law, and the

Dufour effect, in which heat flux is caused by a concentration gradient, *in addition* to the heat flux caused by temperature gradient. Thermodynamic, thermoelectric, and galvano-magnetic effects, are other examples of coupled phenomena. The description of coupled phenomena requires a generalization of the linear law (2.6.1).

Following Bear and Bachmat (1991, p. 100), let us assume that a set of N variables, referred to as *parameters of state*, ψ^n, $n = 1, 2, \ldots, N$, is sufficient to completely define the motion and physico-chemical properties of a phase. These ψ's are specific values of the extensive quantities pertinent to the system, i.e., $\psi^n = dE^n/dm (\equiv e^n/\rho)$. Examples of ψ^n's are the specific mass, $\omega^\gamma (= \rho^\gamma/\rho)$, of a γ-species of a phase, the specific volume, $v (= 1/\rho)$, of a phase, the specific momentum of a phase, $\mathbf{V}$ (= mass weighted velocity = $\rho\mathbf{V}/\rho$), and the specific entropy of a phase, s. These ψ's constitute fields that are functions of the spatial coordinates and of time.

The behavior of a phase is characterized by its *constitutive equations*. The most fundamental one is the *caloric equation of state* that relates the *specific internal energy of a phase*, u, to the complete set of the N parameters of state, ψ^n, through a single valued function which is *independent of time, position, motion, or stress*. We may express this relation in the general form:

$$u = \mathrm{f}(\psi^1, \psi^2, \ldots, \psi^N, \boldsymbol{\xi}),$$

where the ξ_i's are the material coordinates (Sect. 2.1) of a considered element of the system.

The differential increment in the thermodynamic state of an element is expressed by:

$$\begin{aligned} du &= \frac{\partial u}{\partial \psi^1} d\psi^1 + \frac{\partial u}{\partial \psi^2} d\psi^2 + \cdots + \frac{\partial u}{\partial \psi^N} d\psi^N \\ &= \sum_{n=1}^{N} \frac{\partial u}{\partial \psi^n} d\psi^n \equiv \sum_{n=1}^{N} \Phi^n \, d\psi^n, \\ \Phi^n &= \frac{\partial u}{\partial \psi^n} = \Phi^n(\psi^1, \psi^2, \ldots, \psi^N)\Big|_{\xi}, \quad n = 1, 2, \ldots, N, \end{aligned} \qquad (2.6.2)$$

where Φ^n is the increment of internal energy per unit increment in the value of the parameter of state ψ^n. The various Φ^n's are thus *functions of the state* of the system as expressed by the ψ^r's. Accordingly, we refer to the N equations for the Φ^n's appearing in (2.6.2) as *equations of state*.

For example, by comparing (2.6.2) with (2.2.20), we obtain in the latter case:

$$\begin{aligned} d\psi^1 &= ds, \quad d\psi^2 = -dv, \quad d\psi^\gamma = d\omega^\gamma, \\ \Phi^1 &= T, \quad \Phi^2 = p, \quad \Phi^\gamma = \mu^\gamma. \end{aligned}$$

Assuming that the linear relationship of the type (2.6.1) is valid, and continuing to omit the subscript that denotes a phase, we find that the flux of an extensive quantity, E^q, belonging to the coupled subset, is given by

$$j_i^q = -\sum_{j=1}^{3} L_{ij}^{qq} \frac{\partial \Phi^q}{\partial x_j} = -\sum_{r=1}^{\ell} \sum_{j=1}^{3} L_{ij}^{qq} \frac{\partial \Phi^q}{\partial \psi^r} \frac{\partial \psi^r}{\partial x_j}, \tag{2.6.3}$$

with no summation on q.

Defining (Veynik 1961):

$$X_j^r = -\frac{\partial \Phi^r}{\partial \psi^r} \frac{\partial \psi^r}{\partial x_j}, \qquad r = 1, 2, \ldots, \ell, \tag{2.6.4}$$

with no summation on r, as the *thermodynamic force*, which is *conjugate* to the gradient of ψ^r, and with:

$$L_{ij}^{qr} = L_{ij}^{qq} \frac{\partial \Phi^q / \partial \psi^r}{\partial \Phi^r / \partial \psi^r}, \qquad r = 1, 2, \ldots, \ell, \tag{2.6.5}$$

with no summation on r, we rewrite (2.6.3) in the form of the ℓ equations:

$$j_i^q = \sum_{r=1}^{\ell} \sum_{j=1}^{3} L_{ij}^{qr} X_j^r, \quad i = 1, 2, 3; \quad q = 1, 2, \ldots, \ell. \tag{2.6.6}$$

We recall that ℓ *depends on the considered* q.

From (2.6.6) it follows that the flux, $\mathbf{j}^q$ (components j_i^q), of an extensive quantity, E^q, is a single-valued function of *all* the (coupled) thermodynamic forces associated with E^q. Equation (2.6.6), for the various E^q's, are also called the *phenomenological equations* of a system possessing ℓ coupled degrees of freedom. They express linear relationships between fluxes and thermodynamic forces. In general, however, the relationships between fluxes and thermodynamic forces may be nonlinear.

While (2.6.4) serves as a definition for the thermodynamic force, $\mathbf{X}^r$, Eq. (2.6.5) determines the nature of the *phenomenological coefficients*, $\mathbf{L}^{qr}$.

Stokes (1951) postulated that in (2.6.3), the coefficients L_{ij}^{qq} are symmetric with respect to the coordinates i and j, i.e.:

$$L_{ij}^{qq} = L_{ji}^{qq}. \tag{2.6.7}$$

In other words, this coefficient is a *second rank symmetric tensor*.

Of special interest are the *cross coefficients* for $q \neq r$, which give the flux of E^q caused by the force, $\mathbf{X}^r$, associated with the gradient of e^r ($\equiv$ the density of E^r). Employing the *principle of microscopic reversibility of processes*, and methods of statistical mechanics, Onsager (1931) showed that for the linear equation (2.6.6), *and provided a proper choice is made for the fluxes*, $\mathbf{j}^q$, *and conjugated forces*, $\mathbf{X}^r$, *the phenomenological coefficients are also symmetric in* r *and* q, i.e.:

$$L_{ij}^{qr} = L_{ij}^{rq}, \qquad q \neq r, \tag{2.6.8}$$

and L^{rq} is also a second rank symmetric tensor.

Equation (2.6.8) is known as *Onsager's*, or *Onsager–Casimir's, reciprocal relations* (or *Onsager's law*). They express a relationship between any pair of cross-phenomena arising from simultaneously occurring irreversible processes (e.g., heat conduction and molecular diffusion).

Together, the relationships (2.6.7) and (2.6.8) take the form:

$$L_{ij}^{qr} = L_{ji}^{qr} = L_{ij}^{rq} = L_{ji}^{rq}. \tag{2.6.9}$$

According to Onsager, the reciprocal relations (2.6.7), hold under two conditions:

(a) The relationship between each individual flux and its conjugate thermodynamic force is linear.
(b) The fluxes, $\mathbf{j}^q$, and their conjugate forces, $\mathbf{X}^q$, should be selected such that:

$$\dot{\mathbb{S}} = \sum_{q=1}^{\ell} \sum_{i=1}^{3} j_i^q X_i^q, \tag{2.6.10}$$

where $\dot{S}$ is the rate of entropy production of the system, and we recall that by the second law of thermodynamics, (Sect. 2.2.1 B), $\dot{\mathbb{S}} \geq 0$.

Hence, we require that:

$$\sum_{(q,r,i,j)} L_{ij}^{qr} X_j^r X_i^q \geq 0. \tag{2.6.11}$$

A sufficient condition for the validity of (2.6.11) is:

$$L_{ij}^{qq} \geq 0, \qquad \text{for all } i,\, j \tag{2.6.12}$$

$$L_{ij}^{qq} L_{ij}^{rr} \geq \tfrac{1}{4}(L_{ij}^{rq} + L_{ij}^{qr}). \tag{2.6.13}$$

For an isotropic medium, $L_{ij}^{qr} = L^{qr}\delta_{ij}$, and:

$$L^{qq} \geq 0, \qquad L^{qq} L^{rr} \geq \tfrac{1}{4}(L^{qr} + L^{rq}). \tag{2.6.14}$$

• **Example A: Thermo-mechanical system.**

Consider the fluxes of *volume*, $\mathbb{V}$, and *entropy*, $\mathbb{S}$, in a fluid phase possessing two coupled degrees of freedom: one mechanical and the other thermal. The corresponding equations of state are:

$$\left.\begin{aligned} p &= p(s, v), \\ T &= T(s, v), \end{aligned}\right\} \tag{2.6.15}$$

where $v = 1/\rho$ is the specific volume, s is the specific entropy and T is the absolute temperature. For this case, following (2.2.20), Eq. (2.6.2) yields:

$$\Phi^1 = p, \quad \Phi^2 = T,$$
$$\psi^1 = -v, \quad \psi^2 = s,$$

with $\mathbf{j}^1$ denoting volumetric flux, and $\mathbf{j}^2$ denoting entropy flux.

From (2.6.4) through (2.6.6), we obtain:

$$L_{ij}^{11} = L_{ij}^{11} \frac{(\partial p/\partial v)\big|_s}{(\partial p/\partial v)\big|_s} = L_{ij}^{11}, \quad L_{ij}^{12} = L_{ij}^{11} \frac{(\partial p/\partial s)\big|_v}{(\partial T/\partial s)\big|_v},$$

$$X_j^1 = -\frac{\partial p}{\partial v}\bigg|_s \frac{\partial v}{\partial x_j} = -\frac{\partial p}{\partial x_j}\bigg|_s, \quad X_j^2 = -\frac{\partial T}{\partial s}\bigg|_v \frac{\partial s}{\partial x_j} = -\frac{\partial T}{\partial x_j}\bigg|_v,$$

and the flux

$$j_i^1 = L_{ij}^{11} X_j^1 + L_{ij}^{12} X_j^2 = -L_{ij}^{11} \frac{\partial p}{\partial x_j}\bigg|_s - L_{ij}^{12} \frac{\partial T}{\partial x_j}\bigg|_v. \tag{2.6.16}$$

Thus, in (2.6.16) the total volume flux is produced by both a pressure gradient and a temperature gradient.

In a similar way, the flux of entropy is given by:

$$j_i^2 = L_{ij}^{21} X_j^1 + L_{ij}^{22} X_j^2 = -L_{ij}^{21} \frac{\partial p}{\partial x_j}\bigg|_s - L_{ij}^{22} \frac{\partial T}{\partial x_j}\bigg|_v, \tag{2.6.17}$$

where:

$$L_{ij}^{21} = L_{ij}^{22} \frac{(\partial T/\partial v)\big|_s}{(\partial p/\partial v)\big|_s}, \qquad L_{ij}^{22} = L_{ij}^{22} \frac{(\partial p/\partial s)\big|_v}{(\partial p/\partial s)\big|_v} = L_{ij}^{22}. \tag{2.6.18}$$

• **Example B: Thermo-diffusive system**.

For such system, we consider the fluxes of mass and entropy in a fluid phase composed of a solvent and a γ-solute. The equations of state of this system are:

$$\begin{aligned} \mu^\gamma &= \mu^\gamma(s, \omega^\gamma), \\ T &= T(s, \omega^\gamma), \end{aligned} \tag{2.6.19}$$

where μ^γ represents the *chemical potential* of the solute (Sect. 2.2.6), and $\omega^\gamma (\equiv \rho^\gamma/\rho)$ is its specific mass (= mass of γ per unit mass of phase). Bear and Bachmat (1991, p. 106) show that for this case:

$$\Phi^1 = \mu^\gamma, \quad \Phi^2 = T,$$
$$\psi^1 = \omega^\gamma, \quad \psi^2 = s.$$

With $\mathbf{j}^\gamma$ denoting mass flux of the solute, and $\mathbf{j}^2$ denoting entropy flux, they show that:

$$j_i^\gamma = -L_{ij}^{\gamma\gamma}\frac{\partial\mu^\gamma}{\partial x_j}\bigg|_s - L_{ij}^{\gamma s}\frac{\partial T}{\partial x_j}\bigg|_{\omega^\gamma},$$
$$j_i^s = -L_{ij}^{s\gamma}\frac{\partial\mu^\gamma}{\partial x_j}\bigg|_s - L_{ij}^{ss}\frac{\partial T}{\partial x_j}\bigg|_{\omega^\gamma}, \tag{2.6.20}$$

where $L_{ij}^{\gamma s}$ is a coefficient which represents the diffusive mass flux of the solute caused by a temperature gradient, referred to as *thermodiffusion*, or *Soret effect*, and $L_{ij}^{s\gamma}$ is a coefficient which represents the flux of entropy caused by a gradient of the chemical potential, referred to as the *Dufour Effect*. Altogether, we note that the flux of heat is affected also by the gradient of the γ-species, while the flux of γ is also driven by the temperature gradient

Bear and Bachmat (1991, p. 107) rewrite these fluxes in terms of more commonly used coefficients, e.g., thermal conductivity and coefficient of molecular diffusion.

The above discussion on coupled phenomena has been presented at the microscopic level, i.e., we have considered fluxes of extensive quantities at a point is a fluid continuum. However, following the phenomenological approach which underlies the presentation in this book, we can extend the presentation to the macroscopic level, i.e., at a point in a porous medium domain. Nield and Bejan (2013, p. 433) present a discussion with many references on the Soret and Dufour cross-diffusion, or thermodiffusion effects.

Unless otherwise stated, the phenomena of coupling between fluxes of heat and mass of species as described here are neglected in this book.

References

Adamson AW (1982) Physical chemistry of surfaces, 4th edn. Wiley, New York, 664 p

Altunin VV, Sakhabetdinov MA (1972) Viscosity of liquid and gaseous carbon dioxide at 2917 temperatures 220–1300 K and pressures up to 1200 bar. Teploenergetika 8:85–89

Ambrose DA, Walton J (1989) Vapor pressures up to their critical temperatures of normal alkanes and 1-alkanols. Pure Appl Chem 618:1395–1403

Anderson WG (1987) Wettability literature survey-part 4: effects of wettability on capillary pressure. J Petrol Technol 39:1283–1300

Antoine Ch (1888) Tension des vapeurs: nouvelle relation entre les tension et les temperatures. Comptes Rendus 107:681–684, 778–780, 836–837

Appelo CAJ, Postma D (2005) Geochemistry, groundwater and pollution, 2nd edn. CRC Press, Boca Raton, 652 p

Aslyng HC (1963) Soil physics terminology. Int Soc Soil Sci Bull

Bear J, Nitao JJ (1995) On equilibrium and primary variables in transport in porous media. Transp Porous Media 18:151–184

Bear J, Nitao JJ (1996) Potentials and their role in transport in porous media. Water Resour Res 32:225–250

Bear J, Bachmat Y (1991) Introduction to modeling phenomena of transport in porous media. Kluwer Publishing Company, Dordrecht, 553 p

ChemicaLogic, ChemicaLogic Corporation (1999). www.chemicalogic.com, http://www.chemicalogic.com/Pages/DownloadMollierCharts.aspx

Corey AT (1977) Mechanics of heterogeneous fluids in porous media. Water Resources Publications, Colorado, 259 p
de Groot (1963) Thermodynamics of irreversible process, North Holand Publishing Co. Amsterdam
Dekker LW, Ritsema CJ (1994) How water moves in a water repellent sandy soil. 1. Potential and actual water repellency. Water Resour Res 30:2507–2517
Demond AH, Desai FN, Hayes KF (1994) Effect of cationic surfactants on organic liquid water capillary-pressure saturation relationships. Water Resour Res 30:333–342
Denbigh KG (1981) The principles of chemical equilibrium, 4th edn. Cambridge University Press, Cambridge, 494 p
Edelfsen NE, Anderson ABC (1943) Thermodynamics of soil moisture. Hilgardia 15:31–298
Escobedo J, Ali Mansoori G (1996) Surface tension prediction for pure liquids. A.I.Ch.E J 42:1425–1433
Gibbs JW (1873) In: Collected Works published in Trans. Conn. Acad., Vol. 1, Longmans, New York, 1928
Gibbs WJ (1906) The scientific papers of J. Willard Gibbs. In: Thermodynamics, vol 1. Longmans, Greens (reprinted by Dover Publications, 1961)
Girifalco LA, Good RJ (1957) A theory for the estimation of surface and interfacial energies. I. Derivation and application to interfacial tension. J Phys Chem 61(7):904–909
Hassanizadeh SM, Gray WG (1989) Averaging theorems and averaged equations for transport of interface properties in multiphase systems. Int J Multiph Flow 15:81–95
Knudsen M (1934) Reciprocal relations in irreversible processes. Phys Rev 37:405–426
Lyman WJ, Reehl WF, Rosenblatt DH (eds) (1982) Adsorption coefficients for soils and sediments. In: Handbook of chemical property estimation methods. McGraw-Hill, New York, 977 p
Nield DA, Bejan A (2013) Convection in porous media, 4th edn. Springer, Berlin, 778 p
Nitao JJ, Bear J (1996) Potentials and their role in transport in porous media. Water Resour Res 32:225–250
Onsager L (1931) Reciprocal relations in irreversible processes. Phys Rev 37:405–426
Parker JG (1986) Hydrostatics of water in porous media. In: Sparks DL (ed) Soil physical chemistry. CRC Press, Boca Raton
Peng DY, Robinson DB (1976) A new two-constant equation of state. Ind Eng Chem Fundam 15:59–64
Poling BE, Prausnitz JM, O'Connell JP (2000) Properties of gases and liquids, 5th edn. McGraw-Hill, New York
Redlich O, Kwong JNS (1949) On the thermodynamics of solutions. V. An equation of state. Fugacities of gaseous solutions. Chem Rev 44(1):233–244
Schowalter TT (1979) Mechanics of secondary hydrocarbon migration and entrapment. Water Sci Technol 23:467–476
Schramm LL (ed) (2000) Surfactants: fundamentals and applications in the petro-leum industry. Cambridge University Press, Cambridge
Smith JM, Van Ness HC, Abbott MM (2005) Introduction to chemical engineering thermodynamics, 7th edn. McGraw-Hill, New York, 342 p
Soave G (1972) Equilibrium constants from a modified Redlich-Kwong equation of state. Chem Eng Sci 27:1197–1203
Stokes GG (1951) On the effect of internal friction of fluids on the motion of pendulums, Trans Cambridge Phil Soc 9
Truesdell CA, Toupin R (1960) The classical field theories in Flügge's Handbuch der Physik 3(1):226–793. Berlin: Springer-Verlog
Veynik AI (1961) Thermodynamics (in Russian). Izd Min Vysh Obraz USSR, Minsk
Vinet P, Ferrante J, Rose JH, Smith JR (1987) Compressibility of solids. J Geophys Res 92:9319–9325
Wagman DD, Evans WH, Parker VB, Schumm RH, Halow I, Bailey SM, Churney KL, Nuttall RL (1982) The NBS tables of chemical thermodynamic properties: selected values for inorganic and C1 and C2 organic substances in S1 units. J Phys Chem Ref Data Suppl 2, 392 p

Wagner G (1973) New vapour pressure measurements for argon and nitrogen and a new method for establishing rational vapour pressure equations. Cryogenics 13:470–482

Walas SM (1985) Phase equilibria in chemical engineering. Butterworth-Heinemann, London, 671 p

Yaws CL (1999) Chemical properties handbook: chemical, thermodynamic, environmental, transport, safety, and health related properties for organic and inorganic chemicals. McGraw-Hill, New York, 799 p

Yaws CL (ed) (2003) Yaws' handbook of thermodynamic and physical properties of chemical compounds. Gulf Publishing, Houston, 784 p

Chapter 3
Fundamental Balance Equations and Fluxes

There are two major approaches to the mathematical description of flow and transport phenomena in a continuum. The first approach, referred to as the *Lagrangian approach*, or *Lagrangian description of motion*, focuses on a fixed amount of a considered extensive quantity and follows what happens to it as it travels in a considered spatial domain. Both the volume occupied by this amount and the shape of the surface surrounding it vary with time.

In the second approach, called the *Eulerian approach*, still in a continuum domain, we focus on a fixed (in space and time) finite domain, referred to as a *control volume*, of arbitrary shape, bounded by a *fixed* closed surface, and follow what happens *within* that volume with time. In this book we shall use the *Eulerian approach* to develop flow and transport models.

In Sect. 1.4, we identified two levels of description of flow and transport phenomena in porous medium domains. One, called the *microscopic level*, at which we consider and describe what happens at a point inside any of the fluid phases that occupy the void space, or at a point within the solid matrix. The other, the *macroscopic level*, describes what happens at a point, which is the centroid of an REV, in a porous medium domain, regarded as a continuum. When we say 'at a point', we mean 'at every point'. This level, employed in practice for describing and solving flow and transport problems of interest, is the one that we shall use in this book. In Sect. 1.1.6, we have also identified a still higher level, the *megascopic* one, introduced in an effort to cope with heterogeneity at the macroscopic level, say in permeability.

As described in Sect. 1.4, macroscopic models of a transport process undergone by any extensive quantity, can be obtained in a number of ways. One way is to start by modeling the considered phenomenon of transport at the microscopic level, and then make use of one of the averaging techniques described in Sect. 1.4.2 in order to upscale the microscopic level model to the macroscopic one. A second way also starts from the microscopic level model, but derives the macroscopic one by *mathematical homogenization*. In the third approach, the *phenomenological* one (Sect. 1.4.4), the macroscopic model is obtained directly from observations; it does

J. Bear, *Modeling Phenomena of Flow and Transport in Porous Media*,
Theory and Applications of Transport in Porous Media 31,
https://doi.org/10.1007/978-3-319-72826-1_3

not require the formal passage from the microscopic level to the macroscopic one. This is the approach that we use in this book.

We start by discussing the meaning of a *point*, a *particle*, e.g., a fluid particle, the *velocity of an extensive quantity*, and the *flux* of the latter. We then use the phenomenological approach to construct balance equations for a number of extensive quantities. Although we start by writing balance equations at the microscopic level, our main objective is to write balance equations at the macroscopic one, as this is the level at which processes and phenomena with applications and consequences in practice are analyzed and described. All balance equations contain terms that express fluxes, interphase transfers and sources. In the last section of this chapter, we discuss fluxes in general terms, making use of thermodynamic considerations, using the flux of a dissolved species as an example. In later chapters, we shall present and discuss fluxes of other extensive quantities as well as interphase transfer rates and strength of sources of these quantities.

And a few words about symbols. Throughout this book, we are using the term 'flux' for the quantity of an extensive quantity E, *per unit area* and *per unit time*. At the microscopic level, we use the symbol $\mathbf{j}_\alpha^E$ for the flux of E in the α-phase, and the unit area is of the considered phase; the additional subscripts indicate advective (adv) and diffusive (dif) fluxes. In multiphase situations, a subscript (say α) will indicate a considered α-phase. At the macroscopic level, we use the symbol $\mathbf{J}_\alpha^E$ for the same definition of flux, except that now the unit area is of the α-phase occupying part of the void space. At that level, we also use the symbol $\mathbf{q}_\alpha^E$ to denote the quantity of E of an α phase passing through a unit area of porous medium. Thus, $\mathbf{q}_\alpha^E = \theta_\alpha \mathbf{J}_\alpha^E$. However, at the macroscopic level, we shall often use $\phi\mathbf{V}_\alpha$, rather than $\mathbf{q}_\alpha$, to facilitate the option of a variable porosity, say, in a deformable porous medium.

In Sect. 1.1.5, we have introduced the concept or areal porosity, ϕ^A. We have also suggested that in practical cases of interest, this porosity is approximated by the (volumetric) one, ϕ. Henceforth, we shall make this assumption. In fact, we shall continue to make this assumption throughout the remaining chapters of this book.

3.1 Point, Particle, Velocity and Flux

3.1.1 Point and Particle

The continuum at the microscopic level is defined in Sect. 1.1. There, we wrote that a domain behaves as a continuum, *'if values of state variables (e.g., density, pressure, temperature), or phase coefficients (e.g., viscosity, or thermal conductivity), can be assigned to every point within the domain'*. In this definition, the term *point* is used to indicate a *location* in the considered spatial domain. Another important basic definition is *particle*. We use this term to denote a *point in a continuum*. Here, the continuum is of a considered extensive quantity, e.g., the mass of a fluid phase, or of a chemical species dissolved in the fluid, or of the energy of a fluid. While points are (1) fixed in space, and (2) independent of time, the position of a particle may vary with time. Accordingly, we consider two distinct, yet related concepts of points, coordinates, and particles:

- **Spatial (Eulerian) coordinates**: We denote a point in a spatial domain by its *position vector*, $\mathbf{x}$, with respect to a *fixed* coordinate system, serving as a *frame of reference*. We use x_i, $i = 1, 2, 3$, to denote components of $\mathbf{x}$ in a *Cartesian coordinate system*.
- **Material (Lagrangian) coordinates** of a particle: We denote a particle in space by its position vector, $\boldsymbol{\xi}$ (components ξ_i, $i = 1, 2, 3$). These coordinates are assigned, *once and for all*, to a particle of a continuum as it travels in the considered domain. Usually, we use the initial location of the particle as its material coordinates, i.e., $\boldsymbol{\xi} \equiv \mathbf{x}|_{t=0}$.

Having defined a particle as a point in an E-continuum, we have also to consider its corresponding intensive quantity. In fact, in Sect. 1.1, we have introduced two kinds of such quantities: E *per unit mass*, usually referred to as the *specific value* of E, denoted by e, and E *per unit volume* of the phase, referred to as the *density* of E, denoted by e'. We shall also use *molar values* of E, i.e., E per mole, denoted by $\check{e}$.

Over what volume do we take the mass in order to determine the mass density of a particle *at a point in a considered domain*?

Because of the molecular structure of matter, in order to define the density of a moving particle, its volume has to be within a certain range: as small as possible, yet not too small, in order to avoid the effect of motion of individual molecules. This means that if at $t = 0$ we identify a particle of a certain solute mass, at a certain solute concentration, then, as that particle moves, that mass will spread out over a volume which is too large. We then have to 'freeze' the motion, re-define the particle at that point, maintaining the same concentration, and then allow the particle to proceed on its motion as an E-particle.

In what follows, we shall assume that the reader is familiar with the concepts of 'vector' and 'tensor' and with tensor operations. In Sect. 2.3.4, we have presented some introductory remarks on second rank tensors, in connection with stress and strain in porous medium domains.

3.1.2 Velocity

As a *particle of a continuum* of an extensive quantity (e.g., mass) moves, the (*spatial*) coordinates of its position, x_i, vary in time, whereas its *material* coordinates, ξ_i, remain unchanged. Thus, $\mathbf{x}$ is a function of both time, t, and the initial position, $\boldsymbol{\xi}$, of the particle, and the particle's motion can be described by:

$$\mathbf{x} = \mathbf{x}(\xi, t), \quad \text{or} \quad x_i = x_i(\xi_1, \xi_2, \xi_3, t), \quad i = 1, 2, 3. \tag{3.1.1}$$

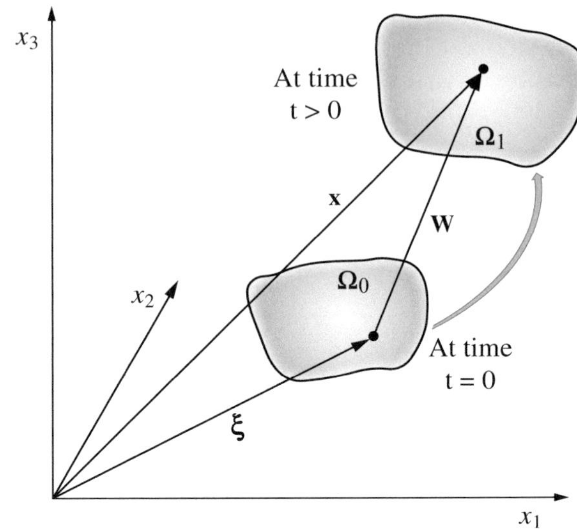

Fig. 3.1 Definition sketch for particles, points and displacements

This description of *motion* is known as the *Lagrangian formulation of motion* (see preamble of this chapter).

Figure 3.1 shows a spatial domain Ω_o occupied at $t = 0$ by a continuum with material coordinates $\boldsymbol{\xi}$. The set of points in Ω_o specifies the initial *configuration* of this continuum. At some later time, $t > 0$, the domain occupied by the same continuum is Ω_1. As $\boldsymbol{\xi}$ runs over the set of points in Ω_o, the value of $\mathbf{x}$ runs over the set of points in Ω_1, and hence, Ω_1 may be regarded as *deformed configuration* of the continuum initially occupying Ω_o. Referring to a continuous sequence of configurations as *motion*, (3.1.1) also describes the motion of any particle (of the continuum contained in Ω_o) initially at $\boldsymbol{\xi}$, i.e., it gives its place, $\mathbf{x}(t)$, as a function of time.

Assuming that (3.1.1) can be inverted to yield the initial position (i.e., material coordinates) of a particle which at time t is at position $\mathbf{x}$, we have:

$$\boldsymbol{\xi} = \boldsymbol{\xi}(\mathbf{x}, t), \qquad \text{or} \quad \xi_i = \xi_i(x_1, x_2, x_3, t), \quad i = 1, 2, 3. \tag{3.1.2}$$

This description of motion is known as the *Eulerian formulation of motion*.

It is important to emphasize that a *particle* here should not be interpreted as a 'small material body'. Instead, it is a point that belongs to a specific continuum of an extensive quantity, which, at some specified (or initial) time, has occupied a certain finite domain. The configuration of the domain (e.g., the shape and the size of the volume corresponding to a unit of mass of a moving fluid) occupied by the extensive quantity may vary with time, but it will always contain the same amount of the extensive quantity. If sources and/or sinks of the extensive quantity are present, i.e., new particles are being created, or existing particles are being removed, (3.1.1) does not hold, since particles exist in the domain only for a short duration and have to be continuously redefined. We note that the concept of a particle as defined above allows particles of different continua (e.g., of mass of a phase and mass of a chemical species) to occupy *the same point*, simultaneously.

The *material derivative* (also called *convected derivative*) of a variable g^E of a particle of a considered E-continuum is the (temporal) rate of change of that variable for the considered particle. The symbol:

$$\left.\frac{\partial g^E(\boldsymbol{\xi}^E, t)}{\partial t}\right|_{\boldsymbol{\xi}^E=\text{const.}} \equiv \frac{D_E g^E\{\mathbf{x}(\boldsymbol{\xi}^E, t), t\}}{Dt}, \tag{3.1.3}$$

i.e., a derivative of g^E with respect to time, keeping $\boldsymbol{\xi}^E$ constant. In other words, $D_E g^E / Dt$ gives *the rate of change of g of a fixed E-continuum particle to an observer situated on that particle*.

With $\mathbf{x}^m$ denoting the position of a particle of a mass (m)-continuum as it is being displaced, and $\boldsymbol{\xi}^m$, denoting its material coordinates, the velocity, $\mathbf{V}^m$, of a mass particle, is given by the rate of change of its position in time:

$$\mathbf{V}^m = \left.\frac{\partial \mathbf{x}^m}{\partial t}\right|_{\boldsymbol{\xi}^m=\text{const.}}. \tag{3.1.4}$$

We may now generalize (3.1.4) to a particle of a continuum of any extensive quantity denoted by E. Its velocity is defined by:

$$\mathbf{V}^E = \left.\frac{\partial \mathbf{x}^E}{\partial t}\right|_{\boldsymbol{\xi}^E=\text{const.}} \tag{3.1.5}$$

Examples for E representing mass as expressed by mass density and by molar density are presented on Sect. 3.1.3.

The material derivative of g^E, which is a Lagrangian concept, can also be expressed in terms of the spatial, or Eulerian, description, using the relationship $g^E(\mathbf{x}, t) = g^E[\mathbf{x}(\mathbf{X}, t), t]$:

$$\begin{aligned}
&\frac{D_E g^E\{\mathbf{x}(\boldsymbol{\xi}^E, t), t\}}{Dt} \\
&= \left.\frac{\partial g^E}{\partial t}\right|_{\mathbf{x}=\text{const.}} + \left.\frac{\partial g^E}{\partial x_k}\right|_{t=\text{const.}} \left.\frac{\partial x_k(\boldsymbol{\xi}^E, t)}{\partial t}\right|_{\boldsymbol{\xi}^E=\text{const.}} \\
&= \frac{\partial g^E}{\partial t} + \frac{\partial g^E}{\partial x_k} V_k^E,
\end{aligned} \tag{3.1.6}$$

where

$$\frac{\partial g^E}{\partial x_k} V_k^E \equiv \sum_{k=1}^{3} \frac{\partial g^E}{\partial x_k} V_k^E \equiv \mathbf{V}^E \cdot \nabla g^E.$$

This abbreviated form of representing a sum of terms by a single, typical, one is known as *Einstein's (double index) summation convention*. Unless otherwise stated, this convention will be used throughout this book. It states that any index (called a *dummy index*) repeated *twice and only twice* in a term is held to be summed over the range of its values.

Obviously, the two expressions in (3.1.2) represent the same motion. Hence, they yield the same equation for the pathline of a particle, as long as the material coordinates of the particle are defined in the same way, with $x_i = \xi_i$, at $t = 0$, and if the two equations are *mutually invertible*, i.e.,

$$\mathcal{J} \equiv \left| \frac{\partial x_i}{\partial \xi_j} \right| \neq 0, \qquad i, j = 1, 2, 3, \tag{3.1.7}$$

where $\mathcal{J}$, referred to as *Jacobian*, is the determinant of a matrix in which the typical element is $\partial x_i / \partial \xi_j$.

For the sake of simplifying notation, henceforth, we shall use the symbol $\mathbf{V}$ to denote the mass weighted velocity, $\mathbf{V}^m$, and $\mathrm{D}E/\mathrm{D}t$ to denote $\mathrm{D}_m E/\mathrm{D}t$.

3.1.3 E-Fluxes, Pathlines and Transport Lines

At the microscopic level, i.e., at a point in a phase continuum, the flux, or total flux, $\mathbf{j}^E$, of an extensive quantity, E, describes the amount of E passing through a unit area of the phase, during a unit time:

$$\mathbf{j}^E = e'\mathbf{V}^E, \tag{3.1.8}$$

in which e' denotes the density of E, and $\mathbf{V}^E$ is the velocity of E. The specific value of E is e, with $e' = \rho e$. In the above equation, the unit area is normal to the direction of $\mathbf{V}^E$. As emphasized in Sect. 1.1.4 D, we often monitor not a quantity per unit volume, but a quantity per unit mass. Then:

$$\mathbf{j}^E = \rho e \mathbf{V}^E, \tag{3.1.9}$$

The total flux of E, can be expressed as the sum of two fluxes:

$$\mathbf{j}^E (\equiv e'\mathbf{V}^E) = e'\mathbf{V} + e'(\mathbf{V}^E - \mathbf{V}) = \mathbf{j}^E_{adv} + \mathbf{j}^E_{dif}, \tag{3.1.10}$$

i.e., the sum of an *advective flux*, $\mathbf{j}^E_{adv}$ $(= e'\mathbf{V})$, and a *diffusive flux* $\mathbf{j}^E_{dif}$ $\big(= e' (\mathbf{V}^E - \mathbf{V})\big)$. The first expresses the flux of E as carried by the fluid moving at the fluid phase (mass-weighted) velocity, $\mathbf{V}$. The second flux expresses the flux of E relative to this advective flux.

The diffusive mass flux of a dissolved chemical species in a fluid phase will be further discussed in Sects. 3.4 and in 7.2.2. *Thermal conduction is the diffusive flux of thermal energy.*

In a multi-species fluid, the *mass-weighted velocity*, $\mathbf{V}(\equiv \mathbf{V}^m)$ is:

$$\mathbf{V}(\equiv \mathbf{V}^m) = \frac{1}{m}\sum_{\gamma=1}^{N} m^\gamma \mathbf{V}^\gamma = \sum_{\gamma=1}^{N} \frac{\rho^\gamma}{\rho}\mathbf{V}^\gamma = \sum_{\gamma=1}^{N} \omega^\gamma \mathbf{V}^\gamma, \tag{3.1.11}$$

where $\omega^\gamma = m^{\gamma/m} = \rho^{\gamma/\rho}$ is the *mass fraction* of the γ-species in the phase, with $\sum_{(\gamma)} \omega^\gamma = 1$, $\rho^\gamma = m^\gamma/V$ and $\rho = m/\mathbb{V}$, with $\mathbb{V}$ denoting the volume of the domain Ω. Note that the symbol $\mathbf{V}$ denotes the velocity of the mass, $\mathbf{V}^m$.

When considering reacting chemical species (see Chap. 7), it is convenient to measure the mass of chemical species in a fluid in terms of *moles* and write the species balance equations also in term of moles, both at the microscopic and macroscopic levels. In such equations, the fluid's velocity is the *molar averaged velocity* $\mathbf{V}^{mol}$, and the diffusive molar flux $\mathbf{j}_{dif}^{\gamma,mol}$ defined in Sect. 7.2.1.

Since momentum of a mass m is $m\mathbf{V}^m$, we may regard $\mathbf{V}^m$ ($\equiv$ 'velocity of the mass') as 'momentum per unit mass'. In this book, unless otherwise specified, we shall use the symbol $\mathbf{V}$ for $\mathbf{V}^m$. This is the *mass averaged velocity* defined in Sect. 1.1.4.

Although our interest in the microscopic level description is only as the starting point for velocity and transport at the macroscopic one, let us add two additional concepts—pathlines and streamlines, as they aid in the understanding of the concept of tortuosity (Sect. 4.2.5) as a porous medium property. We shall follow here the presentation in Bear and Bachmat (1991, p. 62).

A (microscopic level) *pathline* is a curve (or line) along which a *fixed particle* of a continuum moves in the course of time. The term *trajectory* is sometimes used instead of pathline. A pathline is thus a *Lagrangian* concept. Let ξ_i^E, $i = 1, 2, 3$, denote the *material coordinates* of a fixed E-particle. The Lagrangian description of its motion, as given by (3.1.1), is:

$$\mathbf{x} = \mathbf{x}(\boldsymbol{\xi}^E, t), \quad \text{or} \quad x_i = x_i(\xi_1^E, \xi_2^E, \xi_3^E, t), \quad i = 1, 2, 3. \tag{3.1.12}$$

The above equations provide the coordinates of the time-dependent position-vector of the particle. Eliminating the time from a pair of these equations, and repeating this process for a second pair, yields two equations:

$$F_1(x_1, x_2, \boldsymbol{\xi}^E) = 0, \qquad F_2(x_2, x_3, \boldsymbol{\xi}^E) = 0.$$

Each of these equations describes a surface. Together, they define the pathline of the E-particle (coinciding with the intersection of the two surfaces).

In the *Eulerian formulation* of the material derivative, the differential equation of motion of an E-continuum, is given by:

$$\frac{dx_i}{dt} = V_i^E(\mathbf{x}, t),$$

or

$$\frac{dx_1}{V_1^E(\mathbf{x}, t)} = \frac{dx_2}{V_2^E(\mathbf{x}, t)} = \frac{dx_3}{V_3^E(\mathbf{x}, t)} = dt, \tag{3.1.13}$$

in which $\mathbf{V}^E$ denotes the velocity of an E-particle.

The solution of these equations gives the *Eulerian description of motion*, (3.1.2)

$$\xi_i^E = \xi_i^E(\mathbf{x}, t), \qquad i = 1, 2, 3, \tag{3.1.14}$$

where the ξ_i^E's are parameters identifying a particle. By fixing the values of these parameters, one obtains, the pathline of a specific particle.

While a *pathline* is a curve along which a *given particle* moves, a *streamline* is a curve along which a *string of particles* move at a *given instant*. By definition, the tangent to a streamline at each point on it is *collinear* with the velocity vector, $\mathbf{V}^E$, at that point. Accordingly, the mathematical definition of a streamline of an E-continuum at a given instant, say, $t = t_o$, is:

$$dx_i = aV_i^E(\mathbf{x}, t_o), \qquad i = 1, 2, 3,$$

or,

$$\frac{dx_1}{V_1^E(\mathbf{x}, t_o)} = \frac{dx_2}{V_2^E(\mathbf{x}, t_o)} = \frac{dx_3}{V_3^E(\mathbf{x}, t_o)}, \tag{3.1.15}$$

where a is a scalar and the dx_i's are the components of an infinitesimal distance along a streamline. A streamline is thus an *Eulerian concept*.

Once the velocity field, $\mathbf{V}^E(\mathbf{x}, t_o)$, is known, the general solution of the system (3.1.15) yields the family of streamlines, referred to as the *motion pattern*, of the E-continuum at the instant $t = t_o$.

For *unsteady motion* of an E-continuum (i.e., $\partial V_i^E/\partial t \neq 0$), the streamlines may vary from one instant to the next, whereas for a *steady motion* ($\partial V_i^E/\partial t = 0$), the streamlines remain unchanged with time. In the latter case, streamlines and pathlines coincide. For any scalar E-continuum (e.g., mass, mass of a solute, heat), a streamline which is a *vector line* of the velocity field, $\mathbf{V}^E$, is also a vector line of the total flux, $\mathbf{j}^{tE}(= e\mathbf{V}^E)$ of that continuum. This line is defined by:

$$\frac{dx_1}{j_1^{tE}(\mathbf{x}, t_o)} = \frac{dx_2}{j_2^{tE}(\mathbf{x}, t_o)} = \frac{dx_3}{j_3^{tE}(\mathbf{x}, t_o)}. \tag{3.1.16}$$

A line or curve defined by (3.1.16) is called an *E-transport line*, or *curve* of the (scalar) E-continuum. Bear and Bachmat (1991, p. 65) use the above discussion to introduce streamlines and stream-tubes in a two-dimensional domain. In three dimensional transport, a *stream-tube* may also be defined as a control volume bounded by streamlines. Bear (1972, p. 226) discusses stream-tubes in three-dimensional domains. In Sect. 4.2.5, we shall introduce the concept of *tortuosity* which expresses the fact that the length of the microscopic level flow, or transport, through the actual (microscopic) tortuous pathways within the void-space is larger than the length of averaged transport.

So far, we have considered phenomena at the microscopic level continuum. In Chap. 1, we also defined the macroscopic. All the concepts presented above, like velocity and streamline, are also applicable at that level.

3.2 Microscopic Balance Equations for Extensive Quantities

As stated earlier, the core of any model that describes the transport of an extensive quantity is the *balance equation* of that quantity. Some authors refer to it as a 'conservation equation'. In this book we consider mainly five extensive quantities: (1) mass, (2) mass of a chemical species dissolved in a fluid phase, or adsorbed on a solid, (3) linear momentum, (4) energy, and (5) entropy. In certain cases, e.g., when dealing with the energy of a porous medium as a whole, we'll also consider extensive quantities of the solid phase.

We have already explained why, in practice, the required models have to be written at the macroscopic level (Sect. 1.1.3). Accordingly, in this chapter, using the phenomenological approach, we shall develop the macroscopic balance equation for any extensive quantity of a phase and apply it to the five extensive quantities mentioned above. In later chapters, we shall present the same equations in more details by adding specific information on the fluxes, interphase exchange terms and sources, as well as additional information that is required in order to present complete, well posed mathematical models.

3.2.1 *The General Microscopic Balance Equation*

We consider an extensive quantity E, within a domain Ω, of volume $\mathbb{V}$, bounded by a fixed surface $\mathcal{S}$ around some point $\mathbf{x}$ (Fig. 3.2). We can write the balance of E in the verbal form:

$$\underset{\text{(a)}}{\left\{\begin{array}{l}\text{Rate of}\\ \text{accumulation of}\\ E \text{ within } \Omega\end{array}\right\}} = \underset{\text{(b)}}{\left\{\begin{array}{l}\text{Net influx of}\\ E \text{ into } \Omega\\ \text{through } \mathcal{S}\end{array}\right\}} + \underset{\text{(c)}}{\left\{\begin{array}{l}\text{Net rate of}\\ \text{production of}\\ E \text{ within } \Omega\end{array}\right\}},$$

in which:

(a) The rate of increase in the amount of E within the Ω-domain, of volume $\mathbb{V}$, with $e' = dE/d\mathbb{V}$, is expressed by:

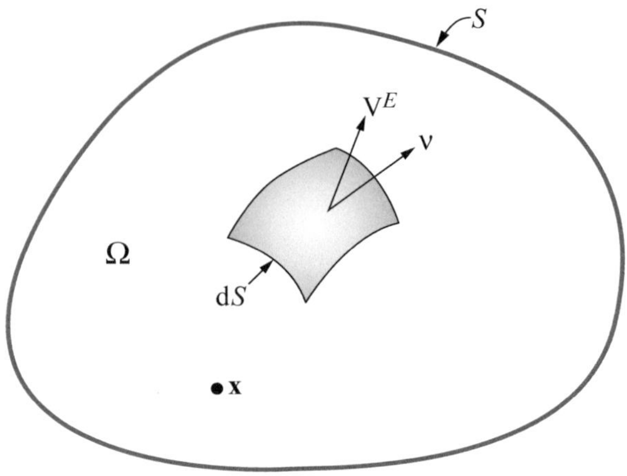

Fig. 3.2 A 3-d domain Ω bounded by a fixed surface $\mathcal{S}$

$$\frac{\partial}{\partial t}\int_{\mathbb{V}} e'\, d\mathbb{V}\left(=\int_{\mathbb{V}}\frac{\partial e'}{\partial t}\, d\mathbb{V}\right), \tag{3.2.1}$$

where e' is the density of E, i.e., the amount of E per unit volume of the phase; the exchange of integration and differentiation is permitted since the boundary $\mathcal{S}$ is fixed.

(b) The net influx (= total influx minus total efflux) of E into Ω, through $\mathcal{S}$, is expressed as:

$$-\int_{\mathcal{S}} e'\mathbf{V}^E\cdot\boldsymbol{\nu}\, d\mathcal{S}, \tag{3.2.2}$$

where $\boldsymbol{\nu}$ is the *outward* normal unit vector on the elemental area (= $d\mathcal{S}$), and $\mathbf{V}^E$ denotes the velocity of E as defined in Sect. 3.1; the product $e'\mathbf{V}^E$ denotes the *flux* of E, i.e., the amount of E passing through a unit area of the phase, per unit time.

(c) The net rate of production of E by sources inside Ω is expressed by:

$$\int_{\mathbb{V}} \rho\Gamma^E\, d\mathbb{V}, \tag{3.2.3}$$

where ρ is the mass density of the phase, and Γ^E denotes the rate of *internal production* of E, per unit mass of the phase. When the considered extensive quantity disappears, e.g., mass of a liquid disappears by change of phase, Γ^E will take on a negative value, representing a sink. We may include in Γ^E both *internal* and *external* sources, making a distinction between E-sources resulting from internal activities, i.e., within the phase, and E-sources that result from external activities: $\Gamma^E = \Gamma^E_{int} + \Gamma^E_{ext}$, each per unit mass of the phase, per unit time. The emphasis in the definition of Γ^E is that the production/disappearance *is not through any interphase boundary*. Gravity, a force that is a source for momentum, radioactive decay of a chemical species, added or disappearance of a species by chemical reactions and heating by microwave may serve as examples. In the macroscopic E-balance equation (Sect. 3.2), we shall introduce a separate term to express sources and sinks that are due to exchange across interphase boundaries.

Altogether, the balance of E within Ω is expressed by:

$$\int_{\mathbb{V}} \frac{\partial e'}{\partial t}\, d\mathbb{V} = -\int_{\mathcal{S}} e'\mathbf{V}^E \cdot \boldsymbol{\nu}\, d\mathcal{S} + \int_{\mathbb{V}} \rho\Gamma^E\, d\mathbb{V}. \tag{3.2.4}$$

Equivalently, in terms of the specific value, e:

$$\int_{\mathbb{V}} \frac{\partial \rho e}{\partial t}\, d\mathbb{V} = -\int_{\mathcal{S}} \rho e\mathbf{V}^E \cdot \boldsymbol{\nu}\, d\mathcal{S} + \int_{\mathbb{V}} \rho\Gamma^E\, d\mathbb{V}. \tag{3.2.5}$$

As emphasized when the concept of e' of a phase was introduced in Sect. 1.1.4, e' must be such that it is additive over volume of that phase, and $e'\mathbf{V}^E$ expresses a flux.

At this point, we introduce the *Gauss Theorem*, the proof of which can be found in any book on vector analysis. We consider a tensorial quantity, $G_{ijk...}$ (of any rank), that is defined and differentiable within a regular *convex spatial domain*, Ω, bounded by a fixed closed surface, $\mathcal{S}$. The surface $\mathcal{S}$ consists of a finite number of parts, with a continuously turning tangent plane.

Gauss' Theorem states that:

$$\int_{\mathbb{V}} \frac{\partial G_{jkl...}}{\partial x_j}\, d\mathbb{V} = \int_{\mathcal{S}} G_{jkl...}\nu_j\, d\mathcal{S}, \tag{3.2.6}$$

in which the *double summation convention* is used. This equation is also called the *Gauss divergence theorem*. Note that throughout this book, we are making use of indicial notation and vector notation, interchangeably.

For the special case in which $G_{jkl...}$ is a vector, G_i, the Gauss theorem takes the form:

$$\int_{\mathbb{V}} \frac{\partial G_i}{\partial x_i}\, d\mathbb{V} = \int_{\mathcal{S}} G_i\nu_i\, d\mathcal{S}, \quad \text{or} \quad \int_{\mathbb{V}} \nabla\cdot\mathbf{G}\, d\mathbb{V} = \int_{\mathcal{S}} \mathbf{G}\cdot\boldsymbol{\nu}\, d\mathcal{S}. \tag{3.2.7}$$

With Gauss' theorem, and with $e'\mathbf{V}^E$ expressing flux, we have:

$$\int_{\mathbb{V}} [\nabla\cdot(e'\mathbf{V}^E)]\, d\mathbb{V} = \int_{\mathcal{S}} (e'\mathbf{V}^E)\cdot\boldsymbol{\nu}\, d\mathcal{S}. \tag{3.2.8}$$

The r.h.s. of (3.2.8) represents the net amount of E leaving the domain Ω of volume $\mathbb{V}$, through its bounding surface $\mathcal{S}$. An interesting consequence is:

$$\nabla\cdot e'\mathbf{V}^E \equiv \lim_{\mathbb{V}\to 0} \frac{1}{\mathbb{V}} \int_{\mathcal{S}} e'\mathbf{V}^E\cdot\boldsymbol{\nu}\, d\mathcal{S}. \tag{3.2.9}$$

This provides a physical interpretation of the *divergence of a flux of E as the excess of efflux over influx of E through a closed surface surrounding a domain, per unit volume, as the latter shrinks to zero around the point*.

Another interesting conclusion of the *Gauss theorem* is that at the microscopic level, for a surface $\mathcal{S}$ enclosing a volume $\mathbb{V}$, we have:

$$\int_{\mathbb{V}} (\nabla p)\, d\mathbb{V} = \int_{\mathcal{S}} p\, d\mathcal{S},$$

with $-\int_{\mathcal{S}} p\, d\mathcal{S}$ expressing the net force acting on the closed surface. Thus $-\nabla p$ expresses a force per unit volume.

Assuming that $e'\mathbf{V}^E$ is differentiable within Ω, we apply *Gauss' theorem*, (3.2.7), with $G_i = e'V_i^E$, to the first term on the r.h.s. of (3.2.4), obtaining:

$$\int_{\mathbb{V}} \left(\frac{\partial e'}{\partial t} + \nabla \cdot e'\mathbf{V}^E - \rho \Gamma^E \right) d\mathbb{V} = 0. \tag{3.2.10}$$

By shrinking the volume $\mathbb{V}$ to zero around an the arbitrary point $\mathbf{x}$, we obtain the microscopic level E-balance equation:

$$\frac{\partial e'}{\partial t} + \nabla \cdot e'\mathbf{V}^E - \rho \Gamma^E = 0, \tag{3.2.11}$$

where all terms refer to the considered point.

In view of (3.1.10), the balance equation (3.2.11) can be rewritten as:

$$\frac{\partial e'}{\partial t} = -\nabla \cdot (\mathbf{j}^E_{adv} + \mathbf{j}^E_{dif}) + \rho \Gamma^E, \quad \mathbf{j}^E_{adv} = e'\mathbf{V}, \tag{3.2.12}$$

in which $\mathbf{j}^E_{adv}$ and $\mathbf{j}^E_{dif}$ denote the *advective* and the *diffusive* fluxes of E, respectively, i.e., the amounts of E passing by advection and by diffusion through a unit area of a cross section through the phase, per unit time. We recall that $\mathbf{V} \equiv \mathbf{V}^m$ is the mass-weighted (also called *barycentric velocity*) of the fluid. In a multi-species fluid, $\mathbf{V} = \sum_\gamma \omega^\gamma \mathbf{V}^\gamma$, $\omega^\gamma = \rho^\gamma/\rho$. It is also the *momentum per unit mass*.

Equation (3.2.12) is the *microscopic level differential balance equation of any extensive quantity, E, in a phase domain considered as a continuum*. It expresses the balance of E over a small volume around any point in the domain, in the limit as this volume is reduced, converging to the point.

The E-balance equation (3.2.12) is written in terms of E-density, i.e., E per unit volume of the considered phase. However, sometimes, E is monitored as *E per unit mass of the considered phase*, denoted as e defined in (1.1.15). Under such conditions, we write the E-balance equation in terms of $e(\equiv e'/\rho)$:

$$\frac{\partial (\rho e)}{\partial t} = -\nabla \cdot (\rho e \mathbf{V} + \mathbf{j}^E_{dif}) + \rho \Gamma^E, \quad \rho e \mathbf{V} \equiv \mathbf{j}^E_{adv}, \tag{3.2.13}$$

recalling that $\mathbf{V}$ is the mass-weighted (*barycentric*) velocity defined in (3.1.10) and (3.1.11). In (3.2.13), $\rho\Gamma^E$ represents a source (which adds E) per unit phase volume per unit time. A negative Γ^E-value means a sink. Some authors make a distinction between internal and external sources.

The above equation can also be written in the form:

$$\frac{D(\rho e)}{Dt} = -\mathbf{V}\cdot\nabla(\rho e) - \nabla\cdot\mathbf{j}^E_{dif} + \rho\Gamma^E. \tag{3.2.14}$$

3.2.2 Particular Cases

Here we present the cases of mass, momentum, energy and entropy balance equations as applications of the general balance equation (3.2.13).

A. E Represents the Mass, m, of a Phase

Here, $e' = \rho$, $e = 1$, $\Gamma^E = \Gamma^m = 0$ (as mass cannot be created), and $\mathbf{V}^E = \mathbf{V}^m \equiv \mathbf{V}$ is the phase velocity defined by (3.1.11). For this case, the *diffusive mass flux*, $\mathbf{j}^m_{dif}$, vanishes, i.e., $\mathbf{j}^m_{dif} \equiv \rho(\mathbf{V}^m - \mathbf{V}) \equiv 0$. Hence, for the mass of a fluid phase, (3.2.12), or (3.2.13) leads to the mass balance equation:

$$\frac{\partial\rho}{\partial t} = -\nabla\cdot\rho\mathbf{V}, \tag{3.2.15}$$

in which:

- On the l.h.s. we have the rate of mass increase per unit volume of the phase.
- On the r.h.s. we have the net flux into the vicinity of the point through the bounding surface of a small domain, per unit volume of the latter, as this domain converges to the point.

Equation (3.2.15) is the *microscopic level balance equation for the mass of a phase*. It may also be written in the form:

$$\left(\frac{\partial\rho}{\partial t} + \mathbf{V}\cdot\nabla\rho \equiv\right)\frac{D\rho}{Dt} = -\rho\nabla\cdot\mathbf{V}, \quad \text{or} \quad \frac{1}{\rho}\frac{D\rho}{Dt} = -\nabla\cdot\mathbf{V}, \tag{3.2.16}$$

where $\mathrm{D}(.)/\mathrm{D}t \equiv \mathrm{D}_m(.)/\mathrm{D}t$ denotes the *total (= material) derivative* defined by (3.1.6), i.e., with respect to an observer moving at the fluid's mass averaged velocity $\mathbf{V}(\equiv \mathbf{V}^m)$. Recall:

- For any *scalar* φ, the gradient is a *vector*,

$$\nabla\varphi = (\partial\varphi/\partial x)\mathbf{1}_x + (\partial\varphi/\partial y)\mathbf{1}_y + (\partial\varphi/\partial z)\mathbf{1}_z,$$

  in which $\mathbf{1}_x$, $\mathbf{1}_y$, $\mathbf{1}_z$ denote unit vectors in the x, y, z directions, respectively.

- For any *vector* **V**, the divergence of the vector is a *scalar*, e.g.,

$$\nabla \cdot \mathbf{V} = \partial V_x/\partial x + \partial V_y/\partial y + \partial V_z/\partial z.$$

- For any vector **V**, the gradient, $\nabla\mathbf{V}$, is a second rank tensor, with 9 components $\partial V_i/\partial x_j$, e.g., $i, j = x, y, z$.

B. E Represents the Mass of a γ-Chemical Species, m^γ, in a Phase

$E = m^\gamma_\alpha$ represents the mass of a γ-species in an α-phase. Then, $e' = \rho^\gamma_\alpha$ is the concentration of the γ-species (= mass of γ-species per unit volume of the phase), $\Gamma^E_\alpha = \Gamma^{m^\gamma}_\alpha \equiv \Gamma^\gamma_\alpha$ expresses a source (= negative sink) of the γ-species in the α-phase, say by production of the mass of the considered γ-species within the phase by chemical reactions, or radioactive decay, per unit mass of the phase, $\mathbf{V}^E \equiv \mathbf{V}^\gamma_\alpha$ denotes the velocity of the mass of the γ-species, and $\mathbf{j}^\gamma_{\alpha,dif} (= \rho^\gamma_\alpha(\mathbf{V}^\gamma_\alpha - \mathbf{V}_\alpha))$ represents the diffusive flux of the mass of the γ-species, with respect to α-phase mass particles. Then, the mass balance equation for the considered species takes the form:

$$\underset{(a)}{\frac{\partial \rho^\gamma_\alpha}{\partial t}} = \underset{(b)}{-\nabla\cdot(\rho^\gamma_\alpha \mathbf{V}_\alpha + \mathbf{j}^\gamma_{\alpha,dif})} + \underset{(c)}{\rho_\alpha \Gamma^\gamma_\alpha}, \tag{3.2.17}$$

where $\sum_{(\gamma)} \rho^\gamma_\alpha = \rho_\alpha$ and $\sum_{(\gamma)} \mathbf{j}^\gamma_{\alpha,dif} = 0$, i.e., the sum of diffusive fluxes for all chemical species is identically zero. In the above equations,

(a) is the rate of added species mass per unit phase volume,
(b) is the rate of net influx of species mass through the surface bounding the domain, per unit domain volume, by advection and diffusion, and
(c) is the rate of produced/destroyed species mass per unit volume, by sources within the latter.

Except for the case of point injection or extraction of γ-species mass, or cases in which the considered species is introduced through domain boundaries, the total mass of the system as a whole is conserved, i.e., the combined mass production of all γ-species must vanish, $\sum_{(\gamma)} \Gamma^\gamma_\alpha = 0$. By summing (3.2.17) for all γ's comprising the fluid phase, we obtain the mass balance equation for the fluid, (3.2.15). It is of interest to note that (3.2.17) may also be written in the *mixed Eulerian–Lagrangian form*:

$$\frac{D\rho^\gamma_\alpha}{Dt} = -\rho^\gamma_\alpha \nabla\cdot\mathbf{V}_\alpha + \nabla\cdot\mathbf{j}^\gamma_{\alpha,dif} + \rho_\alpha \Gamma^\gamma_\alpha. \tag{3.2.18}$$

The diffusive flux of a γ-species is further discussed in Sect. 7.2.2. Chemical reactions will be discussed in Sect. 7.3.3.

The mass balance equation (3.2.17) for a chemical species can also be written in terms of the γ-mass fraction ω^γ,

$$\frac{\partial \rho_\alpha \omega^\gamma_\alpha}{\partial t} = -\nabla\cdot(\rho\omega^\gamma_\alpha \mathbf{V}_\alpha + \mathbf{j}^\gamma_{\alpha,dif}) + \rho_\alpha \Gamma^\gamma_\alpha, \tag{3.2.19}$$

or, in terms of the *molar fraction*, X^γ_α, defined in Sect. 7.1,

$$\frac{\partial \eta_\alpha X^\gamma_\alpha M^\gamma_\alpha}{\partial t} = -\nabla\cdot(\eta_\alpha X^\gamma_\alpha M^\gamma_\alpha \mathbf{V}_\alpha + \mathbf{j}^\gamma_{\alpha,dif}) + \rho_\alpha \Gamma^\gamma_\alpha. \tag{3.2.20}$$

C. E Represents the Linear Momentum, M, of a Fluid Phase

Here, $E = \mathbf{M} \equiv m\mathbf{V}$ denotes linear momentum, and $e = \mathbf{V}$. $e' = \rho\mathbf{V}$. The source of momentum within a unit volume is the *total* force acting on that volume. Here, this force is due to gravity,

$$\Gamma^E \equiv \Gamma^\mathrm{M} = -\mathbf{F}, \tag{3.2.21}$$

in which $\mathbf{F}$ is a body force per unit mass. When $\mathbf{F}$ is due only to gravity, we have $\mathbf{F} = -\rho g \nabla z$, with ∇z denoting a unit vector pointing upward.

The *diffusive flux of linear momentum is the stress*, $\boldsymbol{\sigma}$, which causes energy dissipation:

$$\mathbf{j}^\mathrm{M}_{dif} \equiv \rho\mathbf{V}\left(\mathbf{V}^\mathrm{M} - \mathbf{V}\right) = -\boldsymbol{\sigma}. \tag{3.2.22}$$

The microscopic level momentum balance (3.2.12) takes the form:

$$\underset{(a)}{\frac{\partial \rho\mathbf{V}}{\partial t}} = \underset{(b)}{-\nabla\cdot(\rho\mathbf{V}\mathbf{V} - \boldsymbol{\sigma})} + \underset{(c)}{\rho\mathbf{F}}, \tag{3.2.23}$$

in which $\mathbf{V}\mathbf{V}$ is the *dyadic product* ($\equiv V_i V_j$) of the two vectors, and

(a) is the rate of accumulation of momentum,
(b) is the rate of momentum gained by momentum advection and diffusion,
(c) is the rate of momentum gained by external (e,g., body) force,

and each term expresses the rate at which momentum is added *per unit volume of the phase*. Equation (3.2.23) is the (microscopic level) *differential balance equation of linear momentum of a phase*.

In view of the mass balance (3.2.15), another form of the above momentum balance equation is *Cauchy's equation of motion* (or *Cauchy's first law*):

$$\rho\frac{D\mathbf{V}}{Dt} = \nabla\cdot\boldsymbol{\sigma} + \rho\mathbf{F}, \quad \text{with} \quad \boldsymbol{\sigma} = \boldsymbol{\tau} - \mathrm{p}\boldsymbol{\delta}, \tag{3.2.24}$$

in which $\boldsymbol{\delta}$ denotes the *Kronecker delta*; it is a second rank symmetric tensor, with components δ_{ij}, such that $\delta_{ij} = 1$ for $i = j$, and $\delta_{ij} = 0$ for $i \neq j$. The *Kronecker delta* is often referred to as the *unit tensor* denoted by $\mathbf{I}$. For gravity as the only external force, $\mathbf{F} = \mathbf{g} = -g\nabla z$, with g denoting gravity acceleration and ∇z denoting a unit vector directed upwards.

Equation (3.2.23), combined with the mass balance equation (3.2.15), can also be written in the form of the motion equation:

$$\rho \frac{D\mathbf{V}}{Dt} = \nabla \cdot \boldsymbol{\tau} - \nabla p + \rho \mathbf{F}. \tag{3.2.25}$$

This equation has to be supplemented by an expression that relates $\boldsymbol{\tau}$ to $\mathbf{V}$, according to the nature of the considered fluid. Henceforth in this book, we shall limit the discussion to Newtonian fluids. For such fluids, with $\boldsymbol{\tau}$ denoting the shear stress, we use the constitutive relationship:

$$\boldsymbol{\tau} = \mu(\nabla \mathbf{V} + \nabla \mathbf{V}^T), \tag{3.2.26}$$

where μ is the fluid's dynamic viscosity.

When gravity is the only body force, for $\mathbf{V} = 0$, and for $D\mathbf{V}/Dt = 0$, (3.2.23) reduces to the *equilibrium equation*:

$$\nabla \cdot \boldsymbol{\sigma} + \rho \mathbf{F} = 0, \tag{3.2.27}$$

which is actually an equation that expresses a *(static) balance of forces*.

The momentum balance equation (3.2.25) then takes the form:

$$\rho \frac{D\mathbf{V}}{Dt} = \mu \nabla \cdot (\nabla \mathbf{V}) - \nabla p + \rho \mathbf{F}, \tag{3.2.28}$$

known as the *Navier–Stokes equation* for viscous flow. Recall that *as is common in Fluid Mechanics, pressure is considered negative for compression.*

Note that whenever the dynamic viscosity, μ, appears in a balance equation as the only coefficient that represents the fluid's viscosity, the equation is limited to a Newtonian fluid.

The l.h.s. of (3.2.28) expresses the *inertial effects*, which are a consequence of non-uniform velocity. From the definition of $D\mathbf{V}/Dt$ in (3.1.6), it follows that the inertial effects involve two phenomena: one is due to velocity acceleration at a point, and the other to velocity non-uniformity in space. The first will vanish in steady flow. The second will vanish in uniform flow (but uniform flow in a porous medium does not exist at the microscopic level!).

For $\nabla \cdot \boldsymbol{\tau} = 0$, i.e., viscous effects are negligible, (3.2.28) reduces to

$$\rho \frac{D\mathbf{V}}{Dt} = -\nabla p + \rho \mathbf{F}, \tag{3.2.29}$$

known as *Euler's equation*.

For small Reynolds numbers, $Re(= Vd/\nu) \ll 1$, we are left with:

$$\rho_f \frac{\partial \mathbf{V}}{\partial t} = \mu \nabla^2 \mathbf{V} - \nabla p + \rho \mathbf{F}, \tag{3.2.30}$$

known as *Stokes equation.*

D. E Represents the Energy, $\mathbb{E}$, of a Fluid Phase

The total energy of the phase, $\mathbb{E}$, consists of its *internal energy* (due to thermal agitation and short range intermolecular forces), and its *kinetic energy*. The *potential energy* of the body as a whole (due to gravity) does not appear explicitly in the balance equation, as here we choose to include it in the term that expresses work by the body forces (compare with De Groot and Mazur 1962, p. 17). Accordingly, the specific energy and the energy density are expressed by:

$$e^{\mathbb{E}} = \left(u + \tfrac{1}{2}V^2\right), \quad e'^{\mathbb{E}} = \left(u' + \tfrac{1}{2}\rho V^2\right), \tag{3.2.31}$$

where $\mathbf{V}$ is the velocity ($\equiv$ momentum per unit mass), $V \equiv |\mathbf{V}|$, and u is the *specific internal energy*, (i.e., internal energy per unit mass).

Although we could make use of the general microscopic level balance equation (3.2.13), we shall develop the microscopic level energy balance equation in detail.

Following the phenomenological approach, and referring to Fig. 3.2, we note that energy is supplied to the phase contained in a domain Ω through its total interface with other phases, $\mathcal{S}$:

- by advection, following (3.2.8): $\mathbf{j}^{\mathbb{E}}_{adv} = \int_{\mathcal{S}} \left(u' + \tfrac{1}{2}\rho V^2\right) \mathbf{V}\cdot\boldsymbol{\nu}\, d\mathcal{S}$, and
- by thermal diffusion ($\equiv$ conduction: $\mathbf{j}^{\mathbb{E}}_{dif}$): $\; -\int_{\mathcal{S}} \mathbf{j}^{\mathbb{H}}_{dif}\cdot\boldsymbol{\nu}\, d\mathcal{S}$,

with $\boldsymbol{\nu}$ denoting the outward normal unit vector on $\mathcal{S}$.

Employing Gauss divergence theorem (3.2.7) to transform the sum of the last two integrals into a volume one, we obtain:

$$-\int_{\mathbb{V}} \nabla\cdot\left[(u' + \tfrac{1}{2}\rho V^2)\mathbf{V} + \mathbf{j}^{\mathbb{H}}_{dif}\right] d\mathbb{V}.$$

The rate of energy production within Ω-domain is expressed by $\int_{\mathbb{V}} \rho \Gamma^{\mathbb{H}}\, d\mathbb{V}$, where $\Gamma^{\mathbb{H}}$ is the rate of heat produced within Ω, per unit mass, e.g., by chemical reactions.

Finally, energy is added to the domain Ω by the work of the forces acting on the phase contained in Ω. These include:

(a) $\int_{\mathbb{V}} \mathbf{V}\cdot\rho\mathbf{F}\, d\mathbb{V}$, where $\mathbf{F}$ represents body force per unit mass, expressing the rate of supply of kinetic energy by the body force acting on the phase contained in Ω, and

(b) $-\int_{\mathcal{S}} \mathbf{V}\cdot(-\boldsymbol{\sigma})\cdot\boldsymbol{\nu}\, d\mathcal{S}$, expressing the rate of work done by the surface force acting on the surface $\mathcal{S}$ of Ω, with the tensor $\boldsymbol{\sigma}$ denotes stress. Employing the divergence theorem, this term can be replaced by $\int_{\mathbb{V}} \nabla\cdot(\boldsymbol{\sigma}\cdot\mathbf{V})\, d\mathbb{V}$.

By combining all the above terms, dividing by $\mathbb{V}$ and passing to the limit as $\mathbb{V} \to 0$, we obtain the differential *energy balance equation* in the form:

$$\underset{(a)}{\frac{\partial}{\partial t}\left(u' + \tfrac{1}{2}\rho V^2\right)} = \underset{(b)}{-\nabla\cdot\left[\left(u' + \tfrac{1}{2}\rho V^2\right)\mathbf{V} + \mathbf{j}^{\mathbb{H}}_{dif}\right]} + \underset{(c)}{\mathbf{V}\cdot\rho\mathbf{F} + \nabla\cdot(\boldsymbol{\sigma}\cdot\mathbf{V}) + \rho\Gamma^{\mathbb{H}}}, \tag{3.2.32}$$

where:

(a) is the rate of energy accumulation,
(b) is rate of net energy influx by advection and heat conduction, and
(c) is the rate of energy supplied to the phase by the work of mechanical forces and heat sources (three terms),

and all terms are per unit volume.

Combining (3.2.32) with the mass balance equation (3.2.15) and the momentum balance equation (3.2.23) yields another form of the energy balance equation,

$$\underset{(a)}{\rho\frac{Du'}{Dt}} = \underset{(b)}{\boldsymbol{\sigma} : \nabla\mathbf{V}} - \underset{(c)}{\nabla\cdot\mathbf{j}^{\mathbb{H}}} + \underset{(d)}{\rho\Gamma^{\mathbb{H}}}, \tag{3.2.33}$$

where:

(a) is the material rate of growth of internal energy,
(b) is the rate of increase of internal energy by the work done in producing strain ($= \boldsymbol{\sigma} : \dot{\varepsilon}$, with ε denotes strain),
(c) is the net influx of internal energy by heat conduction, and
(d) is the rate of increase of internal energy from internal sources,

and all terms are per unit phase volume.

In a fluid, $\boldsymbol{\sigma} = \boldsymbol{\tau} - p\boldsymbol{\delta}$, and (3.2.33) can be rewritten as:

$$\underset{(a)}{\rho\frac{Du'}{Dt}} = \underset{(b)}{\boldsymbol{\tau} : \nabla\mathbf{V}} \underset{(c)}{-p\nabla\cdot\mathbf{V}} - \underset{(d)}{\nabla\cdot\mathbf{j}^{\mathbb{H}}} + \underset{(e)}{\rho\Gamma^{\mathbb{H}}}, \tag{3.2.34}$$

where:

(a) is the material rate of growth of internal energy,
(b) is the irreversible rate of increase in internal energy (= heat) due to shear,
(c) is the reversible rate of internal energy gain by compression,
(d) is the net influx of internal energy by heat conduction, and
(e) is the rate of increase of internal energy from internal sources.

For a multi-species system, we should add the work due to the diffusive flux,

$$\sum_{(\gamma)} \mathbf{j}^{\gamma}_{dif} \cdot \mathbf{F}^{\gamma}, \qquad \mathbf{j}^{\gamma}_{dif} = \rho^{\gamma}(\mathbf{V}^{\gamma} - \mathbf{V}),$$

where $\mathbf{F}^{\gamma}$ is the body force acting on the γ-species, provided $\mathbf{F}^{\gamma}$ is different for different chemical species.

It is often more convenient to write the energy balance equation (3.2.34) in terms of the absolute temperature, T, and the *heat capacity*. To do so, we follow the discussion in Sect. 2.2.2, where we have introduced the *specific internal energy* of a phase, u', as a single valued function of a set of specific values (per unit of mass) of extensive state variables, e.g., $u' = u'(s, v, \omega^{\gamma})$, where s' is the *specific entropy*, $v(= 1/\rho)$ is the specific volume, and $\omega^{\gamma}(= \rho^{\gamma}/\rho)$ is the mass fraction of the γ-species, with $\gamma = 1, 2, \ldots$, all of a phase. Based on that discussion, the l.h.s. of (3.2.34) becomes:

$$\rho \frac{Du'}{Dt} = \rho \left(T \left. \frac{\partial p}{\partial T} \right|_{v,\omega^{\gamma}} - p \right) \frac{Dv}{Dt} + \rho c_v \frac{DT}{Dt} + \rho \sum_{(\gamma)} \mu^{\gamma} \frac{D\omega^{\gamma}}{Dt}, \tag{3.2.35}$$

in which c_v (dims: $\mathsf{M}^{-1}\mathsf{L}^2\mathsf{T}^{-2}\Theta^{-1}$, with SI units of J/kg/K) denotes the specific heat at constant volume.

From (3.2.16), or (3.2.34), we also have:

$$\frac{1}{v} \frac{Dv}{Dt} \equiv \rho \frac{D(1/\rho)}{Dt} = -\frac{1}{\rho} \frac{D\rho}{Dt} = \nabla \cdot \mathbf{V}. \tag{3.2.36}$$

By substituting (3.2.36) in (3.2.35), and combining the resulting equation with (3.2.34), we obtain:

$$\rho C_v \frac{DT}{Dt} = \boldsymbol{\tau} : \nabla \mathbf{V} - \nabla \cdot \mathbf{j}^{\mathbb{H}} - T \left. \frac{\partial p}{\partial T} \right|_{v,\omega^{\gamma}} \nabla \cdot \mathbf{V} - \rho \sum_{(\gamma)} \mu^{\gamma} \frac{D\omega^{\gamma}}{Dt} + \rho \Gamma^{\mathbb{H}}. \tag{3.2.37}$$

This is the (microscopic level) differential *equation of heat balance*, written in terms of the temperature of the considered phase.

Another useful form of this equation is:

$$\rho c_p \frac{DT}{Dt} = \boldsymbol{\tau} : \nabla \mathbf{V} - \nabla \cdot \mathbf{j}^{\mathbb{H}} + T \left. \frac{\partial \rho}{\partial T} \right|_{p,\omega^{\gamma}} \nabla \cdot \mathbf{V} - \rho \sum_{(\gamma)} \mu^{\gamma} \frac{D\omega^{\gamma}}{Dt} + \rho \Gamma^{\mathbb{H}}, \tag{3.2.38}$$

where c_p (dims.: $\mathsf{M}^{-1}\mathsf{L}^2\mathsf{T}^{-2}\Theta^{-1}$, with SI units of J/kg/K) is the *specific heat at constant pressure*.

E. E **Represents the Enthalpy,** $\mathbb{H}$**, of a Fluid Phase**

It is often advantageous to use *specific enthalpy*, h (Sect. 2.2.2), of the considered phase, as the dependent variable, especially when dealing with liquids and gases which undergo phase change. We then express the energy balance equation (3.2.34) in the form:

$$\rho\frac{Dh}{Dt} = \boldsymbol{\tau} : \nabla\mathbf{V} - \nabla\cdot\mathbf{j}^{\mathbb{H}} + \frac{Dp}{Dt} + \rho\Gamma^{\mathbb{H}}. \tag{3.2.39}$$

By adding $h(\partial\rho/\partial t + \nabla\cdot\rho\mathbf{V})$ which is equal to zero (i.e., the mass balance equation) to the l.h.s. of the last equation, we obtain another form of the energy balance equation:

$$\frac{\partial\rho h}{\partial t} = -\nabla\cdot(\rho h\mathbf{V} + \mathbf{j}^{\mathbb{H}}) + \boldsymbol{\tau} : \nabla\mathbf{V} + \frac{Dp}{Dt} + \rho\Gamma^{\mathbb{H}}. \tag{3.2.40}$$

F. E **Represents the Entropy,** $\mathbb{S}$**, of a Fluid Phase**

With s denoting entropy ($\mathbb{S}$) per unit mass, the entropy balance takes the form:

$$\frac{\partial\rho s}{\partial t} = -\nabla\cdot(\rho s\mathbf{V} + \mathbf{j}^{\mathbb{S}}) + \rho(\Gamma^{\mathbb{S}}_{int} + \Gamma^{\mathbb{S}}_{ext}), \tag{3.2.41}$$

where $\mathbf{j}^{\mathbb{S}}$ denotes the diffusive entropy flux, $\Gamma^{\mathbb{S}}$ is the rate of production of entropy per unit mass, and we have made a distinction between internal production ($\Gamma^{\mathbb{S}}_{int}$) and external supply ($\Gamma^{\mathbb{S}}_{ext}$).

By the second law of thermodynamics (Sect. 2.2.1 B), $\Gamma^{\mathbb{S}}_{int} \geq 0$.

3.2.3 *Initial and Boundary Conditions*

In order to solve the E-balance equation (3.2.13) for a specified phase domain within the void space, we have to specify: (1) (possibly moving) phase domain boundaries, (2) initial conditions within the phase domain, and (3) conditions on interphase boundaries. However, since our interest is primarily in modeling at the macroscopic level, we shall skip the discussion on microscopic level initial and boundary conditions. These can be found in any text on Continuum Mechanics.

Overlooking for a moment the fact that microscopic level interphase boundaries are unknown (and, in fact, this is one of the reasons for developing macroscopic models), we may describe a (possibly moving) interphase boundary surface by $F(x, y, z, t) = 0$. Although we are considering here boundary surfaces and boundary conditions at the microscopic level (i.e., between phases inside the void space), the material on the moving macroscopic boundary presented in Sect. 5.2.1 D is valid also here, with x, y, z denoting coordinates at the microscopic level.

3.3 Macroscopic Balance Equations for E (1)

In the previous section, we used the phenomenological approach to develop E-balance equations at the microscopic level. Once we have that equation, we can use one of the averaging approaches outlined in Chap. 1 to derive the corresponding macroscopic balance equations. Instead, we shall develop the corresponding macroscopic also by the *phenomenological approach*, recalling that an equation at the macroscopic level describes what happens 'at a point in the porous medium considered as a continuum'. Within a porous medium domain, the considered E may be transported by a single (fluid or solid) phase, or by all fluid and solid phases present in the considered domain. The fluid phase itself is usually composed of a number of chemical species.

3.3.1 The General Macroscopic Balance Equation

We shall generalize the discussion by considering what happens at a point $\mathbf{x}$ (to be interpreted as 'within the REV centered at the point $\mathbf{x}$') at time t, to a fluid α-phase that occupies only part of the void space at *phase saturation*, S_α, or *volumetric fraction*, θ_α, where:

$$\theta_\alpha(\mathbf{x}, t) = \frac{\text{Volume of } \alpha\text{-fluid in REV}}{\text{Volume of REV}}, \qquad 0 \leq \theta_\alpha \leq \phi, \quad \sum_{(\alpha)} \theta_\alpha = \phi,$$

$$S_\alpha(\mathbf{x}, t) = \frac{\text{Volume of } \alpha\text{-fluid in REV}}{\text{Volume of void space in REV}}, \qquad 0 \leq S_\alpha \leq 1, \quad \sum_{(\alpha)} S_\alpha = 1.$$

In both definitions, the sum is over all fluid phases present in the void space. The two definitions are related to each other by:

$$\theta_\alpha = \phi S_\alpha, \tag{3.3.1}$$

where ϕ is the porosity at the point.

When the porous medium is inhomogeneous with respect to porosity, or when it undergoes deformation, which alters the porosity, the fluid saturation should be used to describe the quantity of a fluid in the void space at a point (meaning in the vicinity of the point). This enables a separate treatment of porosity changes.

Within a porous medium domain, consider a domain Ω surrounded by a surface $\mathcal{S}_\Omega$. Within Ω, we have a subdomain Ω_α, occupied by a fluid α-phase, and a subdomain Ω_β occupied by β, i.e., non-α, fluid or solid phases. The macroscopic E_α-balance equation for the porous medium domain Ω_α, bounded by a fixed closed surface $\mathcal{S}_\Omega$, can be written in the form:

$$\left\{\begin{array}{c}\text{Quantity of } E_\alpha\\ \text{accumulating}\\ \text{in } \Omega_\alpha\\ \text{during } \Delta t\end{array}\right\} = \left\{\begin{array}{c}\text{Net quantity of}\\ E_\alpha \text{ entering } \Omega_\alpha\\ \text{through } \mathcal{S}_\Omega\\ \text{during } \Delta t\end{array}\right\}$$

$$+ \left\{\begin{array}{c}\text{Net quantity of}\\ E_\alpha \text{ entering } \Omega_\alpha\\ \text{through } \mathcal{S}_{\alpha\beta}\\ \text{during } \Delta t\end{array}\right\} + \left\{\begin{array}{c}\text{Net production}\\ \text{of } E_\alpha \text{ in } \Omega_\alpha\\ \text{during } \Delta t\end{array}\right\}. \tag{3.3.2}$$

where E_α may be the mass of an α-phase, or the of a γ-species in the α-phase, m^γ_α, and we have taken into account that E_α can leave/enter the α-phase across any $\mathcal{S}_{\alpha\beta}$-surface within Ω. This includes the cases of *phase change* from α to β, and a γ-species diffusing from α to β.

When the above balance equation is written for a small volume in a porous medium domain around a point, and for a small time interval, and then letting the volume and the time interval shrink to zero, the balance equation can be written as a *partial differential equation* (PDE) that expresses the balance of E *at that instant of time* and *that point in space* within the domain:

$$\underset{(a)}{\frac{\partial \theta_\alpha e'_\alpha}{\partial t}} = \underset{(b)}{-\nabla\cdot\theta^A_\alpha \mathbf{J}^E_{\alpha,tot}} \underset{(c)}{+f^E_{\beta\to\alpha}} \underset{(d)}{+\theta_\alpha\rho_\alpha\Gamma^E_\alpha}, \tag{3.3.3}$$

where e'_α denotes the E_α-density, i.e., E_α per unit volume of fluid α-phase, θ^A_α denotes the areal fraction of the α-phase in the cross-section, and $\mathbf{J}^E_{\alpha,tot}$ denotes the *total* macroscopic flux of E, *with* and *in* the moving α-phase, per unit phase area. Or, in terms of the specific value of e_α ($= E_\alpha$ per unit mass of fluid α-phase), and approximating θ^A_α by θ_α:

$$\underset{(a)}{\frac{\partial \theta_\alpha \rho_\alpha e_\alpha}{\partial t}} = \underset{(b)}{-\nabla\cdot\theta_\alpha \mathbf{J}^E_{\alpha,tot}} \underset{(c)}{+f^E_{\beta\to\alpha}} \underset{(d)}{+\theta_\alpha\rho_\alpha\Gamma^E_\alpha}. \tag{3.3.4}$$

The terms appearing in the above balance equations can be interpreted in the following way:

(a) Rate of accumulation of E in the α-phase, per unit volume of porous medium, per unit time, with:

$$\theta_\alpha = \frac{1}{\mathbb{V}_o}\int_{\mathbb{V}_o} \gamma_\alpha d\mathbb{V},$$

in which γ_α denotes the characteristic function of the α-phase, defined by (1.1.9):

$$\theta_\alpha\rho_\alpha(\equiv \theta_\alpha\overline{\rho_\alpha}^\alpha) = \frac{1}{\mathbb{V}_o}\int_{\mathbb{V}_o} \rho_\alpha\gamma_\alpha d\mathbb{V},$$

and:

$$\theta_\alpha \rho_\alpha e_\alpha (\equiv \theta_\alpha \overline{e'_\alpha}^\alpha) = \frac{1}{\mathbb{V}_o} \int_{\mathbb{V}_o} (\rho_\alpha e_\alpha) \gamma_\alpha d\mathbb{V},$$

with $e_\alpha = 1$ for $\mathbb{E}_\alpha = m_\alpha$. Note that, although we are not using different symbols, in the above equation, as in the following ones, the integral is over microscopic level values, while the l.h.s. involves macroscopic values.

(b) Net inflow of E_α, per unit volume of porous medium, per unit time, with $\mathbf{J}^E_{\alpha,tot}$ denoting the total macroscopic flux (per unit area of the α-phase), per unit volume of the α-phase, by advection ($\mathbf{J}^E_{adv} \equiv \overline{e_\alpha}^\alpha \overline{\mathbf{V}_\alpha}^\alpha$), dispersion ($\mathbf{J}^E_{\alpha,dis}$, to be discussed later in this chapter) and diffusion ($\mathbf{J}^E_{\alpha,dif}$), with

$$\theta_\alpha \rho_\alpha \mathbf{V}_\alpha (\equiv \theta_\alpha \overline{\rho_\alpha \mathbf{V}_\alpha}^\alpha) = \frac{1}{\mathbb{V}_o} \int_{\mathbb{V}_o} (\rho_\alpha \mathbf{V}_\alpha) \gamma_\alpha d\mathbb{V}.$$

(c) Rate at which E enters the α-phase from all other (non-α) phases through the microscopic $\mathcal{S}_{\alpha\beta}$-surface that surrounds the α-phase within an REV, per unit volume of porous medium. However, in determining this transfer, we have to take into account the possibility that the microscopic interface between a considered α-phase and any other phase inside the REV *may be moving* (e.g., in two phase flow, or when the solid matrix is deformable, or when phase change occurs). Thus:

$$f^E_{\alpha\to\beta} = -\frac{1}{\mathbb{V}_o} \int_{S_{\alpha\beta}} \left[\rho_\alpha e_\alpha (\mathbf{V}_\alpha - \mathbf{u}_{\alpha\beta}) + \mathbf{j}^E_{dif}\right] \cdot \boldsymbol{\nu}_\alpha dS, \qquad (3.3.5)$$

in which $\mathbb{V}_o$ denotes the volume of an REV, $S_{\alpha\beta}$ is the total α-β interface within the REV, and $\boldsymbol{\nu}_\alpha$ denotes the outward unit vector on this surface, which moves at a velocity $\mathbf{u}_{\alpha\beta}$. When the $\mathcal{S}_{\alpha\beta}$-interface is a *material surface* with respect to E, the first term in the square brackets above vanishes, and interphase transfer is possible only as a diffusive flux:

$$f^\gamma_{\alpha\to\beta} = -\frac{1}{\mathcal{U}_o} \int_{\mathcal{S}_{\alpha\beta}} \mathbf{j}^\gamma_\alpha \cdot \mathbf{n}_\alpha \, dS. \qquad (3.3.6)$$

This means that the chemical species can reach and cross interphase boundaries *only by diffusion*. Note that all symbols in the above integrand denote *microscopic values*, and

$$f^E_{\alpha\to\beta} + f^E_{\beta\to\alpha} \equiv 0,$$

unless we allow E to accumulate on the interface itself (see Sect. 1.4.2 C).

When a change of phase takes place across the $\mathcal{S}_{\alpha\beta}$-interface, *the latter is no more a material surface*. In fact, the first term in the square parenthesis expresses the rate at which the considered E_α crosses the microscopic $\alpha - \beta$ interface as a consequence of the phase change. In fact, in (3.3.6) we note both the effect of change of phase and that of transfer (of the considered E across interphase boundaries. Some authors (e.g., Hassanizadeh and Gray 1990) represent $f_{\alpha\rightarrow\beta}$ in (3.3.5) as two separate terms: one representing the amount of E crossing the interface as a consequence of phase change, and the other due to the diffusive E-flux (or phase exchange).

(d) Rate of production (i.e., a source) of E in the α-phase, per unit volume of porous medium, per unit time:

$$\theta_\alpha \rho_\alpha \Gamma^E_\alpha (\equiv \theta_\alpha \overline{\rho_\alpha \Gamma^E_\alpha}^\alpha) = \frac{1}{\mathbb{V}_o} \int_{\mathbb{V}_o} (\rho_\alpha \Gamma^E_\alpha) \gamma_\alpha d\mathbb{V}.$$

In Chap. 7, we shall consider the case of E_α that represents the mass of a γ chemical species in an α phase. Then, Γ^γ_α will represent the production of the γ-species by chemical reactions. Another example will be the appearance of the mass of a phase by phase change, say, due to heating at a point.

Let us add here a few comments:

(1) First, a comment concerning notation. For the flux of E, we have introduced, and shall be using throughout the book, two different symbols, $\mathbf{j}^E$ and $\mathbf{J}^E$. The first for the microscopic flux and the second for the macroscopic one. However, for the sake of simplicity, except for the flux, we shall not use different symbols for the two levels; the appropriate level for each equation should be inferred from the content of the equation itself, recalling that each equation belongs *only* to one of the two levels, and so do all the terms appearing in it. For example, if an equation includes a porosity, ϕ, or a volumetric fraction, θ_α, or a permeability, k, it is obviously at the macroscopic level.

(2) In view of (1.1.15), e'_α in (3.3.3) represents the intrinsic phase average $\overline{e'_\alpha}^\alpha$, while e_α in (3.3.4) represents the intrinsic mass average $\widetilde{e_\alpha}^\alpha$. In both cases, ρ_α represents $\overline{\rho_\alpha}^\alpha$. In fact, all variables appearing in (3.3.4) are mass averaged ones.

(3) At the macroscopic level, the flux is still per unit area of the considered phase, but the latter occupies only part of the cross-section through the porous medium domain. The flux per unit area of porous medium is obtained by multiplying the $\mathbf{J}$-flux by the areal fraction of the phase, $\theta^A_\alpha (\approx \theta_\alpha)$. This facilitates modeling the case of variable porosity and saturation.

(4) We have not made a distinction between the *porosity*, ϕ (= volume of void space per unit volume of porous medium), and the areal porosity, ϕ^A (= area of void in a planar cross-section, per unit area of cross section), introduced in Sect. 1.1.5. Similarly, in the case of multiple phases, we do not distinguish between θ_α and θ^A_{α}, with $\phi^A = \phi^A(\nu_1, \nu_2, \nu_3)$, in which ν_1, ν_2, ν_3 denote the components of the unit vector, $\boldsymbol{\nu}$, normal to the considered cross-section. Instead, we assume that $\phi^A \approx \phi$ and $\theta^A_\alpha \approx \theta_\alpha$.

The balance equation (3.3.4), obtained *phenomenologically*, will be the basic balance equation to be used for all extensive quantities in this book. It is a *differential macroscopic balance equation of* E in an α-phase. It describes the transport of any extensive quantity in an α-phase that occupies the entire void space, or part of it, in a porous medium domain. Appropriate expressions have to be provided for the flux, the transfer and the source terms. The term 'flux (for $\mathbf{J}^E_{\alpha,tot}$) is used here to denote the quantity of E passing through a unit area of α-phase in a planar cross-section, per unit time

We wish to emphasize that every term in (3.3.4), as in all macroscopic balance equations, expresses what happens in the vicinity of a point (and this means in the vicinity of *every* point) in the porous medium domain, regarded as a continuum, in terms of values of state variables that may be regarded as average values for that vicinity, and coefficients that represent various aspects of the effects of the solid matrix configuration in that vicinity.

The macroscopic balance equations presented below contain terms that express total flux, rate of interphase exchange and sources of the considered extensive quantities. These will be discussed in subsequent chapters.

3.3.2 Particular Cases

The particular cases discussed below are based on (3.3.3).

A. Mass Balance Equation of an α-Phase, $e_\alpha = 1$

We consider the mass of a fluid α-phase ($\mathbb{E} = m_\alpha$, density $e' = \rho_\alpha$, specific mass, $e_\alpha = 1$) that occupies part the void space, at the volumetric fraction θ_α. Assuming no sources or sinks of α-phase mass, i.e., $\Gamma^m_\alpha = 0$, the macroscopic α-fluid mass balance equation (3.3.3) takes the form:

$$\frac{\partial \theta_\alpha \rho_\alpha}{\partial t} = -\nabla \cdot \theta_\alpha \mathbf{J}^m_{\alpha,tot} + f^m_{\beta\rightarrow\alpha}, \tag{3.3.7}$$

where $\rho_\alpha \equiv \overline{\rho_\alpha}^\alpha$, $\mathbf{J}^m_{\alpha,\,tot}$ denotes the total α-mass flux (= mass per per unit time per unit area of α-phase), and $f^m_{\beta\rightarrow\alpha} (\equiv f^m_{\alpha\rightarrow\beta})$ denotes the rate of mass transfer from all β-phases to the α-phase, e.g., by phase change like evaporation, condensation, and dissolution, per unit volume of porous medium. In single-phase flow, $\theta_\alpha \rightarrow \phi$, and $f^m_{\beta\rightarrow\alpha} \equiv 0$. However, we use this term to express the rate at which a solid dissolves in the fluid, or a liquid evaporates. Note that here and in what follows, we have approximated θ^A_α by θ_α in the flux expression, and that as everywhere in this book, we refer to sources, with sinks being equivalent to negative sources.

Examples of two phase flow are air and water, oil and gas, and water and CO_2. Fluxes of specific extensive quantities will be discussed in Sect. 3.4.

B. Balance Equation for the Mass of a Chemical Species, $e = \omega^\gamma_\alpha$

We consider a chemical γ-species present in a fluid α-phase, which occupies part of the void space, at the volumetric fraction θ_α. For this case ($\mathbb{E} = m^\gamma_\alpha$, $e' = \rho^\gamma_\alpha$, $e = \omega^\gamma_\alpha$), Eq. (3.3.4), takes the form:

$$\frac{\partial \theta_\alpha \rho_\alpha \omega^\gamma_\alpha}{\partial t} = -\nabla \cdot \theta_\alpha \mathbf{J}^\gamma_{\alpha,tot} + f^\gamma_{\beta\rightarrow\alpha} + \theta_\alpha \rho_\alpha \Gamma^\gamma_\alpha, \tag{3.3.8}$$

in which $\mathbf{J}^{\gamma}_{\alpha,\,tot}$ denotes the total flux of the γ-mass *with* and *in* the α-phase, i.e., advective, dispersive and diffusive fluxes, $f^{\gamma}_{\beta\rightarrow\alpha}$ denotes the rate of transfer of m^{γ} to the α-phase from the solid and from all non-α fluid phases present in the void space, across α-β interfaces, e.g., by volatilization of γ, and the source Γ^{γ}_{α} may include a source of m^{γ}_{α} due to such phenomena as chemical reactions or radioactive decay *within* the α-phase in which γ is produced (or destroyed). Each term in (3.3.8) expresses added mass of the γ-species in the α-phase, per unit volume of porous medium, per unit time.

The flux, $\mathbf{J}^{\gamma}_{\alpha,tot}$, the rate of interphase transfer, $f^{\gamma}_{\beta\rightarrow\alpha}$, and sources, $\theta_{\alpha}\rho_{\alpha}\Gamma^{\gamma}_{\alpha}$, will be discussed in detail in Chap. 7.

C. Balance Equation for the Momentum of a Fluid Phase, $e = \mathbf{V}$

We consider a single fluid phase (of density ρ) that occupies the entire void space. The case of two fluids will be discussed in Sect. 6.2. Here, $\mathbb{E} = \mathbf{M}$, $e = \mathbf{V}$, $e' = \rho\mathbf{V}$, and the momentum balance equation obtained from (3.3.4) takes the form:

$$\underset{(a)}{\frac{\partial \phi\rho\mathbf{V}}{\partial t}} = \underset{(b)}{-\nabla\cdot\phi\mathbf{J}^{\mathrm{M}}_{tot}} \underset{(c)}{+\mathbf{f}^{\mathrm{M}}_{s\rightarrow f}} \underset{(d)}{+\phi\rho\Gamma^{\mathrm{M}}}, \tag{3.3.9}$$

in which:

(a) denotes the rate of momentum added to the fluid, per unit volume of porous medium,
(b) denotes the net influx of momentum per unit volume of porous medium, as flow takes place, with $\mathbf{J}^{\mathrm{M}}_{tot}$ denoting the total flux of momentum (per unit fluid phase area), i.e., advective, dispersive and diffusive fluxes,
(c) denotes the rate at which momentum is transferred from the solid phase to the fluid, across their common interface, per unit volume of porous medium, and
(d) denotes the rate at which momentum is produced within a unit volume of porous medium.

As we shall see in Sect. 6.2.1, in the case of two-phase (α, β) flow, momentum is transferred to the α-phase also from the β-phase.

In the above equation, the source of momentum, $\boldsymbol{\Gamma}^{\mathrm{M}}$, is due to body (and other) forces (per unit mass of the phase), and the velocity, $\mathbf{V}$, has the meaning of *momentum per unit mass of the phase*.

In the case of a single fluid that occupies the entire void space, by (1) combining the above equation with the mass balance equation (without mass sources), (2) recalling that the total momentum flux is made up of the advective flux, $\rho\mathbf{V}\mathbf{V}$, the diffusive flux, expressed by the stress, $-\boldsymbol{\sigma}_f$, and a dispersive flux, which we neglect here, (3) expressing the solid to fluid momentum transfer per unit volume by $(1/\mathbb{V}_o)\int_{S_{sf}} \boldsymbol{\sigma}\cdot\boldsymbol{\nu}\,dS$, (4) assuming that the only source (= production) of momentum, Γ^{M}, is the external body force per unit mass acting on the phase is $\mathbf{F}$, we can rewrite the above momentum balance equation in the form:

$$\phi\rho\frac{D\mathbf{V}}{Dt} = \nabla\cdot\phi\boldsymbol{\sigma} + \frac{1}{\mathbb{V}_o}\int_{\mathcal{S}_{\alpha s}} \boldsymbol{\sigma}\cdot\boldsymbol{\nu}\, d\mathcal{S}_{\alpha s} + \phi\,\rho_f\mathbf{F}. \tag{3.3.10}$$

The total momentum flux, $\mathbf{J}^{\mathrm{M}}_{tot}$, the rate of interphase transfer, $\mathbf{f}^{\mathrm{M}}_{s\to f}$, and the momentum source, $\phi\rho_f\mathbf{F}$, will be further discussed in Sect. 3.5.

When we are interested in the momentum balance equation for the porous medium as a whole, say in single phase flow, we write one momentum balance equation for the fluid and one for the solid, and add them (see Chap. 9). The momentum exchange terms will then vanish. Actually, in the phenomenological approach, we can write the momentum balance equation directly for the porous medium as a whole.

D. Total Stress, Capillary Pressure and Equilibrium Equation

Let us add some information on the case of two fluid phases that occupy the void space of a deformable porous medium domain (Bear and Pinder 1978): a wetting phase (w), and a non-wetting one (n). These terms are introduced in Sect. 2.4.2. The macroscopic momentum balance equation, (3.3.10), for each of the three phases, with subscript s denoting the solid, is rewritten here in the form:

$$\theta_\alpha\rho_\alpha\frac{D\mathbf{V}_\alpha}{Dt} = \nabla\cdot\theta_\alpha\boldsymbol{\sigma}_\alpha + \frac{1}{\mathbb{V}_o}\int_{\mathcal{S}_{\alpha\beta}} \boldsymbol{\sigma}_\alpha\cdot\boldsymbol{\nu}_\alpha\, d\mathcal{S} + \theta_\alpha\rho_\alpha\mathbf{F}, \quad \alpha,\beta = s, w, n, \tag{3.3.11}$$

where we recall that ρ_α and $\mathbf{V}_\alpha$ represent intrinsic phase averages.

By writing (3.3.11) for each of the three phases, we obtain:

$$\begin{aligned}
\theta_n\rho_n\frac{D\mathbf{V}_n}{Dt} &= \nabla\cdot\theta_n\boldsymbol{\sigma}_n + \theta_n\rho_n\mathbf{F}_n + \frac{1}{\mathbb{V}_o}\int_{\mathcal{S}_{ns}+\mathcal{S}_{nw}} \boldsymbol{\sigma}_n\cdot\boldsymbol{\nu}_n d\mathcal{S},\\
\theta_w\rho_w\frac{D\mathbf{V}_w}{Dt} &= \nabla\cdot\theta_w\boldsymbol{\sigma}_w + \theta_w\rho_w\mathbf{F}_w + \frac{1}{\mathbb{V}_o}\int_{\mathcal{S}_{ws}+\mathcal{S}_{wn}} \boldsymbol{\sigma}_w\cdot\boldsymbol{\nu}_w d\mathcal{S},\\
\theta_s\rho_s\frac{D\mathbf{V}_s}{Dt} &= \nabla\cdot\theta_s\boldsymbol{\sigma}_s + \theta_s\rho_s\mathbf{F}_s + \frac{1}{\mathbb{V}_o}\int_{\mathcal{S}_{sn}+\mathcal{S}_{sw}} \boldsymbol{\sigma}_s\cdot\boldsymbol{\nu}_s d\mathcal{S}.
\end{aligned} \tag{3.3.12}$$

We note that the outward normals are such that $\boldsymbol{\nu}_n \equiv -\boldsymbol{\nu}_s$ on $\mathcal{S}_{ns}$, $\boldsymbol{\nu}_s \equiv -\boldsymbol{\nu}_w$ on $\mathcal{S}_{sw}$ and $\boldsymbol{\nu}_n \equiv -\boldsymbol{\nu}_w$ on $\mathcal{S}_{nw}$. The symbol $\mathrm{D}(\,)/\mathrm{D}t$ denotes the material derivative from the point of view of an observer traveling at the average velocity of the considered phase.

By summing the three equations, we obtain:

$$\begin{aligned}
\sum_{(\alpha=n,w,s)} \theta_\alpha\rho_\alpha\frac{D\mathbf{V}_\alpha}{Dt} &= \nabla\cdot\overline{\boldsymbol{\sigma}} + \overline{\rho\mathbf{F}} + \frac{1}{\mathbb{V}_o}\int_{\mathcal{S}_{nw}} [\![\, \boldsymbol{\sigma} \,]\!]_{n,w}\cdot\boldsymbol{\nu}_n\, d\mathcal{S}\\
&+ \frac{1}{\mathbb{V}_o}\int_{\mathcal{S}_{ws}} [\![\, \boldsymbol{\sigma} \,]\!]_{w,s}\cdot\boldsymbol{\nu}_w\, d\mathcal{S} + \frac{1}{\mathbb{V}_o}\int_{\mathcal{S}_{sn}} [\![\, \boldsymbol{\sigma} \,]\!]_{s,n}\cdot\boldsymbol{\nu}_s\, d\mathcal{S},
\end{aligned} \tag{3.3.13}$$

where the overbar indicates *volume average* as defined by (1.1.12) and we have used the symbol:

$$[\![\, (..) \,]\!]_{1,2} \equiv (..)\big|_{\text{side 1}} - (..)\big|_{\text{side 2}} \tag{3.3.14}$$

to denote the jump from side 1 to side 2 across the microscopic interface. By (1.1.12),

$$\begin{aligned} \overline{\boldsymbol{\sigma}} &= \frac{1}{\mathbb{V}_o} \int_{\mathbb{V}_o} \boldsymbol{\sigma} d\mathbb{V} = \frac{1}{\mathbb{V}_o} \sum_{(\alpha=s,n,w)} \int_{\mathbb{V}_{o\alpha}} \boldsymbol{\sigma}_\alpha d\mathbb{V}_\alpha \\ &= \sum_{(\alpha=s,n,w)} \overline{\boldsymbol{\sigma}_\alpha} \equiv \overline{\boldsymbol{\sigma}_s} + \overline{\boldsymbol{\sigma}_n} + \overline{\boldsymbol{\sigma}_w} \end{aligned} \tag{3.3.15}$$

defines the volume averaged stress, or *total stress*, and

$$\overline{\rho \mathbf{F}} = \overline{\rho_n \mathbf{F}_n} + \overline{\rho_w \mathbf{F}_w} + \overline{\rho_s \mathbf{F}_s} \tag{3.3.16}$$

defines the *total body force* per unit volume of porous medium at a (macroscopic) point.

To facilitate the discussion on fluid stress-strain relationships, we express the fluid's average stress, $\overline{\boldsymbol{\sigma}}$, in the form:

$$\overline{\boldsymbol{\sigma}} = \overline{\boldsymbol{\tau}} - \overline{p}\boldsymbol{\delta}, \tag{3.3.17}$$

where the average *deviator* $\overline{\boldsymbol{\tau}}$ is the *viscous stress tensor*, $\overline{p} = -\frac{1}{3}(\overline{\sigma}_{ii} - \overline{\tau}_{ii}) \equiv -\frac{1}{3}\sum_{(i)}(\overline{\sigma}_{ii} - \overline{\tau}_{ii})$ is the pressure, and $\boldsymbol{\delta}$ (components δ_{ij}) is the Kroenecker delta, defined by:

$$\delta_{ij} = \begin{cases} 1, & \text{when} \quad i = j, \\ 0, & \text{when} \quad i \neq j. \end{cases} \tag{3.3.18}$$

In (3.3.17), the pressure $\overline{p}$ is considered *positive for compression*, while $\overline{\sigma}_{ij}$ and $\overline{\tau}_{ij}$ are considered *positive for tension*.

Using (3.3.17), the total stress in the two phase flow considered here can also be expressed by:

$$\overline{\boldsymbol{\sigma}} = \overline{\boldsymbol{\sigma}_s} + \overline{\boldsymbol{\tau}_n} + \overline{\boldsymbol{\tau}_w} - \overline{p_n}\,\mathbf{I} - \overline{p_w}\,\mathbf{I}. \tag{3.3.19}$$

By examining (3.3.13), we note that the interaction between the phases is accounted for by the three surface integrals. The first integral on the r.h.s. of (3.3.13) describes the interaction across the (microscopic) interface between the nonwetting and wetting fluid phases. In principle, similar interactions also occur at the solid-fluid interfaces, i.e., the second and third integrals. On the other hand, we usually assume continuity of traction, i.e., $[\boldsymbol{\sigma}]_{w,s} \cdot \boldsymbol{\nu}_w = 0$ and $[\boldsymbol{\sigma}]_{s,n} \cdot \boldsymbol{\nu}_s = 0$, and neglect *surface tension* phenomena at fluid-solid interfaces. Hence, the last two surface integrals in (3.3.13) vanish. Surface tension is discussed in Sect. 2.4.1.

Back to the first integral, actually, at every point on the microscopic interface between two immiscible fluids (here the wetting and the nonwetting ones), overlooking their molecular structure, and regarding them as two continua separated by a sharp interface, the following equation expresses *continuity of momentum transfer* (Landau and Lifshitz 1960):

$$[\rho V_i(V_j - u_j) - \sigma_{ij}]_{n,w}\nu_j = \left(\frac{1}{r'} + \frac{1}{r''}\right)\gamma_{wn}\nu_i + \frac{\partial \gamma_{wn}}{\partial x_i}$$
$$= \frac{2}{r^*}\gamma_{wn}\nu_i + \frac{\partial \gamma_{wn}}{\partial x_i}, \tag{3.3.20}$$

where γ_{wn} is the magnitude of the *surface-tension* between the wetting and the non-wetting fluids. This is a concept that introduces the molecular level effects between the two fluids in the form of a force (per unit length) that is tangent to the interface, and r^* is the mean radius of curvature of the latter, with r' and r'' denoting its principal radii of curvature. The l.h.s. of (3.3.20) expresses a jump in the component in the ith direction of the total momentum flux. The r.h.s. may be interpreted as the rate of production of linear momentum per unit area of the interface. In this equation, γ_{nw} may be nonuniform, e.g., because of impurities and temperature variations.

Note that because γ_{wn} exists only in the interphase surface, the gradient of γ_{wn} in (3.3.20) should be interpreted as:

$$\frac{\partial \gamma_{wn}}{\partial x_i} \equiv |\nabla \gamma_{wn}| t_i,$$

in which $t_i = \cos(\nabla\gamma_{wn}, \mathbf{1x}_i)$.

When the $n-w$-interface is a material surface with respect to fluid mass, the advective momentum flux vanishes, i.e., $[\rho\mathbf{V}(\mathbf{V}-\mathbf{u})]_{n,w}\cdot\boldsymbol{\nu}_n = 0$. For a stationary fluid, the viscous stress, $\boldsymbol{\tau}(= \boldsymbol{\sigma} + p\mathbf{I})$, vanishes and the jump, $-[\boldsymbol{\sigma}]_{n,w}\cdot\boldsymbol{\nu}_n$, reduces to $[p]_{n,w}\boldsymbol{\nu}_n$. The same conclusion can be obtained if we assume, as an approximation, that the component of the shear force normal to the interface is much smaller than the force due to pressure, i.e.,

$$|\tau_{ij}\nu_j\nu_i| \ll |p|.$$

If also $\nabla\gamma_{wn} = 0$, Eq. (3.3.20) reduces to:

$$p_n - p_w = \left(\frac{1}{r'} + \frac{1}{r''}\right)\gamma_{wn}, \tag{3.3.21}$$

known as the *Laplace formula*. Since $(p_n - p_w) > 0$, the *pressure is greater in the nonwetting fluid for which the surface is convex.* The difference $p_n - p_w$, called (microscopic) *capillary pressure*, was introduced in Sect. 2.4.3. In Sect. 6.1, we shall introduce the pressure difference at the macroscopic level, $p_c = p_n - p_w = p_c(\theta_w)$ referred to as *macroscopic capillary pressure*. Note that $\overline{p_n}^n \equiv p_n, \overline{p_w}^w \equiv p_w$, recalling that, in general, we are not using the averaging symbols whenever, it is obvious that an equation is at the macroscopic level.

Without fluid traction at the fluid solid-interfaces, (3.3.13) reduces to:

$$\sum_{(\alpha=n,w,s)} \theta_\alpha \rho_\alpha \frac{D\mathbf{V}_\alpha}{Dt} = \nabla\cdot\overline{\boldsymbol{\sigma}} + \boldsymbol{\mathcal{F}}_c + \overline{\rho\mathbf{F}}, \tag{3.3.22}$$

where we recall that ρ_α and $\mathbf{V}_\alpha$ represent intrinsic phase averages, and

$$\boldsymbol{\mathcal{F}}_c = \frac{1}{\mathbb{V}_o}\int_{\mathcal{S}_{nw}} [\boldsymbol{\sigma}]_{n,w}\cdot\boldsymbol{\nu}_n \, d\mathcal{S}. \tag{3.3.23}$$

Without inertial forces, the averaged momentum balance (3.3.13) reduces to:

$$\nabla\cdot\overline{\boldsymbol{\sigma}} + \overline{\rho\mathbf{F}} + \boldsymbol{\mathcal{F}}_c = 0. \tag{3.3.24}$$

In the case of a single fluid phase, or, in the case of multiphase flow, neglecting $\boldsymbol{\mathcal{F}}_c$ as an approximation, (3.3.24) reduces to:

$$\nabla\cdot\overline{\boldsymbol{\sigma}} + \overline{\rho\mathbf{F}} = 0, \tag{3.3.25}$$

in which we note the phase average. The above equation is known as the *equilibrium equation* (see Biot 1941; Verruijt 1969). It relates changes in total stress to body force. In Sect. 9.2, we shall make use of this equation as the starting point for modeling deformation in a porous medium.

E. Energy Balance Equation

For $E = \mathbb{E}$, representing the energy of a fluid that occupies the entire void space, and $e = u + \frac{1}{2}V^2, e' = u' + \frac{1}{2}\rho V^2 = \rho(u + \frac{1}{2}V^2)$, i.e., the sum of the internal energy and the kinetic energy of the fluid, per unit fluid volume, or per unit fluid mass, but not potential energy, the macroscopic energy balance equation (3.3.4), written for energy in terms of u', takes the form:

$$\frac{\partial}{\partial t}\left[\phi\left(u' + \tfrac{1}{2}\rho V^2\right)\right] = -\nabla\cdot\phi\mathbf{J}^{\mathbb{E}}_{tot} + f^{\mathbb{E}}_{s\to f} + \phi\rho\Gamma^{\mathbb{E}}, \tag{3.3.26}$$

or, in terms of u,

$$\frac{\partial}{\partial t}\left[\phi\left[\rho(u + \tfrac{1}{2}V^2)\right]\right] = -\nabla\cdot\phi\mathbf{J}^{\mathbb{E}}_{tot} + f^{\mathbb{E}}_{s\to f} + \phi\rho\Gamma^{\mathbb{E}}, \tag{3.3.27}$$

in which $\phi\mathbf{J}^{\mathbb{E}}_{tot}$ denotes the total energy flux carried *in and by* the fluid, $\phi\rho\Gamma^{\mathbb{E}}$ denotes the source of energy in the fluid, both per unit volume of porous medium and per unit time, from both external and internal sources, and $f^{\mathbb{E}}_{s\to f}$, defined by (3.3.5), denotes the energy transferred from the solid to the fluid across their common interface. Note that the *potential energy* is not included as we assume that gravity is the only body force.

The flux $\mathbf{J}^{\mathbb{E}}_{tot}$, the rate of interphase transfer, $f^{\mathbb{E}}_{s\to f}$, and the sources, $\phi\rho\boldsymbol{\Gamma}^{\mathbb{E}}$, will be discussed in Chap. 8.

Usually, we are interested in the energy balance equation for the porous medium as a whole under the assumption of *thermal equilibrium*. We can then write (3.3.26) once for the fluid (f) and once for the solid (s) and sum up the two equations. The total interphase energy transfer term will vanish, and we obtain:

$$\frac{\partial}{\partial t}\sum_{(\alpha=f,s)}\left[\theta_\alpha\left(\rho_\alpha(u_\alpha+\tfrac{1}{2}V^2_\alpha)\right)\right]=-\nabla\cdot\mathbf{J}^{\mathbb{E}}_{pm,tot}+\sum_{(\alpha=f,s)}\theta_\alpha\rho_\alpha\Gamma^{\mathbb{E}}_\alpha, \tag{3.3.28}$$

in which $\mathbf{J}^{\mathbb{E}}_{pm,tot}=\sum_{(\alpha=f,s)}\theta_\alpha\mathbf{J}^{\mathbb{E}}_\alpha$ denotes the total energy flux, i.e., due to advection, dispersion, and diffusion, through the porous medium as a whole.

* * *

All (macroscopic) balance equations in this section were obtained strictly by phenomenological considerations, without resorting to any formal upscaling technique. They contain fluxes, rates of interphase transfers and sources of the considered E's, which will be discussed in subsequent chapters. Obviously, we could have obtained the same equations also by appropriate averaging (over an REV) of the microscopic level balance equations. Examples of the averaging approach are presented, among others, by Bear and Bachmat (1991, see also Sect. 1.4.2 A), Whitaker (1999, see also Sect. 1.4.2 B), Hassanizadeh and Gray (1979a, b, 1980, see also Sect. 1.4.2 C), and Gray and Miller (2014).

3.4 E-Fluxes

E-fluxes appear in all E-balance equations, both at the microscopic level (Sect. 3.2) and at the macroscopic one (Sect. 3.3).

3.4.1 *Microscopic Advective and Diffusive Fluxes*

At the microscopic level of description, the *total flux* of an extensive quantity E is defined as $\mathbf{j}^E_{tot}=e'\mathbf{V}^E$. This flux, within a phase, is the sum of two fluxes:

$$\mathbf{j}^E_{tot}=e'\mathbf{V}^E=e'\mathbf{V}+e'(\mathbf{V}^E-\mathbf{V})=\mathbf{j}^E_{adv}+\mathbf{j}^E_{dif}, \tag{3.4.1}$$

i.e., the sum of an *advective flux*, $\mathbf{j}^E_{adv}=e'\mathbf{V}$, that carries E at the fluid's velocity, $\mathbf{V}$, and a *diffusive flux*, $\mathbf{j}^E_{dif}=e'(\mathbf{V}^E-\mathbf{V})$, relative to the advective one, resulting from the molecular (Brownian) motion within the fluid phase.

Or, in terms of the specific value of E:

$$\mathbf{j}^{E}_{tot} = \rho e \mathbf{V}^{E} = \rho e \mathbf{V} + \rho e(\mathbf{V}^{E} - \mathbf{V}) = \mathbf{j}^{E}_{adv} + \mathbf{j}^{E}_{dif}, \tag{3.4.2}$$

A. Advective Flux of Mass of a Fluid Phase

In this case, $E = m$, $e' = \rho$, $e = 1$, and the (microscopic level) flux is

$$\mathbf{j}^{m}_{adv} = \rho \mathbf{V}. \tag{3.4.3}$$

We recall that $\rho\mathbf{V}$ is also the fluid's momentum per unit mass.

B. Advective Flux of Mass of a γ-Species

In this case, $E = m^{\gamma}$, $e' = \rho^{\gamma} = \rho\omega^{\gamma}$, $e = \omega^{\gamma}$, and we have,

$$\mathbf{j}^{\gamma}_{adv} = \rho\omega^{\gamma}\mathbf{V}. \tag{3.4.4}$$

C. Advective Flux of Momentum of a Phase

Here, $e' = \rho\mathbf{V}$, $e = \mathbf{V}$, and

$$\mathbf{j}^{\mathbf{M}}_{adv} = \rho\mathbf{V}\mathbf{V}, \tag{3.4.5}$$

where $\mathbf{VV}$ is the *dyadic product* of the two vectors. Recall that a dyadic product of two vectors is a 2nd rank symmetric tensor, with components $V_i V_j$.

D. Advective Flux of Energy of a Phase

In this case, $e = \left(u + \frac{1}{2}V^2\right)$, $e' = \rho\left(u + \frac{1}{2}V^2\right) = \left(u' + \frac{1}{2}\rho V^2\right)$, and

$$\mathbf{j}^{\mathbb{E}}_{adv} = \rho\left(u + \tfrac{1}{2}V^2\right)\mathbf{V}. \tag{3.4.6}$$

E. Diffusive Flux of a γ-Species in a Fluid Phase

Consider a fluid containing only two species: γ and δ. The *diffusive mass flux* (mass of γ per init area per unit time) is expressed by *Fick's law* of molecular diffusion, in the form:

$$\mathbf{j}^{\gamma}_{dif} = -\rho\mathcal{D}^{\gamma\delta}\nabla\omega^{\gamma}, \quad \omega^{\gamma} = \frac{\rho^{\gamma}}{\rho}, \quad \sum_{(\gamma)}\mathbf{j}^{\gamma}_{dif} = 0, \tag{3.4.7}$$

where ω^{γ} is the mass fraction of the γ-species, and the scalar $\mathcal{D}^{\gamma\delta}$ is the coefficient of molecular diffusion (dims. L^2/T) of the γ-species in a fluid that contains only two species, γ and δ. The symbol c^{γ} is often used instead of ρ^{γ}.

The diffusive flux of the other species, δ, is given by $\mathbf{j}^{\delta}_{dif} = -\rho\mathcal{D}^{\delta\gamma}\nabla\omega^{\delta}$, such that $\mathbf{j}^{\gamma}_{dif} + \mathbf{j}^{\delta}_{dif} = 0$, implying that $\mathcal{D}^{\gamma\delta} = \mathcal{D}^{\delta\gamma}$. It is usually assumed that $\mathcal{D}^{\gamma\delta}$ is independent of c^{γ}. However, in general, it is a function of pressure and temperature.

When $\nabla\rho = 0$, i.e., in a homogeneous fluid, or when $\rho\nabla\omega^{\gamma} \approx \nabla\rho^{\gamma}$, i.e., $|\omega^{\gamma}\nabla\rho| \ll |\rho\nabla\omega^{\gamma}|$, we may write Fick's law in terms of ρ^{γ}, as:

$$\mathbf{j}^{\gamma}_{dif} = -\mathcal{D}^{\gamma\delta}\nabla\rho^{\gamma}. \tag{3.4.8}$$

In Chap. 7, we shall consider transport with chemical reactions among dissolved species. Under such conditions, it is more convenient to express the concentration of a γ-species, say in an α-phase, in terms of molar fraction, X^{γ}_{α}. Thus, the advective flux is:

$$\mathbf{J}^{\gamma,mol}_{\alpha,adv} = \phi S_{\alpha}\eta_{\alpha}X^{\gamma}_{\alpha}\mathbf{V}_{\alpha}. \tag{3.4.9}$$

and the diffusive flux (i.e., Fick's law) in terms of X^{γ}_{α} is:

$$\mathbf{j}^{\gamma,mol}_{\alpha,dif} = -\eta_{\alpha}\mathcal{D}^{\gamma}_{\alpha,dif}\nabla X^{\gamma}_{\alpha}. \tag{3.4.10}$$

η_{α} and X^{γ}_{α} are defined in Sect. 7.1.1.

The term *Knudsen flow*, or *Knudsen diffusion* (Knudsen 1934; Malek and Coppens 2003) is used to describe the diffusion of a gas when the *mean free path* of its molecules is not much larger than the characteristic dimension of the flow domain, here the size of a pore.

The *Knudsen number*, Kn, defined as:

$$Kn = \frac{\lambda^{kn}}{\ell_{pm}}, \tag{3.4.11}$$

in which λ^{kn} denotes the *mean free path of the molecules*, and ℓ_{pm} is the characteristic length dimension of the void space, is used to identify the regime in which the Knudsen effect occurs.

Typically, Knudsen diffusion occurs when the pore diameter is in the range 2–50 nm. According to Chambre and Schaaf (1961), Knudsen diffusion occurs when $K_n > 10$, as then collisions between the gas molecules and the pore wall dominates. The range $0.1 \leq K_n \leq 10$, is regarded as transition regime. For $10^{-3} \leq K_n \leq 0.1$, the flow at the wall cannot be neglected, which means slip flow exists and Klinkenberg effects have to be taken into account. For $K_n < 10^{-3}$, collisions between gas molecules and the pore wall can be neglected, and the flow behavior can be described by Darcy's Law (Chap. 4), without any correction.

In the above expressions for the flux of molecular diffusion, we have not taken into account the *Soret effect* presented in Sect. 2.6 according to which we also have molecular diffusion as a result of a temperature gradient.

More on diffusion of a chemical species and Fick's law at the macroscopic level is presented in Sect. 7.2.2.

F. Diffusive Flux of Momentum of a Fluid Phase

From the momentum balance equation (3.2.23), it follows that the *viscous stress tensor*, τ_{ij}, represents the *diffusive flux of linear momentum* across a material surface element, due to velocity gradient *across* the element. As such, it acts as a frictional force between adjacent layers of the fluid moving at different velocities, per unit area.

Within the framework of the same concept, we obtain the transfer of momentum between the fluid and the solid, assuming that the fluid sticks to the solid surface.

A typical component of the velocity gradient, relevant to τ_{ij}, is $\partial V_i/\partial x_j$. Therefore, it is generally assumed that for fluids, $\tau_{ij} = \tau_{ij}(\dot{\varepsilon_{k\ell}})$, where $\dot{\varepsilon}_{k\ell}$ represents the *rate of strain* defined by:

$$\dot{\varepsilon}_{k\ell} = \frac{1}{2}\left(\frac{\partial V_k}{\partial x_\ell} + \frac{\partial V_\ell}{\partial x_k}\right). \tag{3.4.12}$$

For the sake of clarity, we are using here indicial notation.

For sufficiently small values of $\partial V_i/\partial x_j$, the diffusive flux of momentum is expressed by:

$$\tau_{ij} = C^v_{ijk\ell}\,\dot{\varepsilon}_{k\ell}, \tag{3.4.13}$$

where $C^v_{ijk\ell}$ (a fourth rank tensor) denotes the fluid's *viscosity coefficient*. The linear relationship (3.4.13) is called *Newton's law*, and the fluid that obeys this law is called a *Newtonian fluid*. Some introductory remarks about tensors are presented in Sect. 2.3.4.

When the viscous stress generated in a fluid element is independent of the orientation of the latter, i.e., when the molecular structure of the fluid is statistically isotropic (which is the case of all gases and simple liquids, but unlike suspensions and solutions that contain very large molecular chains), the coefficient $C^v_{ijk\ell}$ is an *isotropic tensor* that can be written in the form:

$$C^v_{ijk\ell} = b^v(\delta_{ik}\delta_{j\ell} + \delta_{i\ell}\delta_{jk}) + b^{vv}\delta_{ij}\delta_{k\ell}, \tag{3.4.14}$$

where b^v and b^{vv} are scalar coefficients. Hence, by inserting (3.4.14) into (3.4.13), we obtain the following expression for the diffusive flux of an *isotropic single species Newtonian fluid*,

$$\tau_{ij} = 2\mu\dot{\varepsilon}_{ij} + \lambda^{vv}\dot{\varepsilon}_{kk}\delta_{ij}, \qquad \dot{\varepsilon}_{kk} \equiv \frac{\partial V_k}{\partial x_k}, \qquad \tau_{ij} = \tau_{ji}, \tag{3.4.15}$$

with

$$p = -\tfrac{1}{3}\sigma_{ii} + \left(\lambda^{vv} + \tfrac{2}{3}\mu\right)\dot{\varepsilon}_{kk}, \quad \sigma_{ij} = \tau_{ij} - p\delta_{ij}, \tag{3.4.16}$$

where $\mu = b^v$ is the fluid's *dynamic viscosity* and $\lambda^{vv} = b^{vv}$ is another viscosity coefficient of the fluid. In (3.4.16) p is the thermodynamic pressure and $\dot{\varepsilon}_{ij}$ denotes the rate of deformation. Water is a typical example of a Newtonian fluid that obeys (3.4.15).

For an incompressible Newtonian isotropic single species fluid under conditions of isochoric isothermal mass flow, since $\dot{\varepsilon}_{kk} \equiv \partial V_k/\partial x_k = 0$, (3.4.15) reduces to:

$$\tau_{ij} = 2\mu\dot{\varepsilon}_{ij} = \mu\left(\frac{\partial V_i}{\partial x_j} + \frac{\partial V_j}{\partial x_i}\right), \quad \text{with} \quad \sigma_{ij} = \tau_{ij} - p\delta_{ij}. \tag{3.4.17}$$

Then, the *pressure* p in (3.4.16) reduces to $p = -\frac{1}{3}\sum_{i=1,2,3}\sigma_{ii}$ and is called the *mean normal stress*, or *mechanical pressure*. Equation (3.4.17) is often called the *generalized Newton law*, describing the molecular (= diffusive) flux of linear momentum.

For a fluid at rest, or in uniform flow, $\mathbf{V} = const.$, we always have $\tau_{ij} = 0$, and $\sigma_{ij} = -p\delta_{ij}$. In this case, the mean negative normal stress, p, is the *hydrostatic pressure*.

A *perfect fluid* is a *non-viscous* (or *inviscid*) fluid that cannot sustain a viscous stress, i.e.,

$$\tau_{ij} = 0, \qquad \text{and} \qquad \sigma_{ij} = -p\delta_{ij}. \tag{3.4.18}$$

G. Diffusive Flux of Energy of a Fluid Phase

The diffusive flux of internal energy, $\mathbf{j}^{\mathrm{U}}_{dif}$, is identical to the heat flux, $\mathbf{j}^{\mathrm{H}}$, which is, usually, referred to as *heat conduction*. This (microscopic) flux is expressed by *Fourier's law*,

$$\mathbf{j}^{\mathrm{H}} = -\lambda\nabla T, \tag{3.4.19}$$

in which λ and T denote the *thermal conductivity* and the temperature of the phase, respectively.

In the above expressions for heat flux, we have not taken into account the *Dufour effect*, presented in Sect. 2.6, according to which we also have heat flux as a result of a concentration gradient.

More on Fourier's law is presented in Sect. 8.1.1 A.

3.4.2 Macroscopic Advective and Diffusive Fluxes

At the macroscopic level, with $\mathbf{V}$ denoting the fluid's mass-averaged velocity, the advective flux of any E in a porous medium domain is associated with the movement of the fluid phase through the void space. The diffusive fluxes are associated with the molecular motion within the fluid that occupies the void space.

As defined earlier, the (macroscopic) flux, $\mathbf{J}^E$, expresses the amount of E passing through a *unit area of fluid* in a planar cross-section through the porous medium domain. The amount per unit area of the porous medium cross-section is given by $\phi^A\mathbf{J}^E$ ($\cong \phi\mathbf{J}^E$). We have been avoiding the use of the specific discharge, $\mathbf{q}^E (= \phi\mathbf{V}^E)$ for the various extensive quantities, as the porosity may vary spatially and with time.

A. Advective Flux of the Mass of a Fluid α-Phase

In this case,

$$e = 1, \quad e' = \rho_\alpha, \quad \mathbf{J}^m_{\alpha,adv} = \rho_\alpha\mathbf{V}_\alpha. \tag{3.4.20}$$

We recall that $\mathbf{V}$ is interpreted as *momentum per unit fluid mass*. It is, thus, a state variable like ρ and ρ^γ, and we need to include the momentum balance equation in the set of equations to be solved for these variables.

B. Advective Mass Flux of a γ-Chemical Species in an α-Fluid

The advective flux of the mass of a γ-chemical species carried with the α-fluid moving at the velocity $\mathbf{V}_\alpha$, with $e_\alpha^\gamma = \omega^\gamma$, $e_\alpha^\gamma = \rho_\alpha \omega_\alpha^\gamma = \rho_\alpha^\gamma$, is

$$\mathbf{J}^\gamma_{\alpha,adv} = \rho^\gamma_\alpha \mathbf{V}_\alpha = \rho_\alpha \omega^\gamma_\alpha \mathbf{V}_\alpha. \tag{3.4.21}$$

C. Advective Flux of Momentum of a Fluid α-Phase

Here, the advective flux of momentum of the α-fluid has the form

$$e'^{\mathbf{M}}_\alpha = \rho_\alpha \mathbf{V}_\alpha, \quad e^{\mathbf{M}}_\alpha = \mathbf{V}_\alpha, \quad \mathbf{J}^{\mathbf{M}}_{\alpha,adv} = \rho_\alpha \mathbf{V}_\alpha \mathbf{V}_\alpha. \tag{3.4.22}$$

D. Advective Flux of the Energy of a Fluid α-Phase

For a moving α-fluid, the advective energy flux is given by

$$\mathbf{J}^{\mathbb{E}}_{\alpha,adv} = \rho_\alpha \left(u_\alpha + \tfrac{1}{2} V_\alpha^2\right) \mathbf{V}_\alpha. \tag{3.4.23}$$

E. Diffusive Mass Flux of a Fluid α-Phase

For this case,

$$e = 1, \quad \mathbf{J}^m_{\alpha,dif} = \rho_\alpha (\mathbf{V}_\alpha - \mathbf{V}_\alpha) \equiv 0. \tag{3.4.24}$$

i.e., there is no diffusion of the mass of a fluid phase.

F. Diffusive Flux of the Mass of a γ-Species in an α-Fluid

Phenomenologically, in analogy to Fick's law in a fluid continuum, with a driving force which is the gradient of the (macroscopic) γ-species concentration, expressed by the mass fraction of the species, $\omega^\gamma_\alpha (\equiv \rho^\gamma_\alpha / \rho_\alpha)$, the macroscopic diffusive flux of a γ-species takes the form:

$$\mathbf{J}^\gamma_{\alpha,dif} = -\rho_\alpha \boldsymbol{D}_{\alpha,pm} \nabla \omega^\gamma_\alpha, \quad \boldsymbol{D}_{\alpha,pm} = \mathbf{T}^*_\alpha \mathcal{D}, \quad \mathbf{T}^*_\alpha = \mathbf{T}^*_\alpha (S_\alpha), \tag{3.4.25}$$

in which $\boldsymbol{D}_{pm}$ denotes the *macroscopic coefficient of molecular diffusion*. It is a second rank symmetric tensor expressed as the product of the scalar molecular diffusivity in a fluid continuum, $\mathcal{D}$, and a geometrical property of the void space, $\mathbf{T}^*$, called *tortuosity*, which is a second rank symmetric tensor. This property expresses the fact that the actual path of extensive quantities (here the mass of a chemical species) by advection along (microscopic) stream-tubes is longer than the length between macroscopic points in the porous medium domain. This tensorial tortuosity exists also in anisotropic porous media. A detailed discussion on tortuosity is presented in Sect. 4.2.5. Note that we indicated that in multiphase flow, the phase tortuosity is a function of phase saturation, S_α.

More on molecular diffusion at the macroscopic level, when mass is measured in moles, is presented in Sect. 7.2.2.

G. Diffusive Flux of Momentum of a Fluid α-Phase

Here, $e' = \rho_\alpha \mathbf{V}_\alpha$, $e = \mathbf{V}_\alpha$, and the macroscopic diffusive momentum flux is:

$$\mathbf{J}^{\mathrm{M}}_{\alpha,dif} = \rho_\alpha \mathbf{V}_\alpha (\mathbf{V}^{\mathrm{M}}_\alpha - \mathbf{V}_\alpha) = -\boldsymbol{\sigma}_\alpha, \tag{3.4.26}$$

in which $\boldsymbol{\sigma}_\alpha$ ($\equiv \boldsymbol{\tau}_\alpha - p_\alpha \boldsymbol{\delta}$) denotes the *shear*, or *deviatoric stress*, in the fluid phase, $\boldsymbol{\sigma}_\alpha$ is the stress in the fluid phase, p_α is pressure and $\boldsymbol{\delta}$ denotes the *unit tensor*.

In the momentum balance equation (3.3.9), we can express the first term on the r.h.s. in the form:

$$\nabla\cdot\phi \mathbf{J}^{\mathrm{M}}_{adv+dif} \equiv \nabla\cdot\phi\,(\rho\mathbf{V}\mathbf{V} - \boldsymbol{\tau}) + \nabla\cdot(\phi p \boldsymbol{\delta}), \tag{3.4.27}$$

in which $-\phi\boldsymbol{\tau}$ expresses the (macroscopic) *diffusive flux of momentum*, since only $\boldsymbol{\tau}$ is contributing to the dissipation of energy.

H. Diffusive Energy Flux in a Fluid α-Phase

In a fluid α-phase that occupies the entire void space, $e'_\alpha = (u'_\alpha + \frac{1}{2}\rho_\alpha V^2_\alpha)$, $e_\alpha = (u_\alpha + \frac{1}{2} V^2_\alpha)$, and the diffusive flux is:

$$\mathbf{J}^{\mathbb{E}}_{\alpha,dif} = \rho_\alpha \left(u_\alpha + \tfrac{1}{2} V^2_\alpha\right) (\mathbf{V}^{\mathbb{E}}_\alpha - \mathbf{V}_\alpha). \tag{3.4.28}$$

At the microscopic level, i.e., in a phase continuum, the *diffusive* ($\equiv$ *conductive*) energy flux is expressed by *Fourier's law*, $\mathbf{j}^{\mathbb{E}}_{\alpha,\,dif} \equiv \rho_\alpha e^{\mathbb{E}}_\alpha (\mathbf{V}^{\mathbb{E}}_\alpha - \mathbf{V}) = -\lambda \nabla T$, in which T denotes the temperature, and λ denotes the thermal conductivity of the phase. By the phenomenological approach, at the macroscopic level, in the fluid, the diffusive energy flux, driven by ∇T, takes the form:

$$\mathbf{J}^{\mathbb{E}}_{\alpha,dif} = -\boldsymbol{\lambda}^*_\alpha \cdot \nabla T, \tag{3.4.29}$$

in which the $\boldsymbol{\lambda}^*_\alpha$ denotes the macroscopic thermal conductivity within the fluid α-phase inside the void space.

For the porous medium as a whole, the diffusive energy flux is expressed by:

$$\mathbf{J}^{\mathbb{E}}_{pm,dif} = -\boldsymbol{\lambda}^*_{pm} \cdot \nabla T, \quad \boldsymbol{\lambda}_{pm} = \phi \lambda^*_f + (1-\phi)\lambda^*_s, \tag{3.4.30}$$

in which $\boldsymbol{\lambda}_{pm}$ denotes the thermal conductivity for the porous medium (pm) as a whole, i.e., through the *composite material* composed of the (assume thermally conducting) solid matrix and the (thermally conducting) fluid occupying the void space. Thus, $\boldsymbol{\lambda}_{pm}$ depends on both λ_f and λ_s and on the configuration of the two phases in the porous medium domain.

In two-phase flow, the combined porous medium conductivity, $\boldsymbol{\lambda}_{pm}$, is a complicated function of conductivities and volume fractions of the participating phases, as heat streamlines (see Sect. 3.1.3) may pass through all phases.

* * *

Altogether, we have 3 driving forces: $\nabla\omega^\gamma$ for the diffusive mass of a species, $\nabla\mathbf{V} + (\nabla\mathbf{V})^T$ (see below) for the diffusive flux of momentum, and ∇T for the diffusive flux of energy. In this book, we do not take into account coupled processes (in the Onsager sense, e.g., de Groot and Mazur, 1962, p. 30), although, for the sake of a complete presentation, the topic is presented in Sect. 2.6.

3.4.3 Dispersive Fluxes

A. The Need for a Dispersive Flux

Around the 1950s, mainly in dealing with the quality of groundwater in aquifers, as associated with the movement, accumulation and transformation of dissolved chemical species, and, later, with ground water contamination, it was noted, in field and laboratory experiments (e.g., summary in Bear 1972, Sect. 10.3), that a dissolved species is transported in a porous medium domain, both in the general direction of the flow and also normal to it, *in a way that could not be explained merely by the movement of the fluid at its average velocity, (i.e., advective flux at a velocity described by Darcy's law (Chap. 4)) and by molecular diffusion*. This spreading phenomenon was called *dispersion*. To bridge the discrepancy, an additional flux was introduced *at the macroscopic level*–the *dispersive flux*. It became obvious that this additional flux is not a real flux, like the advective and diffusive fluxes discussed above; *it does not exist at the microscopic level! It is a flux that is added in order to compensate for the fact that the advective transport of the solute at the macroscopic level is described in terms of the volume (or mass) averaged velocity* (and this was done because Darcy's law provides only this velocity), *while, actually, within the void space, the solute is transported at every (microscopic) point by the* local microscopic velocity. All this is in addition to the flux due to molecular diffusion, which is also included in the macroscopic flux. Altogether, the *total macroscopic flux* of a solute is expressed as the sum:

$$\mathbf{J}^{m^\gamma}_{tot} = \mathbf{J}^{m^\gamma}_{adv} + \mathbf{J}^{m^\gamma}_{dis} + \mathbf{J}^{m^\gamma}_{dif}, \tag{3.4.31}$$

i.e., advection, dispersion and diffusion.

Following is a simple way to understand the need for introducing the dispersive flux of any extensive quantity E of a fluid α-phase. This flux is a consequence of the fact that both the fluid's velocity and the intensive quantity, e (e.g., concentration of a solute, when E is the mass of the solute), vary from point to point within a fluid phase that occupies the entire void space, or part of it.

The advective flux of E ($= E$ per unit area of fluid) at a (microscopic) point, $\boldsymbol{\xi}$, within a fluid phase (f) that occupies part of the void space within an REV centered at a point $\mathbf{x}$, is given by $e'(\boldsymbol{\xi}, t; \mathbf{x})\mathbf{V}(\boldsymbol{\xi}, t; \mathbf{x})$. The intrinsic phase average of this flux is $\overline{e'\mathbf{V}}^f(\mathbf{x}, t)$. In order to express this flux in terms of the average values, $\overline{e'}^f(\mathbf{x}, t)$ and $\overline{\mathbf{V}}^f(\mathbf{x}, t)$, the velocity, $\mathbf{V}(\boldsymbol{\xi}, t; \mathbf{x})$, and the value of $e'(\boldsymbol{\xi}, t; \mathbf{x})$, are decomposed into two parts: an intrinsic phase average value and a deviation from that value:

$$\mathbf{V}(\boldsymbol{\xi}, t; \mathbf{x}) = \overline{\mathbf{V}}^f(\mathbf{x}, t) + \mathring{\mathbf{V}}(\boldsymbol{\xi}, t; \mathbf{x}),$$
$$e(\boldsymbol{\xi}, t; \mathbf{x}) = \overline{e'}^f(\mathbf{x}, t) + \mathring{e}'(\boldsymbol{\xi}, t; \mathbf{x}). \tag{3.4.32}$$

Because an average value is constant over the REV, we have $\overline{\overline{(..)}^f}^f = \overline{(..)}^f$. As a consequence,

$$\overline{\mathring{\mathbf{V}}}^f = 0, \qquad \text{and} \qquad \overline{\mathring{e}}^f = 0. \tag{3.4.33}$$

To obtain the average flux (still per unit area of fluid), we write:

$$\overline{e'\mathbf{V}}^f = \overline{(\overline{e'}^f + \mathring{e}')(\overline{\mathbf{V}}^f + \mathring{\mathbf{V}})}^f = \overline{\overline{e'}^f\overline{\mathbf{V}}^f}^f + \overline{\overline{c}^f\mathring{\mathbf{V}}}^f + \overline{\mathring{e}'\overline{\mathbf{V}}^f}^f + \overline{\mathring{e}'\mathring{\mathbf{V}}}^f. \tag{3.4.34}$$

Because the average of the deviations vanishes, the second and third terms on the right-hand side of (3.4.34) vanish, leaving the relationship:

$$\overline{e'\mathbf{V}}^f = \overline{e'}^f\overline{\mathbf{V}}^f + \overline{\mathring{e}'\mathring{\mathbf{V}}}^f. \tag{3.4.35}$$

Thus, the average (= macroscopic) flux of e' at a point in a porous medium domain (= centroid of an REV) is equal to the sum of two *macroscopic* fluxes:

- An advective flux, $\overline{e'}^f\overline{\mathbf{V}}^f$, expressing the mass of the species carried by the fluid at the latter's average velocity, $\overline{\mathbf{V}}^f$.
- A flux, $\mathbf{J}^E_{dis} \equiv \overline{\mathring{e}'\mathring{\mathbf{V}}}^f$, that results from the variation of e' and $\mathbf{V}$ within the REV for which the point, $\mathbf{x}$, serves as a centroid. This is the flux that produces the (often called mechanical) dispersion of E, with $E = m^\gamma$ as an example. We refer to this flux as the *dispersive flux* of E.

Although we have presented the above discussion for the case of the intensive property e' (= E-density = E per unit phase volume), an analogous development can be written with respect to the specific value, e (= E per unit mass of the phase), and to intrinsic mass averaging, $\widetilde{e}^\alpha$, defined in (1.1.15), with a deviation from the average, $\check{e}_\alpha$, defined by (1.1.16). However, in this case, the microscopic advective is $j^E_\alpha = \rho_\alpha e_\alpha \mathbf{V}_\alpha$, and, following (1.4.68), the intrinsic mass average flux is defined by:

$$\overline{\rho_\alpha e_\alpha \mathbf{V}_\alpha}^\alpha = \overline{\rho_\alpha}^\alpha \widetilde{e_\alpha \mathbf{V}_\alpha}^\alpha = \overline{\rho_\alpha}^\alpha \left(\widetilde{e_\alpha}^\alpha \widetilde{\mathbf{V}_\alpha}^\alpha + \widetilde{\check{e}_\alpha \check{\mathbf{V}}_\alpha}^\alpha \right). \tag{3.4.36}$$

Again, we note here the flux term that expresses the effect of velocity deviations.

Over the years, research, using a variety of models, has led to various expressions that describe the dispersive flux of a solute, $E = m^\gamma$, but not much research efforts have been devoted to the dispersion of momentum and energy, although, in principle, this kind of flux should be present in the respective (macroscopic) balance equation models.

The dispersive flux of the various E's will be considered in more details below.

B. Dispersive Flux of a Solute

Starting with Nikolaevski (1959), Bear (1961), Scheidegger (1961), and others, the most commonly used expression for the dispersive flux in single phase flow has been one in which $\nabla\rho^\gamma$, or $\nabla\omega^\gamma$, serves as the driving force, with a coefficient that is proportional to the fluid's average velocity vector, $\mathbf{V}$ (see Sect. 7.2.3 for more details).

The term 'Fickian expression' is often used when referring to the dispersive flux of a solute which is assumed to be proportional to the gradient of latter's concentration, because of its resemblance to Ficks law for molecular diffusion, i.e., a flux that is driven by the gradient of concentration. Over the years, some researchers, mainly (but not exclusively) on the basis of field observations of plume (of a solute) propagation in aquifers, have proposed various 'non-Fickian' alternative approaches to the determination of plume shape. The main, but not sole reason for the non-Fickian expression for dispersion has been field heterogeneity, although Berkowitz et al. (2008, p. 226) have observed non-Fickian behavior also at laboratory rather short sand columns.

In Sect. 7.2.4 A we introduce an approach to modeling solute transport that is based on Fickian dispersion, with averaging over an RMV (defined in Sect. 1.1.6). In Sect. 7.2.4 B we present a field-scale approach, called *Continuous Time Random Walk*, which takes into account domain heterogeneity, employs a *non-Fickian approach* to dispersion and does not involve volume averaging. In Sect. 7.6, we shall present other approaches to solute transport – upscaling – that is attempting to treat the effects of porous medium heterogeneity as encountered under field conditions.

In what follows, we shall employ the phenomenological approach to develop an expression for the dispersive flux for the mass of a solute, based on the observation that this flux must depend on the fluid's velocity, as *there is no dispersion when* $\mathbf{V} = 0$, and on a *driving force*, associated with the spatial distribution of solute concentration in the fluid phase, as there is no dispersion unless this distribution in not uniform.

Although the discussion in this section starts with the dispersive flux of a solute, the same phenomenon of dispersion and the consequence in the form of a dispersive flux occur also in the cases of transport of momentum and of energy.

Dispersion of a solute and of heat will be considered in detail in Sects. 7.2.3, and 8.3.6, respectively.

3.4.4 Non-advective Fluxes

Within the framework of the phenomenological approach to modeling transport in porous media, we shall now derive expressions for the non-advective fluxes of extensive quantities: the diffusive and dispersive fluxes. We shall start with the flux of a solute. The same approach will then be applied to the fluxes of other extensive quantities.

A. Non-advective Solute Flux

Following Bear and Fel (2012), we make use of the generic terms in the *polynomial representation* of the total (macroscopic) flux of a γ-species in a fluid phase. The total non-advective flux, $\mathbf{J}^{\gamma}_{non-adv}$, of a γ-species is produced by two *independent* factors:

- the fluid phase velocity, $\mathbf{V}$, as there is no dispersion when $\mathbf{V} = 0$, and
- the gradient of the mass fraction $\nabla\omega^{\gamma}$, as there is no solute dispersion when $\omega^{\gamma}(\mathbf{x}) = const.$

The second factor can be expressed in the form of a *driving force* represented by $-\nabla\omega^{\gamma}$. Thus, the flux is a smooth function of these two factors, and of the fluid's density,

$$\mathbf{J}^{\gamma} = \mathbf{J}^{\gamma}(\mathbf{V}, \nabla\rho^{\gamma}), \quad \text{or} \quad \mathbf{J}^{\gamma} = \mathbf{J}^{\gamma}(\rho, \mathbf{V}, \nabla\omega^{\gamma}).$$

By developing this functional relationship in a power series, say, up to third order terms, we obtain:

$$\begin{aligned} J_i^{\gamma}(\rho, \nabla\omega^{\gamma}, \mathbf{V}) = {} & A_{ik}\nabla_k\omega^{\gamma} + B_{ikl}\nabla_k\omega^{\gamma}\nabla_l\omega^{\gamma} + C_{ikl}V_l\nabla_k\omega^{\gamma} \\ & + D_{iklm}\nabla_k\omega^{\gamma}\nabla_l\omega^{\gamma}\nabla_m\omega^{\gamma} + E_{iklm}V_k\nabla_l\omega^{\gamma}\nabla_m\omega^{\gamma} \\ & + G_{iklm}V_lV_m\nabla_k\omega^{\gamma}, \end{aligned} \tag{3.4.37}$$

in which the various coefficients (A_{ik}, B_{ikl}, etc.) are *tensorial* coefficients that are associated with fluid and porous medium properties. The latter represent various characteristics of the geometry of the void space configuration. In multiphase flow, and when (3.4.37) is written separately for each fluid phase, these coefficients depend also on fluid saturations. As everywhere in this book, whenever indicial notation is being used, *Einstein's summation convention* is employed (e.g., $A_{ik}\nabla_k\omega^{\gamma} \equiv \sum_{(k)} A_{ik}\nabla_k\omega^{\gamma} \equiv \sum_{(k)} A_{ik}\partial\omega^{\gamma}/\partial x_k$).

In thermodynamics, the *rate of entropy production*, denoted by $\dot{\mathbb{S}}$ (Sect. 2.2.1) is related to the *thermodynamic driving force*, $\mathbf{X}$, and to the *thermodynamic flux*, $\mathbf{Y}$, by De Groot and Mazur (1962, p. 65):

$$\dot{\mathbb{S}} = Y_iX_i. \tag{3.4.38}$$

Furthermore, the rate of entropy production must be positive, i.e., $\dot{\mathbb{S}} \geq 0$. Here, the solute flux, $\mathbf{J}^{\gamma}$ is driven by $-\nabla\omega^{\gamma}$, which acts as a 'driving force'. Thus, in this case, $\mathbf{X} = -\nabla\omega^{\gamma}$, and its conjugate flux is $\mathbf{Y} = -\mathbf{J}^{\gamma}$. Altogether, we have:

$$\begin{aligned} \dot{\mathbb{S}}(\rho, \nabla\omega^{\gamma}, \mathbf{V}) = {} & A_{ik}\nabla_i\omega^{\gamma}\nabla_k\omega^{\gamma} + B_{ikl}\nabla_i\omega^{\gamma}\nabla_k\omega^{\gamma}\nabla_l\omega^{\gamma} + C_{ikl}\nabla_i\omega^{\gamma}\nabla_k\omega^{\gamma}V_l \\ & + D_{iklm}\nabla_i\omega^{\gamma}\nabla_k\omega^{\gamma}\nabla_l\omega^{\gamma}\nabla_m\omega^{\gamma} + E_{iklm}\nabla_i\omega^{\gamma}\nabla_k\omega^{\gamma}\nabla_l\omega^{\gamma}V_m \\ & + G_{iklm}\nabla_i\omega^{\gamma}\nabla_k\omega^{\gamma}V_lV_m. \end{aligned} \tag{3.4.39}$$

Note that we have extended the structure of $\dot{\mathbb{S}}$, as proposed by De Groot and Mazur (1962), for a flux linearly proportional to a driving force, to the nonlinear case considered here.

The requirement that $\dot{\mathbb{S}} \geq 0$, i.e., $\dot{\mathbb{S}}$ be *positive definite*, leaves only the quadratic and the two quadratic terms in (3.4.39):

$$\dot{\mathbb{S}}(\rho, \nabla\omega^\gamma, \mathbf{V}) = A_{ik}\nabla_i\omega^\gamma\nabla_k\omega^\gamma \\ +D_{iklm}\nabla_i\omega^\gamma\nabla_k\omega^\gamma\nabla_l\omega^\gamma\nabla_m\omega^\gamma + G_{iklm}\nabla_i\omega^\gamma\nabla_k\omega^\gamma V_l V_m \geq 0. \tag{3.4.40}$$

In (3.4.40), we note that certain symmetries exist in the three tensors **A**, **D** and **G**:

$$A_{ik} = A^1_{ki}, \quad D_{iklm} = D_{kilm} = \cdots = D_{ilmk} \\ G_{iklm} = G_{kilm} = G_{ikml} = G_{lmik}, \tag{3.4.41}$$

where the tensor D_{iklm} is invariant under every permutation of the *full symmetric group* (Sirotine and Chaskolskaya 1984).

Thus, with (3.4.40), and since

$$\dot{\mathbb{S}} = \langle \mathbf{J}^\gamma, \nabla\omega^\gamma \rangle, \tag{3.4.42}$$

in which $\langle \mathbf{A}, \mathbf{B} \rangle$ denotes the *scalar product* of the vectors **A** and **B**, it follows that the non-advective flux is expressed as

$$J^\gamma_{non-adv,i} = -A_{ik}\nabla_k\omega^\gamma - D_{iklm}\nabla_k\omega^\gamma\nabla_l\omega^\gamma\nabla_m\omega^\gamma - G_{iklm}\nabla_k\omega^\gamma V_l V_m. \tag{3.4.43}$$

The first two terms on the r.h.s. of (3.4.43) do not involve the velocity. They describe *diffusion*. The first term is actually the *diffusive flux expressed by Fick's law*, with $A_{ik} \equiv \rho\mathcal{D}^{*\gamma}_{ij}$. The second term represents a non-linear, or 'non-Fickian' diffusive flux. The last term expresses the *dispersive flux* discussed above, with $-\nabla\omega^\gamma$ as the driving force, but with a proportionality to V^2. This may be still considered a 'Fickian' law, as it is proportional to $\nabla\omega^\gamma$, as in Fick's law, but it is different from the Fickian expression for the coefficient of dispersion presented in Sect. 3.4.3, in that here the flux depends on V^2. It may be interesting to mention that in one of the earliest works on dispersion, the work of Taylor (1953), concerning dispersion in a circular capillary tube, due to the parabolic distribution of velocities within the tube, the dispersive flux was also proportional to V^2 (see Sect. 7.2.3).

A detailed analysis of the tensors **A**, **D** and **G** in a 3-dimensional porous medium domain with a prescribed symmetry (e.g., isotropic, axisymmetric) shows that the increase in symmetry towards isotropy causes a reduction in the number of independent moduli associated with these tensorial coefficients. For an isotropic porous medium —i.e., highest symmetry—we need *four* moduli for the description of the three tensorial coefficients:

$$A_{ik} = a\delta_{ik}, \quad D_{iklm} = \frac{d}{3}\left(\delta_{ik}\delta_{lm} + \delta_{il}\delta_{km} + \delta_{im}\delta_{kl}\right),$$
$$G_{iklm} = g_1\delta_{ik}\delta_{lm} + \frac{g_2}{2}\left(\delta_{il}\delta_{km} + \delta_{im}\delta_{kl}\right), \tag{3.4.44}$$

The corresponding rate of entropy production is then

$$\begin{aligned}\dot{\mathbb{S}} &= \dot{\mathbb{S}}(\rho, \nabla\omega^\gamma, \mathbf{V})\\ &= a(\nabla\omega^\gamma)^2 + d(\nabla\omega^\gamma)^4 + g_1(\nabla\omega^\gamma)^2\mathbf{V}^2 + g_2\langle\nabla\omega^\gamma, \mathbf{V}\rangle^2,\end{aligned} \tag{3.4.45}$$

with the coefficients $a, d, g_1, g_2 > 0$.

Hence, the non-advective flux, presented above as (3.4.43), takes the form:

$$\begin{aligned}J^\gamma_{non-adv,i} &= J^\gamma_{non-adv,i}(\nabla\omega^\gamma, \mathbf{V})\\ &= -a\nabla_i\omega^\gamma - d(\nabla\omega^\gamma)^2\nabla_i\omega^\gamma - \left(g_1\mathbf{V}^2\nabla_i\omega^\gamma + g_2\langle\nabla\omega^\gamma, \mathbf{V}\rangle V_i\right).\end{aligned} \tag{3.4.46}$$

Again, in this equation, the first term on the r.h.s. expresses linear diffusion, the second term expresses non-linear diffusion, and the last term expresses the dispersive flux of γ. In (3.4.46), we note the driving force, $\nabla\omega^\gamma$ and the dependence on V^2 (see comment following (3.4.43).

For *axisymmetric* porous medium domains, with the axis of symmetry indicated by the unit vector **e**, we need *eleven* (2+3+6) moduli to describe these three tensorial coefficients:

$$\begin{aligned}&\dot{\mathbb{S}}(\rho, \nabla\omega^\gamma, \mathbf{V}, \mathbf{e}) =\\ &\quad a_1(\nabla\omega^\gamma)^2 + a_2\langle\nabla\omega^\gamma, \mathbf{e}\rangle^2 + d_1(\nabla\omega^\gamma)^4 + 2d_2(\nabla\omega^\gamma)^2\langle\nabla\omega^\gamma, \mathbf{e}\rangle^2\\ &+d_3\langle\nabla\omega^\gamma, \mathbf{e}\rangle^4 + g_1\mathbf{V}^2(\nabla\omega^\gamma)^2 + g_2\langle\nabla\omega^\gamma, \mathbf{V}\rangle^2 + g_3\mathbf{V}^2\langle\nabla\omega^\gamma, \mathbf{e}\rangle^2\\ &+g_4\langle\mathbf{e}, \mathbf{V}\rangle^2(\nabla\omega^\gamma)^2 + g_5\langle\mathbf{e}, \mathbf{V}\rangle\langle\mathbf{V}, \nabla\omega^\gamma\rangle\langle\nabla\omega^\gamma, \mathbf{e}\rangle + g_6\langle\mathbf{e}, \mathbf{V}\rangle^2\langle\mathbf{e}, \nabla\omega^\gamma\rangle^2,\end{aligned} \tag{3.4.47}$$

with the following thermodynamic constraints imposed on the eleven moduli:

$$a_1, a_2 > 0, \quad d_1, d_3 > 0, \quad d_1 d_3 > (d_2)^2,$$

and, similar to the results obtained in Fel and Bear (2010),

$$g_1 > 0, \quad g_1 + g_2 > 0, \quad g_1 + g_3 > 0, \quad g_1 + g_4 > 0, \quad g_1 + g_2 + g_3 + g_4 + g_5 + g_6 > 0,$$

$$(g_1)^2 + g_1(3g_2 + g_3 + g_4 + g_5 + g_6) + 2g_2(g_3 + g_4 + g_6) > \tfrac{1}{2}(g_5)^2.$$

Altogether, the non-advective γ-species flux is:

$$\begin{aligned}&J^\gamma_{non-adv,i}(\rho, \nabla\omega^\gamma, \mathbf{V}, \mathbf{e}) = a_1\nabla_i\omega^\gamma + a_2\langle\nabla\omega^\gamma, \mathbf{e}\rangle e_i\\ &\quad +d_1(\nabla\omega^\gamma)^2\nabla_i\omega^\gamma + d_2\left((\nabla\omega^\gamma)^2 e_i + \langle\nabla\omega^\gamma, \mathbf{e}\rangle^2\nabla_i\omega^\gamma\right) + d_3\langle\nabla\omega^\gamma, \mathbf{e}\rangle^3 e_i\end{aligned}$$

$$+g_1\mathbf{V}^2\nabla_i\omega^\gamma + g_2\langle\nabla\omega^\gamma, \mathbf{e}\rangle V_i + g_3\mathbf{V}^2\langle\nabla\omega^\gamma, \mathbf{e}\rangle e_i + g_4\langle\mathbf{e}, \mathbf{V}\rangle^2\nabla_i\omega^\gamma$$
$$+g_5\langle\mathbf{e}, \mathbf{V}\rangle\left(\langle\mathbf{V}, \nabla\phi\rangle e_i + \langle\nabla\omega^\gamma, \mathbf{e}\rangle V_i\right) + g_6\langle\mathbf{e}, \mathbf{V}\rangle^2\langle\mathbf{e}, \nabla\omega^\gamma\rangle e_i\ , \quad (3.4.48)$$

in which the first two terms describe the linear diffusive flux, with the tensorial coefficient of diffusion depending on two scalar moduli, a_1, a_2. The next three terms, with moduli d_1, d_2 and d_3, describe the *nonlinear diffusive flux* components. The remaining terms describe the dispersive flux. We note that for this axially symmetric porous medium, the dispersivity coefficient is defined by *six* dispersivity moduli, $g_1, \ldots, g_6$.

More on solute dispersion is presented in Sect. 7.2.3.

B. Non-advective Momentum Flux

The advective momentum flux at the microscopic level is $\mathbf{j}^{\mathrm{M}}_{adv} = \rho\mathbf{V}\mathbf{V}$. Recalling that the macroscopic non-diffusive momentum flux is $\mathbf{J}^{\mathrm{M}}_{non-dif} = \overline{\rho}\overline{\mathbf{V}}\overline{\mathbf{V}}$, and assuming, for the sake of simplicity, that the variations in fluid density are relatively small, i.e., that $|\overline{\rho}| \gg |\mathring{\rho}|$, we may approximate the dispersive momentum flux by $\mathbf{J}^{\mathrm{M}}_{dis} = \overline{\rho}\overline{\mathring{\mathbf{V}}\mathring{\mathbf{V}}}$. Altogether, the non-diffusive momentum flux nay be expressed as the sum of advective and dispersive momentum flux:

$$\mathbf{J}^{\mathrm{M}}_{non-dif} = \overline{\rho}\overline{\mathbf{V}}\overline{\mathbf{V}} + \overline{\rho}\overline{\mathring{\mathbf{V}}\mathring{\mathbf{V}}}. \quad (3.4.49)$$

In view of (3.4.36), the non-advective momentum flux is expressed by $\overline{\rho_\alpha}^\alpha\widetilde{\check{\mathbf{V}}_\alpha\check{\mathbf{V}}_\alpha}^\alpha$.

At the microscopic level, i.e., in a fluid continuum, the deviatoric stress, τ_{ij}, which expresses the dissipative part of the diffusive momentum flux, is related to the driving force $W_{ij}(\equiv \nabla_i V_j + \nabla_j V_i)$, which is a symmetric 2nd rank tensor. With this in mind, at the macroscopic level, the non-advective flux of momentum depends also on the fluid's velocity, $\mathbf{V}$, and on the driving force, $\mathbf{W}$, by the general constitutive relationship:

$$\tau_{ij} = \tau_{ij}(\mathbf{W}, \mathbf{V}) = M_{ijkl}W_{kl} + N_{ijklpstr}W_{kl}W_{ps}W_{tr} + L_{ijklps}W_{kl}V_pV_s, \quad (3.4.50)$$

where $M_{ijkl} = M_{ijkl} = M_{jikl} = M_{ijlk}$, $N_{ijklpstr} = N_{jiklpstr} = \cdots = N_{trijklps}$ and $L_{ijklps} = L_{jiklps} = L_{ijlkps} = L_{klijps} = L_{ijklsp}$ are tensorial coefficients that depend on fluid and void-space properties, and all terms are at the macroscopic level.

The first term on the r.h.s. represents the linear diffusive flux of momentum, with the 4th rank tensor M_{ijkl} standing for the usual fluid viscosity for a Newtonian fluid. The second term, with the 8th rank tensorial coefficient, $N_{ijklpstr}$, is responsible for non-linear viscous effects. The dispersive part of the non-advective momentum flux involves the 6th rank tensorial coefficient L_{ijklps}.

The corresponding rate of entropy production is:

$$\dot{S}(\mathbf{W}, \mathbf{V}) = \tau_{ij}W_{ij}$$
$$= M_{ijkl}W_{ij}W_{kl} + N_{ijklpstr}W_{ij}W_{kl}W_{ps}W_{tr} + L_{ijklps}W_{ij}W_{kl}V_pV_s. \quad (3.4.51)$$

To facilitate the use of (3.4.51) for an isotropic porous medium, we introduce the following notations for operations with the **W**-tensor:

$$\langle \mathbf{W}, \boldsymbol{\delta} \rangle \equiv W_{ii}\ , \quad \langle \mathbf{W}, \mathbf{W} \rangle \equiv W_{ij} W_{ji}\ , \quad \langle \mathbf{W}, \mathbf{W}, \mathbf{W} \rangle \equiv W_{ij} W_{jk} W_{ki},$$
$$\langle \mathbf{W}, \mathbf{W}, \mathbf{W}, \mathbf{W} \rangle \equiv W_{ij} W_{jk} W_{kl} W_{li}, \quad \langle \mathbf{W}\mathbf{V}, \mathbf{W}\mathbf{V} \rangle \equiv W_{ij} V_j W_{ik} V_k, \tag{3.4.52}$$

recalling that the symbol $\langle \mathbf{A}, \mathbf{B} \rangle$ denotes the scalar product of the vectors **A** and **B**. With this notation, we get *eleven* $(2 + 5 + 4)$ viscous moduli:

$$\begin{aligned}\dot{S}(\mathbf{W}, \mathbf{V}) = {} & M_1 \langle \mathbf{W}, \mathbf{W} \rangle + M_2 \langle \mathbf{W}, \boldsymbol{\delta} \rangle^2 + N_1 \langle \mathbf{W}, \mathbf{W} \rangle^2 + N_2 \langle \mathbf{W}, \mathbf{W}, \mathbf{W}, \mathbf{W} \rangle \\ & + N_3 \langle \mathbf{W}, \mathbf{W}, \mathbf{W} \rangle \langle \mathbf{W}, \boldsymbol{\delta} \rangle + N_4 \langle \mathbf{W}, \mathbf{W} \rangle \langle \mathbf{W}, \boldsymbol{\delta} \rangle^2 + N_5 \langle \mathbf{W}, \boldsymbol{\delta} \rangle^4 \\ & + L_1 \langle \mathbf{W}, \mathbf{W} \rangle \mathbf{V}^2 + L_2 \langle \mathbf{W}\mathbf{V}, \mathbf{W}\mathbf{V} \rangle + L_3 \langle \mathbf{V}, \mathbf{W}, \mathbf{V} \rangle \langle \mathbf{W}, \boldsymbol{\delta} \rangle \\ & + L_4 \langle \mathbf{W}, \boldsymbol{\delta} \rangle^2 \mathbf{V}^2. \end{aligned} \tag{3.4.53}$$

When $\langle \mathbf{W}, \boldsymbol{\delta} \rangle = 0$, equivalent to $\nabla \cdot \mathbf{V} = 0$, i.e., *isochoric flow*, and the domain is isotropic, only five $(1 + 2 + 2)$ moduli are left:

$$\begin{aligned}\dot{S}(\mathbf{W}, \mathbf{V}) = {} & M_1 \langle \mathbf{W}, \mathbf{W} \rangle + N_1 \langle \mathbf{W}, \mathbf{W} \rangle^2 + N_2 \langle \mathbf{W}, \mathbf{W}, \mathbf{W}, \mathbf{W} \rangle \\ & + L_1 \langle \mathbf{W}, \mathbf{W} \rangle \mathbf{V}^2 + L_2 \langle \mathbf{W}\mathbf{V}, \mathbf{W}\mathbf{V} \rangle. \end{aligned} \tag{3.4.54}$$

The corresponding non-advective momentum flux is:

$$\begin{aligned} J^{\mathrm{M}}_{non-adv,ij} = {} & M_1 W_{ij} + N_1 \langle \mathbf{W}, \mathbf{W} \rangle W_{ij} + N_2 W_{ik} W_{kl} W_{lj} \\ & + L_1 \mathbf{V}^2 W_{ij} + L_2 W_{ik} V_k V_j. \end{aligned} \tag{3.4.55}$$

The first term expresses the diffusive flux of fluid momentum. The next two terms express the nonlinear momentum flux. The last two terms express the dispersive flux of momentum (proportional to V^2). For a Newtonian fluid, $M_1 \equiv \mu$, i.e., the fluid's viscosity.

C. Non-advective Heat Flux

We consider the entire porous medium domain. At every point, i.e., within an REV centered at the point, we have a rigid stationary solid matrix and a void space occupied by a single fluid. We assume that both the solid and the fluid phases are at thermal equilibrium, i.e., a single temperature T describes the thermal state of both phases at the considered point. The development of the expressions for the non-advective fluxes of heat is similar to those of solute, except that in this case the diffusive flux has to take into account the heat transported in both fluid and solid phases. Altogether, (3.4.43)–(3.4.48) are valid, except that the numerical values of the various coefficients are different, and the thermal diffusivity of the porous medium as a whole depends on the porosity and on the thermal conductivity of the two phases, but not on their densities.

For example, for an isotropic porous medium, we may express the non-advective heat flux in the form (3.4.43), replacing ω^γ by T, A_{ik} by $\lambda^*_{pm,ik}$, and omitting the nonlinear diffusive term. We obtain:

$$J^H_{non-adv,i} = -\lambda^*_{pm,ik}\nabla_k T - G^H_{iklm}V_l V_m \nabla_k T, \tag{3.4.56}$$

i.e., the sum of a diffusive term and a dispersive one.

3.5 Interphase Transfers and Sources

In most cases of flow and transport, regardless of the number of phases involved, a considered extensive quantity is being transported across fluid-fluid and fluid-solid (microscopic) interfaces. In the macroscopic balance equation (3.3.3), we have denoted the rate of such transfer—of an extensive quantity, E, to an α-phase from all other β-phases, including the solid, within a porous medium domain—by $f^E_{\beta\rightarrow\alpha}$. In this section, we shall consider two cases: momentum transfer, $E = \mathbf{M}$, and energy transfer, $E = \mathbb{E}$. The transfer of mass of a γ-species across the interface between a fluid phase and the solid matrix and across the fluid-fluid interface in multiphase flow, will be presented in Chap. 7.

Actually, interphase transfer (as a term in the E-balance equation) does not exist at the microscopic level. To the extent that such transfers exist, they appear in the microscopic level models as *boundary conditions*. During the passage from the microscopic to the macroscopic levels of description, these flux boundary conditions are transformed into interphase exchange terms in the balance equations. Sources and sinks that exist at the microscopic level will be transformed into macroscopic source-sink terms at the macroscopic one. Both cases will be discussed in the next section which deals with the macroscopic balance equations and in the respective chapters where the models of specific extensive quantities are presented.

3.5.1 Fluid to Solid Momentum Transfer

Obviously, at the microscopic level, there is no fluid-solid momentum transfer in the momentum balance equation. However, such transfer appears as a boundary condition at a fluid-solid boundary. It then appears as a term in the macroscopic M-balance equation. Thus, in the macroscopic momentum balance equation (3.3.9) for single phase flow, $\mathbf{f}^{\mathrm{M}}_{f\rightarrow s}$ denotes the rate of momentum transfer from the fluid to the solid matrix, per unit volume of porous medium. To obtain this constitutive relationship, we shall make the following assumptions:

- The fluid in the void space adheres to the solid surface (*no-slip condition*).
- The fluid is *Newtonian*. This means that the momentum transfer from the fluid to the solid, per unit f-s-area, is proportional to the dynamic viscosity, μ, and to the velocity gradient, $(V_{f,j} - V_{s,j})/\Delta \equiv V_{r,j}/\Delta$, at the solid surface, with the hydraulic radius, Δ, of the void-space, $(= \mathbb{V}_{of}/\mathcal{S}_{sf})$ serving as a length characterizing the size of a 'pore'. To obtain the momentum transfer per unit volume of porous medium, we have to multiply the flux at a point by $\mathcal{S}_{fs}/\mathbb{V}_o = (\mathcal{S}_{fs}/\mathbb{V}_{of})(\mathbb{V}_{of}/\mathbb{V}_o) = \phi/\Delta$.
- The solid matrix may be in motion (e.g., due to deformation), at the velocity $\mathbf{V}_s$.

Altogether, *phenomenologically*, the **M**-transfer from the fluid to the solid, per unit area of the (microscopic) solid-fluid interface, due to the fluid's velocity gradient at that s-f surface area is expressed by:

$$\left(f^{\mathrm{M}}_{f\to s}\right)_i = \phi R_{ij}\,\mu\frac{V_{f,j} - V_{s,j}}{\Delta^2} \equiv \phi R_{ij}\,\mu\frac{V_{r,j}}{\Delta^2}. \tag{3.5.1}$$

in which R_{ij}, are components of a second rank symmetric tensor, R, which is a coefficient of proportionality associated with the geometrical configuration of the void space. It is possible to add in the above expression a coefficient of proportionality, say, C_f, to be determined, like all coefficients, experimentally. We prefer not to add this coefficient here, and to envision that it is imbedded either in Δ or R. In fact, R may be envisioned as some kind of tortuosity associated with the momentum transferred from the moving fluid to the solid. It is an extension of Newton's (diffusive) momentum flux to a porous medium, taking into account the directions of the solid-fluid microscopic interface elements.

With the terminology introduced in the preamble to Sect. 2.3, (3.5.1) is a *constitutive relationship*. Later, we'll suggest that this makes Darcy's law also a constitutive relationship.

In Sect. 4.2.4, we use (3.5.1) to derive Darcy's law and other flux expressions as simplified versions of the momentum balance equation. In Sect. 6.2.1 we shall consider the momentum exchange in the case of two phase flow, leading to the corresponding fluid flux expressions.

However, when the velocity difference, $\mathbf{V}_r$ becomes larger, there is no reason to assume that the momentum transfer depends only on the first power of $\mathbf{V}_r$. In fact, all experiments show that as $\mathbf{V}_r$ increases, the relationship between the rate of momentum transfer and the relative velocity is no longer linear, as in (3.5.1). For example, we may write the non-linear constitutive relationship:

$$\left(f^{\mathrm{M}}_{f\to s}\right)_i = A_{ij}V_{r,j} + B_{ijk}V_jV_k + C_{ijkl}V_jV_kV_l + \cdots \tag{3.5.2}$$

In Sect. 4.3.2, we present *Forchheimer's equation* which relates the pressure gradient, ∇p, to the sum of a linear and a quadratic terms in $\mathbf{V}$.

3.5.2 Interphase Energy Transfer

There is no interphase energy transfer when we assume that all phases at a (macroscopic) point are at the same temperature. However, we may encounter cases in which this equilibrium assumption is not valid. One such example is a chemical reaction which is fed by streams of fluids at different temperatures. Another example is at the initiation of steam (or hot water) injection into a (cold) rock formation. The temperature difference may gradually be closed, but it does exist for a while. While it exists, the rate of energy transfer is (1) proportional to the difference in temperature, (2) proportional to the contact area between the solid and the fluid in the void space, and (3) inversely proportional to some length characterising the distance between the void space occupied by the fluid and the solid matrix. The coefficient of proportionality will be an equivalent thermal conductivity which takes into account the thermal conductivity of the fluid, that of the solid matrix and the porosity. The subject is discussed in detail in Sect. 8.3.4.

3.5.3 Sources of Extensive Quantities

In the macroscopic level E-balance equation (3.3.4), this term appears as $\theta_\alpha \rho_\alpha \Gamma^E$.

A. Source of Mass of a Fluid Phase

There are no fluid mass sources within a fluid phase. However, often especially in dealing with flow in ground water aquifers and in oil and gas reservoirs, fluids are injected into the geological formation, or pumped from it through wells. Because a well is small, relative to the extent of the considered geological formation, it is often considered as a point source (or sink) for fluid mass. This subject will be discussed in Sect. 5.1.1.

B. Sources of Mass of a Chemical Species in a Fluid Phase

A source of this kind may be due to (1) decay (e.g., radioactive), or production of a considered chemical species, and (2) by chemical reactions. We shall elaborate on such sources in Chap. 7.

C. Sources of Momentum of a Fluid Phase

We are considering sources only in the fluid due to forces acting on it (but per unit volume of porous medium). In the case of single phase flow considered here, we have two sources of momentum per unit volume of porous medium:

- A source due to body forces, here due to gravity, i.e., $\Gamma^{\mathrm{M}} = -\phi \rho g \nabla z$.
- A source due to the pressure gradient in the fluid, i.e., $-\phi \nabla p$, where we have taken into account that the considered fluid occupies only part of any cross-sectional area through the porous medium.

D. Sources of Energy of a Phase

There are four possible sources of energy:

- Sources, $\phi\rho\Gamma^{\mathbb{E}}_{chem}(\equiv \phi\rho\Gamma^{H}_{chem})$, due to heat generated by chemical reactions in the fluid phase (when such reactions occur).
- A source $\mathbf{V}\cdot(\phi\rho\mathbf{F})$, due to the work, per unit volume of porous medium, by the body force $\mathbf{F}$ $(= -g\nabla z)$.
- A source $-\nabla\cdot[\mathbf{V}\cdot\phi(-\boldsymbol{\tau})]$, resulting from the work of the viscous (shear) stress in the fluid, per unit volume of porous medium.
- A source due to the work of the pressure, per unit volume of the porous medium.

3.6 Macroscopic E-Balance Equations (2)

In this section, we shall insert the expressions for fluxes, rates of transfer and sources, all for E, into the balance equations presented in Sect. 3.3, in order to obtain specific balance equations in terms of the problem variables: p, T, $\mathbf{V}$, ρ^{γ}, $\gamma = 1, 2, \ldots, NC$. We also have $\rho = \sum_{(\gamma)} \rho^{\gamma}$. Altogether we have $NC + 3$ variables, and $NC + 3$ equations: $NC + 2$ balance equations and one constitutive relation,

$$\rho = \rho(p, T, \rho^{\gamma}; \gamma = 1, 2, \ldots, NC).$$

3.6.1 *Mass Balance of a Fluid Phase*

Inserting the advective mass flux into (3.3.7), considering a single fluid phase that occupies the entire void space, with no mass sources, we obtain:

$$\frac{\partial\phi\rho}{\partial t} = -\nabla\cdot(\phi\rho\mathbf{V}), \quad \text{or} \quad \frac{\partial\phi\rho}{\partial t} = -\nabla\cdot(\rho\mathbf{q}), \quad \mathbf{q} = \phi\mathbf{V}, \tag{3.6.1}$$

where $\mathbf{q}$ denotes the *specific discharge* and we have assumed that the advective flux of the fluid mass is much larger than the dispersive one (see Sect. 3.4.3) so that the latter can be neglected. Note that many authors prefer to state this equation in the second form, i.e., in terms of $\mathbf{q}$. However, this is inconvenient when ϕ varies in space and time, or when we consider solid deformation.

3.6.2 *Mass Balance for a γ-Chemical Species*

From (3.3.8), rewritten for a single fluid phase, we obtain

$$\frac{\partial\phi\rho\omega^{\gamma}_{\alpha}}{\partial t} = -\nabla\cdot\phi\left(\rho\omega^{\gamma}\mathbf{V} + \mathbf{J}^{m^{\gamma}}_{dis} + \mathbf{J}^{m^{\gamma}}_{dif}\right) + f^{m^{\gamma}}_{s\rightarrow\alpha} + \phi\rho\Gamma^{m^{\gamma}}, \tag{3.6.2}$$

into which we can now insert appropriate expressions for the non-advective γ-fluxes that appear in Sect. 3.4.4. As an example, for an isotropic case, we shall make use of (3.4.46):

$$\frac{\partial \phi \rho \omega_\alpha^\gamma}{\partial t} = \\ -\nabla\cdot\phi[\rho\omega^\gamma \mathbf{V} - a\nabla\omega^\gamma - 2d(\nabla\omega^\gamma)^2\nabla\omega^\gamma - \big(g_1\mathbf{V}^2\nabla\omega^\gamma + g_2\langle\nabla\omega^\gamma, \mathbf{V}\rangle\mathbf{V}\big)] \\ +f_{\beta\to\alpha}^{m^\gamma} + \phi\rho\Gamma^{m^\gamma}. \tag{3.6.3}$$

We note that the total m^γ-flux is made up of an advective flux, a (linear) diffusive flux, with a representing the (scalar) coefficient of diffusion in a porous medium, $\rho\mathcal{D}_{pm}^*$, a non-linear diffusive flux, and a dispersive flux, which is proportional to the fluid's velocity squared and depends on two (scalar) coefficients that represent the effect of the void space geometry. More on the coefficients of molecular diffusion and the coefficients of dispersion is presented in Chap. 7.

3.6.3 Momentum Balance of a Newtonian Fluid

For a Newtonian fluid, and (1) expressing the advective momentum flux by $\mathbf{J}_{adv}^{\mathrm{M}} \equiv \phi\rho\mathbf{V}\mathbf{V}$, $\mathbf{V} \equiv \mathbf{V}_f$, (2) expressing the momentum diffusive flux, $\mathbf{J}_{dif}^{\mathrm{M}}$, by the dissipative part of the stress, i.e., $\phi\boldsymbol{\tau}$, (3) neglecting the dispersive momentum flux, (4) combining the pressure part with the body force, i.e., $-\phi(\nabla p + \rho g\nabla z)$, where $-g\nabla z$ expresses the body force due to gravity, and (5) using (3.5.1) to express the fluid-to-solid momentum transfer, $\mathbf{f}_{f\to s}^{\mathrm{M}} = (\phi\mu/\Delta^2)\mathbf{R}\cdot(\mathbf{V}_f - \mathbf{V}_s)$, we rewrite the fluid's momentum balance equation (3.3.9) in the form:

$$\phi\rho\frac{D\mathbf{V}_f}{Dt} = \nabla\cdot(\phi\boldsymbol{\tau}) - \phi\,(\nabla p + \rho g\nabla z) - \phi\frac{\mu\mathbf{R}}{\Delta^2}\cdot(\mathbf{V}_f - \mathbf{V}_s). \tag{3.6.4}$$

In the above equation, each term expresses added momentum per unit volume of porous medium per unit time.

If we (1) neglect the nonlinear terms in the series expression for the diffusive flux of momentum, (2) neglect the dispersive flux of momentum, and (3) express the diffusive flux of momentum, $\boldsymbol{\tau}$, by $\phi\mu\nabla\mathbf{V}$, i.e., assuming isochoric flow at the microscopic level, the momentum balance equation (3.6.4) reduces to:

$$\phi\rho\frac{D\mathbf{V}_f}{Dt} = \nabla\cdot\phi\mu\nabla\mathbf{V}_f - \phi(\nabla p + \rho g\nabla z) - \phi\frac{\mu\mathbf{R}}{\Delta^2}\cdot(\mathbf{V}_f - \mathbf{V}_s). \tag{3.6.5}$$

The macroscopic momentum balance equation (3.6.5) may be referred to as the *Navier–Stokes equation for a porous medium*. With a number of special cases, it will be further discussed in Sect. 4.2.4.

3.6.4 Energy Balance

We start from the energy balance equation (3.3.26). For the total energy flux, in the case of an isotropic porous medium, we make use of (3.4.56), i.e., taking into account only the linear diffusive heat flux, and the 'Fickian' dispersive flux. Energy per unit volume of porous medium is added by (1) the work done on the fluid phase by external body forces, e.g., gravity ($\mathbf{V}\cdot\phi\rho\mathbf{F} \equiv -\mathbf{V}\cdot\phi\rho g\nabla z$), (2) work done on the fluid by the stress within the fluid, composed of the work done by the viscous forces ($\nabla\cdot(\mathbf{V}\cdot\phi\boldsymbol{\tau})$) and by the pressure forces, ($\nabla\cdot\mathbf{V}\cdot\phi p\boldsymbol{\delta}$), and (3) heat produced by chemical reactions within the fluid ($= \phi\rho\Gamma^H$). We obtain:

$$\begin{aligned}\frac{\partial}{\partial t}\left[\phi\rho\left(I+\tfrac{1}{2}V^2\right)\right] = &-\nabla\cdot\left[\phi\rho\left(I+\tfrac{1}{2}V^2\right)\mathbf{V}\right] - \nabla\cdot\phi\mathbf{J}^H_{dif} - \nabla\cdot\phi\mathbf{J}^H_{dis}\\ &+\nabla\cdot[\mathbf{V}\cdot(\phi\boldsymbol{\tau})] - \nabla\cdot[\mathbf{V}\cdot(\phi p\boldsymbol{\delta})] - \mathbf{V}\cdot\phi\rho g\nabla z + \phi\rho\Gamma^H,\end{aligned} \tag{3.6.6}$$

in which the diffusive and dispersive fluxes of heat are defined by (3.4.56), and the shear stress, $\boldsymbol{\tau}$, can be expressed by (3.4.50). Note that (1) there is no exchange of heat between solid and fluid, as we have assumed that both are at the same T, and (2) we have assumed heat flux only in the fluid, as if the solid matrix is an insulator.

Other forms of the energy balance equation are presented in Sect. 8.2.

* * *

In Chaps. 5–9, we shall use the E-balance equations presented above as the cores of models that describe mass, energy, and momentum transport. There, to present complete well-posed models, we shall also present and discuss the appropriate initial and boundary conditions for each case.

3.7 Constitutive Equations

So far in this chapter, we have presented the fundamental balance equations for flow and transport in any fluid phase, first at the microscopic level and then at the macroscopic one, i.e., in a porous medium. To simplify the discussion here, let us consider the former case, i.e., the microscopic level. The same conclusions are applicable also to multiple phases at the macroscopic one, i.e., for modeling any problem of transport in a porous medium domain. Since, in what follows, we shall refer to a single fluid continuum only, no subscript will be used.

For any fluid phase at the microscopic level, we have the following system of balance equations:

- **Mass balance equation for a fluid** ($\mathbf{V}_f \equiv \mathbf{V}$). Since there is no source of phase mass:

$$\frac{\partial\rho}{\partial t} = -\frac{\partial(\rho V_i)}{\partial x_i}. \tag{3.7.1}$$

It is a single equation that contains four scalar variables: the density, ρ, and three scalar velocity components, V_i.

- **Linear momentum balance**:

$$\frac{\partial(\rho V_i)}{\partial t} = -\frac{\partial}{\partial x_j}(\rho V_i V_j - \sigma_{ij}) + \rho F_i, \tag{3.7.2}$$

which is a set of six equations for the three components of the (assumed known) body force, F_i, and the six components of the stress, σ_{ij}, $(\equiv \sigma_{ji})$.

- **Internal energy balance**:

$$\rho\frac{\partial u'}{\partial t} = -\rho V_i \frac{\partial u'}{\partial x_i} + \sigma_{ij}\frac{\partial V_i}{\partial x_j} + \frac{\partial(\rho j_i^H)}{\partial t} + \rho\Gamma^H, \tag{3.7.3}$$

where u' denotes internal energy density, $\mathbf{j}_i^H$ is the ith component of heat flux, and Γ^H denotes energy source, and we have taken into account the mass balance (3.7.1).

- **Mass balance for a γ-chemical species in a fluid phase**:

$$\frac{\partial \rho^\gamma}{\partial t} = -\frac{\partial}{\partial x_i}\left(\rho^\gamma V_i + j_{i,dif}^\gamma\right) + \rho\Gamma^\gamma, \tag{3.7.4}$$

and we write one such equation for every γ-species. In the above equation, we have four scalar variables: ρ^γ and $j_{i,dif}^\gamma$ for every γ-species, and Γ^γ is known.

- **Entropy balance**:

$$\begin{aligned}\rho\frac{\partial s}{\partial t} = &- \rho V_i \frac{\partial s}{\partial x_i} + \frac{1}{T}\left(\tau_{ij}\frac{\partial V_i}{\partial x_j} - \frac{\partial j_i^H}{\partial x_i}\right.\\ &\left. + \rho\Gamma^H + \sum_{(\gamma)}\mu^\gamma \frac{\partial j_{i,dif}^\gamma}{\partial x_i} - \rho\sum_{\gamma=1}^{NC}\mu^\gamma\Gamma^\gamma\right),\end{aligned} \tag{3.7.5}$$

with $NC + 8$ additional variables: μ^γ (NC variables), τ_{ij}, T and s ($\equiv$ specific entropy).

Altogether, for an NC species phase, since $\sum_{\gamma=1}^{NC}\rho^\gamma = \rho$, we have $NC + 5(\equiv (NC - 1) + 6)$ independent balance equations in the $17 + 5NC(\equiv 22 + 5(NC - 1))$ variables that are functions of time and position:

$$\rho, V_i, \sigma_{ij}, u', j_i^H, \rho^\gamma, j_i^\gamma, \mu^\gamma, \tau_{ij}, T, s.$$

The above equations are nothing but *universal statements of balance*, which are *valid for any phase continuum*. They contain no information on the specific material (fluid phase, solid, chemical species) involved in any specific investigated case. For example, they say nothing on the internal constitution of the considered fluid or solid, or of how these phases respond to pressure or temperature changes. This observation is important in view of the fact that different fluids and solids behave differently as pressure, temperature, stress, and solute concentrations vary. Moreover, their state of aggregation -solid, liquid, gas- may also vary during a considered case. Furthermore, within each of these states, different materials behave differently.

For any fluid or solid, a total of additional 12 + 4NC relationships are needed in order to complete the above description of transport for any *specific material.* These take the form of a closed set of equations (i.e., number of variables equal to the number of equations) that describe the behavior of the *particular material* that comprises the specific phase and chemical species in a considered problem. We refer to these relations as *constitutive equations*. As we shall see in the examples below, constitutive equations take the forms of equations of state, as well as *stress-strain relationships* and *flux equations*, which may be regarded as *dynamic equations of state*.

Accordingly, let us add six additional equations:

$$\sigma_{ij} = \tau_{ij} - p\delta_{ij} \qquad \text{(6 equations)}, \tag{3.7.6}$$

which decompose the stress components into a deviatoric part τ_{ij}, and a pressure, p, which, in a fluid, is taken *positive for compression*). This equation is a *definition*, valid for any fluid, It contains no coefficient. We now have $11 + NS$ equations, but the total number of variables has been increased to $18 + 5NC$, as we have added the pressure, p, as a variable. The $7 + 4NC$ relationships still required to obtain a *closed set of equations*, take the forms of *equations of state*. These can be expressed in one of the three following forms:

- **Equations of state (EOS)**:

$$\left.\begin{array}{lll} & \rho = \rho(p, \rho^\gamma, T,), & \text{(one equation)}, \\ & s = s(T, v, \omega^\gamma), & \text{(one equation)}, \\ \text{or} & & \\ & u' = u'(T, v, \omega^\gamma), & \text{(one equation)}, \\ & \mu^\gamma = \mu^\gamma(p, \omega^\gamma, T), & (N-1 \text{ equations}), \end{array}\right\} \tag{3.7.7}$$

where $\rho = 1/v$, μ^γ is the chemical potential of the γ-species, and $\omega^\gamma = \rho^\gamma/\rho$. The symbols ρ^γ and ω^γ stand for all the ρ^γ's and ω^γ's of all the γ species present in the system. Under certain circumstances, we may use an expression for the chemical potential $\mu^\gamma = \mu^\gamma(p, \omega^\gamma, T)$.

- **Stress-strain relationships**:

$$\tau_{ij} = \tau_{ij}(V_i), \qquad \text{(6 equations)}. \tag{3.7.8}$$

Under certain circumstances, it may be advantageous to introduce the relationship $\varepsilon_{ij} = \varepsilon(\sigma_{ij}, T)$.

- **Flux equations**:

Heat conduction: $j_i^H = j_i^H(T)$, (3 equations),

Diffusive mass flux: $j_{i,dif}^\gamma = j_{i,dif}^\gamma(\rho^\gamma)$, (3(N − 1) equations).

We wish to emphasize one important difference between the balance equations and the constitutive relationships. Unlike the former, *the latter always contain (scalar or tensorial) coefficients that characterize the specific considered material, phase, or chemical species*. They have to be determined *experimentally*, for any considered phase or chemical species.

So far, the entire discussion, as well as the schematic examples of constitutive equations presented above, are related to a single phase continuum, viz., at the microscopic level. However, the same ideas–balance equations, constitutive relationships and definitions–are also applicable to the macroscopic level. Just as we obtain the macroscopic balance equations by averaging the microscopic ones, the macroscopic constitutive equations for a phase in the multiphase system called 'porous medium' (regarded as a continuum), or for the porous medium as a whole, can be obtained, by averaging the appropriate microscopic equations.

For example, given a microscopic constitutive equation in terms of the variables a and b in the form

$$a = f(b),$$

valid for an α-phase, we assume that by averaging over the α-phase within an REV, we obtain the corresponding macroscopic constitutive relationship in terms of the macroscopic variables $\overline{a}^\alpha$ and $\overline{b}^\alpha$, in the form:

$$\overline{a}^\alpha = f(\overline{b}^\alpha). \tag{3.7.9}$$

Note that this conclusion is not necessarily correct if a constitutive relationship involves derivatives.

Denoting the deviation of b from its mean by $\mathring{b} \equiv b - \overline{b}^\alpha$, we have

$$f(b) \approx f(\overline{b}^\alpha) + \mathring{b}\frac{df}{db}\bigg|_{\overline{b}^\alpha} + (\mathring{b})^2\frac{d^2 f}{db^2}\bigg|_{\overline{b}^\alpha}. \tag{3.7.10}$$

Since, $\overline{\mathring{b}}^\alpha = 0$, we have:

$$\overline{a}^\alpha = \overline{f(b)}^\alpha \approx f(\overline{b}^\alpha) + \overline{(\mathring{b})^2}^\alpha \frac{d^2 f}{db^2}\bigg|_{\overline{b}^\alpha}. \tag{3.7.11}$$

Thus, if the mean-square deviation, $\overline{(\mathring{b})^2}^\alpha$, is sufficiently small within the REV, then the relationship in (3.7.9) holds. This condition is satisfied when the constitutive relationship is close to a linear function. For example, fluid density is almost a linear function of pressure over a large range of pressures. However, for highly nonlinear constitutive relationships, it is necessary to verify that the deviations of the arguments are not too large.

Or, the macroscopic constitutive relationship can be obtained by making use of the *phenomenological approach*. According to this approach, *we assume that the general structure of a macroscopic constitutive equation is the same as that of the microscopic one*, without going through an averaging procedure. It then remains only to determine *experimentally* the values of the coefficients that appear in the macroscopic constitutive relation written in this way. By following this procedure, we lose the interactions across interphase boundaries that appear in the averaging process, *assuming* that they are relatively minor.

We should mention one important difference between the microscopic level and the macroscopic one. It concerns interphase E-transfers, expressed by the term $f^E_{\alpha\rightarrow\beta}$, or $f^E_{\alpha\rightarrow s}$, in the macroscopic E-balance equation. These terms usually involve interphase transfer coefficients.

Obviously, it is essential to select the appropriate constitutive relationship for a given problem (and this means for a given set of fluid and solid materials), and use in them appropriate values of coefficients which are independent of the fluid/solid movement. However, while the equations of state are derived experimentally, and this includes both their general structure as well as the values of the coefficients appearing in them, *they must obey the fundamental laws of physics and thermodynamics*. Gray and Hassanizadeh (1998), Hassanizadeh (2004, p. 187), following Eringen (1980, p. 151), and others, refer to these rules as 'axioms' which constrain the equations of state:

(a) **Axiom of causality**: When investigating the behavior of a considered system, certain properties can be selected as *observable*, or measured. Examples are the location of a fluid particle and the temperature. The remaining properties and quantities are considered as *causes*. The latter are regarded as dependent, or consequences of the former. For example, pressure and temperature are observable quantities.
(b) **Axiom of determinism**: The value of the thermodynamic constitutive functions at a material point within a considered domain, at a specified time, is determined by the history of the motion and temperature of all material points within the considered domain. Thus, it excludes any dependence on points outside the considered body and future times.
(c) **Axiom of equipresence**: All constitutive functions should be expressed in terms of the same list of independent constitutive variables until otherwise deduced.
(d) **Axiom of objectivity**: This axiom stems from the recognition that the behavior of material must be independent of the motion of an observer.
(e) **Axiom of material invariance**: Constitutive relationships must be invariant under certain groups of orthogonal transformations and translations of the material coordinates. These constraints are the result of symmetry conditions implied by these transformations in the material frame of reference. In order to obey this constraint, the constitutive equations are expressed in *tensorial form*.

(f) **Axiom of neighborhood**: It is safe to assume that what happens very far from a considered system has very little effect on the latter.
(g) **Axiom of memory**: Similarly, things that happened in the far past have very little effect on the present behavior of a considered system.
(h) **Axiom of memory**: Similarly, things that happened in the far past have very little effect on the present behavior of a considered system. **Axiom of admissibility**: The constitutive equation should not violate conservation laws (i.e., mass, momentum and energy balance equations) and the second law of thermodynamics.

(see also Classical Field Theories in Truesdell and Toupin 1960.) Some authors present a slightly different list.

These are important axioms. They should be obeyed whenever choosing a constitutive relationship. However, beyond their presentation here, we shall not dwell on them any further, assuming, implicitly, that all constitutive relations mentioned in this book obey these axioms, both at the microscopic level, but also when selecting such relationships within the framework of macroscopic models.

As mentioned earlier, constitutive relationships always include *coefficients* that represent material properties. They are *material dependent* and have to be determined empirically for any considered case. Usually, they are determined at the microscopic level. However, when considering phenomena of transport in porous media, say, with the void space occupied by one or more fluids, we encounter interactions across interphase boundaries (= interphase transfers). These interactions occur as boundary conditions at the microscopic level, but their effects have to be taken into account when we consider their macroscopic (averaged) counterpart. This is done when the averaging approach is employed, but also in the phenomenological approach. Then, what happens at phase boundaries appears as interphase exchange terms. Because we do not know the detailed form of interphase boundaries, constitutive relationships are obtains by conducting experiments on porous medium samples e.g., on a core extracted from a geological formation. Obviously, the above axioms are also applicable to the macroscopic level constitutive relations.

Altogether, when considering phenomena of transport in porous media, we encounter (macroscopic):

- Constitutive relationships that express solid and fluid behavior, e.g., density and enthalpy, in response to changes in pressure, temperature, and solute concentration. These relationships are the same for the microscopic and the macroscopic models, as they involve no interphase boundaries.
- Constitutive relations that express the transfer of extensive quantities, e.g., solute, heat, and momentum, across microscopic interphase boundaries. This is the f^E term appearing in the macroscopic balance equation.
- Flux equations that express the motion of extensive quantities (e.g., mass of a chemical species and heat) in response to driving forces.
- Definitions that express stress-strain relationships for the solid matrix and for the saturated porous medium as a whole.
- Definitions, e.g., $S_\ell + S_g = 1$.

Finally, a few comments on certain specific constitutive equations. The first is Darcy's law, which plays a central role in phenomena of transport in porous media. While Fick's law and Fourier's law, expressing the diffusive fluxes of mass and heat, respectively, exist at both the microscopic and the macroscopic level, Darcy's law exists only at the macroscopic one. It expresses at the macroscopic level the momentum transfer from the fluid to the solid at their microscopic interfaces. Although it is often considered as an approximate form of the momentum balance equation (see Sect. 3.5.1), it is a *constitutive relationship* of the flux type. It expresses the momentum transferred from the fluid to the solid (some refer to it as 'friction'). It may be considered as a macroscopic extension of Newton's law. Henry Darcy discovered it as an expression for the loss of head in an experiment of flow through a sand column. The coefficient-hydraulic conductivity-involves the fluid's viscosity (for a Newtonian fluid) and the geometrical features of the fluid-solid interface. We express the momentum transfer, f^M by (3.5.1) and then express it by the momentum balance equation in which we neglect certain terms.

The second comment is on the capillary pressure. Here we do have a microscopic law, in the form of Laplace formula (Sect. 2.4.12), but we do have also a macroscopic law in the form of (6.1.5). However we should take into account the condition of continuity of momentum transport expressed by (3.3.20).

Equations of state and flux equation for extensive quantities are presented throughout the book. For example, equations of state for fluid and solid phases are presented in Sect. 2.3. Stress-strain relationship for an elastic porous medium is presented in Sect. 2.3.5. Macroscopic flux equation for a dissolved chemical species is presented in Sect. 7.2.2, for heat in Sect. 8.3.4, and for momentum in Sect. 3.5.1.

3.8 The Finite Volume Method

So far in this book, we have emphasized that the core of the mathematical model that describes the flow and transport of any extensive quantity through a porous medium domain is a *partial differential equation* (PDE) that describes the balance of that quantity at a point in the considered domain. The analytical solution of this equation, subject to appropriate initial and boundary conditions, provides information on future values of relevant state variable at *all* points within the considered domain. Unfortunately, although *analytical solutions* are always preferable, they are seldom solvable for problems of practical interest, because of the irregular boundaries of the problem domain, the heterogeneity of the domain, with respect to its physical properties, and, often, because of the nonlinearity of the equations. Instead, in practice, computer-based *numerical techniques* are used for solving such equations.

In most numerical techniques, the mathematical model, written in terms of *continuous* state variables, like $p(\mathbf{x}, t)$ and $c(\mathbf{x}, t)$, is replaced by a numerical model, written in terms of *discrete* variables, such as $p_j^n \equiv p(\mathbf{x}_j, t_n)$, which represent values of state variables at grid points $\mathbf{x}_j$ and times t_n, in space and time, respectively.

Furthermore, in the *finite difference method*, the derivatives that appear in the PDE are replaced by approximate expressions written in terms of values at the grid points.

In all numerical methods, the sought values of state variables are obtained only for specified points in space and time (= the grid points). Information related to other points in the problem space and time domains are obtained by *interpolation*. Numerical methods of solution, or numerical models, are beyond the scope of this book. However, since we have been emphasizing that the core of a mathematical model is the partial differential equation that expresses the balance of the considered extensive quantity, we wish to show that also in numerical models, the core is a statement of balance of the considered extensive quantity, this time taken over a specified *finite domain*, and over a specified time interval. Just a reminder: the partial differential equation that represents the balance of an extensive quantity 'at a point' is obtained by first writing the balance for a finite domain in space and a finite time interval, and then going to the limit as the domain and the time interval are shrunk to a point.

Obviously, seeking the solution in the form of values of state variables at many (often very many) points in space and time, requires an appropriate computer and a computer program, referred to as 'code', or 'software'.

The *finite volume method* is one of the numerical models that are based on writing the relevant balance equation (of a considered extensive quantity) for *specified finite subdomains*, often called 'cell', 'elements', or 'gridblocks'.

As examples, we may mention the TOUGH2 (Pruess et al. 1999) and TOUGH+ (Moridis et al. 2008) codes. In these codes, the balance equation of the considered extensive quantity is presented in its *integro-differential form*, rather than in the form of a partial differential equation. The considered porous medium domain is divided into small volume elements, or cells, and values of relevant state variables and of porous medium properties are assigned to every one of them. A balance of the considered extensive quantity is then written for every cell in terms of these values.

Specifically, Fig. 3.3 shows a porous medium subdomain (= finite volume element, cell) at the macroscopic level. The cell, Ω_n, of finite volume $\mathbb{V}_n$, bounded

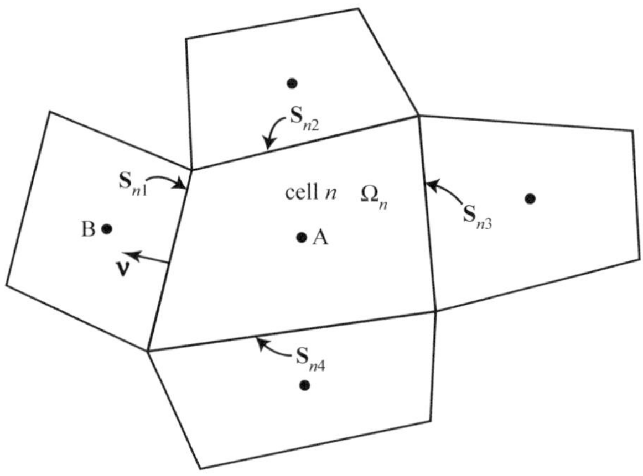

Fig. 3.3 A quadrilateral cell in the finite volume method

Table 3.1 Example 1: Two phases and two components

Phases (α)	Aqueous (w)	Non-aqueous (n)
Components (κ)		
H_2O	✓	✓
CO_2	✓	✓

by a closed surface of area $\mathbb{S}_n, n = 1, \ldots, 4$, with $\boldsymbol{\nu}$ indicating the outward normal unit vector on the latter. The figure also shows the adjacent cells as the considered extensive quantity can cross inter-cell boundaries.

The void space in the cell is occupied by one, two or three α-fluid phases, each at the volumetric fraction $\theta_\alpha (\equiv \phi S_\alpha)$. In addition, we have the solid phase (relevant for heat transport). An extensive quantity E^κ_α (read: the κth-extensive quantity of the α-phase', (specific value e^κ_α) is transported *in* and *with* each of the (one, two or three) fluid α-phases that occupy the void space in this domain and through the solid (e.g., heat, or momentum).

As an example, consider the case of a geological formation saturated by brine into which supercritical CO_2 is injected. As the latter–non-wetting phase– spreads out, displacing the brine (= wetting phase), some CO_2 will dissolve in the brine, while some H_2O will dissolve into the CO_2. The following table represents the phases and components in this example (Table 3.1):

As a second example, consider the case represented by the following table (Falta et al. 1995, p. 6):

This is a case in which the void space is occupied by *three fluid phases*: a gas, an aqueous phase and a Non-aqueous Liquid Phase (NAPL). We declare three components: water, air, and Volatile Organic Compound (VOC). We note that the component water can be a gas or an aqueous phase, while air and VOC can be in any of the three phases. If we consider the extensive quantity mass, and its density, then we have m^κ_α and ρ^κ_α. Note that the 'component' (as defined in Sect. 1.1.1) represents a certain fixed combination of γ-species. However, in non-isothermal transport, heat is also considered as a *(pseudo) component*. For modeling heat transport, we often use for heat: $E^\kappa \to \mathbb{H}$.

Heat and mass transport (of fluid phases and dissolved chemical species) in a petroleum reservoir may serve as a third example. The void space in the reservoir is occupied by three fluid phases: an aqueous wetting phase, a NAPL, which is the non-wetting phase, and an aqueous intermediate wetting phase–the oil. Each of these phases is made up of components; gas, oil and water, In addition, we may consider solutes that are present in the oil and may also dissolve in the water.

As a result of fluxes to and from the neighboring cells, the considered quantity of E^κ_α accumulates in the considered volume element, due to:

(a) The net quantity of E that enters the considered cell from adjacent cells through its bounding surfaces, at the fluxes of E^κ_α that prevail on them.

(b) The net production of E^κ_α by various sources that exist within the considered cell.

The total flux mentioned in (a) above involves advection, diffusion and dispersion. However, we recall that there is no diffusion nor dispersion of the total mass of a phase. The same extensive quantity (e.g., heat) may also be present and transported in the solid (s)-phase, within the Ω_n-cell. In this case, the only flux through the solid is the diffusive (= conductive) heat flux.

Let $\mathfrak{M}^\kappa \equiv \sum_{(\alpha)} \mathfrak{M}^\kappa_\alpha$, denote the total amount of a considered extensive quantity associated with the relevant α-phases, per unit volume of the nth cell, of volume $\mathbb{V}_n$. The rate at which this quantity is increased is expressed by:

$$\frac{\partial}{\partial t}\int_{\mathbb{V}_n} \mathfrak{M}^\kappa \, d\mathbb{V}_n.$$

The total flux is expressed by:

$$\mathfrak{J}^\kappa \equiv \sum_{(\alpha)} \mathbf{J}^\kappa_{\alpha,adv+dis+dif}.$$

Within the phases that occupy the cell, the considered E^κ_α is produced (e.g., by chemical reactions) at a rate $\mathfrak{G}^\kappa = \sum_\alpha \theta_\alpha \rho_\alpha \Gamma^\kappa_\alpha$, per unit volume of porous medium, with Γ^κ_α denoting the rate of production of κ per unit mass of the α-phase. Following (3.2.4), we can write the balance of E^κ within the Ω_n cell, in the form (Pruess et al. 1999)

$$\frac{\partial}{\partial t}\int_{\mathbb{V}_n} \mathfrak{M}^\kappa \, d\mathbb{V}_n = -\int_{\mathbb{S}_n} \mathfrak{J}^\kappa \cdot \boldsymbol{\nu} \, d\mathbb{S}_n + \int_{\mathbb{V}_n} \mathfrak{G}^\kappa \, d\mathbb{V}_n. \tag{3.8.1}$$

Following are four examples. Since our objective here is only to demonstrate how balance equations of considered extensive quantities are written as integro-differential equations, no details will be presented on the actual fluxes and storage coefficients of the considered extensive quantities. These are presented and discussed in the relevant chapters of the book.

A. Mass of a Fluid α-Phase in Multi Phase Flow

We consider the flow of an α-phase at saturation S_α in two-phase flow, say a liquid (ℓ) and a gas (g), in the absence of sources. For this case, $\mathfrak{M}^\kappa = \phi S_\alpha \rho_\alpha$. No production of m_α, i.e., $\Gamma_\alpha = 0$. The mass flux of the α phase is $\mathbf{J}^\kappa \equiv \rho_\alpha \mathbf{q}_\alpha \equiv \phi S_\alpha \rho_\alpha \mathbf{V}_\alpha$ is by advection only, obeying, for example, Darcy's law for multiphase flow. The balance equation (3.8.1) takes the form:

$$\int_{\mathbb{V}_n} \frac{\partial \phi S_\alpha \rho_\alpha}{\partial t} \, d\mathbb{V}_n = \int_{\mathbb{S}_n} \phi S_\alpha \rho_\alpha \mathbf{V}_\alpha \cdot \boldsymbol{\nu} \, d\mathbb{S}_n, \quad \alpha = \ell, g. \tag{3.8.2}$$

Table 3.2 Example 2: Multiple phases and components

Phases (α)	Gas	Aqueous	NAPL –
Components (κ)			
Water	✓	✓	–
Air	✓	✓	✓
VOC	✓	✓	✓

Table 3.3 Example 3: Multiple Phases and components

Phases (α)	Gas (G)	Aqueous (A)	NAPL (N)	Solid ($\bar{S}$)
Components (κ)			–	
Gas (g)	✓	✓	✓	
Oil (o)	✓	✓	✓	
Water (w)	✓	✓	–	
Solutes (γ)		✓	✓	
Solid (s)				✓
Heat ($\mathbb{H}$)	✓	✓	✓	✓

B. Mass of Multiple κ-Components in Multiphase Flow

In this case (Table 3.2), for any of the κ-components:

$$\mathfrak{M}_\alpha^\kappa = \sum_{\alpha=G,A,N} \phi S_\alpha \rho_\alpha \omega_\alpha^\kappa, \quad \kappa = g, o, w, s. \tag{3.8.3}$$

C. Mass of a γ-Solute in Multiphase Flow

We are considering the transport of mass of a γ-chemical species in a fluid phase in multiphase flow, using ω_α^γ to denote the mass fraction of γ in α. For all phases together, we have (Table 3.3):

$$\mathfrak{M}^\kappa = \sum_{(\alpha)} \phi S_\alpha \rho_\alpha \omega_\alpha^\gamma, \quad \gamma = 1, 2, \ldots, NC. \tag{3.8.4}$$

The flux $\mathfrak{J}^\gamma$ is by advection, diffusion and dispersion of γ in all phases:

$$\mathfrak{J}^\gamma = \sum_{(\alpha)} \phi S_\alpha \rho_\alpha \omega_\alpha^\gamma \mathbf{V}_\alpha + \sum_{(\alpha)} \mathbf{J}_{\alpha,dif+dis}^\gamma.$$

The rate of production of γ per unit cell volume is

$$\mathfrak{G}^\gamma = \sum_{(\alpha)} \phi S_\alpha \rho_\alpha \Gamma_\alpha^\gamma.$$

Altogether, the γ-balance equation for the nth cell takes the integro-differential form:

$$\int_{\mathbb{V}_n} \frac{\partial}{\partial t} \sum_{(\alpha)} \phi S_\alpha \rho_\alpha \omega_\alpha^\gamma, \; d\mathbb{V}_n = \int_{\mathbb{S}_n} \mathfrak{J}^\gamma \cdot \boldsymbol{\nu} \, d\mathbb{S}_n + \int_{\mathbb{V}_n} \mathfrak{G}^\gamma \, d\mathbb{V}_n, \quad \gamma = 1, 2, \ldots, NC, \tag{3.8.5}$$

and we have NC such equations–one for every species present in the system.

When the γ-species can be adsorbed on the solid, the term $(1-\phi)\rho_s \Upsilon_s^\gamma$, where Υ_s^γ denotes the mass of adsorbed γ-species per unit mass of solid, is added on the r.h.s. of (3.8.4).

D. Heat in Multiphase Flow

In this case, again with multiple β-fluid phases occupying the void space, $\mathfrak{M}^\kappa$ denotes heat (H) per unit volume of cell, treated as a *pseudo-component*:

$$\mathfrak{M}^H = (1-\phi)\rho_s C_s T + \sum_{(\alpha)} \phi S_\alpha \rho_\alpha u_\alpha,$$

in which T denotes temperature, C_s denotes the heat capacity of the solid (s), and u_α denotes the specific internal energy of the α-phase fluid. Accordingly, the heat balance equation is:

$$\int_{\mathbb{V}_n} \frac{\partial}{\partial t} \left[(1-\phi)\rho_s C_s T + \sum_{(\alpha)} \phi S_\alpha \rho_\alpha u_\beta \right] d\mathbb{V}_n = \int_{\mathbb{S}_n} \mathfrak{J}^H \cdot \boldsymbol{\nu} \, d\mathbb{S}_n + \int_{\mathbb{V}_n} \mathfrak{G}^H \, d\mathbb{V}_n. \tag{3.8.6}$$

Finally, a comment about monitoring. Here, like in any numerical technique (e.g., finite element method, finite differences, etc.), in which the domain is divided into finite volume cells, eventually, we assign the average value of state variables in the elementary cell to the cell's center. Then, by interpolation, we obtain the continuous variation of the variables throughout the domain. These are the values that we compare with values that we monitor in the field, say, in wells or by taking samples.

* * *

This concludes our presentation of the balance equations for the extensive quantities: mass of a fluid phase, mass of a solute in a fluid phase, momentum of a phase, and energy of a phase and of the porous medium as a whole. In subsequent chapters, we shall elaborate on each of these cases, and discuss the various coefficients that appear in the flux and transfer expressions. However, before leaving this chapter, we should refer to the observation that a large number of variables of state are involved in any flow and transport problem.

3.9 Primary Variables and Degrees of Freedom

As will be shown in subsequent chapters, a rather large number of variables is required in order to fully describe flow and transport in a porous medium domain, especially when the considered problem involves multiple phases, dissolved interacting chemical species, non-isothermal conditions, and a deformable solid matrix. However, on the basis of balance and thermodynamic relationships, the number of variables to be solved for in a mathematical model can be significantly reduced, thus simplifying the task of solving transport models. Obviously, although the actual solution of the (usually, partial differential) equations is carried out for a small number of variables, eventually, the values of *all* system variables are obtained.

The number of *degrees of freedom* was introduced in Sect. 2.3.1 as the minimal number of variables that are needed in order to *completely* describe a system, whether the latter is at the microscopic level, or the macroscopic one. All other variables are functions of these degrees of freedom, or of their derivatives. Thus, for a given transport problem, the number of degrees of freedom, NF, is equal to the number of variables, NV, minus the constitutive relationships and constraints, NE, imposed on these variables i.e.,

$$\mathrm{NF} = \mathrm{NV} - \mathrm{NE}. \tag{3.9.1}$$

The degrees of freedom are also referred to as *primary variables*. By definition, the primary variables, cannot solely be algebraically expressed in terms of each other, as, often, they must satisfy partial differential equations, in particular, when the latter are equations of balance of extensive quantities.

The material in his section is based on Bear and Nitao (1995), who presented a comprehensive discussion on the subject of *degrees of freedom* in modeling phenomena of transport in porous media under various conditions.

3.9.1 Degrees of Freedom in Multiphase Flow

Let us start with the case in which all phases and species within a system are *at equilibrium*, or when the rate of transformation of the system from one state to another is such that it can be assumed to be continuously close to equilibrium.

The state of a system composed of NP phases and *NC* chemical components under conditions of equilibrium is fully defined by NF state variables determined by the relationship called *Gibbs phase rule* (3.9.2) (e.g., Denbigh 1981, p. 185). Repeated here for convenience, we write:

$$\mathrm{NF} = \mathrm{NC} - \mathrm{NP} + 2. \tag{3.9.2}$$

degrees of freedom.

For example, consider the case of a single fluid phase composed of a single species, say H_2O, at a density ρ and temperature T. For this system, NC $=$ NP $= 1$, NF $= 2$. This means that the state of such system at equilibrium is fully defined by *two* independent (of each other) variables, say the pressure, p, and the temperature, T. We could, however, select also T and ρ, with the constitutive relation, $\rho = \rho(p, T)$, as long as this relationship can, at least in theory, be solved for p as a function of ρ and T. As a second example, consider two fluid phases: a liquid, consisting of a single-species water, and a gas, composed only of water vapour. This means that we have two phases and one species, and by Gibbs phase rule we have NF $= 1$, i.e., one degree of freedom. Suppose we select T as the independent variable. Since we have here liquid water and water vapour in equilibrium, the gas pressure is determined as a function of T by $p_g = p_{sat}(T)$, where p_{sat} is the saturated vapour pressure of water at which the system can exist at any given temperature. Once we know p and T, we can determine the densities of the phases, or the value of any other thermodynamic property.

Next, let us consider a model that describes the behavior at the *macroscopic* level of a system composed of multiple multi-species fluid phases within a possibly deformable porous medium domain under non-isothermal conditions. Here, the behavior *at a point* means the averaged behavior within an REV centered at the considered point.

Based on balance considerations and on thermodynamic relationships, Bear and Nitao (1995) showed that when conditions of thermodynamic equilibria prevail (or are assumed to prevail as a good approximation) among all phases and species present within a deformable porous medium under non-isothermal conditions, the number of degrees of freedom, NF, in a problem of heat and mass transport, involving NP fluid phases and NC non-reacting species, is given by the relationship

$$\mathrm{NF} = \mathrm{NC} + \mathrm{NP} + 4. \tag{3.9.3}$$

Under conditions of nonequilibrium between the phases, this rule becomes

$$\mathrm{NF} = \mathrm{NC} \times \mathrm{NP} + 2 \times \mathrm{NP} + \mathrm{NC} + 4. \tag{3.9.4}$$

In both cases, when Darcy's law is used to determine the velocities of the fluid phases, NF is reduced by NP. When the solid matrix is non-deformable, NF is reduced by 3, leading to the relationship

$$\mathrm{NF} = \mathrm{NC} + 1. \tag{3.9.5}$$

Under isothermal conditions,

$$\mathrm{NF} = \mathrm{NC}. \tag{3.9.6}$$

The above rules are, thus, extensions of the well known *Gibbs phase rule* to phenomena of flow and transport in porous media. They are applicable to NC *non-reacting* species. For a system with chemical reactions, let NS_r be the number of reacting species and NR_{eq} denote the number of equilibrium reactions. Then, by expressing the reactions in the form of a canonical set of equations, and using of the law of mass action, we find NC = NS_r - NR_{eq}.

As an example, consider the case (Lichtner and Karra 2014, p. 81) of NP fluid phases that move and interact within a geological formation under non-isothermal conditions. In this general case, we denote the fluid phases by subscript α, $\alpha = 1, \ldots, NP$. For example, in the case of CO_2 sequestration, the fluids are H_2O and CO_2. Each fluid involves chemical γ-species that move and interact. The considered problem involves the following variables:

- Temperature, T, assuming thermal equilibrium between all participating phases, independent of their number.
- Pressure, p_α, of the α-phase, with $\alpha = 1, 2, \ldots,$ NP pressures of all the participating phases.
- NC chemical γ-species, $\gamma = 1, 2, \ldots$ NC, in every α-phase, with concentrations measured as *molar fractions* X^γ_α, such that $\sum_{(\gamma\,\mathrm{in}\,\alpha)} X^\gamma_\alpha = 1$. We have here 2 variables for every γ – one in any of the two adjacent phases.
- Two phase saturations, S_α, at every point, with $\sum_{(\alpha)} S_\alpha = 1$, and $S_\alpha - S_\beta = p_c$, where β is a phase which is not α, and p_c denotes the capillary pressure. Note that we have a capillary pressure relationship for every pair of fluid phases:

$$p_c|_{\alpha,\beta} = p_\alpha - p_\beta, \quad \alpha \neq \beta.$$

Altogether, the number of variables at a point is: 1 (for T) + 2NP (for S_α, p_α) + NP×NC concentrations (i.e., X^γ_α's). To solve for these variables, we have the following equations:

- Equality of the chemical potential (Sect. 2.2.4) at every point in space and for every instant, between the two adjacent phases (i.e., "no jump" condition):

$$[\![\; \mu^\gamma_\alpha(p_\alpha, c^\gamma_\alpha, T) \;]\!]_{(\mathbf{x},t)} = 0,$$

  i.e., $\mu^\gamma_\alpha = \mu^\gamma_\beta$ for $\gamma = 1, 2, \ldots,$ NC.
- $p_\alpha - p_\beta = p_c(\mathbf{x}, t)$ at every point, i.e., one more constraint (for the two-phase case).

Altogether, we have:

- Number of variables: T, p_α, S_α, X^γ_α, for a total on 1+ 2NP + NP×NC.
- Number of constraints and equations: 1 + NP + NC(NP-1) + NP - 1.

Thus the number of degrees of freedom, is (Lichtner and Karra 2014, p. 83):

$$\mathrm{NF} = \mathrm{NC} + 1.$$

We note that this number of degrees of freedom is independent of the number of phases present in the system.

The number of degrees of freedom for reactive transport problems was also discussed by Saaltink et al. (1998).

Once the number of degrees of freedom has been determined for a given problem, we select the most convenient variables to be declared as primary ones, and identify the (same number of) balance (partial differential) equations which have to be solved in order to determine the values of these variables. All other variables are, subsequently, determined by using the remaining equations—constitutive relations and definitions.

Under certain conditions, as when dealing with multiple multi-component phases, conditions may develop under which the initially selected primary variables cannot continue to play this role (e.g., when a phase disappears due to evaporation). Under such conditions, the initially selected primary variables have to be changed.

3.9.2 Degrees of Freedom Under Nonequilibrium Conditions

In a porous medium, a system that undergoes changes in time due to motion of the phases can never be strictly in complete/exact thermodynamic equilibrium (Bear and Nitao 1995). Conditions of mechanical non-equilibrium prevail as a consequence of the transfer of momentum from the moving fluid to the solid by viscous forces. This gives rise to pressure gradients at the microscopic level within the REV. Temperature gradients may also occur because of viscous dissipation. If these pressure and temperature gradients are large, the system will be far from chemical and thermal equilibrium. Perhaps, more importantly, flow can transport components into the REV, resulting in non-equilibrium concentrations. For a multi-phase REV, flow can cause some phases to be under non-equilibrium conditions, and some of the phases may be in non-equilibrium with each other. Bear and Nitao (1995) consider also this case.

The phases moving within the void space of a porous medium domain are never in thermodynamic equilibrium. Conditions of mechanical non-equilibrium are the consequence of the transfer of momentum by viscous forces from the moving fluid, or fluids, to the solid. This gives rise to pressure gradients at the microscopic level within the REV. Temperature gradients may also occur because of viscous dissipation. If these pressure and temperature gradients are large, the system will be far from chemical and thermal equilibrium. Furthermore, flow can transport chemical species into the REV, thus resulting in non-equilibrium concentrations. For a multi-phase REV, flow may cause some phases to be under non-equilibrium conditions. In spite of the above statements, we usually assume that changes are slow so that we are considering unsteady conditions f flow and transport, but the at every point, the system is close to equilibrium and we use all kinds of equilibrium laws.

A. Degrees of Freedom Under Non-equilibrium Conditions

Bear and Nitao (1995) consider cases of non-equilibrium conditions in a deformable porous medium domain containing a number of multi-species fluid phases under non-isothermal conditions. Some of the chemical species may adsorb on the solid surface. Actually, they assume approximate equilibrium *within* each phase present in the REV, but not *between* phases. The balance equations provide information on the following state variables:

Kind	Number of variables
$\omega_\alpha^\gamma, \rho_{ad}^\gamma$	NC × (NP + 1)
θ_α, θ_s	NP + 1
T_α, T_s	NP + 1
p_α	NP
$\mathbf{V}_\alpha, \mathbf{V}_s$	NP + 1
$\mathbf{w}_s$	1

where ρ_{ad}^γ denotes the REV-averaged mass density of an adsorbed γ-species, $\mathbf{w}_s$ denotes the displacement in the solid matrix, with:

$$\frac{D_{\mathbf{V}_s}\mathbf{w}_s}{Dt} = \mathbf{V}_s, \quad \text{or approximately} \quad \frac{\partial \mathbf{w}_s}{\partial t} = \mathbf{V}_s,$$

and $\alpha = 1, \ldots, \text{NP}$. The total number of variables is

$$\text{NV} = \text{NC} \times \text{NP} + \text{NC} + 4\text{NP} + 4. \tag{3.9.7}$$

We have not listed the phase internal energies, u_α, and the densities, ρ_α, as they can be related to the other variables through appropriate constitutive relations, assuming macroscopic thermodynamic equilibrium within each phase.

The actual number of independent variables is much smaller due to various constraints. There are NP+1 constraints on these variables due to the following relations:

$$\sum_{\alpha=1}^{\text{NP}} \theta_\alpha + \theta_s = 1, \qquad \text{and} \qquad \sum_{(\gamma)} \omega_\alpha^\gamma = 1. \tag{3.9.8}$$

In addition, we have to take into account the NP-1 capillary pressure relationships,

$$p_1 = p_\alpha - p_{c1\alpha}(\theta_\delta, T_\beta, \omega_\beta^\gamma), \quad \alpha = 2, \ldots, \text{NP}. \tag{3.9.9}$$

Altogether, the total number of constraints is

$$\text{NE} = 2\text{NP}, \tag{3.9.10}$$

and the number of degrees of freedom is equal to

$$\text{NF} = \text{NV} - \text{NE} = \text{NC} \times \text{NP} + \text{NC} + 2\text{NP} + 4. \tag{3.9.11}$$

To solve for these primary variables, we have

$$\text{NB} = \text{NV-NE} = \text{NC} \times \text{NP} + \text{NC} + 2\text{NP} + 4 = \text{NF}$$

three partial differential equations:

Kind	Number of equations
Massbalances	NC × NP + NC + 1
Energybalances	NP + 1
Momentumbalances	NP + 1,

and the definition of $\mathbf{w}_s$ as the 4th equation.

The set of primary variables is, in general, not unique. An example of a set of NF primary variables is:

Kind	Range	Number
ω_α^γ	$\gamma = 1, \ldots, \text{NC} - 1$	$(\text{NC} - 1) \times \text{NP}$
	$\alpha = 1, \ldots, \text{NP}$	
ρ_{ad}^γ	$\gamma = 1, \ldots, \text{NC}$	NC
$\theta_s (\equiv 1 - \phi)$		1
p_α	$\alpha = 1, \ldots, \text{NP}$	NP
T_α	$\alpha = 1, \ldots, \text{NP}$	NP
T_s		1
$\mathbf{V}_\alpha$	$\alpha = 1, \ldots, \text{NP}$	NP
$\mathbf{V}_s$		1
$\mathbf{w}_s$		1

Let us show how the other variables can be solved in terms of these primary variables. For example:

- We first solve for the θ_α $(\alpha = 2, \ldots, \text{NP})$ in terms of the p_α, using the capillary relationships (3.9.9).
- We then solve for θ_1 in terms of the other θ_α and θ_s, using:

$$\sum_{\alpha=1}^{\text{NP}} \theta_\alpha + \theta_s = 1, \qquad \text{and} \tag{3.9.12}$$

- Finally, we solve for ω_α^γ, using $\sum_{(\gamma)} \omega_\alpha^\gamma = 1$.

Low volumetric fractions may not be unique function of pressure, and we may need to use them instead of pressure (see Nitao and Bear 1996). Thus, another possible set of primary variables is obtained by replacing the p_α's by the set, p_1, θ_α

Table 3.4 Degrees of freedom and primary variables for nonequilibrium case (Bear and Nitao 1995)

Deformable solid	Darcy's law	Iso-thermal	NF	Primary variables ($\gamma' = 1, \ldots, $ NC-1; $\gamma = 1, \ldots, $ NC; $\alpha = 1, \ldots, $ NP)
Yes	No	No	NC $\times$ NP $+$ NC $+$ 2NP $+$ 4	$p_\alpha,\ \omega_\alpha^{\gamma'},\ \rho_{ad}^\gamma,\ \mathbf{V}_\alpha,$ $\mathbf{V}_s,\ \mathbf{w}_s,\ \theta_s,\ T_\alpha,\ T_s,$
Yes	Yes	No	NC $\times$ NP $+$ NC $+$ NP $+$ 4	$p_\alpha,\ \omega_\alpha^{\gamma'},\ \rho_{ad}^\gamma,\ \mathbf{V}_s,$ $\mathbf{w}_s,\ \theta_s,\ T_\alpha,\ T_s,$
No	Yes	No	NC $\times$ NP $+$ NC $+$ NP $+$ 1	$p_\alpha,\ \omega_\alpha^{\gamma'},\ \rho_{ad}^\gamma,\ T_s,\ T_\alpha,$
No	No	No	NC $\times$ NP $+$ NC $+$ 2NP $+$ 1	$p_\alpha, \omega_\alpha^{\gamma'},\ \rho_{ad}^\gamma,\ \mathbf{V}_\alpha, T_s,\ T_\alpha,$
Yes	No	Yes	NC $\times$ NP $+$ NC $+$ NP $+$ 1	$p_\alpha,\ \omega_\alpha^{\gamma'},\ \rho_{ad}^\gamma,\ \mathbf{V}_\alpha, \mathbf{V}_s \mathbf{w}_s,\ \theta_s,$
Yes	Yes	Yes	NC $\times$ NP $+$ NC $+$ 3	$p_\alpha,\ \omega_\alpha^{\gamma'},\ \rho_{ad}^\gamma,\ \mathbf{V}_s, \mathbf{w}_s,\ \theta_s,$
No	Yes	Yes	NC $\times$ NP $+$ NC	$p_\alpha,\ \omega_\alpha^{\gamma'},\ \rho_{ad}^\gamma,$
No	No	Yes	NC $\times$ NP $+$ NC $+$ NP	$p_\alpha,\ \omega_\alpha^{\gamma'},\ \rho_{ad}^\gamma,\ \mathbf{V}_\alpha,$

($\alpha = 2, \ldots,$ NP). In this case, the p_α, ($\alpha = 2, \ldots,$ NP) are obtained as functions of the θ_α and p_1, using (3.9.9).

Since, within an REV, the number of phases may change as phases may disappear or appear, the actual number and type of primary variables may change with time.

The primary variables depend also on the type of problem that is being modeled and the simplifications involved, e.g., whether or not the solid is assumed to be deformable. In the latter case, the variables $\mathbf{V}_s$, $\mathbf{w}_s$, and θ_s are not needed, and NF is reduced by three. Other cases are when the problem is isothermal, so that $T_s = T_\alpha =$ initial temperature. In this case, the temperatures are not required as primary variables, and NF is reduced by NP+1. When approximate thermal equilibrium exists between phases, $T_s = T_\alpha$, only the temperature, say T_s, is needed, and then NF is reduced by NP. When Darcy's law is valid, the phase velocities $\mathbf{V}_\alpha$ can be expressed in terms of the other primary variables so that they should not be counted as primary variables. These special cases are summarized in Table 3.4.

If the phases are in thermal equilibrium with each other, then $T_\alpha = T_s$ and the T_α's may be removed from the set of primary variables. NF is then reduced in all the above cases by the amount NP. If the adsorbed chemical species on the solid are in chemical equilibrium with the fluid phases, then the ρ_{ad}^γ's are no longer needed as primary variables and NF is reduced by NC.

B. Degrees of Freedom for Approximate Chemical–Thermal Equilibrium

Suppose that at every (macroscopic) point within the domain, and at every instant of time, the system is assumed to be in *approximate thermodynamic equilibrium*. Bear and Nitao (1995) also consider the case in which the exchange of certain extensive quantities between NPE phases (1 < NPE < NP) occurs sufficiently fast, so as to establish equilibrium between them, with respect to the considered quantities, while the other NP-NPE phases are not in equilibrium with respect to the same quantities.

Under equilibrium, we assume that adsorption isotherms take the form:

$$\rho_{ad}^{\gamma}/\rho_b = f_\alpha(\omega_\alpha^\gamma \rho_\alpha), \tag{3.9.13}$$

where ρ_b is the bulk density of the solid matrix (assumed constant), with:

$$\rho_b \equiv \frac{1}{\mathcal{U}_o} \int_{\mathcal{U}_{so}} \rho_s d\mathcal{U}. \tag{3.9.14}$$

We start with the same NV=NC×NP+NC+4NP+4 variables as above. We have the following constraints:

$\theta_s + \sum_{(\alpha)} \theta_\alpha = 1$	1
$\sum_{(\gamma)} \omega_\alpha^\gamma = 1$	NP
$T_s = T_\alpha$	NP
$\mu_\alpha^\gamma = \mu_\beta^\gamma$	(NP-1) × NC
$\rho_{ad}^{\gamma}/\rho_b = f_\alpha(\omega_\alpha^\gamma \rho_\alpha)$	NC
capillary pressure	NP − 1

The number of constraints is NE=NP×NC+3NP. The number of degrees of freedom is

$$\text{NF} = \text{NV} - \text{NE} = \text{NC} + \text{NP} + 4. \tag{3.9.15}$$

The mass balance equations for all components may be summed over all the phases to give NC component balance equations. The exchange fluxes cancel since $f_{\alpha\rightarrow\beta}^\gamma = -f_{\beta\rightarrow\alpha}^\gamma$. This leaves NC mass balance equations for the γ component and one mass balance equation for the solid matrix. Similarly, the exchange fluxes cancel when the phase energy balance equations are summed over all the phases within an REV, resulting in a single energy balance equation.

We now count the number of balance equations.

Mass balance	NC + 1
Energy balance	1
Momentum balance	NP
Definition of $\mathbf{w}_s$	1

The total number of equations is

Table 3.5 Degrees of freedom and primary variables for equilibrium case (Bear and Nitao 1995)

Deformable solid	Darcy's law	Iso-thermal	NF	Primary variables ($\gamma = 1, \dots,$ NC-NP; $\alpha = 1, \dots,$ NP)
Yes	No	No	NC+NP+4	$p_\alpha,\ \omega_1^\gamma,\ \mathbf{V}_\alpha,\ \mathbf{V}_s,\ \mathbf{w}_s,\ \theta_s,\ T_s,$
Yes	Yes	No	NC+4	$p_\alpha,\ \omega_1^\gamma,\ \mathbf{V}_s,\ \mathbf{w}_s,\ \theta_s,\ T_s,$
No	Yes	No	NC+1	$p_\alpha,\ \omega_1^\gamma,\ T_s,$
No	No	No	NC+NP+1	$p_\alpha,\ \omega_1^\gamma,\ \mathbf{V}_\alpha,\ T_s,$
Yes	No	Yes	NC+NP+1	$p_\alpha,\ \omega_1^\gamma,\ \mathbf{V}_\alpha,\ \mathbf{V}_s,\ \mathbf{w}_s,\ \theta_s,$
Yes	Yes	Yes	NC+3	$p_\alpha,\ \omega_1^\gamma,\ \mathbf{V}_s,\ \mathbf{w}_s,\ \theta_s,$
No	Yes	Yes	NC	$p_\alpha,\ \omega_1^\gamma,$
No	No	Yes	NC+NP	$p_\alpha,\ \omega_1^\gamma,\ \mathbf{V}_\alpha.$

$$\text{NB} = \text{NC} + \text{NP} + 4. \tag{3.9.16}$$

This is also equal to the number of degrees of freedom, NF.

A possible set of primary variables is p_α ($\alpha = 1, .., \text{NP}$), ω_1^γ ($\gamma = 1, .., \text{NC-NP}$), $T_s, \mathbf{V}_\alpha$ ($\alpha = 1, .., \text{NP}$), $\mathbf{V}_s, \mathbf{w}_s, \theta_s$. Solution of the θ_α in terms of the primary variables is described above for the nonequilibrium case. It is obvious that the T_α's are obtained through $T_\alpha = T_s$. The values of ω_α^γ ($\alpha = 2, \dots, \text{NP}$; $\gamma = 1, \dots, \text{NC-NP}$) are found through solving the equations:

$$\mu_\alpha^\gamma(\omega_\alpha^\kappa, p_\alpha, T_\alpha) = \mu_1^\gamma(\omega_1^\kappa, p_1, T_1). \tag{3.9.17}$$

The remaining ω_α^γ ($\alpha = 1, \dots, \text{NP}$; $\gamma = \text{NC-NP+1}, \dots, \text{NC}$) are found by solving

$$\sum_{\gamma=\text{NC}-\text{NP}+1}^{\text{NC}} \omega_\alpha^\gamma = 1 - \sum_{\gamma=1}^{\text{NP}-\text{NC}} \omega_\alpha^\gamma, \tag{3.9.18}$$

where the right-hand side is known and the left-hand side contains the variables to be solved for. The ρ_{ad}^γ's are obtained from (3.9.13). A number of particular cases are considered in Table 3.2.

It is obvious that for isothermal problems, T_s is no longer a primary variable and that NF is reduced by one. For a nondeformable, isothermal system with Darcy's law, we have NF = NC (Table 3.5).

3.9.3 *Partial Phase Equilibrium*

It is possible that the exchange of certain extensive quantities between NPE phases (1 < NPE < NP) occurs sufficiently fast, so as to establish equilibrium between them, with respect to the considered quantities. The other NP-NPE phases are not in equilibrium with respect to the same quantities. This means that the potentials associated with the considered quantities are approximately constant over the NPE phases. It is then clear that the number of degrees of freedom is smaller than in the case where all phases are not in equilibrium with each other. In particular, consider the case where NPE phases are in equilibrium with each other with respect to the mass and energy of the NC components. Then, we have the (NPE-1)×NC constraints from the equality of chemical potentials between the NPE phases, and NPE-1 constraints for the equality of temperatures, for a total of NPE×NC-NC+NPE-1 additional constraints. Adding these to the NE (= 2NP) constraints that correspond to the case of complete nonequilibrium, we obtain:

$$\mathrm{NE} = 2\mathrm{NP} + \mathrm{NPE} \times \mathrm{NC} - \mathrm{NC} + \mathrm{NPE} - 1$$

constraints. Since the number of variables is NV=NC×NP+NC+4NP+4, the number of degrees of freedom is:

$$\mathrm{NF} = \mathrm{NV} - \mathrm{NE} = \mathrm{NC} \times (\mathrm{NP} - \mathrm{NPE}) - \mathrm{NPE} + 2\mathrm{NC} + 2\mathrm{NP} + 5.$$

A possible set of primary variable is: $\omega^{\delta}_{\alpha_o}$ ($\delta = 1, \ldots,$ NC-NPE), $\omega^{\gamma}_{\alpha'}$ ($\gamma = 1, \ldots,$ NC-1), ρ^{γ}_{ad} ($\gamma = 1, \ldots, NC$), θ_s, p_α, $T_{\alpha'}$, T_s, $\mathbf{V}_\alpha$, $\mathbf{w}_s$, where $\alpha = 1, \ldots,$ NP, $\alpha' = 1, \ldots,$ NP-NPE denotes all the phases that are not in equilibrium, and α_o is one of the NPE phases.

3.10 Dimensionless Numbers and Non-dominant Effects

In each of the balance equations presented throughout this book, we see the effect of a number of phenomena. In fact, we note these phenomena already in the verbal form of the general balance equation (3.3.2). When all these phenomena occur simultaneously, it is possible that under certain conditions, the effect of some phenomena are much larger than others, to the extent that by neglecting the less dominant phenomena, or the *non-dominant phenomena*, we obtain a simpler, albeit approximate model. The latter model is, usually, much easier to solve, yet it provides a solution which is not much different from that obtained by solving the full model.

A simple approach is to define dimensionless numbers and express flux and balance equations in terms of these number. For example. Let us define the following dimensionless numbers:

$$\text{Reynolds number}: \quad Re = \frac{q_c \ell_c}{\nu_c}, \qquad \text{Euler number} \quad Eu = \frac{(\Delta p)_c}{\rho_c q_c^2},$$

$$\text{Froude number}: \quad Fr = \frac{q_c^2}{g \ell_c}, \qquad \text{Darcy number} \quad Da = \frac{k}{\ell_c^2}.$$

in which subscript c denotes a characteristic value.

A number of dimensional numbers often used in dealing with phenomena of transport in prous media will be added below.

Following Bear and Bachmat (1991, p. 268), let us suggest a methodology that leads to the derivation of dimensionless numbers. We shall make use of two examples. In these examples, the emphasis will be on the identification of non-dominant effects, and not on the specific case described by these equations. Other examples are scattered throughout the book.

Another subject, associated with the evaluation of dominance of effects is that of *nondimensionalization*. Because each equation that appears in a mathematical model expresses a physical phenomenon, it must be *dimensionally homogeneous*. Hence, it can always be *nondimensionalized*, i.e., written in terms of dimensionless quantities. This goal is achieved by the following three steps:

(a) By introducing for each dependent variable, independent variable and coefficient, a dimensionless variable that expresses the ratio between the considered (dimensional) quantity and a corresponding *intrinsic reference quantity* of the same dimension (e.g., length, time and force).
(b) By inserting the dimensionless ratios into the original equation, we obtain an equation in which each term is a product of the dimensionless ratio and a dimensional coefficient made up only of the reference quantities.
(c) By dividing all terms appearing in the equation by one of the coefficients, we obtain the *dimensionless form* of the original equation.

Let us demonstrate the method through a number of examples (see also Bear and Bachmat 1991, p. 274):

Example 1: Consider the macroscopic mass balance equation (3.3.7), written for the case of saturated flow ($\theta \to \phi$) in a rigid solid matrix ($\partial\phi/\partial t = 0$), with $\mathbf{J} = \rho\mathbf{V}$, and $f^m_{\beta\to\alpha} = 0$:

$$\phi\frac{\partial\rho}{\partial t} = (-\nabla\cdot\rho\mathbf{q}) = -\rho\nabla\cdot\mathbf{q} - \mathbf{q}\cdot\nabla\rho. \tag{3.10.1}$$

We introduce the following dimensionless variables (indicated by an asterisk) and *scales*, or characteristic values (indicated by subscript c):

$$\phi^* = \frac{\phi}{\phi_c}, \quad t^* = \frac{t}{t_c^{(\rho)}}, \quad \rho^* = \frac{\rho}{\rho_c}, \quad \rho_c = \rho_{max}, \quad q_c = |\mathbf{q}|_{max}$$

$$\left(\frac{\partial\rho}{\partial t}\right)^* \equiv \frac{\partial\rho^*}{\partial t^*} = \frac{\partial\rho}{\partial t}\Big/\left(\frac{\partial\rho}{\partial t}\right)_c, \quad \left(\frac{\partial\rho}{\partial t}\right)_c = \left|\frac{\partial\rho}{\partial t}\right|_{max},$$

$$(\Delta t)_c^{(\rho)} = \frac{\rho_c}{(\partial\rho/\partial t)_c} = \frac{\rho_{max}}{|\partial\rho/\partial t|_{max}},$$

$$\mathbf{q}^* = \frac{\mathbf{q}}{|\mathbf{q}|_{max}}, \quad (\nabla\cdot\mathbf{q})^* = \nabla^*\cdot\mathbf{q}^* = \frac{\nabla\cdot\mathbf{q}}{(\nabla\cdot\mathbf{q})_c}, \quad (\nabla\cdot\mathbf{q})_c = \left|\nabla\cdot\mathbf{q}\right|_{max},$$

$$(\nabla\rho)^* \equiv \nabla^*\rho^* = \frac{\nabla\rho}{(\nabla\rho)_c}, \quad (\nabla\rho)_c = \left|\nabla\rho\right|_{max}, \quad \nabla^* = L_c\nabla,$$

$$L_c^{(q)} = \frac{|\mathbf{q}|_{max}}{|\nabla\cdot\mathbf{q}|_{max}}, \quad L_c^{(\rho)} = \frac{\rho_{max}}{|\nabla\rho|_{max}}.$$

With these substitutions, we rewrite (3.10.1) and divide all terms by $\rho_{max}|\mathbf{q}|_{max} \equiv \rho_c q_c$. We obtain

$$\mathrm{St}^{(\rho)}\phi^*\frac{\partial\rho^*}{\partial t^*} + \frac{L_c^{(q)}}{L_c^{(\rho)}}\rho^*\nabla^*\cdot\mathbf{q}^* + \mathbf{q}^*\cdot\nabla^*\rho^* = 0, \tag{3.10.2}$$

where

$$\mathrm{St}^{(\rho)} = \frac{L_c^{(\rho)}/V_c}{t_c^{(\rho)}}$$

is the Strouhal number of this example. We have here two characteristic length scales: $L_c^{(q)}$, which characterizes the length over which a significant change in $\mathbf{q}$ takes place, and $L_c^{(\rho)}$, which indicates the length over which the same change is in ρ.

It is interesting to understand the interpretation of the Strouhal number in this case. Verbally, for the general case of any E, it relates the added E per unit volume and unit time to the added E per unit volume and unit time by advection. With $E = m$ and $e = \rho$, we obtain

$$\mathrm{St}^{(e)} = \frac{|\partial e/\partial t|_c}{|\nabla\cdot e\mathbf{V}|_c} = \frac{L_c^{(e)}/V_c}{t_c^{(e)}}. \tag{3.10.3}$$

Another interpretation of $\mathrm{St}^{(e)}$ is that it expresses the reciprocal of the time during which significant changes in values of e are induced, compared to that required for these changes to spread out out throughout the domain by advection.

Consider $E = m$, $e = \rho$. Unless we have very rapid changes in ρ, the time increment, $(\Delta t)_c^\rho$ for a local change in ρ is large compared to the travel time required to obtain the same spatial change in ρ, and, therefore, $\mathrm{St}^{(\rho)} \ll 1$. Then, the first term (3.10.2) is much smaller than the third, and may be deleted. Often, $L_c^{(q)} \ll L_c^{(\rho)}$, so that the third term may also be neglected. Then, (3.10.2) reduces, *as an approximation*, to the dimensionless equation:

$$\nabla^*\cdot\mathbf{q}^* = 0. \tag{3.10.4}$$

Example 2: Consider the balance equation for a chemical species (concentration c) that undergoes radioactive decay. At the microscopic level, under certain conditions, this equation takes the form

$$\frac{\partial c}{\partial t} = -\frac{\partial}{\partial x_i}\left(cV_i - \mathcal{D}_{ij}\frac{\partial c}{\partial x_j}\right) - \lambda c, \tag{3.10.5}$$

where the last term on the r.h.s. expresses the rate (per unit fluid volume) at which the concentration of the considered species is reduced by radioactive decay; λ denotes the decay constant.

The proposed methodology will be presented as a number of steps:

Step 1: For every dependent variable, independent variable, and coefficient that appears in the considered equation, we introduce a *dimensionless variable* (denoted by an asterisk) that represents the ratio between the considered (dimensional) quantity and a corresponding *characteristic quantity*, of the same dimension, denoted by subscript c. In the example considered here,

$$\frac{\partial c}{\partial t} = \left(\frac{\partial c}{\partial t}\right)^* \frac{(\Delta c)_c}{(\Delta t)_c} \equiv \frac{\partial c^*}{\partial t^*}\frac{(\Delta c)_c}{(\Delta t)_c}, \tag{3.10.6}$$

$$\frac{\partial c V_i}{\partial x_i} \equiv c\frac{\partial V_i}{\partial x_i} + V_i\frac{\partial c}{\partial x_i} = c^*\frac{\partial V_i^*}{\partial x_i^*}\frac{(\Delta V)_c(\delta c)_c}{L_c^{(v)}} + V_i^*\frac{\partial c^*}{\partial x_i^*}\frac{V_c(\Delta c)_c}{L_c^{(c)}}, \tag{3.10.7}$$

$$\frac{\partial}{\partial x_i}\left(\mathcal{D}_{ij}\frac{\partial c}{\partial x_j}\right) = \frac{\partial}{\partial x_i^*}\left(\mathcal{D}_{ij}^*\frac{\partial c^*}{\partial x_j^*}\right)\frac{\mathcal{D}_c(\Delta c)_c}{L_c^{(c)2}}, \tag{3.10.8}$$

$$\lambda c = \lambda^* c^* \lambda_c (\Delta c)_c, \quad x_i = x_i^* L_c^{(.)}, \tag{3.10.9}$$

where $(\Delta t)_c$, is a characteristic time interval, $L_c^{(c)}$ and $L_c^{(v)}$ are lengths characterizing the spatial changes in c and $\mathbf{V}$, respectively, and $\mathcal{D}_c$ is a characteristic coefficient of dispersion. We may relate these lengths to the gradients in the respective quantities, e.g.,

$$\frac{(\Delta V)_c}{L_c^{(v)}} = \max\left|\frac{dV}{dx}\right|, \qquad \text{viz.,} \quad L_c^{(v)} = \frac{|V|_{\max}}{|dV/dx|_{\max}}. \tag{3.10.10}$$

We can also use some characteristic length of the domain. Usually, we assume a single common characteristic length, associated with all the domain's dimensions. We shall do so here, denoting it by L_c. This is not essential, as we could proceed with different characteristic lengths. A similar discussion applies to the characteristic time, which may take on different values, depending on the particular transported quantity and mode of transport. Here, $(\Delta t)_c$ denotes the characteristic time for a change in concentration at a point in the domain. We note that the characteristic *rate* of a process is inversely proportional to its characteristic time. The maximum velocity within a considered domain may be taken as the characteristic velocity.

In principle, there may be different characteristic times for different processes, and they need not be equal to each other.

Step 2: When we insert these relationships into the considered balance equation, we obtain:

$$\begin{aligned}\frac{\partial c^*}{\partial t^*}\frac{(\Delta c)_c}{(\Delta t)_c} &= -c^*\frac{\partial V_i^*}{\partial x_i^*}c_c\frac{(\Delta V)_c}{L_c} - V_i^*\frac{\partial c^*}{\partial x_i^*}\frac{V_c(\Delta c)_c}{L_c}\\ &\quad + \frac{\partial}{\partial x_i^*}\left(\mathcal{D}_{ij}^*\frac{\partial c^*}{\partial x_j^*}\right)\frac{\mathcal{D}_c(\Delta c)_c}{L_c^2} - \lambda^* c^* \lambda_c (\Delta c)_c.\end{aligned} \tag{3.10.11}$$

Step 3: We rewrite the original balance equation in one of the three dimensionless forms:

$$\mathrm{St}\frac{\partial c^*}{\partial t^*} = -\frac{\partial}{\partial x_i^*}\left(c^* V_i^* - \frac{1}{\mathrm{Pe}}\mathcal{D}_{ij}^*\frac{\partial c^*}{\partial x_j^*}\right) - \mathrm{Dm}'\lambda^* c^*, \tag{3.10.12}$$

$$\frac{1}{\mathrm{Fo}}\frac{\partial c^*}{\partial t^*} = -\frac{\partial}{\partial x_i^*}\left(\mathrm{Pe}\, c^* V_i^* - \mathcal{D}_{ij}^*\frac{\partial c^*}{\partial x_j^*}\right) - \mathrm{Dm}''\lambda^* c^*, \tag{3.10.13}$$

$$\frac{\mathrm{St}}{\mathrm{Pe}}\frac{\partial c^*}{\partial t^*} = -\frac{1}{\mathrm{Pe}}\frac{\partial}{\partial x_i^*}\left(c^* V_i^* - \frac{1}{\mathrm{Pe}}\mathcal{D}_{ij}^*\frac{\partial c^*}{\partial x_j^*}\right) - \mathrm{Dm}'''\lambda^* c^*, \tag{3.10.14}$$

in which, with c_c denoting $(\Delta c)_c$, V_c denoting $(\Delta V)_c$, and $\Delta t_c \equiv t_{c,accum}$, we have made use of the following *dimensionless numbers*:

$$\begin{aligned}
\mathrm{St} &\equiv \frac{L_c}{V_c t_c} = \frac{L_c/V_c}{t_c} = \frac{t_{c,adv}}{t_c} = \text{Strouhal number},\\
\mathrm{Pe} &\equiv \frac{L_c V_c}{\mathcal{D}_c} = \frac{L_c^2/\mathcal{D}_c}{L_c/V_c} = \frac{t_{c,dif}}{t_{adv}} = \text{Peclet number},\\
\mathrm{Fo} &\equiv \frac{t_c}{L_c^2/\mathcal{D}} = \frac{t_c}{t_{c,dif}} = \frac{1}{\mathrm{St\,Pe}} = \text{Fourier number},\\
\mathrm{Dm}' &\equiv \frac{L_c/V_c}{1/\lambda} = \frac{t_{c,adv}}{t_{c,react}} = \text{1st kind Damkohler number},\\
\mathrm{Dm}'' &\equiv \frac{L_c^2/\mathcal{D}}{1/\lambda} = \frac{t_{c,dif}}{t_{c,react}} = \mathrm{Pe\,Dm}' = \text{2nd kind Damkohler number},\\
\mathrm{Dm}''' &\equiv \frac{\mathrm{Dm}'}{\mathrm{Pe}} = \frac{\mathrm{Dm}''}{\mathrm{Pe}^2} = \frac{\lambda\mathcal{D}}{V_c^2}\\
&= \frac{t_{c,adv}}{t_{c,dif}}\frac{t_{c,adv}}{t_{c,react}} = \text{3rd kind Damkohler number}.
\end{aligned} \tag{3.10.15}$$

Note that

$$t_{c,adv} \equiv \frac{L_c}{V_c}, \qquad t_{c,dif} \equiv \frac{L_c^2}{\mathcal{D}_c}, \qquad t_{c,react} \equiv \frac{1}{\lambda}, \tag{3.10.16}$$

denote the characteristic times of advection, diffusion, and chemical reactions, respectively:

Since, if the reference (or characteristic) values are properly selected, the asterisk'ed terms are always of *of order one* (Bear and Bachmat 1991, p. 271), the dominance of a term is determined by *the magnitude of the dimensionless numbers that appear in that term*. We note that in (3.10.12), the rate of accumulation by advection (= inverse of characteristic time of advection) is used as a reference. In (3.10.13), the rate of accumulation by diffusion (= inverse of characteristic time of diffusion) is used as a reference time.

Step 4: We consider the ratio between any two terms in (3.10.5), which we wish to compare with each other. For simplicity, consider a one-dimensional case. For example,

$$\frac{|\partial c/\partial t|}{|\partial(cV)/\partial x|} = \mathrm{St}\frac{\partial c^*/\partial t^*}{\partial(c^*V^*)/\partial x^*}. \tag{3.10.17}$$

Since every term with an asterisk is of order one, the *Strouhal number*, St, indicates the ratio between two characteristic time intervals: that required for a significant change in concentration to spread throughout the considered domain by advection, and the other ($t_c \equiv t_{c,accum}$) that is required for local changes in concentration to take place.

Example 3: Let us compare advective and diffusive fluxes:

$$\frac{|cV|}{|\mathcal{D}\partial c/\partial x|} = \mathrm{Pe}\frac{c^*V^*}{\mathcal{D}^*\partial c^*/\partial x^*}. \tag{3.10.18}$$

From the above equation, it follows that the *Peclet number expresses the ratio between the advective and diffusive fluxes*. For Pe $\gg$ 1, advection dominates over diffusion. For Pe $\ll$ 1, diffusion dominates.

It is possible to interpret the Peclet number also as a ratio between two time scales: one ($t_{c,dif} = L_c^2/\mathcal{D}$) that is required for spreading by diffusion, the other ($t_{c,adv} = L_c/V_c$) for spreading by advection. When the former time scale is smaller than the latter, diffusion dominates over advection.

Example 4: Let us compare the rate of production of a source (here, a sink due to radioactive decay) with the rate of accumulation (or spreading out) by advection. This is expressed by the *Damköhler number* of the first kind defined above. Again, this number may be *interpreted as the ratio between two characteristic times: that of advection, and that of production* ($= 1/\lambda$). It is also possible to define another (second kind) Damköhler number, by replacing the time required for spreading by advection, by the time required for spreading by diffusion, or by comparing the source term in the balance equation ($|\lambda c|$), with that expressing accumulation by diffusion ($|\nabla\cdot\mathcal{D}\nabla c|$). We obtain

$$\mathrm{Dm}'' = \frac{L_c^2/\mathcal{D}}{1/\lambda} = \mathrm{Pe}\,\mathrm{Dm}'. \tag{3.10.19}$$

Altogether, the two Damköhler numbers are defined as the ratio between the reaction rate and the rates of mass transport by advection and by diffusion.

Note that Dm$''$ is independent of the characteristic velocity, and that Dm$'''$ is not a ratio between two characteristic times. All three are referred to as 'Damköhler numbers', as they involve the characteristic time of reaction.

By examining whether a dimensionless number is much smaller or much larger than unity, we may learn the relative significance of various transport processes. Following are some examples:

- When $\mathrm{Dm}' \ll 1$, advection dominates over the source term. In other words, the time required for transport by advection is much smaller than that required for production or removal by the source. Conversely, when $\mathrm{Dm}' \gg 1$, the source term dominates over advection.
- When $\mathrm{Dm}'' \ll 1$, the spreading of the contaminant by diffusion dominates over the source term. In other words, the time required for spreading by diffusion is much smaller than that required for production, or removal by the source.

From (3.10.12)–(3.10.14), it follows that:

- When $\mathrm{Pe} \ll 1$, transport by diffusion dominates over that by advection. If also:
  - $\mathrm{Dm}'' \gg 1$, the reaction is referred to as a *fast reaction*.
  - $\mathrm{Dm}'' \ll 1$, the reaction is referred to as a *slow reaction*.
- When $\mathrm{Pe} \gg 1$, transport by advection dominates over that by diffusion. If also:
  - $\mathrm{Dm}' \gg 1$, the reaction is referred to as a *fast reaction*.
  - $\mathrm{Dm}' \ll 1$, the reaction is referred to as a *slow reaction*.
- When $\mathrm{Pe} \gg 1$, and also $\mathrm{Dm}'''' \gg 1$, we have a process dominated by advection and reaction. If also
  - $\mathrm{Dm}' \gg 1$, the reaction is referred to as a (relatively) *fast reaction*.
  - $\mathrm{Dm}' \ll 1$, the reaction is referred to as a (relatively) *slow reaction*.
- When $\mathrm{Pe} \gg 1$, and $\mathrm{Dm}' \ll 1$, the situation implies $\mathrm{Dm}''' \ll 1$; the process is dominated by advection and reaction.
- When $\mathrm{Pe} \ll 1$, and $\mathrm{Dm}' \ll 1$, the situation implies $\mathrm{Dm}''' \gg 1$; the process is dominated by diffusion and reaction.

Let us introduce two additional useful dimensionless numbers:

Fourier number:
$\mathrm{Fo}^{(E)}$, for any E:

$$\mathrm{Fo}^E = \frac{|\nabla\cdot(\mathbf{D}^{(E)}\cdot\nabla e)|_c}{|\partial e/\partial t|_c} = \frac{t_c^E}{L_c^E/D_c^E}. \tag{3.10.20}$$

It expresses the *ratio between the time interval required for the introduction of changes in the density of an extensive quantity* (E) *into a system, say, in the vicinity of a point, and that required for these changes to spread throughout the system by the dispersion–diffusion process*.

Another kind of Fourier number, associated with the fluid's kinematic viscosity, can be defined as:

$$\mathrm{Fo}^{\nu} = \frac{t_c^{(q)}}{(L_c^{(q)})^2/\nu_c}. \tag{3.10.21}$$

It expresses the ratio between the time interval during which a significant change in velocity (= momentum per unit mass!) occurs and the time required for smoothing out spatial velocity differences by molecular transfer of momentum.

References

Bear J (1972) Dynamics of fluids in porous media. American Elsevier, Amsterdam, p 764 (also published by Dover Publications, 1988; translated into Chinese)
Bear J (1961) On the tensor form of dispersion. J. Geophys. Res., vol. 66, pp, 1185–1197
Bear J, Bachmat Y (1991) Introduction to modeling phenomena of transport in porous media. Kluwer Publ. Co, Dordrecht, p 553
Bear J, Fel L (2012) A phenomenological approach to modeling transport in porous media. Transp. Porous Media 92(3):649–665
Bear J, Nitao JJ (1995) On equilibrium and primary variables in transport in porous media. Transp. Porous Media 18:151–184
Bear J, Pinder GF (1978) Porous medium deformation in multiphase flow. Proc. A.S.C.E. 1(04):891–894
Berkowitz B, Dror I, Yaron B (2008) Contaminant geochemistry: interactions and transport in the subsurface environment. Springer, Heidelberg, p 412
Biot MA (1941) General theory of three-dimensional consolidation. J. Appl. Phys. 12:155–164
Chambre PL, Schaaf SA, (1961) Flow of Rarified Gases. Princeton Univ. Press., 66p
De Groot SR, Mazur P (1962) Non-equilibrium thermodynamics. North-Holland Pub. Co., Amsterdam, The Netherlands, 510 pp
Denbigh KG (1981) The Principles of Chemical Equilibrium, 4th edn. Cambridge University Press, Cambridge, p 494
Eringen AC (1980) Mechanics of Continua, 2nd edn. Krieger Publishing Company, Malabar, p 592
Falta RE, Pruess K, Finsterle S, Battistelli A (1995) T2VOC User's Guide, LBNL 36400, 158 pp
Fel LG, Bear J (2010) Dispersion and dispersivity tensors in saturated porous media with uniaxial symmetry. Trans. Porous Media 85(1):259–268
Gray WG, Hassanizadeh SM (1998) Macroscale continuum mechanics for multiphase porous-media flow including phases, interfaces, common lines, and common points. Adv. Water Res. 21:261–281
Gray WG, Miller CT (2014) Introduction to the thermodynamically constrained averaging theory for porous medium systems. Springer, Berlin, p 582
Hassanizadeh SM (2004) Continuum description of thermodynamic processes in porous media: fundamentals and applications. In: Kubik J, Kaczmarek M, Murdoch I, (eds) Modellimg coupled phenomena in saturated porous materials. Polish Academy of Science, IPPT, pp 179–223
Hassanizadeh M, Gray W (1979a) General conservation equations for multiphase systems: 1. Averaging procedure. Adv. Water Res. 2:131–144
Hassanizadeh M, Gray W (1979b) General conservation equations for multiphase systems: 2. Mass, momentum, energy and entropy equations. Adv. Water Res. 2:191–203
Hassanizadeh SM, Gray WG (1980) General conservation equations for multi-phase systems, 3. Constitutive theory for porous media flow. Adv. Water Res. 3:25–40
Hassanizadeh SM, Gray WG (1990) Mechanics and thermodynamics of multiphase flow in porous media including interphase boundaries. Adv Water Res. vol. 13, pp. 169–186
Knudsen M (1934) The kinetic theory of gases. Methuen, London
Landau L, Lifshitz EM (1960) Fluid mechanics. Addison-Wesley, Reading
Lichtner, P.C. Karra, S. Modeling Multiscale-Multiphase-Multicomponent Reactive Flows in Porous Media: Application to CO_2 Sequestration and Enhanced Geothermal Energy using PFLOTRAN, Chap. 3 in *Computational Models for CO2 Geo-sequestration and Compressed Air Energy Storage* by Al-Khoury. R. and Bundschuh, J. (Edts.), CRC Press, 31pp., 2014,
Malek K, Coppens M-O (2003) Knudsen self- and Fickian diffusion in rough nanoporous media. J. Chem. Phys. 119:2801–2828
Moridis GJ, Kowalsky M, Pruess K (2008) TOUGH+HYDRATE v1.0 Users manual: a code for the simulation of system behavior in hydrate-bearing geologic media, LBNL-00149E
Nikolaevski VN (1959) Convective diffusion in porous media. J. Appl. Math. Mech. (P.M.M.) 2(3):1042–1050

Nitao JJ, Bear J (1996) Potentials and their role in transport in porous media. Water Resour Res 32:225–250

Pruess K, Oldenburg C, Moridis G (1999) TOUGH2 User's Guide, Version 2.0, Lawrence Berkeley National Laboratory Report LBNL-43134. Berkeley, CA

Saaltink MW, Ayora C, Carrera J (1998) A mathematical formulation for reactive transport that eliminates mineral concentrations. 34:1649–1656

Scheidegger AE (1961) General theory of dispersion in porous media. J. Geophys. Res. 66:3273–3278

Sirotine, Y. and Chaskolskaya, M. *Fondaments de la physique des crystaux*. Edition Mir, 680 p., (Russian Ed., 1975) 1984

Taylor GI (1953) Dispersion of soluble matter in solvent flowing slowly through a tube. Proc. R. Soc. A 219(1137):186–203

Truesdell C, Toupin RA (1960) Handbuch der Physik. Springer, Berlin, p 293

Verruijt A (1969) Elastic storage in aquifers. In: De Wiest, R.J.M. (ed) Flow through porous media. Academic Press, New York, pp 331–376

Whitaker S (1999) The method of volume averaging. Kluwer Academic Publishers, Dordrecht, p 219

Chapter 4
Momentum Balance and Motion Equation

Darcy's law, often referred to as the 'motion equation of a fluid in a porous medium' is the law that governs the flux of a fluid that occupies the entire void space, or part of it (in multiphase flow). It relates the macroscopic (i.e., averaged) velocity of the fluid, or its *specific discharge*, to the forces that produce the fluid's motion, primarily pressure gradient and gravity.

In this chapter we consider the motion equation for *saturated flow*, i.e., when a single fluid occupies the entire void space. In Sect. 6.2 we shall consider the motion equations when two or three fluid phases occupy the void space simultaneously.

Here, as in the entire book, we are using the following five terms at the macroscopic level:

- **Fluid's velocity**, $\mathbf{V}_\alpha$: It is the distance traveled by a fluid α-phase 'particle' per unit time, or the volume of an α-fluid passing through a unit *area of fluid* in a planar cross-section through a porous medium domain, per unit time.
- **Solid's velocity**, $\mathbf{V}_s$: It is the distance traveled by a point of a solid matrix per unit time, or the volume of solid passing through a unit area of solid in a planar cross-section through a porous medium domain, per unit time.
- **Relative fluid velocity**, $\mathbf{V}_{\alpha r}$: It is the velocity of the α-fluid relative to that of the solid matrix: $\mathbf{V}_{\alpha r} \equiv \mathbf{V}_\alpha - \mathbf{V}_s$.
- **Fluid's flux**, $\mathbf{J}_\alpha$: Volume of fluid passing through a unit area of α-fluid (in a planar cross-section through a porous medium domain), per unit time.
- **Fluid's specific discharge**, $\mathbf{q}_\alpha$ $(= \theta_\alpha \mathbf{V}_\alpha)$: Volume of an α-fluid passing through a *unit area of porous medium* (in a planar cross-section through a porous medium domain), per unit time.

J. Bear, *Modeling Phenomena of Flow and Transport in Porous Media*, Theory and Applications of Transport in Porous Media 31, https://doi.org/10.1007/978-3-319-72826-1_4

An important comment should be introduced here. In Sect. 1.1.7 A, we have introduced the concept of *effective porosity*, stemming from the observation that flow (may) take place only through part of the void space. Accordingly, in single phase flow, the specific discharge, $\mathbf{q}$ should be expressed by $\mathbf{q} = \phi_{eff}\mathbf{V}$. Henceforth, in this book, whenever $\phi_{eff} < \phi$, the specific discharge of a fluid should be expressed by $\phi_{eff}\mathbf{V}$. Furthermore, in most applications, especially in the case of a geological formation, it is practically impossible to determine/measure both effective porosity and fluid velocity. It is much more convenient to determine the fluid's specific discharge, $\mathbf{q}$. As a consequence, in practice, it is much more convenient to refer to $\mathbf{q}$ as the variable that describes fluid motion in a porous medium—volume of fluid per unit area of a porous medium per unit time.

In the continuum approach employed here, all macroscopic variables are assigned to a point within the porous medium domain, regarded a continuum.

4.1 Some Historical Notes

With no effort to present a comprehensive review, let us review some of the ways used for developing the motion equation, as defined in Chap. 3.

4.1.1 Obtaining the Law Experimentally

Actually, this is how, in 1856, Henri Darcy, the water engineer of Dijon, a city in the southern part of France, obtained the linear law named after him. He conducted experiments in the vertical sand column shown in Fig. 4.1. In his experiments, Darcy (1856) varied the piezometric head (see Sect. 4.2.1) difference acting on a sand column, and measured the resulting water discharge. Based on these experiments, he concluded that the discharge through the column is:

- proportional to the cross-sectional area of the column, $\mathcal{A}$,
- proportional to the difference in water level elevations, h_1 and h_2, at the inflow and outflow reservoirs of the column, respectively, and
- inversely proportional to the column's length, L.

When combined, these conclusions give the famous *Darcy's formula* (or *law*):

$$Q = K\mathcal{A}\frac{h_1 - h_2}{L}, \tag{4.1.1}$$

in which L, is the length of the sand column, Q is the (constant) discharge through the column, $\mathcal{A}$ denotes the column's cross-section, and h_1 and h_2 denote the piezometric heads at the column's inlet and outlet reservoirs (see Fig. 4.2).

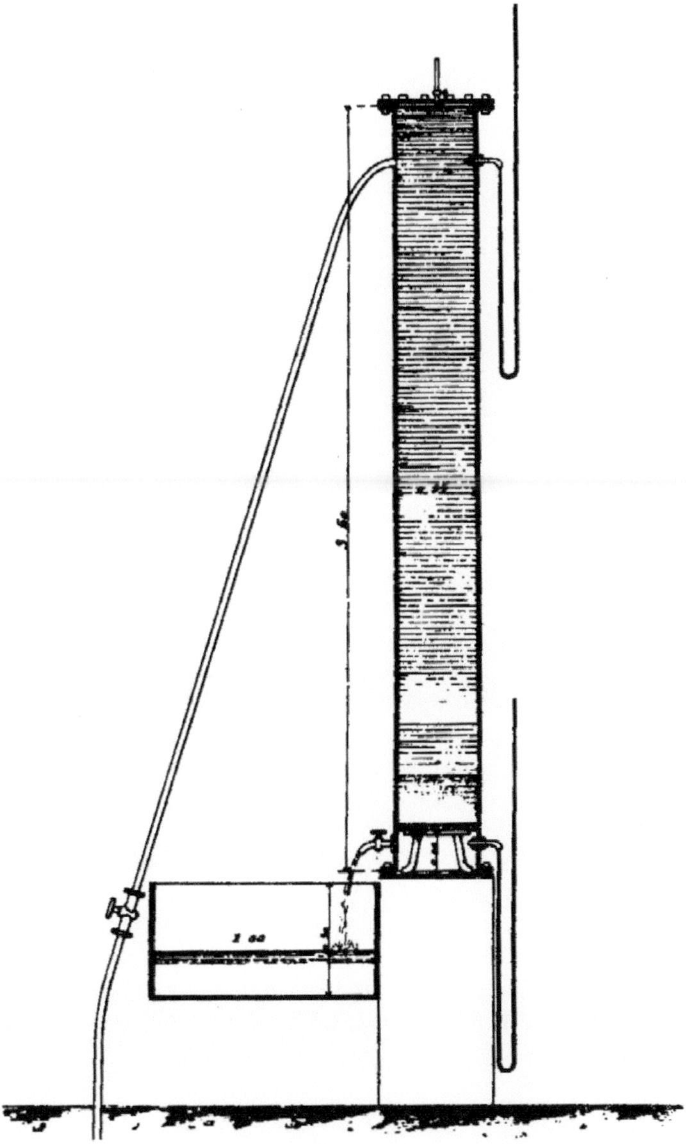

Fig. 4.1 Darcy's column experiment (after Darcy 1856)

The coefficient of proportionality, K, called *hydraulic conductivity* (dims. L/T), is discussed in detail in Sect. 4.2.3. With the definition of *specific discharge* introduced earlier, Darcy's law (for the 1-D case considered here) can also be written in the form:

$$q = K\,\frac{h_1 - h_2}{L}. \tag{4.1.2}$$

4.1.2 Analogy to Flow Through Capillary Tubes

As discussed in Chap. 1, it is (practically) impossible to analyze and mathematically solve the detailed (microscopic level) flow of a fluid through the void space of most porous media, by solving, simultaneously, the mass and momentum balance equations, say the Navier–Stokes equation, for the fluid occupying the void space. Instead, we replace the actual porous medium domain by a fictitious, simpler one. This is the one regarded in Chap. 1 as a *continuum*. In this domain, the flow problem reduces to one that is amenable to exact mathematical treatment. Of course, this approach is justified only if the real system and the simpler fictitious one—the continuum at the macroscopic level—exhibit the same major features that control the flow through the void space. By analyzing the flow in the simpler system, we obtain a relationship between the flux and the relevant driving force for that system. We then *assume* that the same (e.g., linear) relationship exists also in a real porous medium, but with a different coefficient that represents the effect of the (microscopic) geometrical features of the considered void space. In a real porous medium, such coefficients have, anyway, to be determined experimentally. In fact, the only way to validate the model obtained from the simplified (conceptual) model is through experiments.

We start by considering flow of a homogeneous fluid through a parallelepiped body of porous medium of length L in the s-direction, and a cross-section of unit area normal to this direction. The simplest physical model that represents the flow through such body is that of laminar flow (or flow at small *Reynolds number*, Re; see Sect. 4.1) through a single straight circular capillary tube of diameter δ (see Bear 1972, p. 161) and length L in the s-direction. The *Hagen–Poisseuille law* that governs such flow is:

$$Q_s = -\frac{\pi\delta^4}{128}\frac{\rho g}{\mu}\frac{\partial h}{\partial s}, \quad \text{or} \quad q_s\left(\equiv \frac{Q_s}{\pi\delta^2/4}\right) = -\frac{\delta^2}{32}\frac{\rho g}{\mu}\frac{\partial h}{\partial s}, \tag{4.1.3}$$

where Q_s is the total discharge through the tube, ρ and μ are the density and viscosity of the fluid, respectively, q_s denotes the fluid's specific discharge in the s-direction, p denotes pressure and for constant density, i.e., when the fluid is homogeneous and incompressible, and the flow is under isothermal conditions, we define the *piezometric head*:

$$h = z + \frac{p}{\rho g}, \tag{4.1.4}$$

where z is the elevation of the point at which the piezometric head is being considered above some datum level, p and ρ are the fluid's pressure and mass density, respectively, and g is the gravity acceleration. The piezometric head is a concept often used in groundwater hydrology; the term *potential* is also often used. The quotient $p/\rho g$ is called *pressure head*. The Hagen–Poisseuille law (4.1.3) is the steady state solution of the Navier–Stokes equation for a capillary tube.

Next we represent the same porous medium domain by N identical capillary tubes of diameter δ each. Equation (4.1.3) is then replaced by:

$$Q_s = -N\frac{\pi\delta^4}{128}\frac{\rho g}{\mu}\frac{\partial h}{\partial s}, \quad q_s = \frac{Q_s}{N\pi\delta^2/4} = -\frac{\phi\delta^2}{32}\frac{\rho g}{\mu}\frac{\partial h}{\partial s}, \tag{4.1.5}$$

or:

$$q_s = -k\frac{\rho g}{\mu}\frac{\partial h}{\partial s}, \quad k = \frac{\phi\delta^2}{32}, \tag{4.1.6}$$

which is *Darcy's law*, with the permeability, k, related to ϕ and δ. Bear (1972, p. 163) presents also the case in which there are N_i pipes of diameter δ_i per unit area of the cross-section. In Sect. 4.2.5, we consider tortuous capillary tubes, leading to the concept of *tortuosity*.

Note that the use of of piezometric head, h, is limited to cases of constant density only, i.e., when the fluid is homogeneous and incompressible, and the flow is isothermal.

For variable density, e.g., a compressible fluid, $\rho = \rho(p)$, the *pressure head* is expressed by the integral $\int_{p_o}^{p}(dp/\rho(p)g)$, where p_o is a reference pressure (Muskat 1946, p. 129).

In petroleum engineering, the concept of *pseudopotential* Φ (Hubbert 1940, 1956) is often used. Thus, *Hubbert's potential* for a compressible fluid, $\rho = \rho(p)$, is defined by:

$$\Phi = \int_{p_0}^{p}\frac{dp}{\rho(p)g} + z, \tag{4.1.7}$$

Darcy's Law for the fluid's flux can then be rewritten in the form:

$$\mathbf{q} = -\frac{k}{\mu}\rho g\nabla\Phi. \tag{4.1.8}$$

With the piezometric head, h, defined in (4.1.4), Darcy's law takes the form:

$$\mathbf{q} = -K\nabla h, \quad \text{where} \quad K = \frac{k\rho g}{\mu}. \tag{4.1.9}$$

Scheidegger (1953, 1960) related the distribution of diameters of the capillary tubes to the pore size distribution.

For flow through a single narrow fracture of aperture width b, it is easy to show (e.g., Bear 1972, p. 165) that:

$$q_s = \phi \frac{b^2}{12} \frac{\rho g}{\mu} \frac{\partial h}{\partial s}, \quad \text{where} \quad k = \frac{\phi b^2}{12}, \tag{4.1.10}$$

i.e., the permeability is proportional to the square of the aperture, b. Irmay (1955) analyzed the flow through narrow parallel capillary fissures of aperture b, spaced a distance a apart, and obtained an expression which is analogous to Darcy's law, i.e., the flux is linearly proportional to the head gradient, with a permeability proportional to b^2.

In all these models, laminar flow takes place through narrow channels, and it is easy to recognize the similarity between the flux expressions and Darcy's law. This law is nothing but a statement that the flux is linearly proportional to a driving force, which is the gradient of the piezometric head. The latter is the sum of the pressure gradient and a force due to gravity. In (4.1.3), $(\delta^2/32)(\rho g/\mu)$ is analogous to the hydraulic conductivity, K, in a porous medium.

We note that the hydraulic conductivity is made up of two parts: one that expresses the geometry of the flow domain; specifically, it is proportional to the square of the tube's diameter. The other is a property of the fluid (actually, $\rho/\mu \equiv 1/\nu$). In fact, the same expression appears also in all other models mentioned above, except that in each case the length that characterizes the capillary opening, e.g., tube diameter, or width of aperture, may be different. Obviously, the similarity between all above cases stems from the observation that they are all based on solving the (linear) momentum balance equation (in this case, the Navier–Stokes equation) for small Reynolds numbers, i.e., neglecting all inertial terms. A widely used expression for the permeability of a porous medium is the one proposed by Kozeny (1927) and later modified by Carmen (1937, 1956; see Bear 1972, p. 165). Actually, Blake (1922) has derived the same expression for permeability earlier.

Kozeny (1927) also treated a porous medium as a bundle of capillary tubes of equal length, but not necessarily circular, neglecting the velocity normal to a tube's axis. By solving the (linear) momentum balance equations for a Newtonian fluid, i.e., the Navier–Stokes equations, simultaneously for all capillary tubes passing through a cross-section which is normal to the flow, he obtained the flux through the fictitious porous medium in the form:

$$q_s = -\frac{c_o \phi^3}{\mu M_s^2} \frac{\partial p}{\partial s}, \quad \text{where} \quad k = \frac{c_o \phi^3}{M_s^2}, \tag{4.1.11}$$

where M_s denotes the specific surface of the tubes, and c_o is a numerical coefficient referred to as Kozeny's constant; it varies slightly according to the shape of the individual tubes. The above equation is referred to as Kozeny's equation. Carman (1937) suggested $c_o = 0.2$, and lead to the Kozeny–Carman equation:

$$k = \frac{1}{5M_s^2} \frac{\phi^3}{(1-\phi)^2}, \tag{4.1.12}$$

in which M_s is the specific surface (= surface of solid exposed to fluid per unit volume of solid), and d_m is the mean diameter of the spheres comprising the porous medium. Recall that for a sphere of diameter d_m, the surface area per unit volume of sphere is $6/d_m$.

With $\mathbb{V}_o$ and $\mathbb{V}_{so}$ denoting volume of a representative elementary volume, and the volume of solid in the latter, it is interesting to note that with:

$$\phi = \mathbb{V}_{vo}/\mathbb{V}_o, \quad 1-\phi = \mathbb{V}_{so}/\mathbb{V}_o,$$

$$M_s = S_{so}/\mathbb{V}_{so} = S_{so}/(\mathbb{V}_o - \mathbb{V}_{vo}) = S_{so}/\mathbb{V}_o(1-\phi),$$

Equation (4.1.12) can be written as $k = \phi\Delta^2 T^*$, in which $\Delta(=\mathbb{V}_{vo}/S_{vo})$ is the *hydraulic radius* of the void space, and $T^* = 1/5$ plays the role of *tortuosity*. This can be compared with (4.2.26).

While (4.1.12) is a relatively simple relationship, it by no means represents the full spectrum of the k–ϕ relationships presented in the literature for natural geological porous media, nor for the kind of porous media used in reactors in the chemical industry. Actually, there is no universal convenient equation relating k to ϕ; a wide variety of equations have been proposed.

Some authors construct models that take into account a statistical distribution of the capillary tubes in space. For example, de Josselin de Jong (1969; Bear 1972, p. 173) used a model composed of a network of interconnected capillary channels (Fig. 4.3). He assumed that the probability of a fluid particle choosing a direction between θ and $\theta + \Delta\theta$ is proportional to the total fluid's discharge in that direction. His analysis lead to a linear relationship between the flux vector and the hydraulic gradient vector that took the form of a second rank symmetric tensor. Some introductory remarks about tensors are presented in Sect. 2.3.4 (Fig. 4.2).

4.1.3 Models Based on Resistance to Flow Around Spheres

When a fluid is flowing relative to a solid surface, say bounding a solid grain, it exerts a force on the latter. This force is due to (1) pressure variations on the surface and (2) shear stress resulting from the velocity gradient at the surface, since we assume that the fluid which is adjacent to the solid sticks to the latter (i.e., a *no slip condition*). The sum of these forces, integrated over the surface, gives the resultant force exerted by the fluid on the considered solid grain. The component of this force in the direction of the velocity (relative to the solid, as the latter may be moving too) is called *drag*. The resistance to flow, or drag on the solid surface, has been used by various researchers to derive expressions for the permeability of a porous medium.

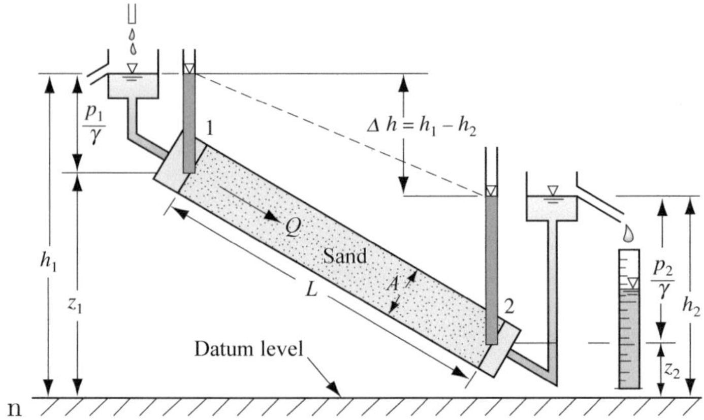

Fig. 4.2 Flow in an inclined porous medium column

Fig. 4.3 Interconnected channels. (de Josselin de Jong, 1958)

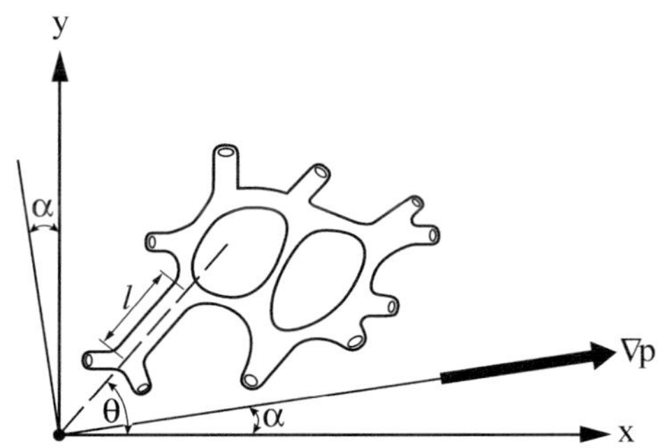

The simplest case is that of laminar flow around a single stationary sphere of diameter d, representing the case of drag acting on a solid particle of a porous medium (Bear 1972, p. 167): In slow fluid motion, this drag, D, can be expressed by:

$$D = 3\pi d\mu V, \tag{4.1.13}$$

i.e., proportional to the first power of the specific discharge, or:

$$D = C_D \frac{\rho V^2}{2} A_{ref} = \frac{24}{Re}\frac{\rho V^2}{2}\frac{\pi d^2}{4}; \quad C_D = \frac{24}{Re}, \quad Re = \frac{\rho q d}{\mu}, \tag{4.1.14}$$

where A_{ref} is a chosen reference surface area. This equation is known as the *Stokes equation*, with the *drag coefficient*, C_D valid for $Re < 1$. Various authors suggested values of C_D for high values of Re, where inertia cannot be neglected.

The general macroscopic linear momentum balance equation was presented as (3.3.9) in Sect. 3.3.2C. That equation includes terms that represent momentum flux, momentum transfer and momentum sources, expressed in terms of state variables that are more relevant to the considered transport problem, leading to the fundamental macroscopic momentum balance equation for a fluid phase that occupies the entire void space. We shall then show that under certain conditions, certain phenomena associated with momentum transport may be neglected, and the macroscopic momentum balance equation can be simplified. In fact, one of these simplified forms is *Darcy's law*. In Chap. 6, we shall discuss the case in which two or three fluid phases occupy the void space, simultaneously.

The development of Darcy's law as a simplified form of the momentum balance equation explains why we have chosen to present the discussion on the macroscopic momentum balance equation (in this chapter) before the chapter on the mass balance equation (next chapter); the former will provide the required expression for fluid flux (or fluid velocity, in the more general case), which is required for the mass balance equation.

Another application of the macroscopic momentum balance equation will be in the modeling of deformation and wave propagation in porous media (Sect. 9.4).

4.2 Darcy's Law

Although in the previous subsection, we have presented Darcy's law as an approximate form of the momentum balance equation (Case A), we shall now present this law as an empirical phenomenological law. For the sake of simplicity, we shall overlook the possible motion of the solid skeleton, as did Darcy.

4.2.1 The Empirical Law

In Sect. 4.1.1, we have introduced Darcy's law, as proposed by Darcy (1856) on the basis of column experiments, say, in the form of (4.1.2). In this equation, q denotes the specific discharge, i.e., the discharge (= volume of fluid per unit time) per unit area of porous medium in a planar cross-section perpendicular to the flow direction. Figure 4.2 provides also the meaning of the symbol h that expresses the mechanical energy per unit weight of fluid due to elevation and pressure; it does not include the kinetic energy, expressed as $\beta V^2/2g$, where β is a coefficient that is associated with the fact that V is the average velocity. We may interpret the head difference $h_1 - h_2$ as the loss of mechanical energy due to friction in the flow through the narrow tortuous pathways in the porous medium; changes in the kinetic energy have been neglected as being much smaller than those in the piezometric head.

Although, originally, Darcy's law (4.1.2) was derived from experiments on a finite length column, we can extend Darcy's conclusion to what happens *at a point* along a column, or streamtube, by considering flow in a segment of a (not necessarily straight) streamtube in a three-dimensional space, aligned in a direction indicated by the unit vector **1s** (Fig. 4.4).

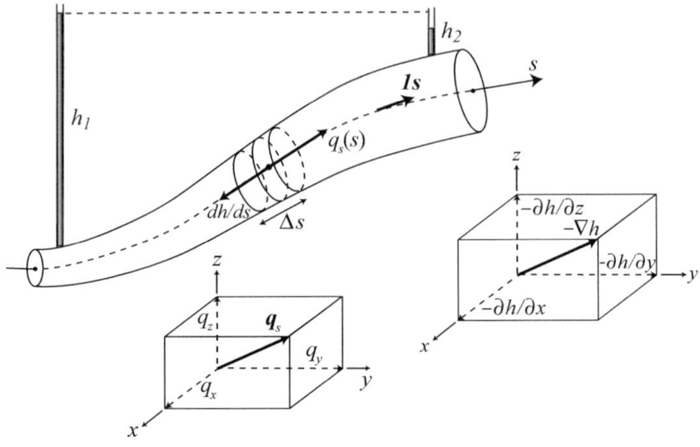

Fig. 4.4 A streamtube in three-dimensional space

The piezometric head varies along the stream-tube i.e., $h = h(s)$. We consider a segment of length Δs along the s-axis, between the coordinates $s - \Delta s/2$, and $s + \Delta s/2$. With this notation, (4.1.2) takes the form:

$$q_s(s) = K \frac{h\big|_{s-\frac{\Delta s}{2}} - h\big|_{s+\frac{\Delta s}{2}}}{\Delta s}, \tag{4.2.15}$$

where the subscript s in q_s indicates that the flow is in the s-direction. In the limit, as $\Delta s \to 0$, we obtain:

$$\lim_{\Delta s \to 0} \frac{h\big|_{s-\frac{\Delta s}{2}} - h\big|_{s+\frac{\Delta s}{2}}}{\Delta s} = -\frac{dh}{ds}, \tag{4.2.16}$$

and (4.2.15) reduces to:

$$q_s = -K \frac{dh}{ds}, \tag{4.2.17}$$

where q_s is considered positive in the positive direction of the s-axis. This is the 1-d *differential form of Darcy's law at a point*. In a heterogeneous domain, $K = K(x, y, z)$, i.e., a function of the location of the considered point.

The derivative dh/ds in (4.2.17) expresses the slope of the *piezometric curve*, $h = h(s)$, with a positive value indicating a rising function, $h = h(s)$, and a negative value indicating that h decreases with s. We refer to $-dh/ds$ as the *hydraulic gradient*. The use of the piezometric head, h, in Darcy's law is permitted only for a fluid of constant density. When the fluid's density varies, because of variations in pressure, concentration of dissolved matter, or temperature, the hydraulic gradient should not be used as a driving force. Later in this section, we shall present the motion equation (which we usually still refer to as Darcy's law) for a variable density fluid.

So far, we have been discussing the fluid's specific discharge. However, when considering the transport of a solute dissolved in the moving fluid (Chap. 7), we need to know the fluid's *velocity*. In fact, the entire discussion in Sect. 4.2.4 on the momentum balance equation and its simplified forms, often referred to as *motion equations*, was presented in terms of the fluid's (average) velocity as the variable of state.

At the microscopic level, flow takes place only through part of the cross-sectional area of the porous medium domain, the remaining part being occupied by the *solid matrix*, or *solid skeleton*. Because it can be shown (Bear and Bachmat 1991, p. 37) that the *average areal porosity*, ϕ^A, *equals the volumetric one*, ϕ, the portion of the area $\mathcal{A}$ available to flow is $\phi^A\mathcal{A} \approx \phi\mathcal{A}$, where ϕ and ϕ^A denote the porosity (= volume of voids per unit volume of a porous medium sample) and areal porosity (= area of voids per unit area of a planar cross-section through the porous medium). Accordingly, the *average velocity*, V, of a fluid flowing through a porous medium is given by:

$$V = \frac{Q}{\phi\mathcal{A}} = \frac{q}{\phi}. \tag{4.2.18}$$

As explained in Sect. 3.1.3, the velocity as defined above is the fluid's *mass-weighted velocity*.

Sometime, part of the void space is unavailable to fluid flow, or almost so, due to dead-end pores in which the fluid is (practically) immobile. We then define an *effective porosity*, ϕ_{eff}, and use it to determine the velocity:

$$V = \frac{Q}{\phi_{eff}\mathcal{A}} = \frac{q}{\phi_{eff}}. \tag{4.2.19}$$

4.2.2 Extension to Three Dimensions

Our next step is to extend Darcy's law, say (4.2.17) to three dimensions. We envision any flow pattern in a 3-d porous medium domain (recalling that we are regarding this domain as an isotropic continuum at the macroscopic level). Then, using a cartesian coordinate system, at every point (x, y, z), and any instant of time, t, the value of the piezometric head is $h(x, y, z, t) = z + p(x, y, z, t)/\rho g$. At that point and at that instant of time, the flux vector $\mathbf{q}(x, y, z, t)$, is driven by the *hydraulic gradient*, $-\nabla h$ (which is the 3-d extension of $-dh/ds$), having the components $-\partial h/\partial x$, $-\partial h/\partial y$, $-\partial h/\partial z$. Thus, the cartesian components of the flux vector are:

$$q_x = -K(x, y, z)\frac{\partial h}{\partial x}, \quad q_y = -K(x, y, z)\frac{\partial h}{\partial y}, \quad q_z = -K(x, y, z)\frac{\partial h}{\partial z}. \tag{4.2.20}$$

In the above equation, $K = K(x, y, z)$ denotes the hydraulic conductivity at the point (x, y, z). In this way, (4.2.20) is valid also for heterogeneous porous medium domains. Darcy's law (4.2.20) can also be written in the compact vector form:

$$\mathbf{q} = -K(x, y, z)\nabla h, \tag{4.2.21}$$

or making use of indicial notation, in the form:

$$q_i = -K(x, y, z)\frac{\partial h}{\partial x_i}. \tag{4.2.22}$$

Note that when written in terms of the piezometric head, h, Darcy's law, e.g., (4.2.22), is applicable only to a constant density fluid.

4.2.3 Hydraulic Conductivity and Permeability

The coefficient of proportionality, K, appearing in Darcy's law (4.1.1) is called the *hydraulic conductivity* of the porous medium. In an *isotropic* porous medium, (4.1.1) may be used to define it as the *specific discharge per unit hydraulic gradient*. It is a scalar that expresses the ease with which a fluid flows through the narrow tortuous pathways comprising the void space. It is, therefore, a coefficient that depends on both matrix and fluid properties. The relevant fluid properties are the density, ρ, and the *dynamic viscosity*, μ (or in the combined form of the *kinematic viscosity*, ν $(= \mu/\rho)$). The relevant solid matrix property is the permeability, denoted by k. It is related to the geometrical features of the void space, which, in turn, depend on grain- or pore-size distribution, shape of grains, or pores, tortuosity of passages, specific surface, roughness of solid surface, porosity, and other void-space characteristics. Following Bear and Bachmat (1991), we shall use (in saturated flow) the *hydraulic radius* of the fluid-filled void space as the (microscopic) length that characterizes the void-space, and, hence, is the main feature that determines the permeability.

Accordingly, the hydraulic conductivity, K, can be expressed as:

$$K = k\frac{\rho g}{\mu} \equiv \frac{k\,g}{\nu}, \tag{4.2.23}$$

where g is the gravity acceleration, μ and ν are the dynamic and kinematic viscosities of the fluid, respectfully, and k (dims. L^2) is the *permeability* of the porous medium. It is a coefficient that depends solely on the properties of the configuration of the void-space and not on those of the fluid. For water at 20 °C, $\mu = 1.002\times 10^{-3}$ Pa.s.

Various units are used in practice for the hydraulic conductivity, K. Some hydrologists prefer the unit m/d (meters per day). Soil scientists and geotechnical engineers often use cm/s (centimeters per second). The recommended system is, of course, the

$-\log_{10} \cdot K$(cm/sec)	–2	–1	0	1	2	3	4	5	6	7	8	9	10	11
Permeability	Pervious				Semipervious				Impervious					
Aquifier	Good					Poor				None				
Soils	Clean gravel		Clean sand or sand and gravel			Very fine sand, silt, loess, loam, solonetz								
					Peat		Stratified clay			Unweathered clay				
Rocks					Oil rocks			Sand-stone		Good limestone, dolomite		Breccia, granite		
$-\log_{10} \cdot k$(cm²)	3	4	5	6	7	8	9	10	11	12	13	14	15	16
$\log_{10} k$(md)	8	7	6	5	4	3	2	1	0	–1	–2	–3	–4	–5

Fig. 4.5 Representative values of hydraulic conductivity (for water at 20 °C) and permeability for selected soils (Bear et al. 1968)

SI system of units, where m/s (meters per second) is used. Representative values of hydraulic conductivity are given in Fig. 4.5.

In the SI system, the permeability, k, is measured in m^2. Petroleum engineers use the unit *Darcy*, suggested by Muskat (1937) to honor Henry Darcy. The unit 1 darcy is defined as that permeability which allows the discharge of 1 cm^3/s of fluid of viscosity 1 centipoise (abbrev. cp) through an area of 1 cm^2, under a pressure gradient of 1 atm/cm:

$$1 \text{ darcy} = \frac{1\,\text{cm}^3/\text{s}/\text{cm}^2 \;\times\; 1 \text{ centipoise}}{1\,\text{atm}/\text{cm}}. \tag{4.2.24}$$

where we recall that 1 darcy is equivalent to 9.869233×10^{13} m^2, or 0.9869233 μm^2. This conversion is usually approximated as 1 μm^2. Groundwater hydrologists, for water at 20°, use 1 darcy $= 9.613 \times 10^{-4}$ cm/s, or 0.831 m/day.

Although the *darcy* as a unit for permeability is not an SI unit, it is very commonly used, especially by reservoir engineers.

Numerous formulæ that relate permeability to various geometric properties of the void space are presented in the literature. Some are purely empirical, as, for example:

$$k = Cd^2, \tag{4.2.25}$$

where C is a dimensionless coefficient and d is an effective grain diameter, say, d_{10} (i.e., 10% of the grains by weight are smaller than this diameter). Krumbein and Monk (1943) suggest $C = 6.17 \times 10^{-4}$ for k and d expressed in cm^2 and cm, respectively. Although this is an empirical formula; the dependence on the square of a characteristic length of the void-space can be justified by a theoretical analysis (see, for example, Bear and Bachmat 1991, p. 174).

Another example is the Fair and Hatch (1933) formula, developed from dimensional considerations, and verified experimentally:

$$k = \frac{1}{\beta}\left[\frac{(1-\phi)^2}{\phi^3}\left(\frac{\alpha}{100}\sum_{(m)}\frac{P_m}{d_m}\right)^2\right]^{-1}, \tag{4.2.26}$$

where β is a *packing factor*, found experimentally to be 5, α is a sand *shape factor*, varying from 6 for spherical grains to 7.7 for angular ones, P_m is the weight percentage of sand held between adjacent sieves, and d_m is the geometric mean diameter of the adjacent sieves.

An often used formula for permeability is the Kozeny–Carman equation:

$$k = C_o\frac{\phi^3}{(1-\phi)^2(\Sigma_{vs})^2}, \tag{4.2.27}$$

where Σ_{vs} is the specific surface area of the solid (defined per unit volume of solid matrix), and C_o is a coefficient for which Carman (1937) suggested the value 0.2. Often, $1/(\Sigma_{vs})^2$ is replaced by d^2, with d = mean grain size, or by Δ^2, with Δ denoting the hydraulic radius of the void-space.

It is interesting to note that in all permeability expressions, e.g., (3.6.5), (4.2.16), (4.2.26) and (4.2.32), the permeability is proportional to the porosity, to the square of a length characterizing a pore, and to a tortuosity. For an anisotropic porous medium, the latter is a second rank symmetric tensor. The tortuosity is discussed in the next subsection.

Under certain conditions, the permeability may vary with time. Such a change may be caused by compaction of a layer due to external loads. Clogging by precipitation or dissolution of minerals, filtration of fine-grained solids, or swelling of clay may also produce changes in the structure and texture of the solid matrix.

4.2.4 Simplified Macroscopic Momentum Balance

Although we have already suggested (in Sect. 3.5.1) that Darcy's law is a *constitutive* type of law, we shall elaborate now on how Darcy's law and other flux laws are obtained as simplified forms of the momentum balance equation for a fluid in a porous medium domain. However, in all these laws, we note the term that expresses the transfer of momentum from the fluid to the solid. This actually makes these laws constitutive ones (see preamble to Chap. 2).

Let us start from the macroscopic momentum balance equation for a Newtonian fluid, (3.6.5), rewritten here, for convenience, as:

$$\phi\rho\frac{D\mathbf{V}}{Dt} = \nabla\cdot\phi\left(\mathbf{J}^{\mathrm{M}}_{dif} + \mathbf{J}^{\mathrm{M}}_{dis}\right) - \phi\left(\nabla p + \rho g\nabla z\right) - \phi\frac{\mu\mathbf{R}}{\Delta^2}\cdot\mathbf{V}_r, \tag{4.2.1}$$

in which $\mathbf{V} \equiv \mathbf{V}_f$, $\mathbf{V}_r = (\mathbf{V} - \mathbf{V}_s)$, and the specific discharge relative to the solid matrix is expressed by $\mathbf{q}_r = \phi \mathbf{V}_r$.

In what follows, we shall:

(a) neglect the nonlinear diffusive flux of momentum,
(b) assume a Newtonian fluid and express the diffusive flux of momentum, $\mathbf{J}^{\mathrm{M}}_{dif}$, by $\mathbf{J}^{\mathrm{M}}(\equiv \boldsymbol{\tau}) = \phi\mu\nabla\mathbf{V}$, i.e., assuming *isochoric flow at the microscopic level*, $\nabla\cdot\mathbf{V} = 0$, and
(c) neglect the dispersive flux of momentum, $\mathbf{J}^{\mathrm{M}}_{dis}$.

Under these assumptions, the momentum balance (4.2.1) reduces to:

$$\phi\rho\frac{D\mathbf{V}}{Dt} = \nabla\cdot\phi\mu\nabla\mathbf{V} - \phi(\nabla p + \rho g\nabla z) - \phi\frac{\mu\mathbf{R}}{\Delta^2}\mathbf{V}_r. \tag{4.2.2}$$

Sometimes, different viscosities are used in the first and last terms on the r.h.s. of the above equation.

We shall later show (Case A below) that the product $\phi\Delta^2\mathbf{R}^T$ denotes the *permeability tensor*, $\mathbf{k}$, of the considered porous medium. The symbol $(.)^T$ denotes the *transpose* of (.).

Equation (4.2.2) may be regarded as a sufficiently generalized *motion equation* for a Newtonian fluid. Let us consider a number of simplified cases.

We wish to investigate the conditions under which the magnitude of the fluid's velocity and of the fluid's temporal and spatial velocity variations are such that the viscous force, resisting the flow, due to the transfer of momentum at the fluid-solid interface, is much larger than both the inertial force and the viscous resistance to the flow. Mathematically, we are looking for the conditions under which:

$$\left|\phi\rho\frac{\partial\mathbf{V}}{\partial t}\right| \ll |\phi^2\mu\mathbf{k}^T\mathbf{V}_r|, \tag{4.2.3}$$

$$\left|\phi\rho\mathbf{V}\cdot\nabla\mathbf{V}\right| \ll |\phi^2\mu\mathbf{k}^T\mathbf{V}_r|, \tag{4.2.4}$$

$$|\nabla\cdot\phi\mu\nabla\mathbf{V}| \ll |\phi^2\mu\mathbf{k}^T\mathbf{V}_r|, \tag{4.2.5}$$

where $\mathbf{k} = \phi\Delta^2\mathbf{R}^T$ is the permeability tensor. With $\mathbf{V}_s \ll \mathbf{V}_f \equiv \mathbf{V}$, we follow the discussion is Sect. 3.10 on dominance of effects (represented as terms in the balance equation), we can investigate the conditions under which (4.2.3)–(4.2.5) are valid. We rewrite the above equations in the form:

$$\frac{\phi_c\rho_c V_c}{(\Delta t)_c}\left|\phi^*\rho^*\frac{\partial V_i^*}{\partial t^*}\right| \ll \phi_c^2\mu_c k_c^{-1}V_c|\phi^{*2}\mu^* k_{ij}^{*T}V_{ri}^*|, \tag{4.2.6}$$

$$\frac{\phi_c\rho_c V_c^2}{L_c}\left|\phi^*\rho^* V_j^*\frac{\partial V_i^*}{\partial x_j^*}\right| \ll \phi_c^2\mu_c k_c^{-1}V_c|\phi^{*2}\mu^* k_{ij}^{*T}V_{ri}^*|, \tag{4.2.7}$$

$$\frac{\phi_c\mu_c V_c}{L_c^2}\left|\frac{\partial}{\partial x_j^*}\phi^*\mu^*\frac{\partial V_i^*}{\partial x_j^*}\right| \ll \phi_c^2\mu_c k_c^{-1}V_c|\phi^{*2}\mu^* k_{ij}^{*T}V_{ri}^*|. \tag{4.2.8}$$

In the above inequalities, all symbols with subscript c indicate characteristic values, while those with an asterisk ($*$) denote the (dimensionless) ratio between a value and its characteristic counterpart.

Let us introduce three dimensionless numbers which are related here to flow in porous media:

• The **Darcy number**:

$$\mathrm{Da} = \frac{k_c/\phi_c}{(L_c^V)^2}, \quad \text{where} \quad L_c^V = \frac{|V|_{\max}}{|dV/dx|_{\max}}, \tag{4.2.9}$$

which expresses the ratio between the square of a characteristic length of the void space, i.e., at the microscopic level, and the square of the characteristic length over which the fluid's velocity varies significantly. Note that the characteristic length at the microscopic level is represented here by $\sqrt{k_c/\phi_c}$. The hydraulic radius of the pore space may serve as the characteristic length of the void space. Bear and Bachmat (1991, p. 276) define the microscopic characteristic length as $\sqrt{k_c/\phi_c T_c^*}$, in which T^* denotes the characteristic (dimensionless) *tortuosity* (see Sect. 4.2.5) of the porous medium. In most cases of flow through porous media, $Da \ll 1$.

• The **Reynolds number**, which relates the inertial to the viscous forces acting on the fluid:

$$\mathrm{Re} = \frac{V_c\sqrt{k_c/\phi_c}}{\nu_c}, \tag{4.2.10}$$

• The **Strouhal number** (also Sect. 3.10), which is defined here as the ratio between two time intervals: the travel time L_c^V/V_c, required in order to encounter a significant spatial change in velocity, and the time interval Δt_c^V required to encounter the same change in velocity *at a point*:

$$\mathrm{St} = \frac{L_c^V}{(\Delta t)_c^V V_c}, \quad \text{with} \quad (\Delta t)_c^V = \frac{|V|_{\max}}{|\partial V/\partial t|_{\max}}. \tag{4.2.11}$$

A 'significant change' may mean, for example, a change from zero to the characteristic velocity, V_c. The latter may be selected as $|V_{max}|$.

With the above dimensionless numbers, (4.2.6)–(4.2.8) can be rewritten as:

$$\mathrm{Re}\mathrm{Da}^{\frac{1}{2}}\mathrm{St}\left|\phi^*\rho^*\frac{\partial V_i^*}{\partial t^*}\right| \ll |\phi^{*2}\mu^* k_{ij}^{*T} V_{ri}^*|, \tag{4.2.12}$$

$$\mathrm{Re}\mathrm{Da}^{\frac{1}{2}}\left|\phi^*\rho^* V_j^*\frac{\partial V_i^*}{\partial x_j^*}\right| \ll |\phi^{*2}\mu^* k_{ij}^{*T} V_{ri}^*|, \tag{4.2.13}$$

$$\mathrm{Da}^{\frac{1}{2}}\left|\frac{\partial}{\partial x_j^*}\phi^*\mu^*\frac{\partial V_i^*}{\partial x_j^*}\right| \ll |\phi^{*2}\mu^* k_{ij}^{*T} V_{ri}^*|, \tag{4.2.14}$$

Since all terms indicated by an asterisk ($*$) are *of order one*, the validity of the inequalities (4.2.3)–(4.2.5) depends on the values of the relevant dimensionless numbers.

Case A

When, in a considered case, Reynolds number, Darcy number and the Strouhal number are such that:

$$\mathrm{Da} \ll 1, \quad \mathrm{Re}\,\mathrm{Da}^{\frac{1}{2}} \ll 1, \quad \mathrm{St} \sim O(1), \tag{4.2.15}$$

which is the case in most groundwater and petroleum reservoir engineering cases, the momentum balance equation (4.2.1) reduces to:

$$\mathbf{q}_r \equiv \phi(\mathbf{V}_f - \mathbf{V}_s) = -\frac{\mathsf{k}}{\mu}\left(\nabla p + \rho g \nabla z\right), \quad \mathsf{k} = \phi\Delta^2\mathbf{R}^T, \tag{4.2.16}$$

which is Darcy's law, with k denoting the permeability. Thus, *Darcy's law, which is usually regarded as a flux equation, is nothing but a simplified form of the momentum balance equation*. The product $\phi\mathbf{V}_f$ is usually referred to as the *specific discharge* of the fluid. However, we note that on the l.h.s. of the above equation, we have the fluid flux relative to the (possibly moving) solid matrix.

The permeability, k is a second rank tensor, as it relates the velocity vector to the driving force vector composed of the pressure gradient and the force due to gravity.

It is interesting to compare the expressions for the permeability tensor, k, as defined in (4.2.16), in (4.2.26) and in (4.2.32).

Darcy's law (4.2.16) is the form most commonly used in modeling, obviously, within the limits of the constraints underlying it. We note that in it, the density and the viscosity depend on the temperature and the concentration of dissolved species.

Let us express this driving force (for a constant density fluid) as the gradient of the piezometric head, $h = z + p/\rho g$ (valid for $\rho =$ const.). Then, we make use of (3.4.38), with $X_i \equiv q_i$, where $q_i \equiv \phi V_i$, $V_i \approx V_{fi} \gg V_{si}$, and $Y_i \equiv \partial h/\partial x_i \equiv \nabla_i h$. In this case, the, $\dot{\mathcal{S}}$ is related to $\mathbf{X}$ and $\mathbf{Y}$, by:

$$\dot{\mathcal{S}} \equiv X_i Y_i = \left(-K_{ij}\frac{\partial h}{\partial x_j}\right)\left(-\frac{\partial h}{\partial x_i}\right) \geq 0.$$

Hence, the matrix K_{ij} is symmetric and definite positive; *the permeability is a second rank symmetric tensor*.

Note that when the driving force in Darcy's law is expressed in terms of pressure and gravity, like in (4.2.16), the law is applicable to a variable density fluid.

We note, that although we have started from a momentum balance equation, the final form, contains a coefficient—the fluid's viscosity,

Case B

We assume that (4.2.13) and (4.2.14) are valid, i.e., that we may neglect the inertial effects, but we maintain the internal viscous friction expressed by the first term on the r.h.s. of (4.2.1). Then, with:

$$\text{Da} \sim O(1), \quad \text{Re}\,\text{Da}^{\frac{1}{2}} \ll 1, \quad \text{St} \sim O(1), \tag{4.2.17}$$

the momentum balance equation (4.2.1) reduces to the simplified form:

$$\nabla\cdot\phi\tilde{\mu}\nabla\mathbf{V} - \phi(\nabla p + \rho g\nabla z) - \phi\frac{\mu\mathsf{R}}{\Delta^2}\cdot(\mathbf{V} - \mathbf{V}_s) = 0, \tag{4.2.18}$$

known as the *Brinkman equation* (Brinkman 1948). In it, we note two viscosity symbols: one which is due to the momentum transfer from the solid to the fluid, and the other, which is associated with the fluid's velocity gradient, unless the (macroscopic) velocity is uniform everywhere, i.e., $\mathbf{V} = \text{const}$. The Brinkman equation is usually employed to describe saturated flow with a high velocity gradient, e.g., when the porous medium domain is bounded by a body of free water.

Case C

When the local acceleration, $\partial\mathbf{V}/\partial t$, cannot be neglected, e.g., when flow starts from rest, or in oscillatory flow, we have:

$$\text{Da} \sim O(1), \quad \text{Re}\,\text{Da}^{\frac{1}{2}} \ll 1, \quad \text{St} \gg O(1), \tag{4.2.19}$$

and the momentum balance equation takes the form:

$$\phi\rho\frac{\partial\mathbf{V}}{\partial t} = \nabla\cdot\phi\mu\nabla\mathbf{V} - \phi(\nabla p + \rho g\nabla z) - \phi\frac{\mu\mathsf{R}}{\Delta^2}\cdot(\mathbf{V}_f - \mathbf{V}_s). \tag{4.2.20}$$

Focussing our attention on the three momentum balance equations (4.2.16), (4.2.18) and (4.2.20), it is obvious that if, in a given experiment, we try to use any of these equations to determine the value of the coefficient R/Δ^2, we shall obtain different values. This means that the permeability, defined as $\phi\Delta^2\mathsf{R}^T$ will take on different values in the three cases. Cheng (2016, p. 486) uses the term "dynamic permeability" for k obtained from (4.2.20).

Let us focus on the last two terms on the r.h.s. of the momentum balance equation (4.2.20), recalling that $k_{ij} = \phi\Delta^2 R^T_{ij}$. It is implicit that these two terms are of the same order of magnitude, i.e.,

$$O\left(\phi_c\rho_c g\right) \sim O\left(\frac{\phi_c^2\mu_c L_c}{k_c t_c}\right), \quad O\left(p_c\right) \sim O\left(\rho_c g L_c\right).$$

This leads to the conclusion that in all three cases considered above:

$$\frac{\text{Fr}^2}{\text{ReDa}^{\frac{1}{2}}} \sim O(1), \quad \Rightarrow \quad Fr \ll 1, \quad Fr = \frac{V_c}{\sqrt{gL_c}}, \tag{4.2.21}$$

where Fr denotes the Froude number associated with flow through porous media. We recall that in fluid dynamics, the Froude number is used for the ratio of inertial to gravity forces.

4.2.5 *Tortuosity*

The term *tortuosity* has been introduced already in Sect. 3.4.2, in connection with the macroscopic expression for molecular diffusion, and in connection with fluid-to-solid momentum transfer in Sect. 3.5.1. In fact, the concept of tortuosity is associated with the macroscopic *diffusive fluxes* of mass, heat and momentum in a porous medium. The latter—that of momentum—is the extension of the microscopic Newton's law to the transfer momentum of momentum from the fluid to the solid matrix mentioned above. Eventually, Darcy's law is obtained as a simplified case of the momentum balance equation. Let us elaborate on these concepts in connection with Darcy's law and porous medium permeability.

Ghanbarian et al. (2012) present an extensive critical review and discussion on tortuosity in saturated and unsaturated flow, making a distinction between geometric, hydraulic, electrical, and diffusive tortuosities. They emphasize that the proposed tortuosity models are distinct and thus may not be used interchangeably. Especially, they suggest making a distinction between tortuosity with and without chemical reactions. They also review a number of expressions for tortuosity.

In this subsection, the discussion on macroscopic fluxes in based on the phenological approach.

A. Tortuosity in Single-phase Flow

In Sect. 4.1.2, we used a bundle of straight capillary tubes to represent the flow through a porous medium domain. We showed how a single straight capillary tube, or a bundle of such tubes, can represent a parallelepiped block of porous medium.

Bear (1972, p. 110) followed Carman (1937), who made use of Poisseuille's law for the average velocity, V_x, in a straight capillary tube in the x-direction:

$$V_x = -\frac{g\rho}{\mu}\varpi\Delta^2\frac{\Delta h}{L_x}, \tag{4.2.22}$$

where Δ is the *hydraulic radius* of the tube, ϖ is a factor that depends on the shape of the tube's cross-section, and $\Delta h/L_x$ is the average hydraulic gradient along the tube. By using Δ and ϖ, we account for a non-circular cross-section. By averaging over many such elementary tubes of different cross-sections, we obtain the average velocity:

$$\overline{V}_x = \frac{\rho g}{\mu} \overline{\varpi(\Delta^2)} \frac{\Delta h}{L_x}, \tag{4.2.23}$$

However, a bundle of straight parallel capillary tubes does not really represent the flow through the void-space of a porous medium domain. The real flow takes place through *tortuous stream-tubes* that fill up the entire void-space. Along each stream-tube, the direction of the flow varies continuously, say, with respect to a fixed x, y, z coordinate system. The cross-section also varies along each stream-tube. Thus, we start from the single tortuous stream-tube, with length $L_s(> L_x)$ and constant cross-section. The axial average axial velocity in an elementary straight stream-tube is expressed by:

$$V_s = -\frac{\rho g}{\mu} \overline{\varpi(\Delta^2)} \frac{\Delta h}{L_s}. \tag{4.2.24}$$

Then, we replace $\Delta h/L_s$ by $(\Delta h/L_x)(L_x/L_s)$ and V_s by $(L_s/L_x)V_x$, leading to:

$$\overline{V}_x = -\frac{\rho g}{\mu} \overline{\varpi(\Delta^2)} \left(\frac{L_x}{L_s}\right)^2 \frac{\Delta h}{L_x} = -\frac{\rho g}{\mu} \overline{\varpi(\Delta^2)} T^* \frac{\Delta h}{L_x}, \tag{4.2.25}$$

where $T^* = (L_x/L_s)^2$ may be regarded as the porous medium's tortuosity (e.g., Corey 1977, p. 91; Bear 1972, p. 110). It is an elementary void-space property that takes into account the fact that the (actual) microscopic stream-tubes are tortuous and longer through the void-space than the macroscopic ones; they do not coincide with the direction of the piezometric head gradient, which is the driving force. Bear and Bachmat (1967, see also Bear 1972, p. 105) obtained this coefficient when deriving Darcy's law by averaging over tortuous stream-tubes within an REV, taking into account also variations in their cross-sectional area. Bear (1972, p. 107) defines tortuosity of a porous medium as a "non-random porous medium operator (property) that transforms the average components of an external force acting at a physical point of a porous medium into the average components (in the x_i system) of its projections along the streamlines". Actually, we could define the tortuosity as obtained from (4.2.25) by $\varpi(L_x/L_s)^2$, as m may also vary with direction. Other authors suggest different definitions.

Henceforth, the symbol $\mathbf{T}^*$ will be used to denote (flow) *tortuosity*. In an anisotropic porous medium, the lengths of the tortuous stream-tubes per unit length of porous medium, and their cross-sectional area, vary with direction. Hence, in an anisotropic porous medium, the tortuosity is a *second rank symmetric tensor*, $\mathbf{T}^*$ (components T^*_{ij}).

Altogether, we can write (4.2.25) in the form:

$$q_i \equiv \phi V_i = -\frac{\rho g}{\mu} k_{ij} \frac{\partial h}{\partial x_j}, \quad k_{ij} = \phi \varpi \overline{\Delta^2} T^*_{ij}, \tag{4.2.26}$$

where k_{ij}, the porous medium's permeability, is related to the porosity, to the average of the square of the hydraulic radius of the void-space, and to some numerical coefficient that is related to the shape of the voids. We note the role played by the (second rank tensor) tortuosity. It is interesting to compare the tortuosity T^*_{ij} with R^T_{ij}, in which the superscript T denotes transpose, by comparing (4.2.26) with (4.1.12) and (4.2.2), and, especially, with (4.2.16).

Bear and Bachmat (1991, p. 126), as part of their discussion on the rule that relates the average of a gradient to the gradient of an average, present another definition of the tortuosity of a porous medium domain in the form of (1.4.15), which expresses the *total static moment of the oriented elementary surfaces comprising the $\mathcal{S}_{\alpha\alpha}$-surface, with respect to planes passing through the centroid of the* REV, *per unit volume of the α-phase within $\mathbb{V}_o$*. This macroscopic coefficient depends only on the microscopic configuration of the α-phase within the REV. For an isotropic porous medium, they suggest:

$$T^*_{\alpha,ij} = \frac{\theta^S_\alpha}{\theta_\alpha}\delta_{ij}, \tag{4.2.27}$$

where θ_α denotes the volumetric fraction of the α-phase (i.e., the void space), and θ^S_α is the fraction of the $\alpha - \alpha$ surface on the surface of the REV.

B. Tortuosity in Multi-phase Flow

In multi-phase flow (Chap. 6), each fluid occupies only part of the void space, with this part depending on the phase saturation. It is obvious that the concept of tortuosity is applicable also to the flow of each of the phases, with the effective permeability of each phase related to the tortuosity of that phase. Thus, the tortuosity of each phase will be a function of that phase's saturation. Effective permeability is discussed in Sect. 6.2.2.

* * *

In what follows we shall introduce additional cases in which tortuosity is defined and employed. We shall consider the following cases: (1) mass of a solute diffusing through the fluid occupying the void space, with no adsorption, obeying Fick's law, (2) heat conduction through a fluid saturating the void space, with a non-thermally-conductive solid, obeying Fourier law, and (3) electric current through a fluid saturating the void space, obeying Ohm's law, with a non-conductive solid. In all these cases, the microscopic flux is proportional to the gradient of a potential and stream-tubes of the considered extensive quantity can be identified. These tortuous stream-tubes underlie the concept of tortuosity.

C. Tortuosity in Molecular Diffusion

We consider the case of molecular diffusion, i.e., the case in which the transported extensive quantity is the mass of a dissolved chemical species. At the microscopic level, the flux of molecular diffusion is described by *Fick's law*, presented in Sect. 7.2.2 A. This law states that the flux of a solute is driven by the latter's concentration gradient. The discussion in that subsection introduces also the tortuosity associated with solute transport by molecular diffusion at the macroscopic level.

Treated there are also the cases of molecular diffusion in multi-phase flow and the effect of adsorption of the solute on the solid surface, discussed in Sect. 7.4.1.

D. Tortuosity in Heat Conduction

The diffusive flux of heat, or heat conduction, is described by *Fourier's law* of heat conduction (8.1.4). This law states that this flux is proportional to the temperature gradient; the latter acts as a driving force. Similar to molecular (mass) diffusion, described by Fick's law, we can envision the transport of heat through tortuous heat-carrying stream-tubes. Indeed, Bear and Bachmat (1991, p. 62) extended the concepts of stream-tubes and *stream function* to any extensive quantity. However, unlike mass diffusion, in the absence of adsorption, *heat can cross interphase boundaries*, including the fluid-solid interface, unless the solid is a thermal insulator. Let us focus first on the case in which a single fluid occupies the entire void space, while the solid behaves as a thermal insulator. We can follow the discussion presented earlier in this subsection on mass flow in the void space, noting that the actual travel distance along a heat stream-tube is much longer than the distance between equal temperature surfaces under a macroscopic temperature gradient. This will lead to the definition of a tortuosity in a way similar to that defined for flow and for molecular diffusion. Then, using λ_α to denote the thermal conductivity of the α-phase at the microscopic level, the corresponding macroscopic thermal flux law and the thermal conductivity of the fluid occupying the entire void space and a *non-conducting solid* is:

$$\mathbf{q}^H_{pm} = \phi \mathbf{J}^H_\alpha = -\phi \boldsymbol{\lambda}^*_\alpha \cdot \nabla T, \quad \boldsymbol{\lambda}^*_\alpha = \lambda_\alpha \mathbf{T}^*_\alpha, \tag{4.2.28}$$

in which $\mathbf{T}^*_\alpha$ denotes the tortuosity of the fluid in the void space. In fact we note the similarity between the above equation and (4.2.26). Thus, (4.2.28) describes heat conduction in the fluid occupying the void space in an anisotropic porous medium, with a non-conducting solid.

When, in single phase flow, the solid matrix is also heat conducting, we have to take into account the possibility that the tortuous heat stream-tubes refract at the solid-fluid interfaces and pass also through the solid. Altogether, following the same analysis that led to (4.2.26), we obtain:

$$\begin{aligned} \mathbf{q}^H_{pm} &= \phi \mathbf{J}^H_f + (1-\phi)\mathbf{J}^H_s = -\Lambda^{*H}_{pm} \cdot \nabla T. \\ \text{with} \quad \boldsymbol{\Lambda}^{*H}_{pm} &= \phi \boldsymbol{\lambda}^*_f + (1-\phi)\boldsymbol{\lambda}^*_s \end{aligned} \tag{4.2.29}$$

denoting the thermal conductivity of the saturated porous medium. We note the two tortuosities—one for the fluid saturated void space and the other for the solid matrix.

In Sect. 8.1.1 A we shall present another approach for deriving (4.2.29), making use of the derivation presented by Bear and Bachmat (1991, p. 128).

E. Formation Factor in Reservoir Engineering

The discussion presented earlier, on the tortuous pathways in the void space, etc., is also valid for the flow of electricity through the void-space occupied by an electrically conducting fluid, assuming a non-conducting solid. The flux of electricity in the (conducting) fluid is described by *Ohm's law*. The *formation factor*, F, is a concept used in reservoir engineering. It is defined (Archie 1942) as the resistivity (defined as the electrical resistance of a porous medium cube having unit length sides measured with uni-directional electrical current flow entering one face and leaving the opposite one) of a porous medium saturated with an ionic solution to the bulk resistivity of the same solution. It is used in the interpretation of electric logs, providing information on the permeability of the formation. Assuming that the solid matrix is an insulating material, F provides information on the geometry of the void space. Bear (1972, p. 114), following Cornell and Katz (1953) and performing an analysis similar to that leading to the definition of tortuosity above, suggested that:

$$F = \frac{1}{\phi T^*}. \tag{4.2.30}$$

Thus, the formation factor is another way of introducing the actual (tortuous) pathways of extensive properties transport through porous medium domains.

4.2.6 Range of Validity of Darcy's Law

Column experiments, similar to those conducted by Darcy, indicate that as the specific discharge increases, its relationship to the hydraulic gradient gradually deviates from the linear relationship expressed by Darcy's law (4.2.17). Figure 4.6a shows this deviation.

In fluid mechanics, the (dimensionless) *Reynolds number*, Re, which expresses the ratio between inertial and viscous forces acting on a moving fluid, is used as a criterion for distinguishing between *laminar flow* occurring at low velocities and turbulent flow occurring at higher ones (see any textbook on fluid mechanics). In

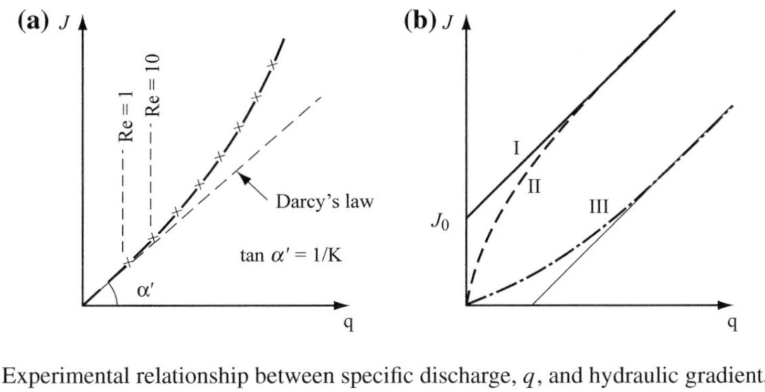

Fig. 4.6 Experimental relationship between specific discharge, q, and hydraulic gradient, $\mathcal{J}$

pipes, the critical value Re ≈ 2000, indicates the transition from laminar to turbulent flow, although widely varying values may apply under special conditions. By analogy, a Reynolds number is defined for flow through porous media, as:

$$\mathrm{Re} = \frac{q\,d}{\nu}, \tag{4.2.31}$$

where d is some representative (microscopic) length characterizing the void space and ν is the fluid's kinematic viscosity. Although, by analogy with pipe flow, d should be a length characterizing the cross-section of an elementary channel of the porous medium, it is customary, for unconsolidated porous media, to employ for d some characteristic length of the grains, probably because it is more easily measured.

Often, the mean grain diameter is used for d in (4.2.31). Sometimes, d_{10} is mentioned in the literature as the representative grain diameter used for d. Collins (1961) suggested $d = (k/\phi)^{\frac{1}{2}}$, or $k = \phi d^2$, where k is the permeability and ϕ is the porosity, and d serves as the representative length. Bear and Bachmat (1991), on the basis of theoretical analysis, suggested the *hydraulic radius* of the void space (defined as the ratio of the volume of the void space to the area of solid-fluid interface) as the characteristic length. In their analysis, they define a Reynolds number and a *Darcy number*:

$$\mathrm{Re} = \frac{V\Delta}{\nu}, \quad \mathrm{Da} = \left(\frac{\Delta}{L^{(v)}}\right)^2, \quad \Delta = \sqrt{\frac{k}{\phi T^*}}, \tag{4.2.32}$$

where Δ denotes the characteristic hydraulic radius, $L^{(v)}$ is the length characterizing spatial variations in the fluid's velocity, V, and T^* is the *tortuosity* (Sect. 4.2.5). They suggest that Darcy's law be used for $\mathrm{ReDa}^{1/2} \ll 1$. Following their analysis, we suggest replacing (4.2.31) by:

$$\mathrm{Re} = \frac{V\sqrt{k/\phi T^*}}{\nu}. \tag{4.2.33}$$

In spite of the various definitions for the characteristic length used in (4.2.31), practically, all evidence indicates that *Darcy's law is valid as long as* Re *does not exceed a value of about* 1 (but sometimes as high as 10). Most saturated groundwater flows occur in this range, except in the very close vicinity of high-rate pumping or recharging wells, or large (point) springs. High Reynolds number flows may also be observed in very porous aquifers, such as cavernous limestone, where the hydraulic radius is large.

Curve I in Fig. 4.6b indicates the existence of some minimum hydraulic gradient, $J = J_o$, below which there is, practically, no flow. On this figure, Curves II and III indicate cases in which the growth is q slower and faster, respectively as J increases. Curve I shows a case when no flow takes place below a threshold gradient. Bear and Verruijt (1987, p. 34) present a number of explanations to these phenomena: (1) pores are very small such that water molecules in them are strongly influenced by the double layer effects of clay particles. Because water molecules are *polar*, water near the electrically charged clay particles has a more crystalline structure which causes the viscosity to be higher than ordinary water. Under such conditions, a minimum hydraulic gradient is required to produce water movement, (b) the effect of *streaming potential*. As water moves near the clay surface it carries along some cations in the diffuse layer Tne cations are electrically attached to the clay particles, a fact that produces a resistance to the movement of the cations. This, in turn, produces a drag on the moving water. The potential difference due to this migration of cations is called *streaming potential*; it acts in the direction opposite to that of the flow, (3) non-Newtonian behavior of the fluid in capillary passages, and (4) electroosmosis counterflow.

Liu (2017, p. 9) presents a detailed literature survey and discussion on the flux law for low permeability materials.

The case of flow at high Re is discussed in Sect. 4.3.2.

4.2.7 Darcy's Law in an Anisotropic Porous Medium

A porous medium domain is said to be *isotropic at a point*, with respect to a considered property, if that property does not vary with direction at that point. It is said to be *anisotropic at a point*, with respect to a property, if that property varies with direction at that point. A typical porous medium property that exhibits anisotropy is the permeability, $\mathbf{k}$. In Sect. 4.2.4, Case A, we have shown that the permeability of a porous medium is a *second rank symmetric tensor*. A symmetric second rank tensor has a set of *three mutually orthogonal principal axes*, with three corresponding real *principal values*. If the principal values differ from each other, then there exists *only one set of principal axes*. In general, the principal values differ from each other, and they correspond to only one set of *three mutually orthogonal principal directions* (see any textbook on tensors).

Consider the case in which the permeability tensor, $\mathbf{k}$, has three principal values, k_1, k_2 and k_3, and three corresponding principal directions indicated by the unit vectors $\mathbf{e}^{(1)}$, $\mathbf{e}^{(2)}$ and $\mathbf{e}^{(3)}$. We can express the tensor's components in any Cartesian coordinate system, in the form:

$$k_{ij} = k_1 e_i^{(1)} e_j^{(1)} + k_2 e_i^{(2)} e_j^{(2)} + k_3 e_i^{(3)} e_j^{(3)}. \tag{4.2.34}$$

If the principal axes are chosen as coordinate axes, (4.2.34) reduces to:

$$k_{ij} = k_1 \delta_{1i} \delta_{1j} + k_2 \delta_{2i} \delta_{2j} + k_3 \delta_{3i} \delta_{3j}. \tag{4.2.35}$$

when $k_1 \neq k_2 \neq k_3$, we say that the permeability represented by these coefficient is *anisotropic*.

In certain cases, a tensorial property of a porous material may exhibit the same response in some, but not all directions. Such behavior indicates that the *microstructure* of the solid matrix is such that it possesses certain features of *macroscopic symmetry*. The existence of such symmetries simplifies the mathematical form of the corresponding coefficient, and, hence, also the description of the process under consideration.

When $k = k_1 = k_2 = k_3$, Eq. (4.2.35) reduces to the simple form:

$$k_{ij} = a\delta_{ij}, \tag{4.2.36}$$

which represents an *isotropic second rank tensor*. It is characterized by a single *scalar*, k.

When $\mathbf{k}$ represents a tensorial coefficient that has one principal direction, $\mathbf{e}$, with a corresponding principal value, k_1, while all directions in a plane normal to $\mathbf{e}$ are principal directions, with a common principal value, k_2, the components k_{ij} can be expressed by (4.2.34), reduces to the form:

$$k_{ij} = k_1 e_i e_j + k_2 \delta_{ij}. \tag{4.2.37}$$

A tensor $\mathbf{k}$ that satisfies (4.2.37) is said to be an *anisotropic tensor with axial symmetry*.

The components of $\mathbf{K}$ in a three-dimensional space, can be written in the matrix form:

$$\mathbf{K} = \begin{bmatrix} \mathsf{K}_{xx} & \mathsf{K}_{xy} & \mathsf{K}_{xz} \\ \mathsf{K}_{yx} & \mathsf{K}_{yy} & \mathsf{K}_{yz} \\ \mathsf{K}_{zx} & \mathsf{K}_{zy} & \mathsf{K}_{zz} \end{bmatrix}, \tag{4.2.38}$$

In two dimensions, the transformation of K-components from any coordinate system, (x, y) to the principal coordinate system, (x', y'), is given by the relationship:

$$\begin{matrix} \mathsf{K}'_{x'x'} \\ \mathsf{K}'_{y'y'} \end{matrix} = \frac{\mathsf{K}_{xx} + \mathsf{K}_{yy}}{2} \pm \left[\left(\frac{\mathsf{K}_{xx} - \mathsf{K}_{yy}}{2} \right)^2 + \mathsf{K}_{xy}^2 \right]^{1/2}. \tag{4.2.39}$$

The angle of rotation needed to reach the principle axes is given by:

$$\alpha = \frac{1}{2} \tan^{-1} \frac{2\mathsf{K}_{xy}}{\mathsf{K}_{xx} - \mathsf{K}_{yy}}. \tag{4.2.40}$$

When the principal values $\mathsf{K}'_{x'x'}$ and $\mathsf{K}'_{y'y'}$ are given, and xy are the cartesian coordinates rotated clockwise by an angle α, with respect to $x'y'$, we have:

$$\begin{aligned} \begin{matrix} \mathsf{K}_{xx} \\ \mathsf{K}_{yy} \end{matrix} &= \frac{\mathsf{K}'_{x'x'} + \mathsf{K}'_{y'y'}}{2} \pm \frac{\mathsf{K}'_{x'x'} - \mathsf{K}'_{y'y'}}{2} \cos 2\alpha, \\ \mathsf{K}_{xy} &= -\frac{\mathsf{K}'_{x'x'} - \mathsf{K}'_{y'y'}}{2} \sin 2\alpha. \end{aligned} \tag{4.2.41}$$

When the principal directions are aligned with a selected coordinate system, (4.2.38) may be represented in matrix form as:

$$[\mathsf{K}] = \begin{bmatrix} \mathsf{K}_{xx} & 0 & 0 \\ 0 & \mathsf{K}_{yy} & 0 \\ 0 & 0 & \mathsf{K}_{zz} \end{bmatrix}, \tag{4.2.42}$$

so that Darcy's law reduces to:

$$q_x = \mathsf{K}_{xx}\mathcal{J}_x, \qquad q_y = \mathsf{K}_{yy}\mathcal{J}_y, \qquad q_z = \mathsf{K}_{zz}\mathcal{J}_z. \tag{4.2.43}$$

In an anisotropic porous medium, since the direction of the flow ($\mathbf{q}$) does not coincide with the direction of the driving force (a vector, say, $\vec{\mathcal{J}} = -\nabla h$), it is possible to consider the permeability in a specific direction in the porous medium domain. However, it is interesting to note that we may define *two* kinds of permeability for the same direction in space. For the sake of simplicity, consider the flow of a homogeneous fluid in such a porous medium. With θ denoting the angle between the two directions,

$$\cos\theta = \frac{\mathbf{q}\cdot\vec{\mathcal{J}}}{\mathcal{J}},$$

and, for the sake of simplicity, referring to the hydraulic conductivity, K, we distinguish between two cases (Bear 1972, p. 143):

(a) Permeability, K_q, in a direction that coincides with that of the flow,

$$K_q = \frac{q}{\mathcal{J}\cos\theta}.$$

(b) Permeability, K_J, in a direction that coincides with that of the hydraulic gradient,

$$K_J = \frac{q\cos\theta}{\mathcal{J}},$$

with $K_J/K_q = \cos^2\theta < 1$.

Altogether, the basic expression that describes the flux of a variable density fluid (relative to the (possibly moving) solid) takes the form:

$$\mathbf{q}_r \equiv \phi\left(\mathbf{V}_f - \mathbf{V}_s\right) = -\frac{\mathsf{k}}{\mu}\cdot(\nabla p + \rho g \nabla z), \tag{4.2.44}$$

or, in indicial notation:

$$q_{ri} \equiv \phi\left(V_{fi} - V_{si}\right) = -\frac{k_{ij}}{\mu}\left(\frac{\partial p}{\partial x_j} + \rho g \frac{\partial z}{\partial x_j}\right). \tag{4.2.45}$$

This equation will be used throughout this book as the fundamental single flux equation for saturated flow in the range in which Darcy's law is applicable. However, we usually (but not always) overlook the fact that the above equation involves the fluid's velocity relative to the solid, i.e., and assume $\mathbf{V}_s \equiv 0$.

Note that in the above equation, ρ and μ may vary with p and T, and concentrations of dissolved species. To understand the gravity effect, we may rewrite (4.2.45) in the form:

$$q_{ri} = -\frac{k_{ij}}{\mu}\left(\frac{\partial p}{\partial x_j} + \rho_o g \frac{\partial z}{\partial x_j}\right) - (\rho - \rho_o)\frac{k_{ij}}{\mu}. \tag{4.2.46}$$

4.3 Non-Darcy Flux Laws

In Sect. 4.2.4, we have presented an expression for fluid flux, Darcy's (linear) law, as a simplified, or approximate, version of the fluid's momentum balance equation, where the simplification is based on neglecting certain terms in the general momentum balance equation. We started from (4.2.1) and lead to Darcy's linear law (4.2.16). However, under certain circumstances, the assumptions leading to the linear flux law (4.2.16) are not valid and other simplified versions of the momentum balance equation have to be used as 'motion equations'.

4.3.1 Brinkman's Equation

Brinkman's law (Brinkman 1949) was already presented as (4.2.18). With the permeability defined in (4.2.16), we can rewrite it in the form:

$$\nabla\cdot\phi\mu\nabla\mathbf{V} - \phi(\nabla p + \rho g\nabla z) - \phi\frac{\mu\mathbf{R}}{\Delta^2}\cdot(\mathbf{V}_f - \mathbf{V}_s) = 0. \tag{4.3.1}$$

Brinkman's equation is used, primarily, when it is not possible to ignore the viscous shearing stresses acting on the fluid (Brinkman 1949). While such simplification is acceptable under the conditions of relatively low permeability, for which Darcy's Law is valid, it introduces significant errors at the boundaries between low-permeability subdomains and subdomains where resistance to flow is limited (e.g., at the boundary of a wellbore or a fracture in contact with a porous medium). Although the use of Brinkman's equation allows an accurate description of flow over the entire spectrum of permeabilities within a domain, it has found limited application because of its complexity.

4.3.2 Forchheimer's and Other High Re Flux Laws

A. Forchheimer's Flux Law

As shown in Sect. 4.2.6, Darcy's law, which involves a *linear relationship* between the fluid's flux and the hydraulic gradient, is valid only for flow at low Reynolds' numbers, e.g., Re $\leq$ 1–10. Experimentally, as Re increases, or, equivalently, as the fluid's velocity increases, we observe a growing deviation from the linear relationship between the pressure gradient and the flux. In the range of validity of Darcy's law, i.e., Re $<$ 1–10, the viscous forces that resist flow are predominant. As the flow velocity increases, a gradual transition is observed (Fig. 4.6) from (microscopically) laminar flow, where viscous forces are predominant, to, still essentially laminar flow, but with inertial forces gradually taking over. Often, the value of Re $=100$ is mentioned as the upper limit of this transition region in which Darcy's linear law is no longer valid. The reason for this deviation from the linear law is that *at the microscopic level*, as velocities increase, local *separation* of the flow from the walls of the solid matrix occurs at an increasing number of places where the flow curves or diverges. Local vortices and countercurrent flow regimes are caused by inertial and viscous forces along portions of the solid. It may be interesting to note that the use of the range 1–10 for the limiting value of Re results for the use of different values for the characteristic microscopic length: mean grain diameter, mean pore diameter, hydraulic radius of pore, etc. For this range, the motion equation, known as the *Darcy–Forchheimer equation*, is used. It was suggested by Dupuit (1848, 1863) and Forchheimer (1901)on the basis of experiments of flow at high Re in a sand column. In his 1901 paper, Forchheimer suggested the relationship:

$$\frac{\Delta h}{\Delta \ell} = aV + bV^2, \tag{4.3.2}$$

in which the l.h.s. denotes the hydraulic gradient.

Barak and Bear (1981) suggested the following form of the motion equation:

$$\mathcal{J}_i = \frac{\nu}{g} w_{ij} q_j + \beta'_{ijkl} q_j q_k q_l + \frac{1}{g} \beta''_{ijk} q_j q_k, \tag{4.3.3}$$

as a good approximation for a constant density Newtonian fluid. In this equation, $\mathcal{J}_i$ denotes the ith component of the hydraulic gradient, w_{ij}, β''_{ijk} and β'_{ijkl} are tensors of the second, third and fourth orders, respectively, which represent matrix properties only. If the resistance to flow in any direction is the same as in the opposite direction, the second term must vanish. At low Re, the last two terms on the r.h.s. vanishes and we obtain Darcy's law. The last term expresses the effect of matrix asymmetry.

Nield and Bejan (1998, pp. 9–12) present reviews and discussions on Forchheimer's equation. They present the expression suggested by Joseph et al. (1982):

$$\nabla p = -\frac{\mu}{k}\mathbf{q} - c_F k^{-1/2} \rho |\mathbf{q}| \mathbf{q}, \tag{4.3.4}$$

in which a, b are coefficients and c_F is a dimensionless form-drag constant, k is the permeability and μ is the fluid's viscosity. At low Re, the second term on the r.h.s. which expresses the average of the microscopic inertial effects, becomes negligible.

Hassanizadeh and Gray (1987) concluded that:

$$\phi\left(\frac{\partial p}{\partial x_k}+\rho g\nabla z\right)=\left(a+b|V_k^d|+c|V_k^d|^2\right)V_k^d, \tag{4.3.5}$$

where V_k^d denotes the kth component of the fluid's velocity relative to the solid, and a, c, c are coefficients which depend on density and porosity.

Altogether, for a rigid porous medium, neglecting the effects of inertia *at the macroscopic level*, leads to the flux equation in the form:

$$q_i=-\frac{k_{ij}}{\mu}\left(\frac{\partial p}{\partial x_j}+\rho g\frac{\partial z}{\partial x_j}\right)-\frac{\rho\beta_{ij}}{\mu}qq_j, \tag{4.3.6}$$

where $q=|\mathbf{q}|$, and β_{ij}, sometimes referred to as the *non-Darcy flow coefficient*, or the *Forchheimer inertial resistance coefficient* (dims. L^{-1}) is a second rank tensorial coefficient that is related to the configuration of the void space, thus related also to the permeability of the void-space. In multiphase flow, β_{ij} is also related to phase saturation. In an isotropic porous medium, this coefficient is a scalar. At low Re, the second term on the right-hand side, which expresses the average of the *microscopic inertial effects*, becomes negligible. It is important to emphasize that although we have here flow at high Re, this is not turbulent flow, which in a porous medium appears at a much higher Re.

Whitaker (1996), starting from the Navier–Stokes equation at a point within the void space, employs his averaging method described in Sect. 1.4.2 B, to develop the Forchheimer equation by averaging the Navier–Stokes equation. His result takes the form:

$$\langle\mathbf{V}_\alpha\rangle=-\frac{\mathsf{k}}{\mu_\alpha}\cdot(\nabla\langle p_\alpha\rangle_\alpha-\rho_\alpha\mathbf{g})-\mathsf{F}_\alpha\cdot\langle\mathbf{V}_\alpha\rangle, \tag{4.3.7}$$

where $\langle\mathbf{V}_\alpha\rangle$ is the superficial average velocity, $\langle e_\alpha\rangle$ and $\langle e_\alpha\rangle_\alpha$ represent the phase average and the intrinsic phase average, respectively, of e_α, and F_α is a tensorial correction coefficient.

Finsterle (2001) expresses the β-coefficient for a phase in multiphase flow in an isotropic porous medium,in the form:

$$\beta=A_1\frac{1}{k_{eff}^{A_2}}\frac{1}{\theta^{A_3}}\frac{1}{\tau^{A_4}}, \tag{4.3.8}$$

where the A's are coefficients, θ is the fluid's content, assuming that the Forchheimer equation applies to both liquid and gas flow, and τ is the tortuosity for multiphase flow, defined by Millington (1959) as:

$$\tau = \phi^{1/3} S_\alpha^{10/3}. \tag{4.3.9}$$

Finsterle (2001) provides a table in which the values of various coefficients are presented, as appearing in various publications.

B. Friction Factor and Ergun's Equation

The concept of *friction factor* and Ergun's equation are commonly used in Chemical Engineering (see Appendix A). When fluid flows relative to a submerged solid particle, it exerts a force, or *drag* on the surface of the latter. This force is a consequence of (1) shear stresses due to the fluid's viscosity and to velocity gradients, which produces forces that are tangential to surface, and (2) pressure which produces a force normal to the surface. Together, the (vector) sum of these forces, integrated over the entire surface area of the particle, produces a *resultant force*. The component of the latter in the direction of the fluid's velocity is called the *drag force*, or just *drag*. It is also referred to as *surface resistance*. The force component normal to direction of the relative velocity is referred to as *lift force*. The two force component include the effects of both pressure and friction.

Each of these forces can be expressed as a product of three factors: a coefficient, the kinetic energy of the fluid and some suitably selected characteristic area of the solid:

$$D_{fric} = C_{fric}\,\rho \frac{V_{o,f}^2}{2} A_{fs}, \quad D_{pres} = C_{pres}\,\rho \frac{V_{o,f}^2}{2} A_{ps} \tag{4.3.10}$$

It is common to choose the frontal area for A_{fs} and the horizontal projection are for A_{ps}. When the solid is not stationary, we have to consider the velocity relative to the solid.

In principle, when we consider the flow of a fluid through the void space of a porous medium domain, the same drag and lift phenomena occur, but the geometry is more complicated. Still, we may consider the same two types of forces: one associated with the flow, i.e., with the velocity of the fluid, and the other associated with the pressure (which is present even in the absence of flow).

Let F_k denote the force exerted by the moving fluid acting on a characteristic area A. We can write:

$$F_k = K \times A \times f, \tag{4.3.11}$$

in which K is a coefficient, A is the area normal to the flow direction and f is a coefficient, referred to as *friction factor* that depends on the properties of both the flow and the porous medium, e.g., $f = f(Re)$. For flow in a porous medium column of length L, packed with solid particles of diameter d_p, it is common to write (e.g., Bird et al. 1960, p. 199):

$$\frac{p_o^* - p_L^*}{\frac{1}{2}\rho q_o^2} = \frac{L}{d_p} \cdot 4f, \quad p^* = p + \rho g z, \tag{4.3.12}$$

where q denotes the fluid's specific discharge, f is the friction factor, and p^* (for $\rho = const.$) may be referred to as a *modified pressure*. Actually, $p^* = \rho g(p/\rho g + z) = \rho g h$, where h denotes the piezometric head.

Following the discussion presented by Bird et al. (1960, p. 197), let us consider 1-d average laminar flow in a narrow tube of radius R. Making use of the Hagen–Poisseuille law (4.1.3), we write;

$$q = \frac{(p_o^* - p_L^*)R^2}{8\mu L}. \tag{4.3.13}$$

When applied to laminar flow in column of length L, packed with a porous medium, visualized as a network of capillary tubes of non-uniform length and cross-section, we may rewrite the above equation in the form:

$$q = \frac{(p_o^* - p_L^*)\Delta^2}{2\mu L}, \qquad \Delta = \text{hydraulic radius}, \tag{4.3.14}$$

where Bird et al. (1960, p. 197), following the work of Kozeny (see (4.1.12)), the *hydraulic radius* (Δ) can be related to the porosity and to the specific surface (M_s) by:

$$\Delta = \frac{\phi}{M_s}, \qquad M_s = \frac{6}{d_p}, \tag{4.3.15}$$

where d_p denotes the mean particle diameter (sphere).

Altogether, by combining the *Hagen–Poiseuille formula* (4.1.3) with the above considerations, we obtain:

$$q = \frac{p_o^* - p_L^*}{L} = \frac{d_p^2}{150\mu}\frac{\phi^3}{(1-\phi)^2}. \tag{4.3.16}$$

According to Bird et al. in the above equation, referred to as the *Blake–Kzeny equation*, the coefficient 150 is a consequence of taking into account the tortuosity of the individual pathways within a real porous medium domain.

Following similar consideration for flow at higher Re, Ergun (1952) suggested the following flux law for both low and high Re 1-d flow in an isotropic *packed beds* of length L:

$$\frac{\Delta p^*}{L} = 150\frac{\mu}{d_p^2}\frac{(1-\phi)^2}{\phi^3}q + 1.75\frac{\rho}{d_p}\frac{1-\phi}{\phi^3}q^2, \tag{4.3.17}$$

where 180 and 1.75 are experimentally derived coefficients, $\Delta p^*/L$ includes both the pressure gradient and the effect of gravity, and d_p denotes particle size. Ergun's equation is commonly used in chemical engineering to calculate pressure drop in packed bed reactors. Slightly different values for the numerical coefficients are also mentioned in the literature. Irmay (1958) derived the same equation, but without the numerical values of the coefficients.

C. Rapid Velocity Changes

When local acceleration may not be neglected, especially at the onset of flow and in oscillatory flows, but the advective acceleration, $\mathbf{V}\nabla\mathbf{V}$, and the internal friction, $\phi\mu\nabla\mathbf{V}$, in the fluid may be neglected, the motion equation takes the form:

$$\frac{\rho}{\mu}(T^*_{im})^{-1}k_{mj}\frac{\partial V_j}{\partial t} + \frac{k_{ij}}{\mu}\left(\frac{\partial p}{\partial x_j} + \rho g\frac{\partial z}{\partial x_j}\right) + q_i = 0, \qquad (4.3.18)$$

where T^*_{im} denote components of the *tortuosity tensor*, defined in Sect. 4.2.5.

Although, for the sake of completeness, we have introduced equations applicable to cases in which Darcy's law is not applicable, conditions that justify their application are seldom encountered in problems of flow and contaminant transport in the subsurface.

4.3.3 The Klinkenberg Effect in Gas Flow

The general flux law for a fluid was introduced in the form of Darcy's law (4.2.44). When applied to a gas, the gravity term in that flux law is often neglected. One example is the case where gas flows through a relatively thin gas-bearing reservoir. Another example is when considering the flow of air in the unsaturated zone, just below ground surface. Obviously, the gravity term cannot be neglected when the flow is gravity-driven, e.g., when a volatile organic compound is present and diffusing in the air that occupies part of the void space in the unsaturated zone.

However, when (4.2.44) is applied to a gas, this flux law requires special attention. As presented, this law is applicable to the flow of any fluid phase–liquid, or gas—through a porous medium domain. However, there are some significant differences between liquid and gas flow that stem mainly from the very strong relationship between the gas pressure and certain properties of the solid matrix, as compared to those of liquids. Focussing first on a single-component gaseous phase, the main difference stems from the assumption that liquid *adheres* to the solid's (microscopic) surface, i.e., from the assumption of a "no-slip" condition at the microscopic solid–liquid interface. On the other hand, at a solid–gas interface, this assumption is no more valid and we have "slippage" of the gas relative to the solid surface. This phenomenon is amplified as the area of this interface (per unit volume of porous medium) becomes larger as voids (or grains) become smaller, i.e., in low permeability media. This phenomenon of gas slippage is well known in flow though capillary tubes when the diameter of the latter becomes smaller, particularly approaching the mean free path of the gas. It is referred to as the *Klinkenberg effect* (Klinkenberg 1941).

As the mean free path of the molecules is inversely proportional to the gas pressure, Klinkenberg's experiments with gas flow in a glass capillary tube showed that the measured permeability and the reciprocal mean pressure can be expressed by:

$$k_g = k_\ell \left(1 + 4c\frac{\lambda}{R}\right), \tag{4.3.19}$$

in which k_g and k_ℓ are the permeabilities to gas and to liquid (or gas at very high pressure), respectively, λ denotes the mean free path of the gas molecules under the mean pressure p, $c \approx 1$ is a proportionality factor, and R is the radius of the tube. He concluded that for a porous medium, the rule should be:

$$k_g = k_\ell \left(1 + \frac{b}{p_g}\right), \tag{4.3.20}$$

in which b (dim. $ML^{-1}T^{-2}$) is a coefficient for the gas-solid system that depends on the mean free path of the gas (and, hence, on its pressure) and on the size of the openings in the porous medium. Since k_ℓ is related to this size, b is also a function of k_ℓ. Note that in the literature k_ℓ may also be referred to as k_∞, i.e., the gas-phase permeability to gas at very high pressures, at which the Klinkenberg effect becomes negligible (Wu et al. 1998). Examples of b suggested by the *American Petroleum Institute* are: $b = 3, 0.5, 0.1$ for $k_\ell = 3 \times 10^{-18}, 5 \times 10^{-18}$, and $2 \times 10^{-13}\,\text{m}^2$ (0.03, 0.05 and 200 millidarcy (or mD), respectively).

When the permeability is sufficiently large, e.g., $> 10^{-13}\,\text{m}^2$ ($= 100\,\text{mD}$), the dependence of k_g on pressure, associated with the slippage (or Klinkenberg) effect, can safely be ignored (Aronofsky 1954).

Wu et al. (1998) suggested $b = 3.95 \times 10^3$, 4.75×10^4 and 7.60×10^5 for $k_\ell = 10^{-12}, 10^{-15}$ and $10^{-18}\,\text{m}^2$ (1000, 1 and 0.001 mD), respectively. Jones (1972) expressed the k - b relationship by the equation:

$$b = k_\ell^{-0.36}. \tag{4.3.21}$$

This relationship can then be used to estimate b from a reference medium with known properties:

$$\frac{b}{b_o} = \left(\frac{k_\ell}{k_{\ell o}}\right)^{-0.36}, \tag{4.3.22}$$

where the subscript o denotes a reference medium.

Wang et al. (2014) suggested the relationship:

$$b = \frac{16c\nu}{w}\sqrt{\frac{2RT}{\pi M}}, \tag{4.3.23}$$

where c is a constant (usually taken as 0.9), ν is the kinematic viscosity, M is the molecular weight of the gas, R is the universal gas constant, w is the average flow path width, and T is the absolute temperature.

We recall that Darcy's law is based on the *assumption of no slip* of the fluid with respect to the pore's wall. This is similar to the assumption of no slip of the fluid in a capillary tube, which leads to Poiseuille's equation. The introduction and quantification of the Klinkenberg effect, or the *gas slippage effect* , is based on the concept of a fluid layer that is thinner than the *molecular mean free path,* i.e., the average distance between two consecutive molecular collisions (sometimes called *Knudsen layer*), which is in contact with the pore walls. In this layer, only molecules-to-wall collisions are taken into account, while collisions among molecules are ignored. Thus, the slippage velocity, described by the Klinkenberg permeability correction, represents the contribution of molecule-wall interactions. It yields the Poiseuille velocity profile in a capillary tube, based on the no slip assumption. The Klinkenberg effect is important when $k_g < 10^{-18}\,\mathrm{m}^2$, and can be used to estimate the liquid permeability from gaseous permeability measurements. Klinkenberg applied this slippage condition to the flow of a gas in a porous medium to derive a first order correction for the gas slippage.

Knudsen flow characterizes gas flow through the void space of ultra-low permeability media. In such media, the mean molecular path length is of the order of magnitude of the size of the very small pores size and the permeability is controlled by them. A typical case is gas flow in shale reservoirs. This is a rather unconventional gas resource whose exploitation began mainly around 2005 (with the introduction of hydraulic fracturing (fracking) techniques) and soon turned into a significant gas source, particularly in the U.S.A. Knudsen flow occurs also at very low pressure; the gas is then referred to as "rare gas".

Altogether, the assumption underlying Darcy's law fails as the flow apertures become very small. The dimensionless *Knudsen number*:

$$\mathrm{Kn} = \frac{\lambda}{\Delta}, \tag{4.3.24}$$

in which Δ is the length characterizing the size of the void space, for example, its hydraulic radius, and λ denotes the mean free path of the gas molecules, is used to determine the kind of flow regime prevailing in gas flow through a porous medium domain.

When the Knudsen number is near or greater than one, the mean free path of a gas molecule is comparable to a length scale of the aperture, and the continuum assumptions underlying fluid mechanics is no longer valid. For $\mathrm{Kn} < 0.01$, the mean free path of the gas molecules is negligible compared to the characteristic dimension of the flow geometry and Darcy's law (including all the assumptions underlying it) is valid. As pressure drops, the mean free path and the Knudsen number increase, and Darcy's law is no more valid.

The Klinkenbergs approach ignores the transition flow region, in which both molecule-molecule and molecule-wall interactions are significant. The Klinkenberg approach and equations are sufficiently accurate to describe the problem of gas flow in tight organic media for a wide range of practical purposes. More accurate estimates that account for this transition zone (the *Knudsen layer*) can be obtained by considering *Knudsen diffusion* which is valid in the range Kn $<$ 10 in which the gas molecules collide with the void space boundaries more often than among themselves.

In *shale reservoirs*, the characteristic length of the void space is close to the mean free path of the gas and Darcy law fails to describe gas flow.

Kuila et al. (2010) suggested that gas flow in the transition, or intermediate, regime falls between Darcy flow and Knudsen flow. The term *Knudsen flow* (reference) is used here to describe the flow of a gas when the mean free path of the gas molecules is lager than the characteristic length of the flow domain; here, this is the characteristic length dimension of a pore. Thus, the Klinkenberg effect and the equation for permeability describe what happens when we have Knudsen flow in a porous medium with very small apertures and/or very low gas pressure.

The subject of gas flow, with and without the Klinkenberg effect is presented in Sect. 5.1.7.

References

Archie GE (1942) The electrical resistivity log as an aid in determining some reservoir characteristics. Trans AIME 146:54–61

Aronofsky JS (1954) Effect of a gas slip on unsteady flow of a gas through porous media. J Appl Physics 25(1):48–53

Barak AZ, Bear J (1981) Flow at high reynolds numbers through anisotropic porous media. Adv Water Resour 4(2):54–66

Bear J (1972) Dynamics of fluids in porous media. Elsevier, Amsterdam, 764 p (also published by Dover Publications, 1988; translated into Chinese)

Bear J, Bachmat Y (1991) Introduction to modeling phenomena of transport in porous media. Kluwer Publishing Company, Dordrecht, 553 p

Bear J, Verruijt A (1987) Modeling groundwater flow and pollution. D. Reidel Publishing Company, Dordrecht, 414 p

Bear J, Zaskavsky D, Irmay S (1968) Physical Principles of Water Percolation and Seepage, UNWSCO Paris, pp 465

Bird RB, Stewart WE, Lightfoot EN (1960) Transport phenomena. Wiley, New York, 780 p

Blake FC (1922) The resistance of packing to fluid flow. Trans Am Inst Chem Eng 14:415–422

Brinkman HC (1948) Calculations of the flow of heterogeneous mixture through porous media. Appl Sci Res 2:81–86

Brinkman HC (1949) A calculation of the viscous force exerted by a flowing fluid on a dense swarm of particles. Appl Sci Res 1(27):27–34

Carman PC (1937) Fluid flow through granular bed. Trans Inst Chem Eng (London) 15:150–156

Carman PC (1956) Flow of gases through porous media. Butterworth, London, 152 p

Cheng AH-D (2016) Poroelascicity. Springer, Berlin, 877 p

Collins RE (1961) Flow of fluids through porous media. Reinhold, New York, 270 p

Corey AT (1977) Mechanics of heterogeneous fluids in porous media. Water Resources Publications, Fort Collins, 259 p

Cornell D, Katz DL (1953) Flow of gases through consolidated porous media. Ind Eng Chem 45:2145–2152

Darcy H (1856) Les Fountaines Publiques de la Ville de Dijon. Dalmont, Paris

De Josselin de Jong G (1958) Longitudinal and transverse diffusion in granular deposits. Trans Am Geophys Union 39:67–74

De Josselin de Jong G (1969) The tensor character of dispersion coefficient in anisotropic porous media. 1st IAHR Symp. Fundamentals of Transport in Porous Media, pp 259–277, Haifa

Dupuit J (1863) Études Théoriques et Pratiques sur les Mouvementdes des Eaux dans les Cannaux Découverts et à Travers les Terrains Perméables, 2nd edn. Dunod, Paris, 304 p 1st edn. (1948)

Ergun S (1952) Fluid flow through packed columns. Chem Eng Progr 48:89–94

Fair GM, Hatch LP (1933) Fundamental factors governing the streamline flow of water through sand. J Am Water Works Assoc 25:1551–1556

Finsterle F (2001) Implementation of the forchheimer equation in iTOUGH2. Project report, LBNL, 38 p

Forchheimer P (1901) Wasserbewegung durch boden. Z Ver Deutsch Ing 45:1782–1788

Hassanizadeh M, Gray W (1987) High velocity flow in porous media. Trans Phen Porous Media 2:521–531

Hubbert MK (1940) The theory of ground water motion. J. Geol., 48: 785–944

Hubbert MK (1956) Darcy law and the field equations of the flow of underfround fluids, rans. Amer. Inst. Min. Metal. Eng., 207: 222–239

Irmay S (1955) Flow of liquid through cracked media, Bull. Res. Council of Israel, 5A(1), 84 p

Irmay S (1958) On the theoretical derivatin of Dacy's and Forchheimer formulas. Trans Am Geophys Union 39:702–707

Jones SC (1972) A rapid accurate unsteady-state Klinkenberg parameter, SPE Journal, pp 383–397

Joseph DD, Nield DA, Papanicolaou G (1982) Nonlinear equation governing flow in a saturated porous medium. Water Resour Res 18:1049–1105

Klinkenberg LJ (1941) The permeability of porous media to liquids and gases. American Petroleum Institute, New York, pp 200–213

Kozeny J (1927) Ueber kapillare leitung des wassers im boden. Sitzungsber Akad Wiss wien 136(2a):271–306

Krumbein WC, Monk GD (1943) Permeability as a function of the size parameters of unconsolidated sands. Trans Inst Min Met Eng 15:153–163

Kuila U, Prasad M, Batzle N (2010) Pore size distribution and ultrasonic velocities of compacted Na-montmorillonite clays. Paper presented at 8th biennial international confernce and exposition on petroleum geophysics, 143 p

Liu H-H (2017) Fluid flow in the subsurface. Springer, Berlin, 230 p

Millington RJ (1959) Gas diffusion in porous media. Science 130:100–102

Muskat M (1946) The flow of homogeneous fluids through porous media. J.W. Edwards Inc., Ann Arbor, 763 p (1st Ed. 1937)

Nield DA, Bejan A (1998) Convection in porous media, 2nd edn. Springer, Berlin, 546 p

Scheidegger AE (1953) Theoretical models of porous matter. Prod Mon 10(17):17–23

Scheidegger AE (1960) The physics of flow through porous media, 2nd edn. University of Toronto Press, Toronto

Wang G, Res T, Wang K, Zhou A (2014) Improved apparent permeability models of gas ow in coal with klinkenberg ekect. Fuel 128:41–53

Whitaker S (1996) The forchheimer equation: a theoretical development. Transp Porous Media 25:27–61

Wu YS, Pruess K, Persoff P (1998) Gas flow in porous media with Klinkenberg effects. Transp Porous Media 32(1):117–137

Chapter 5
Modeling Single-Phase Mass Transport

In the previous chapter, we have presented the law that governs the mass flux of a fluid phase that occupies the entire void space. We noted that this law contains at least two variables: the flux and the pressure, or the piezometric head (in the case of constant density). This means that to obtain a complete model, we need at least one additional equation—the mass balance equation. This equation will be discussed in the current chapter for the case of a single fluid phase that occupies the entire void space, i.e., saturated flow. We shall make use of the general microscopic and macroscopic balance equations for any extensive quantity, presented in Sects. 3.2, 3.3 and 3.6, and apply it to cases in which the considered extensive quantity is the mass of a fluid. In the next chapter, we shall consider the case of multiple fluid phases that together occupy the void space.

The ultimate objective of this chapter is to develop and present complete, well-posed models of single phase mass flow in a porous medium domain. We shall also develop flow models based on the approximations of constant density fluid and *essentially horizontal flow*, commonly used to describe flow in aquifers. The entire discussion here is under the assumption of isothermal conditions. Non-isothermal conditions will be discussed in Chap. 8.

Two additional subjects will be discussed in this chapter. One is flow in a deformable porous medium, primarily as associated with storage of water in aquifers. The general subject of flow and other phenomena of transport in deformable porous media will be discussed in detail in Chap. 9. The other subject is an introduction to flow in fractured porous medium domains.

The presentation in this chapter is at the macroscopic level, although, as is common in this book, we shall not use any special symbol to indicate this fact.

J. Bear, *Modeling Phenomena of Flow and Transport in Porous Media*,
Theory and Applications of Transport in Porous Media 31,
https://doi.org/10.1007/978-3-319-72826-1_5

We shall make use of the concepts of *stress* and *shear* which are *second rank tensors*, assuming that the reader is familiar with these concepts (or will seek information in any good text on the subject). Some introductory remarks about stress and strain are presented in Sect. 9.1.

The last section in this chapter is an introduction to modeling flow in fractured rock domains. In this book, we do not cover the cases of solute transport, nor of energy transport in fractured rock domains (see, for example, Bear et al. 1993).

5.1 Mass Balance Equation for a Deformable Porous Medium

In principle, fluids are *compressible* and porous media are *deformable*, i.e., their solid matrix deforms under applied stress. In connection with phenomena of transport in porous media, we encounter such deformation or its consequences in many forms:

- One form is as solid's velocity, $\mathbf{V}_s$. We have encountered $\mathbf{V}_s$ in Chaps. 3 and 4, when considering Darcy's law. We have shown that Darcy's law really expresses the velocity of the fluid relative to the solid. In most cases of interest, $|\mathbf{V}_s| \ll |\mathbf{V}_f|$, so that $\mathbf{V}_s$ may be neglected relative to $\mathbf{V}_f$. However, in the case of pressure waves, the strain produced in a porous medium domain as a result of the application of stresses, strongly affects the propagation of pressure waves, and we have to take into account the solid's velocity. This possibility is of interest in geophysical engineering.
- A stress acting on a porous medium domain will cause the solid matrix to deform. When the domain is a geological formation, the strain produced by the applied stress may damage the formation. Cracks and fissures may develop. *Land subsidence*, or *consolidation*, in response to loading at ground surface, and land upheaval, in the case of recharge of fluids into a confined formation, may serve as examples. These phenomena are of interest to geotechnical engineers. This subject is discussed in detail in Chap. 9. There, we shall also show the relationship between the solid's velocity, the strain and the displacement of points within the solid matrix.
- The deformation produced by changes in stress within a formation manifests itself also as change in porosity. This means that the storage of fluid(s) within the formation is changing. This change in storage, of interest to groundwater hydrologists and to petroleum reservoir engineers, is considered here. In the current chapter, we shall also develop the concept of *specific storativity*, encountered when modeling flow in confined geological formations, e.g., groundwater aquifers.
- Using injection of fluids at high pressure to produce cracks and fractures in deep geological rock formations is a technique used in reservoir engineering to enhance fluid recovery.
- In a deformable porous medium, we have: (1) a time-dependent porosity, i.e., $\partial\phi/\partial t \neq 0$, and (2) a moving solid matrix, i.e., $\mathbf{V}_s \neq 0$. We recall that *Darcy's*

law, e.g., (4.2.16), *expresses the fluid's flux relative to the (possibly moving) solid.* Hence, when considering fluid (f) flow through a deformable porous medium, we have to take into account $\mathbf{V}_s \neq 0$, i.e., $\mathbf{q}_r = \phi(\mathbf{V}_f - \mathbf{V}_s)$ in Darcy's law.

5.1.1 The Basic Fluid's Mass Balance Equation

The basic macroscopic mass balance equation for a single fluid phase that occupies the entire void space was already presented in Sect. 3.6.1. With subscript f denoting the fluid phase, $\theta_f \equiv \phi$, and with a fluid source term, this balance equation takes the form:

$$\frac{\partial \phi \rho_f}{\partial t} = -\nabla \cdot \rho_f \, \mathbf{q}_f + \rho_f \Gamma^f, \tag{5.1.1}$$

where: $\mathbf{q}_f (= \phi \mathbf{V}_f)$ is the *specific discharge* ($\equiv$ flux) of the fluid, and the source, Γ_f, denotes added fluid mass per unit volume of porous medium per unit time. Usually, as a practical approximation, $\mathbf{q}_f$ is expressed by Darcy's law, but we have to recall that, actually, this law expresses the fluid's flux *relative to the solid*, i.e., $\mathbf{q}_r$ and not $\mathbf{q}_f$, with $\mathbf{q}_r \equiv \phi(\mathbf{V}_f - \mathbf{V}_s)$. As we shall see below, since we added $\mathbf{V}_s$ as an additional variable, we shall have to add the balance equation for the solid's mass as an additional equation.

It may be interesting to note the special case in which the porous medium domain is homogeneous, isotropic and rigid, and the fluid is a liquid of constant density. For the case $\rho, \phi =$ constant, and $\Gamma_f = 0$, the mass balance equation reduces to:

$$\nabla \cdot \mathbf{q} = 0, \tag{5.1.2}$$

known as the *Boussinesq equation*. In this case, we may use the piezometric head, $h(x, y, z, t)$, defined by (4.1.4), instead of pressure as a variable of state. The mass balance equation that describes steady flow (ϕ and ρ_f are constant) in a homogeneous isotropic porous medium domain, in the absence of liquid sources and sinks, reduces to the *Laplace equation*:

$$\nabla^2 h = 0, \quad \text{or} \quad \frac{\partial^2 h}{\partial x^2} + \frac{\partial^2 h}{\partial y^2} + \frac{\partial^2 h}{\partial z^2} = 0. \tag{5.1.3}$$

When the considered fluid is extracted from the void space of a geological formation through a well, considered as a point sink, or injected into the porous medium domain through a well, considered as a point source, we may replace Γ^f on the right hand side of (5.1.1) by specific expressions that describe these activities, per unit volume of porous medium. For example, when fluid extraction takes the form of point sinks located at points $\mathbf{x}_m$, where the fluid's densities are ρ_m, and the rates are $Q_m(\mathbf{x}, t)$ (dims. L^3/T), we use the *Dirac delta-function*, $\delta(\mathbf{x} - \mathbf{x}_m)$ (dims. L^{-3}), formally defined by:

$$\delta(\mathbf{x}-\mathbf{x}_m) = \lim_{a\to 0} \begin{cases} 1/a^3, & \text{if } |\mathbf{x}-\mathbf{x}_m| < a, \\ 0, & \text{if elsewhere,} \end{cases} \tag{5.1.4}$$

where a is a small length, to express the combined mass withdrawal of a number of sources in the form: $\sum_{(m)} \rho_m Q_m(\mathbf{x}_m, t)\delta(\mathbf{x}-\mathbf{x}_m)$. Actually, Q_m represents both sinks (with $Q_m < 0$) and sources (with $Q_m > 0$). The mass balance equation (5.1.1) may then be written in the form:

$$\frac{\partial \phi\rho}{\partial t} = -\nabla\cdot\rho\mathbf{q} + \sum_{(m)} \rho_m Q_m(\mathbf{x}_m, t)\delta(\mathbf{x}-\mathbf{x}_m), \tag{5.1.5}$$

where subscript f has been omitted. Note that ρ_m represents the density of the pumped water when the later is withdrawn and that of the injected water in the case of injection wells. The l.h.s. of the above equation expresses the mass added to a unit volume of porous medium, per unit time, because of changes in both porosity and fluid density. In the next section, we shall focus on these changes, as the fluid's pressure undergoes changes.

5.1.2 *Mass Balance Equation for the Solid Matrix*

Because $\mathbf{V}_s$ is an additional variable, we have to consider also the solid's mass balance equation. To do so, we make use of the general macroscopic balance equation (1.4.30), with $E = m_s$, $e' = \rho_s$, $\theta \to (1-\phi)$, $\mathbf{V} = \mathbf{V}_s$, $\mathbf{j}^E = 0$, $\Gamma^E = 0$, and assume that *at the microscopic level* the solid-fluid interface is a *material surface* with respect to the solid's mass, i.e., $(\mathbf{V}_s - \mathbf{u}) \cdot \boldsymbol{\nu} = 0$, where $\mathbf{u}$ denotes the speed of displacement of this surface and $\boldsymbol{\nu}$ is the unit outward vector on the latter. Under such conditions, (1.4.30) reduces to the solid's macroscopic mass balance equation:

$$\frac{\partial}{\partial t}[(1-\phi)\rho_s] = -\nabla\cdot[(1-\phi)\rho_s\mathbf{V}_s], \tag{5.1.6}$$

where we have omitted all averaging symbols, as it is obvious that the equation is at the macroscopic level.

Introducing the *material* (or *total*) *time derivative* for the solid phase, defined by:

$$\frac{D_s(..)}{Dt} = \frac{\partial(..)}{\partial t} + \mathbf{V}_s \cdot \nabla(..), \tag{5.1.7}$$

Equation (5.1.6) can be rewritten in the form:

$$\frac{1}{1-\phi}\frac{D_s(1-\phi)}{Dt} + \frac{1}{\rho_s}\frac{D_s\rho_s}{Dt} = -\nabla\cdot\mathbf{V}_s, \tag{5.1.8}$$

or:

$$\frac{1}{1-\phi}\frac{D_s\phi}{Dt} = \frac{1}{\rho_s}\left(\frac{D_s\rho_s}{Dt} + \rho_s\nabla\cdot\mathbf{V}_s\right). \tag{5.1.9}$$

When we assume $D_s\rho_s/Dt = 0$, the above equation reduces to:

$$\frac{1}{1-\phi}\frac{D_s\phi}{Dt} = \nabla\cdot\mathbf{V}_s. \tag{5.1.10}$$

The solid (as a phase, not the solid matrix !) is usually assumed to be *volume preserving*, i.e., deformation of the solid matrix is caused only by change of shape or rearrangement of particles, with the solid itself assumed incompressible. This means that *at the microscopic level*, both $\nabla\cdot\mathbf{V}_s = 0$, and $D_s\rho_s/Dt = 0$. Hence, (5.1.8) reduces to:

$$\frac{1}{1-\phi}\frac{D_s(1-\phi)}{Dt} = -\nabla\cdot\mathbf{V}_s, \quad \Longrightarrow \quad \frac{\partial\phi}{\partial t} = \nabla\cdot[(1-\phi)\mathbf{V}_s]. \tag{5.1.11}$$

The l.h.s. of the first equation in (5.1.11) may be interpreted as 'the relative rate of expansion of the volume occupied by the solid phase'. However, this does not necessarily mean that the porosity does not undergo changes.

5.1.3 Mass Balance Equation for the Fluid

We now rewrite the fluid (f) mass balance equation (5.1.1) as:

$$\begin{aligned}\frac{\partial\phi\rho_f}{\partial t} &= -\nabla\cdot\phi\rho_f\mathbf{V}_f + \rho_f\Gamma^f = -\nabla\cdot\phi\rho_f(\mathbf{V}_f - \mathbf{V}_s) - \nabla\cdot(\phi\rho_f\mathbf{V}_s) + \rho_f\Gamma^f\\ &= -\nabla\cdot(\rho_f\mathbf{q}_r) - \mathbf{V}_s\cdot\nabla(\phi\rho_f) - \phi\rho_f\nabla\cdot\mathbf{V}_s + \rho_f\Gamma^f\\ &= -\nabla\cdot(\rho_f\mathbf{q}_r) - \mathbf{V}_s\cdot\nabla(\phi\rho_f) + \phi\rho_f\frac{1}{1-\phi}\frac{D_s(1-\phi)}{Dt} + \rho_f\Gamma^f,\end{aligned} \tag{5.1.12}$$

or:

$$\phi\frac{D_s\rho_f}{Dt} + \rho_f\frac{1}{1-\phi}\frac{D_s\phi}{Dt} = -\nabla\cdot(\rho_f\mathbf{q}_r) + \rho_f\Gamma^f. \tag{5.1.13}$$

In terms of the material derivative with respect to the fluid phase, this equation takes the form:

$$\phi\frac{D_f\rho_f}{Dt} + \rho_f\frac{1}{1-\phi}\frac{D_s\phi}{Dt} = -\rho_f\nabla\cdot\mathbf{q}_r + \rho_f\Gamma^f, \tag{5.1.14}$$

where:

$$\frac{D_f(..)}{Dt} = \frac{\partial(..)}{\partial t} + \mathbf{V}_f\cdot\nabla(..). \tag{5.1.15}$$

For a *stationary* ($\mathbf{V}_s = 0$) *nondeformable solid matrix* ($D_s(1-\phi)/Dt = -D_s\phi/Dt = 0$), Eq. (5.1.13) reduces to (5.1.1). Assuming that in a deformable porous medium:

$$\left|\frac{\partial \rho_f}{\partial t}\right| \gg \left|\mathbf{V}_s \cdot \nabla \rho_f\right|, \quad \left|\frac{\partial \phi}{\partial t}\right| \gg \left|\mathbf{V}_s \cdot \nabla \phi\right|, \tag{5.1.16}$$

i.e., assuming that the spatial variations are much smaller than the corresponding temporal ones, (5.1.13) reduces to the mass balance equation for a fluid in a deformable porous medium:

$$\phi\frac{\partial \rho_f}{\partial t} + \rho_f \frac{1}{1-\phi}\frac{\partial \phi}{\partial t} = -\nabla\cdot(\rho_f \mathbf{q}_r) + \rho_f \Gamma^f. \tag{5.1.17}$$

A detailed analysis of the deformation of a porous medium requires the introduction of the latter's (macroscopic) *volumetric strain*, or *dilatation*, ε_s. Denoting the (macroscopic) *displacement* vector of the porous medium's solid skeleton by $\mathbf{w}_s$, the porous medium's volumetric strain is expressed by:

$$\varepsilon_s = \nabla\cdot\mathbf{w}_s. \tag{5.1.18}$$

Then, with the assumption $|\partial \mathbf{w}_s/\partial t| \gg |\mathbf{V}_s \cdot \nabla \mathbf{w}_s|$, we obtain:

$$\mathbf{V}_s \equiv \frac{D_s \mathbf{w}_s}{Dt} \approx \frac{\partial \mathbf{w}_s}{\partial t}. \tag{5.1.19}$$

Equation (5.1.11) then becomes:

$$\frac{\partial \varepsilon_s}{\partial t} = -\frac{1}{1-\phi}\frac{D_s(1-\phi)}{Dt}, \tag{5.1.20}$$

and the mass balance equation (5.1.13) is replaced by:

$$\phi\frac{D_s \rho_f}{Dt} + \rho_f \frac{\partial \varepsilon_s}{\partial t} = -\nabla\cdot(\rho_f \mathbf{q}_r) + \rho_f \Gamma^f. \tag{5.1.21}$$

The mass balance equation (5.1.17) can be rewritten as:

$$\phi\frac{\partial \rho_f}{\partial t} + \rho_f \frac{\partial \varepsilon_s}{\partial t} = -\nabla\cdot(\rho_f \mathbf{q}_r) + \rho_f \Gamma^f. \tag{5.1.22}$$

Finally, if we assume that:

$$\left|\phi\frac{\partial \rho_f}{\partial t}\right| \gg \left|\mathbf{q}_r \cdot \nabla \rho_f\right|, \tag{5.1.23}$$

which may be interpreted as stating that the temporal rate of density change at a point is much larger than the spatial one, we may approximate $\nabla\cdot(\rho_f \mathbf{q}_r)$ in (5.1.17),

(5.1.21), and (5.1.22) by $\rho_f \nabla\cdot\mathbf{q}_r$. For example, the mass balance equation for a fluid, (5.1.17), then reduces to:

$$\phi\frac{\partial\rho_f}{\partial t} + \rho_f\frac{1}{1-\phi}\frac{\partial\phi}{\partial t} = -\rho_f\nabla\cdot\mathbf{q}_r + \rho_f\Gamma^f. \tag{5.1.24}$$

For a compressible fluid, $\rho_f = \rho_f(p)$, the above equation takes the form:

$$\phi\beta\frac{\partial p}{\partial t} + \frac{\partial\varepsilon_s}{\partial t} = -\nabla\cdot\mathbf{q}_r + \Gamma^f, \tag{5.1.25}$$

in which β is the coefficient of fluid compressibility, defined by:

$$\beta = \frac{1}{\rho_f}\frac{d\rho_f}{dp_f}. \tag{5.1.26}$$

Equation (5.1.25) is often referred to as the *storage equation*.

Note that the above two forms of the fluid mass balance equation involve the relative specific discharge, $\mathbf{q}_r$, while in the mass balance equation (5.1.5) it is the specific discharge, $\mathbf{q}$.

Expressing $\mathbf{q}_r$ by (4.2.16), we obtain a single equation in the four variables p, ρ, ϕ and ε_s. We recall that $\rho = \rho(p)$, and that we have shown earlier that changes in ε_s are associated with changes in ϕ. The second term on the l.h.s. of (5.1.24) expresses the temporal rate of change in the *volume strain* of the solid skeleton. It has to be expressed in terms of the variable(s) of the problem, e.g., in terms of the rate of change in fluid pressure. To achieve this goal, we shall introduce the concept of *effective stress* in the next subsection. Then we shall define the *specific storativity* of a deformable saturated porous medium, which leads to the derivation of a model that describes saturated flow and deformation in a three-dimensional porous medium domain.

In the next section, we shall present a different form of (5.1.25), which describes the mass balance for a fluid in a deformable porous medium—an equation written in terms of a single variable, pressure, or piezometric head.

5.1.4 Effective Stress

In simple terms, a fluid mass balance equation means that for a given spatial domain, the total fluid mass inflow minus the total mass outflow plus the quantity of fluid mass added from all sources, is being stored in that domain. The mechanisms that enable such storage are the compressibility of the fluid and of the solid matrix. In this subsection, we shall discuss these mechanisms and develop a coefficient that expresses their effect in the mass balance equation.

Although Chap. 9 is devoted to poro-mechanics, we have introduced the discussion on *effective stress* already here, as we need this concept in the discussions on *specific storativity* in 3-d flow in a porous medium and on *aquifer storativity* in the mass balance equation for essentially horizontal flow in aquifers. These are two basic concepts in modeling flow in aquifers. The material on effective stress presented below will also serve the discussion in Chap. 9, where we consider the general case of poro-mechanics. Note that the terms "storativity" and "effective storativity" are used mainly by groundwater hydrologists in connection with flow and water storage in aquifers. They are not common in reservoir engineering. It may also be interesting to note that the concept of effective stress was originally developed and initially employed in connection with soil loaded by structures and in connection with what happens in aquifers and petroleum reservoirs as fluids are extracted or injected. Here, we shall extend the concept to any stressed porous medium domain.

Following Bear and Bachmat (1991, p. 153), we consider the particular case of a *single fluid phase* (subscript f) that *fully occupies the void space*. Then, neglecting the shear stress in the fluid, or when the system is at rest, i.e., $\boldsymbol{\tau}_f \equiv 0$, we have, from (3.3.19):

$$\overline{\boldsymbol{\sigma}} = (1 - \phi)\boldsymbol{\sigma}_s - \phi p_f \mathbf{I}, \tag{5.1.27}$$

in which $\boldsymbol{\sigma}_s$ and p_f are intrinsic phase averages of the stress in the solid and the pressure in the fluid, respectively, and $\overline{\boldsymbol{\sigma}}$ denotes the volume average stress defined by (3.3.17).

Our objective is to determine the stress that produces the strain in the solid matrix. Knowledge and understanding of this strain is required in problems of flow through deformable porous media.

Terzaghi (1925), while investigating the delay in the deformation caused by the slow drainage of water from the void space in a low permeability soil, when a compressive load is applied, introduced the concept of *effective stress*, or *intergranular stress* in Soil Mechanics (see any text on Soil Mechanics, e.g., Verruijt 2010, p. 28). Essentially, this concept assumes that as the solid comprising the solid matrix (e.g., grains in a granular material) is (*almost*) completely surrounded by an ambient fluid, the latter's pressure (actually stress, unless we neglect $\boldsymbol{\tau}_f$) acting on the solid-fluid interface produces a stress of equal magnitude in the solid (say, within each individual grain), without contributing to the solid matrix' deformation, which is produced mainly by forces that are transmitted (in a granular material) from grain to grain at contact points. Thus, the *strain producing stress*, or *inter-granular stress*, is obtained by subtracting the pressure in the fluid from the stress in the solid, where both pressure and stress are average values. We recall that in this book stress is positive for tension, while pressure is positive for compression.

Terzaghi's concept of effective stress is based on the assumption that soil grains are very rigid compared to the soil as a whole, and on the assumption that the interparticle contact areas are very small compared to the surface area of the grains. These are reasonable assumptions for granular materials or soils, but are not valid for porous or fractured rocks. In the latter, rock compressibility must be taken into account.

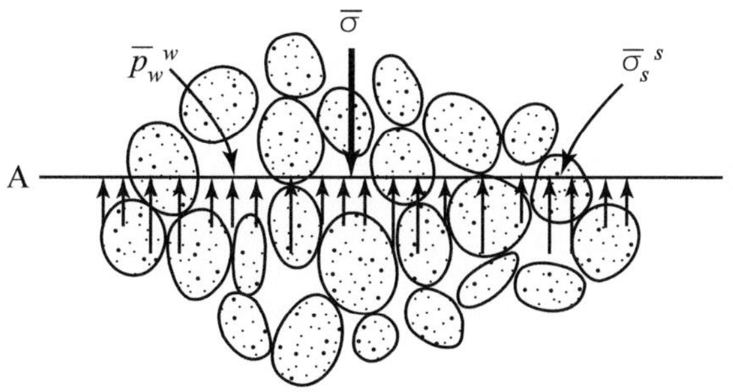

Fig. 5.1 Nomenclature for the definition of Terzaghi's effective stress

Verruijt (2010, p. 29) comments that Terzaghi's effective stress principle is often quoted as total stress equals effective stress plus pore pressure, but that it should be noted that "this applies only to the normal stresses. Shear stresses can be transmitted by the grain skeleton only".

Essentially, the concept of effective stress stems from the observation that the deformation of a granular material is much larger than can be explained by the compression of the solid material itself. This suggests, at least in a granular material, that deformation is produced mainly by the rearrangement of grains, with localized slipping and rolling, implying that deformation is governed by the transmission of localized normal and shear forces at contact points. As these forces are not affected by the pressure in the fluid, a change in fluid pressure, accompanied by an equal change in total stress, produces no deformation and, hence, should produce no change in effective stress.

Accordingly, when the (macroscopic) shear stress in the fluid, $\overline{\boldsymbol{\tau}_f}^f$, is neglected, we write (3.3.17) in the form:

$$\begin{aligned}\overline{\boldsymbol{\sigma}} &= (1-\phi)\overline{\boldsymbol{\sigma}_s}^s - \phi\overline{p_f}^f\boldsymbol{\delta}\\ &= (1-\phi)\{\overline{\boldsymbol{\sigma}_s}^s + \overline{p_f}^f\boldsymbol{\delta}\} - (1-\phi)\overline{p_f}^f\boldsymbol{\delta} - \phi\overline{p_f}^f\boldsymbol{\delta},\end{aligned}$$

or,

$$\overline{\boldsymbol{\sigma}} = \overline{\boldsymbol{\sigma}'_s} - \overline{p_f}^f\boldsymbol{\delta}, \qquad (5.1.28)$$

where $\overline{p_f}^f$ is the (intrinsic phase average) pressure in the fluid, $\boldsymbol{\delta}$ is the *unit tensor*, and the effective stress, $\overline{\boldsymbol{\sigma}'_s}$, is expressed by:

$$\overline{\boldsymbol{\sigma}'_s} = (1-\phi)\{\overline{\boldsymbol{\sigma}_s}^s + \overline{p_f}^f\boldsymbol{\delta}\}. \qquad (5.1.29)$$

Figure 5.1 illustrates the above stress balance in a simple way. The figure is limited to *vertical forces only*, and considers a horizontal area, AB, in a vertical cross-section through a saturated porous medium.

In the theory of poroelasticity, (5.1.29) is replaced by:

$$\overline{\boldsymbol{\sigma}'_s} = (1-\phi)\{\overline{\boldsymbol{\sigma}_s}^s + \alpha_B\overline{p_f}^f\boldsymbol{\delta}\}, \qquad (5.1.30)$$

in which α_B, with $0 \le \alpha_B \le 1.0$, a coefficient introduced by Biot (1941), is called the *Biot effective stress coefficient*. This coefficient depends on the nature of the solid matrix only, independent of the nature of the fluid that occupies the void space. Cheng (2016, p. 61) interprets α_B as the ratio of the fluid volume gained (or lost) in a porous sample to the volume change of that sample, when the pore pressure is returned to its initial state, with $\alpha_B = 1$ in the case of an incompressible solid.

Terzaghi's work was related, primarily, to one-dimensional case. Biot (1941) extended poroelasticity theory to three dimensions. The effective stress defined by (5.1.30) is sometimes called *Biot's effective stress*. Over the years, Biot's theory has been extended (e.g., Geertsma 1957; Gambolati et al. 2000; Bishop 1973; Verruijt 1995) to include the compressibility of the fluid and that of the soil particles. This last extension is especially important because it means that the theory can also be used for time-dependent deformation of such materials as rocks (in reservoir engineering) and in bone structures. Biot (1935, 1941), who was the first to develop a linear theory of poroelasticity, suggested that the α_B-coefficient be related to the compressibility of the solid (say grains), C_s, and to that of the porous medium as a whole (C_{pm}). The upper limit of α_B is 1, when the solid is incompressible. The lower limit is associated with the porosity. It can approach zero only when the porosity goes to zero (i.e., we have a pure solid). In between, it depends on the porosity and the compressibility of the solid. The usual assumption (e.g., Verruijt 2014; Jha and Juanes 2014; Kim et al. 2013) is that $\alpha_B = 1 - C_{dr}/C_s$, where C_{dr} denotes the drained bulk modulus (= inverse of compressibility) and C_s is that of the solid material. Often it is assumed that $\alpha_B = 1$.

A comprehensive review of soil consolidation, including the above developments can be found in a comprehensive review on Soil Consolidation, published by Verruijt (2005).

From (5.1.29) it follows that the effective stress is made up of two parts: one is an average stress (*positive for tension*) within the solid matrix, and the other is an average pressure (*positive for compression*) in the fluid occupying the void space. The stresses $\overline{\boldsymbol{\sigma}}$ and $\overline{\boldsymbol{\sigma}'_s}$ are forces *per unit area of porous medium cross-section*. In soil mechanics, the minus sign in (5.1.28) is usually replaced by a plus sign, i.e., both $\boldsymbol{\sigma}$ and $\overline{\boldsymbol{\sigma}'_s}$ are positive for compression.

When $\boldsymbol{\tau}_f$ within the fluid cannot be neglected, we obtain:

$$\overline{\boldsymbol{\sigma}} = \overline{\boldsymbol{\sigma}'_s} + \overline{\boldsymbol{\sigma}_f}^f, \tag{5.1.31}$$

with:

$$\overline{\boldsymbol{\sigma}'_s} = (1 - \phi)(\overline{\boldsymbol{\sigma}_s}^s - \overline{\boldsymbol{\sigma}_f}^f). \tag{5.1.32}$$

5.1.5 Specific Storativity in Single Phase Flow

The term on the l.h.s. of (5.1.1) represents the *mass of fluid added to a unit volume of porous medium per unit time*. This term can also be written as:

$$\frac{\partial}{\partial t}(\phi\rho_f) = \phi\frac{\partial\rho_f}{\partial t} + \rho_f\frac{\partial\phi}{\partial t}. \tag{5.1.33}$$

We note the two effects that contribute to the added fluid mass: fluid compressibility and porous medium deformability.

In general, the fluid's density depends on its pressure, p, solute concentration, c, and temperature, T, i.e., $\rho_f = \rho_f(p, c, T)$. Thus,

$$\frac{\partial\rho_f}{\partial t} = \frac{\partial\rho_f}{\partial p}\frac{\partial p}{\partial t} + \frac{\partial\rho_f}{\partial c}\frac{\partial c}{\partial t} + \frac{\partial\rho_f}{\partial T}\frac{\partial T}{\partial t}. \tag{5.1.34}$$

Note that c represents c^γ, $\gamma = 1, 2, \ldots, NC$. In this section, we restrict the discussion to the case in which the fluid's density depends on pressure only, i.e., $\rho_f = \rho_f(p)$. Then,

$$\frac{\partial\rho_f}{\partial t} = \frac{\partial\rho_f}{\partial p}\frac{\partial p}{\partial t} = \rho_f\beta\frac{\partial p}{\partial t}, \tag{5.1.35}$$

where β is the *fluid's compressibility*, defined in (5.1.26).

To develop the second term on the r.h.s. of (5.1.33), we start from the assumption (already introduced earlier) that the solid's density, ρ_s, *not of the solid matrix*, remains unchanged as the porosity undergoes changes. Given a fixed mass of solid matrix, m_s, this means that $\partial\mathbb{V}_s/\partial t = 0$, where $\mathbb{V}_s$ denotes the solid's volume. We recall that the entire discussion here is at the macroscopic level.

The total stress in a three-dimensional saturated porous medium is expressed by (5.1.28), repeated here for convenience as:

$$\overline{\boldsymbol{\sigma}} = \overline{\boldsymbol{\sigma}'_s} - \overline{p_f}^f\boldsymbol{\delta}, \tag{5.1.36}$$

in which $\boldsymbol{\delta}$ denotes the Kronecker delta (tensor).

It is interesting to recall here the *equilibrium equation* (3.3.25), repeated without the averaging symbols:

$$\nabla\cdot\boldsymbol{\sigma} + \rho\mathbf{F} = 0. \tag{5.1.37}$$

With $\rho\mathbf{F}$ including only the total external force per unit mass due to gravity, i.e.,

$$\rho\mathbf{F} \equiv -g[(1-\phi)\rho_s + \phi\rho_f]\nabla z \equiv -g\rho_{pm}\nabla z,$$

and total stress related to effective stress, $\boldsymbol{\sigma}'_s$, by (5.1.30), the equilibrium equation (5.1.37) takes the form:

$$\nabla\cdot\boldsymbol{\sigma}'_s - \alpha_B\nabla p_f + (\rho_{pm}\mathbf{F}) = 0, \quad \mathbf{F} = -g\nabla z. \tag{5.1.38}$$

Stresses in the soil below ground surface are caused by the weight of the soil itself, i.e., solid matrix + water, as well as by any external load applied to ground surface. For a homogeneous soil and a horizontal ground surface, the vertical stress, σ_{zz} at a depth d is obtained from:

$$\sigma_{zz} = \int_0^d g\rho_{pm}(z)dz, \quad \rho_{pm} = \phi\rho_f + (1-\phi)\rho_s. \tag{5.1.39}$$

If a water table is present at some depth below ground surface, we assume that the soil above it is dry. The effect of any additional load (positive, or negative for excavation) at ground surface should also be taken into account (see Bear 1972, p. 187).

When changes take place, either in the external load (producing changes in the total stress distribution, $\overline{\boldsymbol{\sigma}}$), or in the fluid pressure, as a result of changes in flow conditions, e.g., by pumping or injection, we have, omitting the averaging symbol, as the effective stress is a macroscopic concept only,

$$d\boldsymbol{\sigma} = d\boldsymbol{\sigma}'_s - \alpha_B dp_f \boldsymbol{\delta}, \tag{5.1.40}$$

in which α_B is the Biot coefficient.

We shall now limit the discussion to the case of *vertical stresses only*. The statement that the solid is not deformable, is expressed by:

$$\frac{\partial \mathbb{V}_s}{\partial \sigma'_{s,z}} = 0, \tag{5.1.41}$$

where $\sigma'_{s,z}$ is the vertical effective stress. With $\mathbb{V}_{\text{pm}}$ $(= \mathbb{V}_s/(1-\phi))$ denoting the porous medium volume containing $\mathbb{V}_s$, we rewrite (5.1.41) as:

$$\frac{\partial \mathbb{V}_s}{\partial \sigma'_{s,z}} \equiv (1-\phi)\frac{\partial \mathbb{V}_{\text{pm}}}{\partial \sigma'_{s,z}} + \mathbb{V}_{\text{pm}}\frac{\partial(1-\phi)}{\partial \sigma'_{s,z}} = 0. \tag{5.1.42}$$

Hence, in view of (5.1.36), *written for the vertical direction only*, and *assuming no change in the total stress*, i.e., $d\sigma = 0$, and $d\sigma'_s = dp$, we have:

$$\frac{1}{\mathbb{V}_{\text{pm}}}\frac{\partial \mathbb{V}_{\text{pm}}}{\partial \sigma'_{s,z}} = \frac{1}{1-\phi}\frac{\partial \phi}{\partial \sigma'_{s,z}} = \frac{1}{1-\phi}\frac{\partial \phi}{\partial p}. \tag{5.1.43}$$

At this point, we assume that we deal with relatively small volume changes, and that the porous medium may be assumed to behave as an *elastic material*. The *coefficient of porous medium compressibility*, α_{pm}, is defined for this case of vertical stresses only, as:

$$\alpha_{pm} = \frac{1}{\mathbb{V}_{\text{pm}}}\frac{\partial \mathbb{V}_{\text{pm}}}{\partial \sigma'_{s,z}} = \frac{1}{1-\phi}\frac{\partial \phi}{\partial p}. \tag{5.1.44}$$

The coefficient α_{pm} (dimensions of reciprocal of stress) can be determined in a laboratory experiment with a fixed mass of porous medium, and a *representative volume* of porous medium.

We now return to the second term on the right-hand side of (5.1.33). Making use of (5.1.44), we obtain

$$\frac{\partial \phi}{\partial t} = (1-\phi)\alpha_{pm}\frac{\partial p}{\partial t}. \tag{5.1.45}$$

With the above equation, we can now rewrite (5.1.33) in the form:

$$\frac{\partial}{\partial t}(\phi\rho_f) = \rho_f\left[\phi\beta + (1-\phi)\alpha_{pm}\right]\frac{\partial p}{\partial t} \equiv S_{op}^{m*}\frac{\partial p}{\partial t}. \tag{5.1.46}$$

Recalling the physical interpretation of the l.h.s. of the above equation, S_{op}^{m*} can be interpreted as the *specific mass storativity*, here for a saturated porous medium. It is defined as the *mass of fluid released from (or added to) storage in a unit volume of a deformable porous medium per unit decline (or rise) in fluid pressure*. We have used the superscript m and subscript p to indicate that this is a specific *mass* storativity associated with *pressure* changes, as several other types of storativity will be defined below. Through the dependence of ρ_w and ϕ on pressure, it follows that the specific storativity is also pressure dependent.

Groundwater hydrologists, who deal with water (w) as the dominant fluid, define a *specific storativity* (for saturated flow),

$$S_o^* \equiv gS_{op}^{m*} = \rho_w g[\phi\beta + (1-\phi)\alpha_{pm}], \tag{5.1.47}$$

as the *volume of water (w) released from (or added to) storage in a unit volume of porous medium, per unit decline (or rise) in the piezometric head* (e.g., Bear 1972, p. 204),

$$S_o^* = \frac{\Delta \mathbb{V}_w}{\mathbb{V}_{\text{pm}}\Delta h}. \tag{5.1.48}$$

Following the above discussion, we may now rewrite the mass balance equation (5.1.1), say, for water (w), in the form:

$$S_{op}^{m}\frac{\partial p}{\partial t} = -\nabla\cdot(\rho_w\mathbf{q}_w) + \rho_f\Gamma^f, \quad S_{op}^{m} = \rho_w(\phi\beta + \alpha_{pm}). \tag{5.1.49}$$

We can also write the mass balance equation for water in saturated flow, in terms of the piezometric head, h, in the form:

$$\rho_w S_o\frac{\partial h}{\partial t} = -\nabla\cdot(\rho_w\mathbf{q}_w) + \rho_w\Gamma^w, \quad S_o = \rho_w g(\phi\beta + \alpha_{pm}), \tag{5.1.50}$$

in which S_o is another form for the *specific (volume) storativity*. Note that in order to express $\mathbf{q}_w$ by Darcy's law, we have to assume that $\mathbf{V}_s \approx 0$. Then, when expressing $\mathbf{q}_w$ in terms of the piezometric head, h, i.e., $\mathbf{q}_w = -\mathbf{K}\cdot\nabla h$, the above mass balance equation contains only a single variable, h, to be solved for. We recall that the use of the piezometric head as a variable is permitted only when the fluid's density is constant.

Equation (5.1.50) is a partial differential equation of the parabolic type, often referred to the *heat equation*.

We note the difference between the expressions for S_o and S_o^* defined in (5.1.47). This difference is explained by the difference between $\mathbf{q}_w(\equiv \phi\mathbf{V}_w)$ and $\mathbf{q}_r(\equiv \phi(\mathbf{V}_w - \mathbf{V}_s))$ appearing in the divergence term in the mass balance equation. In Soil Mechanics, an *undrained test* is one in which a stress applied to a saturated soil sample produces deformation (of the sample), but *no fluid is allowed to drain out of the deforming porous medium sample*. The fluid is (practically) stationary relative to the solids. Such conditions can occur when the sample is bounded by impervious boundaries, or when the permeability is very low, or when the external stress is applied very quickly and the low permeability does not allow any significant outflow of water. *The deformation of the sample is due mainly to the rearrangement of particles*.

Under *drained conditions* of a saturated soil sample, as the stress is applied to the sample, producing changes in the fluid's pressure, fluid can enter or leave the sample.

Accordingly, S_o $(= \rho_w(\alpha_{pm} + \phi\beta))$ may be considered to be another definition for specific storativity, this time under conditions equivalent to those prevailing in an undrained test. Thus, S_o is applicable to a coordinate system that moves with the solid phase, while S_o^* is appropriate for a reference frame in which solid and fluids are allowed to move freely.

When we assume $|\phi\partial\rho_w/\partial t| \gg |\mathbf{q}_w \cdot \boldsymbol{\nabla}\rho_w|$, the mass balance equation (for constant ρ) (5.1.50) is simplified to the form:

$$S_o\frac{\partial h}{\partial t} = -\nabla{\cdot}\mathbf{q}_w + \Gamma^w, \quad \text{or} \quad \mathbf{q}_{\mathrm{w}} = -\mathsf{K}{\cdot}\nabla \mathrm{h}. \tag{5.1.51}$$

This equation, with specific storativity, $S_o \equiv S_o^*$, defined by (5.1.48), i.e., with the same verbal definition, is the one commonly used when considering (saturated) groundwater flow in aquifers. The variable solved for is $h = h(x, y, z, t)$. Actually, groundwater hydrologists use (5.1.51) with S_o defined by the right-hand side of (5.1.47), and with $\mathbf{q}_w$ expressed by Darcy's law.

We wish to emphasize again that underlying (5.1.51) is the assumption that water density is assumed constant, *except* in the expression for the specific storativity, S_o, where we do take water compressibility into account.

5.1.6 Three-Dimensional Flow with Deformation

Following Verruijt (1969), we separate the total stress, $\boldsymbol{\sigma}$, the effective stress, $\boldsymbol{\sigma}'_s$, the pressure p, and the body force $\mathbf{f}(\equiv \rho\mathbf{F})$, appearing in the equilibrium equation (5.1.37), into initial steady-state values, $\boldsymbol{\sigma}^o$, $\boldsymbol{\sigma}'^{o}_s$, p^o and $\mathbf{f}^o$, and deformation-producing increments, $\boldsymbol{\sigma}^e$, $\boldsymbol{\sigma}'^{e}_s$, p^e and $\mathbf{f}^e$. We obtain

$$\boldsymbol{\sigma}^o = \boldsymbol{\sigma}'^o_s - p^o\mathbf{I}, \qquad \text{and} \quad \boldsymbol{\sigma}^e = \boldsymbol{\sigma}'^e_s - p^e\mathbf{I}. \tag{5.1.52}$$

We can, of course, modify the above equations by including the effect of the Biot coefficient (Sect. 5.1.4). We shall continue without this coefficient.

As a good approximation, we assume that the body force, $\mathbf{f}$, remains unchanged, although ϕ and ρ_w do vary, i.e., $\mathbf{f}^e = 0$. Then, the equilibrium equation for the initial steady state, is

$$\nabla \cdot \boldsymbol{\sigma}'^o_s + \mathbf{f}^o - \nabla p^o = 0. \tag{5.1.53}$$

For the incremental (deformation producing) effective stress and pressure, we have

$$\nabla \cdot \boldsymbol{\sigma}^e \equiv \nabla \cdot \boldsymbol{\sigma}'^e_s - \nabla p^e = 0. \tag{5.1.54}$$

We now make the assumptions that the solid matrix is isotropic (although the same approach can also be applied to an anisotropic solid matrix) and, for the relatively small excess effective stresses considered here, is made of a perfectly elastic material that obeys the macroscopic strain-stress relationship (2.3.73), with $\mathbf{w}(\equiv \mathbf{w}_s)$ denoting the macroscopic solid's *displacement vector*. Thus, the constitutive relationship is:

$$(\sigma'^e_s)_{ij} = \mu''_s\left(\frac{\partial w_i}{\partial x_j} + \frac{\partial w_j}{\partial x_i}\right) + \lambda''_s\left(\frac{\partial w_k}{\partial x_k}\right)\delta_{ij} = 2\mu''_s\varepsilon_{ij} + \lambda''_s\varepsilon\delta_{ij}. \tag{5.1.55}$$

as only the incremental effective stress causes displacement.

By inserting (5.1.55) into (5.1.54), we obtain

$$\frac{\partial}{\partial x_i}\left[\mu''_s\left(\frac{\partial w_i}{\partial x_j} + \frac{\partial w_j}{\partial x_i}\right) + \lambda''_s\frac{\partial w_k}{\partial x_k}\delta_{ij}\right] - \frac{\partial p^e}{\partial x_i}\delta_{ij} = 0, \tag{5.1.56}$$

to be used for determining $\mathbf{w}$.

The mass balance equation (5.1.22) may also be rewritten as two balance equations, one representing the initial steady state (with variables denoted by superscript o), and the other, involving the pressure increment (denoted by superscript e) that produces displacement. Thus, the second equation may be written in the form:

$$\nabla \cdot \left(\rho_w \mathbf{q}^e_r\right) + \phi\rho\beta\frac{\partial p^e}{\partial t} + \rho\frac{\partial \varepsilon_{sk}}{\partial t} = 0, \tag{5.1.57}$$

where $\varepsilon^e_{sk} \equiv \varepsilon_{sk}$ since $\varepsilon^o_{sk} \equiv 0$. For the isotropic porous medium considered here, Darcy's law takes the form:

$$\mathbf{q}^e_r = -\frac{k}{\mu}(\nabla p^e + \rho_w g \nabla z). \tag{5.1.58}$$

In writing (5.1.57) and (5.1.58), we have introduced the approximations

Table 5.1 Balance equations and constitutive relations for Darcian, saturated flow of a compressible Newtonian fluid in an isotropic linearly elastic porous medium (Bear and Bachmat 1991)

Equations		Additional dependent variables
Mass balance equation for the fluid, (5.1.22)–(1 Eq.)	$\nabla \cdot \rho_w \mathbf{q}_r^e + \phi \frac{\partial \rho_w}{\partial t} + \rho_w \frac{\partial \varepsilon_{sk}}{\partial t} = 0$	$\rho_w,\ \phi,\ \mathbf{q}_r^e,\ \varepsilon_{sk}$ (6 vars.)
Equation of motion for the fluid (5.1.58)–(3 Eqs.)	$\mathbf{q}_r^e = -\frac{k}{\mu}(\nabla p^e + \rho_w g \nabla z)$	p^e (1 var.)
Equilibrium relationships (5.1.54)–(3 Eqs.)	$\nabla \cdot \boldsymbol{\sigma}_s'^e - \nabla p^e = 0$	$\boldsymbol{\sigma}_s'^e$ (6 vars.)
Stress-strain relationships for the solid matrix (5.1.55)–(6 Eqs.)	$\boldsymbol{\sigma}_s''^e = \mu_s'' \left[\nabla \mathbf{w} + (\nabla \mathbf{w})^T\right] + \lambda_s''(\nabla \cdot \mathbf{w})\mathbf{I}$	$\mathbf{w}$ (3 vars.)
Dilatation-displacement relations (5.1.18)–(1 Eq.)	$\varepsilon_{sk} = \nabla \cdot \mathbf{w}$	(none)
Equation of state for the fluid–(1 Eq.)	$\rho_w = \rho_w(p)$	(none)
Dilatation-porosity relation (5.1.20)–(1 Eq.)	$\dot{\varepsilon}_{sk} = \frac{1}{1-\phi}\dot{\phi}$	(none)
Total: 16 equations		16 (scalar) variables

$$\rho_w = \rho_w^o + \rho_w^e \simeq \rho_w^o, \quad \rho_w^e \ll \rho_w^o, \quad \mu^e \ll \mu^o, \quad \phi = \phi^o + \phi^e \simeq \phi^o, \quad \phi^e \ll \phi^o.$$

We also assume that the permeability, k, remains unchanged in spite of the deformation that takes place.

By inserting the expression for $\mathbf{q}_r^e$ into (5.1.57), we obtain the mass balance equation for a compressible fluid phase in a deformable, isotropic and linearly elastic porous medium, in the form:

$$-\nabla \cdot \left[\rho_w \frac{k}{\mu}(\nabla p^e + \rho_w g \nabla z)\right] + \phi \rho_w \beta \frac{\partial p^e}{\partial t} + \rho_w \frac{\partial \varepsilon_{sk}}{\partial t} = 0. \tag{5.1.59}$$

This is a single equation in two variables p^e and ε_{sk}. We need a second PDE.

The complete set of equations describing the flow of a single compressible Newtonian fluid (ρ, μ) in a deformable isotropic porous medium consists now of the equations and relationships summarized in Table 5.1.

From this table, it follows that we have a sufficient number of equations to solve for the various dependent variables involved. In principle, this is the model introduced by Biot (1941), except that the Biot coefficient has been omitted, bu can be added. We note that this model also yields the displacement vector, $\mathbf{w}$. It can be used, for example, for determining soil consolidation and land subsidence.

As an example, consider a homogeneous isotropic porous medium, with λ_s'', $\mu_s'' =$ const. We rewrite (5.1.56) in the form of the three equations:

$$\mu_s'' \frac{\partial^2 w_i}{\partial x_j \partial x_j} + (\lambda_s'' + \mu_s'') \frac{\partial \varepsilon_{sk}}{\partial x_i} - \frac{\partial p^e}{\partial x_i} = 0, \qquad i, j = 1, 2, 3. \tag{5.1.60}$$

By differentiating each of these equations with respect to the corresponding x_i, and adding the resulting three equations, we obtain the single equation (Verruijt 1969):

$$(\lambda_s'' + 2\mu_s'')\nabla^2 \varepsilon_{sk} - \nabla^2 p^e = 0, \tag{5.1.61}$$

which, together with (5.1.25) and (4.2.44), often simplified for a homogeneous isotropic porous medium to the form:

$$-\frac{k}{\mu}\nabla^2 p^e + \phi\beta \frac{\partial p^e}{\partial t} + \frac{\partial \varepsilon_{sk}}{\partial t} = 0, \tag{5.1.62}$$

constitute two equations in the variables p^e and ε_{sk}.

Following Verruijt (1969), we integrate (5.1.61) over the vertical, and obtain:

$$(\lambda_s'' + 2\mu_s'')\varepsilon_{sk} = p^e + \Pi(\mathbf{x}, t), \tag{5.1.63}$$

where Π is a function of position and time that for every value of time, t, satisfies

$$\nabla^2 \Pi = 0. \tag{5.1.64}$$

When $\Pi \equiv 0$ (see below), we may insert:

$$\varepsilon_{sk} = \frac{p^e}{\lambda_s'' + 2\mu_s''} \tag{5.1.65}$$

in (5.1.62), and obtain:

$$\frac{k}{\mu}\nabla^2 p^e = \left(n\beta + \frac{1}{\lambda_s'' + 2\mu_s''}\right)\frac{\partial p^e}{\partial t} \equiv (n\beta + \alpha_{pm})\frac{\partial p^e}{\partial t}, \tag{5.1.66}$$

which is a simple (diffusion-type) mass balance equation commonly employed in hydraulics of groundwater for determining the pressure distribution. In the above equation, we have

$$\alpha_{pm} = \frac{1}{\lambda_s'' + 2\mu_s''}, \tag{5.1.67}$$

which may be interpreted as a *coefficient of porous medium compressibility*. This is the same coefficient α (defined in (5.1.44)) that appears in the definition of specific storativity, (5.1.47).

As pointed out by Verruijt (1969, p. 348), the function Π 'describes the deviation of the simplified Terzaghi-Jacob theory from the Biot theory', where the former assumes vertical consolidation only, while the latter takes into account the three-dimensional nature of consolidation. Here, Π expresses the deviation in the integrated approach to aquifer consolidation. In principle, however, horizontal displacements do take place. Their effect in hydrology may be negligible, but as part of consolidation, their damage may be significant. A discussion on modeling land subsidence is presented in Sect. 9.3.2.

5.1.7 Balance Equation for Gas Flow

Equation (5.1.1) is valid also for gas flow. Repeated here for convenience, without the source term, it takes the form:

$$\frac{\partial \phi \rho_g}{\partial t} = -\nabla \cdot \rho_g \mathbf{q}_g, \tag{5.1.68}$$

or:

$$\rho_g \frac{\partial \phi}{\partial t} + \phi \frac{\partial \rho_g}{\partial t} = -\nabla \cdot \frac{k_g}{\mu} \left(\nabla p_g + \rho_g g \nabla z \right), \tag{5.1.69}$$

where we can express changes in ϕ and ρ_g as in terms of variations in p_g.

In general, Darcy's law, say (4.2.44), is applicable to the flow of both gas and liquid at low Reynolds numbers. However, as discussed in Sect. 4.3.3, when considering the flow of gas at low pressure, or through a porous medium in which the void space is comprised of very small pores, the permeability in Darcy's law has to be modified, say in the form of (4.3.20).

Let us now consider a *real gas*. The density of a real gas in a single phase system is presented in Sect. 2.3.3 A, where the dependence of the gas density, ρ_g, on pressure is presented, for example, in the form of (2.3.40), repeated here for convenience, as:

$$\rho_g = \rho_g(p, T) = \frac{M}{RT} \frac{p}{Z(p, T)}, \tag{5.1.70}$$

in which $Z = Z(p, T)$ is the *compressibility factor*. For an ideal gas, $Z = 1$. Conversely, for a large permeability, e.g., in excess of 10^{-12} m^2($\simeq 1\ D$) and/or when gas velocity is high, Darcy's law is not applicable. Instead, a non-linear flux equation, such as Forchheimer's equation (Sect. 4.3.2), may be required in order to express the flux.

Consider gas flow under the following conditions:

(a) We take into account the Klinkenberg effect (Sect. 4.3.3), such that:

$$k_g = k_g(p).$$

(b) Flow is under isothermal conditions.
(c) We neglect the effect of gravity in Darcy's law. This assumption is valid when the reservoir's thickness is small and pressure is low.
(d) The solid matrix is slightly deformable. For example, we may use the approximation:

$$\phi = \phi(p) = \phi_o[1 + C_{pm}(p - p_o)], \quad \text{or} \quad \mathrm{C_{pm}} = \frac{1}{\partial \phi_o} \frac{\partial \phi}{\partial p}.$$

Altogether, the flow of gas is governed by the mass balance equation:

$$\frac{\partial}{\partial t} \frac{\phi(p)p}{Z(p)} = \nabla \cdot \frac{p}{Z(p)} \frac{k_g(p)}{\mu(p)} \nabla p. \tag{5.1.71}$$

If, further, we assume (1) that k_g is homogeneous over the domain, and is pressure-independent, and (2) that the solid matrix is incompressible, then $\partial\phi/\partial p = 0$ (a reasonable assumption when pressure changes are small, given the very large difference between the compressibility of the matrix and that of the gas phase) and (5.1.71) is simplified to the form:

$$\frac{\phi\mu(p)c_g(p)}{k} \frac{\partial p^2}{\partial t} = \nabla^2 p^2 - \frac{d\{\ln[\mu(p)Z(p)]\}}{dp^2} (\nabla p^2)^2, \tag{5.1.72}$$

where:

$$c_g = \frac{1}{p} - \frac{1}{Z(p)} \frac{dZ(p)}{dp}. \tag{5.1.73}$$

Equation (5.1.72) is rather cumbersome, because it involves both ln and $(\nabla p^2)^2$ components. However, the second term on the r.h.s. of (5.1.72) may be ignored when

$$\nabla^2 p^2 >> \frac{d\{\ln[\mu(p)Z(p)]\}}{dp^2} (\nabla p^2)^2,$$

a condition which is valid only if both the pressure gradient in the system and the variations in $[\mu(p)Z(p)]$ are small compared to the initial p.

The difficulty in dealing with the second term on the l.h.s. of (5.1.72) is alleviated by using the concept of *pseudo-pressure* (Al-Hussainy et al. 1966), which is defined as:

$$\psi_g = \psi_p(p) = 2 \int_{p_o}^{p} \frac{p}{\mu(p)\, Z(p)}. \tag{5.1.74}$$

We then obtain the simpler more robust mass balance equation:

$$\frac{\phi\, \mu(p)\, c_g(p)}{k} \frac{\partial \psi_p(p)}{\partial t} = \nabla \cdot \left[\nabla \psi_p(p)\right]. \tag{5.1.75}$$

This equation is applicable over the entire range of pressures. Despite its non-linearity, (5.1.75) can be solved, leading to an analytical solutions with significant engineering applications, especially in the petroleum industry (e.g., Fraim and Wattenbarger 1986, 1987) for well test analysis and for reservoir evaluation.

Obviously, there is no need for many of these approximations and simplifications if the problem of gas flow through porous media is solved by numerical techniques, as such solutions can handle the non-linearities of the pressure-dependence appearing in the flow equations.

5.2 Complete Flow Models

The (macroscopic) partial differential equations (5.1.1) and (5.1.51) are different forms of the mass balance equations of a fluid that fully occupies the void space of a porous medium domain. Each equation is associated with a different set of underlying assumptions. In subsequent chapters, we shall see additional equations, e.g., ones that describe the mass balance of a chemical component in a fluid phase and that of energy. All these balance equations contain no information related to any particular flow or solute transport problem, because a balance equation contains no information on the boundaries of the problem domain, and how the external world interacts with phenomena within the considered domain across such boundaries, nor information on the behavior of the specific materials (solid matrix, fluids, chemical species) involved.

Accordingly, for a balance equation, or a set of such equations, to fully describe a *particular* case of interest, it has to be supplemented by the following information:

- *The constitutive equations* that provide information on the behavior of the specific solid and fluid phases involved in the considered case.
- The numerical values of all the coefficients that appear in the constitutive equations and in the source terms. In Sect. 7.6, we introduce some comments on techniques aimed at the treatment of cases for which the information concerning model coefficients is insufficient.
- Functions that represent the rate of (positive or negative) production of the various extensive quantities that are relevant to the considered case.
- The configuration of the boundaries of the domain within which the considered transport phenomena take place.
- A description of the initial state of the considered domain (= *initial conditions*) in terms of the considered state variables.
- A description of the interaction of the domain under consideration with its environment, i.e., conditions on the specified boundaries. These conditions are referred to as *boundary conditions*.

The number of variables may be large. However, we have to solve *partial differential (balance) equations* only for a small number of these variables, referred to as *primary variables*. All other variables can then be obtained from these variables by solving *algebraic equations* and relationships. The latter contain definitions and constitutive relationships. In practice, rather than solve PDE's, we solve their numerical equivalent, e.g., the integro-differential equations discussed in Sect. 3.8. A further discussion on this subject is presented in Sect. 3.9.

When all this information is put together, we obtain a *closed set of equations*, i.e., a set in which the number of equations equals the number of variables to be solved for. The solution provides the future spatial distributions of the value(s) of the considered state variable(s) within the considered domain.

Different boundary conditions result in different solutions. Hence, it is important to select them, as part of the formulation of the conceptual model of the problem, in a way that reflects the actual physical conditions of the problem on hand. These may be current conditions, as actually observed in the field, or conditions anticipated (or assumed) to prevail in the future.

Altogether, solving a problem of transport in a specified domain means determining the spatial and temporal distributions of certain dependent variables that satisfy a given (1) set of equations, (2) initial conditions at all points *within* the considered domain, and (3) conditions specified *on* its boundary.

We shall start with a discussion on the conceptual model of a (sharp) boundary surface. We shall then discuss the general boundary condition that is based on the continuity of fluxes of extensive quantities across a boundary, and present a number of the conditions that are more commonly encountered in problems of flow in aquifers. In the current section, we focus on boundary and initial conditions required for modeling fluid mass flow. In subsequent chapters, we shall add conditions associated with solute transport, heat transport and stresses in deforming porous media.

Prior to presenting examples of types of initial and boundary conditions that may be encountered, we wish to emphasize again that no mathematical model can be solved unless appropriate initial and boundary conditions are specified. If we do not know them, but we still wish to solve the problem, we have to *assume, or guess*, possible conditions.

5.2.1 Boundary Surface

Although this section deals with macroscopic boundaries, the basic ideas presented in it apply also to microscopic interphase boundaries in a porous medium domain.

Any closed surface may serve as a boundary of an investigated domain, *provided* we can state the conditions that prevail on it. It is, therefore, convenient, but not mandatory, in groundwater hydrology and in reservoir engineering, to select natural boundaries for a considered problem domain, e.g., an impervious geological formation, an aquifer in contact with a hydrocarbon reservoir, a lake or a river in contact with a groundwater aquifer. An infinite domain may be visualized as an idealization of a very large finite one.

A boundary may also coincide with a *surface of discontinuity* in any (macroscopic) parameter characterizing the solid matrix, e.g., porosity. In the strict continuum sense, a *sharp boundary* that separates a porous medium domain from its environment, or that delineates a subdomain of different porous media at the macroscopic level, does not exist. By taking averages over REVs at points (= centers of these REVs) located along a line normal to a boundary between two different media, e.g., between two porous media (with $\phi_1 \neq \phi_2$), between a porous medium and an adjacent domain of solid without voids ($\phi = 0$), or between a porous medium and a body of fluid ($\phi = 1$), we obtain a *gradual transition* in the averaged solid matrix properties. Usually, no information is available on how the averaged values of a considered property vary within this transition zone. However, we recall that in Sect. 1.1.2, in defining a porous medium, we required: (a) the existence of an REV the size of which is much smaller than the size of the considered domain, and (b) that the variation of any macroscopic quantity (e.g., porosity) over the REV be *linear*, or approximately so. If these conditions are satisfied, the actual variation in porosity across the transition zone may be replaced by an *idealized boundary in the form of a surface across which an abrupt change in porosity takes place*. The sharp boundary may be arbitrarily located at any point within the transition region. For convenience, however, we usually locate this surface at the point corresponding to the mean value of the considered property between the two adjacent regions. These considerations are applicable also to an impervious boundary, i.e., when the external domain is impervious ($\phi = 0$).

The sharp boundary surface introduced in this way, divides the entire space into two parts: the bounded investigated domain and the external world. Conditions are imposed on this boundary. They represent the way the exterior world interacts with and constrains what happens on the interior side of the boundary. Rigorously, the behavior close to the surface that serves as a (sharp) boundary of an investigated domain, say, within a boundary domain of thickness equal to half the size of an REV, cannot be described by the continuum approach, as we do not have *within* such domain the REV required for obtaining averaged values. In principle, a porosity cannot be defined within this boundary layer. In most cases, we extrapolate the value of ϕ from the interior of the domain. Some authors (e.g., Beavers et al. 1973) regard the situation within some distance from a rigid wall as one of *variable porosity*. In the continuum approach as presented in this book, the expression "porosity at a point within a porous medium domain" means the "fraction of the void space within an REV centered at the point". Therefore, this definition cannot be applied to any point within half the size of an REV from the boundary. According to the definition of an REV, the width of the boundary domain must be much smaller than the size of the domain itself, so that the effect (on the solution) of the error resulting from extrapolating the value of porosity from the interior to the boundary should be negligible. The only way to study what happens *within* the boundary layer is to do so at the *microscopic level*.

By *hypothesizing* the existence of sharp boundaries, we obtain regular continuum domains for all phases present in the system *up to the boundary surface*. Boundary conditions have, then, to be specified; they describe the interactions between the interior and the exterior domains, across these boundaries.

In addition to the boundary that delineates a considered porous medium domain from its environment, we may encounter some special cases of surfaces that are often considered as boundaries:

A. Boundary Between Two Miscible Fluids

We consider 'two miscible fluids' separated by an *assumed* sharp surface. This may be a case with either two different fluid phases, or the same fluid, but with significantly different concentrations of some dissolved species (e.g., fresh water and sea water in an aquifer), possibly leading to two different viscosities and densities. This is an approximation of the real situation where a transition zone always occurs between two fluids, whether miscible or not.

Even if initially the two fluids are separated by a sharp interface, as the two fluids move, a transition zone is created between them because of dispersion and diffusion phenomena (Chap. 7). The concentrations vary gradually across this *transition zone*. However, when the latter is narrow, relative to the dimensions of each of the two fluid domains of interest, we may *approximate* the boundary surface between the two fluids, across which the concentration changes abruptly from that of one fluid to that of the other, as a sharp boundary. This approximation was often used for modeling seawater intrusion into coastal aquifers. Nowadays, there is no need to use this approximation, as numerical solutions, run on powerful computers, can treat the fluids' motion with a transition zone of variable concentration, density and viscosity.

B. Boundary Between Two Immiscible Fluids

Here, even if the two fluids are initially separated by a sharp interface, as the fluids move, due to capillary effects (Sect. 6.1), the saturation of each fluid varies gradually across a transition zone between the two fluids. If this zone is narrow, relative to the domains of interest on its two sides, it may be *approximated* as a sharp boundary across which a jump in the saturation of the considered fluids is stipulated. The phreatic surface (Sect. 5.2.4 E) may serve as an example; the two fluids are air and water, and we *assume* that only water is present in the void space below this surface, while only air is present in the void space above it. Here also, numerical solutions with fast computers can handle the moving/widening transition zone. Nowadays, there is no need for the "sharp interface approximation".

C. Boundary Between States of Aggregation

Under certain conditions, a substance in the void space undergoes a change of phase (= change in state of aggregation). Evaporation, condensation, freezing, thawing and melting of water, may serve as examples. When this change takes place within a relatively narrow zone in a porous medium domain, we introduce, *as an approximation*, a macroscopic boundary surface across which the phase change is *assumed* to take place. We assume that the void space on each side of such a boundary is completely occupied by a uniform state of aggregation. The interface boundary may move as a result of the change of phase.

Transport problems having a moving boundary as discussed above are called *Stefan problems*.

D. Shape of a Moving Sharp Boundary

In general, a boundary surface may be stationary or moving. It may also be material or non-material with respect to any considered extensive quantity.

Let $F(x, y, z, t) = 0$ represent the equation that describes a possibly moving (macroscopic) boundary surface between two fluids. If the surface is between a porous medium domain and an impervious domain, then fluid particles on this boundary stay on it; neither fluid particles, nor dissolved species can cross this boundary (but heat can, unless the impervious domain is also an insulator). If this boundary is between two fluid domains, then, fluid particles cannot cross it, but molecules of species dissolved in the fluids may (and, in fact, do) cross this interface. The speed, $\mathbf{u}$, of boundary displacement should not be mixed up with the velocities of the fluids present on its two sides.

As the surface moves, its shape may change, but the equation describing it, $F(\mathbf{x}, t) = 0$, remains unchanged. The quantity F is, thus, a *conservative property* of the points on the surface, for which the total derivative vanishes, i.e.,

$$\frac{DF}{Dt} \equiv \frac{\partial F}{\partial t} + \mathbf{u} \cdot \nabla F = 0, \quad \text{hence} \quad \mathbf{u} \cdot \nabla F = -\frac{\partial F}{\partial t}. \tag{5.2.1}$$

The unit vector, $\boldsymbol{\nu}$, normal to the surface $F = 0$, is expressed by

$$\boldsymbol{\nu} = \frac{\nabla F}{|\nabla F|}. \tag{5.2.2}$$

The component of $\mathbf{u}$ normal to the surface is then given by

$$u_\nu \equiv \mathbf{u} \cdot \boldsymbol{\nu} = -\frac{\partial F/\partial t}{|\nabla F|}. \tag{5.2.3}$$

5.2.2 *Initial Conditions*

Initial conditions specify the value of the (macroscopic) dependent variable, e.g., p, at all points within the modeled domain at some initial time, usually taken as $t = 0$. For example, in terms of p,

$$p(x, y, z, 0) = f(x, y, z), \tag{5.2.4}$$

where $f(x, y, z)$ is a known function.

5.2.3 General Boundary Conditions

Although the current chapter deals with a single fluid that occupies the entire void space, we shall begin with the more general case of two-phase flow. The subsequent discussion on specific types of boundary conditions will focus on single-phase flow.

In general, there exist two kinds of conditions that have to be satisfied on a boundary surface:

- Continuity in the (macroscopic) value of the considered intensive quantity, $\overline{e'_\alpha}^\alpha$ (or $\widetilde{e_\alpha}^\alpha$) across a boundary surface (sides 1 and 2):

$$[\![\ \overline{e'_\alpha}^\alpha \ ]\!]_{1,2} = 0, \tag{5.2.5}$$

where $e' = \rho e$ and $[\![\ .. \ ]\!]_{1,2}$ denotes a jump from side 1 to side 2. This *no-jump condition* is a consequence of the continuity in the microscopic value as any microscopic boundary is crossed. A jump in $\overline{e'_\alpha}^\alpha$ or $\widetilde{e_\alpha}^\alpha$ would lead to an infinite gradient which, in turn, would create an infinite flux that will instantly eliminate the jump.
- In the absence of (1) sources and sinks on a boundary, and (2) accumulation of a considered extensive quantity on the boundary, the total amount of any considered extensive quantity that is transferred by all phases present in the porous medium domain must be conserved as it is being transported across a boundary. This condition arises from the balance of that quantity as it is transported across a considered boundary.
However, it is possible that a certain extensive quantity can accumulate on the boundary. This would lead to a model of a different kind (see Sect. 1.4.2 C).

With the above considerations in mind, for any extensive quantity, E, in the absence of sources and sinks of E on the boundary surface, and with no accumulation of E on the boundary, the boundary condition may be stated in the form (Bear and Bachmat 1991, p. 238):

$$\sum_{(\delta=\alpha,\beta,s)} [\![\ \theta_\delta(\overline{e'_\delta}^\delta(\overline{\mathbf{V}_\delta}^\delta - \mathbf{u}) + \mathbf{J}^{*E}_\delta) \ ]\!]_{1,2} \cdot \mathbf{n} = 0, \tag{5.2.6}$$

where, assuming that E can be transferred from one phase to another within the surface, the sum is over all the phases present in the domain (including the solid), with $\sum_{(\delta)} \theta_\delta = 1$, $\mathbf{J}^{*E}_\delta$ denotes the sum of the dispersive and (macroscopic) diffusive fluxes of E, and $\mathbf{u}$ denotes the velocity of the (possibly moving) boundary. Equation (5.2.6) represents the *general macroscopic boundary condition for any extensive* quantity, E, *in a porous medium*. It is often referred to as the *no-jump* condition, meaning no jump in the normal component of the total flux across the boundary. We note that it expresses the notion that E does not accumulate (or disappear) on the boundary. If it does, we have a "jump condition", and the value of the jump is specified.

To be used as a boundary condition in a transport problem, *we must know the value of the variable, or the total flux, on the external side of the boundary*. By using this information, the no-jump condition becomes a constraint that serves as a boundary condition in the transport problem.

However, although we have stated above that the considered E does not have sources or sinks and does not accumulate on the boundary, it is possible to envision conditions under which (1) E does accumulate on the boundary, and (2) sources and sinks of the considered E do exist on the boundary surface. We may think of a boundary between two states of aggregation, or the accumulation of surfactant. Under such conditions, and allowing the exchange among phases at a (macroscopic point) within the surface, (5.2.6) is replaced by:

$$\sum_{\delta=\alpha,\beta,s} \frac{\partial \theta_\delta \rho_{sur,\delta} e_{sur,\delta}}{\partial t} = \sum_{\delta=\alpha,\beta,sur} [\![\ \theta_\delta(\rho_\delta e_\delta(\mathbf{V}_\delta - \mathbf{V}_{sur}) + \mathbf{J}^E_{\delta,dif})\ ]\!]_{1,2} \cdot \boldsymbol{\nu} + \sum_{\delta=\alpha,\beta,sur} \theta_\delta \rho_{sur,\delta} \Gamma_{sur}, \tag{5.2.7}$$

in which $\rho_{\delta,sur}$ denotes, the mass density of the surface (= mass per unit area), $e_{\delta,sur}$ denotes the density of E (= E per unit δ-mass), $\mathbf{V}_\delta$ denotes the velocity of the δ-phase, and $\boldsymbol{\nu}$ denotes the unit vector normal to the surface.

If local thermodynamic equilibrium is assumed, then thermodynamic variables, such as p_α, T, and concentration, say, expressed as the mass fraction ω^γ_α, must be continuous across the boundary. Some non-thermodynamic variables, such as θ_α and S_α, which do not exist at the microscopic scale, are not necessarily continuous across boundary surfaces; they, do not necessarily satisfy the condition of no-jump across a considered boundary surface.

Gray and Hassanizadeh, in a number of publications (e.g., Gray and Hassanizadeh 1989; Hassanizadeh and Gray 1979a) also present E-balance equation for surfaces. However, they deal with inter-phase surfaces (see Sect. 1.4.2 C for a description of the Hassanizadeh and Gray approach to modeling transport in porous medium domains).

In practice, the boundary conditions used in models of flow and transport in porous medium domains, although a consequence of the no-jump conditions (5.2.6), take forms that specify values of variables, or of their derivatives, on the boundary.

The kind of PDE that describes the mass balance, e.g., (5.1.50), requires only *one* condition on each boundary segment, and we prefer a condition based on flux continuity, if such information is available. If not, we base the condition on available information on values of scalar variables, e.g., a known pressure. Sometimes, approximations concerning the continuity in fluxes produce a jump in the values of the variables; we have to accept this consequence.

Because the momentum balance equation has been reduced to a flux expression in the form of Darcy's law, there is no need to solve the momentum balance equation. To predict flow in a given domain, we have to solve only any of the flow equations (presented in the previous subsection) that expresses the mass balance equation for the considered fluid. By inserting Darcy's law into the mass balance equation, we obtain a single linear second order PDE in terms of either pressure (p), or piezometric head (h). In Sect. 5.2.4 G, we shall introduce a case in which the motion equation takes the form of Brinkman's equation, rather than Darcy's law. The reason for the need to use the Brinkman equation is that the momentum transfer due to the fluid's velocity gradient cannot be neglected, at least in part of the considered domain.

The type of partial differential mass balance equation considered here, e.g., (5.1.51), requires just *one* condition. We satisfy this requirement using the condition of no-jump in mass flux. In multiphase flow, the pressure is related to saturation. The selection of the condition to be used in any particular case depends on the kind of information we have concerning what happens on the external side of the boundary.

In the next subsection, we present boundary conditions for single phase flow models, without presenting the details of their development from the general boundary condition stated above. The type of boundary condition to be used in any particular case depends on the available data concerning the actual or anticipated behavior in the field.

5.2.4 *Particular Boundary Conditions*

Following are some of the more commonly encountered boundary conditions for saturated flow. In each case, the entire surface bounding an investigated domain is divided into segments on each of which we may specify a different boundary condition. The latter should be stated in terms of the relevant state variable of the problem. The boundary surface is described by $F(\mathbf{x}, t) = 0$ (Sect. 5.2.1). The discussion is at the macroscopic level, and we shall use the symbol $\mathcal{B}$ to denote a boundary segment. We recall that all boundary conditions are based on the no-jump condition presented in Sect. 5.2.3.

A. Boundary of Prescribed Pressure, or Piezometric Head

In this case, the boundary condition takes the form:

$$p = f_1(x, y, z, t), \quad \text{or} \quad h = f_2(x, y, z, t), \quad \text{on} \quad \mathcal{B}, \tag{5.2.8}$$

where f_1 and f_2 are known functions, and $\mathcal{B}$ denotes the boundary.

Actually, the value of p is seldom known on the boundary, except when a porous medium domain is bounded by a body of water (e.g., a pond). In such case, the pressure along the pond's bounding surface is dictated by the water level in the pond. Whenever the density, ρ_w, is constant, the piezometric head, h, may also be prescribed on such a boundary.

A boundary condition that specifies the value of a state variable (here, p, or h) along a boundary segment is called *boundary condition of the first type*, or *Dirichlet boundary condition*.

B. Boundary of Prescribed Flux

This case occurs when fluid at a *known flux* enters a domain through its boundary. This includes the case of 'no-flow' through such a boundary. We shall assume that such boundary is always a *material surface* with respect to the solid, i.e.:

$$(\mathbf{V}_s - \mathbf{u})\big|_{\text{side 1}} \cdot \boldsymbol{\nu} = (\mathbf{V}_s - \mathbf{u})\big|_{\text{side 2}} \cdot \boldsymbol{\nu} = 0. \tag{5.2.9}$$

When $[\![\ \rho_f \ ]\!]_{1,2} = 0$, the general boundary condition, (5.2.6), for such a surface takes the form:

$$[\![\ \phi(\mathbf{V} - \mathbf{u}) \ ]\!]_{1,2} \cdot \boldsymbol{\nu} = 0. \tag{5.2.10}$$

With (5.2.9), Eq. (5.2.10) reduces to the form:

$$[\![\ \mathbf{q}_r \ ]\!]_{1,2} \cdot \boldsymbol{\nu} = 0, \quad \text{or} \quad \mathbf{q}_r\big|_1 \cdot \boldsymbol{\nu} = \mathbf{q}_r\big|_2 \cdot \boldsymbol{\nu}. \tag{5.2.11}$$

To serve as a boundary condition, information must be available on what happens on the external side of the boundary, say, side 2. Obviously, the relative specific flux, $\mathbf{q}_r$, has to be expressed by an appropriate flux equation, written in terms of p, or h.

For an *impervious boundary*, say, a pervious side 1 and an impervious side 2, Eq. (5.2.11) reduces to:

$$\mathbf{q}_r \cdot \boldsymbol{\nu} = 0. \tag{5.2.12}$$

Note that this equation constrains only the normal component of the flux. The tangential components may take on any value; we may have *slippage* along such a boundary.

Let $\mathbf{N}$ denote the prescribed flux on the external side of a stationary boundary ($\mathbf{u} = 0$) described by $F = F(\mathbf{x})$, with $\boldsymbol{\nu} \equiv \nabla F / |\nabla F|$ denoting the unit outward normal vector to it. We assume that the fluid density obeys $[\![\ \rho_f \ ]\!]_{1,2} = 0$, and that ρ_f is a constant. Then, the prescribed flux boundary condition takes the form:

$$\mathbf{q}_r \cdot \nabla F = N|\nabla F|, \qquad N = \mathbf{N} \cdot \boldsymbol{\nu}. \tag{5.2.13}$$

In this equation, $\mathbf{q}_r$ can be expressed by any of the motion equations presented in Chap. 4. For example, we can rewrite (5.2.13) in the form:

$$-(\mathbf{K} \cdot \nabla h) \cdot \boldsymbol{\nu} = \mathbf{N} \cdot \boldsymbol{\nu}, \tag{5.2.14}$$

where $\mathbf{K}$ denotes the hydraulic conductivity.

A condition that specifies the gradient of a scalar variable on the boundary, here ∇h, is called a *boundary condition of the second kind* or a *Neumann boundary condition*.

The condition of prescribed flux provides no explicit information on the values of the state variables, say, p or h, at (i.e., just inside) the boundary. These values will adjust themselves to accommodate the specified rate of flow through the boundary.

C. Semipervious Boundary

A layer of fine sediments on the bottom of a pond may serve as an example of a semi-pervious membrane that resists the movement of water through it (Fig. 5.2).

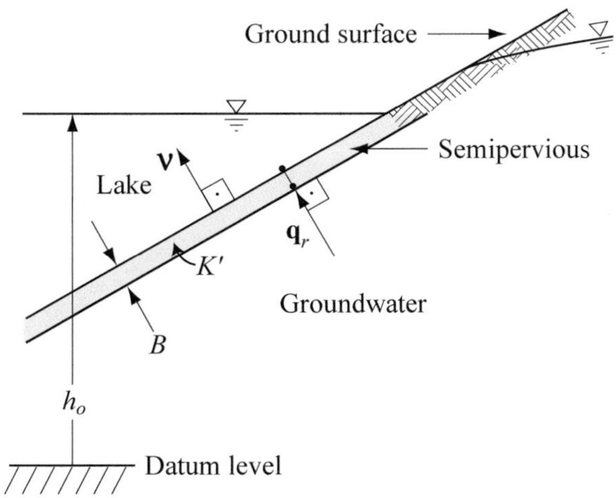

Fig. 5.2 A semipervious boundary

We assume that this 'membrane' is saturated when present at the bottom of an active water pond.

Let us assume that water is ponded on the upper side of this 'membrane' such that a piezometric head h_o is specified there. Let us denote the *resistance* of the semipervious membrane by c_r (= thickness of the membrane, B, divided by its hydraulic conductivity, K', i.e., the reciprocal of the leakance), and the piezometric head on the lower side of the membrane by h. Then, the flux through the membrane is expressed by:

$$\mathbf{q}_r \cdot \boldsymbol{\nu} = \frac{h - h_o}{c_r}, \quad c_r = \frac{B}{K'}, \tag{5.2.15}$$

where $\mathbf{q}_r$ may be expressed by any of the flux equations. The reason for writing $\mathbf{q}_r$ and not $\mathbf{q}$ is that, as explained earlier, $\mathbf{V}_s \cdot \boldsymbol{\nu} \equiv 0$ on the boundary. This is a *third type boundary condition*, or a *Robin boundary condition*.

D. Boundary Between Different Porous Media

Figure 5.3 shows a boundary between two regions of different permeability. In principle, in a mathematical model, we should avoid modeling of discontinuities within a modeled domain, e.g., discontinuity in the values of coefficients. When an investigated domain does include such discontinuities, it is useful to divide the domain into sub-domains along the surfaces of discontinuity, in order to obtain sub-domains without discontinuities. We then write a complete model for each sub-domain. On each of the common boundary segments, we need *two* boundary conditions—one for each side: these are the continuity of flux and the continuity of pressure. Because both flux and pressure are unknown a priori, we have to write these conditions in terms of the state variables for both sides, and solve for all the sub-domains simultaneously.

As the boundary is approached from within each side, the continuity of pressure (or piezometric head), is expressed as:

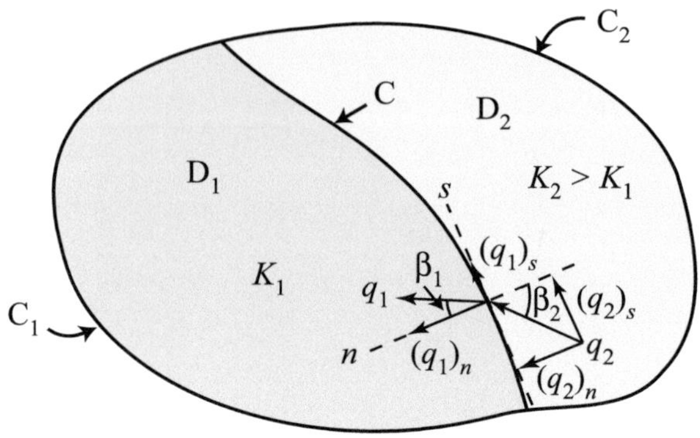

Fig. 5.3 Boundary between regions of different hydraulic conductivities

$$p\big|_{\text{side 1}} = p\big|_{\text{side 2}}, \quad \text{or} \quad h\big|_{\text{side 1}} = h\big|_{\text{side 2}}, \tag{5.2.16}$$

and the continuity of flux, following the discussion leading to (5.2.11), takes the form:

$$\mathbf{q}_r\big|_{\text{side 1}} \cdot \boldsymbol{\nu} = \mathbf{q}_r\big|_{\text{side 2}} \cdot \boldsymbol{\nu}. \tag{5.2.17}$$

Note that in view of, (5.2.9), we have expressed the boundary condition in terms of the relative flux, $\mathbf{q}_r$, expressed by Darcy's law.

Although Fig. 5.3 is presented in two-dimension, (5.2.16) and (5.2.17) are valid also on a boundary in a three-dimensional domain. The explicit expression (5.2.17), in terms of h_1 in D_1, and h_2 in D_2, depends on the nature of the materials occupying the two sub-domains (also, with respect to isotropy or anisotropy).

Thus, for the case shown in Fig. 5.3, the two boundary conditions to be satisfied on C are (5.2.16) and (5.2.17). Since each of these equations includes both h_1 and h_2, the two problems (for D_1 and D_2) must be solved simultaneously.

From (5.2.16), it follows that $\partial h_1/\partial s = \partial h_2/\partial s$, where s is a distance measured along the tangent to C (in Fig. 5.3). This can also be expressed as:

$$\frac{(q_s)_1}{K_1} = \frac{(q_s)_2}{K_2}, \tag{5.2.18}$$

where both K_1, and K_2 are isotropic. By combining (5.2.17) with (5.2.18), we obtain:

$$\frac{K_1}{\tan\beta_1} = \frac{K_2}{\tan\beta_2}, \quad \tan\beta_1 = \frac{(q_s)_1}{(q_\nu)_1}, \quad \tan\beta_2 = \frac{(q_s)_2}{(q_\nu)_2}, \tag{5.2.19}$$

where β_1 and β_2 are the angles which $\mathbf{q}_1$ and $\mathbf{q}_2$ make with the normal to the boundary C. This means that along such a boundary, the incident streamline is refracted. Equation (5.2.19) is the *law of refraction of streamlines for two-dimensional flow*, when both sub-domains are isotropic.

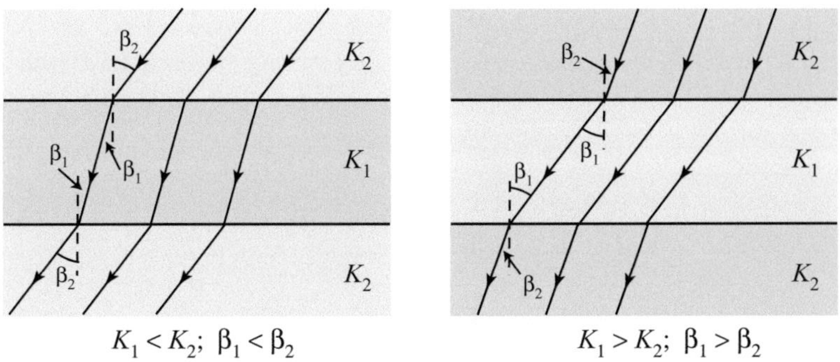

Fig. 5.4 Refraction of streamlines at an interface between different hydraulic conductivities

Bear (1972, p. 263) discusses the laws of refraction of streamlines and of equipotentials also for three-dimensional flows and for cases where the two sub-domains are anisotropic.

From (5.2.19), it follows that when $K_1 \gg K_2$, then $\beta_1 \gg \beta_2$, and the refracted streamline approaches the normal to the common boundary upon passing from a more pervious to a less pervious medium. When (Fig. 5.4) $K_1 \ll K_2$, then $\beta_1 \ll \beta_2$, and the refracted streamline tends to become almost parallel to the common boundary upon passing from a less pervious (e.g., semi-pervious) to a more pervious medium. This justifies the assumption of 'essentially horizontal flow' in a leaky aquifer.

E. Phreatic Surface

A (possibly moving) phreatic surface, or free surface, may serve as the upper boundary of a saturated zone below ground surface. This surface is defined as *the locus of all points at which the pressure in the liquid/water phase is equal to the gas/air pressure* (e.g., Bear 1972, p. 252; Bear 1979, p. 98). Below this surface, the soil is saturated. Above it, the void space is occupied by both water and air. The subject of two phase flow is discussed in Chap. 6. Thus, in principle, a solution of the problem of flow below ground surface requires the statement of the problem as one of two phase flow, or, *as an approximation*, as one of flow of two fluid phases separated by an *assumed sharp interface*, discussed in Sect. 5.2.6.

In Soil Physics, it is often assumed that the air in the unsaturated zone is (practically) immobile and under atmospheric conditions. This approximation is also assumed here. This leads to a simplified problem in which we need to consider only the flow of water below the phreatic surface, although we do consider accretion reaching the water table from above.

Actually, for some small distance, $h_c^{\rm cr}$, above the phreatic surface, the soil is still saturated, but the water there is at a (small) pressure less than atmospheric (which is the air pressure in the unsaturated void space above this zone). The value of $p_c^{\rm cr}(= h_c^{\rm cr}/\rho g)$ is called *bubbling pressure* or *air entry pressure* (see Sect. 6.1 B).

As for every boundary, we have to specify for the phreatic surface both the *shape* of the boundary surface and the *condition* to be satisfied on it.

Usually, the shape of the phreatic surface, say, expressed by the equation $F(x, y, z, t) = 0$, is a priori unknown. In fact, as we have already emphasized earlier, in many flow problems, determining the shape and (possibly time-dependent) position of this surface is the very objective of model investigations. However, once we have a solution, say, in the form of $p = p(x, y, z, t)$, or $h = h(x, y, z, t)$, whether in the unsaturated flow domain, or in the saturated one underlying it, since on the phreatic surface

$$p\Big|_{\text{sat}} = p\Big|_{\text{unsat}} = 0,$$

the shape of the phreatic surface boundary is given by:

$$F(x, y, z, t) \equiv p(x, y, z, t) = 0. \tag{5.2.20}$$

Let us assume that the water density, ρ_w, remains unchanged. This is a valid approximation in the vadose zone of an unconfined groundwater system. The elevation of points on the phreatic surface, denoted as $\zeta = \zeta(x, y, t)$, can be found from the requirement that $h_{sat}(x, y, z, t)|_{z=\zeta} = h_{unsat}(x, y, z, t)|_{z=\zeta} = \zeta$. Thus, we can define the *shape* of the phreatic surface as either:

$$z = \zeta(x, y, t), \quad \text{or} \quad F(x, y, z, t) = z - \zeta(x, y, t) = 0, \tag{5.2.21}$$

or as:

$$F(x, y, z, t) \equiv h(x, y, z, t) - z = 0. \tag{5.2.22}$$

The *condition* on the phreatic surface boundary is that of continuity of the normal water flux across it. Usually, when we consider a phreatic surface as a boundary, the underlying assumption is that the moisture saturation above this surface is at its irreducible level, θ_{wr}. The concept and definition of irreducible moisture saturation is discussed in Sect. 6.1 D. This 'sharp interface approximation' is valid as long as the thickness of the capillary fringe is small relative to either the thickness of the unsaturated zone, or of the saturated one. Such a sharp front may exist in the case of very permeable media (such as gravels) However, under certain circumstances (e.g., in fine soils), the capillary fringe, or the transition from the water table to the zone of irreducible water saturation may be several meters, or tens of meters thick. Under such conditions, *the 'sharp interface approximation' is no more valid.*

When the sharp interface approximation is justified, say, for a phreatic surface, the condition to be satisfied on it, assuming no change in density as water crosses this surface, is expressed in the form:

$$\phi(\mathbf{V}_w - \mathbf{u})\Big|_{\text{sat}} \cdot \boldsymbol{\nu} = \theta_{rw}(\mathbf{V}_w - \mathbf{u})\Big|_{\text{unsat}} \cdot \boldsymbol{\nu}, \tag{5.2.23}$$

where $\boldsymbol{\nu}$ denotes the unit vector normal to that surface, pointing away from the saturated zone, and $\mathbf{u}$ is the speed of the moving phreatic surface. They are related to the shape of the surface, $F(x, y, z, t) = 0$, by (5.2.1) and (5.2.2).

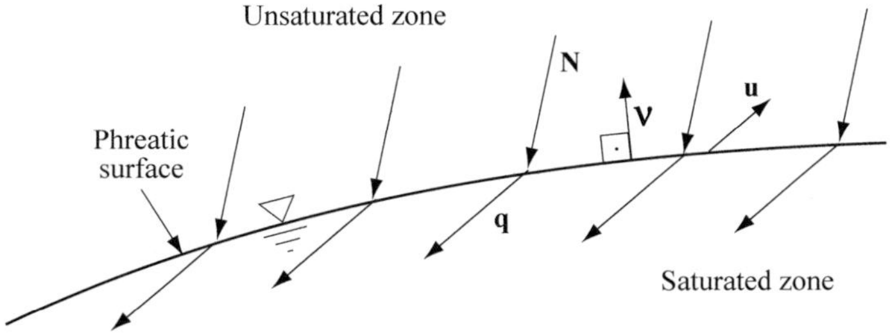

Fig. 5.5 Phreatic surface with accretion

Let us consider the case of flow in a phreatic aquifer, encountered in groundwater hydrology. Here, the details of flow in the unsaturated zone are of no interest. Instead, we assume that accretion, **N** (e.g., from precipitation), takes place on the upper side *of the phreatic surface* (Fig. 5.5).

The rate at which water travels from the unsaturated zone to the saturated one through the phreatic surface may be expressed by:

$$\theta_{rw}(\mathbf{V}_w - \mathbf{u})\Big|_{\text{unsat}} \cdot \boldsymbol{\nu} \equiv (\mathbf{N} - \theta_{rw}\mathbf{u})\Big|_{\text{unsat}} \cdot \boldsymbol{\nu}, \qquad (5.2.24)$$

where $\mathbf{N} = \theta_{rw}\mathbf{V}_w\big|_{\text{unsat}}$. For a vertically downward accretion at a rate N, we use $\mathbf{N} = -N\nabla z$. Equation (5.2.24) is the sought boundary condition. Let us rewrite it in a number of equivalent forms.

We have used the term 'accretion' to denote the rate at which water is added to, or removed from the phreatic surface, independent of the movement of the latter and of any moisture (if present) in the void space above it. However, it should be emphasized that $\mathbf{N} \cdot \boldsymbol{\nu}$ is not the rate at which water actually crosses the phreatic surface and augments (or reduces) the quantity of water in the saturated zone. This net rate depends also on the movement of the phreatic surface.

We can rewrite (5.2.23), or (5.2.24), in the form:

$$(\mathbf{q}_w\big|_{\text{sat}} - \phi\mathbf{u}) \cdot \boldsymbol{\nu} = (\mathbf{N} - \theta_{rw}\mathbf{u}) \cdot \boldsymbol{\nu}, \qquad (5.2.25)$$

or:

$$(\mathbf{q}_w\big|_{\text{sat}} - \mathbf{N}) \cdot \boldsymbol{\nu} = (\phi - \theta_{rw})\mathbf{u} \cdot \boldsymbol{\nu}. \qquad (5.2.26)$$

In view of (5.2.2) and (5.2.3), we may rewrite (5.2.26) as:

$$(\mathbf{q}_w\big|_{\text{sat}} - \mathbf{N}) \cdot \nabla F = -(\phi - \theta_{rw})\frac{\partial F}{\partial t}. \qquad (5.2.27)$$

Making use of (5.2.22), we may rewrite (5.2.27) in terms of the piezometric head, h, in the form:

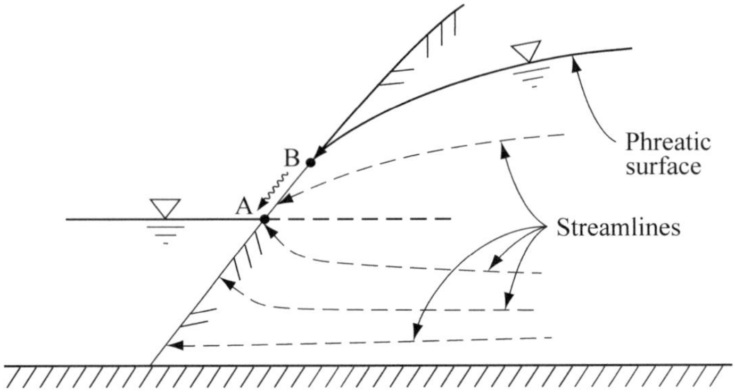

Fig. 5.6 The seepage face, AB

$$(\mathbf{q}_w\big|_{\text{sat}} - \mathbf{N}) \cdot \nabla(h - z) = -(\phi - \theta_{rw})\frac{\partial h}{\partial t}. \tag{5.2.28}$$

By inserting $\mathbf{q}_w\big|_{\text{sat}} \equiv \mathbf{q}_r = -\mathbf{K} \cdot \nabla h$ into this equation, we obtain:

$$(\mathbf{K}_w \cdot \nabla h + \mathbf{N}) \cdot \nabla(h - z) = \phi_{\text{eff}}\frac{\partial h}{\partial t}, \tag{5.2.29}$$

where $\phi_{\text{eff}} \equiv \phi - \theta_{rw}$. We wish to reiterate that this equation expresses nothing but the continuity of fluid flux across the phreatic surface boundary. For an isotropic porous medium, the above condition can be written in the form:

$$K\left[\left(\frac{\partial h}{\partial x}\right)^2 + \left(\frac{\partial h}{\partial y}\right)^2 + \left(\frac{\partial h}{\partial z}\right)^2\right] - (K + N)\frac{\partial h}{\partial z} + N = \phi_{\text{eff}}\frac{\partial h}{\partial t}. \tag{5.2.30}$$

Although we have presented here the boundary condition of a phreatic surface, assuming that the rate of accretion, **N**, is known, in most cases of practical interest, this rate is actually unknown. It is certainly not the rate of rainfall; among other factors, its value depends both on the rainfall and on the moisture conditions of the soil at ground surface. We shall discuss this issue in detail in Sect. 6.4.2.

F. Seepage Face

This kind of boundary appears when a phreatic surface approaches a body of open water, a river or a lake, which serves as part of the boundary of a flow domain (Fig. 5.6). In such cases, the phreatic surface will *always* terminate on that (known) boundary at a point (Point B in Fig. 5.6) located at some elevation *above* the water surface of the body of open water (Point A). The segment AB is called the *seepage face*. Through it, water seeps out of the porous medium domain. Along the seepage face, the water will seep out of the formation and flow as a thin layer along the AB slope.

The reason for the existence of a seepage face is that otherwise (i.e., if point B would coincide with A), the velocity at that point would be infinite. This is an impossible situation (Muskat 1946, p. 303; Bear 1972, p. 260, 288).

Since on a seepage face, which is exposed to the atmosphere, the pressure in the water is $p = 0$ (assuming atmospheric pressure is $p_a = 0$), the boundary condition is:

$$p(\mathbf{x}, t) = 0, \quad \text{or} \quad h(\mathbf{x}, t) = z, \tag{5.2.31}$$

i.e., the head at every point of the seepage face is specified to be equal to its *known* elevation. The geometry of the seepage face is known, except for the location of its end point, B, which is also a point on the (a priori unknown) phreatic surface. It is interesting to note that the component (tangential to the slope) of the seepage flow is a constant $-K(\partial h/\partial \ell) = -K \sin\beta$, with β and ℓ denoting the slope of the seepage face and length along the slope, respectively.

G. Boundary with a Body of Flowing Water

In Sect. 5.2.4 A, we discussed the boundary condition of a porous medium in contact with a static body of water; in Sect. 5.2.4 C, the presence of a semi-pervious layer separating open water and a porous medium domain was discussed. In this subsection, we shall examine the conditions on the boundary between a porous medium domain and a body of *flowing water*.

In a body of water, the flow is governed by the Navier–Stokes equation (= momentum balance equation for a Newtonian fluid), or by some simplified version of it, together with the mass balance equation. To have a well-posed boundary value problem, either the velocity components, or the stresses, but not both, have to be specified on all parts of the boundary (Ladyzhenskaya 1963). For example, for the two-dimensional flow shown in Fig. 5.7a, the boundary condition on the solid surface is that the fluid must adhere there to the solid; this is referred to as the 'no-slip' condition. In the case of a stationary solid surface, the boundary conditions are $V_x = 0$ and $V_z = 0$. On the other hand, on a free surface, the shear stress and pressure are specified as $\tau_{zx} = 0$ and $p = 0$.

In principle, flows in a porous medium, and in a fluid continuum, are governed by the same physical laws—momentum balance and mass balance (and solute balance and heat balance, in the cases of solute and heat transport). However, for flow in a porous medium domain, as a consequence of the process of averaging or homogenization, these two fundamental equations are often expressed in different forms, containing *coefficients*, such as permeability; and the microscopic geometry of the void space no longer appears explicitly. Furthermore, the porous medium equations are often simplified, because the effects of certain terms appearing in these equations are negligible.

Let us start by assuming that a considered flow in the porous medium domain is governed by Darcy's law, as discussed as Case A in Sect. 4.2.4. This law is a simplified form of the homogenized Navier–Stokes equation. When Darcy's law is combined with the mass balance equation, we obtain a single governing, or flow, equation, expressed, say for a constant density fluid, in terms of a single scalar

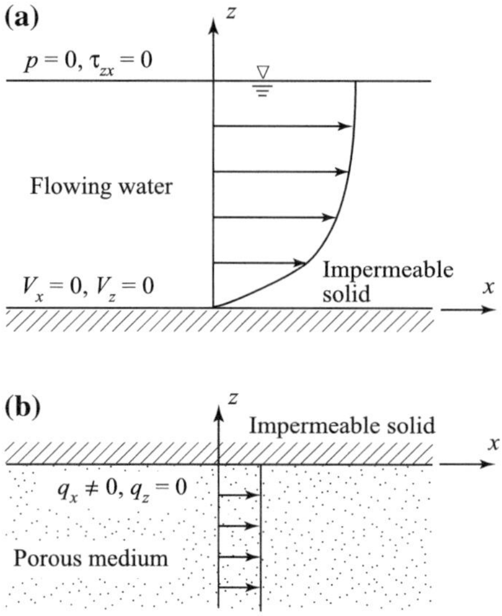

Fig. 5.7 Boundary conditions for: **a** Flowing water with impermeable boundary, **b** Porous medium with impermeable boundary, and **c** Common boundary between flowing water and porous medium

variable—the piezometric head, h. As a consequence, one and only one boundary condition, either the normal flux or the head, needs to be specified on any part of the domain's boundary. As illustrated in Fig. 5.7b, $q_z = 0$ on a horizontal impervious boundary. Particularly, we notice that q_x cannot be specified and must be solved for; hence, $q_x \neq 0$ on the impervious boundary, i.e., we have a 'slip'. We recall that $\mathbf{q}$ is the macroscopic average of the microscopic velocity over the REV, and that the microscopic velocity *does not slip* on a solid surface.

The discussion above serves to illustrate that although the physical principles need to be obeyed, the averaging process makes the two sets of equations, one based on Navier–Stokes equation in the free-flowing water, and the other based on the homogenized Darcy's law in the porous medium, *incompatible* on a shared boundary. Hence, a coupled solution of the two domains is not possible.

This incompatibility between governing equations may be resolved if we assume that flow in the porous medium domain is governed by *Brinkman's equation* (Case B in 4.2.4). When the two flow domains are in contact with each other, that is, the flowing (viscous) fluid is bounded from below by a porous medium saturated with the same fluid (Fig. 5.7c), we need to consider the conditions on the common boundary carefully, in order to determine the set of necessary and sufficient boundary conditions that will ensure the existence of a solution of the problem.

First, consider the interface between two immiscible flowing fluids with different densities and viscosities. A total of four conditions are needed on the interface (for two-dimensional flow): (1) velocity continuity, $V_x|_{z=0^+} = V_x|_{z=0^-}$; (2) $V_z|_{z=0^+} = V_z|_{z=0^-}$; (3) pressure continuity, $p|_{z=0^+} = p|_{z=0^-}$; and (4) shear stress continuity $\tau_{zx}|_{z=0^+} = \tau_{zx}|_{z=0^-}$. A similar situation exists at the interface between a free-flowing fluid and a porous medium (Kohr and Sekhar 2007). Two of these interface conditions are obvious: (1) velocity continuity normal to the interface (mass conservation), $V_z|_{z=0^+} = q_z|_{z=0^-}$, and pressure continuity $p|_{z=0^+} = \gamma h|_{z=0^-}$. The other two conditions: the relation between the horizontal velocity of the flowing fluid, $V_x|_{z=0^+}$ and the flux $q_x|_{z=0^-}$, and between the two shear stresses, are not so obvious, as the quantities in the porous medium are homogenized ones.

Bear and Bachmat (1991, p. 245) considered the case of free flowing water overlaying a porous medium (Fig. 5.7c) containing an incompressible, Newtonian fluid. They showed that *when we assume no jump in pressure, and no jump in effective stress, there should not be a jump in the normal (to the common interface) component of the shear stress, $\boldsymbol{\tau}$, across the interface*, i.e.,

$$[\![\ \boldsymbol{\tau}_f \ ]\!]_{1,2} \cdot \boldsymbol{n} = 0, \tag{5.2.32}$$

where $\boldsymbol{n}$ is the unit normal vector. In order to express (5.2.32) in terms of fluid velocities, we need an appropriate constitutive relation. For $\boldsymbol{\tau}_f\big|_2$, i.e., in the free-flowing fluid, we use the constitutive relationship for an incompressible single component Newtonian fluid,

$$\tau_{ij} = \mu\left(\frac{\partial V_i}{\partial x_j} + \frac{\partial V_j}{\partial x_i}\right). \tag{5.2.33}$$

We assume that this relationship is valid also for the porous medium at the macroscopic level, but with an apparent viscosity μ^* that takes into consideration the added porous medium resistance (Shavit et al. 2004). Different studies have derived and used different values of μ^*, $\mu^* >, =, < \mu$ (Nield and Bejan 2013; Koplik et al. 1983; Kim and Russel 1985). In the case under consideration here, we shall neglect the velocity gradient terms $\partial V_z/\partial x|_{z=0^+}$ and $\partial q_z/\partial x|_{z=0^-}$ in (5.2.33), to obtain the condition for shear stress continuity on the interface,

$$\left.\frac{\mu^*}{\phi}\frac{\partial q_x}{\partial z}\right|_{z=0^-} = \left.\mu\frac{\partial V}{\partial z}\right|_{z=0+}. \tag{5.2.34}$$

This can be used as the condition on the interface between the two domains in the coupled boundary value problems. A more thorough examination of the condition of shear stress compatibility can be found in Kubik (2004), who suggested that the horizontal velocities tangential to the interface, $V|_{z=z^+}$ and $q_x|_{z=0^-}$ are not continuous. Their relation should be determined from the continuity of both momentum and energy near the interface. More discussion about the interface condition between free-flowing fluid and porous medium can be found in Rosenzweig and Shavit (2007).

Finally, let us examine a well-known condition at the interface between a free flowing water domain and a saturated porous medium domain, known as the *Beavers–Joseph condition* (Beavers and Joseph 1967). This condition was motivated by the observation that in open channel flow, with a porous channel bottom, the discharge tends to be slightly greater than the one bounded by an impermeable bottom. The reason is that the porous interface condition allows the velocity at the channel bottom to *slip*, as shown in Fig. 5.7c. The Beavers–Joseph condition approximates the velocity gradient on the left side of (5.2.34) in the form:

$$\frac{\alpha^M}{\phi}\left(V\big|_{z=0^+} - q_o\right) = \frac{\partial V}{\partial z}\bigg|_{z=0^+}, \tag{5.2.35}$$

where α^M (dims. L^{-1}) is a momentum transfer coefficient that depends only on porous medium properties, such as permeability and porosity, and q_o is the uniform specific discharge in the porous medium starting from a certain distance away from the 'velocity boundary layer' in the vicinity of the interface. Beavers and Joseph (1967) proposed $\alpha^M = C^M\phi/\sqrt{k}$, where the dimensionless coefficient, C^M, which depends only on ϕ and k, has to be determined experimentally. We note that (5.2.35) is a third type boundary condition for the free-flowing water, with empirical coefficients α^M and q_o. As the equation does not contain head or specific discharge information, it is not a boundary condition for the porous medium flow. More discussion on the Beavers–Joseph type boundary condition can be found in Ochoa-Tapia and Whitaker (1995), Nield and Bejan (2013), and Jager and Mikelic (2000).

5.2.5 Complete 3-D Mathematical Flow Model

We now have all the elements required in order to formulate the complete mathematical model of a problem of forecasting the flow of a single fluid phase (saturated flow) in a porous medium domain. The objective of this subsection is to review the standard content of any such model.

Note that although this subsection refers to the particular case in which the transported extensive quantity is the mass of a single fluid phase, the same model content is applicable also to the model of any other extensive quantity.

A. A Well-Posed Problem

The solution of the mathematical model of a problem takes the form of temporal and spatial distributions of the state variables of interest within the problem's prescribed time and space domains.

From the mathematical point of view, given a model composed of one or more partial differential equations, not every set of conditions imposed on the boundaries of the problem domain is satisfactory. This is even more so because, often, we have to resort to estimates of coefficients and simplifications of the mathematical models in specifying the boundary conditions.

A mathematical model that represents a *physical reality* (and only such cases are considered in this book) is said to be *well-posed* if it satisfies the following requirements (e.g., Courant and Hilbert 1962):

- A solution of the problem exists (*existence*).
- The solution is unique (*uniqueness*).
- The solution is stable (*stability*).

The first requirement simply states that at least one solution exists. The second one stipulates completeness of the problem statement, with no ambiguity. There exists no other solution that satisfies the stated problem. The third requirement means that small variations in data (e.g., initial and boundary conditions, and/or values of model coefficients) should lead to small changes in the resulting solution. If small errors in the data do not lead to correspondingly small errors in the solution, then the mathematical model is *ill-posed*. This last requirement is of particular interest, as all our observations have always some measurement errors. A model will be meaningless if these small errors will significantly affect the solution.

Thus, once a complete mathematical model has been stated, the next step is to ensure that it is well-posed. Only then should a solution be sought.

The models developed and presented in this book, since they are based on a thorough analysis of the physical reality and on its description, albeit with certain simplifying assumptions, are implicitly *assumed* to be always well-posed. Therefore, they should provide unique, stable solutions. We shall not go into the mathematical analysis of whether a model developed here is well-posed, or not, although, as stated above, such an analysis is an essential step in the modeling process. The techniques used for such analysis can be found in appropriate mathematical texts on partial differential equations.

B. Standard Content of a Flow Model

A complete, well-posed mathematical model consists of the following items:

(a) Definition of the geometry of the flow domain's boundaries. The boundary surface must form a closed surface.
(b) The primary variable that describes the state of the system, e.g., the pressure (or the piezometric head for $\rho_w = const.$).
(c) Partial differential flow equation that expresses the mass balance equation.
(d) An expression for the fluid's flux (e.g., Darcy's law).
(e) Constitutive equations (and equations of state) for the phases involved, including, if necessary, the solid matrix. For the case of saturated flow, we may need the relationships between density and pressure, and between porosity and effective stress.
(f) Information on the various sources and sinks of fluid mass. Sometimes, these take the form of functions of the problem's state variables.
(g) Formulation of the conditions that prevail everywhere within the considered domain at some initial time, in terms of the problem's state variables.
(h) Formulation of the conditions that prevail on the domain's boundaries, specified in item (a) above, during the period of interest. In many cases, the delineation of a boundary segment and the conditions on it have to be considered simultaneously, i.e., we select boundaries such that we can specify the conditions on them.

(i) Numerical values, or functional relations, of all the coefficients and parameters that appear in the model's equations.

The set of equations (mass balance equations, motion equations, and constitutive relations) must constitute a *closed* one, i.e., it should contain a sufficient number of equations to enable the simultaneous solution for all state variables of the problem.

After writing the closed set of equations, we use the methodology discussed in Sect. 3.9 to determine the number of *degrees of freedom*, or *primary variables* of the problem and to select the most convenient ones. We then identify an equal number of (partial differential) balance equations that have to be solved in order to determine the values of these variables (whether they appear explicitly in the equations or not). All the remaining equations and relationships, including partial differential equations, are then employed in order to determine the remaining variables. Initial and boundary conditions are specified *only* for the partial differential equations that have to be solved. In practice, in most cases, the PDE's are replaced by equivalent numerical equations, e.g., the integro-differential equations discussed in Sect. 3.8, which have to be solved by appropriate software.

We have presented the discussion above in a somewhat generalized form in order to facilitate the discussion in Chaps. 6 and 7, where we shall be discussing more than one fluid phase and dissolved chemical species. The situation described above becomes much simpler in the case of saturated (or single phase) flow considered in this chapter, for which we have to solve only a single partial differential equation–the mass balance equation—for the single variable pressure, or piezometric head.

5.2.6 Two Phases Separated by a Sharp Interface

Although this chapter covers modeling of flow of a single fluid only, we include in it also this subsection that considers two fluid phases. However, as we shall see below, we actually deal simultaneously with two domains, each containing a single phase only, except that the two domains are interacting through a common boundary.

A sharp interface does not really exist between two immiscible fluids, because of capillarity. In the case of two miscible fluids, a transition zone will always occur between them because of diffusion and dispersion. This subject is discussed in Chap. 7. In both cases, under certain conditions, when the *transition zone* between the two fluids is narrow, relative to the zones occupied by the considered fluids, we may replace it by a *sharp interface* between the two fluids. This approximation was common prior to the era of computers, e.g., when considering the phreatic surface, or the interface in coastal aquifers. Nowadays, there is seldom a justification for this *sharp interface approximation*.

Actually, we have already introduced this approximation in Sect. 5.2.1 A. There, because of capillarity, a transition zone occurs. Across the latter, water saturation varies from 100% to an irreducible saturation value. Nevertheless, we often replace this transition zone by a sharp air-water interface, referred to as the *water table*, or

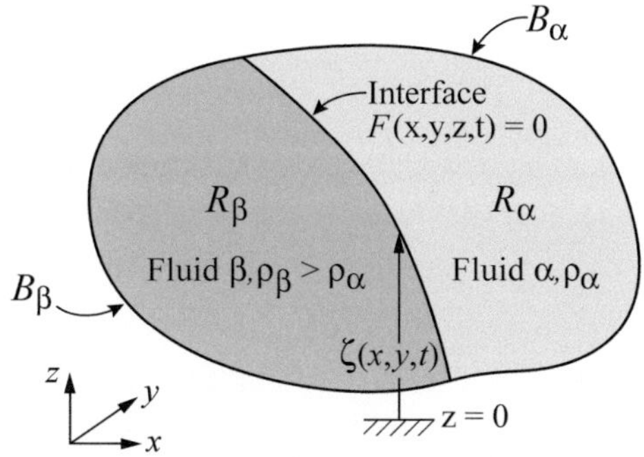

Fig. 5.8 Two immiscible liquid zones separated by a sharp interface

phreatic surface. We have presented the condition on this surface in Sect. 5.2.4 E, where we have also assumed that the air is at a constant pressure.

Another example is the boundary between two fluids of different densities, e.g., fresh water and seawater in a coastal aquifer. These are two miscible fluid zones: a fresh groundwater zone and a zone of sea water, with a transition zone between them due to diffusion/dispersion of dissolved salts. However, when each of these zones is much thicker than the width of the transition zone, it is possible to regard the latter as a *sharp interface* (e.g., Bear 1979, Sect. 9.5).

Figure 5.8 shows two fluid zones separated by a sharp interface. The two fluids are characterized by their densities and viscosities.

In order to determine the flow regime in the two zones, and the time-dependent location of the interface, we have to solve an appropriate mathematical model that is composed of two zones: an α-fluid zone, with density ρ_α and viscosity μ_α, and a β-fluid zone, with $\rho_\beta > \rho_\alpha$ and $\mu_\beta \neq \mu_\alpha$. We write a complete model for each zone. It consists of a mass balance equation, a flux equation, and conditions on the zone's boundaries, including on the surface separating the two zones. However, the (time-dependent) location of the latter is a-priori unknown. In fact, in most cases, this location is what we are seeking by solving this problem.

Let us express the location and shape of the interface considered here in the form $F = F(x, y, z, t) = 0$, with F describing an a priori unknown (until the problem is solved) surface. Denoting the elevation, z, of points on the interface by $\zeta = \zeta(x, y, t)$ (Fig. 5.8), the relationship satisfied by F at points on the moving interface is:

$$z = \zeta(x, y, t), \quad \text{or} \quad F(x, y, z, t) \equiv z - \zeta(x, y, t) = 0. \tag{5.2.36}$$

The pressure at a point on the interface, $p(x, y, z, t)$, is the same when the point is approached from both sides. Because ρ is constant within each of the two zones, the use of *piezometric head* as a variable is permitted. Hence, from the definitions of h_α and h_β (see (4.1.4)), we obtain:

$$\rho_\alpha(h_\alpha - \zeta) = \rho_\beta(h_\beta - \zeta), \tag{5.2.37}$$

or:

$$\zeta = h_\beta \frac{\rho_\beta}{\rho_\beta - \rho_\alpha} - h_\alpha \frac{\rho_\beta}{\rho_\beta - \rho_\alpha} = h_\beta(1+\delta) - h_\alpha \delta, \qquad (5.2.38)$$

where h_α and h_β take on their respective values on the interface, and:

$$\delta = \frac{\rho_\alpha}{\rho_\beta - \rho_\alpha}. \qquad (5.2.39)$$

If we can solve the appropriate PDE's for $h_\alpha = h_\alpha(x, y, z, t)$ and for $h_\beta = h_\beta(x, y, z, t)$, in their respective domains, (5.2.38) will be the sought *equation describing the shape of the (possibly moving) interface*. We can rewrite it in the form:

$$F \equiv z - h_\beta(1+\delta) + h_\alpha \delta = 0. \qquad (5.2.40)$$

As this relationship is valid only for points z on the interface, i.e., $z \equiv \zeta$, we recall that $h_\alpha = h_\alpha(x, y, \zeta, t)$, and $h_\beta = h_\beta(x, y, \zeta, t)$.

Once we have the location of the boundary ($\equiv$ the interface), the boundary conditions on it—one for each side—are obtained from the fact that the interface is a *material surface* with respect to the mass of each of the liquids; no liquid mass crosses it. The two conditions are:

$$(\mathbf{V}_\alpha - \mathbf{u}) \cdot \boldsymbol{\nu} = 0, \quad (\mathbf{V}_\beta - \mathbf{u}) \cdot \boldsymbol{\nu} = 0 \qquad (5.2.41)$$

in which $\mathbf{V}_\alpha$ and $\mathbf{V}_\beta$ are the velocities of the respective fluids, $\mathbf{u}$ is the speed of displacement of the interface F, and $\boldsymbol{\nu}$ denotes the outward unit vector on F, with:

$$\frac{DF}{Dt} \equiv \frac{\partial F}{\partial t} + \mathbf{u} \cdot \nabla F, \quad \boldsymbol{\nu} = \frac{\nabla F}{|\nabla F|}, \quad \mathbf{u} \cdot \nabla F = -\frac{\partial F}{\partial t}. \qquad (5.2.42)$$

In addition, the interface is also a material surface with respect to the solid, and hence,

$$(\mathbf{V}_{\text{solid}} - \mathbf{u}) \cdot \mathbf{n} = 0. \qquad (5.2.43)$$

From the above two equations, it follows that on the interface, which serves as a common boundary to both subdomains, we have:

$$(\mathbf{q}_{r\alpha} - \phi \mathbf{u}) \cdot \mathbf{n} = 0, \quad \alpha = f, s. \qquad (5.2.44)$$

Making use of Darcy's law and (5.2.42), we obtain the two conditions on the interface, for the R_α and R_β sub-domains, respectively, in the form:

$$\phi\delta \frac{\partial h_\alpha}{\partial t} - \phi(1+\delta)\frac{\partial h_\beta}{\partial t} = \mathbf{K}_\alpha \cdot \left[\nabla z - (1+\delta)\nabla h_\beta + \delta \nabla h_\alpha\right] \cdot \nabla h_\alpha, \qquad (5.2.45)$$

$$\phi\delta \frac{\partial h_\alpha}{\partial t} - \phi(1+\delta)\frac{\partial h_\beta}{\partial t} = \mathbf{K}_\beta \cdot \left[\nabla z - (1+\delta)\nabla h_\beta + \delta \nabla h_\beta\right] \cdot \nabla h_\beta. \qquad (5.2.46)$$

In principle, we can find solutions for h_α and h_β in their respective domains by solving the governing equations for each domain, together with the boundary (interface) conditions, (5.2.44)–(5.2.46). Unfortunately, the interface conditions, (5.2.45) and (5.2.46), are nonlinear coupled partial differential equations in the variables h_α and h_β. It is practically impossible to directly solve the coupled system analytically, in order to determine the shape and position of the interface. Instead, numerical methods can be employed to approximately find the interface location.

5.3 Modeling 2-D Flow in an Aquifer

In principle, flow always takes place in a three-dimensional domain. However, sometimes the domain has one dimension which is much larger than the other two, so that a one-dimensional model seems a good simplification. Another case is when one dimension is much smaller than the other two such that a simplified two-dimensional flow model is called for. We refer to such domain as a *thin* one. An example is the case of regional flow in an aquifer. The main feature of an aquifer is that it is an essentially horizontal flow domain, characterized by a thickness that is much smaller than its horizontal extent of interest; hence, the vertical variations in piezometric head are, usually, much smaller than the horizontal ones. Under such circumstances, flow in an aquifer may be *conceptually* modeled (albeit *as an approximation*) as taking place in a horizontal two-dimensional domain. We refer to this approximation as 'essentially horizontal flow' approximation, or 'the hydraulic approach'. The transformation of the three-dimensional mathematical model into a horizontal two-dimensional one is performed by integrating (or averaging) the former along the vertical coordinate axis (see next subsection). Obviously, although we have referred here to an aquifer, the same approach is applicable to any reservoir for which the *essentially horizontal flow approximation* is valid.

5.3.1 Deriving 2-D Balance Equations by Integration

Consider the piezometric head, $h = h(x, y, z, t)$. Its average over the vertical thickness of an aquifer, $B(x, y, t)$, is defined by

$$\hat{h}(x, y, t) = \frac{1}{B(x, y, t)} \int_{B(x,y,t)} h(x, y, z, t)\, dz. \tag{5.3.1}$$

In terms of this averaged variable, $\hat{h} = \hat{h}(x, y, t)$, the flow equation is reduced to a two-dimensional one in the horizontal, xy-plane.

The assumption of 'essentially horizontal flow in an aquifer', usually referred to as the *Dupuit assumption*, was introduced by Dupuit (1863) in connection with flow in phreatic aquifers. Dupuit suggested that in such an aquifer, we assume, as a good approximation, that flow is horizontal, equipotentials are vertical, and, equivalently, the vertical pressure distribution is hydrostatic. Although we shall apply here the Dupuit assumption to water flow in an aquifer, both confined and phreatic, the presented material may be extended to the transport of any extensive quantity in any relatively thin domain (Bear and Bachmat 1991, p. 481).

We start by developing the *integrated balance equation* for *any* macroscopic extensive quantity, E, having a density e (= amount of E per unit volume of the phase). The general *macroscopic balance equation* that describes the transport of E in a three-dimensional domain is given by (3.3.3), rewritten here in the form:

$$\frac{\partial}{\partial t}(\phi e) + \nabla \cdot \left(e\mathbf{q} + \phi \mathbf{J}_h^E\right) - \Gamma'' = 0, \tag{5.3.2}$$

where ϕ denotes the porosity, $\mathbf{q}$ denotes the specific discharge of the phase, $\phi\mathbf{J}_h^E$ denotes the sum of dispersive and diffusive fluxes (Sect. 3.4.4 A) of the extensive quantity, per unit area of porous medium, and Γ'' denotes the total source (= rate of production) of E, due to both internal production and influx across the (microscopic) surface that bounds the phase, per unit volume of porous medium.

The methodology of the hydraulic approach calls for the integration of (5.3.2) along the vertical (possibly varying) thickness of the aquifer. Let the aquifer be bounded from above and below by (possibly moving) surfaces whose elevations are at $z = b_1(x, y, t)$ and $z = b_2(x, y, t)$, respectively, with $b_2 - b_1 = B$. Another way of expressing the geometry of these boundary surfaces is by (Fig. 5.9):

$$\begin{aligned} F_1 &\equiv F_1(x, y, z, t) = z - b_1(x, y, t) = 0, \\ F_2 &\equiv F_2(x, y, z, t) = z - b_2(x, y, t) = 0, \end{aligned} \tag{5.3.3}$$

where $F_i(\mathbf{x}, t) = 0$ represents the equation of a boundary surface, or a segment of it. Time is introduced to allow for the possibility of a moving boundary, e.g., a phreatic surface, with $\mathbf{u}$ denoting the speed of displacement of such a boundary. Since the quantity F is a conservative quantity of the points on the surface, we have $DF_i/Dt = 0$, $i = 1.2$, and hence:

$$\boldsymbol{\nu} = \frac{\nabla F}{|\nabla F|}, \qquad u_\nu = \mathbf{u} \cdot \boldsymbol{\nu} = -\frac{\partial F/\partial t}{|\nabla F|}, \tag{5.3.4}$$

where $\boldsymbol{\nu}$ denotes the unit vector normal to the surface $F = 0$.

With the above definitions of the two F-surfaces, we have from (5.3.3):

$$\nabla F_i = \nabla(z - b_i) \qquad \text{and} \qquad \frac{\partial F_i}{\partial t} = -\frac{\partial b_i}{\partial t}, \qquad i = 1, 2. \tag{5.3.5}$$

For example, for a horizontal F_1-surface, ∇F_1 is directed upward, normal to the surface. Also, for any surface $F_i = F_i(x, y, z, t)$, we have:

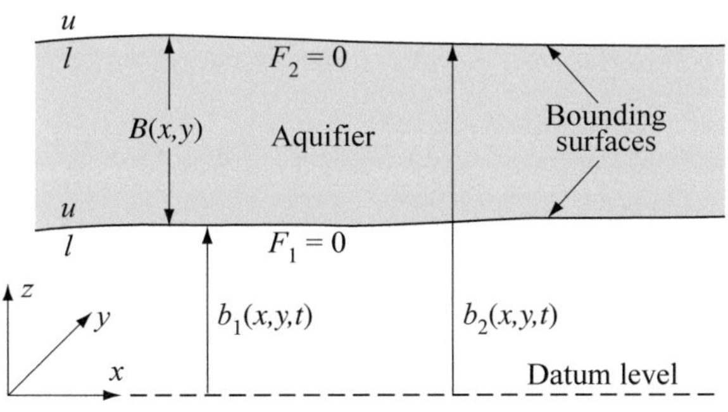

Fig. 5.9 Nomenclature for integration over the thickness of an aquifer

$$\frac{\partial F_i}{\partial t} + \mathbf{u} \cdot \nabla F_i = 0, \quad \text{or} \quad \frac{\partial b_i}{\partial t} - \mathbf{u} \cdot \nabla (z - b_i) = 0, \qquad i = 1, 2. \tag{5.3.6}$$

For a stationary boundary:

$$b_i = b_i(x, y), \quad \frac{\partial F_i}{\partial t} = 0, \quad i = 1, 2. \tag{5.3.7}$$

By integrating (5.3.2) along the thickness, B, we obtain:

$$\int_{b_1}^{b_2} \frac{\partial \phi e}{\partial t}\, dz + \int_{b_1}^{b_2} \nabla \cdot (e\mathbf{q} + \phi \mathbf{J}_h^E)\, dz - \int_{b_1}^{b_2} \Gamma''\, dz = 0. \tag{5.3.8}$$

Since we have here integrals of derivatives, with integration boundaries that are space-(and possibly time-)dependent, we have to introduce a certain rule for taking integrals of derivatives. This rule is based on *Leibnitz' rule* for a derivative of an integral with respect to a variable upon which the boundaries of the latter depend (see any textbook on Calculus). Here, we rewrite this rule in the form:

$$\frac{\partial}{\partial r} \int_{b_1}^{b_2} \mathbf{A}\, dz = \int_{b_1}^{b_2} \frac{\partial \mathbf{A}}{\partial r}\, dz + \mathbf{A}\Big|_{b_2} \frac{\partial b_2}{\partial r} - \mathbf{A}\Big|_{b_1} \frac{\partial b_1}{\partial r}, \tag{5.3.9}$$

where $\mathbf{A} = \mathbf{A}(x, y, z, t)$ is any tensor field, and r stands for x, y, z, or t.

Let us define the symbol $\widehat{\mathbf{A}}$ as:

$$\widehat{\mathbf{A}}'(x, y, t) = \frac{1}{B(x, y, t)} \int_{b_1(x,y,t)}^{b_2(x,y,t)} \mathbf{A}(x, y, z, t)\, dz, \tag{5.3.10}$$

in which the prime symbol denotes a vector (or vector operator) in the two-dimensional (xy)-plane only, viz.,

$$\mathbf{A}' = A'_x \mathbf{1x} + A'_y \mathbf{1y}, \qquad \nabla'(..) = \frac{\partial}{\partial x}(..)\, \mathbf{1x} + \frac{\partial}{\partial y}(..)\, \mathbf{1y},$$

with $\mathbf{1}x$ and $\mathbf{1}y$ denoting unit vectors in the x- and y-directions, respectively.

Making use of Leibnitz rule, we may write for any vector, $\mathbf{A}$:

$$\begin{aligned}\int_{b_1(x,y,t)}^{b_2(x,y,t)} \nabla\cdot\mathbf{A}\,dz &= \int_{b_1}^{b_2}\left(\nabla'\cdot\mathbf{A}' + \frac{\partial A_z}{\partial z}\right)dz\\ &= \nabla'\cdot\int_{b_1}^{b_2}\mathbf{A}'\,dz - \mathbf{A}'\big|_{b_2}\cdot\nabla' b_2 + \mathbf{A}'\big|_{b_1}\cdot\nabla' b_1 + A_z\big|_{b_2} - A_z\big|_{b_1}\\ &= \nabla'\cdot B\widehat{\mathbf{A}'} + \mathbf{A}\big|_{b_2}\cdot\nabla(z-b_2) - \mathbf{A}\big|_{b_1}\cdot\nabla(z-b_1)\\ &= \nabla'\cdot B\widehat{\mathbf{A}'} + \mathbf{A}\big|_{F_2}\cdot\nabla F_2 - \mathbf{A}\big|_{F_1}\cdot\nabla F_1.\end{aligned} \tag{5.3.11}$$

Here, and henceforth, $\big|_{F_i}$ stands for $\big|_{F_i=0}$.

For any scalar, $A(x, y, z, t)$, with $b_1 = b_1(x, y, t)$, $b_2 = b_2(x, y, t)$, we have:

$$\begin{aligned}\int_{b_1}^{b_2}\frac{\partial A}{\partial t}\,dz &= \frac{\partial}{\partial t}\int_{b_1}^{b_2} A\,dz - A\big|_{b_2}\frac{\partial b_2}{\partial t} + A\big|_{b_1}\frac{\partial b_1}{\partial t}\\ &= \frac{\partial}{\partial t}B\widehat{A} + A\big|_{F_2}\frac{\partial F_2}{\partial t} - A\big|_{F_1}\frac{\partial F_1}{\partial t}.\end{aligned} \tag{5.3.12}$$

By applying (5.3.6), (5.3.12), (5.3.11), and (5.3.8), we obtain

$$\begin{aligned}&\frac{\partial}{\partial t}B\widehat{\phi e} + \nabla'\cdot B\left(\widehat{e\mathbf{q}'} + \widehat{\phi\mathbf{J}'^E_h}\right) + \left[\phi e(\mathbf{V}-\mathbf{u}) + \phi\mathbf{J}^E_h\right]\big|_{F_2}\cdot\nabla F_2\\ &\quad - \left[\phi e(\mathbf{V}-\mathbf{u}) + \phi\mathbf{J}^E_h\right]\big|_{F_1}\cdot\nabla F_1 - B\widehat{\Gamma''} = 0,\end{aligned} \tag{5.3.13}$$

in which $\mathbf{q}'$ denotes the vector of specific discharge in the horizontal xy-plane. This is the averaged, two-dimensional (in the horizontal plane) balance equation for any E in an aquifer. The dependent variables and fluxes, $\widehat{\phi e}$, $\widehat{e\mathbf{q}'}$, and $\widehat{\phi\mathbf{J}'^E_h}$, are functions of x, y, and t only.

In (5.3.13), the terms

$$\left[\phi e(\mathbf{V}-\mathbf{u}) + \phi\mathbf{J}^E_h\right]\Big|_{F_2}\cdot\nabla F_2 \quad\text{and}\quad \left[\phi e(\mathbf{V}-\mathbf{u}) + \phi\mathbf{J}^E_h\right]\Big|_{F_1}\cdot\nabla F_1$$

represent the total flux of E through the (possibly moving) boundaries $F_2 = 0$ and $F_1 = 0$, which bound the aquifer from above and below, respectively. In other words, these terms represent *boundary conditions* on these surfaces. We note that while these terms are boundary conditions for the three-dimensional balance equation, (5.3.2), they appear as *source terms* in the averaged, two-dimensional equation (5.3.13). Our next task is to express these conditions.

The general condition that must be satisfied at any point on a boundary $F = 0$, for any of the α-phases present in a system, in the absence of sources and sinks of a considered quantity E on the boundary, is that of continuity of the normal component of the total flux of E in any of the phases. However, if there exist on the boundary portions that provide direct contact between phases, continuity exists only for the flux through all phases together, rather than for each phase separately. In what follows, we shall refer to the latter case. Using subscripts ext and int to denote the *external* and *internal* sides of the boundary $F(x, y, z, t) = 0$, respectively, we may rewrite this condition in the form:

$$\sum_{(\alpha=s,f)} \left[\!\left[\theta_\alpha \left\{e'_\alpha(\mathbf{V}_\alpha - \mathbf{u}) + \mathbf{J}^{E_\alpha}_{h,\alpha}\right\}\right]\!\right]_{\text{ext,int}} \cdot \boldsymbol{\nu} = 0, \tag{5.3.14}$$

where the symbols f and s denote the fluid and solid phases, respectively, $[\![\ (..)\ ]\!]_{1,2}$ denotes the jump in (..) from side 1 (here, external) of the boundary to side 2 (here, internal), and $\mathbf{J}^E_{h,\alpha}$ denotes the non-advective flux of E_α. When the microscopic interphase boundary is a *material boundary* with respect to the considered quantity, i.e., there is no exchange of that quantity among the phases, the *no-jump condition* (5.3.14) may be written separately for each phase.

In what follows, we shall assume that the top and bottom surfaces that bound a confined or a leaky aquifer are material surfaces with respect to the solid mass. Hence, on these surfaces, $(\mathbf{V}_s - \mathbf{u}) \cdot \boldsymbol{\nu} = 0$, and, therefore:

$$\begin{aligned} \phi(\mathbf{V}_f - \mathbf{u}) \cdot \boldsymbol{\nu} &= \phi(\mathbf{V}_f - \mathbf{u}) \cdot \boldsymbol{\nu} - \phi(\mathbf{V}_s - \mathbf{u}) \cdot \boldsymbol{\nu} \\ &= \phi(\mathbf{V}_f - \mathbf{V}_s) \cdot \boldsymbol{\nu} \equiv \mathbf{q}_r \cdot \boldsymbol{\nu}. \end{aligned} \tag{5.3.15}$$

Recall that $\mathbf{q}_r$ denotes the specific discharge of the fluid relative to the solid; it is expressed by Darcy's law.

We can now use (5.3.14) to replace the terms in (5.3.13) that express the flux conditions on the 'internal sides' of the boundaries by terms that involve (*known*) information on the corresponding 'external sides.'

Let us develop the condition for an upper boundary, $F_2 = 0$, and for the specific case of the mass of water phase, $e = \rho$, in saturated flow. Since we have assumed that the boundary is a material surface with respect to the solid, we have on it:

$$\rho_s(\mathbf{V}_s - \mathbf{u}) \cdot \nabla F_2 = 0. \tag{5.3.16}$$

In addition, for the F_2-surface, (5.3.14) reduces to:

$$[\phi\rho(\mathbf{V}_f - \mathbf{u})]\big|_{\text{ext}} \cdot \nabla F_2 = [\phi\rho(\mathbf{V}_f - \mathbf{u})]\big|_{\text{int}} \cdot \nabla F_2, \tag{5.3.17}$$

where $\rho \equiv \rho_f$, or

$$(\rho\mathbf{q}_r)\big|_{\text{ext}} \cdot \nabla F_2 = (\rho\mathbf{q}_r)\big|_{\text{int}} \cdot \nabla F_2. \tag{5.3.18}$$

For an impervious boundary, $\mathbf{q}_r\big|_{\text{ext}} \cdot \nabla F_2 = \mathbf{q}_r\big|_{\text{ext}} \cdot \boldsymbol{\nu} = 0$, so that (5.3.18) reduces to:

$$(\rho\mathbf{q}_r)\big|_{\text{int}} \cdot \nabla F_2 = 0. \tag{5.3.19}$$

At a leaky boundary, i.e., a boundary through which fluid mass can enter or leave the aquifer at a *known* rate, $\rho\mathbf{q}_{\text{leak}}$, the condition is

$$(\rho\mathbf{q}_r)\big|_{\text{int}} \cdot \nabla F_2 = (\rho\mathbf{q}_{\text{leak}})\big|_{\text{ext}} \cdot \nabla F_2. \tag{5.3.20}$$

The term $\mathbf{q}_{\text{leak}}\big|_{\text{ext}}$ represents the leakage into (or out of) the aquifer on the *external* side of the latter. It can now be expressed in terms of the state variables of a considered problem.

Let the surface F_2 serve as the upper boundary for the saturated domain of a phreatic aquifer with accretion. The condition on such a boundary takes the form:

$$[\phi\rho(\mathbf{V}_f - \mathbf{u})]\big|_{\text{int}} \cdot \nabla F_2 = \rho_N(\mathbf{N} - \theta_{rf}\mathbf{u})\big|_{\text{ext}} \cdot \nabla F_2, \tag{5.3.21}$$

or, by rearranging terms, and using (5.3.4):

$$(\rho_N\mathbf{N} - \rho\mathbf{q}) \cdot \nabla F_2 = (\phi\rho - \theta_{rf}\rho_N)\frac{\partial F_2}{\partial t}. \tag{5.3.22}$$

Here, $\mathbf{N}$ denotes the rate of accretion of water of density ρ_N, and θ_{rf} denotes the irreducible moisture content that is assumed to prevail above the phreatic surface.

For downward accretion at a rate N, we introduce $\mathbf{N} = -N\nabla z$, and (5.3.22) becomes:

$$(\rho_N N\nabla z + \rho\mathbf{q}) \cdot \nabla F_2 = -(\phi\rho - \theta_{rf}\rho_N)\frac{\partial F_2}{\partial t}. \tag{5.3.23}$$

Let us now rewrite the mass balance equation (5.3.13), making use of the following approximations:

- The macrodispersive flux (Sect. 7.2.4) of the total mass, $\widehat{\mathring{\rho}\mathring{\mathbf{q}}'}$, due to vertical variations in the horizontal flux, $\mathbf{q}'$, and in ρ, may be neglected, i.e.,

$$\widehat{\rho\mathbf{q}'} = \widehat{\rho}\widehat{\mathbf{q}'} + \widehat{\mathring{\rho}\mathring{\mathbf{q}}'} \approx \widehat{\rho}\widehat{\mathbf{q}'}, \tag{5.3.24}$$

  where the symbol $(\mathring{..})$ denotes deviation of (..) from its average, $\widehat{(..)}$, over the vertical, B.
- The average of the sum of the components of the dispersive and diffusive fluxes of the total mass is much smaller than the advective mass flux at the averaged level, i.e.,

$$|\widehat{\phi\mathbf{J}^{m\prime}_{fh}}| \ll |\widehat{\rho}\widehat{\mathbf{q}'}|.$$

With these approximations, (5.3.13) can be rewritten in the form:

$$\frac{\partial}{\partial t}(B\widehat{\phi\rho}) + \nabla' \cdot (B\widehat{\rho}\widehat{\mathbf{q}'}) + \left[\rho(\mathbf{q} - \phi\mathbf{u})\right]\big|_{F_2} \cdot \nabla F_2$$
$$- \left[\rho(\mathbf{q} - \phi\mathbf{u})\right]\big|_{F_1} \cdot \nabla F_1 - B\widehat{\Gamma''} = 0. \tag{5.3.25}$$

in which we note the use of the specific discharge, $\widehat{\mathbf{q}'}$, rather than the specific discharge relative to the moving solid, $\widehat{\mathbf{q}'_r}$.

Let us now rewrite the last balance equation for specific types of aquifer.

A. Confined Aquifer

Both the upper and lower bounding surfaces are now impervious. On such boundaries, $\mathbf{u} \cdot \nabla F = \mathbf{V}_s \cdot \nabla F$; hence, two conditions prevail: (5.3.19) and a similar one for $F_1 = 0$. By inserting these conditions into (5.3.25), we obtain:

$$\frac{\partial (B\widehat{\phi\rho})}{\partial t} + \nabla' \cdot (B\widehat{\rho}\widehat{\mathbf{q}'}) - B\widehat{\Gamma''} = 0. \tag{5.3.26}$$

This is the *(integrated) balance equation for flow in a confined aquifer*. Pumping and artificial recharge may serve as examples of distributed sinks and sources expressed by $B\widehat{\Gamma''}$.

In order to express the mass balance equation (5.3.26), which applies to a confined aquifer with a compressible fluid, in terms of a single state variable, $\hat{h}$, we make use of (5.3.12), and introduce the approximations:

$$\frac{\partial (B\widehat{\phi\rho})}{\partial t} \approx B\frac{\partial (\widehat{\phi\rho})}{\partial t} \approx \widehat{\rho} B S_o \frac{\partial \widehat{h^*}}{\partial t}, \tag{5.3.27}$$

$$(\phi\rho)\big|_{F_1} \approx (\phi\rho)\big|_{F_2} \approx \widehat{\phi\rho}, \tag{5.3.28}$$

where h^* is Hubbert's potential defined by (4.1.7), and neglect averages of products of fluctuations over the thickness. We also assume that:

$$h^*\big|_{F_1} \approx h^*\big|_{F_2} \approx \widehat{h^*} \approx \widehat{h} \equiv h, \tag{5.3.29}$$

$$\nabla' \cdot (B\widehat{\rho}\widehat{\mathbf{q}'}) \approx -\nabla' \cdot (\widehat{\rho} B\widehat{\mathbf{K}'} \cdot \nabla'\widehat{h^*}), \tag{5.3.30}$$

$$\mathbf{q} \approx \mathbf{q}_r, \tag{5.3.31}$$

and:

$$\left| B\widehat{\phi}\frac{\partial\widehat{\rho}}{\partial t} \right| \gg \left| B\widehat{\mathbf{q}'} \cdot \nabla'\widehat{\rho} \right|. \tag{5.3.32}$$

These approximations lead to the averaged mass balance equation

$$\widehat{\rho} S\frac{\partial h}{\partial t} = \nabla' \cdot (\widehat{\rho}\mathbf{T} \cdot \nabla' h) + B\widehat{\Gamma''}, \quad h \equiv \hat{h}, \tag{5.3.33}$$

in which the *aquifer storativity* or storage coefficient, S, is:

$$S = \int_{b_1}^{b_2} S_o \, dz, \qquad S = B\widehat{S_o}, \tag{5.3.34}$$

where $\widehat{S_o}$ represents the average value of S_o along B. In words, S is defined as the *volume of water release from (or added to) storage in a confined aquifer per unit area of aquifer per unit decline (or rise) of the piezometric head.*

The term $-\mathbf{T} \cdot \nabla' \widehat{h}$, which we denote as $\mathbf{Q}'$, expresses the horizontal discharge through the entire thickness of the aquifer per unit width, with

$$\mathbf{T} = \mathbf{T}(x, y) = \int_{b_1}^{b_2} \mathbf{K} \, dz \equiv B\widehat{\mathbf{K}}' \tag{5.3.35}$$

denoting *aquifer transmissivity*. It may be interesting to note that for a *stratified aquifer*, say an aquifer composed of N layers, B_i, K_i, $\widehat{\mathbf{K}}$ is the harmonic mean hydraulic conductivity of the layers, and the transmissivity, obtained by integration, takes the form $\mathbf{T} = \sum_N \mathbf{K}_i B_i$. Note that we have assumed above that the layers are anisotropic, but with x, y, z principal directions. The resulting transmissivity, $\mathbf{T}$, is then also anisotropic (in 2-d), with x, y as principal directions.

In the *Dupuit assumption*, equipotentials are approximated as vertical, i.e., $\widehat{h}(x, y) \approx h(x, y, b_1) \approx h(x, y, b_2)$, so that:

$$\mathbf{Q}' = -\mathbf{T} \cdot \nabla' h, \qquad \mathbf{T} = \widehat{\mathbf{K}}(x, y) B(x, y). \tag{5.3.36}$$

with $\mathbf{Q}$, $\mathbf{T}$ and $h \equiv \hat{h}(x, y, t)$.

In (5.3.33), $B\widehat{\Gamma''}/\widehat{\rho}$ denotes a water source (dims. $L^3/T/L^2$). For wells, we express the sources by using the *Dirac delta function*, $\delta(x - x^i, y - y^i)$, modified to two dimensions from the three-dimensional definition (5.1.4).

In groundwater hydrology, it is often assumed that $\widehat{\rho} =$ const. in (5.3.33). Thus, for water of constant density, with pumping wells at points (x^m, y^m), with pumping rates, P^m, we obtain the confined aquifer flow equation:

$$S\frac{\partial h}{\partial t} = \nabla' \cdot (\mathbf{T} \cdot \nabla' h) - \sum_{(m)} P^m \delta(x - x^m, y - y^m), \quad h \equiv \hat{h}. \tag{5.3.37}$$

B. Leaky Aquifer

In this case, (5.3.20) serves as a boundary condition on the upper bounding surface (a similar expression can be written for the lower surface). We assume that the upper semipervious boundary can be approximated as a thin membrane through which water leaks out of the aquifer into an overlying aquifer. Then, the rate of leakage through the upper surface of the aquifer can be expressed by:

$$\mathbf{q}_{\text{leak}}\big|_{F_2} \cdot \nabla F_2 = K_\ell \frac{\widehat{h} - h\big|_{\text{ext}}}{B_\ell} |\nabla F_2|, \tag{5.3.38}$$

where $h\big|_{ext}$ denotes the piezometric head above the upper semipervious layer, and K_ℓ and B_ℓ denote the hydraulic conductivity and thickness of the semipervious layer. This term is sometimes called *leakance*. The term B_ℓ/K_ℓ is referred to as the *resistance* (dims. T) of the semipervious layer. In writing (5.3.38), we have also assumed that water density is constant, and is, thus, the same on both sides of the semipervious layer.

We now use (5.3.13), making the same approximations used to derive (5.3.33) and (5.3.36) for the confined aquifer, defining $q_{leak}\big|_{F_i} \equiv \mathbf{q}_{leak}\big|_{F_i} \cdot \nabla F_i \; i = 1, 2$, to obtain the following relationships from (5.3.20):

$$\mathbf{q}\big|_{F_1} \cdot \frac{\nabla F_1}{|\nabla F_1|} = q_{leak}\big|_{F_1} \quad \text{and} \quad \mathbf{q}\big|_{F_2} \cdot \frac{\nabla F_2}{|\nabla F_2|} = q_{leak}\big|_{F_2},$$

which represent the conditions on the top and the bottom boundaries. With $R(x, y, t) \equiv B\widehat{\Gamma''}/\rho$, we then obtain:

$$\nabla' \cdot (T \cdot \nabla'\widehat{h}) + R(x, y, t) - q_{leak}\big|_{F_2}|\nabla F_2| + q_{leak}\big|_{F_1}|\nabla F_1| = \widehat{S}\frac{\partial \widehat{h}}{\partial t}. \qquad (5.3.39)$$

This is the *(integrated) mass balance equation for a leaky aquifer*. The leakage terms express (possible) loss of water to the overlying and underlying aquifers.

C. Phreatic Aquifer

Let the lower boundary be impervious, so that (5.3.19) is valid there. Then, using (5.3.21) with (5.2.3), and employing the same assumptions as introduced above, including constant water density, (5.3.25) becomes:

$$\frac{\partial (B\widehat{\phi})}{\partial t} + \nabla' \cdot (B\widehat{\mathbf{q}'}) + \left(\mathbf{N} \cdot \nabla F_2 + \theta_{rw}\frac{\partial F_2}{\partial t}\right) - B\frac{\widehat{\Gamma''}}{\rho} = 0, \qquad (5.3.40)$$

where we have assumed $b_2 \approx \hat{h}$ so that $B = \hat{h} - b_1$. This is the *(integrated) mass balance equation for a phreatic aquifer*.

Assuming that $|B(\partial\widehat{\phi}/\partial t)| \ll |(\phi|_{\hat{h}}\, \partial\hat{h}/\partial t)|$, and since $F_1 = z - b_1(x, y)$, and $F_2 = z - \hat{h}(x, y, t)$, the first term of the left-hand side of (5.3.40) becomes:

$$\begin{aligned}\frac{\partial B\widehat{\phi}}{\partial t} &\equiv \frac{\partial}{\partial t}\int_{b_1}^{\widehat{h}} \phi \, dz = \int_{b_1}^{\widehat{h}} \frac{\partial \phi}{\partial t}\, dz + \phi\Big|_{\widehat{h}}\frac{\partial \widehat{h}}{\partial t} \\ &= B\frac{\partial \widehat{\phi}}{\partial t} + \phi\Big|_{\widehat{h}}\frac{\partial \widehat{h}}{\partial t} \approx \phi\Big|_{\widehat{h}}\frac{\partial \widehat{h}}{\partial t}. \end{aligned} \qquad (5.3.41)$$

With $\boldsymbol{\nu} = -N\nabla z$, by inserting (5.3.41) into (5.3.40), we now obtain:

$$S_y\frac{\partial \widehat{h}}{\partial t} + \nabla' \cdot [(\widehat{h} - b_1)\widehat{\mathbf{q}'}] - N - R + P = 0, \qquad (5.3.42)$$

Fig. 5.10 Relation between specific yield and grain size (from Conkling 1934, as modified by Davis and de Wiest 1966)

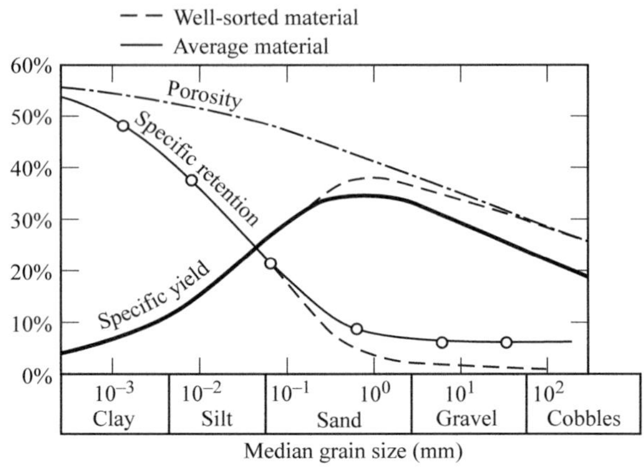

in which we have assumed that $\nabla F_2 \approx -\nabla z$, i.e., the water table is approximately horizontal with respect to **N**, and the net withdrawal from the aquifer, i.e., pumping minus artificial recharge, is denoted by $P - R \equiv -(\widehat{h} - b_1)\widehat{\Gamma''}/\rho_w$.

The symbol $S_y \equiv \phi_{\text{eff}} = \phi - \theta_{rw}$ is the *specific yield*, which is equivalent to the storativity of a phreatic aquifer. It is defined as (Fig. 6.5) the *volume of water, $\Delta\mathcal{U}_w$, released from storage in a phreatic aquifer, per unit area and per unit decline in the water table elevation*:

$$S_y = \frac{\Delta\mathcal{U}_w}{\mathcal{A}\Delta h}. \tag{5.3.43}$$

In this definition of S_y, we have assumed that $(\widehat{h} - b_1)S_o \ll \phi_{\text{eff}}$, i.e., the effect of elastic storativity is negligible.

Figure 5.10 shows the dependence of the specific yield, S_y, on grain- (actually pore-) size. We note that for clay, S_y is very small, although the porosity is relatively large. This behavior stems from the fact that the size of the pores in clays is very small, so that capillary forces are very large, and so is the irreducible moisture content, θ_{wr}.

Equation (5.3.42) is the balance equation commonly employed for a phreatic aquifer. It is often referred to by groundwater hydrologists as the *Boussinesq equation*. In (5.3.42), $\mathbf{Q}' (\equiv (\widehat{h} - b_1)\widehat{\mathbf{q}'})$ denotes the total discharge through the saturated thickness, $\widehat{h} - b_1$, per unit width of aquifer.

Let us use vertical integration to derive an expression for $\mathbf{Q}'$ in terms of $\widetilde{h}$. Assuming that $\mathbf{K} = \mathbf{K}(x, y)$, we obtain:

$$\begin{aligned}\mathbf{Q}' &= \int_{b_1(x,y)}^{\widehat{h}(x,y,t)} \mathbf{q}\, dz = -\mathbf{K} \cdot \int_{b_1}^{\widehat{h}} \nabla h\, dz \\ &= -\mathbf{K} \cdot \left\{ \nabla' \left[(\hat{h} - b_1)\hat{h} \right] - h\Big|_{x,y,z=\widehat{h}} \nabla' \widehat{h} + \widehat{h}\Big|_{x,y,z=b_1} \nabla' b_1 \right\},\end{aligned} \tag{5.3.44}$$

where $h\Big|_{(x,y,z=h)} = \widehat{h}$.

By invoking now the Dupuit assumption, i.e., $\widehat{h} \approx h\big|_{(x,y,z=h)} \approx h\big|_{(x,y,z=b_1)}$, we obtain:

$$\mathbf{Q}' = -(\widehat{h} - b_1)\mathbf{K} \cdot \nabla' \widehat{h}. \tag{5.3.45}$$

Equation (5.3.42) can then be written as:

$$S_y \frac{\partial \widehat{h}}{\partial t} + \nabla \cdot [(\widehat{h} - b_1)\mathbf{K} \cdot \nabla' \widehat{h}] - N - R + P = 0. \tag{5.3.46}$$

When recharge and pumping is implemented through wells ($\equiv$ point sources and sinks), we may, symbolically, use the notation:

$$\begin{aligned} &R(x, y, t) - P(x, y, t) \equiv \\ &\quad \sum_{(i)} R^i(t)\delta(x - x^i, y - y^i) - \sum_{(j)} P^j(t)\delta(x - x^j, y - y^j), \end{aligned} \tag{5.3.47}$$

where $\delta(x - x^i, y - y^i)$ is the *Dirac delta function* at (x^i, y^i).

In (5.3.46), the product $(\hat{h} - b_1)\mathbf{K}$ plays the role of *transmissivity* of a phreatic aquifer. However, here the transmissivity may be time dependent, because $\tilde{h} = \hat{h}(x, y, t)$. As a result, the equation for flow in a phreatic aquifer is *nonlinear*.

In principle, the non-linear balance equation (5.3.46) can be solved numerically. However, often this equation is approximated by linearization prior to being solved numerically. Commonly, linearization is achieved by replacing $\widehat{h}$ in the product $(\widehat{h} - b_1)\mathbf{K}$, which represents the aquifer transmissivity, by some mean (in time!) value $\widehat{\widehat{h}}$, assuming $|\widehat{h} - \widehat{\widehat{h}}| \ll \widehat{\widehat{h}}$. Equation (5.3.46) then becomes

$$S_y \frac{\partial \widehat{h}}{\partial t} + \nabla \cdot [(\widehat{\widehat{h}} - b_1)\mathbf{K} \cdot \nabla' \widehat{h}] - N - R + P = 0, \tag{5.3.48}$$

which is now linear in $\widehat{h} = \widehat{h}(x, y, t)$. The introduction of an average thickness of the saturated zone is justified whenever fluctuations in the water table elevations are much smaller than the thickness itself.

5.3.2 Initial and Boundary Conditions

As in the 3-dimensional case discussed above, to present a well-posed problem in a 2-d domain, we need to specify appropriate initial and boundary conditions for every flow equation. We recall that we are considering here flow equations based on the *essentially horizontal flow* assumption. Hence, the flow domain, $\Omega(x, y)$, which is in the horizontal plane, is bounded by a *closed boundary* Ω, composed of curved and straight line segments, Ω_i, with $F_i = F_i(x, y) = 0$ representing the equation of the i-th segment. Note that only stationary boundaries are considered here. We shall regard the piezometric head, $h = h(x, y, t)$, as the variable in the flow equation.

A. Initial Conditions

Initial conditions take the form:

$$h = h(x, y, t) = f(x, y), \quad \text{on} \quad \Omega, \tag{5.3.49}$$

where $f = f(x, y)$ is a known function.

B. Boundary Conditions

Several types of boundary conditions may be encountered:
(a) *Boundary of prescribed piezometric head.* This condition takes the form:

$$h = f_1(x, y, t), \quad \text{on} \quad \Omega_1, \tag{5.3.50}$$

where $f_1 = f_1(x, y, t)$ is a known function.
(b) *Boundary of prescribed flux.* Along such a boundary:

$$Q'_\nu = \mathbf{Q}' \cdot \boldsymbol{\nu} = f_2(x, y, t), \quad \text{on} \quad \Omega_2, \tag{5.3.51}$$

where $f_2 = f_2(x, y, t)$ is a known function.
(c) *Semi-pervious Boundary.* As in three-dimensional flow, this kind of boundary condition occurs, for example, when a partly clogged river-bed (e.g., by a thin layer of silt or clay) serves as a boundary of a flow domain. Because of the resistance to the flow offered by the semi-pervious layer, the piezometric head in the aquifer, next to this layer, is different from that on its external side.

With h_o denoting the piezometric head on the external side of the boundary, continuity of flux through the entire thickness of the aquifer requires that:

$$(Q'_n \equiv \mathbf{Q}' \cdot \mathbf{n} =) - (\mathbf{K}h \cdot \nabla' h) \cdot \mathbf{n} = h \frac{h_o - h(x, y, t)}{c_r}, \tag{5.3.52}$$

where $c_r = B'/K'$, with K' and B' denote the hydraulic conductivity and the thickness of the semi-pervious layer, respectively.

C. Complete Model Statement

The standard content of a two-dimensional flow model include the following items:

- Delineation of the closed curve that bounds the problem domain. This means that all boundaries are really vertical surfaces extending through the entire thickness of the aquifer.
- Specification of the state variable, usually the average piezometric head h, for which a solution is sought.

- Statement of the partial differential equation that represents the mass balance of water in the aquifer. In the case of multiple leaky aquifers, we need a variable of state and a mass balance equation for each aquifer.
- Specification of all the (storage and transport) coefficients that appear in the balance equation(s).
- Statement of initial conditions that the state variables have to satisfy.
- Statement of boundary conditions for each balance equation.

5.4 Heterogeneity and Monte Carlo Simulations

In Sect. 1.1.6, we discussed the conditions that have to be satisfied in order to justify the modeling of phenomena of transport in a geological formation regarded as a continuum. However, we have also emphasized that *most geological formations*, e.g., petroleum reservoirs, groundwater aquifers and geothermal reservoirs, *are highly heterogeneous* to the extent that it is not obvious that an REV, or an RMV can be identified. In fact, even if an REV can be found, or *assumed to exist*, the spatial variability of coefficients, like porosity and permeability, obtained from cores, from a pumping test, or by some inverse technique, is significant and very irregular. Usually, there is not enough information concerning their spatial variability. Interpolation does not solve the problem as it leaves a high level of uncertainty as to the values of the considered coefficients between points at which actual data are available.

In spite of the above severe conclusion concerning the lack of sufficient information about coefficients, in the real world, decisions (based on predicting excitation-response relations) have still to be made with respect to flow and transport in such (highly heterogenous) formations. This led to efforts to find techniques for coping with the lack of sufficient information concerning model coefficients, obviously at the cost of a certain degree of uncertainty in the predicted values.

In fact, the above introductory remarks focus on the assumption that the basic mass balance equation and Darcy's law remain unchanged, with the same set of coefficients, e.g., S and T for an aquifer, but that we have insufficient information concerning the spatial distribution of the formation's coefficients. However, this is not necessarily true, as strong heterogeneity may lead to additional phenomena that are not represented by the set of coefficients that correspond to a homogeneous, or a slightly heterogeneous formation. The term "upscaling" is often used in the literature to describe modeling of transport in highly heterogeneous formation.

In the current section, we shall introduce a method–Monte carlo Simulations–often used to cope with the lack of information on the spatial distribution of coefficients in heterogeneous domains.

We introduce the Monte Carlo method by presenting the case of flow through a heterogenous isotropic confined aquifer described by the mass balance equation (5.3.37), rewritten here in the form:

$$S\frac{\partial h}{\partial t} = \nabla' \cdot (T \cdot \nabla' h) - R(x, y, t), \quad h = h(x, y, t), \tag{5.4.1}$$

where $h = h(x, y, t)$ is the piezometric head, ∇' denotes the gradient operator in the 2-d (x, y) domain, $R(x, y, t)$ denotes sources, and $S(x, y)$ and $T(x, y)$ denote the aquifer's transmissivity and storativity, respectively. These two coefficients, S and T, are known at a small number of points throughout the aquifer, say from pumping tests. Thus, using interpolation to derive values for all points throughout the domain is an unacceptable approach.

The Monte Carlo method deals with the uncertainty concerning the values coefficients (here, S and T) as a probability issue. It uses the available measured (thus, assumed reliable) information on these coefficients to produce their statistical characteristics (e.g., mean, standard deviation, covariance), and then uses the latter to produce a large number of *realizations, each of which is a possible manifestation of the unknown reality*. All these realizations obey the same statistical characteristics as the actually measured information. Each of these (equally likely to occur) realizations, is used as input to the considered (deterministic) model, producing one possible model prediction as output. In this way, a large number of simulations is conducted, each making use (as input) of one of the realizations of the spatial distribution of model coefficients. Each of the produced outputs contains detailed information on the distributions of the sought model variables, say, $h(x, y, t)$. It is a sample taken from the ensemble space. The ensemble statistics can then be applied to this information. In this way, instead of a single deterministic prediction, say $h(x, y, t)$, obtained by solving the given mathematical model with known coefficients, we obtain *many* solutions, one for each realization of the coefficients' distributions. From them, we obtain the statistical characterization of the solution. For example, for a specified location and a specified point in time, we obtain the mean value of h, its standard deviation, as well as its correlation length and correlation time. The correlation length tells us the persistence of a value in a spatial direction and in time. In fact, the Monte Carlo method produces much more than merely statistical information. The large number of (artificially produced) samples allows us to construct a histogram of any output prediction and provides the probability distribution of that quantity.

The Monte carlo method requires the (usually numerical) solution of a very large number, often many thousands of cases, each with a different set of the considered coefficients. However, with the power of computing nowadays, this is not a serious obstacle.

By applying a statistical analysis to these many 'equally likely to occur' outcomes, we can provide quantitative, albeit probabilistic, answers to questions like, 'what is the probability that a water table elevation, h, at a certain location at a certain time will be below or above a certain specified value, in response to certain specified values of natural replenishment, R'. Thus, the results take the form of a probability distribution for h at a desired location. We have to ensure that we run a sufficient number of cases so as to ensure convergence of the results. Even then, convergence is not always ensured.

Although the Monte Carlo procedure as described above seems simple and straightforward, it is actually not. Dagan (1989, p. 182) and Bear and Cheng (2010, p. 668) discuss some of the difficulties associated with the implementation of this technique.

Let us discuss some issues associated with it:

(1) To generate the many realizations of the spatial distribution of transmissivity and storativity, we need to determine their probability distributions, i.e., their *probability density functions* (abbreviated: pdf). In most fields, however, the amount of actually measured data is insufficient for such purpose. Fortunately, based on a few studies in which a large quantity of data was indeed available for a given site, or could be compiled from different similar sites, it was concluded that the hydraulic conductivity is *log-normally* distributed, $Y = \ln T$ (e.g., Freeze 1975; Hoeksema and Kitanidis 1985a, b; Gelhar 1986, 1993). The same conclusion may be extended to transmissivity and storativity. With this assumption, the full determination of their pdf's is reduced to determining only two statistical moments: the mean and the standard deviation of the required distributions.

(2) To determine these statistical moments of the pdf, a large number of samples in the ensemble space is needed. Unfortunately, such ensemble space does not exist in groundwater modeling, as each investigated hydrogeological field is unique. This obstacle is overcome by making the assumption of *ergodicity*. The ergodicity hypothesis allows us to conduct the spatial statistical analysis by using data collected at different locations of the same field, and using the results as the sought ensemble statistics. However, as emphasized earlier, the amount of spatial data is often insufficient for producing a statistical model with a spatial trend. In other words, we may not be able to reliably obtain a mean that varies from location to location. Most likely, we can only obtain a statistically homogeneous mean. This means that a constant mean is obtained for the entire field.

(3) For the second moment, e.g., the covariance, the amount of data is usually insufficient for determining its spatial trend. Hence, the covariance is almost always assumed to be statistically homogeneous; in other words, we assume that the variance and the correlation length are everywhere the same. Due to the lack of sufficient data for the construction of a detailed *empirical statistical model*, the simulated results are often questionable.

(4) The generation of random realizations, required in the Monte Carlo simulation, calls for the generation of a sequence of random numbers. For this purpose, we use a computerized *pseudo random number generator*. This is based on a mathematical algorithm, programmed for a computer, that can generate a seemingly random sequence of numbers, say, between 0 and 1, with a certain precision. Actually, the process is only *pseudo-random*, because, given the same 'seed number', the same sequence of numbers will be generated every time. Using an algebraic transformation, meaning replacing one variable by another, defined by a functional relation, this sequence of random numbers, can be mapped onto a Gaussian (normal) probability distribution. With another transformation, the normal probability distribution can be mapped onto a log-normal probability distribution (Press et al. 2007).

With the random number generator and a probability distribution, the random fields (of parameters, such as storativity and transmissivity) required as input for the Monte Carlo simulations of the considered mathematical model can now be generated. We divide the modeled inhomogeneous domain into a number of small cells, each assumed to be homogeneous. Selecting one cell to start from, we can 'randomly' assign to it a parameter value, say, the value of transmissivity. We then move to the next cell and assign to it another random transmissivity value. We continue this process, until the entire transmissivity field is defined. To obtain these transmissivity values, we start by inserting a 'seed number' (between zero and one) into a random number generator. The random number generator then produces a sequence of pseudo-random (i.e., almost random) numbers, uniformly distributed between zero and one. Based on the assumed pdf (for example, log-normal distribution), and the provided mean and standard deviation, these numbers are then transformed into the sought transmissivity values. Note that by conducting a statistical analysis on these random transmissivity values, we should return to the same mean and standard deviation originally used for their generation. These transmissivity values can then be assigned, cell by cell, as described above. This process, however, generates an *incoherent* random parameter field that does not exhibit a spatial correlation, as a real field should. The value selected for one cell bears no correlation with those assigned to neighboring cells, and the resulting field is rougher than it should be. In fact, the smaller the cell, the rougher is the field. In reality, however, if the value in one cell is higher than average, there is a high probability that the values at neighboring cells will also be higher than the average. Thus, the values at neighboring cells should be *conditionally* generated, on the basis of known information concerning the covariance.

(5) Once the transmissivity has been generated for all cells, we need to generate the spatial distribution of storativity. If we allow the random number generator to arbitrarily select a storativity value for individual cells, that value may not be physically feasible, as it has been observed that a large transmissivity is often associated with a low storativity (Freeze 1975; Dagan 1979). The conclusion is that the selection of transmissivity and storativity must be based on a *joint probability density function* for these two parameters. This function can be obtained by compiling the storativity and transmissivity data pairs on a histogram that sorts these values into ranges.

The above discussion (in which we have used transmissivity and storativity just as examples of two porous medium parameters) indicates that the generation of a multidimensional, multivariate, and spatially correlated random field of the relevant model parameters, to be used as input in a Monte Carlo simulation, is a complicated process; special algorithms are required (e.g., Mantoglou and Wilson 1982; Mantoglou and Gelhar 1987; Tompson et al. 1989; Tompson and Gelhar 1990; Robin et al. 1993; Bellin and Rubin 1996).

Although we have presented here the Monte Carlo method for the solution of the aquifer flow equation, it can be applied also to solute transport and to any problem of transport.

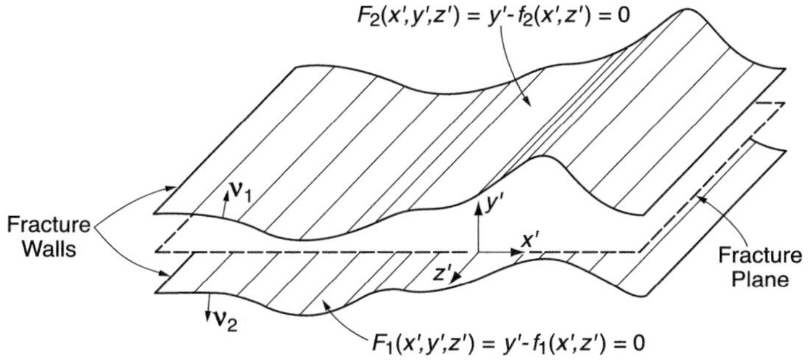

Fig. 5.11 Nomenclature for a fracture geometry

5.5 Flow in Fractured Domains

Fractured rock domains and fractured porous rock formations were introduced in Sect. 1.3. In many parts of the world, such formations constitute important fresh water aquifers, petroleum reservoirs and sources of geothermal energy. In Sect. 1.3.4 we have mentioned fracking and acidizing as techniques that are employed to increase the permeability of fractured rocks, thus enhancing the production of oil and gas.

In principle, as long as an REV can be identified for a fractured domain, the latter can be regarded as a porous medium domain and the mass, momentum and energy transport models presented in this book are applicable, at least in principle. The objective of this section is to highlight certain features that are typical to fractured domains.

5.5.1 Flow in a Single and Multiple Fractures

As already emphasized in Sect. 4.2.4, Case A, the flux equation in the form of Darcy's law can be obtained as a simplified version of the Navier–Stokes (momentum balance) equation. This will be our starting point for expressing the flux of a fluid in a fracture. We shall start by introducing some definitions related to the geometry of a fracture. The discussion will be confined to a non-deformable fracture, with a *fracture aperture* that may vary in space in such a way that a fracture axis surface can be defined throughout the fracture. To obtain this surface, we consider a point on one of the walls that bound the fracture and determine the largest sphere that (1) can be placed inside the fracture, and (2) is tangent to the wall at that point. The centroid of this sphere is a point on the axis surface. We repeat this procedure for all points on the selected wall. The centroid of all spheres form the *axis surface*. Often, we approximate this surface as a plane (or as composed, piecewise, of planar segments). Figure 5.11 shows a single fracture with a variable aperture.

At every point on a fracture's axis plane, an orthogonal coordinate system, x', y', z', can be defined, with x', z' denoting a point in the axis plane, and y' is the coordinate normal to the axis plane. The stationary walls of the fracture can be described by the equations:

$$F_1(x', y', z') \equiv y'_1 - f_1(x', z') = 0, \quad F_2(x', y', z') \equiv y'_2 - f_1(x', z') = 0, \tag{5.5.1}$$

in which y'_1, y'_2 are the values of y' on the walls of the fracture. From the above equation it follows that the normal outwardly directed unit vector at a point on a fracture wall, is defined by:

$$\boldsymbol{\nu}_m = \frac{\nabla F_m}{|\nabla F_m|}, \qquad m = 1, 2, \tag{5.5.2}$$

and the fracture aperture, $b = b(x', z')$, at any point on the axis surface is expressed by:

$$b(x', z') = f_2(x', z') - f_1(x', z'). \tag{5.5.3}$$

A. Flow in a Single Fracture

Assuming negligible pressure variations across the fracture's width (based on the assumption that the latter is much smaller than its extent in the direction of the fracture surface), the point (i.e., microscopic) balance equations for the fluid's mass and linear momentum can be averaged over the fracture width to produce averaged (or integrated) equations for two-dimensional flow in the fracture plane (see Sect. 5.3.1).

The three-dimensional balance equation for the linear momentum of an incompressible constant density fluid in a fracture, when combined with the mass balance equation, takes the form:

$$\rho \frac{\partial \mathbf{V}}{\partial t} = -\rho \nabla \cdot (\mathbf{V}\mathbf{V}) + \mu \nabla^2 \mathbf{V} - \nabla p + \rho \mathbf{g}, \tag{5.5.4}$$

where ρ and μ denote the fluid's density and dynamic viscosity, respectively, p is pressure, $\mathbf{V}$ is the fluid's velocity, t denotes time, and $\mathbf{g} = -g\nabla z$ denotes gravity acceleration.

For a constant density, we can use the piezometric head, $h = z + p/\rho g$ as the state variable, so that (5.5.4) can be rewritten in the form:

$$\rho \frac{\partial \mathbf{V}}{\partial t} = -\rho \nabla \cdot (\mathbf{V}\mathbf{V}) + \mu \nabla^2 \mathbf{V} - \rho g \nabla h. \tag{5.5.5}$$

Our next step is to average (5.5.5) over the width of the aperture, b, normal to the latter's axis. With the nomenclature of Fig. 5.11, we write:

$$\int_{f_1(x',z')}^{f_2(x',z')} \left[\frac{\partial \rho}{\partial t} + \rho(\mathbf{V}\mathbf{V}) + \rho g \nabla h - \mu \nabla^2 \mathbf{V} \right] dy' = 0. \tag{5.5.6}$$

Making use of Leibnitz rules (5.3.9) and (5.3.12), we obtain (Bear 1993, p. 13):

$$\begin{aligned}\rho\frac{\partial b\widetilde{\mathbf{V}}}{\partial t} &+ \rho\nabla'\cdot\left(b\widetilde{\mathbf{V}}\widetilde{\mathbf{V}}\right) + \rho\nabla'\cdot\left(b\widetilde{\mathring{\mathbf{V}}\mathring{\mathbf{V}}}\right)\\ &- \rho\mathbf{V}\mathbf{V}|_{f_2}\cdot\nabla F_2 + \rho\mathbf{V}\mathbf{V}|_{f_1}\cdot\nabla F_1\\ &+ \rho g\nabla'(b\tilde{h}) - \rho gh|_{f_2}\nabla F_2 + \rho gh|_{f_1}\nabla F_1\\ &+ \mu\nabla'^2(b\widetilde{\mathbf{V}}) - \mu\nabla'\cdot(\mathbf{V}|_{f_2}\nabla F_2 - \mathbf{V}|_{f_1}\nabla F_1)\\ &- \mu(\nabla\mathbf{V}|_{f_2}\cdot\nabla F_2 - \nabla\mathbf{V}|_{f_1}\cdot\nabla F_1) = 0, \end{aligned} \tag{5.5.7}$$

In the above equation, ∇' involves differentiation only with respect to coordinates lying in the fracture (x', z')-plane, and we have used:

$$\widetilde{A}(x', z') \equiv \frac{1}{b}\int_{f_1}^{f_2} A(x', y', z')dy', \tag{5.5.8}$$

with $\mathbf{V}(x', y', z') = \widetilde{\mathbf{V}}(x', z') + \mathring{\mathbf{V}}(x', y', z')$, so that

$$\widetilde{\mathbf{V}\mathbf{V}} = \widetilde{\mathbf{V}}\widetilde{\mathbf{V}} + \widetilde{\mathring{\mathbf{V}}\mathring{\mathbf{V}}}, \quad \mathbf{V}(x', y'z') \equiv \widetilde{\mathbf{V}}(x', z') + \mathring{\mathbf{V}}(x', y', z'), \quad \widetilde{\mathring{\mathbf{V}}} \equiv 0. \tag{5.5.9}$$

In the absence of sources and sinks, the microscopic level mass balance equation at a point in a fracture is:

$$\frac{\partial\rho}{\partial t} = -\nabla\cdot(\rho\mathbf{V}). \tag{5.5.10}$$

By integrating over the aperture,

$$\int_{f_1(x',z')}^{f_2(x',z')} \left[\frac{\partial\rho}{\partial t} + \nabla\cdot(\rho\mathbf{V})\right] dy' = 0, \tag{5.5.11}$$

using Leibnitz rules (5.3.9) and (5.3.12), we obtain the averaged mass balance equation:

$$\frac{\partial b\widetilde{\rho}}{\partial t} + \nabla'\cdot(b\widetilde{\rho\mathbf{V}}) - (\rho\mathbf{V})|_{f_2}\nabla F_2 + (\rho\mathbf{V})|_{f_1}\nabla F_1 = 0. \tag{5.5.12}$$

With a constant density and stationary, non-deformable fracture walls, the above equation reduces to:

$$\nabla'\cdot(b\tilde{\rho}) = \mathbf{V}|_{f_2}\cdot\nabla F_2 - \mathbf{V}|_{f_1}\cdot\nabla F_1. \tag{5.5.13}$$

Substituting (5.5.13) into (5.5.7), we obtain:

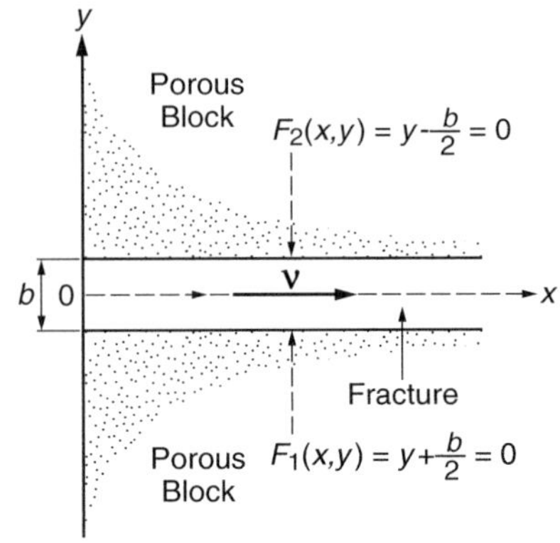

Fig. 5.12 Fracture-porous block geometry in a one-dimensional case

$$
\begin{aligned}
\rho\frac{\partial b\widetilde{\mathbf{V}}}{\partial t} &= -\rho b\widetilde{\mathbf{V}}(\nabla'\cdot\widetilde{\mathbf{V}}) - \rho\nabla'\cdot(b\widetilde{\mathring{\mathbf{V}}\mathring{\mathbf{V}}})\\
&\quad+\rho\mathbf{V}\mathbf{V}|_{f_2}\cdot\nabla F_2 - \rho\mathbf{V}\mathbf{V}|_{f_1}\cdot\nabla F_1\\
&\quad-\rho g\nabla'(b\tilde{h}) + \rho gh|_{f_2}\nabla F_2 - \rho gh|_{f_1}\nabla F_1\\
&\quad-\mu\nabla'^2(b\widetilde{\mathbf{V}}) + \mu\nabla'\cdot(\mathbf{V}|_{f_2}\nabla F_2 - \mathbf{V}|_{f_1}\nabla F_1)\\
&\quad+\mu(\nabla\mathbf{V}|_{f_2}\cdot\nabla F_2 - \nabla\mathbf{V}_{f_1}\cdot\nabla F_1)\\
&\quad-\rho\widetilde{\mathbf{V}}(\mathbf{V}|_{f_2} - \mathbf{V}|_{f_1}\nabla F_1).
\end{aligned}
\tag{5.5.14}
$$

To analyze (5.5.14), we consider the simple case of steady, unidirectional flow through a two-dimensional fracture bounded by the planar, parallel walls defined in Fig. 5.12.

Furthermore, we assume that (1) the dispersive momentum flux is much smaller that the advective one, i.e., $|\rho\widetilde{\mathring{\mathbf{V}}\mathring{\mathbf{V}}}| \ll |\rho\widetilde{\mathbf{V}}\widetilde{\mathbf{V}}|$, and (2) across any aperture, the piezometric heads at the fracture walls satisfy $h|_{f_1} \simeq h|_{f_2}$. The stronger condition $h|_{f_1} \simeq h|_{f_2} \simeq \tilde{h}$ is required when the fracture walls are not assumed parallel. Under these assumptions, and for steady flow, (5.5.14) reduces for the x-direction to:

$$
\rho gb\frac{\partial\tilde{h}}{\partial x} = \mu\left(\left.\frac{\partial V_x}{\partial y}\right|_{f_2} - \left.\frac{\partial V_x}{\partial y}\right|_{f_1}\right). \tag{5.5.15}
$$

The assumption of steady flow is not required if from the outset, i.e., already at the microscopic level, we would have made the reasonable assumption that the inertial effects are negligible.

For fracture walls that are stationary and impervious, and under a *no-slip condition*, $\mathbf{V} = 0$, on the walls, the velocity distribution across the fracture width is *parabolic* and symmetric about the fracture's axis (Lamb 1945), with:

$$V_x(y) = \frac{6\widetilde{V}_x}{b^2}\left(\frac{b}{2}+y\right)\left(\frac{b}{2}-y\right), \quad -\frac{b}{2} \leq y \leq \frac{b}{2}. \tag{5.5.16}$$

By differentiating (5.5.16) and substituting the result into (5.5.15), we obtain the average velocity in a fracture:

$$\widetilde{V}_x = -\frac{\rho g}{\mu}\frac{b^2}{12}\frac{\partial h}{\partial x}, \tag{5.5.17}$$

and a similar expression in the z-direction. The above equation can also be written in the form:

$$\widetilde{V}_i = -K_{fr}\frac{\partial h}{\partial x_i}, \quad x_1 \equiv x, \quad x_2 \equiv z, \quad K_{fr} = \frac{\rho g}{\mu}\frac{b^2}{12} \equiv \frac{\rho g}{\mu}k_{fr}, \quad k_{fr} = \frac{b^2}{12}, \tag{5.5.18}$$

where K_{fr} and k_{fr} are referred to as the *hydraulic conductivity* and *permeability of a fracture*. Altogether, the discharge through an individual fracture is:

$$\mathbf{Q}'_{fr} = b\widetilde{\mathbf{V}} = -\frac{\rho g}{\mu}\frac{b^3}{12}\nabla'\tilde{h} = -T_{fr}\nabla'\tilde{h}, \tag{5.5.19}$$

in which the prime indicates a vector in the two dimension, x, z, and T_{fr} denotes the *transmissivity of a fracture*. The last equation is often referred to as the 'cubic law' (Witherspoon et al. 1980).

When fluid can pass from fracture to adjacent rock block, or from the rock block to the fracture, we have to modify the above analysis. Let us return to (5.5.14) and consider the case of fracture walls that are permeable. Then, with the other assumptions that led to (5.5.15), we obtain:

$$\begin{aligned}\rho g b\frac{\partial \tilde{h}}{\partial x} &= \mu\left(\left.\frac{\partial V_x}{\partial y}\right|_{f_2} - \left.\frac{\partial V_x}{\partial y}\right|_{f_1}\right) - \rho\widetilde{V}_x\left(V_y|_{f_2} - V_y|_{f_1}\right)\\ &\quad +\mu\frac{\partial}{\partial x}\left(V_y|_{f_2} - V_y|_{f_1}\right).\end{aligned} \tag{5.5.20}$$

Assuming that the leakage into or out of the fracture is equal on both fracture walls, and is uniform over the fracture length, i.e., $V_y|_{f_2} = V_y|_{f_1} = q_\ell$, Eq. (5.5.15) reduces to:

$$\rho g b\frac{\partial \tilde{h}}{\partial x} = \mu\left(\left.\frac{\partial V_x}{\partial y}\right|_{f_2} - \left.\frac{\partial V_x}{\partial y}\right|_{f_1}\right) - 2\rho\widetilde{V}_x q_\ell. \tag{5.5.21}$$

We now assume that the velocity distribution across a fracture is described by (5.5.16), and that the no-slip condition exists also in the case of flow in a fracture with leakage through the walls. In fact, this assumption is valid only when the flux through the fracture is much larger than the leakage through the fracture walls.

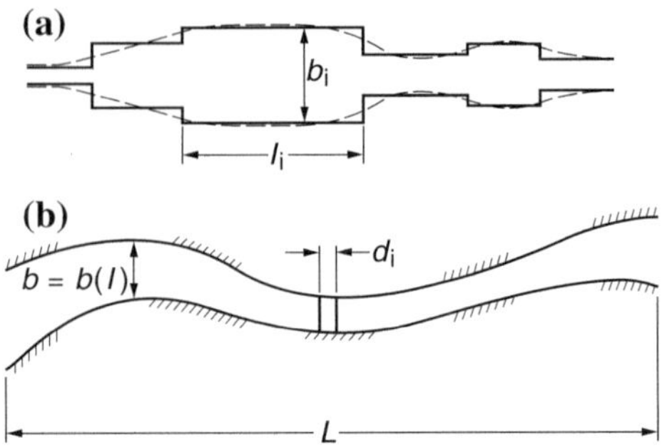

Fig. 5.13 Fracture of variable aperture: **a** Discrete aperture variation, and **b** Continuous aperture variation

Differentiating (5.5.16), and substituting the result into (5.5.21), yields a'modified' average velocity in the fracture:

$$\widetilde{V}_x = -\frac{\rho g b^2}{12\mu + 2b\rho q_\ell}\frac{\partial \tilde{h}}{\partial x}, \tag{5.5.22}$$

to be compared with (5.5.17). Thus, for a fracture imbedded in a porous host rock, the hydraulic conductivity in a fracture is also a function of the magnitude of the leakage through the fracture walls. When $\mu \gg b\rho q_\ell/6$, the permeability in the fracture can be approximated by $k_{fr} = b^2/12$.

In addition to the observation made above, that fracture walls are not smooth, the aperture of a fracture may vary in size. The ideal model of parallel plates does not exist in reality. Moreover, at points and areal segments within a fracture, the aperture may disappear altogether, as adjacent blocks come into direct contact.

Several authors have studied the effects of a variable aperture on the flow in a fracture. For a fracture approximated as a series of m discrete segments with different apertures (Fig. 5.13a), Wilson and Witherspoon (1974) expressed the 'effective aperture', b_{eff}, by:

$$b^3_{eff} = \frac{\sum_{n=1}^{n=m} \ell_n}{\sum_{n=1}^{n=m} (\ell_n/b^3_n)}, \tag{5.5.23}$$

in which ℓ_n denotes the length of a fracture segment of aperture b_n.

It is also possible to introduce a statistical distribution of apertures. Then:

$$T_{fr} = \frac{1}{12}\frac{\rho g}{\mu} L \int_0^\infty b^3 f(b) db, \tag{5.5.24}$$

where $f(b)$ is the frequency distribution of b. For flow parallel (Q'_1) and normal (Q'_2) to these changes, we have:

$$Q'_1 = -\frac{b^3_{eff}}{12}\frac{\rho g}{\mu}\frac{\partial h}{\partial x_1}, \qquad Q'_2 = -\frac{b^3_{eff}}{12}\frac{\rho g}{\mu}\frac{\partial h}{\partial x_2}, \tag{5.5.25}$$

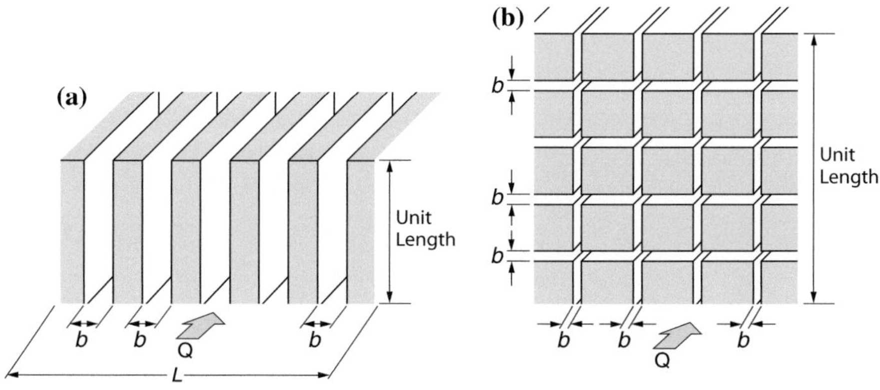

Fig. 5.14 Multiple fracture systems: **a** One family of parallel fractures, and **b** Two mutually orthogonal families of parallel fractures

B. Flow in an Ordered Fractured Rock Domain

With the result of the analysis presented above for a single fracture, it is possible to construct flow expressions for variety of systems composed of multiple parallel fractures. For example, for the case of parallelepiped fractured rock domain intersected by a single family of m fractures of equal aperture b (Fig. 5.14a), oriented parallel to the direction of flow, the total discharge Q and the specific discharge q, through a cross section of this domain, having a width L and a height of unit length normal to the flow direction, is:

$$\mathbf{Q} = -m\frac{b^3}{12}\frac{\rho g}{\mu}\nabla h, \quad \mathbf{q} \equiv \frac{\mathbf{Q}}{L} = -\frac{m}{L}\frac{b^3}{12}\frac{\rho g}{\mu}\nabla h \equiv -K_{fr}\nabla h, \tag{5.5.26}$$

$$K_{fr} = \frac{mb^3}{12L}\frac{\rho g}{\mu} = \phi_{fr}\frac{b^2}{12}\frac{\rho g}{\mu}, \quad \phi_{fr} = \frac{mb}{L}, \tag{5.5.27}$$

denoting the hydraulic conductivity and the fractured rock porosity, respectively.

When the parallel fractures are of varying apertures, b_i, $i = 1, 2, \ldots, m$, then

$$\mathbf{Q} = qL = -\frac{1}{12}\frac{\rho g}{\mu}\left(\sum_{i=1}^{i=m} b_i^3\right)\nabla h, \quad \mathbf{q} \equiv \frac{\mathbf{Q}}{L} = -\frac{m}{L}\frac{b^3}{12}\frac{\rho g}{\mu}\nabla h \equiv -K_{fr}L\nabla h, \tag{5.5.28}$$

and:

$$K_{fr} = \frac{1}{12L}\frac{\rho g}{\mu}\sum_{i=1}^{i=m} b_i^3 = \frac{\phi_{fr}}{\sum_{i=1}^{i=m} b_i}\frac{1}{12}\frac{\rho g}{\mu}\sum_{i=1}^{i=m} b_i^3. \tag{5.5.29}$$

Next, we consider the case of a fracture system composed of two orthogonal families of parallel fractures, in the cross-sectional area $F \times L$, normal to the flow direction, shown in Fig. 5.14b. The two families of fractures are denoted by $b_i = 1, 2, \ldots, m_i$, and b_j, $j = 1, 2, \ldots, m_2$. The total discharge and hydraulic conductivity are determined by superposition:

$$Q = Q_{m_1} + Q_{m_2} = qL^2 = \frac{1}{12}\frac{\rho g}{\mu}\left(\sum_{i=1}^{m_1} b_i^3 + \sum_{j=1}^{m_2} b_j^3\right)\nabla h \equiv K_{fr} L \nabla h, \quad (5.5.30)$$

with an hydraulic conductivity of the fracture network defined by:

$$K_{fr} = \frac{1}{12L^2}\frac{\rho g}{\mu}\left(\sum_{i=1}^{i=m_1} b_i^3 + \sum_{j=j}^{j=m_2} b_j^3\right), \qquad \phi_{fr} = \frac{\sum_{i=1}^{i=m_1} b_i + \sum_{j=1}^{j=m_2} b_j}{L^2}, \quad (5.5.31)$$

with ϕ_{fr} denoting the porosity due to the void space of the fractures. We note that in (5.5.30) and (5.5.31), the flow through the fracture junctions is counted twice. However, it seems reasonable to assume that this will have only a very small effect on the calculated discharge and porosity.

Let us add here the option that the blocks are porous ($\equiv$ a double porosity rock domain), with k_{fr} denoting the permeability of the porous blocks. Then, for a fractured porous rock composed of parallel fractures of constant aperture, the total flux is expressed by:

$$q_{fr} = -\frac{\rho g}{\mu L}\left(\frac{mb^3}{12} + k_{pb}(L - mb)\right)\nabla h. \quad (5.5.32)$$

Since $mb \ll L$, we have:

$$q_{fr} \approx -\frac{\rho g}{\mu}\left(\frac{mb^3}{12L} + k_{pb}\right)\nabla h = -\frac{\rho g}{\mu}\left(\frac{\phi_{fr} b^2}{12} + k_{pb}\right)\nabla h. \quad (5.5.33)$$

Note that in the above development, it is implicitly assumed that the fracture walls are impervious, or that the leakage across them, between the fractures and the porous blocks, is negligible.

Bear (in Bear et al. 1993, pp. 19–21) discusses the general case when the fracture network is made up of randomly oriented fracture segments having different orientations within a fractured rock domain.

5.5.2 Flow in a Fractured Porous Medium Domain

The underlying assumption in this subsection is that the rock blocks surrounded by fractures are porous. We have already introduced such a fractured rock domain as a *fractured porous rock*, or, for obvious reason, as a *double porosity domain*. Accordingly, we have already defined two porosities: one of the fractures and the other of the porous blocks. The entire void space may contain one or more fluids. In the current section we deal only with a single fluid. According to our approach in this book, we envision such domain as three overlapping continua at the macroscopic level: one of the network of fractures, the other of the porous blocks, and the third one of the *fractured porous rock* as a whole. The first two continua may exchange fluid mass between them at every macroscopic point within the considered domain, recalling that "at a point" means "within the REV centered at the point", and the REV always contains both fractures and porous blocks. Obviously, to regard the considered system as overlapping continua requires that a *common* REV exists for both of them. The transport of other extensive quantities, e.g., mass of a chemical species dissolved in the fluid(s), or heat, may also take place.

Many authors developed models of fluid flow based on the two-continua (or double-porosity) approach introduced by Barenblatt and Zheltov (1960). Among many others, we may mention Warren and Root (1963), Odeh (1965), Kazemy et al. (1969) and Streltsova (1976).

Let us consider the mass balance of a single fluid that occupies the void space of both the fractures and the porous blocks. We shall use subscripts s, fr and pb to denote the solid, the fractures and the porous blocks, respectively. Needless to add that we require that a common REV can be found for the network of fractures and for the porous medium in the blocks.

Equation (5.5.10) is the microscopic mass balance equation for a fluid in the void space of any porous medium domain without sources and sinks. Let us rewrite this equation, once for the fluid in the fractures (subscript fr) and once for the fluid in the void space of the porous blocks (subscript pb), in the form:

$$\frac{\partial \rho_{fr}}{\partial t} = -\nabla\cdot(\rho_{fr}\mathbf{V}_{fr}), \tag{5.5.34}$$

$$\frac{\partial \rho_{pb}}{\partial t} = -\nabla\cdot(\rho_{pb}\mathbf{V}_{pb}). \tag{5.5.35}$$

The corresponding averaged mass balance equations for the fluid in the fractures and in the porous blocks, are:

$$\frac{\partial \phi_{fr}\overline{\rho_{fr}}^{fr}}{\partial t} = -\nabla\cdot\phi_{fr}\overline{\rho_{fr}}^{fr}\overline{\mathbf{V}_{fr}}^{fr} - f^{m}_{fr\to pb}, \tag{5.5.36}$$

and:

$$\frac{\partial \phi_{pb}\overline{\rho_{pb}}^{pb}}{\partial t} = -\nabla\cdot\phi_{pb}\overline{\rho_{pb}}^{pb}\overline{\mathbf{V}_{pb}}^{pb} - f^m_{pb\to fr}, \tag{5.5.37}$$

where we note the terms that express mass exchange between the two continua. (Barenblatt et al. 1960) developed the above mass balance equations phenomenologically (from mass balance considerations) and suggested expressions for the mass transfer terms. A review of expressions representing the transfer between the phases is presented by Abushaikha and Gosselin (2008).

Let us assume that the fluid densities obey $\overline{\rho_{pb}}^{pb} \equiv \overline{\rho_{fr}}^{fr} \equiv \rho$. Then, omitting the averaging symbols, as it is obvious that the equations are at the macroscopic level, we can write flux equations for the fluids in the fractures and in the blocks, each regarded as a continuum:

$$\phi_{pb}\mathbf{V}_{pb} = -\frac{k_{pb}}{\mu}\left(\nabla p_{pb} + \rho g \nabla z\right), \tag{5.5.38}$$

and

$$\phi_{fr}\mathbf{V}_{fr} = -\frac{k_{fr}}{\mu}\left(\nabla p_{fr} + \rho g \nabla z\right). \tag{5.5.39}$$

For the mass transfer terms, Barenblatt et al. (1960) suggested:

$$f^m_{fr\to pb} = C_{frpb}\frac{K_{pb}\rho A^2}{\mu}\left(p_{fr} - p_{pb}\right), \tag{5.5.40}$$

where A denotes the area of fracture-porous block interface, per unit volume of fractured porous medium ($\approx 2\theta_{fr}/ < b >$), with $< b >$ representing the average fracture aperture, and C_{frpb} denoting a dimensionless shape factor of the fractured porous medium.

For a compressible fluid, Barenblatt et al. (1960) approximated the density-pressure dependence by the linear relationship:

$$\rho = \rho_o\left(1 + \beta_p(p - p_o)\right), \tag{5.5.41}$$

where β_p denotes the fluid's compressibility at constant temperature and p_o denotes a reference pressure.

For a deformable fractured porous rock, for which:

$$\phi_{fr} = \phi_{fr}(p_{fr}, p_{pb}), \qquad \phi_{pb} = \phi_{pb}(p_{fr}, p_{pb}),$$

we write:

$$d\phi_{fr} = \alpha_{11}dp_{fr} - \alpha_{12}dp_{pb}, \quad d\phi_{pb} = \alpha_{21}dp_{fr} - \alpha_{22}dp_{pb},$$

where:

$$\alpha_{11} = \frac{\partial\phi_{fr}}{\partial p_{fr}}, \quad \alpha_{12} = \frac{\partial\phi_{fr}}{\partial p_{pb}}, \quad \alpha_{21} = \frac{\partial\phi_{pb}}{\partial p_{fr}}, \quad \alpha_{22} = \frac{\partial\phi_{pb}}{\partial p_{pb}},$$

are coefficients of porous rock compressibility.

In spite of the deformation of the fracture porous rock as a whole, we assume that the volume of solids, per unit volume of fracture porous rock remains unchanged, viz., $1 - \phi_{fr} - \phi_{pb} = const$. Hence, $\alpha_{11} = -\alpha_{21}, \quad \alpha_{12} = -\alpha_{22}$. Under these conditions, and for an isotropic fractured porous rock, we insert the flux equations into the mass balance equations, and assume that the effect of gravity is much smaller than that of the pressure gradient. We obtain:

- For flow in the fracture network,

$$\frac{k_{fr}}{\mu}\nabla^2 p_{fr} = (\alpha_{11} + \beta_p\phi_{fr})\frac{\partial p_{fr}}{\partial t} + \alpha_{12}\frac{\partial p_{pb}}{\partial t} - C_{frpb}\frac{k_{pb}A^2}{\mu}(p_{fr} - p_{pb}) - \frac{\beta_p k_{fr}}{\mu}(\nabla p_{fr})^2. \quad (5.5.42)$$

- For flow in the porous blocks,

$$\frac{k_{pb}}{\mu}\nabla^2 p_{pb} = (\alpha_{22} + \beta_p\phi_{pb})\frac{\partial p_{pb}}{\partial t} + \alpha_{21}\frac{\partial p_{fr}}{\partial t} - C_{frpb}\frac{k_{pb}A^2}{\mu}(p_{fr} - p_{pb}) - \frac{\beta_p k_{pb}}{\mu}(\nabla p_{pb})^2. \quad (5.5.43)$$

The last term on the r.h.s. of the above two equations is very small (relative to the other terms) and may be neglected. Assuming that:

$$|\alpha_{12}dp_{pb}| \ll |\alpha_{11}dp_{fr}|, \qquad |\alpha_{21}dp_{fr}| \ll |\alpha_{22}dp_{pb}|,$$

$$|\nabla\cdot\rho\phi_{pb}\mathbf{V}_{pb}| \ll \left|\frac{\partial(\phi_{pb}\rho)}{\partial t}\right|, \qquad |\nabla\cdot\rho\phi_{fr}\mathbf{V}_{fr}| \gg \left|\frac{\partial(\phi_{fr}\rho)}{\partial t}\right|,$$

we obtain the approximate mass balance equation for flow in the network of fractures in the form:

$$k_{fr}\nabla^2 p_{fr} + C_{frpb}k_{fr}A^2\left(p_{fr} - p_{pb}\right) = 0. \quad (5.5.44)$$

We note that in this equation we have only flow in the fractures and exchange of fluid between fractures and porous blocks.

For porous blocks, the approximate mass balance equation takes the form:

$$(\alpha_{22} + \beta_p \phi_{pb}) \frac{\partial p_{pb}}{\partial t} - C_{frpb} \frac{k_{fr} A^2}{\mu} \left(p_{fr} - p_{pb} \right) = 0. \tag{5.5.45}$$

Here we note only fluid exchange between fractures and porous blocks, and elastic storage in the porous blocks.

By eliminating p_{pb} from (5.5.44) and (5.5.45), we obtain another approximate mass balance equation for the fluid in the fractures:

$$\frac{\partial p_{fr}}{\partial t} - \frac{k_{fr}}{k_{pb}} \frac{1}{C_{frpb} A^2} \frac{\partial}{\partial t} \left(\nabla^2 p_{fr} \right) = \chi \nabla^2 p_{fr}, \quad \chi = \frac{k_{fr}}{\mu(\alpha_{22} + \beta_p \theta_{pb})}. \tag{5.5.46}$$

The expression $\alpha_{22} + \beta_p \theta_{pb}$ plays here the role of *specific storativity* of the porous block, where most of the water is stored (to be compared with the same definition as used in flow through ordinary porous media, say, (5.1.50).

If we assume that flow takes place only in the fractures, while (compressibility) storage is mainly in porous blocks, the governing mass balance is:

$$\beta_p \phi_{pb} \frac{\partial p_{fr}}{\partial t} = \frac{k_{fr}}{\mu} \nabla^2 p_{fr}. \tag{5.5.47}$$

Warren and Root (1963) employed a model which is essentially the same as (5.5.44) and (5.5.45), except that they did not neglect the effect of fluid storage in the fractures, related to $\partial p_{fr}/\partial t$ in (5.5.42). They presented an analytical solution for flow to a well producing at a constant rate from an infinite, or finite, horizontal reservoir. Kazemy et al. (1969) solved a similar problem. They solved for the spatial as well as the temporal variation in pressure. Da Prat et al. (1981) solved the problem of flow to a well producing at a constant pressure.

Without presenting their actual solution, it may be interesting to summarize the main features of the Warren and Root (1963) solution:

- Initially, fluid is removed primarily from fractures, because their permeability is much higher than that of the porous blocks. The case of impervious blocks reduces to that of a homogeneous single continuum with fracture permeability and the pressure decline in a well producing at a constant rate is plotted as a straight line against the logarithm of time.
- Gradually flow at an increasing rate takes place from the porous blocks to the fractures. This appears as a curve of variable slope on the, mentioned above plot.
- Then equilibrium is reached between the two continua. This appears on the above mentioned plot as a straight line having the same slope as that describing the pressure decline during the initial period, but now the line is displaced parallel to itself. This indicates that the entire fractured porous rock behaves as an equivalent single homogeneous continuum. The equality of the slopes during the first and

third stages indicates that the permeability 1n the latter is primarily that of the fractures.

The solution derived by Warren and Root (1963) requires the knowledge of the coefficient C_{frpb} appearing in the transfer function. This coefficient depends on the configuration of the blocks. Warren and Root assumed that C_{frpb} is a constant that can be determined from tests. Crawford et al. (1976) describe such tests. In a non-homogeneous material, the magnitude of this coefficient varies in space. However, Braester (1984) showed that even when this coefficient is constant (as in the case of uniform regular porous blocks), the pressure decline is insensitive to block size. He, therefore, questions the usefulness of the solution.

An important conclusion is that flow in a fractured porous rock differs from that in an ordinary porous medium only during the initial stages of flow and only in the vicinity of a producing well, After that, the flow regime is identical to that of a single continuum; it can be described by a single mass balance equation for the fluid in the void space. We can obtain such equation by regarding the fractured-porous-rock as a single continuum that has a porosity and a permeability, like an 'ordinary' porous medium. We can obtain this equation also by writing separate equations for the fractures and for the porous blocks and adding them. The terms expressing the exchange of mass between fractures and blocks will then be eliminated.

Usually we assume that the two sub-systems– the fractures and the porous blocks– are in equilibrium–the same pressure at a point in both systems. There is no net exchange of fluid between them.

* * *

In this book, we do not consider modeling of reactive transport, nor energy transport in fractured rock domains. Material on these subjects were presented by Bear in Bear, Tsang and de Marisly (1993). In this book, among others, Smith and Schwartz (in Tsang and de marsily 1993, pp. 129–168) discuss solute transport through fracture networks, Kazemi and Gilman (pp. 267–324) discuss multiphase flow in fractured petroleum reservoirs, and Wang and Narasimhan (pp. 325–395) discuss unsaturated flow in fractured media. Bear and Braester (1972) discuss the simultaneous flow of immiscible liquids in a fractured medium.

References

Abushaikha A, Gosselin O (2008) Matrix-fracture transfer function in dual-medium flow simulation: review, comparison and validation. In: SPE-113890, proceedings of the SPE EUROPEC, Rome, 9–12 June 2008

Al-Hussainy R, Ramey HJ Jr, Crawford PB (1966) The flow of real gases through porous media. J Pet Technol 624

Barenblatt GI, Zheltov IP, Kochina IN (1960) Basic concepts in the theory of seepage of homogeneous liquids in fissured rocks. J Appl Math Mech (PMM) 24:852–864

Barenblatt GI, Zheltov IP (1960) Fundamental equations of filtration of hpmgeneous Liquids in fissured rocks, Soviet Dokl Akad Nauk 13:545–548
Bear J (1972) Dynamics of fluids in porous media. Elsevier, Amsterdam, 764 p (also published by Dover Publications, 1988; translated into Chinese)
Bear J (1979) Hydraulics of groundwater. McGraw-Hill, New York, 569 p (also published by Dover Publications, 2007; translated into Chinese)
Bear J (1993) Modeling flow and contaminant transport in fractured rocks. In: Bear J, Tsang CF, de Marsily G (eds) Flow and contaminant transport in fractured rock. Academic Press, London, pp 1–37 (560 p)
Bear J, Braester C (1972) On the flow of two immiscible fluids in fractured porous media. In: Proceedings of the 1st international IAHR symposium on the fundamentals of transport phenomena in porous media, Haifa, Israel, pp 177–202
Bear J, Bachmat Y (1991) Introduction to modeling phenomena of transport in porous media. Kluwer Publishing Co., Dordrecht, 553 p
Bear J, Tsang CF, de Marsily G (eds) (1993) Flow and contaminant transport in fractured rock. Academic Press Inc., London, 560 p
Bear J, Cheng AH-D (2010) Modelinmg Goundwater Flow and Contaminant Transport. Springer, 834 pp
Beavers GS, Joseph DD (1967) Boundary conditions at a naturally permeable wall. J Fluid Mech 30:197–207
Beavers GS, Sparrow EM, Rodenz DE (1973) Influence of bed size on the flow characteristics and porosity of randomly packed beds of spheres. J Appl Mech 40:655–660
Bellin A, Rubin Y (1996) HYDRO GEN: a spatially distributed random field generator for correlated properties. Stoch Hydrol Hydraul 10:253–278
Biot MA (1935) Le problème de la consolidation des matières agileuses sous une charge. Ann Soc Sci Brux Ser B 55:110–113
Biot MA (1941) General theory of three-dimensional consolidation. J Appl Phys 12:155–164
Bishop AW (1973) The influence of an undrained change in stress on the pore pressure in porous media of low compressibility. Geotechnique 23:435–442
Braester C (1984) Influence of block size on the transition curve for a drawdown test in a naturally fractured reservoir. J Soc Pet Eng 494–504
Cheng AH-D (2016) Poroelasticity. Springer, Berlin, 877 p
Conkling G (1934) Ventura County Investigations, Calif Div of Water Resources Bull 6, 244 pp
Courant R, Hilbert D (1962) Methods of mathematical physics. Wiley Interscience, N. Y, 560 p
Crawford CE, Hagedom NR, Pierce NE (1976) Analysis of pressure build-up test in a naturally fractured reservoir. J Pet Technol 1295–1300
Da Prat G, Cinco-Ley H, Remey JJ Jr (1981) Decline curve analysis using type-curve for two-porosity system. J Soc Pet Eng 354–362
Dagan G (1979) Models of groundwater flow in statistically homogeneous porous formation. Water Resour Res 15:47–63
Dagan G (1989) Flow and transport in porous formations. Springer, New York, 465 p
Davis SN, de Wiest RJM (1966) Hydrogeology. Wiley, New York, 463 p
Dupuit J (1863) Études Théoriques et Pratiques sur les Mouvementdes des Eaux dans les Cannaux Découverts et à Travers les Terrains Perméables, 2nd edn. Dunod, Paris, 304 p (1st edn. 1948)
Fraim ML, Wattenbarger RA (1986) Gas reservoir decline-curve analysis using type curves with real gas. In: SPE 15028 presented at the permian basin oil & gas recovery conference, Midland, Texas, 13–14 March 1986
Fraim ML, Wattenbarger RA (1987) Gas reservoir decline curve analysis using type curves with real gas pseudopressure and normalized time. SPEFE 24:671–682
Freeze RA (1975) A stochastic-conceptual analysis of one-dimensional groundwater flow in nonuniform homogeneous media. Water Resour Res 11:725–741
Gambolati G, Teatini P, Bau D, Ferronato M (2000) Importance of poroelastic coupling in dynamically active aquifers of the Po river basin, Italy. Water Resour Res 36:2443–2459

Geertsma J (1957) The effect of fluid pressure decline on volumetric changes of porous rocks. Trans AIME 210:331–340
Gelhar LW (1986) Stochastic subsurface hydrology from theory to applications. Water Resour Res 22:135S–145S
Gelhar LW (1993) Stochastic subsurface hydrology. Prentice-Hall, Englewood Cliffs
Gray WG, Hassanizadeh SM (1989) Averaging theorems and averaged equations for transport of interface properties in multiphase systems. Int J Multiph Flow 15:81–95
Hassanizadeh M, Gray W (1979a) General conservation equations for multiphase systems: 2. Mass, momentum, energy and entropy equations. Adv Water Resour 2:191–203
Hoeksema RJ, Kitanidis PK (1985a) Analysis of spatial variability of properties of selected aquifers. Water Resour Res 21:563–572
Hoeksema RJ, Kitanidis PK (1985b) Comparison of Gaussian conditional mean and kriging estimation in the geostatistical approach to the inverse problem. Water Resour Res 21:825–836
Jager W, Mikelic A (2000) On the interface boundary condition of Beavers, Joseph, and Saffman. SIAM J Appl Math 60:1111–1127
Jha B, Juanes R (2014) Coupled multiphase flow and poromechanics: a computational model of pore-pressure effects on fault slip and earthquake triggering. Water Resour Res 50(5):3776–3808
Kazemy H, Seth MS, Thomas GW (1969) The interpretation of interference tests in naturally fractured reservoirs, with uniform fracture distribution. Soc Pet Eng J 4(4):463–472
Kim S, Russel WB (1985) Modeling of porous-media by renormalization of the Stokes equation. J Fluid Mech 154:269–286
Kim J, Tchelepi HA, Juanes R (2013) Rigorous coupling of geomechanics and multiphase flow with strong capillarity. SPE J 18(6):1123–1139
Kohr M, Sekhar GPR (2007) Existence and uniqueness result for the problem of viscous flow in a granular material with a void. Quart Appl Math 65:683–704
Koplik J, Levine H, Zee A (1983) Viscosity renormalization in the Brinkman equation. Phys Fluids 26(10):2864–2870
Kubik J (2004) Elements of constitutive modelling of saturated porous materials. In: Kubik J, Kaczmarek M, MurdochModelling I (eds) Coupled phenomena in saturated porous materials, Institute of Fundamental Technological Research, Polish Academy of Sciences, Warsaw, Poland, pp 279–347
Ladyzhenskaya OA (1963) The mathematical theory of viscous incompressible flow. Gordon and Breach, New York
Lamb H (1945) Hydrodynamics. Dover, New York, 602 p
Mantoglou A, Wilson JL (1982) The turning bands method for simulation of random-fields using line generation by a spectral method. Water Resour Res 18:1379–1394
Mantoglou A, Gelhar LW (1987) Stochastic modeling of large-scale transient unsaturated flow systems. Water Resour Res 23:37–46
Muskat M (1946) The flow of homogeneous fluids through porous media. J.W. Edwards Inc., Ann Arbor, 763 p (1st edn. 1937)
Nield DA, Bejan A (2013) Convection in porous media, 4th edn. Springer, Berlin, 778 p
Ochoa-Tapia JA, Whitaker S (1995) Momentum-transfer at the boundary between a porous-medium and a homogeneous fluid. 2. Comparison with experiment. Int J Heat Mass Transf 38:2647–2655
Odeh AS (1965) Unsteady-state behavior of naturally fractured reservoirs. Soc Pet Eng 17(3):245
Press WH, Teukolsky SA, Vetterling WT, Flannery BP (2007) Numerical recipes, the art of scientific computing, 3rd edn. Cambridge University Press, Cambridge, 1256 p
Robin MJL, Gutjahr AL, Sudicky EA, Wilson JL (1993) Cross-correlated random-field generation with the direct Fourier-transform method. Water Resour Res 29:2385–2397
Rosenzweig R, Shavit U (2007) The laminar flow field at the interface of a Sierpinski carpet configuration. Water Resour Res 43:w10402
Shavit U, Rosenzweig R, Assouline S (2004) Free flow at the interface of porous surfaces: a generalization of the Taylor brush configuration. Transp Porous Media 54:345–360

Streltsova TD (1976) Hydrodynamics of groundwater flow in a fractured formation. Water Resour Res 12(3):504–514
Tompson AFB, Gelhar LW (1990) Numerical-simulation of solute transport in 3-dimensional, randomly heterogeneous porous-media. Water Resour Res 26:2541–2562
Tompson AFB, Ababou R, Gelhar LW (1989) Implementation of the 3-dimensional turning bands random field generator. Water Resour Res 25:2227–2243
Verruijt A (1969) Elastic storage in aquifers. In: De Wiest RJM (ed) Flow through porous media. Academic Press, New York, pp 331–376
Verruijt A (1995) Computational Geomechanics. Kluwer Academic Publ 584 pp
Verruijt A (2005) Consolidation of soils. In: Anderson MG (ed) Encyclopedia of hydrological sciences. Wiley, New York, 17 p
Verruijt A (2010) Soil mechanics. Delft University of Technology, 330 p
Verruijt A (2014) Theory and problems of poroelasticity. Delft University of Technology, 266 p
von Terzaghi K (1925) Erdbaumechanik auf Bodenphysikalische Grundlage. Franz Deuticke, Leipzig, 399 p
Warren JE, Root PJ (1963) The behavior of naturally fractured reservoirs. J Soc Pet Eng 3:245–255
Wilson CR, Witherspoon PA (1974) Steady state flow in rigid networks of fractures. Water Resour Res 10(2):328–335
Witherspoon PA, Wang JSY, Iwai K, Gale JE (1980) Validity of cubic law for fluid flow in a deformable rock fracture. Water Resour Res 16(6):1016–1024

Chapter 6
Modeling Multiphase Mass Transport

In the previous chapter, we have presented models that describe flow of a single fluid that fills up the entire void space. Here, we consider cases in which the void space is occupied, simultaneously, by more than one fluid phase. The fluids are referred to as *immiscible fluids*. Actually, we mean "practically immiscible fluids", as all fluids are miscible in each other to some extent. They occupy disjoint (microscopic) subdomains that together fill up the entire void space. Because of *surface tension phenomena* (Sect. 2.4.1), one of the fluids, called *wetting fluid*, tends to adhere to the solid, while the other, called *nonwetting fluid*, finds itself farther from the solid surface. This means that a wetting fluid always coats the *entire* solid's (microscopic) surface.

It is important to emphasize that although we are considering here two phases, with a visible interface at the microscopic level, it is possible that certain chemical species present in these phases do cross this microscopic interface. This possibility is discussed in Sect. 7.4.

The term drainage is often used to describe the situation in which a nonwetting fluid displaces a wetting one. An example is air displacing water.

When a porous medium, initially saturated by a wetting fluid, is drained, a small amount of that fluid will always remain on the surface of the solid matrix in the form of a very thin *film*, with a thickness of some tens of molecules, that adheres to the wall by strong molecular forces and cannot be easily displaced. Even if initially a pore is occupied by a nonwetting fluid, a wetting fluid that invades the void space will tend to spread on the solid surface by *imbibition*, gradually displacing the nonwetting fluid.

It is interesting to note that in many chemical reactors, mass of chemical species is exchanged between a gas in the void space and a (moving) wetting liquid that covers the solid surfaces.

J. Bear, *Modeling Phenomena of Flow and Transport in Porous Media*,
Theory and Applications of Transport in Porous Media 31,
https://doi.org/10.1007/978-3-319-72826-1_6

One example of two fluid phases that occupy the void space is the *unsaturated zone* below ground surface, where the two fluids are water and air. Following a spill of a non-aqueous contaminant at ground surface, the same zone will be occupied also by the downward moving contaminant—a third fluid phase. Special attention is devoted to this zone.

Another example is a petroleum reservoir, where the void space is occupied by two (water and oil), or three (water, oil and gas) fluid phases. Each of these three fluids has a different wetting behavior.

The material in this chapter should be of interest to those who deal with the unsaturated zone in the subsurface, e.g., in connection with irrigation and drainage in agricultural engineering. It should also be if interest to reservoir engineers, to those who plan the disposal of CO_2 in deep brine-containing geological formations, or to those who plan the injection of air or natural gas into depleted oil and gas reservoirs. As will be discussed in Appendix A, many types of chemical reactors involve flow of two (practically) immiscible fluids (two liquids or a liquid and a gas).

We shall start by presenting flux equations for individual phases, leading to well-posed mathematical models based on mass balance equations.

6.1 Macroscopic Capillary Pressure

The concept of capillary pressure was introduced in Sect. 2.4.3. There, the discussion focussed on the pressure jump at the *microscopic* level, i.e., at a point on the interface (= meniscus) between two immiscible fluids–a wetting fluid and a non-wetting one–inside the void space. The *Laplace formula* (2.4.12) was presented as an expression for the capillary pressure at that point. This equation expresses a condition to be satisfied at every point on the interface between the fluids. Here, our interest is in the capillary pressure at the *macroscopic* level, i.e., at every point in a porous medium domain (regarded as a single continuum) in which the void space is occupied by two or three fluid phases. In fact, we consider the wetting fluid, the non-wetting one and the porous medium as a whole as *three overlapping continua*. It seems reasonable to assume that for the many menisci within an REV at such a point, the average capillary pressure will depend on some average radius of the menisci, and, hence, on the saturation (defined in Sect. 1.1.7 B) at the macroscopic point.

Following the phenomenological approach, which serves as the basis for this book, and in view of the definition of capillary pressure at a microscopic point, we assign to every point within a porous medium domain, the average pressure of each of the fluids that occupies the void space, in the vicinity of the point. Each of these fluids is regarded as a continuum that occupies the entire porous medium domain at a certain saturation. The latter may vary from point to point and with time. The pressure of a fluid at a point in a porous medium domain is the average over the fluid present in the REV centered at that point. In analogy to the definition of capillary pressure at the microscopic level, we then define for every point in a porous medium domain a *macroscopic capillary pressure*, p_c, as the difference between the average pressure of the non-wetting fluid (n) and that of the wetting one (w) present at that point:

$$p_c(\mathbf{x}, t) = \overline{p_n}^n(\mathbf{x}, t) - \overline{p_w}^w(\mathbf{x}, t). \tag{6.1.1}$$

Note that $p_c(\mathbf{x}, t)$ here is at the macroscopic level. It is important to note the kind of average, namely, the *intrinsic phase average*, used above. In fact, the above equation serves as a *definition* for the macroscopic capillary pressure, and *not the average of the microscopic capillary pressure*.

Hassanizadeh and Gray (1990, 1993) employed rational thermodynamic averaging approach to derive a different theory of two-phase flow in porous media. Their approach is also based on the behavior of the microscopic fluid-fluid and solid-fluid interfaces. However, as explained in Sect. 1.4.2 C, they write microscopic level E-balance equations also for the (very thin) interphase domains and average these balances (per unit volume of porous medium) to obtain macro-scale description of two-phase flow *within a surface*. By doing so, they obtain also macro-scale quantities. Thus, in addition to the usual averages of mass density, fluid velocity, etc., of the fluid phases involved, they also obtain macroscopic averaged variables for interphase surfaces, e.g., average interfacial mass density, interfacial velocity, average interfacial tension, and specific interfacial area.

In the unsaturated (water-air) zone below ground surface, the wetting fluid is water and the nonwetting one is air. Soil physicists often *assume* that the air is at a constant atmospheric pressure, taken as zero, i.e., $\overline{p_n}^n = \overline{p_a}^a = 0$. Then:

$$p_c = -\overline{p_w}^w. \tag{6.1.2}$$

Under such conditions, especially when considering water in the unsaturated zone, we often introduce the definition of (macroscopic) *capillary pressure head*, ψ (dims. L), also called (macroscopic) *suction*, or *tension*, or *matric suction*:

$$\psi = -\frac{\overline{p_w}^w}{g\overline{\rho_w}^w}. \tag{6.1.3}$$

The symbol h_c is also used for ψ. Note that ψ may be employed only when the (macroscopic) water density, $\overline{\rho_w}^w$, is constant. The same definition is extended to cases where $\overline{p_a}^a \neq 0$, viz.,

$$\psi = \frac{\overline{p_a}^a - \overline{p_w}^w}{g\overline{\rho_w}^w} = \frac{p_c}{g\overline{\rho_w}^w}. \tag{6.1.4}$$

In Soil Science, the unit pF is defined as the logarithm of negative pressure head in the water, measured in cm. Thus, pF = 4 indicates a suction of 10^4 cm (of water).

In the remaining part of this chapter, the averaging symbol, $\overline{(..)_\alpha}^\alpha$ for $(..)_\alpha$, will be omitted when referring to the average of $(..)_\alpha$. Unless otherwise stated, the term capillary pressure, and the symbol p_c will be used for the difference between the macroscopic pressures defined by (6.1.1).

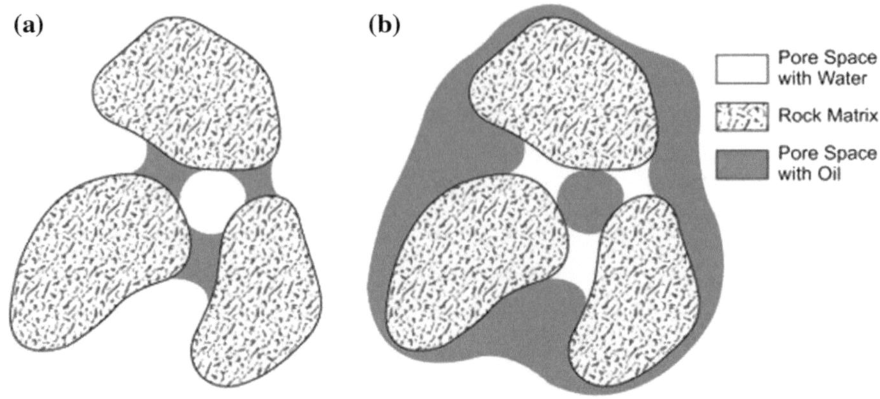

Fig. 6.1 Oil and water distributions as pendular rings when the solid grains are: **a** Water-wet, and **b** oil-wet

A. Drainage and Imbibition

In (2.4.12), we note the relationship between the (microscopic) capillary pressure and the radius of the meniscus at a point on the latter. It seems reasonable to assume that for the many menisci within an REV, the macroscopic capillary pressure will depend on some *average radius of the menisci*, and, hence, on the saturation. In what follows, we shall present the relationship between the (average) capillary pressure and the saturations of two fluid phases that, together, occupy the void space in a domain. The unsaturated (air-water) zone, and a petroleum (oil-water) reservoir, will serve as examples.

A typical example is shown in Fig. 6.1. It shows a number of solid grains and two fluids that occupy the void space: oil and water. In Fig. 6.1a the solid grains are *oil wet*, while Fig. 6.1b shows the case in which the solid grains are *water wet*. In both cases, as a result of surface tension phenomenon, the wetting fluid at low saturations occupies domains in the form of rings around the grain contact points (or grains that are very close). These liquid domains are called *pendular rings*. Note that the figure does not show the thin film of wetting fluid that covers the water-wet solid.

As a second typical example, consider the distribution of air, as an example of a nonwetting phase, and water, as an example of a wetting phase, within the void space of an unsaturated zone. At low saturation, water takes the form of *pendular rings* at contact points (Fig. 6.2a). The air-water interface has the shape of a saddle. A number of adjacent pendular rings may coalesce. We observe how water touches the solid at the *contact angle*. When the grains are close, but not touching each other, the water, or, in general, the wetting fluid, takes the form of a 'bridge', connecting close grains.

In a water-air system, at low water saturation, the pendular rings are isolated and do not form a continuous water phase, except for the very thin film of adsorbed water on the solid surfaces. Figure 6.2b shows a pendular ring between two spheres.

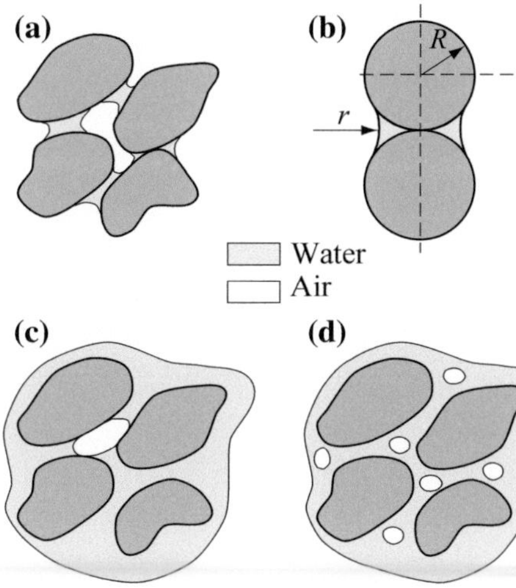

Fig. 6.2 Air and water distributions at various saturations: **a** Pendular saturation, **b** pendular ring between two spheres; **c** funicular water saturation, and **d** insular air saturation

As water saturation increases, the pendular rings expand and coalesce, until a continuous water phase is formed. Above this critical saturation, the 'bulk' water forms a continuous phase, and its saturation is called *funicular*; flow of 'bulk' water is possible (Fig. 6.2c). Both the water and the air phases are continuous. As water saturation increases, a situation develops in which the air is no longer a continuous phase; it breaks into individual bubbles (globules, blobs, ganglia) lodged in the larger pores (Fig. 6.2d). The air is then said to be in a state of *insular air saturation*. An air globule can move only if a pressure difference is applied to it by the surrounding water that is sufficient to squeeze it through the constriction. In the absence of air in the void space, we have complete water saturation. Similarly, a NAPL ganglion may be trapped. Residual NAPL, say in the form of a ganglion, will remain entrapped unless displacement pressure exceeds entry pressure.

The above discussion on a water-air system is, of course, valid for any wetting-nonwetting pair of fluids.

Depending on the pore size distribution, the above stages, say from pendular to funicular water, do not have to occur simultaneously across the entire unsaturated domain. It is possible to envision that at low saturations, part of the water forms a continuous phase, while the remaining part is still in a pendular state, with a gradual transition as saturation increases. A similar transition may occur as saturation is reduced. This may give rise to situations in which part of the water in the porous medium is mobile, while the other part is *immobile*.

In the course of time, the volume of air at insular saturation may decrease due to air solubility in water. Similarly, the volume of water in pendular rings may decrease with time due to evaporation.

With these definitions, let us now follow the changes in water saturation as an initially-saturated porous medium sample is gradually drained from its bottom, with air introduced at its top. Figure 6.3 shows several successive stages of drainage (stages

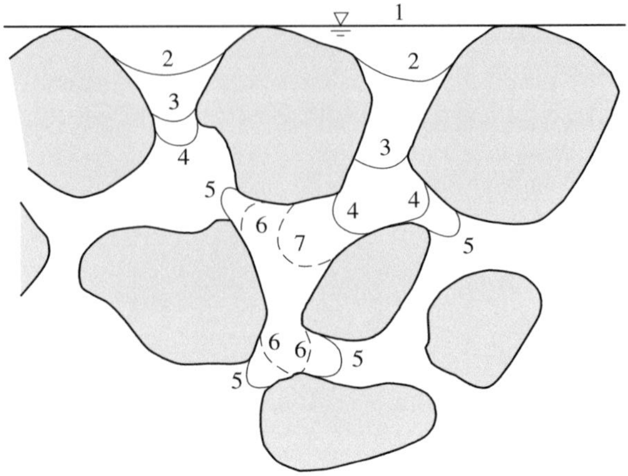

Fig. 6.3 Gradual drainage and rewetting in the unsaturated zone

1 through 5) and rewetting (stages 6 and 7) in the unsaturated zone. Each state corresponds to a certain volume of air occupying a certain portion of the void space at a corresponding saturation. As water drainage progresses, the water-air interface retreats into channels which support a curvature of still smaller radius (e.g., interface 4). The wetting fluid will continue to retreat until the local interfaces have taken up positions of equilibrium in channels which are sufficiently narrow to support interfaces with smaller radius of curvature. Obviously, if all channels are equal and large, at a given capillary pressure, no equilibrium can be maintained any longer, and a sudden, almost complete, drainage of water from the entire porous medium sample will be observed. We use the word 'almost', because some water will always remain as isolated pendular rings and as a film adsorbed to the solid surface. Within the pendular rings, the pressure is independent of that in the remaining, continuous water phase in the void space. However, the pressure there is related to pressure in the gaseous phase (which is a mixture of air and water vapour) by the capillary pressure relationship. As water evaporates, the volume of water in a pendular ring decreases, the radius of curvature of the meniscus decreases, and the capillary pressure increases.

At every stage, the largest capillary pressure that can be maintained by a local interface corresponds to the smallest radius of curvature that can be accommodated in a pore, or channel, through which the interface is being withdrawn. Therefore, the smallest radius of curvature occurs in the narrowest pores that correspond to the prevailing air volume (e.g., interface 3 in Fig. 6.3).

In general, pores have different dimensions and shapes. Therefore, they will not all empty at the same capillary pressure. The large pores (or those with larger channels, or *throats* of entry) will empty at low capillary pressures, while those with narrow channels of entry, supporting interfaces of smaller radius of curvature, will empty at higher capillary pressures.

If, at some point in time, the drainage of the wetting fluid at the bottom of a sample is stopped, overlooking evaporation and air solubility processes, an equilibrium will be established, with no further motion of either fluid. The pressure distribution within each fluid will be hydrostatic, while satisfying the pressure jump condition at every point of the (microscopic) interfaces. At every such point, the pressure jump will correspond to the radius of curvature of the interface at that point. In this way, equilibrium is established between surface tension and gravity.

Let us now reverse the process and begin to refill the pore space with water. In Fig. 6.3, this is shown as transition from stage 5 to stages 6 and 7. The interfaces' radii of curvature become progressively larger.

From the above description, it is obvious that at each stage of a drainage process, the quantity of water remaining in the void space, say, within an REV centered at a point, takes on a certain (microscopic) configuration. The latter is related to the distribution of the (microscopic) interface geometry within that REV. As a consequence, the quantity of water remaining in the void space depends on the (macroscopic) capillary pressure defined by (6.1.1). The capillary pressure increases as the water saturation decreases, which, in turn, corresponds to a decrease in interfacial surface area.

Finally, a comment concerning the effect of wettability is appropriate here. Figure 6.3 and the discussion of drainage describe what happens in an air-water system. However, when we consider *two liquids*, once a solid comes in contact with the wetting liquid, even when the adjacent pore is drained of that liquid, the solid surface will remain covered by a layer, albeit a thin one, of the wetting liquid. Special means are required to completely remove that layer.

B. Capillary Pressure Relationship

Equation (2.4.12) relates the capillary pressure at a point on the water-air interface to the mean radius of curvature of the latter. In Fig. 2.13 we note how the capillary pressure in a tube is related to the radius of curvature of the meniscus, which, in turn, is related to the radius of the tube. Visualizing a porous medium domain as a random assembly of tubes of various radii, the above relations can be interpreted as indicating the fraction of these tubes that will drain at any given capillary pressure (as long as there is a continuous passage for the drained water to reach the external boundary of the sample). We may conclude that a strong relationship exists between the macroscopic capillary pressure and the fraction of the void space occupied by water (or, in general, by the wetting phase). In other words, the *capillary pressure is a function of the water saturation*.

The relationship between the quantity of water present "at a point" in the void space (within an REV centered at the point), in terms of its saturation, to the prevailing capillary pressure, is recorded as a *capillary pressure curve*, $p_c = p_c(S_w)$. In unsaturated (air-water) flow, the $h_c(S_w)$-curve is called *retention curves*, as it shows how much water is retained in the soil by the capillary pressure (Bear et al. 1968, p. 43).

We recall that at every microscopic point on a meniscus, the capillary pressure depends on the surface tension, γ_{wn}, between the wetting and nonwetting fluids. The concept of surface tension, γ_{wn}, introduced in Sect. 2.4.1, depends on the two fluids, the temperature, $T (\equiv T_w \equiv T_n)$, and the concentrations of dissolved species at the interface. For the sake of simplicity, in this chapter we shall continue to assume that

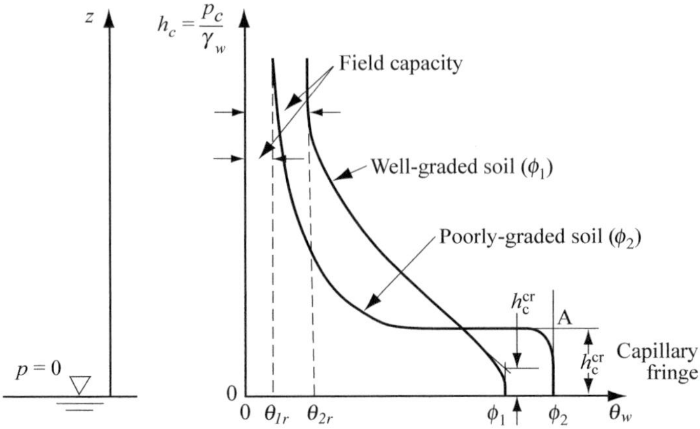

Fig. 6.4 Typical capillary pressure curves during drainage of water from the unsaturated zone

for a given pair of fluids, $p_c = p_c(S_w)$ only. We shall also assume that the porous medium/soil is water-wet.

Figure 6.4 shows two typical examples of capillary pressure head curves, $h_c = h_c(\theta_w)$, during and after cessation of drainage of water in an air-water system above the phreatic surface. Point A in Fig. 6.4 indicates the *threshold capillary pressure head*, h_c^{cr}, corresponding to the largest pore size. If we start from a soil sample that is fully saturated by water, we can produce the capillary head, h_c^{cr}, by draining a very small quantity of water. Practically, no air will penetrate the sample, until the *critical capillary head* is reached. The corresponding pressure, p_c^{cr}, is called the *critical pressure*, or *threshold pressure*. When expressed in terms of pressure, the critical value is also called the *bubbling pressure*, p_b, or *air entry pressure*. As the magnitude of the capillary pressure head, h_c, is increased, an initial small reduction in θ_w, associated with the retreat of the air-water menisci into the pores at the external surface of the sample, is observed. Then, at the critical pressure head value h_c^{cr}, air enters the larger pores and they begin to drain. Recall that the solid is assumed to be *water-wet*.

The shape of the capillary pressure curve, and, hence, also the threshold pressure, depends on the distributions of pore sizes and shapes. The two curves in Fig. 6.4 correspond to *well graded* and *poorly graded* granular porous media, respectively.

The concept of threshold pressure, or pressure entry value, is valid for any porous medium and a pair of immiscible fluids, such that one of them is the wetting fluid, with respect to that solid, and the other is the non-wetting one. A certain minimal pressure is always required such that the non-wetting fluid can enter a wetting fluid saturated porous medium. Similar to air entering a water-saturated zone, a NAPL (= Non Aqueous Phase Liquid) accumulates on a low permeability layer, e.g., clay, and form a pool until the latter's depth will create a pressure in excess of entry pressure.

The threshold pressure plays an essential role in creating petroleum reservoirs. When a water-saturated low permeability (i.e., small pores) layer (also referred to as *caprock*) overlies a layer fully or partly saturated by oil, i.e., a nonwetting fluid, the latter cannot penetrate into the low permeability layer unless its pressure is higher than the pressure entry value of the water saturated caprock. This caprock, although it has (albeit low) permeability, acts as a barrier to the upward migration of the oil.

The concept of *threshold pressure* also serves as the basis for the Mercury Injection Capillary Pressure (MICP) technique for evaluating petroleum reservoir lithology, cap seals (e.g., in the case of CO_2 injection in a CCS project), and determining pore size distribution in a porous medium domain. In this technique, the non-wetting fluid (mercury) is injected into a porous medium, under a gradual step-wise increase of pressure. The volume of injected mercury is recorded as a function of the accumulated injected fluid, until the sample is fully saturated by the injected fluid. This pressure-volume relationship is based on (6.1.5).

Let a fully saturated soil column, $S_w = 1$, be drained through its bottom until no water leaves the column. The drainage will never be complete. Some water (= *irreducible water*) will always remain in the column (against gravity) in the form of *pendular rings* and relatively immobile *thin films* that cover the microscopic solid-void surface. If we now refill the column with water (*imbibition*), displacing the air, the column will never return to full water saturation. Eventually, some air will remain in the sample in the form of isolated bubbles. A detailed discussion on the drainage and of a soil sample is presented in the next subsubsection.

The capillary pressure curve is strongly related the pore-size distribution of the porous medium. The *Laplace formula* at the macroscopic level, as an analog to (2.4.12) and (2.4.13), is:

$$p_c \equiv p_n - p_w = \frac{2}{r^*}\gamma_{wn}, \quad \text{or} \quad p_c \equiv p_n - p_w = \frac{2}{R}\gamma_{wn}\cos\theta_{wn}. \tag{6.1.5}$$

Here r^* is the mean radius of curvature of the microscopic interfaces between the wetting and nonwetting fluids (in our case, water and air) inside an REV, and θ_{wn} is the *contact angle*. We usually write:

$$p_c = p_c(S_w). \tag{6.1.6}$$

For nonisothermal situations and with the effects of dissolved matter, c_w^γ, we have:

$$\gamma_{wn} = \gamma_{wn}(T, c_w^\gamma), \quad \text{leading to} \quad p_c = p_c(S_w, T, c_w^\gamma). \tag{6.1.7}$$

Note that if we assume equilibrium and no gravity effects, all air-water menisci at a (macroscopic) point, i.e., within the REV centered at that point, must have exactly the same radius of curvature. Under such conditions, the microscopic value of capillary pressure and its macroscopic counterpart become identical. Recall that in (2.4.13) we have replaced (the local) r^* by $R/\cos\theta_{wn}$, with R regarded as a measure of the size of the pores occupied by the wetting fluid. Also, with the above discussion, we may conclude that $r^* = r^*(S_w)$ and $R = R(S_w)$. We shall later see that because the above functional relationships are not unique, the capillary pressure relationship, $p_c = p_c(S_w)$, exhibits the phenomenon of hysteresis.

Two dimensionless numbers may be mentioned in connection with entrapment of a non-wetting (nw) fluid ganglion by a surrounding wetting (w) fluid moving at a velocity V_w:

Often, it is useful to use (for an isotropic domain) the dimensionless *capillary number* Ca_α, when considering the flow of an α-phase in two-phase flow. For example, for gas flow in an oil-gas system:

$$\mathrm{Ca}_g = \frac{\mu_g q_g}{\gamma_{og}}. \tag{6.1.8}$$

It expresses the ratio between the viscous force, or drag, tending to move the gas, and the surface tension, or capillary forces, tending to entrap it. The capillary number is small when capillary forces dominate the flow processes

Some authors define the capillary number as:

$$\mathrm{C}_{\mathrm{an}} = \frac{k_\alpha \nabla p_n}{\gamma_{nw}}. \tag{6.1.9}$$

with $\mathrm{C}_{an} < 10^{-5}$ often mentioned as the range for which the flow is capillarity dominated.

It is interesting to mention here the work on the relationships between capillary pressure, p_c, and relative permeability, k_{rw} and interfacial areas published by Joekar-Niasar et al. (2008), following the work of Hassanizadeh and Gray (1990). They conducted experiments on two pore-network models: one, composed of a network of interconnected tubes, that has only pore throats, while the other, of tubes and pores, that has both pore bodies and pore throats. They concluded that the latter kind of network model, namely, the one composed of pore throats and pore bodies, is required to produce hysteresis. Their main conclusion was that the capillary pressure and relative permeability, say k_{rw}, depend not only on saturation, say S_w, but also on the relevant *specific interfacial area* (= surface area per unit volume of porous medium), $a_{\alpha\beta}$, $\alpha\beta = ws, ns, wn$, of the w-s, n-s, and *w-n microscopic interfaces*. Thus, they showed that the $w - n$ specific interfacial area may be considered as an essential variable in the description of multiphase flow. They have also investigated the effects of the specific interfacial areas on hysteresis. Similar results were obtained by Reeves and Celia (1996). They have also investigated the effect of a_{w-n}.

For a given rigid porous medium, neglecting any effect of fluid composition on the structure of the pores, the effect of the pore- or grain-size distribution on capillary pressure curves is the same, regardless of the nature of the two fluids. The effect of fluid properties may be stated by the general expression:

$$S_w(p_c|_{\text{fluids } n_1, w_1}) = S_w(\beta_{12} p_c|_{\text{fluids } n_2, w_2}), \tag{6.1.10}$$

where:

$$\beta_{12} = \frac{\gamma_{n_1,w_1} \cos \theta_{n_1,w_1}}{\gamma_{n_2,w_2} \cos \theta_{n_2,w_2}} \tag{6.1.11}$$

is a *scaling factor*, and $\{n_1, w_1\}$ and $\{n_2, w_2\}$ represent pairs of nonwetting and wetting fluids. When the contact angles remain unchanged, then:

$$\beta_{12} = \frac{\gamma_{n_1,w_1}}{\gamma_{n_2,w_2}}. \tag{6.1.12}$$

The two immiscible fluids may be different, or they may be the same two fluids under different temperature and slightly different chemical compositions. Therefore, using fluids n_1 and w_1 as reference fluids, we can use the above relationships to obtain the capillary pressure curve for any other pair of immiscible fluids.

Let us add a few comments on the concept of *specific yield* introduced in Sect. 5.3.1 C. This concept, defined as the *volume of water added to or released from storage in a phreatic aquifer per unit decline or rise of the phreatic surface*, is one used mainly when hydrologists consider essentially horizontal flow in a phreatic aquifer.

After a prolonged period of time without accretion, the moisture distribution above a phreatic surface takes the form of the capillary pressure curve that corresponds to the relevant soil. Figure 6.5a shows two such steady-state moisture distribution curves that correspond to two water table depths: at time t' and at time t'', with the latter water table lower by Δh. When the initial and final water tables are sufficiently deep below ground surface, and sufficient time has elapsed, so that a new equilibrium moisture distribution has been reached, the curves θ'_w and θ''_w are identical in shape, with one being shifted vertically with respect to the other. The volume of water drained is indicated by the shaded area in the figure. Then, per unit horizontal area, as in the definition of S_y in Sect. 5.3.1 C,

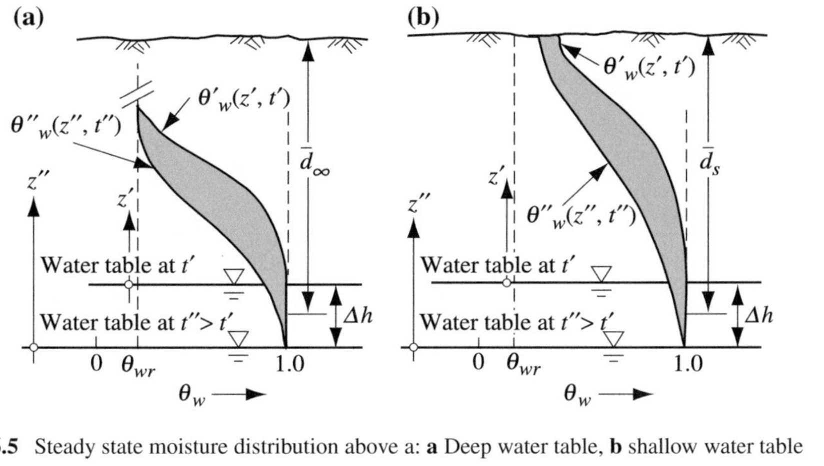

Fig. 6.5 Steady state moisture distribution above a: **a** Deep water table, **b** shallow water table

$$S_y(\bar{d}_\infty, t) = \frac{\text{volume of water drained}}{\Delta h}$$
$$= \frac{1}{\Delta h}\left[\phi(\Delta h) + \int_{z'=0}^{z'=d'} \theta'_w(z', t)dz - \int_{z'=0}^{z'=d''} \theta''_w(z'', t)dz\right]. \quad (6.1.13)$$

When the water table is at a shallow depth below ground surface (Fig. 6.5b), the specific yield, θ_y, is a function of both water table depths, at t' and at t''. The same is true when the soil is inhomogeneous (e.g., layered).

When changes in the water table elevations are slow, the corresponding changes in moisture distribution have sufficient time to adjust continuously, and the lag between the lowering of the water table and the total volume of water drained practically vanishes.

C. Experimental and Analytical Expressions for $p_c = p_c(S_w)$

The capillary pressure relationship, $p_c = p_c(S_w)$, is, usually, obtained by conducting laboratory experiments in which measured static saturation-capillary pressure data are obtained for soil cores. A typical experimental apparatus for determining capillary pressure in a gas-liquid (= the wetting fluid) system, is shown, schematically, in Fig. 6.6. The porous medium sample is placed in a cell on a 'capillary barrier', or 'porous plate', which itself is a porous material, ceramic, sintered metal, or fritted glass. The grain (or pore) size of this barrier is selected such that it is sufficiently small to prevent the gas from invading the sample under the capillary pressures to be applied during the course of the experiment. To achieve this goal, the *gas entry pressure* of the porous plate should exceed the range of capillary pressures that are planned for the experiment. However, if the pores in the barrier are too small, a long time will be required for equilibrium to be reached at every stage of the experiment. The sample (= core) is initially saturated with the liquid, e.g., water, with a zero

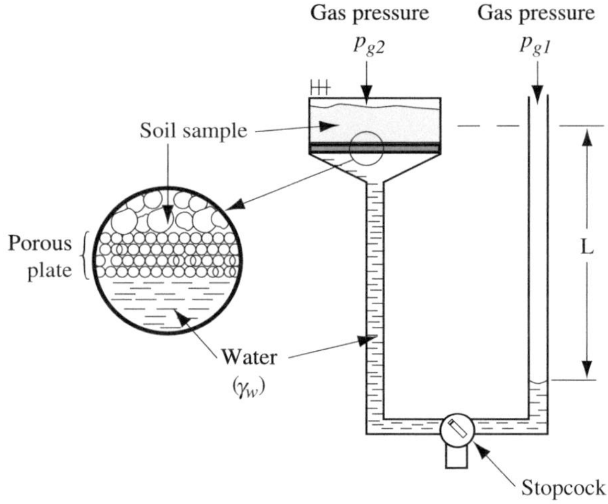

Fig. 6.6 Apparatus for determining the capillary pressure curve of a core

capillary head maintained by adjusting the liquid level in the tube connected to the bottom of the cell. The top of the sample is in contact with gas at a known pressure. The capillary pressure in the sample is increased incrementally, by producing a negative (gauge) liquid pressure at the bottom of the core (e.g., by lowering the tube), or by increasing the gas pressure in contact with the top of the core. Either way, the capillary pressure is increased and liquid will leave the core through the capillary barrier. After equilibrium is attained, following each incremental change in the capillary pressure, the volume of liquid outflow is measured and the new liquid saturation is computed. This procedure is repeated step-wise to generate points on the drainage capillary pressure curve.

Figure 6.7 shows results of a typical drainage-imbibition experiment with hysteresis (to be discussed in Sect. 6.1 E below). In this experiment, the liquid (= wetting fluid) can be removed by drainage as long as the remaining liquid, at least in part, constitutes a continuous phase, i.e., above the *irreducible wetting liquid saturation*, S_{wr}. In the case of water, this saturation is also referred to as *connate water saturation*. When the liquid reaches S_{wr}, it becomes discontinuous everywhere, in the form of isolated *pendular rings*, ganglia, or very thin liquid films on the solid surface in pores from which most of the liquid has been evacuated. Under such conditions, the liquid's effective permeability vanishes, and further drainage by *liquid flow* cannot be produced by a pressure gradient and gravity. The above statement is not completely accurate as some water flow may take place even under saturations below S_{wr} in the form of film flow (see Dullien, 1992). The experiment is terminated when the air entry value (see Sect. 6.1 B) of the porous plate is approached.

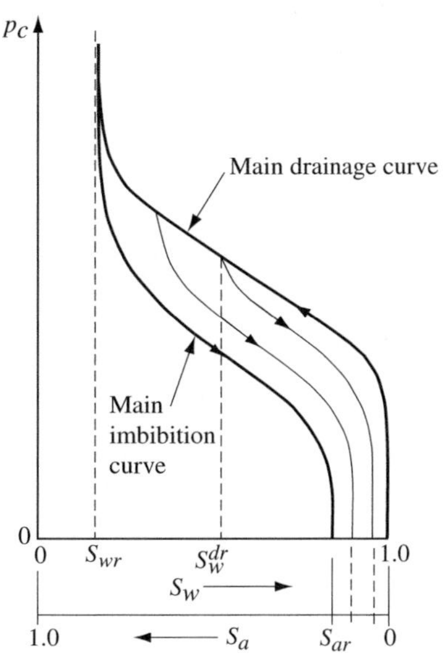

Fig. 6.7 Drainage and imbibition capillary pressure curves

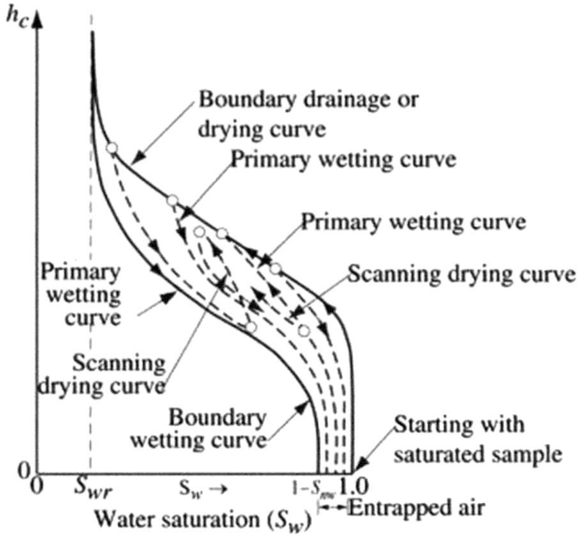

Fig. 6.8 Hysteresis in a capillary pressure-head curve

An imbibition p_c-curve is obtained by reversing the process. We note that once a sample saturated by a wetting phase has been drained, it can never return to full saturation (by flow); part of the void space will always be occupied by some residual non-wetting fluid.

Figure 6.8 shows the relationship between *capillary pressure head*, h_c $(= p_c/\rho_w g)$, and wetting fluid (e.g., water) saturation, $h_c = h_c(S_w)$, as a soil sample, initially at *irreducible wetting fluid saturation*, S_{wr}, is being wette4d. It also shows how a sample, initially at the *residual non-wetting fluid saturation*, S_{nr}, is being wetted. Note that such wetting cannot reach full saturation.

Consider a fully saturated sample, $S_w = 1$. As the sample is being drained, a non-wetting fluid (e.g., air) enters the void space. we move along the *boundary drying curve*. We can drain the sample of the wetting fluid until we reach a saturation, $S_w = S_{wr}$, called *irreducible wetting fluid saturation*, at which the wetting fluid is in the form of isolated menisci, and ganglia. As such, it has no permeability and drainage cannot be continued. We note the asymptotic shape of the curve as it approaches $S_w = S_{wr}$. At this stage, the wetting fluid can further be removed only by evaporation.

We now reverse direction and start wetting the sample (= imbibition), i.e., increasing S_w. We start from a sample at S_{wr} and wet it. As shown on Fig. 6.8, we move now along the *boundary wetting curve*. However, once a saturated sample has been drained, and this means that a nonwetting fluid has entered the void space, the sample saturation because of the phenomenon of *residual nonwetting saturation*, $S_n = S_{nr}$. At this point $S_w = 1 - S_{nr}$. It is called (= *entrapped air* when air is the nonwetting phase); it cannot be removed by flow.

For a given capillary pressure, a higher wetting fluid saturation is obtained during drainage than imbibition. If. during wetting or drying, the direction is reversed, the change in $h_c = h_c(S_w)$ follows wetting or drying *scanning curves*. The scanning curves are shown as dashed lines in the figure. In this way, the macroscopic $h_c = h_c(S_w)$-relationship, described by the capillary pressure head curve, depends also on the wetting-drying history of the particular sample under consideration.

Over the years, methods have also been developed for estimating soil hydraulic properties from grain size distribution data. Such methods are appealing, since grain size distribution data can more easily be obtained than hydraulic data. Some methods for estimating parameters appearing in saturation-capillary pressure models are based on statistical analyses of measured data (e.g., McCuen et al. 1981; Campbell 1985; Rauls and Brakensiek 1985; Carsel and Parrish 1988), while others employ quasi-physical models (e.g., Arya and Paris 1981; Mishra et al. 1989).

Various authors have proposed analytical expressions for the general shape of capillary pressure curves. Each of the proposed expressions involves a number of coefficients that must be determined by solving the inverse problem, i.e., by fitting the analytical expression to measured experimental data.

- Brooks and Corey (1964, 1966) proposed the relationship for an air-water (w) system:

$$S_{we} = \begin{cases} \left(\dfrac{p_b}{p_c}\right)^{\lambda} & \text{for} \quad p_c \geq p_b, \\ 1 & \text{for} \quad p_c < p_b, \end{cases} \tag{6.1.14}$$

where:

$$S_{we} = \frac{S_w - S_{wr}}{1.0 - S_{wr}}, \tag{6.1.15}$$

is called the *effective*, or *reduced* wetting fluid *saturation*, In (6.1.14), λ is called *pore size distribution index*, and p_b is the *bubbling pressure*. This is, approximately, the minimum value of p_c on a drainage capillary pressure curve at which a continuous air phase exists in the void space. The coefficients λ, S_{wr} and S_{nr} (= *residual non-wetting fluid saturation*; see Fig. 6.7 for $a \equiv n$) are fitting parameters whose values may vary, depending on the conditions under which the saturation-capillary pressure data are measured. The 1.0 in the denominator of (6.1.15), is, sometimes, replaced by the maximum wetting fluid saturation, $1 - S_{nr}$, where S_{nr} denoting the residual nonwetting fluid saturation.

- Brutsaert (1966), also for an air-water system, proposed the relationship

$$S_{we} = \begin{cases} \dfrac{1}{1 + (A\psi)^B} & \text{for} \quad \psi \geq 0, \\ 1 & \text{for} \quad \psi < 0, \end{cases} \tag{6.1.16}$$

where A and B are positive curve fitting coefficients.

- Vauclin et al. (1979) introduced the relationship

$$S_{we} = \begin{cases} \dfrac{1}{1+(A \ln \psi)^B} & \text{for } \psi \geq 1 \text{ cm}, \\ 1 & \text{for } \psi < 1 \text{ cm}, \end{cases} \tag{6.1.17}$$

where A and B are positive curve fitting coefficients.

- van Genuchten (1980), for unsaturated flow, proposed the relationship:

$$S'_{we} \equiv \frac{S_w - S_{wr}}{1 - S_{ar} - S_{wr}} = \begin{cases} \left[\dfrac{1}{1+(A\psi)^B}\right]^C & \text{for } \psi \geq 0, \\ 1 & \text{for } \psi < 0, \end{cases} \tag{6.1.18}$$

where A, B and $C\ (= 1 - 1/B)$ are positive curve fitting coefficients.

- van Genuchten (1980), for the more general case of two-phase flow, proposed:

$$S''_{we} \equiv \frac{S_w - S_r}{S_{w,max} - S_r} = \left[1 + \left(\frac{p_c}{p'_c}\right)^n\right]^{-m}, \quad \text{or} \quad p_c = p'_c\left(\frac{1}{(S''_{we})^{1/m}} - 1\right)^{1/n}, \tag{6.1.19}$$

where $S_{w,max}$ denotes the maximum saturation, not necessarily 1.0, and p'_c, n and m are porous medium coefficients that have to be determined experimentally.

The Brooks and Corey capillary expression (6.1.14) is a limiting form of the van Genuchten relation for $p_c/p'_c \gg 1$.

D. Leverett Function

Using dimensional analysis and a semi-empirical approach, Leverett (1941) introduced the dimensionless function $\mathcal{L} = \mathcal{L}(S_w)$, called the *Leverett function* for an isotropic porous medium:

$$\mathcal{L}(S_w) = \frac{p_c}{\gamma}\sqrt{\frac{k}{\phi}}, \quad p_c = p_c(S_w), \quad k = k(S_w), \tag{6.1.20}$$

in which the quotient $\sqrt{k/\phi}$ represents the hydraulic radius of the void space. Leverett showed that the curves $\mathcal{L} = \mathcal{L}(S_w)$ of a number of unconsolidated sands reduce to a common curve.

E. Hysteresis in Capillary Pressure

The concept of capillary pressure as the difference $p_c = p_n - p_w$ is motivated by the pore-scale interfacial pressure difference between the two fluids. However, strictly, the Laplace formula (2.4.12) is valid only for a spherical meniscus. Marle (1981) noted that this formula is strictly valid for a conical capillary in which the interface is spherical. Under such conditions, we obtain a unique relationship between capillary pressure and saturation. However, in a real porous medium, such relationship does not really exist due to the complexity of the shape of the channels constituting the void space. A single length scale characterizing the void space can no more be identified for a given porous medium. Capillary hysteresis, to be discussed below, as well as the void space heterogeneity lead to hysteresis effects in the capillary pressure-saturation relationship. Furthermore, the Laplace formula does not take into account dynamic effects, i.e., dynamic phenomena that occur when changes in saturation occur when the fluids are in motion (see, for example, Stauffer 1978; Das and Mizrahei 2012; Cueto-Felgueroso and Juanes 2012. Thus, several authors describe capillary pressure under dynamic conditions by expressions of the kind (e.g., Dahle et al. 2005)

$$p_n - p_w = p_c(S_w) + \mathcal{F}\left(S_w, \frac{\partial S_w}{\partial t}\right), \tag{6.1.21}$$

where $p_c(S_w)$ is any of the published formula, without the dynamic effect, while $\mathcal{F}$ denotes a functional relationship that introduces the latter.

A *dynamic capillarity theory* was developed by Hassanizadeh and Gray (1990) using a thermodynamic approach. A review of dynamic capillary effects can be found in Hassanizadeh et al. (2002)

The dependence of the capillary pressure curve on the history of drainage and wetting of a sample, i.e., at a point in a porous medium domain, is an observed phenomenon called *hysteresis*. It is attributed to a number of causes. One, called the *ink-bottle effect*, results from the shape of the pore space, with interchanging narrow (throats) and wide passages (see Bear and Cheng 2010, p. 280). During drainage and rewetting, menisci having the same radius of curvature occur at different pores, thus yielding the same capillary pressure for different wetting fluid saturations. As water is drained, the radius of curvature of the $w - n$ meniscus diminishes. At the narrowest part of the throat, the curvature of the meniscus cannot continue to increase gradually; instead, the meniscus abruptly retreats to a nearby throat. This sudden change is called *Haines jump* (Haines 1930). A similar phenomenon occurs during wetting. Altogether, the drying curve depends on the narrow throats (small radii of meniscus curvature), while wetting depends on the maximum diameter of the large pores. The hysteresis effect is more significant in coarse-textured porous medium, in the low-suction range, where pores may empty at an appreciably higher suction than that at which they fill-up (Hillel 1980).

Another effect, called the *raindrop effect* (see Bear and Cheng 2010, p. 280), is due to the fact that the contact angle is larger at the advancing trace of a $w - n$ interface on a solid than at the receding one, because of impurities, possible variability in the minerals that compose the surface, solid roughness, and gravity. Also, when a fluid is *polar*, as is water, the contact angle depends on whether the solid surface has been previously wetted by the fluid, or not.

A third cause for hysteresis is the entrapment of the non-wetting fluid, as an initially wetting fluid saturated sample is drained and then rewetted. Also, during imbibition, the displacing wetting fluid is continuous, while during drainage (of the wetting fluid) part of the wetting fluid may be by-passed by the non-wetting one, and become immobile. Altogether, we observe a higher capillary pressure after drainage than after imbibition.

Finally, consolidation, swelling, and shrinkage of the solid matrix as it is dried and wetted may also contribute to hysteresis in the capillary pressure curve, especially in fine, unconsolidated porous media.

Theoretical analyses of hysteresis in the air-water capillary pressure curve have been presented by Poulovassilis (1962), Topp (1969, 1971), Mualem (1974, 1976, 1984), Kool and Parker (1987), Luckner et al. (1989), and others. Hysteresis in oil-water and gas-oil systems have also been studied by numerous researchers (e.g., Naar and Henderson 1961; Snell 1962; Land 1968; Schneider and Owens 1970; Parker and Lenhard 1987).

Nowadays, the most commonly used theory that explains and describes soil water hysteresis is the *independent domain theory* (Poulovassilis 1962). Mualem (1973) suggested a similarity hypothesis, according to which the bivariate domain density distribution function is represented as a product of two univariate distribution functions. The resulting model significantly reduced the amount of data required for calibration. In subsequent years, Mualem (1974, 1977) introduced the *universal hysteresis model* based on a non-dimensional formulation (Mualem, 1979). Extension of the *domain theory* to the prediction of hysteresis in unsaturated hydraulic conductivity has been successfully initiated by Mualem (1976).

Joekar-Niasar et al. (2013) summarize a number of studies on trapping and hysteresis in two-phase flow in porous media and use a pore-network model to study how the fluids in the (simulated) pore-space. They investigated how the topology of the fluids changes during drainage and imbibition including first, main and scanning capillary curves. They found a strong hysteretic behavior in the relationship between disconnected nonwetting fluid saturation and the wetting fluid saturation in a water-wet medium. They noted how the invading nonwetting phase coalescence with the existing disconnected nonwetting phase and how this behavior depends critically on the presence (or lack) of a connected nonwetting phase at the beginning of the drainage process as well as on the pore geometry. This concluded that this dependence involves a mechanism that they called reversible corner filling.

A method for obtaining up to two scanning drainage and imbibition curves is described by Finsterle et al. (1998) and Doughty (2007, 2013).

6.2 Advective Fluxes in Multiple Phases

Basically, like Darcy's law for a single fluid phase that occupies the entire void space, the advective flux of a phase that occupies only part of the void space in multiphase flow is also a simplified version of the momentum balance equation for that phase. Thus, the starting point is the momentum balance equation of the considered phase.

6.2.1 Two Fluid Phases

Let us return to Figs. 6.1, 6.2 and 6.3. In all these figures we note the interface between a wetting fluid and a nonwetting one, as well as well as on interface between the wetting fluid and the solid. In principle, there is no interface between the nonwetting fluid and the solid. In reality the wetting fluid that separates the solid from the nonwetting fluid is very thin and for the sake of simplicity its effects are neglected. Thus, the total surface surrounding the wetting fluid is made up of (1) a wetting fluid-solid part (disregarding the presence of the thin film) and (2) a wetting fluid-nonwetting fluid one. Similarly, the nonwetting fluid is in contact with both the solid (across the film) and the wetting fluid, through relevant portions of the total surface surrounding it.

The momentum that can be exchanged between the two fluids across their common interphase boundary must be taken into account when writing the macroscopic (= averaged) momentum balance equation for each of the fluids. Thus, for the wetting fluid, the (macroscopic) momentum transfer at a point in the porous medium domain must be expressed by two terms: one related to the surface ($\mathcal{S}_{ws}$) between the wetting fluid and the solid, and the other related to the surface ($\mathcal{S}_{wn}$) between the wetting fluid and the nonwetting one. In this way, two expressions are required in order to express interphase momentum transfer: one for each of these surfaces, in terms of averaged velocities, viscosities and coefficients that represent the configuration of the phase within the REV. This conceptual model serves as a basis for the derivation of flux expressions that will exhibit *coupling* between the two adjacent immiscible fluids, due to momentum transfer across the microscopic interfaces that separate them. As a consequence, the pressure gradient in one fluid will also cause movement in the other fluid.

As a starting point, we consider the macroscopic momentum balance equation for single phase flow, (3.6.4), repeated here for convenience as:

$$\phi\rho\frac{D\mathbf{V}}{Dt} = \nabla\cdot(\phi\mu\nabla\mathbf{V}) - \phi(\nabla p + \rho g\nabla z) - \phi\frac{\mu\mathbf{R}}{\Delta^2}\cdot(\mathbf{V} - \mathbf{V}_s). \qquad (6.2.1)$$

We recall that the last term on the r.h.s. of the above equation expresses momentum transfer from the solid to the fluid, per unit volume of porous medium per unit time, as presented and discussed in Sect. 3.5.1.

Following this phenomenological approach, we assume that the same momentum balance equation applies also to the wetting phase that occupies only part of the void space, at the volumetric fraction θ_w ($\equiv \phi S_w$), except that we have to take into account momentum transfer across both w-n and $w - s$ interfaces. As usual, we use $f^{\mathrm{M}}_{s\rightarrow\alpha}$ to denote the momentum transfer from the solid to the α-fluid, per unit volume of porous medium. We shall use the characteristic length Δ both for representing the length characterizing the hydraulic radius of a phase at the point (e.g., $\Delta_{\alpha s} = \mathbb{V}_{o\alpha}/\mathcal{S}_{\alpha s}$) and the characteristic distance from the interior of the α-phase to the solid surface. Thus:

$$f^{\mathbf{M}}_{s\to\alpha} = \mathbf{R}_{\alpha,s}\mu_\alpha \cdot \frac{\mathbf{V}_s - \mathbf{V}_\alpha}{\Delta_{\alpha s}} \frac{\mathcal{S}_{\alpha s}}{\mathbb{V}_o} \frac{\mathbb{V}_{o\alpha}}{\mathbb{V}_{o\alpha}} = \theta_\alpha \mathbf{R}_{\alpha,s}\mu_\alpha \cdot \frac{\mathbf{V}_s - \mathbf{V}_\alpha}{\Delta^2_{\alpha s}},$$

$$\theta_\alpha = \frac{\mathbb{V}_{o\alpha}}{\mathbb{V}_o}, \quad \Delta_{\alpha s} = \frac{\mathbb{V}_{o\alpha}}{\mathcal{S}_{\alpha s}},$$

where the second rank symmetric tensor $\mathbf{R}$ is a coefficient of proportionality.

Accordingly, for the wetting phase, we write the momentum balance equation:

$$\begin{aligned}\theta_w \rho_w \frac{D\mathbf{V}_w}{Dt} &= \nabla\cdot(\theta_w \mu_w \nabla \mathbf{V}_w) - \theta_w(\nabla p_w + \rho_w g \nabla z)\\ &- \theta_w \frac{\mu_{wn}\mathbf{R}_{w,n}}{\Delta^2_{wn}}(\mathbf{V}_w - \mathbf{V}_n) - \theta_w \frac{\mu_w \mathbf{R}_{w,s}}{\Delta^2_{ws}}(\mathbf{V}_w - \mathbf{V}_s), \qquad (6.2.2)\end{aligned}$$

and for the non-wetting phase:

$$\begin{aligned}\theta_n \rho_n \frac{D\mathbf{V}_n}{Dt} &= \nabla\cdot\theta_n \mu_n \nabla \mathbf{V}_n - \theta_n(\nabla p_n + \rho_n g \nabla z)\\ &- \theta_n \frac{\mu_{nw}\mathbf{R}_{n,w}}{\Delta^2_{nw}}(\mathbf{V}_n - \mathbf{V}_w) - \theta_n \frac{\mu_n \mathbf{R}_{n,s}}{\Delta^2_{ns}}(\mathbf{V}_n - \mathbf{V}_s). \qquad (6.2.3)\end{aligned}$$

Note that each term in the above momentum balance equation expresses added momentum per unit volume of fluid per unit time. We have introduced the fluid viscosities μ_{nw} and μ_{wn} (assumed equal) for the thin fluid domains on both sided of the (microscopic) $n - w$ interfaces.

Again, in the above momentum balance equations, Δ_{ws}, and Δ_{ns} denote characteristic distances from the interior of the w-phase and that of the n-phase, to the $\mathcal{S}_{ws}$ and $\mathcal{S}_{ns}$ surfaces, respectively, and $\Delta_{nw}(\equiv \Delta_{wn})$ denotes a length characterizing the distance between the interiors of the two phases. The $\mathbf{R}_{\alpha,s}$ and $\mathbf{R}_\alpha$, β symbols denote second rank symmetric tensors. Later, we shall suggest how they are related to the *tortuosities* of the α and β fluids, each occupying a portion of the void-space. Both the $\mathbf{R}$-coefficients and the Δ-lengths are functions of the fluids' saturations. Some introductory remarks about tensors are presented in Sect. 9.1.

When considering an REV, there exist two additional surfaces on the external boundary of an REV across which momentum is transmitted by advection: a wetting fluid-wetting fluid surface, and nonwetting fluid-nonwetting fluid one. The possibility that a solid surface will coincide with the boundary of an REV is assumed negligible.

Neglecting inertial effects and the momentum flux resulting from $\nabla\mathbf{V}$, we obtain for the wetting phase,

$$\theta_w(\nabla p_w + \rho_w g \nabla z) = -\theta_w \frac{\mu_{wn}\mathbf{R}_{wn}}{\Delta^2_{wn}}(\mathbf{V}_w - \mathbf{V}_n) - \theta_w \frac{\mu_w \mathbf{R}_{ws}}{\Delta^2_{ws}}(\mathbf{V}_w - \mathbf{V}_s), \quad (6.2.4)$$

and for the non-wetting phase,

$$\theta_n(\nabla p_n + \rho_n g \nabla z) = -\theta_n \frac{\mu_{nw}\mathbf{R}_{nw}}{\Delta^2_{nw}}(\mathbf{V}_n - \mathbf{V}_w) - \theta_n \frac{\mu_n \mathbf{R}_{ns}}{\Delta^2_{ns}}(\mathbf{V}_n - \mathbf{V}_s). \qquad (6.2.5)$$

At this point, let us simplify the discussion by changing the notation:

$$\mathbf{k}_{ws} = \theta_w \Delta^2_{ws} \mathbf{R}^T_{ws}. \qquad (6.2.6)$$

Recall that the μ_{wn} $(= \mu_{nw})$ denotes the fluids' viscosity in the vicinity of the w-n interface, on both sides of the latter.

We can now rewrite the simplified momentum balance equation for the wetting phase in the form:

$$\theta_w(\nabla p_w + \rho_w g \nabla z) = -\theta_w \frac{\mu_{wn}}{k_{wn}}(\mathbf{V}_w - \mathbf{V}_n) - \theta_w \frac{\mu_w}{k_{ws}}(\mathbf{V}_w - \mathbf{V}_s), \qquad (6.2.7)$$

and for the non-wetting phase,

$$\theta_n(\nabla p_n + \rho_n g \nabla z) = -\theta_n \frac{\mu_{nw}}{k_{nw}}(\mathbf{V}_n - \mathbf{V}_w) - \theta_n \frac{\mu_n}{k_{ns}}(\mathbf{V}_n - \mathbf{V}_s). \qquad (6.2.8)$$

Dividing (6.2.7) and (6.2.8) by the θ's, with

$$\nabla F_w = \nabla p_w + \rho_w g \nabla z, \quad \nabla F_n = \nabla p_n + \rho_n g \nabla z,$$

$$x_1 = V_w - V_n,\, x_2 = V_w - V_s;\, x_3 = V_n - V_s, \text{ with } \; x_3 = -x_1 + x_2,$$

$$A_{wn} = \mu_{wn}/k_{wn},\; B_{ws} = \mu_w/k_{ws},\; A_{nw} = \mu_{nw}/k_{nw},\; B_{ns} = \mu_n/k_{ns},$$

we obtain:

$$\begin{aligned} -\nabla F_w &= A_{wn}x_1 + B_{ws}x_2. \\ -\nabla F_n &= -(A_{nw} + B_{ns})x_1 + B_{ns}x_2. \end{aligned} \qquad (6.2.9)$$

By multiplying 1st equation by $(A_{nw} + B_{ns})$ and 2nd equation by A_{wn}, we obtain,

$$\begin{aligned} -(A_{nw} + B_{ns})\nabla F_w &= A_{wn}(A_{nw} + B_{ns})x_1 + (A_{nw} + B_{ns})B_{ws}x_2, \\ -A_{wn}\nabla F_n &= -(A_{nw} + B_{ns})A_{wn}x_1 + B_{ns}A_{wn}x_2. \end{aligned} \qquad (6.2.10)$$

By adding the two equations, we obtain

$$-(A_{nw} + B_{ns})\nabla F_w - A_{wn}\nabla F_n = x_2[(A_{nw} + B_{ns})B_{ws} + B_{ns}A_{wn}],$$

or

$$x_2 = -\frac{(A_{nw} + B_{ns})}{(A_{nw} + B_{ns})B_{ws} + B_{ns}A_{wn}}\nabla F_w - \frac{A_{wn}}{(A_{nw} + B_{ns})B_{ws} + B_{ns}A_{wn}}\nabla F_n.$$

Similarly,

$$x_1 = -\frac{B_{ns}}{A_{wn}B_{ns} + (A_{nw} + B_{ns})B_{ws}}\nabla F_w - \frac{B_{ws}}{(A_{nw} + B_{ns})B_{ws} - A_{ws}B_{ns}}\nabla F_n.$$

One may express this viscous coupling, say between the flux of a wetting phase (w) and of a nonwetting one (n), both with respect to the solid, in the form:

$$\mathbf{q}_{rw} = -\frac{\mathsf{k}_w^w(S_w)}{\mu_w}\cdot(\nabla p_w + \rho_w g\nabla z) - \frac{\mathsf{k}_w^n(S_n)}{\mu_n}\cdot(\nabla p_n + \rho_n g\nabla z), \quad (6.2.11)$$

$$\mathbf{q}_{rn} = -\frac{\mathsf{k}_n^w(S_w)}{\mu_w}\cdot(\nabla p_w + \rho_w g\nabla z) - \frac{\mathsf{k}_n^n(S_n)}{\mu_n}\cdot(\nabla p_n + \rho_n g\nabla z). \quad (6.2.12)$$

Recall that the subscript r above means "relative to the solid".

In this way, at least in principle, the motion in each of the two phases is *coupled* to that of the other; a pressure gradient in one fluid produces flow *also* in the other fluid.

We note the coupling (between the two phases) coefficients in the above equations. Coupled two-phase flow in homogeneous, isotropic porous media has been studied by many authors (e.g., Rose 1972, 1988; Sanchez-Palencia 1980; Whitaker 1986a; Kalaydjian 1987; Auriault et al. 1989). The significance of this coupling has been also extensively debated in the literature, starting in the 1950s (e.g., Yuster 1951; Odeh 1959; Bentsen and Manai 1993; Goode and Ramakrishnan 1993; Lasseux et al. 1996). Rose (1972, 1988, 1990, 1997) and Rose and Rose (2005) suggested a relationship between the two cross-permeability coefficients. Avraam and Payatakes (1995) discussed this topic and reported on experimental investigations. Unfortunately, relatively few experiments (e.g., Liang and Lohrenz 1994; Dullien and Dong 1996) have been conducted to determine the significance of coupling that takes place in multiphase flow due to momentum transfer across fluid-fluid interfaces.

Rose (1972) attributed the coupling to Onsager's theory of coupled processes (Sect. 2.6), and hence suggesting that $K_w^n/\mu_w = K_n^w/\mu_n$, although the macroscopic-level coupling suggested here is not due to molecular-level phenomena, but to macroscopic-level momentum transfer across interphase boundaries.

In case of temperature and concentration gradients, Bear and Bachmat (1991, p. 186) show that an additional term appears in each of the above equations, due to gradients in (averaged) surface tension between the two fluid phases.

By neglecting the momentum exchange across the (microscopic) $w - n$ interfaces, the resulting averaged momentum balance equation, written separately for each fluid phase, is *identical in form* to that written for that phase when it occupies the entire void space. However, since the shape and size of the solid-fluid surfaces and of the volumes occupied by these phases within an REV vary with the saturation of the considered phase, the resistance to the flow of each fluid phase depends also on its saturation. As presented earlier, the permeability of a fluid that occupies the entire void space of a porous medium depends on (1) certain geometric features of the fluid-solid interface, (2) a length (hydraulic radius) that characterizes the distance between the interior of the volume occupied by the considered phase (within the REV) and the fluid-solid interface, and (3) the porosity. When a fluid occupies only part of the void space, the geometrical features, e.g., the tortuosity, become functions of the fluid's saturation, while the porosity is replaced by the volumetric fraction of the void space occupied by the fluid. The conclusion is that *the permeability of a considered fluid phase is a function of the saturation of that fluid.*

With the above considerations in mind, we can now write the macroscopic equations that describe the simultaneous motion of two (assumed) immiscible fluids (w, n) each occupying part of the void space, in the form:

$$\mathbf{q}_{rw} = -\frac{\mathbf{k}_w(S_w)}{\mu_w} \cdot (\nabla p_w + \rho_w g \nabla z), \qquad q_{rw} \equiv \theta_n (\mathbf{V}_w - \mathbf{V}_s), \tag{6.2.13}$$

$$\mathbf{q}_{rn} = -\frac{\mathbf{k}_n(S_n)}{\mu_n} \cdot (\nabla p_n + \rho_n g \nabla z), \qquad q_{rn} \equiv \theta_n (\mathbf{V}_n - \mathbf{V}_s), \tag{6.2.14}$$

with $\mathbf{V}_s$ denoting the velocity of the (possibly moving) solid, $\mathbf{k}_w$ and $\mathbf{k}_n$ denoting the effective permeability tensors, and S_w and S_n denoting the respective fluid saturations. Note that the above equations are written for the general case of an anisotropic porous medium.

It is interesting to note that already in the early 30s of the 20th century, Wyckoff and Botset (1936) and Muskat (1946) have also proposed the use (6.2.13) and (6.2.14), as a generalization of Darcy's law for single phase flow, in connection with two phase flow (oil-water) in reservoir engineering.

We recall that in two phase flow, the pressures p_w and p_n are not independent of each other, as the difference between them is the capillary pressure which, in turn, depends on the saturations of the two phases.

When the density of the two fluids remains constant, the above flux equations can be expressed in terms of the piezometric head, $h_\alpha = z + p_\alpha/\rho_\alpha g$, $\alpha = w, n$, and in terms of the hydraulic conductivity, $K_\alpha = k_\alpha \rho_\alpha g/\mu_\alpha$, instead of the permeability, k_α.

In modeling flow and transport in the unsaturated zone (= air-water system), soil physicists often, but not always, assume that the air in the unsaturated zone is stationary (actually, the correct assumption is that $p_a \approx const.$), and use the concept of *suction head*, defined by (6.1.3), as the variable. They express the water flux by Darcy's law in terms of ψ_w,

$$\mathbf{q}_{rw} = \mathbf{K}_w(\psi_w) \cdot \nabla(\psi_w - z), \tag{6.2.15}$$

where we recall that $\rho_w = const.$, and the (mass) balance equation, without the source term, in the form or *Richard's equation*:

$$\frac{\partial \theta_w}{\partial t} = \nabla \cdot [\mathbf{D}_w(\theta_w) \cdot \nabla \theta_w] + \nabla \cdot [\mathbf{K}_w(\theta_w) \cdot \nabla z]. \tag{6.2.16}$$

If the effect of gravity, represented by the second term on the right-hand side of (6.2.16), is neglected, or the flow is horizontal, (6.2.16) reduces to

$$\frac{\partial \theta_w}{\partial t} = \nabla \cdot [\mathbf{D}_w(\theta_w) \cdot \nabla \theta_w], \tag{6.2.17}$$

known as *moisture diffusivity equation.*

When the porous medium is deformable, $\partial\phi/\partial t \neq 0$, and the flow equation should be written in terms of saturation as a state variable.

The (tensor) coefficients k_w and k_n denote the *effective permeabilities* to the wetting and to the nonwetting phases, respectively, discussed in Sect. 6.2.2. Equations (6.2.13) and (6.2.14) are written for the general case of an *anisotropic* porous medium and for fluid densities that may depend on pressure, concentration of dissolved components, and temperature. These two equations are not independent of each other. They are linked by the condition $S_w + S_n = 1$, and by the relationship between the saturation and the (macroscopic) capillary pressure, $p_c = p_c(S_w)$, discussed in Sect. 6.1.

6.2.2 *Effective Permeability*

The (macroscopic) coefficients k_w and k_n are properties of the (microscopic) geometrical configuration of the portion of void space occupied by each fluid phase. Since, for each phase, this configuration depends on the phase saturation, the effective permeability also depends on phase saturation, i.e.,

$$\mathsf{k}_w = \mathsf{k}_w(S_w) \qquad \text{and} \qquad \mathsf{k}_n = \mathsf{k}_n(S_n).$$

For an anisotropic porous medium, each of these effective permeabilities is a *second rank symmetric tensor*; otherwise, they are scalars. In component notation, we write $k_{wij}(S_w)$ and $k_{nij}(S_n)$ to emphasize that each of the ij-components, of either k_w, or k_n, may have a *different functional relationship to saturation* (Bear and Verruijt 1987; Stephens and Heermann 1988; McCord et al. 1991; and Friedman and Seaton 1996).

For an isotropic porous medium, a *relative permeability* to the w-fluid and to the n-fluid may be defined by (Bear and Verruijt 1987)

$$k_{rw}(S_w) = \frac{k_w(S_w)}{k_{\text{sat}}} \qquad \text{and} \qquad k_{rn}(S_n) = \frac{\mathsf{k}_n(S_n)}{k_{\text{sat}}}, \tag{6.2.18}$$

where k_{sat} is the permeability at full saturation ($S_w = 1$). The relative permeability is a dimensionless number in the range $0 \leq k_{rw} \leq 1$, and $0 \leq k_{rn} \leq 1$. It is a convenient and commonly used concept when considering the permeability in multiphase flow, mainly in reservoir engineering. There, the concept is often applied also to anisotropic

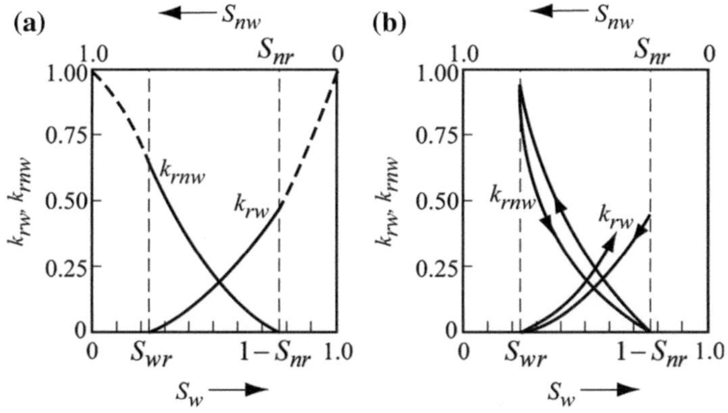

Fig. 6.9 Typical relative permeability curves: **a** Without hysteresis, and **b** with hysteresis

porous media, *assuming*, implicitly, that the same value of relative permeability is applicable to all components of the effective permeability tensor.

Figure 6.9 shows typical relative permeability curves for the unsaturated zone, where water and air are the wetting and the non-wetting phases, respectively. Consider the unsaturated zone below ground surface. Starting the drainage from a water saturated soil sample, we note a rapid decline in k_{rw} as the larger pores are drained first, and flow of water takes place through the smaller pores. This means that a smaller cross-sectional area is available for flow. This trend is expressed by (4.3.9). When the water saturation is below the *irreducible water saturation* S_{wr} (Sect. 6.1 D), the water remaining in the soil is in the form of isolated *pendular rings* (Sect. 6.1 A), and very thin films that cover the solid surface in the larger pores from which water has already been drained. In this form, the water constitutes a discontinuous, immobile phase that cannot transmit pressure. Thus, $k_{rw} = 0$ for $S_w \leq S_{wr}$. In reality, given enough time, the wetting phase will continue to drain by gravity in the form of films, reducing the saturation to below the irreducible saturation (Dullien 1992). For $S_w = 1$, temporarily overlooking the meaning of the dashed portion of the k_{rw}-curve, we have $k_{rw} = 1$, i.e., $k_{rw}(S_w)\big|_{S_w=1} \equiv k_{\text{sat}}$.

When a *nonwetting* fluid is being displaced by a wetting one to below a critical saturation value, S_{nr}, referred to as the *residual nonwetting fluid saturation*, the latter breaks down into isolated blobs, or *globules*. Usually, these remain immobile under the pressure gradient that drives the wetting fluid. The value of S_{nr} is determined by properties of the nonwetting fluid and of the solid. This phenomenon can easily be explained by the concept of capillary pressure discussed in Sect. 6.1. The shape of the globules within the void space is established in response to the capillary forces. These forces establish a pressure gradient within each globule that opposes that in the mobile fluid around it. Following the Laplace formula (2.4.12), the capillary pressure is of the order $2\gamma_{wn}/r$, where r denotes some characteristic radius of a pore. Thus, in an immobile globule, the menisci configurations adjust themselves to maintain a pressure equilibrium with the mobile wetting fluid, as long as the pressure gradient in that fluid is not too high. As this pressure gradient increases, at some point, equilibrium can no longer be maintained, and the globule will be displaced, until equilibrium is re-established.

This explains why the nonwetting fluid becomes immobile below the critical saturation S_{nr}. This phenomenon is of major significance in reservoir (petroleum) engineering and contaminant hydrology, as large quantities of oil remain immobile in the form of globules; they cannot be mobilized by the displacing water (which is the wetting fluid). The same phenomenon occurs also in an air-water system, as water displaces air.

If, at the some value of S_w, the drainage process is stopped and wetting of the sample begins, the latter cannot be brought back to full saturation. The air saturation of the sample cannot be lowered to below the *residual air saturation*, S_{ar}, because of *entrapped air*. The amount of entrapped air, in the form of air bubbles and air-filled portion of the void space completely surrounded by water in the larger pores, grows gradually as the sample is rewetted. Sometimes, this means that effective permeability never rises back to more than 0.5 k_{sat}, especially in view of the fact that for many soils, the slope of the effective permeability-saturation-curve becomes steeper as full saturation is approached.

The relative permeability curve for air is also shown in Fig. 6.9. Again, we note that $k_{ra} = 0$ for $S_a < S_{ar}$, and that due to the irreducible water saturation, k_{ra} cannot rise above $k_{ra}\big|_{S_a=1-S_{wr}}$, unless we start by wetting an initially dry sample at $S_a = 1$.

We emphasize that relative permeability curves have to be determined experimentally for each particular soil. However, various investigators have suggested analytical expressions for the relationship between relative permeability (or relative hydraulic conductivity) and saturation. These expressions were usually obtained by analyzing simplified models of porous media, such as a bundle of parallel capillary tubes, or a network of such tubes. The results, while highlighting the main features of the sought relationship, always contain numerical coefficients that characterize the considered model. They cannot be used for soils that have a much more complicated irregular structure. For a particular soil, the numerical values of the coefficients have to be determined by fitting the analytical expression to experimental curves. Analytical expressions (as compared with tables of experimental results) have the advantages that they can be used in analytical or semi-analytical solutions, and can more easily be used as input to numerical models.

Following are a few examples, all for isotropic porous media.

- Gardner (1958) suggested, for unsaturated flow, the expression

$$K_w = \frac{a}{b + |\psi|^m}, \tag{6.2.19}$$

  where a, b and m are constants, with $m \approx 2$ for heavy clay soil, and $m \approx 4$ for sand. He also suggested the exponential model

$$K_w = K_{\text{sat}}\, e^{-\alpha\psi}, \tag{6.2.20}$$

where α is a soil index parameter, related to pore size distribution, or, according to Raats (1976), to the reciprocal of a macroscopic capillary length scale.

- Childs and Collis-George (1950), for the flow of water in the unsaturated zone, introduced the expression

$$K_w = B\frac{\theta_w^3}{\Sigma_{vs}}, \tag{6.2.21}$$

where Σ_{vs} is the specific surface area of the soil, and B is a coefficient.

- Irmay (1954), for unsaturated flow, suggested

$$K_w(S_w) = K_{\text{sat}}\left(S_{we}\right)^3, \quad S_{we} = \frac{S_w - S_{wr}}{1 - S_{wr}}, \tag{6.2.22}$$

where S_{we} is the *effective water saturation* defined in Sect. 6.1.

- Corey (1957), for unsaturated flow, suggested a relationship proportional to $\left(S_{we}\right)^4$.
- Brooks and Corey (1964, 1966) suggested

$$k_w = \begin{cases} k_{\text{sat}} & \text{for} \quad p_c < p_b, \\ k_{\text{sat}}\left(\frac{p_b}{p_c}\right)^{\frac{2+\lambda}{\lambda}} & \text{for} \quad p_c \geq p_b, \end{cases} \tag{6.2.23}$$

where p_b is the *bubbling pressure*, or *air entry pressure*, related to the largest pore size forming a continuous network of water occupied channels within the porous medium, and λ is an index of the pore-size distribution of the porous medium. In this equation, k_{sat} is the permeability at $S_{we} = 1$, and not at $S_w = 1$. When combined with (6.1.14), Brooks and Corey (1964) obtained

$$k_w(S_{we}) = k_{\text{sat}}\left(S_{we}\right)^{\epsilon}, \qquad \epsilon = \frac{2 + 3\lambda}{\lambda}, \tag{6.2.24}$$

and

$$k_a(S_{we}) = k_{\text{sat}}\left(1 - S_{we}\right)^2\left[1 - (S_{we})^{\gamma}\right], \quad \gamma = \frac{2 + \lambda}{\lambda}, \tag{6.2.25}$$

where S_{we} is less than some maximum value (usually $\approx$0.85) at which k_a still exists.

- Combined with the work of Mualem (1976), van Genuchten (1980) work leads to the relationship:

$$k_w(S_{we}) = k_{\text{sat}}\, S_{we}^{\frac{1}{2}}\left[1 - \left(1 - S_{we}^{\frac{1}{m}}\right)^m\right]^2, \tag{6.2.26}$$

in which because water is immobile at saturations in the range $S_w \leq S_{wr}$, we made use of the *effective water saturation*, S_{we}, defined in (6.2.22).

- A combination of Burdine's expression for permeability with Mualem's (1976) work, leads to:

$$k_w(S_{we}) = k_{\text{sat}}\, S_{we}^2 \left[1 - \left(1 - S_{we}^{\frac{1}{m}}\right)^m\right]. \tag{6.2.27}$$

Effective permeability, or effective hydraulic conductivity, may also be presented as a function of the pressure head, ψ. However, the relationship $k_w(\psi)$ shows much more hysteresis than $k_w(S_w)$, probably due to the large hysteresis in the function $\psi(S_w)$. Hysteresis in $k_w(S_w)$ is generally ignored because the function $p_c(S_w)$ usually exhibits far greater hysteretic effects and because the values of the parameters required to describe hysteresis in $k_w(S_w)$ are highly uncertain.

6.2.3 *Relative Permeability-Capillary Pressure Relationship*

It is of interest to note the relationship between the capillary pressure, $p_c = p_c(S_w)$ and the effective permeabilities, $k_w = \mathsf{k}_w(S_w)$ and $k_a = k_a(S_a)$. The discussion is limited to an isotropic porous medium, and to water and air as typical wetting and non-wetting fluids. This relationship is called *Burdine's equations* (Burdine 1953).

Following Wylie and Gardner (1958), we consider the flow of water in a porous medium visualized as a bundle of capillary tubes with radii, r, in the range $r_{min} \leq r \leq r_{max}$ (corresponding to $S_{wr} \leq S_w \leq 1$, overlooking the presence of entrapped air).

To make the bundle of capillaries better resemble an actual porous medium, let the bundle be cut into a large number of short pieces. Then, the resulting pieces of capillary tubes are rearranged randomly, and re-assembled.

As explained earlier in this section, the drainage process may be simplified by assuming that, for a given capillary pressure, p_c, air occupies all the pores (here, capillary tubes) that are larger than a size (= radius of tube) r', defined by

$$r' = \frac{2\gamma_{wa}}{p_c}, \tag{6.2.28}$$

where γ_{wa} denotes the water-air surface tension. The effective water saturation, S_{we}, can then be expressed by

$$S_{we} = \frac{\int_{r_{min}}^{r'} \pi r^2 \alpha(r)\, dr}{\int_{r_{min}}^{r_{max}} \pi r^2 \alpha(r)\, dr}, \tag{6.2.29}$$

where $\alpha(r)$ is a probability distribution function describing the probability that a capillary tube has a radius between r and $r + dr$. In principle, $\alpha(r)$ can be determined from the $p_c = p_c(S_{we})$ curve. In any slice of total (bundle) area A, the area $\phi(S_w - S_{wr})A \equiv \phi S_{we}(1 - S_{wr})A \equiv \phi \equiv \hat{S}_w A$ is occupied by water occupying pores of radii between r_{min} and r'. An equal area is occupied by the water in neighboring slices.

However, portions of these areas are not connected, due to the random distribution of the 'pores' in the model.

Considering a point on the interface between adjacent slices, the probability that it lies in the water portion of the cross-section in one of two adjacent slices is $\phi \hat{S}_w$. Hence, the probability that it lies in the water simultaneously in both slices is $(\phi \hat{S}_w)^2$. Using similar reasoning, Wylie and Gardner (1958) show that since the probability of a water-filled pore in one slice is $\phi \hat{S}_w$, the area common to a *single* pore of cross-sectional area πr^2 in one slice, and *all* the water-filled pores in a neighboring slice is, therefore, $\pi r^2 \phi \hat{S}_w$. Thus, the passage of water takes place from an area πr^2 to a constricted area $\pi r^2 \phi \hat{S}_w$. One may visualize the constricted area as a pore of smaller radius r'', such that $\pi r''^2 = \pi r^2 \phi (S_w - S_{wr})/\lambda$, where $\lambda\,(\geq 1)$ is a numerical coefficient that reflects the size distribution of the total interconnected pore area; it also depends on $\alpha(r)$.

We now use the Hagen–Poiseuille law (e.g., Bear 1972, p. 162) to describe the discharge, Q_{cap}, through a capillary tube in the model

$$Q_{cap} = \frac{\pi r''^4 \rho_w g \beta}{8\mu} \mathcal{J} = \frac{\pi \phi^2 \hat{S}_w^2 r^4 \rho_w g \beta}{8\mu\lambda^2} \mathcal{J}, \tag{6.2.30}$$

where β is a coefficient that accounts for the nonuniformity of tube diameters and $\mathcal{J}$ is the hydraulic gradient, which is assumed to be approximately uniform over all tubes. For the entire bundle, per unit area A, we obtain

$$q = \frac{\phi^3 \hat{S}_w^2 \beta \rho_w g}{8\mu\lambda^2} \mathcal{J} \frac{\int_{r_{min}}^{r'} \pi r^4 \alpha(r)\, dr}{\int_{r_{min}}^{r_{max}} \pi r^2 \alpha(r)\, dr}, \tag{6.2.31}$$

where

$$\phi A = \int_{r_{min}}^{r_{max}} \pi r^2 \alpha(r)\, dr,$$

and we have assumed that both λ and β are independent of $\alpha(r)$. By differentiating (6.2.29) with respect to r', we obtain

$$d\hat{S}_w = \frac{\pi r'^2 \alpha(r')\, dr'}{\int_{r_{min}}^{r_{max}} \pi r^2 \alpha(r)\, dr}. \tag{6.2.32}$$

Eliminating r' from (6.2.31), using (6.2.28) and (6.2.32), we obtain

$$q = \frac{\phi^3 \hat{S}_w^2 \beta \rho_w g \gamma_{wa}^2}{2\mu\lambda^2} \mathcal{J} \int_0^{S_{we}} \frac{dS'_{we}}{p_c^2(S'_{we})}. \tag{6.2.33}$$

Hence, the effective permeability for the water is

$$k_w(S_{we}) = \frac{\phi^3 \hat{S}_w^2 \beta \gamma_{wa}^2}{2\lambda^2} \int_0^{S_{we}} \frac{dS'_{we}}{p_c^2(S'_{we})}. \qquad (6.2.34)$$

The saturated permeability is found by setting $S_{we} = 1$,

$$k_{sat} = \frac{\phi^3 (1 - S_{wr})^2 \beta \gamma_{wa}^2}{2\lambda^2} \int_0^1 \frac{dS'_{we}}{p_c^2(S'_{we})}. \qquad (6.2.35)$$

The relative permeability to water then takes the form:

$$k_{rw} = S_{we}^2 \int_0^{S_{we}} \frac{dS'_{we}}{p_c^2(S'_{we})} \bigg/ \int_0^1 \frac{dS'_{we}}{p_c^2(S'_{we})}, \qquad (6.2.36)$$

where it is assumed that β/λ^2 is not sensitive to changes in S_w. Similarly, for the air

$$k_{ra} = (1 - S_{we})^2 \int_{S_{we}}^1 \frac{dS'_{we}}{p_c^2(S'_{we})} \bigg/ \int_0^1 \frac{dS'_{we}}{p_c^2(S'_{we})}. \qquad (6.2.37)$$

Thus, given a $p_c(S_{we})$-curve, estimates of k_w and k_a can be obtained.

For an anisotropic porous medium, each of the effective permeability components, k_{wij} and k_{aij}, may have a *different* functional relationship with respect to S_w. Hence, the concept of a relative permeability is not permissible. Bear et al. (1987) have shown, by using computer experiments to simulate an anisotropic porous medium with a three-dimensional network of capillary tubes of random diameters, that the ratio between the principal values of $\mathbf{k}_w$ in the x and y principal directions, $k_{w,xx}/k_{w,yy}$, is a function of saturation and not a constant. For example, if $\mathsf{k}_{w,xx} > k_{w,yy}$ is due to larger pores (or larger pore cross-sections) oriented in the x-direction, then as saturation is reduced, and since the larger pores drain first, we reach a point where $\mathsf{k}_{w,xx} = k_{w,yy}$. At still lower saturation, $\mathsf{k}_{w,xx} < k_{w,yy}$.

The subject of *hysteresis* in the retention curve, which stems from the difference in the configuration of the water-occupied portion of the void space during the two processes–imbibition and drainage–will be discussed later in this subsection. This difference in configuration leads also to a certain degree of hysteresis in the relationships $\mathbf{k}_w(S_w)$ and $\mathbf{k}_a(S_w)$. Figure 6.9 shows hysteresis in typical relative permeability curves.

In Fig. 6.9, we note that upon rewetting back to zero capillary pressure, the permeability is less than at full saturation due to entrapped air. As explained above, the amount of entrapped air and its effect on reducing permeability, is a function of the drainage-imbibition history.

6.3 Specific Storativity

In Sect. 5.1.5, in the discussion on the mass balance equation for flow of a single fluid that occupies the entire void space, we have introduced the effect of fluid and solid matrix compressibility through the concepts of *storativity* and *specific storativity* for flow in a confined aquifer. The same concepts can be extended to two or more fluid phases that, simultaneously, occupy the void space. In fact, we have also introduced there the concept of *effective stress*, which can also be extended to multiphase flow.

Thus, our next step is to focus on the l.h.s. of the mass balance equations (6.4.3), with

$$\frac{\partial}{\partial t}\phi S_w \rho_w = \phi S_w \frac{\partial \rho_w}{\partial t} + S_w \rho_w \frac{\partial \phi}{\partial t} + \phi \rho_w \frac{\partial S_w}{\partial t}. \tag{6.3.38}$$

For the case of two fluids (w and n) that together fill up the void space, with negligible shear stress in both fluids, we define an average pressure in the two fluids (subscript and superscript v) that fill up the void space, by

$$\overline{p_v}^v = S_w \overline{p_w}^w + S_n \overline{p_n}^n. \tag{6.3.39}$$

We can now return to the fundamental definition of effective stress in (5.1.28) and extend it to the case of two phase flow, replacing the single fluid pressure by the average fluid pressure. We obtain the more general case:

$$\overline{\boldsymbol{\sigma}} = \overline{\boldsymbol{\sigma}'_s} - \overline{p_v}^v \mathbf{I}, \tag{6.3.40}$$

in which we recall that $\overline{\boldsymbol{\sigma}'_s}$ denotes the (macroscopic) effective stress and $\overline{p_v}^v$ is the average pressure in the void-space.

Thus, for two fluid phases, with (6.3.39), Eq. (5.1.28) takes the form:

$$\begin{aligned}\overline{\boldsymbol{\sigma}} &= (1-\phi)\overline{\boldsymbol{\sigma}_s}^s - \theta_w \overline{p_w}^w \mathbf{I} - \theta_n \overline{p_n}^n \mathbf{I}\\ &= (1-\phi)\{\overline{\boldsymbol{\sigma}_s}^s + \overline{p_v}^v \mathbf{I}\} - (1-\phi)\overline{p_v}^v \mathbf{I}\\ &\quad -\theta_w \overline{p_w}^w \mathbf{I} - \theta_n \overline{p_n}^n \mathbf{I} = \overline{\boldsymbol{\sigma}'_s} - \frac{\theta_w}{\phi}\overline{p_w}^w \mathbf{I} - \frac{\theta_n}{\phi}\overline{p_n}^n \mathbf{I},\end{aligned} \tag{6.3.41}$$

and the effective stress is defined by

$$\overline{\boldsymbol{\sigma}'_s} = (1-\phi)\{\overline{\boldsymbol{\sigma}_s}^s + \overline{p_v}^v \mathbf{I}\}. \tag{6.3.42}$$

In *unsaturated flow* (= air-water flow), the nonwetting fluid is air while the wetting one is water. When we assume the air to be everywhere at atmospheric pressure, i.e., $\overline{p_a}^a \equiv 0$, (6.3.41) reduces to

$$\overline{\boldsymbol{\sigma}} = \overline{\boldsymbol{\sigma}'_s} - S_w \overline{p_w}^w \mathbf{I}. \tag{6.3.43}$$

In determining the average void pressure, $\overline{p_v}^v$, we have taken a volume average of the pressure in the various fluids occupying the void space. Other weights in determining $\overline{p_v}^v$ would lead to equations that are different from (6.3.41) and (6.3.43). For example, some authors (e.g., Aitchison and Donald 1956) use for air - water flow (with $\overline{p_a}^a = 0$)

$$\overline{\boldsymbol{\sigma}} = \overline{\boldsymbol{\sigma}'_s} - \chi(S_w)\overline{p_w}^w\mathbf{I}, \tag{6.3.44}$$

where $\chi(S_w)$ is some function of the moisture content, θ_w. Bear et al. (1984) used (6.3.44) with $\chi(S_w) = S_w$.

Making use of $\chi(S_w)$, we obtain for two-phase flow:

$$\frac{\partial \phi}{\partial t} = (1-\phi)\alpha\left(\chi(S_w)\frac{\partial p_w}{\partial t} + p_w\frac{\partial \chi}{\partial S_w}\frac{\partial S_w}{\partial t}\right). \tag{6.3.45}$$

To obtain an expression for the third term on the r.h.s. of (6.3.38), we apply the *chain rule of differentiation* to the term $\partial S_w/\partial t$, and noting that $S_w = S_w(p_c)$, we obtain

$$\frac{\partial S_w}{\partial t} = \frac{\mathrm{d}S_w}{\mathrm{d}p_c}\frac{\partial p_c}{\partial t} = \frac{\mathrm{d}S_w}{\mathrm{d}p_c}\left(\frac{\partial p_a}{\partial t} - \frac{\partial p_w}{\partial t}\right). \tag{6.3.46}$$

We define the *water (moisture) capacity*, C_w, by

$$C_w = -\phi\frac{\mathrm{d}S_w}{\mathrm{d}p_c}. \tag{6.3.47}$$

Then,

$$\frac{\partial S_w}{\partial t} = \frac{C_w}{\phi}\left(\frac{\partial p_w}{\partial t} - \frac{\partial p_a}{\partial t}\right). \tag{6.3.48}$$

If p_a is constant, $\partial p_a/\partial t = 0$, and

$$\frac{\partial S_w}{\partial t} = \frac{C_w}{\phi}\frac{\partial p_w}{\partial t}. \tag{6.3.49}$$

In this case, we may also write $S_w = S_w(p_w)$; therefore,

$$\frac{\partial S_w}{\partial t} = \frac{dS_w}{dp_w}\frac{\partial p_w}{\partial t}, \quad \text{and} \quad C_w = \phi\frac{\mathrm{d}S_w}{\mathrm{d}p_w}. \tag{6.3.50}$$

Altogether, we obtain

$$\begin{aligned}\frac{\partial}{\partial t}(\phi S_w \rho_w) = \rho_w\Bigg\{&\phi S_w\beta_w + \phi\frac{\mathrm{d}S_w}{\mathrm{d}p_w}\\ &+ S_w(1-\phi)\alpha\left[\chi(S_w) + p_w\frac{\mathrm{d}\chi}{\mathrm{d}S_w}\frac{\mathrm{d}S_w}{\mathrm{d}p_w}\right]\Bigg\}\frac{\partial p_w}{\partial t},\end{aligned} \tag{6.3.51}$$

in which the coefficients α and $\chi(S_w)$ are defined by (5.1.44) and (6.3.44), respectively.

Equation (6.3.51) can be rewritten in a form similar to (5.1.46), with a specific mass storativity that takes into account also the change in saturation. For the saturated zone, $S_w = 1$, $\chi(S_w) = 1$, and (6.3.51) reduces to (5.1.46).

With the above developments, the mass balance equation for water in a deformable porous medium, (6.4.1), takes the form:

$$\rho_w \left\{ \phi S_w \beta_w + \phi \frac{\mathrm{d}S_w}{\mathrm{d}p_w} + S_w \alpha \left[p_w \frac{\mathrm{d}\chi}{\mathrm{d}S_w} \frac{\mathrm{d}S_w}{\mathrm{d}p_w} + \chi(S_w) \right] \right\} \frac{\partial p_w}{\partial t} \\ = -\nabla \cdot (\rho_w \mathbf{q}_{rw}) + \rho_w \Gamma'^w. \tag{6.3.52}$$

We can write this equation also in the form (5.1.49), repeated here for convenience:

$$S^m_{op} \frac{\partial p_w}{\partial t} = -\nabla \cdot (\rho_w \mathbf{q}_{rw}) + \rho_w \Gamma'^w, \tag{6.3.53}$$

in which the coefficient is defined by (6.3.52). Note that S^m_{op} is different from the coefficient S^{m*}_{op} defined by (6.3.51), and that in the divergence term on the r.h.s. we have $\mathbf{q}_{rw}$ rather than $\mathbf{q}_w$. The explanation is given following (5.1.50).

6.4 Mass Balance Equations and Complete Model

As in the case of single phase flow, the core of a two-phase flow model (NP = 2) is the two mass balance equations of the phases. We assume that each of the fluid phases is a single component phase, and remains so. We added the adjective "complete" to emphasize that we mean both the mass balance equation and the initial and boundary conditions. In Chap. 7 we consider multi-species phases.

6.4.1 *The Flow Model*

Our starting point is (3.3.7) which expresses the mass balance equation for a fluid that occupies part of the void space at the volumetric fraction θ_α. Although there are no (microscopic level) mass sources *within* any of the phases, macroscopic level sources and sinks, in the form of point sources (= wells) may exist. We may then represent the macroscopic Γ by a dirac-delta function (see discussion in Sect. 5.1). Recalling that there is no molecular diffusion nor dispersion of the total mass, and assuming no mass transfer of the phases across interphase boundaries, the macroscopic mass balance equations take the form:

$$\frac{\partial}{\partial t}(\phi S_w \rho_w) = -\nabla\cdot(\rho_w \mathbf{q}_w) + \rho_w \Gamma'_w, \tag{6.4.1}$$

$$\frac{\partial}{\partial t}(\phi S_n \rho_n) = -\nabla\cdot(\rho_n \mathbf{q}_n) + \rho_n \Gamma'_n, \tag{6.4.2}$$

where $\mathbf{q}_\alpha (= \theta_\alpha \mathbf{V}_\alpha)$ denotes the flux, or specific discharge (= volume of the α-phase per unit time per unit area of porous medium), we have assumed that the areal fraction of the phase is equal to its volumetric fraction. The symbol Γ'_α denotes a macroscopic level source of the α-phase (= volume of fluid withdrawn from the void space per init time per unit volume of porous medium); a negative value means fluid withdrawal. Note that the above equations involve $\mathbf{q}_\alpha$ and not $\mathbf{q}_{r\alpha}$. Since Darcy's law provides information on the relative specific discharge $\mathbf{q}_r$, either we deal with a case of $\mathbf{V}_s \approx 0$, or we have to consider $\mathbf{V}_s (= \mathbf{q}_s/\theta_s)$ as an additional variable of the problem, i.e., a deformable porous medium.

Note that in the above two mass balance equations, we have made the assumption (based on the concept of *immiscible fluids*) that there is no (or no significant) fluid mass exchange ($w \rightarrow nw$, $nw \rightarrow s$) across the microscopic interfaces, say, due to phase change. Otherwise, we should add mass transfer terms, e.g., $f^w_{w\rightarrow nw}$-term, and $f^{nw}_{nw\rightarrow w}$-terms, in the mass balance equations, to denote phase change (e.g., evaporation and condensation), respectively of the respective fluids.

The complete statement of a two-phase flow problem contains more information. For example, the complete set of equations that describes the simultaneous isothermal flow of two compressible single species fluid phases (n, w), in a stationary rigid porous medium (i.e., $\mathbf{V}_s = 0$, $\partial\phi/\partial t = 0$, $\mathbf{q}_r \rightarrow \mathbf{q}$), in the absence of interphase mass transfers and external sources, are:

- **Mass balances**:

$$\frac{\partial \phi S_w \rho_w}{\partial t} = -\nabla\cdot(\rho_w \mathbf{q}_w), \qquad \frac{\partial \phi S_n \rho_n}{\partial t} = -\nabla\cdot(\rho_n \mathbf{q}_n). \tag{6.4.3}$$

In the case of *phase change*, e.g., by adding heat at a point by *microwave energy*, $f_{w\rightarrow n}$-type phase transfer terms should be added to the above mass balance equations. Of course, Γ-type source terms can also be added when necessary.

Mass balance equations that consider production of chemical species by chemical reactions and mass transfer across (microscopic) interphase boundaries (expressed as macroscopic terms in the mass balance equations) are discussed in Chap. 7. Phase changes, like aporation/condensation, are considered in Chap. 8.

- **Flux equations**:

$$\mathbf{q}_w = -\frac{\mathbf{k}_w(S_w)}{\mu_w}\cdot(\nabla p_w + \rho_w g \nabla z), \qquad \mathbf{q}_n = -\frac{\mathbf{k}_n(S_n)}{\mu_n}\cdot(\nabla p_n + \rho_a g \nabla z). \tag{6.4.4}$$

- **Capillary pressure and density**:

$$p_c = p_c(S_w) = p_n - p_w, \qquad \rho_w = \rho_w(p_w), \qquad \rho_n = \rho_n(p_n). \tag{6.4.5}$$

• **Sum of saturations**:

$$S_w + S_n = 1. \tag{6.4.6}$$

Altogether, 8 equations to be solved simultaneously for the 8 variables:

$$p_w, \quad p_n, \quad S_w, \quad S_n, \quad \rho_w, \quad \rho_n, \quad \mathbf{q}_w, \quad \mathbf{q}_n.$$

However, following the discussion in Sect. 3.9, we have here 2 components: H_2O and air (all the gases together), i.e. NC = 2, and two phases (liquid and gas), i.e., NP = 2. Hence, using (3.9.1) we find that this problem involves only *two degrees of freedom*. This means that this problem involves only *two independent* variables, e.g., p_w and S_n, or p_n and S_w, and we have to solve only two partial differential equations for these variables. In three-dimensional domains, we may read this problem as involving 12 scalar variables, 2 degrees of freedom and 2 PDE's to be solved for these variables. More on the number of degrees of freedom is presented in Sect. 3.9.

However, under certain conditions, in two (and more so in three) phase flow, the fact that $S_\alpha \leq 1$, may leads too the situation that at some point in the course of solution, the initially selected set of primary variables is no more permissible, and we have to switch to a new set of primary variables. This is part of the routine in numerical solutions of multiphase flow problems.

Comment on *primary variable switching* (PVS) are presented in Sects. 6.5.3 and 7.5.4

6.4.2 Initial and Boundary Conditions

The need for initial and boundary conditions and their role in models has already been presented and discussed in Sect. 5.2, in connection with single phase flow. There, we have also discussed the general concept, the representation of boundary surfaces, and the principles that serve as the basis for determining boundary conditions. Here, we are discussing the simultaneous flow of two (assumed immiscible) fluids, like water and air, or water and oil, with no sharp (macroscopic) interface between them. Instead, there always exists a transition zone across which the saturations of both fluids vary in space and time. In Sect. 5.2.4, we have presented a number of boundary conditions that are commonly encountered in single phase flow. Here, we shall discuss conditions that are usually encountered when modeling two phase flow. The case of flow in3 the unsaturated (air-water) zone in the subsurface will also be presented and discussed as a special example of interest.

As discussed in Sect. 6.4.1, the model describing the flow of two (assumed) immiscible fluids has two degrees of freedom. This means that the flow is described by two independent variables for which we have to solve two partial differential mass balance equations. The kind of PDE describing these mass balance requires only *one* condition on each boundary segment. We should prefer a condition based on flux continuity, if such information is available. If not, we base the boundary condition on available information on values of a scalar variable, e.g., pressure or saturation. Sometimes, approximations concerning the continuity of fluxes produce a jump in the values of the variables; we have to accept this consequence.

A. Initial Conditions

Initial conditions specify the values of the two selected independent variable, e.g., p_w, p_a, S_w, θ_w, or h_w, at all points within the modeled domain at some initial time, usually taken as $t = 0$. For example, in terms of p_w, initial conditions may take the form (5.2.4), in which we replace p by p_w.

B. General Boundary Condition

We note that (5.2.6) is already written for the general case of multiple fluid phases. However, we have to be careful in the case of a boundary of discontinuity between two different porous media. The pressure in each fluid must undergo no jump, but, *there is always a jump in saturation* across the boundary between two porous media. This is acceptable, because the average pressure is not a *thermodynamic variable*.

In two phase flow, e.g., in an oil-water or gas-water reservoir, we encounter the same kinds of boundary conditions that were presented in Sect. 5.2.4, where we have considered the case of a single phase. Specifically, for each of the phases, we specify the pressure, the saturation, or the normal flux on the boundary. Most of the conditions presented below, in connection with the unsaturated zone, are also applicable to any case of two-phase flow.

Following are some of the more commonly encountered boundary conditions for the special case of unsaturated (i.e., air-water) flow. In each case, the boundary condition is stated in terms of the relevant state variable of the problem. The geometry of the boundary surface is stated by the equation $F(x, y, z, t) = 0$, discussed in Sect. 5.2.1.

C. Boundary of Prescribed Saturation

In this case, the external domain imposes a certain saturation on the domain's boundary. In practice, this kind of boundary seldom occurs, except in the case of full saturation, $S_w = 1$, such as when the considered domain is in contact with a body of water (a lake, a river, or a pond). For example, $S_w = 1$ is prescribed on the bottom of a water pond, dictating there a surface at full saturation (even in the limiting case, when a very thin layer of water is present in the pond). Similarly, under the assumption of completely dry soil, the condition $S_a = 1$ is prescribed at ground surface (overlain by the atmosphere) that serves as a boundary to the unsaturated domain.

When ground surface without ponding serves as the upper boundary for the unsaturated zone, neither the water pressure on it, nor the water saturation are known. The only information that we have is the rate of water infiltration through such boundary (including the case of no-infiltration). This information is then used to specify the boundary condition (see below).

D. Boundary of Prescribed Pressure

As in the case of prescribed saturation, the value of p_w is seldom known on a boundary, except when a porous medium domain is bounded by a body of water (e.g., a pond). In the latter case, the pressure along the pond's bottom is dictated by the depth of water in the pond.

In air-water flow, we use atmospheric conditions at ground surface to specify conditions of known air pressure, $p_a = p_{\text{atm}}$.

Another type of boundary condition at ground surface, for air-water flow in the unsaturated zone, is $p_a = p_{atm}$ for the air phase, and a known flux for the water phase. The boundary condition under a pond, is known water pressure, and no flux of air.

The boundary condition that specifies the value of S_α, or p_α, $\alpha = w, n$, along a boundary segment is a *boundary condition of the first type*, or *Dirichlet boundary condition*.

E. Boundary of Prescribed Flux

When the flux, $\mathbf{q}_\alpha = \mathbf{q}_\alpha(\mathbf{x}, t)$, is known along the external side of a boundary segment, e.g., $f(\mathbf{x}, t)$, then the condition on the boundary is:

$$\mathbf{q}_\alpha \cdot \boldsymbol{\nu} = f(\mathbf{x}, t), \tag{6.4.7}$$

Thus condition is based on the *no-jump condition* presented in Sect. 5.2.3.

A special case of interest of such boundary is when water, e.g., from rainfall, infiltrates through the unsaturated zone (= vadose zone) *at a known rate* through ground surface, which serves as a boundary to the unsaturated zone. This includes the case of no-flow (i.e., $f(\mathbf{x}, t) = 0$) through such a boundary.

Because ground surface is a *material surface* with respect to the solid, (5.2.9) is applicable, i.e.,

$$(\mathbf{V}_s - \mathbf{u})\big|_{\text{side 1}} \cdot \boldsymbol{\nu} = (\mathbf{V}_s - \mathbf{u})\big|_{\text{side 2}} \cdot \boldsymbol{\nu} = 0. \tag{6.4.8}$$

Since the microscopic water-solid and air-solid interfaces are material with respect to fluid mass, (5.2.6) can be written separately for each fluid phase.

With $[\![\ \rho_f\ ]\!]_{1,2} = 0$, and replacing $(\mathbf{V} - \mathbf{u})$ by $(\mathbf{V}_\alpha - \mathbf{V}_s) + (\mathbf{V}_s - \mathbf{u})$, the general boundary condition (5.2.6) for such a surface takes the form:

$$[\![\ \theta_\alpha(\mathbf{V}_\alpha - \mathbf{u})\ ]\!]_{1,2} \cdot \boldsymbol{\nu} = 0. \tag{6.4.9}$$

Thus, with (5.2.9), for an α-fluid phase, $\alpha = w, a$, the last equation reduces to the form of (5.2.11), repeated here for convenience as

$$[\![\ \mathbf{q}_{r\alpha}\ ]\!]_{1,2} \cdot \boldsymbol{\nu} = 0, \quad \text{or} \quad \mathbf{q}_{r\alpha}\big|_1 \cdot \boldsymbol{\nu} = \mathbf{q}_{r\alpha}\big|_2 \cdot \boldsymbol{\nu}, \tag{6.4.10}$$

in which one of the sides, say side 2, is the external (atmospheric) one. The relative specific flux, $\mathbf{q}_{r\alpha}$, is expressed by an appropriate motion equation.

For an *impervious boundary*, say, a pervious side 1 and an impervious side 2, Eq. (5.2.11) reduces to (5.2.12). Note that this equation constrains only the normal component of the flux. The tangential components may take on any value, meaning that we may have, and usually have, *slip* along the boundary.

F. Ground Surface as a Boundary

Consider the unsaturated zone, bounded from above by ground surface, and let **N** denote the prescribed flux (say, upward for evaporation and downward for infiltration) on the external side of this stationary ($\mathbf{u} = 0$) boundary. This boundary is described by $F = F(\mathbf{x})$, with $\boldsymbol{\nu} \equiv \nabla F/|\nabla F|$ denoting the unit outward normal vector to it. We assume that the water density, ρ_w, on the external side is the same as within the unsaturated zone, and that it is a constant. Then, the prescribed flux boundary condition takes the form of (5.2.13), repeated here for convenience as

$$\mathbf{q}_{r\alpha} \cdot \nabla F = N|\nabla F|, \qquad N = \mathbf{N} \cdot \boldsymbol{\nu}. \tag{6.4.11}$$

In this equation, $\mathbf{q}_{r\alpha}$ can be expressed by any of the motion equations presented in Sect. 6.2. For example, in terms of ψ, we use (6.2.15) and (5.2.13) becomes

$$[\mathsf{K}_w(\psi) \cdot \nabla(\psi - z)] \cdot \boldsymbol{\nu} = \mathbf{N} \cdot \boldsymbol{\nu}. \tag{6.4.12}$$

As this condition specifies the gradient of a scalar variable on the boundary, it is a *boundary condition of the second kind*, or a *Neumann boundary condition*.

For a horizontal ground surface that serves as the upper boundary of the unsaturated zone, $\boldsymbol{\nu} \equiv \nabla z$. If the soil is isotropic and the infiltration is vertically downward, i.e., $\mathbf{N} = -N\nabla z$, we obtain from (6.4.12):

$$N = -K_w(\psi)\frac{\partial \psi}{\partial z} + \mathsf{K}_w(\psi). \tag{6.4.13}$$

The boundary condition (6.4.13) specifies a constraint that involves both ψ and $\nabla\psi$. This is a *boundary condition of the third type*, or a *Robin boundary condition*.

The condition of prescribed flux provides no explicit information on the values of the state variables, say ψ, at (i.e., just inside) the boundary. These values will adjust themselves (thus modifying also the values of the effective hydraulic conductivity, $\mathsf{K}_w(\psi)$) to accommodate the specified rate of flow through the boundary. The flux from rainfall through ground surface that serves as the upper boundary of the unsaturated zone, requires special attention.

G. Infiltration and Evaporation at Ground Surface

This type of boundary always occurs when ground surface serves as the upper boundary of a modeled unsaturated domain (= vadose zone). For the air, we can specify atmospheric pressure as a known value. However, for the water, neither the pressure nor the saturation is known. The boundary condition to be used in such a case is that of specified flux (due to infiltration or evaporation), with the special case of zero flux when no infiltration from precipitation or from irrigation takes place. There are two approaches that one can take: (1) specify the net infiltration flux (= precipitation minus evaporation), if it is known, or (2) model the evaporation by specifying the relative humidity, h_r, in the air close to ground surface, and use Kelvin's law in (2.5.3) to specify the resulting suction in the water that occupies the void space close to ground surface. The specified suction is then used as a first type boundary condition. The second approach usually requires that thermal effects be also modeled.

To gain some understanding about what is involved in determining infiltration, let us start with a few remarks on infiltration. More information on this subject may be found in standard texts on hydrology (e.g., Bras 1990; Maidment 1993).

Depending on the local conditions during a rainfall event, e.g., rate of precipitation, type of soil, vegetation cover, surface topography, climatic conditions, antecedent soil moisture, etc., part of the precipitation reaching ground surface *infiltrates* through the latter and continues to percolate downward towards an underlying water table. The remaining part will either pond above ground surface, or become *surface runoff*. When the supply of water to ground surface exceeds the rate of infiltration, which is commensurate with the properties of the soil and its moisture content, the infiltration rate is the maximum possible one, under the prevailing conditions. This rate is referred to as *infiltration capacity*. The processes of infiltration and percolation are accompanied by changes in saturation within the unsaturated zone. The redistribution of moisture in the subsurface will continue for some time after cessation of infiltration. This process, its relationship to the rate of infiltration, and to the rate at which water is applied at ground surface, are of major interest in the design of irrigation systems.

The issue of what happens to the air in the unsaturated zone between the water table (i.e., the surface in the saturated zone where $p = p_{atmos}$) and ground surface during a period of infiltration requires special attention. Under conditions of heavy infiltration, the soil zone close to groundwater may become saturated with a downward infiltrating water, and the entrapped air is compressed, unless it has the opportunity to escape sidewise. At the increased air pressure. air may locally escape through the downward advancing saturated zone.

In principle, the rate of (actual) infiltration, should be derived by solving a model of flow in the unsaturated zone, subject to appropriate boundary conditions, especially at ground surface. The properties of the soil and the precipitation characteristics will be represented in such a model. However, the rate of infiltration to be used as a boundary condition *is not known*, as it seldom equals the rate at which water is applied at ground surface. This fact, combined with the difficulties inherent in solving such a problem, has made this approach impractical, at least until recent years. Instead, in the practice of hydrology, various empirical formulas have been employed. In recent years numerical models, solved by appropriate software, have been been developed. Obviously, appropriate boundary conditions have to be specified within the framework of such models. An exact analytical solution of a model that describes flow in a vertical soil column, from ground surface to an underlying water table. subject to recharge conditions at ground surface is not possible, due the nonlinearity of such model. In a series of papers, Philip (1957a, b, c, d, e, 1958a, b, 1969) proposed an approximate solution for the distribution of moisture in the subsurface in the form of an infinite series. Based on this solution, the infiltration capacity, $I_c(t)$, may be estimated from the formula:

$$I_c = I_{c\infty} + \frac{\mathfrak{S}}{2\sqrt{t}}, \tag{6.4.14}$$

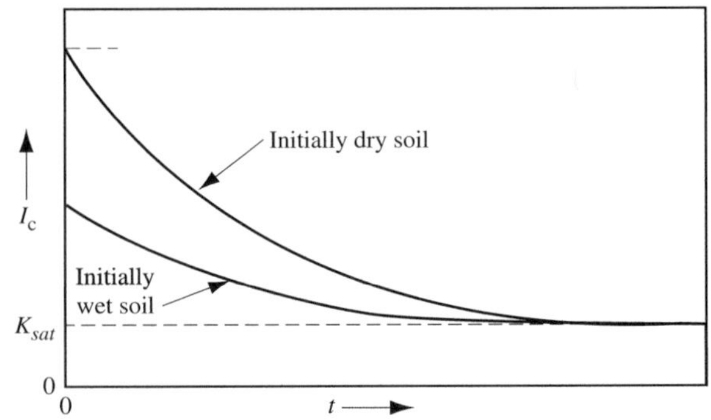

Fig. 6.10 Effect of initial soil moisture on the rate of infiltration

in which $\mathfrak{s}$ is a coefficient that depends on the initial soil moisture. We note that initially, $I_c = \infty$. In the limit, as $t \to \infty$, $I_c(t) \to I_{c\infty}$. Figure 6.10 shows the effect of the initial soil moisture on the infiltration capacity.

It can easily be shown that as infiltration proceeds, in the absence of ponding and air entrapment, the soil close to ground surface becomes saturated, and the rate of infiltration approaches the hydraulic conductivity at saturation. At the same time, the saturation gradient *at* ground surface vanishes asymptotically. The combination of these two processes leaves gravity as the only driving force, with the result that the flux becomes equal to $K\big|_{\text{sat}}$.

With the above introductory remarks, we can now proceed to discuss the infiltration boundary condition.

Let $R(\mathbf{x}, t)$ (≥ 0) denote the rate at which water is applied in a downward direction to ground surface by precipitation, or irrigation, with $\mathbf{x}$ denoting points on ground surface. We start by assuming that this rate is also equal to the rate of (vertically downward) infiltration, i.e., $I(\mathbf{x}, t) = R(\mathbf{x}, t)$. We are faced with two important questions:

- Under what conditions is this assumption valid?
- What do we do when it is not valid, i.e., we know R, but not I?

Consider the case of rainfall at a rate R over a horizontal ground surface, resulting in infiltration at a rate I; the soil is assumed to be isotropic. The boundary condition, say (6.4.13), may then be rewritten as:

$$I = -K_w(\psi)\frac{\partial \psi}{\partial z} + \mathsf{K}_w(\psi). \tag{6.4.15}$$

or, in terms of θ_w:

$$I = D_w(\theta_w)\frac{\partial \theta_w}{\partial z} + \mathsf{K}_w(\theta_w). \tag{6.4.16}$$

From our discussion so far, it follows that at any instant, depending on soil properties and on the prevailing moisture content and distribution, the soil just below ground surface can transmit only a certain flux of water, provided such water quantity is applied at ground surface. If the rate of application is higher, the difference will pond on ground surface, or produce surface runoff. The behavior above and below ground surface is, thus, coupled by the common condition at ground surface. To avoid modeling what happens above ground surface, the verbal constraint is often added that 'no ponding of water is allowed to take place above ground surface'. In reality, because of ground surface roughness, some ponding may take place before surface runoff actually begins. The no-ponding constraint limits the rate of infiltration at every instant to what the soil can transmit at the prevailing saturation and saturation gradient conditions. This constraint has to be incorporated in the statement of the boundary conditions at ground surface.

From (6.4.15), it follows that as the soil just below ground surface approaches full saturation, overlooking entrapped air, $\theta_w \rightarrow \phi$ (or $\theta_w \rightarrow \phi - \theta_{ao}$, if entrapped air is considered), and $\partial\theta_w/\partial z \rightarrow 0$, the first term on the right-hand side vanishes, while the second one approaches the value of the saturated hydraulic conductivity, $K\big|_{\text{sat}}$ (or, with entrapped air, $\mathsf{K}\big|_{\phi-\theta_{ao}}$). In an anisotropic porous medium, it will approach the value of the vertical component, $K_z\big|_{\text{sat}}$. In the current discussion on infiltration, for brevity, we shall use $K_z\big|_{\text{sat}}$ to mean $\mathsf{K}\big|_{\phi-\theta_{ao}}$, when entrapped air may be present.

Let us consider two situations. In both cases, we shall assume that R is constant, and that, initially, the soil is relatively dry, say at *field capacity*. We shall assume that the water table is sufficiently deep so that its influence on soil moisture near ground surface can be neglected.

CASE A. Precipitation is applied at a constant rate $R > K_z\big|_{\text{sat}}$. At first, even at low saturations, the soil can absorb the incoming water at a very high rate, as the gradient in moisture content that is produced at ground surface is very high. Theoretically, this rate is infinite at the initial time. Thus, for a certain period, we have $I = R$. We use this value in the boundary condition (6.4.16). During this initial period, as a wetting front advances downward, the infiltrating water produces two phenomena:

- At ground surface, and just below it, the water content in the soil gradually increases. In the limit, full saturation may be reached (i.e., water content equals porosity). The increase in water content is accompanied by an increase in effective hydraulic conductivity, up to the limiting value of $K_z\big|_{\text{sat}}$, corresponding to a saturated soil.
- As water percolates downward, the gradient in the soil's water content close to ground surface decreases with time. In the limit, this gradient at ground surface approaches the value of zero, so that gravity remains the only driving force there.

The initial period continues until the soil at ground surface reaches a point at which the combination of saturation (and effective hydraulic conductivity) and saturation gradient are such that the soil can no longer transmit water at the rate applied at ground surface. This occurs when the pressure in the water occupying the pore space at ground surface approaches atmospheric pressure. With $p_w = p_a\big|_{\text{atm}}$, we also have full water saturation (or practically so) at ground surface. During this period, the infiltration rate remains constant, equal to the (assumed constant) rate of accretion.

Actually, because of the (discontinuous, i.e., composed of isolated air bubbles) *entrapped air* that remains in the void space, the soil will never reach full saturation. Although the volumetric fraction of entrapped air depends on the wetting-drainage history, we shall approximate it as a constant, θ_{ao}. Thus, upon rewetting, the moisture content of water cannot exceed $\phi - \theta_{ao}$.

In our model, the initial period continues as long as $\theta_w < \phi - \theta_{ao}$, or, equivalently, as long as $p_w < p_a\big|_{\text{atm}}$. Once full saturation is reached ($\theta_w = \phi - \theta_{ao}$), we have to replace (6.4.16) by the first type boundary condition

$$p_w = p_a\big|_{\text{atm}}, \tag{6.4.17}$$

(where we often assume $p_a\big|_{\text{atm}} = 0$ as the datum) corresponding to a condition just below a state of zero ponding depth.

Next, we calculate the rate of infiltration, $I(t)(< R)$ by substituting the solutions for $\theta_w(t)$ in (6.4.16). We would then observe a gradual reduction in $I(t)$, approaching the limiting value of $K\big|_{\phi-\theta_{ao}}$.

Let us now consider the time-varying case $R = R(t)$. The discussion presented so far remains valid. With the resulting value of $I(t)$, we should keep track of whether the calculated rate of infiltration, I, is less than that of application, $R(t)$. As soon as we reach the situation of $I \geq R$, we should switch back to the condition (6.4.16), with $I = R$.

It is important to emphasize again that initially, and for some time (which may be significant when considering irrigation, or an individual storm event), the rate of infiltration will exceed the limiting value of $\mathsf{K}_z\big|_{\text{sat}}$. The latter is approached *from above*, provided the rate of application of water to ground surface remains larger than the rate of infiltration.

CASE B. Precipitation is applied at a constant rate $R < K_z\big|_{\text{sat}}$. Initially, the saturation gradient at ground surface will be very high, and the soil will absorb all incoming water. However, rather rapidly, as infiltration continues, and saturation at ground surface increases, the saturation gradient there will decrease. Eventually, asymptotically, a saturation level is reached with a zero saturation gradient, so that the rate of infiltration (which equals to the rate of application) becomes equal to the effective hydraulic conductivity at the prevailing saturation. Under such conditions, the only force driving infiltration at ground surface is gravity. Capillarity still plays a role in the part of the wetting front that is ahead of the (practically) saturated zone.

For $R = R(t) < K_z\big|_{\text{sat}}$, the same phenomena, as described above, will occur, viz., the rate of infiltration will equal that of accretion, but saturation at ground surface will vary, without leveling off at any asymptotic value.

These phenomena have been known to hydrologists for many years. Here, we have expressed them as constraints associated with the boundary conditions at ground surface.

We may summarize the discussion of the above two cases as follows:

(a) If $p_w(t) = p_a\big|_{\text{atm}}$, then,

- If $R(t) \geq I(t)$, use (6.4.17).
- If $R(t) < I(t)$, use (6.4.16) with $I = R(t)$.

(b) If $p_w(t) < p_a\big|_{\text{atm}}$, use (6.4.16) with $I = R(t)$.

Let us comment about the case of evaporation, or evapotranspiration, produced by solar radiation reaching ground surface. Also in this case, the boundary conditions must be based on the equality of water and water vapour fluxes across ground surface. In this case, as water leaves the soil in the form of vapour, the soil dries out. As the saturation at (i.e., just below) ground surface reaches the irreducible saturation level, S_{wr}, water effective permeability reduces to zero. In the isothermal flow models discussed here, it is usually assumed that when the soil at ground surface reaches S_{wr}, it can no longer transmit liquid water. The boundary condition has to be switched to one of no-flow at ground surface until the water saturation rises above S_{wr}. In more sophisticated, say, nonisothermal models, which are beyond the scope of this book, the water saturation can reduce to below the irreducible value. Obviously, as saturation drops to *below* the irreducible one, e.g., by evaporation and/or root uptake, the effective permeability to water vanishes. Under non-isothermal conditions, the situation may be more complicated, as a *drying front* may move up and down below ground surface.

H. Ponding Above Ground Surface

Finally, let us consider the possibility of ponding. This occurs when the liquid's pressure at ground surface satisfies the condition $p_w > p_a\big|_{\text{atm}}$. Suppose we allow ponding up to a maximum depth that can be specified as $p_w\big|_{\max}/\rho_w g$. Instead of condition (6.4.17) at $S_w = 1$, we treat $p_w|_{z=0}$ as an unknown, and introduce the condition

$$R(t) - I(t) \equiv R(t) - \frac{\mathsf{k}_z\big|_{\text{sat}}}{\mu_w}\left(\frac{\partial p_w}{\partial z} + \rho_w g\right) = \frac{1}{\rho_w g}\frac{\partial}{\partial t}\left(p_w\big|_{z=0} - p_a\big|_{\text{atm}}\right), \tag{6.4.18}$$

allowing $p_w\big|_{z=0}$ to rise up to $p_w\big|_{\max} + p_a\big|_{\text{atm}}$. We usually assume $p_a\big|_{\text{atm}} = 0$.

We do not allow p_w to rise above the specified maximum. If, as a result of $R(t)$, the water level tends to rise higher, it is set at the maximum value. As it drops and reaches zero, or atmospheric gas pressure, we switch to the condition (6.4.17), as long as $S_w = 1$.

We may also assume that ponding occurs *within* an external soil domain that may have different soil characteristics. Defining a specific yield of such a soil by S_y (similar to its definition for a phreatic aquifer), the boundary condition with ponding may be rewritten as

$$R(t) - I(t) \equiv R(t) - \frac{k_z\big|_{\text{sat}}}{\mu_w}\left(\frac{\partial p_w}{\partial z} + \rho_w g\right) = S_y\frac{1}{\rho_w g}\frac{\partial}{\partial t}\left(p_w\big|_{z=0} - p_a\big|_{\text{atm}}\right). \tag{6.4.19}$$

6.4.3 Pressure-Saturation Form of the Balance Equation

Let us rewrite the mass balance equations for two-phase flow in a form that will better expose their nature, using an approach that is often used in petroleum engineering. We shall refer to the two fluids as water (w) and oil (o). The considered domain is assumed anisotropic.

By summing the two equations in (6.4.3), with $\phi =$ const. and $S_w + S_o = 1$, we obtain the balance equation for the total mass of fluids in the void space:

$$\phi \frac{\partial}{\partial t}[S_w \rho_w + (1 - S_w)\rho_o] = -\nabla \cdot (\rho_w \mathbf{q}_w + \rho_o \mathbf{q}_a). \tag{6.4.20}$$

Next, we define a *total specific discharge*:

$$\mathbf{q}_t = \mathbf{q}_w + \mathbf{q}_o, \tag{6.4.21}$$

and an average pressure of the fluids filling up the void space:

$$\mathcal{P} = \tfrac{1}{2}(p_w + p_o), \tag{6.4.22}$$

with

$$p_w = \mathcal{P} - \tfrac{1}{2} p_c, \qquad p_o = \mathcal{P} + \tfrac{1}{2} p_c. \tag{6.4.23}$$

We introduce two *mobility tensors* for the respective phases:

$$\boldsymbol{\mathcal{M}}_w = \mathsf{k}_w / \mu_w, \qquad \boldsymbol{\mathcal{M}}_o = \mathsf{k}_o / \mu_o, \tag{6.4.24}$$

and a *total mobility*, defined by:

$$\boldsymbol{\mathcal{M}}_t = \boldsymbol{\mathcal{M}}_w + \boldsymbol{\mathcal{M}}_o, \tag{6.4.25}$$

where $\boldsymbol{\mathcal{M}}$ indicates a second rank tensor (for an anisotropic porous medium).

Substituting (6.4.23) and (6.4.24) into the two flux equations (6.4.4), leads to a restatement of the two flux equations in the form:

$$\mathbf{q}_w = -\boldsymbol{\mathcal{M}}_w \cdot (\nabla \mathcal{P} - \tfrac{1}{2} \nabla p_c + \rho_w g \nabla z), \tag{6.4.26}$$

$$\mathbf{q}_o = -\boldsymbol{\mathcal{M}}_o \cdot (\nabla \mathcal{P} + \tfrac{1}{2} \nabla p_c + \rho_o g \nabla z). \tag{6.4.27}$$

Substituting these equations into (6.4.20) leads to another form of the mass balance equation,

$$\phi\frac{\partial}{\partial t}[S_w\rho_w + (1-S_w)\rho_o] = \nabla\cdot[(\rho_w\boldsymbol{\mathcal{M}}_w + \rho_o\boldsymbol{\mathcal{M}}_o)\cdot\nabla\mathcal{P}]$$
$$-\tfrac{1}{2}\nabla\cdot[(\rho_w\boldsymbol{\mathcal{M}}_w - \rho_o\boldsymbol{\mathcal{M}}_o)\cdot\nabla p_c] + g\nabla\cdot[\rho_w^2\boldsymbol{\mathcal{M}}_w + \rho_o^2\boldsymbol{\mathcal{M}}_o)\cdot\nabla z], \tag{6.4.28}$$

in terms of the three variables, S_w, p_o and p_w. Equation (6.4.5) can be used to express fluid densities in terms of the fluids' pressures. If densities are assumed constant, or are assumed to be functions of $\mathcal{P}$, then the above equation is written only in terms of S_w and $\mathcal{P}$.

In reservoir engineering, (6.4.28), with $\mathcal{P}$ as the dependent variable, is called the *pressure equation*. The term on the left-hand side is often quite small.

Equation (6.4.28) can be solved only if S_w, p_o and p_w are known, because the coefficients appearing in it depend on S_w, p_o, or p_w.

Consider the total specific discharge, $\mathbf{q}_t$, which may now be written as

$$\mathbf{q}_t = -\boldsymbol{\mathcal{M}}_t\cdot\nabla\mathcal{P} - \tfrac{1}{2}(\boldsymbol{\mathcal{M}}_w - \boldsymbol{\mathcal{M}}_o)\cdot\nabla p_c - g(\rho_w\boldsymbol{\mathcal{M}}_w + \rho_a\boldsymbol{\mathcal{M}}_o)\cdot\nabla z. \tag{6.4.29}$$

Multiplying (6.4.26) by $\boldsymbol{\mathcal{M}}_w^{-1}$ and (6.4.27) by $\boldsymbol{\mathcal{M}}_o^{-1}$, and subtracting the resulting equations, yields

$$\boldsymbol{\mathcal{M}}_w^{-1}\cdot\mathbf{q}_w - \boldsymbol{\mathcal{M}}_o^{-1}\cdot\mathbf{q}_o = \nabla p_c - (\rho_w - \rho_o)g\nabla z. \tag{6.4.30}$$

Then, using (6.4.21) to eliminate $\mathbf{q}_o$ from (6.4.30), we obtain

$$\boldsymbol{\mathcal{M}}_w^{-1}\cdot\mathbf{q}_w - \boldsymbol{\mathcal{M}}_o^{-1}\cdot\mathbf{q}_t + \boldsymbol{\mathcal{M}}_o^{-1}\cdot\mathbf{q}_w = \nabla p_c - (\rho_w - \rho_o)g\nabla z, \tag{6.4.31}$$

which can be rearranged to the form:

$$\mathbf{q}_w = \boldsymbol{\mathcal{M}}_{eff}\cdot\boldsymbol{\mathcal{M}}_o^{-1}\cdot\mathbf{q}_t + \boldsymbol{\mathcal{M}}_{eff}\cdot\nabla p_c - (\rho_w - \rho_o)g\boldsymbol{\mathcal{M}}_{eff}\cdot\nabla z. \tag{6.4.32}$$

The coefficient

$$\boldsymbol{\mathcal{M}}_{eff} = \left(\boldsymbol{\mathcal{M}}_w^{-1} + \boldsymbol{\mathcal{M}}_o^{-1}\right)^{-1} \tag{6.4.33}$$

is called the *effective mobility*.

The equation for S_w is obtained from (6.4.3) and (6.4.32), in the form:

$$\phi\frac{\partial(S_w\rho_w)}{\partial t} = -\nabla\cdot(\rho_w\boldsymbol{\mathcal{M}}_{eff}\cdot\boldsymbol{\mathcal{M}}_o^{-1}\cdot\mathbf{q}_t) - \nabla\cdot(\rho_w\boldsymbol{\mathcal{M}}_{eff}\cdot\nabla p_c)$$
$$+\nabla\cdot[(\rho_w - \rho_o)g\boldsymbol{\mathcal{M}}_{eff}\cdot\nabla z], \tag{6.4.34}$$

which must be solved in conjunction with appropriate density-pressure relationships. The above equation is the sought equation, written in terms of pressure and saturation.

For simplicity, let us assume that the fluids are incompressible, and the solid matrix nondeformable, so that ϕ, ρ_w, ρ_o μ_w, and μ_o are constants. We may then make the following interpretation of (6.4.34). First, we define:

$$\mathbf{F}_w = \boldsymbol{\mathcal{M}}_{eff}\cdot\boldsymbol{\mathcal{M}}_o^{-1}, \quad \mathbf{G}_w = -\boldsymbol{\mathcal{M}}_{eff}, \quad \mathbf{H}_w = (\rho_w - \rho_o)g\boldsymbol{\mathcal{M}}_{eff}. \tag{6.4.35}$$

We note that $\mathbf{F}_w$ can also be written as $\mathbf{F}_w = M_w(M_w + M_o)^{-1}$. If M_w and M_o are scalars, $\mathbf{F}_w$ is called the *fractional flow function for the water phase*.

Then, (6.4.34) can be written as:

$$\phi\frac{\partial S_w}{\partial t} = -\nabla\cdot(\mathbf{F}_w\cdot\mathbf{q}_t) + \nabla\cdot(\mathbf{G}_w\cdot\nabla S_w) + \nabla\cdot(\mathbf{H}_w\cdot\nabla z). \tag{6.4.36}$$

Expanding the second term of this equation as

$$\nabla\cdot(\mathbf{F}_w\cdot\mathbf{q}_t) = \mathbf{q}_t\cdot(\nabla\cdot\mathbf{F}_w) + (\nabla\mathbf{q}_t)\cdot\mathbf{F}_w, \tag{6.4.37}$$

and, noting that $\mathbf{F}_w = \mathbf{F}_w(S_w)$, we may write:

$$\nabla\cdot\mathbf{F}_w = \frac{d\mathbf{F}_w}{dS_w}\cdot\nabla S_w. \tag{6.4.38}$$

Thus, Eq. (6.4.36) can be rewritten in the form:

$$\begin{aligned}\phi\frac{\partial S_w}{\partial t} &= -\left(\frac{d\mathbf{F}_w}{dS_w}\cdot\mathbf{q}_t\right)\cdot\nabla S_w + \nabla\cdot(\mathbf{G}_w\cdot\nabla S_w)\\ &\quad + \nabla\cdot(\mathbf{H}_w\cdot\nabla z) - (\nabla\mathbf{q}_t)\cdot\mathbf{F}_w.\end{aligned} \tag{6.4.39}$$

We note that (6.4.39) is now a single equation in the single variable S_w. Often $\mathbf{G}_w\cdot\nabla S_w$ is small compared to other terms on the right-hand side.

The standard equations derived and used in reservoir engineering, differ from (6.4.28) and (6.4.34) presented here. The main difference results from the treatment of the effective mobilities, $\boldsymbol{\mathcal{M}}_w$ and $\boldsymbol{\mathcal{M}}_o$. Reservoir engineers often limit the analysis to isotropic porous media in which the effective permeability reduces to a *scalar*. They then introduce the definition of *relative permeability*—also a scalar—for each of the phases, with $k_w = k\mathsf{k}_{rw}$ and $k_o = kk_{ro}$, in which permeabilities are defined as *scalars*. This assumption of scalar permeabilities simplifies the results because $\mathcal{M}_w\mathcal{M}_o = \mathcal{M}_o\mathcal{M}_w$, whereas in the general case of an anisotropic medium, $\boldsymbol{\mathcal{M}}_w\cdot\boldsymbol{\mathcal{M}}_o \neq \boldsymbol{\mathcal{M}}_o\cdot\boldsymbol{\mathcal{M}}_w$. This affects the development of $\mathbf{q}_w$ and, hence, the resulting saturation equation.

For completeness, we present the pressure and saturation equations for an isotropic medium. Equation (6.4.28) becomes:

$$\begin{aligned}\phi\frac{\partial}{\partial t}(S_w\rho_w+S_a\rho_o)=&\ \nabla\cdot[(\rho_w\mathcal{M}_w+\rho_o\mathcal{M}_o)\nabla\mathcal{P}]\\&-\tfrac{1}{2}\nabla\cdot[(\rho_w\mathcal{M}_w-\rho_a\mathcal{M}_o)\nabla p_c]\\&+g\nabla\cdot[(\rho_w^2\mathcal{M}_w+\rho_o^2\mathcal{M}_a)\nabla z],\end{aligned}\tag{6.4.40}$$

in which the mobilities are now scalars. Equation (6.4.34) becomes:

$$\begin{aligned}\phi\frac{\partial}{\partial t}(S_w\rho_w)=&-\nabla\cdot\{\rho_w\mathcal{M}_{eff}\mathcal{M}_o^{-1}\left[\mathbf{q}_t-\mathcal{M}_o(\rho_a-\rho_w)g\nabla z\right]\}\\&-\nabla\cdot\left(\rho_w\mathcal{M}_{eff}\nabla p_c\right),\end{aligned}\tag{6.4.41}$$

where $\mathcal{M}_{eff}$ is defined in (6.4.33).

6.4.4 Linear Displacement and Fingering

In Sect. 5.2.1 A and B we have introduced the "sharp interface" approximation that separates two fluids from each other as they move together within a considered domain. However, earlier in the current chapter, we have introduced the concept of (macroscopic) capillary pressure which produces a gradual saturation transition zone between the two fluids. Here, we wish to discuss an important phenomenon that affects the saturation distribution as an initially sharp interface that separates the two moving fluids is being displaced. This instability, referred to as "viscous fingering" (e.g., Bear 1972, p. 544), is strongly dependent on the *mobility ratio* between the two fluids, where *mobility* is defined by (6.4.24).

Consider the horizontal 1-d flow in the direction $+x$ of two incompressible fluids in a homogeneous porous medium domain of length L. Initially, the two fluids are separated by an interface in the form of a plane whose normal is in the direction of $+x$. When flow is initiated, say by a pressure gradient, the interface advances: fluid 1 which is a *wetting fluid*, say water, displaces fluid 2 which is a *non-wetting fluid*, say oil. At some time t the interface reaches a distance x_f from the origin at $x=0$. We can describe the location of this interface by the equation:

$$F(x,t)\equiv x-x_f(t)=0.\tag{6.4.42}$$

For the horizontal flow considered here, we shall express the model in terms of pressure: we wish to determine the pressure distributions: $p_w=p_w(x,t)$ and $p_{nw}=p_{nw}(x,t)$. and the displacement, $x_f(t)$, such that:

$$\begin{aligned}&\frac{\partial^2 p_w}{\partial x^2}=0,\quad 0\le x\le x_f(t),\\&\frac{\partial^2 p_{nw}}{\partial x^2}=0,\quad x_f(t)\le x\le L,\end{aligned}\tag{6.4.43}$$

$$p_w = p_{nw}(\text{neglecting } p_c) \text{ on } x_f = x_f(t), \tag{6.4.44}$$

$$\frac{k_{wo}}{\mu_w}\frac{\partial p_w}{\partial x} = \frac{k_{nwo}}{\mu_{nw}}\frac{\partial p_{nw}}{\partial x} = -q_x, \quad on \quad x_f = x_f(t), \tag{6.4.45}$$

in which k_{wo} denotes the wetting fluid permeability at residual non-wetting fluid saturation, and k_{nwo} denotes the non-wetting fluid permeability at irreducible wetting fluid saturation.

The conditions at the boundaries are

$$x = 0, \quad p_w = p_{wo}, \quad x = L, \quad p_{nw} = p_{nwL}. \tag{6.4.46}$$

By integrating (6.4.43) and using conditions (6.4.44) through (6.4.46), we obtain:

$$p_w = \frac{p_O - p_L}{M_r L + (1 - M_r)x_f} x + p_{wo}, \quad p_{nw} = \frac{p_O - p_L}{M_r L + (1 - M_r)x_f} \mathcal{M}(L - x) + p_{nwL}, \tag{6.4.47}$$

in which $M_r = (k_{wo}/\mu_w)/(k_{nwo}/\mu_{nw})$ is called the *mobility ratio*. Hence:

$$q_x = \frac{k_{wo}}{\mu_w}\frac{p_{wo} - p_{nwL}}{M_r L + (1 - M_r)x_f}. \tag{6.4.48}$$

Since $\partial F/\partial t = -\partial x_f/\partial t$ and $VV{\cdot}\nabla F = V_x \partial F/\partial x = V_x, = q_x/\phi$, we obtain:

$$-\frac{\partial x_f}{\partial t} + \frac{k_{wo}}{\mu_w \phi(1 - S_{wo} - S_{nwo})}\frac{p_{wo} - p_{nwL}}{M_r L + (1 - M_r)x_f} = 0. \tag{6.4.49}$$

Replacing $\partial x_f/\partial t$ by dx_f/dt and integrating, we obtain:

$$t = \frac{\mu_w \phi\,(1 - S_{wo} - S_{nwo})\,L^2}{k_{wo}(p_{wo} - p_{nwL})}\left[M_r\left(\frac{x_f}{L}\right) + \frac{1}{2}(1 - M_r)\left(\frac{x_f}{L}\right)^2\right] \tag{6.4.50}$$

for the time required for the front to advance from $x = 0$ to some distance $x = x_f(t)$. From the last equation it follows that the front either accelerates or decelerates, depending on whether $M_r > 1$ or $M_r < 1$. In the special case of $N_r = 1$, the front moves at a constant speed.

Figure 6.11 shows (6.4.50) in a graphical form (following Collins 1961, p. 175)

Contrary to the above development, which is based on a "sharp interface approximation", reality is different. An important phenomenon has been overlooked when assuming that "a sharp interface separates the two moving fluids". This is the phenomenon of *instability* referred to as *viscous fingering* (e.g., Bear 1972, p. 544). Figure 6.12 shows the instability that occurs when two fluids, initially separated by a sharp interface, displace each other in the void space.

To explain the process that produces instability, consider the case of an advancing front as described above, but due to the irregular shape of the void space at the

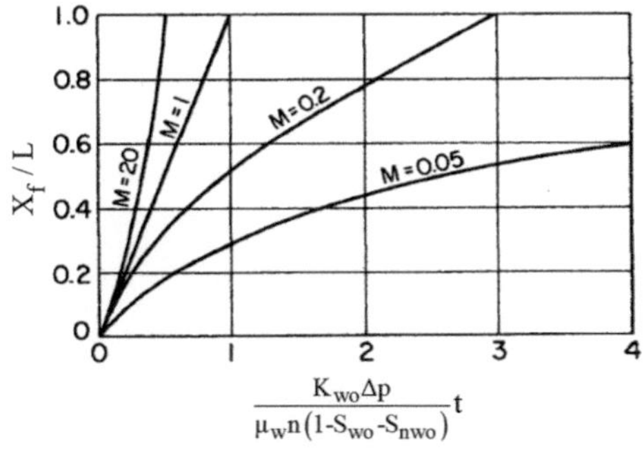

Fig. 6.11 Displacement of a sharp front (Bear 1972, p. 528)

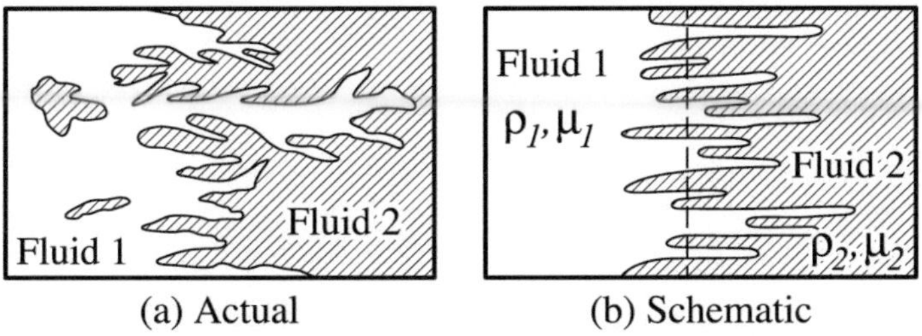

Fig. 6.12 Instability at a moving interface (flow is from left to right)

microscopic level, a small "bump", of length $\epsilon(\ll x_f)$ starts on the advancing front between the two fluids. Collins (1961, p. 197) writes (6.4.49) in the form:

$$\frac{d\epsilon}{dt} = -\frac{k_w}{\phi\mu_w(1 - S_{wr} - S_{nr})} \frac{(1 - \mathcal{M}_{wn})\epsilon}{[\mathcal{M}_{wn}L + (1 - \mathcal{M}_{wn})x_f]^2}(p_{w,0} - p_{w,L}), \quad (6.4.51)$$

concluding that after a bump has been initiated, if $\mathcal{M}_{wn} > 1$, then ϵ will grow exponentially with time; when $\mathcal{M}_{wn} < 1$, the bump will decay exponentially with time. This simple example shows that if the displacing fluid is more mobile than the displaced one, any small perturbation of the advancing front produces fast growing irregularities in the form of "fingers", or "viscous fingers" that extend from the advancing front.

According to Scheidegger (1960a, 1960b), a finger will develop only if its formation will consume less energy than the corresponding stable front displacement of the front. He also reaches the conclusion that a finger will develop when $\mathcal{M}_{wn} > 1$.

The above analysis does not include the effects of gravity and capillarity on fingering at an advancing sharp interface. Bear (1972, p. 547), following Marle (1965, p. 150), discussed the effect of gravity by investigating the vertical displacement of horizontal front shown in Fig. 6.13. In the stable case, i.e., without a finger, we write:

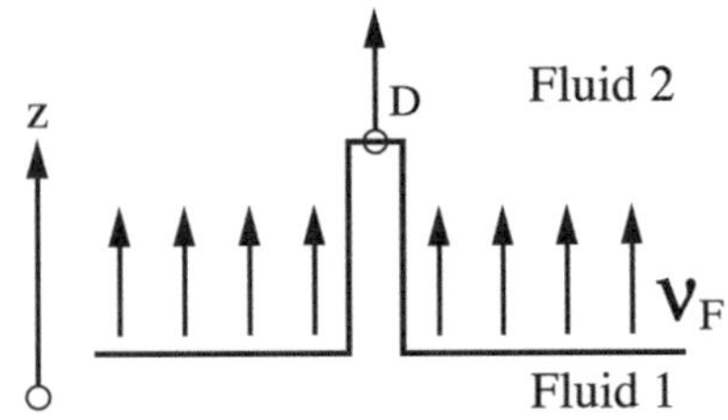

Fig. 6.13 Gravity effect on instability at a moving interface

$$q_1 = -\frac{k_1}{\mu_1}\left(\frac{\partial p_1}{\partial z} + \rho_1 z\right), \quad q_2 = -\frac{k_2}{\mu_2}\left(\frac{\partial p_2}{\partial z} + \rho_2 z\right). \tag{6.4.52}$$

The rising front moves at the velocity $V_F = q_1/\phi^* = q_2/\phi^*$, where ϕ^* represents the porosity for complete separation between the fluids, and for $\phi(1 - S_{wo} - S_{nwo})$ in the case of incomplete separation. and k_1 and k_2 are effective permeabilities. When a finger is formed, let V_D, q_D denote the velocity of the advancing "finger" (point D) in Fig. 6.13. Within this finger:

$$q_D = -\frac{k_1}{\mu_1}\left(\frac{\partial p_2}{\partial z} + \rho_2 z\right) = \phi V_D. \tag{6.4.53}$$

In the above equation, $\partial p_2/\partial z$ is used because the finger is completely surrounded by fluid 2, which imposes its pressure on the fluid within the finger. $V_D > V_F$ means that the finger tends to grow, and vice versa. From the last two equations, it follows that:

$$V_D - V_F = -\frac{k_1}{\phi\mu_1}\left(\frac{\partial p_2}{\partial z} - \frac{\partial p_1}{\partial z}\right), \tag{6.4.54}$$

or,

$$V_D - V_F = -\frac{k_1}{\phi\mu_1}\left[\phi V_F\left(\frac{\mu_1}{k_1} - \frac{\mu_2}{k_2}\right) + g(\rho_1 - \rho_2)\right], \tag{6.4.55}$$

or:

$$V_D - V_F = (\mathcal{M}_{1,2} - 1)(V_F - V_C), \quad V_C = -\frac{k_1 g}{\phi\mu_1}(\rho_1 - \rho_2)(1 - \mathcal{M}_{1,2}) \tag{6.4.56}$$

where $\mathcal{M}_{1,2} = (k_1/\mu_1)/(k_2/\mu_2)$ denotes the *mobility ratio*, with subscripts 1 and 2 denoting the displacing and displaced fluids, respectively.

Marle (1965, p. 150) summarizes the conditions for the development of fingers, making use of Fig. 6.13, which shows a vertical cross-section, with $V_f > 0$ denoting the velocity of the upward moving front, and V_D denoting the velocity at which the finger (Point D) advances:

- If $\mu_2/k_2 < \mu_1/k_1$ i.e., $\mathcal{M}_{1,2} < 1$, and $\rho_1 > \rho_2$, $V_c < 0$, then always, $V_D < V_f$, and the front is stable.
- if $\mu_2/k_2 < \mu_1/k_1$ i.e., $\mathcal{M}_{1,2} < 1$, and $\rho_1 < \rho_2$, always $V_c > 0$, but there are two possibilities: if $V_f > V_c$, the front is stable; if $V_f < V_c$, the front is unstable, where $V_c = -(k_1/\phi\mu_1)g(\rho_1 - \rho_2)(1 - \mathcal{M}_{1,2})$ is a critical velocity that determines the stability of the front.
- If $\mu_2/k_2 > \mu_1/k_1$, i.e., $\mathcal{M}_{1,2} > 1$, and $\rho_1 < \rho_2$, the displacement is stable if $V_f < V_c$, and unstable if $V_f > V_c$.
- If $\mu_2/k_2 > \mu_1/k_1$, i.e., $\mathcal{M}_{1,2} > 1$, and $\rho_1 < \rho_2$, we cannot have $V_f < V_c$, because $V_c > 0$, and the front is always unstable,

Although the subject of fingering was introduced above in connection with two immiscible fluid phases, it is possible that fingering due to density difference will occur also at the interface between two miscible fluids, e.g., domains that contain the same fluid, but with a large difference in the concentration of dissolved solutes. Salt water above fresh water may serve as an example. Natural convection of CO_2 dissolved in water is another example.

Note that a difference in density may be produced not only by solute concentration but also by different temperatures. Nield and Bejan (2013) considered this case.

6.4.5 *The Buckley Balance Equation*

Consider a model of one-dimensional flow along the z-axis (positive upward) of constant density (w and o)-fluids in a stationary nondeformable porous medium. Then, (6.4.41) takes the form:

$$\phi\frac{\partial S_w}{\partial t} + \frac{\partial}{\partial z}\left\{\mathcal{M}_{eff}\left[\frac{q_t}{\mathcal{M}_o} + \frac{\partial p_c}{\partial z} - (\rho_w - \rho_o)g\right]\right\} = 0. \tag{6.4.57}$$

Note that

$$-\frac{\partial p_w}{\partial z} = \rho_w g + \frac{r_w}{\mathcal{M}_w}q_t, \qquad r_w \equiv \frac{q_w}{q_t}, \tag{6.4.58}$$

$$-\frac{\partial p_o}{\partial z} = \rho_o g + \frac{r_o}{\mathcal{M}_o}q_t, \qquad r_o \equiv \frac{q_o}{q_t}, \tag{6.4.59}$$

where, $\mathcal{M}_o$ and $\mathcal{M}_w$ denote the o and w mobilities.

By subtracting (6.4.59) from (6.4.58), and using $r_w + r_o = 1$, we obtain:

$$r_w = \mathcal{M}_{eff}\left\{\frac{1}{\mathcal{M}_o} + \frac{1}{q_t}\left[\frac{\partial p_c}{\partial z} - (\rho_w - \rho_o)g\right]\right\}. \tag{6.4.60}$$

The balance equation for total mass, $\partial q_t/\partial z = 0$, implies that q_t $(\equiv q_w + q_o)$ $=$ const., or possibly $q_t = q_t(t)$. Note that since p_c, $\mathcal{M}_o$, and $\mathcal{M}_w$ are functions of water saturation, we have $r_w = r_w(S_w, t)$. As a result, (6.4.57) reduces to the form:

$$\frac{\partial S_w}{\partial t} + \frac{q_t(t)}{\phi}\frac{\partial r_w}{\partial z} = 0. \tag{6.4.61}$$

Even in this simplified case, obtaining solutions of the nonlinear governing equations is difficult and, generally, requires computer-based methods. Buckley and Leverett (1942) proposed an analytic solution for cases in which the effects of capillary pressure and gravity may be neglected. Their solution is often used in reservoir engineering to investigate horizontal one-dimensional displacement of oil by water in a reservoir.

Let two fluids, water (wetting fluid, w) and oil (nonwetting fluid, n), flow in a one-dimensional horizontal reservoir. If the effects of capillary pressure may be neglected, relative to the other terms in (6.4.61), i.e., when $|\mathrm{d}p_c/\mathrm{d}x|(k_{rnw}/\mu_{nw}) \ll 1$, the saturation equation becomes

$$\frac{\partial S_w}{\partial t} + \frac{q_t(t)}{\phi}\frac{\mathrm{d}r_w}{\mathrm{d}S_w}\frac{\partial S_w}{\partial x} = 0, \tag{6.4.62}$$

with which (6.4.60) reduces to the *fractional flow function*

$$r_w = \frac{\mathcal{M}_{eff}}{\mathcal{M}_{nw}} = \frac{1}{1+\frac{\mathcal{M}_{nw}}{\mathcal{M}_w}} = \frac{1}{1+\frac{k_{rnw}}{k_{rw}}\frac{\mu_w}{\mu_{nw}}}. \tag{6.4.63}$$

Neglecting the effect of capillary pressure is justified only in certain regions. When saturation gradients are large, this assumption is not valid, and a numerical solution is required.

We note that r_w is a function of S_w only, through the dependence on k_w and k_{nw}. Figure 6.14 shows $r_w = r_w(S_w)$ for this case.

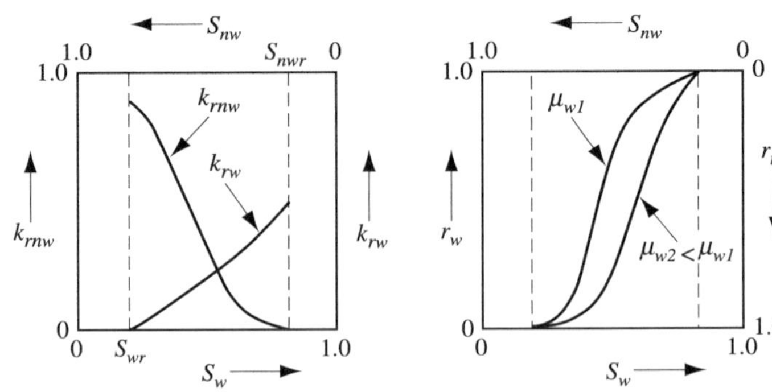

Fig. 6.14 Fractional flow function, $r_w = r_w(S_w)$

Equation (6.4.62) is known as the *Buckley–Leverett equation*. This is a quasi-linear, homogeneous, first-order partial differential equation, which can be solved by the *method of characteristics* (e.g., Marle 1981).

From $S_w = S_w(x,t)$, it follows that

$$\frac{\mathrm{d}S_w}{\mathrm{d}t} = \frac{\partial S_w}{\partial x}\frac{\mathrm{d}x}{\mathrm{d}t} + \frac{\partial S_w}{\partial t}.$$

If $x = \xi(t)$ is chosen to coincide with an advancing point of fixed water saturation, say, $S_w = S$, then $\mathrm{d}S_w/\mathrm{d}t = 0$ at this point, and its velocity is

$$u_w = \left.\frac{\mathrm{d}\xi}{\mathrm{d}t}\right|_{S_w=S} = -\frac{\partial S_w/\partial t}{\partial S_w/\partial x}. \tag{6.4.64}$$

By combining (6.4.62) with (6.4.64), we obtain:

$$u_w = \left.\frac{q_t(t)}{\phi}\frac{\mathrm{d}r_w}{\mathrm{d}S_w}\right|_{S_w=S}. \tag{6.4.65}$$

Then, integrating (6.4.65), the position of the point with *specified water saturation*, $S_w = S$, at time t is given by

$$x\big|_{S_w=S}(t) - x\big|_{S_w-S}(0) = \left.\frac{\mathbb{V}(t)}{\phi}\frac{\mathrm{d}r_w}{\mathrm{d}S_w}\right|_{S_w=S}, \tag{6.4.66}$$

where

$$\mathbb{V}(t) = \int_0^t q_t(t)\,dt$$

denotes the volume of the wetting phase that passes through the horizontal column (per unit area normal to the flow). For q_t equal to a constant, we obtain

$$x\big|_{S_w=S}(t) = x\big|_{S_w=S}(0) + \left.\frac{q_t t}{\phi}\frac{\mathrm{d}r_w}{\mathrm{d}S_w}\right|_{S_w=S}. \tag{6.4.67}$$

Figure 6.15 illustrates the function $\mathrm{d}r_w/\mathrm{d}S_w$. Figure 6.16 shows the solution in graphical form. Note that both small and large values of S_w travel at a velocity that is smaller than that of intermediate S_w-values. As a consequence, it is possible that at a given point in space and time we shall have *three* saturation values. Obviously, this is a non-physical solution. The reason for this situation is that we have neglected the effect of capillary pressure. Mathematically, the triple saturation points in the solution at a given x,t is a result of replacing a second order partial differential equation by a first order one.

Let us demonstrate how to overcome this situation by considering a case in which the nonwetting fluid is injected into a column (representing one-dimensional flow

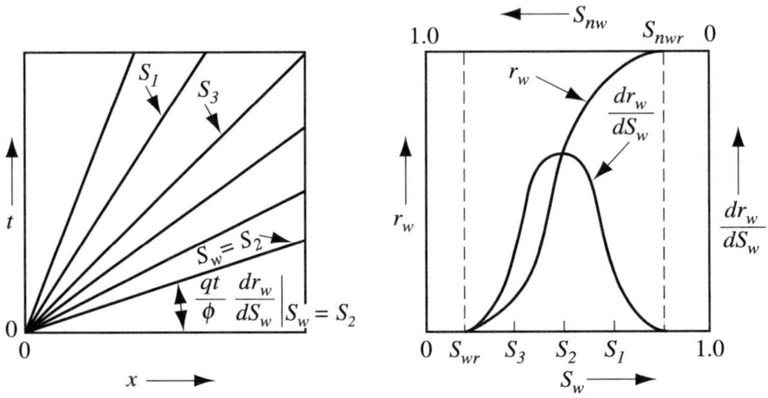

Fig. 6.15 The function dr_w/dS_w

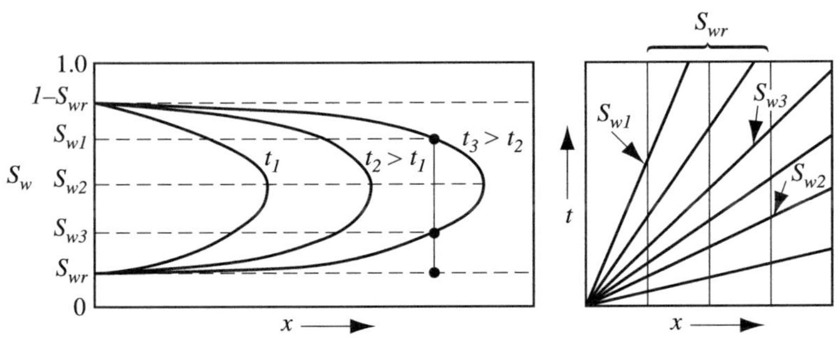

Fig. 6.16 Triple saturations in a Buckley–Leverett solution

in a reservoir) initially at residual water saturation, with the end maintained at the maximum value, i.e.,

$$t = 0, \quad 0 < x < L, \quad S_w = S_{wr},$$

$$x = 0, \quad t > 0, \quad S_w = 1 - S_{nwr}.$$

By injecting water at $x = 0$ at the constant rate, q_t, the saturation at $x = 0$ is raised to $1 - S_{nwr}$. To avoid the non-physical solution, when the flow rate is sufficiently high, we introduce a *discontinuity*, or *shock*, i.e., a sudden spatial change in the saturation profile. Let us denote the saturation at the advancing shock or *front* by S_{wf}. This shock travels along the column as $S_{wf} = S_{wf}(t)$. Obviously, *mass must still be conserved*.

Figure 6.17 shows this shock. In this figure, we note that the rectangular area is: $x_f(S_{wf} - S_{wr})$. With $x(0) = 0$ in (6.4.66), this area must equal

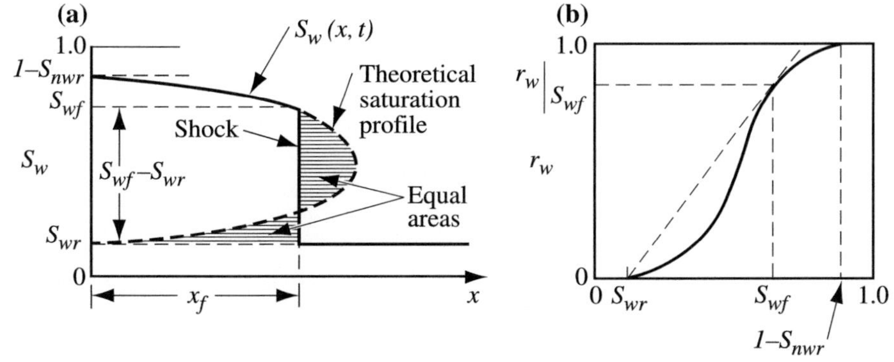

Fig. 6.17 Introducing a saturation discontinuity

$$\int_{S_{wr}}^{S_{wf}} x\,\mathrm{d}S_w \equiv \int_{S_{wr}}^{S_{wf}} \frac{u(t)}{\phi}\frac{\mathrm{d}r_w}{\mathrm{d}S_w}\mathrm{d}S_w$$
$$= \frac{u(t)}{\phi}\int_{r_{wr}}^{r_{wf}} \mathrm{d}r_w = \frac{u(t)}{\phi}\left.\frac{\mathrm{d}r_w}{\mathrm{d}S_w}\right|_{S_{wf}}(S_{wf}-S_{wr}),$$

or

$$\left.\frac{\mathrm{d}r_w}{\mathrm{d}S_w}\right|_{S_{wf}} = \frac{r_w|_{S_{wf}} - r_w|_{S_{wr}}}{S_{wf}-S_{wr}} = \frac{r_w|_{S_{wf}}}{S_{wf}-S_{wr}}. \tag{6.4.68}$$

Thus, the velocity of the front is determined by

$$u_f \equiv \frac{\mathrm{d}x}{\mathrm{d}t} = \frac{q_t}{\phi}\frac{[\![\ r_w\ ]\!]_{1,2}}{[\![\ S_w\ ]\!]_{1,2}}, \tag{6.4.69}$$

where $[\![\ (..)\ ]\!]_{1,2}$ denotes the jump in (..) from side 1 to side 2 across the shock. The location of this moving front is

$$x = \frac{u(t)}{\phi} = \frac{r|_{S_{wf}}}{S_{wf}-S_{wr}}.$$

This relationship, along with Fig. 6.17b, serve as the basis for the graphical determination of x_f and S_{wf} suggested by Welge (1952).

Figure 6.18 shows an advancing shock for the case where initially the column is at $S_w = S_{w1}$, $S_n = 1 - S_{w1}$. By injecting the wetting fluid at $x = 0$, at a constant rate, $q_t r_{w2}$, and the nonwetting fluid at a constant rate $q_t(1 - r_{w2})$, the saturation at $x = 0$ rises to S_{w2}, and stays at that saturation level. We note how from point (S_{w1}, r_{w1}), we draw a *tangent* to the curve $r_w = r_w(S_w)$ to determine the location of the point (S_{wf}, r_{wf}), thus determining the value S_{wf} from the relationship:

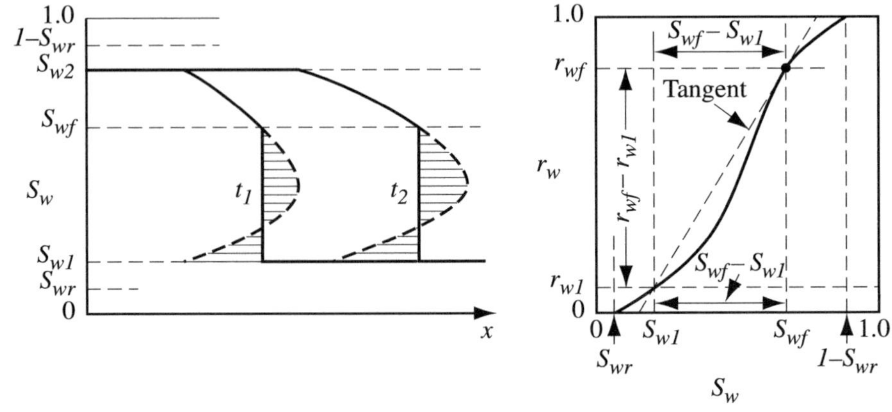

Fig. 6.18 Graphical solution for S_{wf}; initially, $S_w \neq S_{wr}$

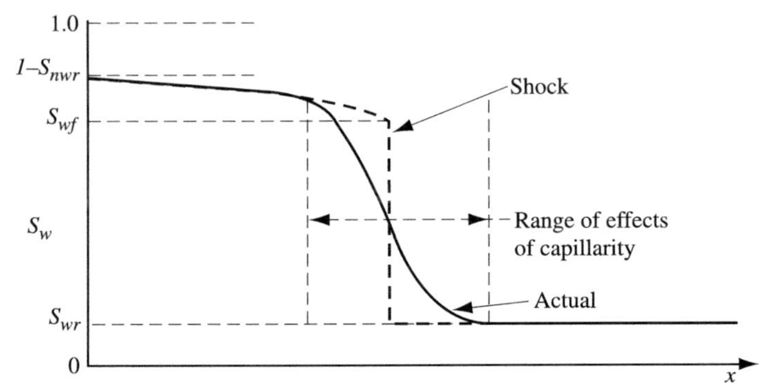

Fig. 6.19 Effect of capillarity

$$\left.\frac{\mathrm{d}r_w}{\mathrm{d}S_w}\right|_{S_{wf}} = \frac{r_w|_{S_{wf}} - r_w|_{S_{w1}}}{S_{wf} - S_{w1}}.$$

As the $r_w(S_w)$-curve shifts to the right, the shock saturation, S_{wf}, becomes higher, while the shock velocity (i.e., the slope of the curve) becomes smaller. Even at high shock saturations, we observe, in experiments and in the field, the effect of the neglected capillary pressure on smearing the shock front (Fig. 6.19).

Obviously, the Welge method described above is valid only when the fractional flow function has an inflection point. Otherwise, we may have one of the two situations shown in Fig. 6.20

So far, in applying the Buckley–Leverett approach to horizontal flow, we have neglected the effect of gravity. This effect is represented by the term $(\rho_w - \rho_n)g$ in (6.4.60). For an inclined system, this term is replaced by $(\rho_w - \rho_n)g \sin\alpha$. When $\alpha \neq 0$, we should take this effect into account when calculating r_w. As a result, the r_w-curve will shift to the right when the ratio $(\rho_w - \rho_n)g \sin\alpha/q_t\mu_w$ increases (i.e.,

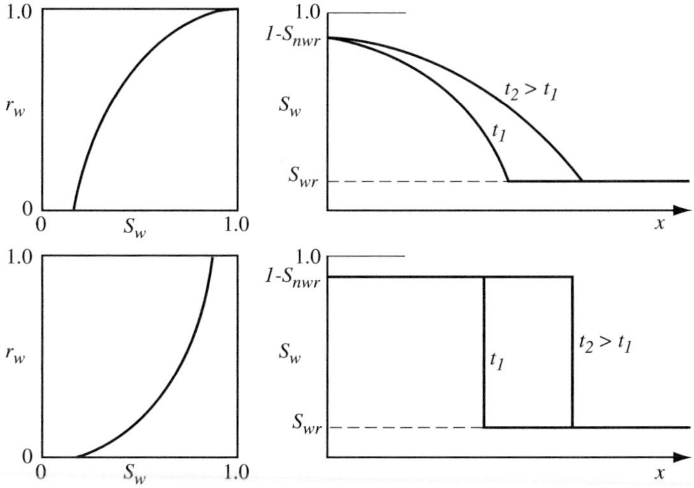

Fig. 6.20 No inflection point in the r_w-curve

at low q_t in updip displacements ($\alpha > 0$), and at high q_t in a downdip displacements ($\alpha < 0$)).

Finally, for a column of length L, breakthrough (i.e., appearance of the injected water at $x = L$) will occur at time t_b, given by

$$t_b = \frac{L\phi}{q_t (\mathrm{d}r_w/\mathrm{d}S_w)\big|_{S_{wf}}}. \tag{6.4.70}$$

Note that $L\phi/q_t$ expresses the mean residence time of the injected water in the column. The breakthrough time, t_b, increases as the r_w-curve shifts to the right.

We may also estimate the volume of produced nonwetting fluid at $x = L$, from:

$$\mathbb{V}_{produced} = A\int_0^t q_{nw}\mathrm{d}t = A\int_0^t q_t(1 - r_w\big|_{x=L})\mathrm{d}t. \tag{6.4.71}$$

Figure 6.21 shows some of these relationships.

The multi-dimensional Buckley–Leverett problem can be derived by taking the constant density form of (6.4.41), and ignoring the ∇S_w term. More on the Buckley–Leverett solution can be found in the reservoir engineering literature, e.g., Marle (1981), and Barenblatt et al. (1990).

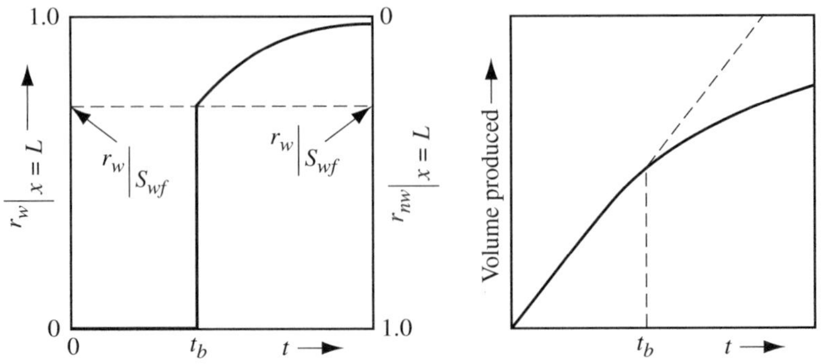

Fig. 6.21 Volume of produced nonwetting fluid

6.4.6 Two Phases with Interphase Mass Transfer

Often, in multiphase flows, components are exchanged among the fluid phases. Air dissolved in water and evaporation of water in the unsaturated zone, may serve as examples. Gas dissolution, retrograde condensation, and vaporization and condensation of injected gases, encountered in oil recovery operations, may serve as additional examples. In fact, terms that express the transfer of an extensive quantity, E, from the wetting phase to the nonwetting one, or vice versa, appear in the fundamental E-balance equations (3.3.4). This equation will serve as the basic mass balance equation for the following presentation.

Let us introduce the subject through three examples.

- **Example 1: Air solubility and water vaporization**

In this example, the wetting phase is an aqueous liquid, denoted by subscript ℓ, and the nonwetting phase is a gas, denoted by subscript g. The liquid is made up of liquid water and dissolved air. The gas phase is made up of air (a) and water vapor (w). Isothermal conditions are assumed.

The mass balance equation for water (H_2O) **in the liquid phase**:

$$\frac{\partial \phi S_\ell \rho_\ell^w}{\partial t} = -\nabla \cdot \phi S_\ell (\rho_\ell^w \mathbf{V}_\ell + \mathbf{J}_{h\ell}^w) + f_{g\to\ell}^w + \phi S_\ell \rho_\ell \Gamma_\ell^w, \tag{6.4.72}$$

where ϕ is the porosity, which under certain circumstances may vary in space (heterogeneity) and time (in a deformable porous medium) and $\mathbf{J}_{h\ell}^w$ denotes the w-flux by hydrodynamic dispersion (dispersion + diffusion) in the ℓ-phase.

- **The mass balance equation for water vapor in the gaseous phase**:

$$\frac{\partial \phi S_g \rho_g^w}{\partial t} = -\nabla \cdot \phi S_g (\rho_g^w \mathbf{V}_g + \mathbf{J}_{hg}^w) + f_{\ell\to g}^w + \phi S_g \rho_g \Gamma_g^w, \tag{6.4.73}$$

where $\mathbf{J}_{hg}^{w}$ denotes the w-flux by hydrodynamic dispersion (dispersion + diffusion) in the g-phase.

- **For dissolved air in the liquid phase**:

$$\frac{\partial \phi S_\ell \rho_\ell^a}{\partial t} = -\nabla\cdot\phi S_\ell(\rho_\ell^a \mathbf{V}_\ell + \mathbf{J}_{h\ell}^a) + f_{g\to\ell}^a + \phi S_\ell \rho_\ell \Gamma_\ell^a. \tag{6.4.74}$$

- **For air in the gaseous phase**:

$$\frac{\partial \phi S_g \rho_g^a}{\partial t} = -\nabla\cdot\phi S_g(\rho_g^a \mathbf{V}_g + \mathbf{J}_{hg}^a) + f_{\ell\to g}^a + \phi S_g \rho_g \Gamma_g^a. \tag{6.4.75}$$

- **Advective fluxes of liquid and gas phases**:

$$\mathbf{q}_\ell \equiv \phi S_\ell \mathbf{V}_\ell = -\frac{\mathbf{k}_\ell}{\mu_\ell}\cdot(\nabla p_\ell + \rho_\ell g \nabla z), \tag{6.4.76}$$

$$\mathbf{q}_g \equiv \phi S_g \mathbf{V}_g = -\frac{\mathbf{k}_g}{\mu_g}\cdot(\nabla p_g + \rho_g g \nabla z), \tag{6.4.77}$$

in which the effective permeabilities are known functions of the respective S_α's.

- **The capillary pressure**:

$$p_g - p_\ell = p_c(S_\ell), \tag{6.4.78}$$

in which $p_c(S_\ell)$ is a known function that relates the capillary pressure to phase saturation.

- **Sum of saturations**:

$$S_\ell + S_g = 1. \tag{6.4.79}$$

- **Concentrations of components related to phase densities**:

$$\rho_\ell = \rho_\ell^w + \rho_\ell^a, \quad \rho_g = \rho_g^a + \rho_g^w. \tag{6.4.80}$$

At this point, we have 14 scalar equations in terms of 18 scalar variables:

$$\rho_\ell^w, \rho_\ell^a, \rho_g^a, \rho_g^w, \mathbf{V}_\ell, \mathbf{V}_g, p_\ell, p_g, S_\ell, S_g, \rho_\ell, \rho,$$

and the two rates of phase changes (= interphase transfers):

$$f_{\ell\to g}^w (= -f_{g\to\ell}^w), \quad f_{g\to\ell}^a (= -f_{\ell\to g}^a).$$

The total fluxes, $\mathbf{J}_\ell$, $\mathbf{J}_g$ have not been counted as variables, because they can easily be related to the advective phase fluxes and to concentration gradients. The source functions, Γ^a_ℓ, Γ^a_g, etc., are also assumed known, and so are the constitutive relations: $\mu_\ell(p_\ell, \rho^w_\ell, \ldots)$, and $\mu_g(p_g, \rho^a_g, \ldots)$. The temperature, T is assumed here a known constant.

To eliminate the rates of phase change from the component balance equations, we sum up, for each chemical substance, the corresponding component balance equations for the two phases. The result is a single balance equation for the considered substance in the porous medium.

Thus, the two mass balance equations for the chemical substances, obtained for the porous medium as a whole, are:

- **Mass balance equation for pure water**

This equation is obtained by summing (6.4.72) and (6.4.73):

$$\frac{\partial}{\partial t}(\phi S_\ell \rho^w_\ell + \phi S_g \rho^w_g) = -\nabla\cdot(\phi S_\ell \rho^w_\ell \mathbf{V}_\ell + \phi S_g \rho^w_g \mathbf{V}_g) \\ - \nabla\cdot(\phi S_\ell \mathbf{J}^w_{h\ell} + \theta_g \mathbf{J}^w_{hg}) + \phi S_\ell \rho_\ell \Gamma^w_\ell + \phi S_g \rho_g \Gamma^w_g. \quad (6.4.81)$$

- **Mass balance equation for dry air**

This equation is obtained by summing (6.4.74) and (6.4.75):

$$\frac{\partial}{\partial t}(\phi S_\ell \rho^a_\ell + \phi S_g \rho^a_g) = -\nabla\cdot(\phi S_\ell \rho^a_\ell \mathbf{V}_\ell + \phi S_g \rho^a_g \mathbf{V}_g) \\ - \nabla\cdot(\phi S_\ell \mathbf{J}^a_{h\ell} + \phi S_g \mathbf{J}^a_{hg}) + \phi S_g \rho_g \Gamma^a_g + \phi S_\ell \rho_\ell \Gamma^a_\ell. \quad (6.4.82)$$

In this way, we have eliminated the rates of phase change, but now each of the substance balance equations involves component concentrations in both phases.

Altogether, we now have 12 scalar equations for the 16 scalar variables. The required additional equations must express *thermodynamic relationships* between components in the two phases. This calls for the introduction of two additional variables, the partial pressures for the gaseous phase, p^a_g and p^w_g, with the relationship

$$p^a_g + p^w_g = p_g. \quad (6.4.83)$$

Now there are 13 equations and 18 state variables to be solved for. Following are 5 additional relationships.

The relationship between component concentrations, densities, and pressures:

$$\rho^a_\ell = \rho^a_\ell(p_g, p_\ell, T), \quad (6.4.84)$$

which relates air solubility in liquid water to air and water pressures. Alternatively, we may use *Henry's law*, which relates the mole fraction of a gas component (= the solute) in a dilute liquid solution (= the solvent) to its partial pressure in the solution.

Assuming that water vapor and dry air components in the gaseous phase behave as *ideal gases*, we write for the gaseous (air) phase:

$$p_g^w = \frac{RT}{M^w}\rho_g^w, \qquad p_g^a = \frac{RT}{M^a}\rho_g^a, \tag{6.4.85}$$

in which $R(= 8.1347\,\text{J/mole}^\circ\text{K})$ is the *universal gas constant*, T is the absolute temperature (in °K), and $M^w(= 18\,\text{g/mole})$ and $M^a(= 29\,\text{g/mole})$ are the molar masses of 'pure' water and 'dry' air, respectively.

The *relative humidity*, $h_r(= (\rho_g^w/\rho_g^w|)_{sat})$, in which $\rho_g^w\big|_{sat}$, is the vapor's concentration (= density) at saturation. It is given by Edelfsen and Anderson (1943). Let us rewrite it here for convenience in the form:

$$\frac{\rho_g^w}{(\rho_g^w)\big|_{sat}} = \exp\left\{-\frac{p_g - p_\ell}{\rho_\ell}\frac{M^w}{RT}\right\}. \tag{6.4.86}$$

Finally, the liquid phase mass density is related to its pressure, to the temperature and to the amount of air dissolved in it by:

$$\rho_\ell = \rho_\ell(p_\ell, T, \rho_\ell^a). \tag{6.4.87}$$

Altogether we now have 18 scalar equations in 18 scalar variables. In principle, with appropriate initial and boundary conditions, a solution can be obtained.

Example 2: Black - oil model

This model is often employed in reservoir engineering. Phase behavior is presented in Sect. 2.3.

The main features introduced here are:

(a) The distinction between the density of a fluid phase when under the conditions of pressure and temperature prevailing in a reservoir, recalling that the balance equations are actually written for these conditions, and the density of that phase, as determined when it is brought to ground surface and exposed to pressure and temperature under atmospheric or stock tank conditions.
(b) The possibility of a gas dissolving in the oil.

We shall make use of the definition of *phase formation volume factor*, B_α, of an α-phase ($\alpha = w, n$, or $\alpha = o, w, g$ for oil, water and gas), presented in Sect. 2.3.1, but express it here in a slightly different form:

$$B_\alpha = \frac{\text{Volume of phase under RC}}{\text{Volume of phase under SC}} = \frac{\rho_{\alpha,SC}}{\rho_{\alpha,RC}}, \tag{6.4.88}$$

where subscripts RC and SC denote *reservoir conditions* and *standard, or stock tank, conditions*, respectively. We note that a liquid phase under reservoir conditions may also include dissolved gas. Its density is affected by this fact. Similarly, a liquid phase may be volatile, so that the gas may contain also a vapor of the liquid.

Models of multiple multicomponent phases, the so called *compositional model*, are discussed in Chap. 6. Here, we shall introduce a simpler form of this model.

Consider the case of three phases: a wetting phase (water, w), an intermediate wetting phase (oil, o) and a nonwetting one (gas, g). For these phases, we have:

$$B_o = \frac{\mathbb{V}_{\text{oil},RC} + \mathbb{V}_{\text{dis.gas},RC}}{\mathbb{V}_{\text{oil},SC}} = B_o(p_o).$$

$$B_w = \frac{\mathbb{V}_{\text{water},RC}}{\mathbb{V}_{\text{water},SC}} = B_w(p_w).$$

$$B_g = \frac{\mathbb{V}_{\text{dis.gas},RC}}{\mathbb{V}_{\text{gas},SC}} = B_g(p_g).$$

Within the reservoir *gas solubility in the oil* is expressed by:

$$R_o^g = \frac{\mathbb{V}_{\text{dis.gas},SC}}{\mathbb{V}_{\text{oil},SC}} = R_o^g(p_o),$$

i.e., the ratio of the volume of gas dissolved in oil to the volume of the latter, both under standard conditions.

Similarly, the *gas solubility in water* is given by:

$$R_w^g = \frac{\mathbb{V}_{\text{dis.gas},SC}}{\mathbb{V}_{\text{water},SC}} = R_w^g(p_w).$$

The mass of (liquid) oil in a reservoir is made up of the mass of oil component in the liquid and that of gas dissolved in the latter, as they are separated on the ground surface, i.e.

$$m_{\text{oil},RC} = m_{\text{oil},SC} + m_{\text{gas},SC}.$$

Expressing this relationship in terms of volumes and densities, we obtain:

$$\begin{aligned}\rho_{o,RC}\mathbb{V}_{\text{oil},RC} &= \rho_{o,SC}\mathbb{V}_{\text{oil},SC} + \rho_{g,SC}\mathbb{V}_{\text{dis.gas},SC}\\ &= \rho_{o,SC}\mathbb{V}_{\text{oil},SC} + R_o^g\rho_{g,SC}\mathbb{V}_{\text{oil},SC}\\ &= \mathbb{V}_{\text{oil},SC}(\rho_{o,SC} + R_o^g\rho_{g,SC}),\end{aligned}$$

or:

$$\rho_{o,RC}(\mathbb{V}_{\text{oil}} - \mathbb{V}_{\text{dis.gas}})_{RC} = \mathbb{V}_{\text{oil},SC} + (\rho_o + R_o^g\rho_g)_{SC}.$$

or:

$$\rho_{o,RC}(\mathbb{V}_{\text{oil}} - \mathbb{V}_{\text{dis.gas}})_{RC} = \mathbb{V}_{\text{oil},SC} + (\rho_o + R_o^g\rho_g)_{SC}.$$

Hence, the density of the oil under reservoir conditions is expressed as:

$$\rho_o \equiv \rho_{o,\mathrm{RC}} = \frac{1}{B_o}(\rho_{o,\mathrm{SC}} + R_o^g \rho_{g,\mathrm{SC}}). \tag{6.4.89}$$

Similarly, for the water:

$$\rho_w \equiv \rho_{w,\mathrm{RC}} = \frac{1}{B_w}\rho_{w,\mathrm{SC}}, \tag{6.4.90}$$

and for the gas:

$$\rho_g \equiv \rho_{g,\mathrm{RC}} = \frac{1}{B_g}\rho_{g,\mathrm{SC}}. \tag{6.4.91}$$

Making use of the above definitions, we can write the balance equation for oil (as a component) in the oil phase in the reservoir, in the form:

$$\frac{\partial}{\partial t}\left(\frac{\phi S_o}{B_o}\right) = -\nabla\cdot\left(\frac{\mathbf{q}_o}{B_o}\right) - \frac{Q_o}{\rho_{o,\mathrm{SC}}}, \tag{6.4.92}$$

where Q_o is the rate of oil withdrawal (in mass per unit volume of porous medium per unit time). Each term expresses volume of oil under stock tank conditions (SC), per unit volume of porous medium per unit time.

The corresponding balance equation for the water phase, takes the form:

$$\frac{\partial}{\partial t}\left(\frac{\phi S_w}{B_w}\right) = -\nabla\cdot\left(\frac{\mathbf{q}_w}{B_w}\right) - \frac{Q_w}{\rho_{w,\mathrm{SC}}}, \tag{6.4.93}$$

To write the gas balance equation under reservoir conditions, we note that gas is present both as free gas and as gas dissolved in the oil. Hence:

$$\frac{\partial}{\partial t}\left\{n\left(\frac{R_o^g S_o}{B_o} + \frac{S_g}{B_g} + \frac{R_w^g S_w}{B_w}\right)\right\}$$
$$= -\nabla\cdot\left(\frac{R_o^g}{B_o}\mathbf{q}_o + \frac{1}{B_g}\mathbf{q}_g + \frac{R_w^g}{B_w}\mathbf{q}_w\right) - \frac{Q_g}{\rho_{g,\mathrm{SC}}} - \frac{R_o^g Q_o}{\rho_{o,\mathrm{SC}}}. \tag{6.4.94}$$

The appropriate specific discharge expressions for oil, gas and water have to be introduced into all the relevant balance equations. We note that these expressions contain, in gravity term, densities which have also to be expressed in terms of SC-densities, using phase formation volumes.

Example 3: Solution gas drive

We introduce this case as an example, encountered in reservoir engineering, in which the liquid, ℓ, and the gas, g, are multicomponent phases, for which the densities depend on pressure, component composition and temperature:

$$\rho_\ell = \rho_\ell(p_\ell, X_\ell^\gamma, T), \qquad \rho_g = \rho_g(p_g, X_g^\gamma, T), \tag{6.4.95}$$

and so are the viscosities:

$$\mu_\ell = \mu_\ell(p_\ell, X_\ell^\gamma, T), \qquad \mu_g = \mu_g(p_g, X_g^\gamma, T), \tag{6.4.96}$$

where X_α^γ $(\equiv (\rho_\alpha^\gamma/M^\gamma)/\sum_{j=1}^N(\rho_\alpha^j/M^j))$ denotes the *molar fraction* of the γ-component in the α-phase, with $\gamma = 1, 2, \ldots, N$, and $\alpha = \ell, g$.

For the sake of simplicity, we shall assume that the solid matrix is nondeformable, and that conditions are isothermal. The effect of temperature changes and solid matrix compressibility, can always be added.

The mass balance equation for a γ-component, this time expressed in terms of moles of γ, per unit volume of porous medium, in *both* phases, is given by

$$\phi\frac{\partial}{\partial t}\left(X_\ell^\gamma\frac{\rho_\ell}{M_\ell}S_\ell + X_g^\gamma\frac{\rho_g}{M_g}S_g\right) = -\nabla\cdot\left(X_\ell^\gamma\frac{\rho_\ell}{M_\ell}\mathbf{q}_\ell + X_g^\gamma\frac{\rho_g}{M_g}\mathbf{q}_g\right), \tag{6.4.97}$$

where component dispersive and diffusive fluxes, as is common in reservoir engineering, have been neglected, $M_\alpha = \sum_{\gamma=1}^N X_\alpha^\gamma M^\gamma$ is the molecular weight of the α-phase, with M^γ denoting the molecular weight of the γ-component, and the saturations satisfy

$$S_\ell + S_g = 1. \tag{6.4.98}$$

Assuming that the two phases are at equilibrium, (2.3.7) is applicable, with g and ℓ replacing V and L.

Altogether, we have $3N$+12 variables to solve for:

$$X_g^\gamma,\ X_\ell^\gamma\ X^\gamma,\ X_\ell,\ X_g,\ \rho_g,\ \rho_\ell,\ \mu_g,\ \mu_\ell,\ p_g,\ p_\ell,\ \mathbf{q}_g,\ \mathbf{q}_\ell,\ S_g,\ S_\ell.$$

In order to achieve this goal, we have:

- N balance equations, (6.4.97), one for every component in every phase.
- Two flux equations, one for each phase.
- Two equations of state for phase density.
- Two equations of state for phase viscosity.
- The capillary pressure that expresses the difference in pressure in the two phases.
- The sum of saturations (6.4.98).
- N Eq. (2.3.7), one for each component.
- N equations $X^\gamma = X_\ell^\gamma X_\ell + X_g^\gamma X_g$ (see Sect. 2.2.5).
- Two equations $\sum_{\gamma=1}^N X_\alpha^\gamma = 1, \qquad \alpha = \ell, g.$
- One equation $\sum_{\gamma=1}^N X^\gamma = 1.$
- One equation $X_\ell + X_g = 1.$

6.5 Three Fluid Phases

The possibility that the void space is occupied by three (NP = 3) rather than two fluid phases has already been presented in Fig. 2.11c. As examples, we may mention petroleum reservoirs in which the void-space is occupied simultaneously by water, oil and gas. Another example is the case where the unsaturated zone below ground surface is contaminated by a spill (at ground surface) of a *nonaqueous* fluid phase (NAPL; or DNAPL, when the nonaqueous phase is heavier than water). Although the NAPL dissolves in water, air dissolves in water, water and NAPL evaporate, etc., the three fluid phases, water, NAPL and gas(air) maintain rather 'sharp' visible microscopic interfaces between them. In what follows, we shall briefly shows how the two-phase flow and transport processes discussed earlier can be extended to three fluid phases.

6.5.1 Statics

Let the entire void space be occupied by three fluid phases: a wetting phase (w), e.g., an aqueous phase, an intermediate wetting phase, e.g., an oil-phase (o), and a gas, which is a nonwetting phase (g). Each of the three phases may be composed of a number of chemical species.

Let S_α, $\alpha = w, n, g$, denote the saturation of the three fluid phases that together occupy the void space, with

$$S_w + S_n + S_g = 1. \tag{6.5.1}$$

The discussion below is an extension of Sect. 6.1 on two fluid phases.

A. Capillary Pressure

The concept of wettability, introduced for two fluid phases in Sect. 2.4.2, is applicable also to three fluid phases. Usually, the wetting phase is in the immediate contact with the solid, while the nonwetting phase is occupying pore space domains which are away from the solid. However, oil-wet, or mixed oil-water-wet soils (= solids) may be encountered, e.g., soils with high content of organic matter, or in cases where mineral surfaces exhibit natural organic coatings. This phenomenon has a strong influence on the behavior of fluid phases within the void space. In what follows, we shall consider the more common case of soils for which water is the wetting phase, oil is the *intermediate wetting phase*, and air (= gaseous phase) is the nonwetting phase.

A schematic example of the three fluid phases in a typical cross-section of a 'pore' is shown in Fig. 6.22 for the phases: water (the wetting fluid), oil (the intermediate wetting fluid), and air (the nonwetting fluid). The three phases are separated from each other by two fluid-fluid (assumed sharp) interfaces: an intermediate wetting-nonwetting interface, and a wetting-nonwetting one. Actually, it is difficult to precisely define a 'pore' and the 'radius of a pore' in a porous medium. Intuitively, let us

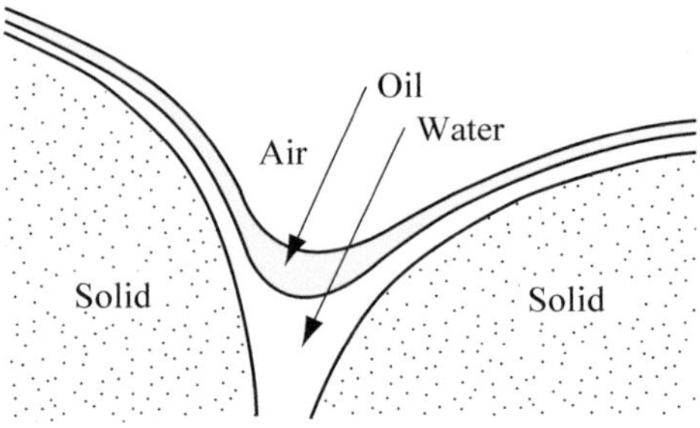

Fig. 6.22 Schematic pore cross-section with three fluid phases

define a 'radius of a pore' (or effective radius) as the radius of the largest sphere that can be placed in a considered portion of the void space, with the solid being tangent to the sphere at least at two points. Due to the assumed order of wettability of the three fluids, the wetting phase occupies primarily pores with the smallest effective radii, the non-wetting phase occupies pores with the largest radii, and the intermediate wetting phase occupies intermediate size pores. Accordingly, the mean radius of curvature of the wetting-intermediate wetting interfaces will always be smaller than that of non-wetting-intermediate wetting interfaces.

The concept of *macroscopic capillary pressure*, introduced in Sect. 6.1 for two fluid phases, can be extended to three phases that occupy the void space. The interface curvature is related to the respective capillary pressures by a generalization of Laplace's formula (6.1.5). When the three phases are water (w, the wetting fluid), oil (o, the intermediate wetting fluid) and gas (g, the non-wetting fluid), it takes the form:

$$p_{cow} \equiv p_o - p_w = \frac{2}{r^*_{ow}}\gamma_{ow}, \qquad p_{cgo} \equiv p_g - p_o = \frac{2}{r^*_{go}}\gamma_{go}, \tag{6.5.2}$$

where p_{cow} and p_{cgo} are oil-water and gas-oil capillary pressures, respectively, r^*_{ow} and r^*_{go} are the average radii of oil-water and gas-oil fluid interfaces, respectively, and γ_{ow} and γ_{go} are the respective interfacial tensions. The effect of the respective contact angles, appearing as factors in $\cos\theta_{LG}$, can be included in (6.5.2), but are commonly neglected in the ensuing analysis. The average radii of curvature of each of the fluids are functions of the respective fluid saturations:

$$r^*_{ow} = r^*_{ow}(S_w), \quad r^*_{go} = r^*_{go}(S_\ell), \tag{6.5.3}$$

where $S_\ell = (S_w + S_o)$ is the total liquid saturation. We note that r^*_{ow} is a function of S_w only, since all pores with radii smaller than r^*_{ow} are assumed to be occupied by water only. However, all pores with radii smaller than r^*_{go} are assumed to be occupied by *both* water and oil, having a combined saturation of S_ℓ $(= 1 - S_g)$. The main assumption here is that, with respect to gas (which is the most nonwetting fluid), the two liquids behave as a *single* wetting fluid.

Since surface tension depends on temperature and concentration of dissolved matter, e.g., expressed by the mass fractions ω_α^γ, we could express the capillary pressure curves in the general forms:

$$p_g - p_o \equiv p_{cgo} = p_{cgo}(S_\ell, \gamma_{go}(T, \omega_o^\gamma)), \tag{6.5.4}$$

$$p_o - p_w \equiv p_{cow} = p_{cow}(S_w, \gamma_{ow}(T, \omega_w^\gamma)), \tag{6.5.5}$$

in which the superscript γ represents all dissolved components.

On the basis of the discussion on the difference between capillary pressure curves and retention curves in two-phase flow, we can also introduce here the *retention curves*

$$p_g - p_o \equiv r_{cgo} = r_{cgo}(S_\ell, T, \omega_o^\gamma), \tag{6.5.6}$$

$$p_o - p_w \equiv r_{cow} = r_{cow}(S_w, T, \omega_w^\gamma). \tag{6.5.7}$$

By extending (6.5.1) and (6.5.3) to three fluid phases, we obtain

$$p_{cnw}(S_w) = \frac{2}{r^*(S_w)}\gamma_{ow}, \qquad p_{cgn}(S_\ell) = \frac{2}{r^*(S_\ell)}\gamma_{go}. \tag{6.5.8}$$

Again, we may replace the surface tension by its product with the cosine of the contact angle.

This implies that, for a given pair of fluids, S_w is a function of p_{cow} only, and S_ℓ is a function of p_{cgo} only. Let us use the superscripts II and III denote two- and three-phase systems, respectively. Based on our previous assumptions, for a given value of p_{cow}, the resulting saturation, $S_w^{\text{III}}(p_{cow})$, in a three-fluid phase system at equilibrium should be identical, or almost identical, to the saturation $S_w^{\text{II}}(p_{cow})$ for a two-phase, oil-water system, except for the influence of the $\cos\theta$'s.. Similarly, at a given value of p_{cow}, the saturation $S_\ell^{\text{III}}(p_{cgo})$ for a three phase system should be identical, or nearly identical, to the saturation $S_\ell^{\text{II}}(p_{cgo})$ for a two-phase (air-oil) system. Put succinctly, we can write

$$S_w^{\text{III}}(p_{cow}) = S_w^{\text{II}}(p_{cow}), \qquad S_\ell^{\text{III}}(p_{cgo}) = S_\ell^{\text{II}}(p_{cgo}). \tag{6.5.9}$$

This protocol, based on the assumption that gas does not touch the solid in the presence of water and oil, was first proposed on theoretical grounds by Leverett (1941), and verified experimentally by Lenhard et al. (1989).

The phenomenon of *hysteresis* in the relationship between capillary pressure and saturation, discussed in Sect. 6.1 for two-phase systems, occurs also in three-phase ones. Again, the reasons are nonwetting fluid entrapment, contact angle hysteresis, and the ink bottle effect.

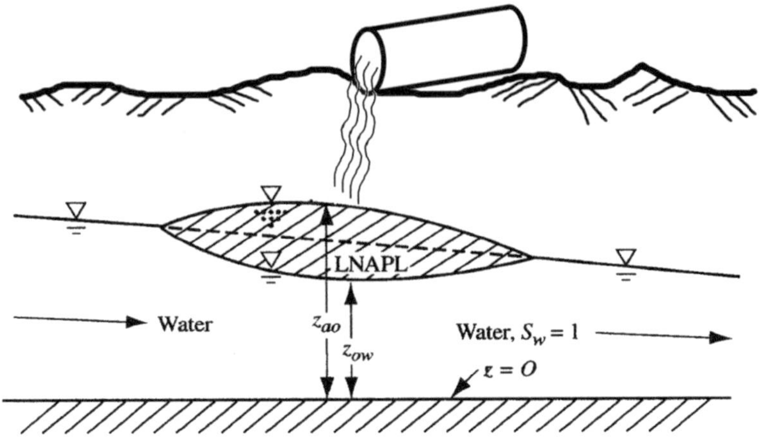

Fig. 6.23 Schematic cross-section of an LNAPL lens above a sloping water table

B. Vertical Equilibrium Saturation Distributions

Consider a sufficiently large spill of LNAPL just above ground surface, such that the percolating LNAPL (n) will reach and accumulate on an underlying water table in the form of a floating lens that spreads out laterally (Fig. 6.23), primarily in the direction of the downward sloping water table. The plume will move in the direction of the water table slope. As it moves, LNAPL will dissolve in the water and a plume of dissolved NAPL will develop.

The vertical distribution of LNAPL saturation, S_n, under the LNAPL source at ground surface depends on the spilled volume, such that the infiltrating LNAPL will become immobile when all the LNAPL is reduced to *residual LNAPL saturation*, S_{nr}. A sufficiently larger spill will create the lens described above. Figure 6.24 shows the vertical distribution of LNAPL in the subsurface resulting from spills of increasing volumes ($\mathbb{V}_1 < \mathbb{V}_2 < \mathbb{V}_3, \ldots, \mathbb{V}_8$).

Because of the essentially horizontal movement of the lens, we may assume that at any instant, the LNAPL in the subsurface is *hydrostatically distributed* along the vertical, and, hence, vertical flow (of all phases) is negligible. This is often called *"vertical-equilibrium (VE-)hypothesis"*, which is equivalent to stating that the *vertical pressure distribution within each phase is hydrostatic*. The vertical distribution *within* the LNAPL source area is shown schematically in Fig. 6.25. Outside the source area, a similar distribution occurs, except that the upper extent of the lens will be where the LNAPL saturation reaches the residual level. Above this point, no LNAPL will be present.

To determine the vertical distributions of the three fluids, say air, water and LNAPL, in the subsurface, under equilibrium conditions, assuming that fluid density of each phase is constant, we define an *equivalent piezometric head* for each phase. Taking water (w, wetting fluid) density as the reference density for all fluids, i.e., $\rho_{\text{ref}} = \rho_w$, the equivalent piezometric heads, $h_{\text{ref},\alpha}$, are defined by

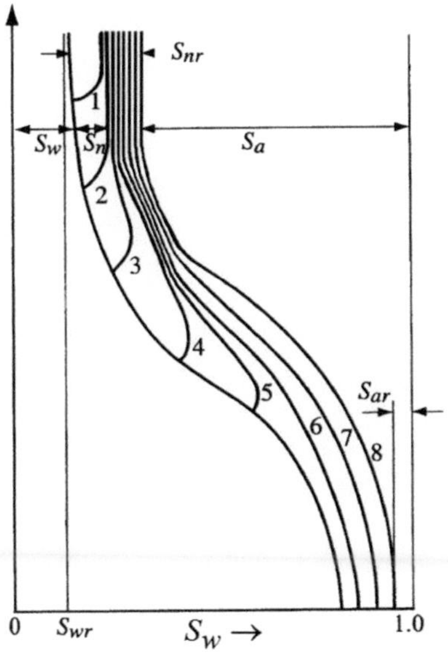

Fig. 6.24 Equilibrium LNAPL distributions in a three-fluid system for various spill volumes

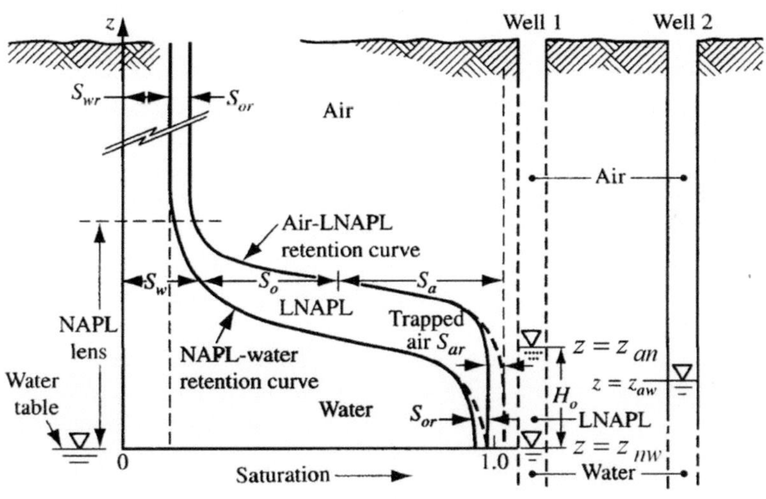

Fig. 6.25 Equilibrium fluid distributions in a three-fluid system with LNAPL

$$h_{\text{ref},w} = \pi_w + z, \qquad h_{\text{ref},n} = \pi_n + z\frac{\rho_n}{\rho_w}, \qquad h_{\text{ref},a} = \pi_a + z\frac{\rho_a}{\rho_w}, \tag{6.5.10}$$

where $\pi_\alpha = p_\alpha/\rho_w g$ denotes the equivalent pressure head in the α-phase (i.e., the height of an equivalent column of water that produces the pressure p_α), and z denotes the elevation above an arbitrary datum.

To facilitate the understanding of the distributions of the three fluid phases in the subsurface, we consider the two wells shown in Fig. 6.25: Well 1, which is screened along its entire length, and Well 2, which is screened only in the water-saturated zone. We note fluid-fluid interfaces in both wells. Well 1 has an air-LNAPL (n) interface at the elevation z_{an}, and an LNAPL-water interface at the elevation z_{nw}. Well 2 has only one air-water interface at the elevation z_{aw}. Pressure is continuous (i.e., no jumps) across these interfaces. Let us refer to the atmospheric pressure in the air at (i.e., immediately above) the air-LNAPL interface inside Well 1 as zero pressure ($p_a = 0$). This is, then, also the pressure in the LNAPL at that point. At the LNAPL-water interface, $p_n = p_w$ so that $p_{cnw} = 0$ there. Note that $S_w = 1$ occurs at the LNAPL-water interface, where the LNAPL-water capillary pressure, p_{cnw}, is zero, while the soil remains fully water saturated for some distance above this elevation. The reason is that LNAPL cannot enter pores until a certain capillary pressure is exceeded. Inside Well 2, the air-water interface, where the pressure is atmospheric, is at an elevation z_{aw}. We note that $S_\ell = 1$ occurs at the air-LNAPL interface, while air does not enter pores for some distance above this elevation. The thickness of the *capillary fringe* above the air-LNAPL interface is smaller than that for the LNAPL-water one, because the air-LNAPL capillary pressure increases with elevation in proportion to the LNAPL's specific gravity, ρ_n/ρ_w, while the LNAPL-water capillary pressure increases in proportion to $(1 - \rho_n/\rho_w)$.

Hydrostatic conditions require that $\partial h_{\text{ref},\alpha}/\partial z = 0$, i.e., the reference piezometric head, defined in (6.5.10), is a constant, independent of z, for $\alpha = a, n, w$. Since $\rho_a/\rho_w \simeq 0$, the pressure gradient in the air may be assumed to be negligible, or, equivalently, the pressure in the air may be taken as approximately constant, equal to $p_{\text{atm}} = 0$. As a consequence, we shall take the reference pressure head in the air, $p_{\text{atm}}/\rho_w a = \pi_a = 0$. For the LNAPL, the value of the constant for h_{ref} is determined by noting that at $z = z_{an}$, $p_a = p_n = 0$, and hence $h_{\text{ref},n}\big|_{z=z_{an}} = z_{an}(\rho_n/\rho_w)$. Accordingly, within the LNAPL, $h_{\text{ref},n} = z_{an}(\rho_n/\rho_w)$. For the water, the value of $h_{\text{ref},w}$ is determined by noting that at $z = z_{aw}$, $p_w = p_a = 0$. Hence, $h_{\text{ref},w} = z_{aw}$. Altogether (Fig. 6.25):

$$\pi_w = z_{aw} - z, \qquad \pi_n = (z_{an} - z)\frac{\rho_n}{\rho_w}, \tag{6.5.11}$$

for the water and for the LNAPL, respectively.

Since $p_w = p_n$ at $z = z_{nw}$, we have

$$p\big|_{z=z_{nw}} = \rho_w g(z_{aw} - z_{nw}) = \rho_n g(z_{gn} - z_{nw}). \tag{6.5.12}$$

It follows that the various interface elevations are related to each other by

$$z_{aw} - z_{nw} = H_n \frac{\rho_n}{\rho_w}, \qquad H_n = z_{an} - z_{nw}, \tag{6.5.13}$$

where H_n is the thickness of the layer of LNAPL inside Well 1, as long as the fluids in the well are in equilibrium with those in the soil. Stipulating any two of the three interface elevations, completely defines the three phase static vertical head distributions within the surrounding soil.

To determine the fluid saturation distributions, we recall that water saturation is controlled by the gas-LNAPL capillary pressure, while total liquid saturation is controlled by the nonwetting-water capillary pressure. Because of the different densities of the three fluids, we introduce here *equivalent capillary pressure heads*, defined, respectively, by

$$\pi_{an} = \pi_a - \pi_n = \frac{p_{can}}{\rho_w g}, \qquad \pi_{nw} = \pi_n - \pi_w = \frac{p_{cnw}}{\rho_w g}. \tag{6.5.14}$$

In view of (6.5.11) and (6.5.13), we write:

$$\pi_{an} = (z - z_{ao}) \frac{\rho_o}{\rho_w}, \qquad \pi_{aw} = (z - z_{aw}) \left(1 - \frac{\rho_a}{\rho_w}\right). \tag{6.5.15}$$

The two equations in (6.5.15) express the relationships between the equivalent capillary pressure and the elevation in the soil for the LNAPL and for the water. On the other hand, following the discussion on capillary pressure presented above, since the three phase capillary pressure relationships, $\pi_{gn}(S_\ell)$ and $\pi_{nw}(S_w)$, between these equivalent pressure heads and the liquid saturation (i.e., combined water and LNAPL) and water, respectively, are known, we may readily compute the sought saturation distributions along the vertical, $S_\ell(z)$ and $S_w(z)$. For example, if hysteresis is disregarded, and the two-phase van Genuchten capillary pressure model, (6.1.18), is used, we obtain

$$S_w = (1 - S_{wr}) \left[1 + (A\beta_{nw}\pi_{nw})^B\right]^{-C} + S_{wr}, \tag{6.5.16}$$

$$S_\ell = (1 - S_{wr}) \left[1 + (A\beta_{an}\pi_{an})^B\right]^{-C} + S_{wr}, \tag{6.5.17}$$

where A, B and $C = 1 - 1/B$ are van Genuchten parameters for the soil, S_{wr} is the irreducible water saturation, and β_{an} and β_{nw} are fluid-dependent scaling factors.

6.5.2 *Motion Equations*

Continuing to assume *no momentum transfer across the microscopic interfaces between two fluid phases* that jointly occupy the void space, the motion equations for three fluid phases, say water (w), oil (o), and gas (g), are similar to (6.2.13) and (6.2.14), except that an equation is required also for the third phase:

$$\mathbf{q}_{rw} = -\frac{\mathbf{k}_{w(on)}(S_w)}{\mu_w} \cdot (\nabla p_w + \rho_w g \nabla z), \tag{6.5.18}$$

$$\mathbf{q}_{ro} = -\frac{\mathbf{k}_{o(wn)}(S_w, S_n)}{\mu_o} \cdot (\nabla p_o + \rho_o g \nabla z), \tag{6.5.19}$$

$$\mathbf{q}_{rn} = -\frac{\mathbf{k}_{n(ow)}(S_n)}{\mu_n} \cdot (\nabla p_n + \rho_n g \nabla z). \tag{6.5.20}$$

In these equations, $\mathbf{k}_{w(on)}$ is the effective permeability to the wetting phase, in the presence of the intermediate and nonwetting phases. Similar definitions apply to $\mathbf{k}_{o(wn)}$ and $\mathbf{k}_{n(ow)}$.

Based on studies by Corey et al. (1956) and Snell (1962), although they actually studied only relative permeabilities of *isotropic* porous media, we assume that

$$\mathbf{k}_{w(on)} = \mathbf{k}_w(S_w), \qquad \mathbf{k}_{n(ow)} = \mathbf{k}_n(S_n), \qquad \mathbf{k}_{o(wn)} = \mathbf{k}_o(S_w, S_n).$$

Following are some key interpretations of these equations:

- The effective permeabilities to the wetting phase and to the nonwetting one *in a two-phase system*, are functions of their respective saturations only, i.e.,

$$\mathbf{k}_{w(n)} = \mathbf{k}_{w(n)}(S_w), \qquad \mathbf{k}_{n(w)} = \mathbf{k}_{n(w)}(S_n).$$

- In a *three-phase system* the effective permeabilities to the wetting (w), and nonwetting (n),phases are the same functions of their respective saturations as they are in a two-phase one, i.e.,

$$\mathbf{k}_{w(on)}(S_w) = \mathbf{k}_{w(o)}(S_w), \qquad \mathbf{k}_{n(ow)}(S_n) = \mathbf{k}_{n(o)}(S_n).$$

- The effective permeability to the intermediate wetting phase, (o), is a function of both the wetting and the nonwetting saturations, i.e.,

$$\mathbf{k}_{o(wn)} = \mathbf{k}_{o(wn)}(S_w, S_n).$$

It is rather difficult to obtain, experimentally, the effective permeabilities as functions of the various saturations (even for isotropic porous media). Practical approaches in petroleum reservoir engineering (where the three phases are hydrocarbon gas, liquid hydrocarbon, and aqueous solution, for the nonwetting, intermediate wetting, and wetting phases, respectively) are based on the *estimation* of three-phase effective (or relative) permeabilities. Two sets of two phase data are used: $\mathbf{k}_{o(w)} = \mathbf{k}_{o(w)}(S_w)$, which is the effective permeability to the o-phase in an o-w-system, and $\mathbf{k}_{o(n)} = \mathbf{k}_{o(n)}(S_n)$ in an o-n-system. The same approach is valid when the intermediate wetting phase is NAPL and the nonwetting phase is air. The under-

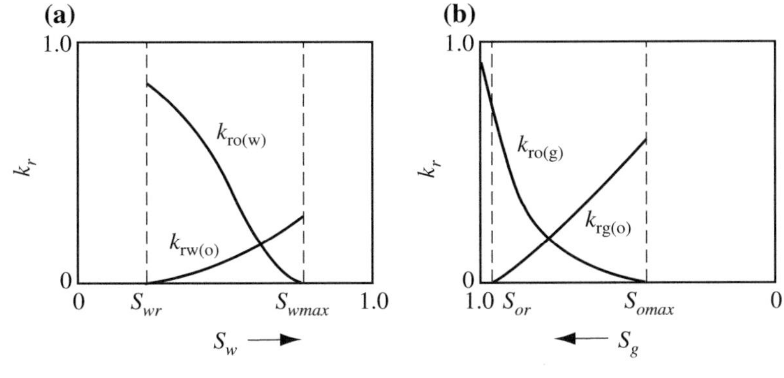

Fig. 6.26 Two phase relative permeability curves: **a** NAPL-water, (o, w), and **b** gas-NAPL, (n, o), in a three phase system

lying conceptual model, say, for a water-NAPL-air system, is that for the water, *both* the NAPL and the air may be considered as *more nonwetting* phases, while for the air, both the water and the NAPL are regarded as *more wetting* phases.

Figure 6.26a, b show relative permeability curves for three phases (w, o, n) in an isotropic porous medium. The point where $k_{ro} = 0$ corresponds to $S_o = 1 - S_{w\,\max}$, where $S_{w\,\max}$ is the maximum value occurring in the NAPL-water system, rather than to the residual o-saturation, S_{or}, in a NAPL-water-air system. The latter saturation can be further reduced by increasing air saturation.

Stone (1970, 1973), Aziz and Settari (1979) and Aleman and Slattery (1985) proposed methods and equations for determining three-phase relative permeabilities. As in the case of two-phase flow, hysteresis is also exhibited in three-phase flow.

6.5.3 Compositional Model–three Multicomponent Phases

Compositional models are ones that track and describe what happens to individual *components*, rather than individual phases or chemical species, during a transport problem. This kind of modeling is very common in reservoir engineering, where the phases are oil, gas and water. Components are chemical substances, or chemical species, that have uniquely defined properties. Thus, CO_2, H_2O, CH_4 are components, but an aqueous phase is not a component, because it may contain H_2O, dissolved gases, etc. Oil is generally not a component, as it usually contains many chemical species that behave differently, say, in interphase exchange, and phase change. However, in many cases, to simplify the model, a mixture of several hydrocarbons with similar properties is considered a *pseudo-component*. Air may be considered a component if we assume that the O_2 to N_2 proportion does not change. However, if we attempt to be accurate in tracking O_2 to N_2 relations (e.g., because O_2's solubility in H_2O is much higher than that of N_2), then we cannot use "air" as a component. Instead, we have to track its two constituent components: O_2 and N_2, separately. This type of model is to be compared with the models considered in Chap. 7.

As an example, we consider the flow of three fluid phases: an aqueous liquid, a nonaqueous liquid, and a gas. Water, oil and gas that together occupy the entire void space, may serve as an example. Each phase is made up of a number of components. The components, or at least some of them, can move from any phase to an adjacent one by such mechanisms as dissolution, volatilization, condensation, adsorption etc.

We shall simplify the discussion by assuming that the solid matrix is rigid and stationary. The term *compositional model*, originating in reservoir engineering, is often used for such a model.

The considered solid and fluid phases are:

- A nonaqueous phase, N, containing a volatile component, d, that can dissolve in the aqueous phase, and the rest of the phase, regarded as a second (non-volatile) component, r. We shall assume that the d-component can evaporate from both the aqueous and the nonaqueous phases, to become a component in the gaseous one.
- An aqueous phase, A, that contains 'pure water' as a component, w, and the d-component as a solute.
- A gas, g, composed of two components: 'dry air', a, and the volatile d-component.
- A solid on which the d-component can be adsorbed, but only from the A-phase.

The three phases, their components, and the interphase transfer rates, are shown, schematically, in Fig. 6.27. No other transfers will be considered.

To facilitate the presentation, let us introduce the *balance operator*, $\mathcal{B}$:

$$\mathcal{B}_\alpha^\gamma(\omega_\alpha^\gamma) \equiv \frac{\partial}{\partial t}(\theta_\alpha \omega_\alpha^\gamma \rho_\alpha) + \nabla\cdot\theta_\alpha \left[\omega_\alpha^\gamma \rho_\alpha \mathbf{V}_\alpha - \rho_\alpha \boldsymbol{\mathcal{D}}_\alpha^{*\gamma}\cdot\nabla\omega_\alpha^\gamma - \mathbf{D}_\alpha\cdot\nabla\omega_\alpha^\gamma \rho_\alpha\right].$$

Note that the mass fractions, ω_α^γ, are used rather than the usual mass concentrations, c_α^γ. One reason is that, from thermodynamics, the density function of a fluid phase is of the form $\rho_\alpha = \rho_\alpha(p, T, \omega_\alpha^\gamma)$. If the mass concentration is used, we are left with the circular result, $\rho_\alpha = \rho_\alpha(p, T, c_\alpha^\gamma/\rho_\alpha)$.

In some cases, one or more of the modes of transport (advection, dispersion, or diffusion), appearing in $\mathcal{B}_\alpha^\gamma$, may not exist.

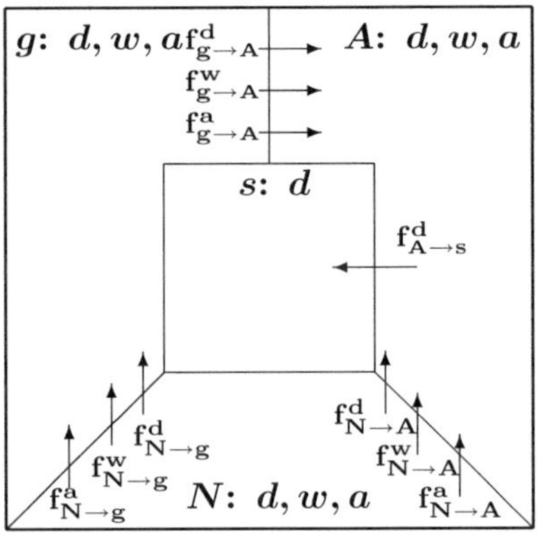

Fig. 6.27 Schematic diagram for component transport in three phase flow

Employing this operator, we can write the mass balance equation for each of the components in all the phases. We shall assume no external sources and sinks of phases and components, no decay and growth, and no chemical interactions. The only sources and sinks of components are due to adsorption and volatilization. Moreover, we assume that the temperature is uniform throughout the domain.

The mass balance equations for the three components in the A-phase are:

$$\mathcal{B}^w_A(\omega^w_A) = f^w_{N\to A} + f^w_{g\to A}, \quad \mathcal{B}^a_A(\omega^a_A) = f^a_{N\to A} + f^a_{g\to A}, \tag{6.5.21}$$

$$\mathcal{B}^d_A(\omega^d_A) = f^d_{N\to A} + f^d_{g\to A} + f^d_{s\to A}. \tag{6.5.22}$$

The mass balance equations for the three components in the g-phase are:

$$\mathcal{B}^w_g(\omega^w_g) = f^w_{N\to g} - f^w_{g\to A}, \quad \mathcal{B}^a_g(\omega^a_g) = f^a_{N\to g} - f^a_{g\to A}, \tag{6.5.23}$$

$$\mathcal{B}^d_g(\omega^d_g) = f^d_{N\to g} - f^d_{g\to A}, \tag{6.5.24}$$

and, for the N-phase, they are:

$$\mathcal{B}^w_N(\omega^w_N) = -f^w_{N\to A} - f^w_{N\to g}, \quad \mathcal{B}^a_N(\omega^a_N) = -f^a_{N\to A} - f^a_{N\to g}, \tag{6.5.25}$$

$$\mathcal{B}^d_N(\omega^d_N) = -f^d_{N\to A} - f^d_{N\to g}. \tag{6.5.26}$$

The mass balance equation for the d-component adsorbed on the s-surface is

$$\mathcal{B}^d_{ads}(c^d_{ads}) = -f^d_{s\to A}, \qquad c^d_{ads} \equiv (1-\phi)\rho_s F^d, \tag{6.5.27}$$

where the balance operator is defined by:

$$\mathcal{B}^d_{ads}(c^d_{ads}) \equiv \frac{\partial c^d_{ads}}{\partial t}.$$

In order to eliminate the interphase transfers, we sum up the mass balance equations for the d-component in all phases, i.e.,

$$\mathcal{B}^d_A(\omega^d_A) + \mathcal{B}^d_N(\omega^d_N) + \mathcal{B}^d_g(\omega^d_g) + \mathcal{B}^d_{ads}(c^d_{ads}) = 0.$$

We then obtain:

$$\begin{aligned}
&\frac{\partial}{\partial t}\big[\theta_A\omega^d_A\rho_A + \theta_N\omega^d_N\rho_N + \theta_g\omega^d_g\rho_g + (1-\phi)\rho_s F^d\big] \\
&+ \nabla\cdot\big(\theta_A\omega^d_A\rho_A\mathbf{V}_A + \theta_N\omega^d_N\rho_N\mathbf{V}_N + \theta_g\omega^d_g\rho_g\mathbf{V}_g\big) \\
&- \nabla\cdot\big(\theta_A\rho_A\mathcal{D}^{*d}_A\cdot\nabla\omega^d_A + \theta_N\rho_N\mathcal{D}^{*d}_N\cdot\nabla\omega^d_N + \theta_g\rho_g\mathcal{D}^{*d}_g\cdot\nabla\omega^d_g\big) \\
&- \nabla\cdot\big(\theta_A\mathbf{D}_A\cdot\nabla\omega^d_A\rho_A + \theta_N\mathbf{D}_N\cdot\nabla\omega^d_N\rho_N + \theta_g\mathbf{D}_g\cdot\nabla\omega^d_g\rho_g\big) = 0.
\end{aligned} \tag{6.5.28}$$

In a similar way, we obtain the mass balance for the water component:

$$\begin{aligned}
&\frac{\partial}{\partial t}\left[\theta_A\omega_A^w\rho_A + \theta_N\omega_N^w\rho_N + \theta_g\omega_g^w\rho_g\right]\\
&\quad + \nabla\cdot\left(\theta_A\omega_A^w\rho_A\mathbf{V}_A + \theta_N\omega_N^w\rho_N\mathbf{V}_N + \theta_g\omega_g^w\rho_g\mathbf{V}_g\right)\\
&\quad - \nabla\cdot\left(\theta_A\rho_A\boldsymbol{\mathcal{D}}_A^{*w}\cdot\nabla\omega_A^w + \theta_N\rho_N\boldsymbol{\mathcal{D}}_N^{*w}\cdot\nabla\omega_N^w + \theta_g\rho_g\boldsymbol{\mathcal{D}}_g^{*w}\cdot\nabla\omega_g^w\right)\\
&\quad - \nabla\cdot\left(\theta_A\mathbf{D}_A\cdot\nabla\omega_A^w\rho_A + \theta_N\mathbf{D}_N\cdot\nabla\omega_N^w\rho_N + \theta_g\mathbf{D}_g\cdot\nabla\omega_g^w\rho_g\right) = 0,
\end{aligned} \tag{6.5.29}$$

and the mass balance equation for the air component:

$$\begin{aligned}
&\frac{\partial}{\partial t}\left[\theta_A\omega_A^a\rho_A + \theta_N\omega_N^a\rho_N + \theta_g\omega_g^a\rho_g\right]\\
&\quad + \nabla\cdot\left(\theta_A\omega_A^a\rho_A\mathbf{V}_A + \theta_N\omega_N^a\rho_N\mathbf{V}_N + \theta_g\omega_g^a\rho_g\mathbf{V}_g\right)\\
&\quad - \nabla\cdot\left(\theta_A\rho_A\boldsymbol{\mathcal{D}}_A^{*a}\cdot\nabla\omega_A^a + \theta_N\rho_N\boldsymbol{\mathcal{D}}_N^{*a}\cdot\nabla\omega_N^a + \theta_g\rho_g\boldsymbol{\mathcal{D}}_g^{*a}\cdot\nabla\omega_g^a\right)\\
&\quad - \nabla\cdot\left(\theta_A\mathbf{D}_A\cdot\nabla\omega_A^a\rho_A + \theta_N\mathbf{D}_N\cdot\nabla\omega_N^a\rho_N + \theta_g\mathbf{D}_g\cdot\nabla\omega_g^a\rho_g\right) = 0.
\end{aligned} \tag{6.5.30}$$

We note that the mass balance equations for any γ-component have the form:

$$\begin{aligned}
&\frac{\partial}{\partial t}\left[\sum_{\alpha=A,N,g}\theta_\alpha\omega_\alpha^\gamma\rho_\alpha + (1-\phi)\rho_s F_s^\gamma\right] + \nabla\cdot\left[\sum_{\alpha=A,N,g}\theta_\alpha\omega_\alpha^\gamma\rho_\alpha\mathbf{V}_\alpha\right]\\
&\quad - \nabla\cdot\left[\sum_{\alpha=A,N,g}\theta_\alpha\rho_\alpha\boldsymbol{\mathcal{D}}_\alpha^{*\gamma}\cdot\nabla\omega_\alpha^\gamma\right] - \nabla\cdot\left[\sum_{\alpha=A,N,g}\theta_\alpha\mathbf{D}_\alpha\cdot\nabla\omega_\alpha^\gamma\rho_\alpha\right] = 0,\\
&\quad \gamma = d, w, a; \quad F_s^w = F_s^a = 0.
\end{aligned} \tag{6.5.31}$$

In these equations, the advective fluxes: $\theta_A\mathbf{V}_A$, $\theta_N\mathbf{V}_N$, and $\theta_g\mathbf{V}_g$, are given by the motion equations, which, neglecting momentum exchange between adjacent fluid phases, are:

$$\theta_\alpha\mathbf{V}_\alpha = -\frac{\mathbf{k}_\alpha}{\mu_\alpha}\cdot(\nabla p_\alpha + \rho_\alpha\nabla z), \quad \alpha = A, N, g. \tag{6.5.32}$$

We assume that in these balance equations, the following variables:

$$\phi, \rho_s, \rho_A, \rho_g, \rho_N, \mu_A, \mu_g, \mu_N,$$

$$\mathbf{k}_A, \mathbf{k}_g, \mathbf{k}_N, \boldsymbol{\mathcal{D}}_A^{*d}, \boldsymbol{\mathcal{D}}_g^{*d}, \boldsymbol{\mathcal{D}}_N^{*d}, \mathbf{D}_A, \mathbf{D}_g, \mathbf{D}_N,$$

are either known constants or functions of the thermodynamic state, as defined by the pressure, temperature, and mass fractions in the appropriate phase. The remaining variables, or unknowns, are:

$$p_A, p_g, p_N, \theta_A, \theta_g, \theta_N,$$

$$c_s^d, \omega_A^d, \omega_g^d, \omega_N^d, \omega_A^w, \omega_g^w, \omega_N^w, \omega_A^a, \omega_g^a, \omega_N^a.$$

The following constraints, on the volumetric fractions:

$$\theta_A + \theta_g + \theta_N = \phi, \tag{6.5.33}$$

and on the phase pressures through the capillary pressure relations:

$$p_g - p_N = p_{cgN}(\theta_g), \qquad p_N - p_A = p_{cNA}(\theta_A), \tag{6.5.34}$$

allow us to eliminate some of the unknown variables. For example, we can eliminate θ_N in (6.5.33), by using the equation $\theta_N = \phi - \theta_A - \theta_g$, leaving θ_A and θ_g. Using (6.5.34), we may eliminate two of the phase pressures, say p_N and p_A, leaving only p_g.

So far, after eliminating the above variables, we are left with the following thirteen unknown variables:

$$p_g, \theta_A, \theta_g, c_s^d, \omega_A^d, \omega_g^d, \omega_N^d, \omega_A^w, \omega_g^w, \omega_N^w. \omega_A^a, \omega_g^a, \omega_N^a.$$

To further reduce the number of unknown variables, we now use thermodynamic relationships to relate the concentrations of the components in adjacent phases to each other. For example, we may use the linear isotherm for the adsorption on the solid:

$$K_d^d = \frac{F^d}{\omega_A^d \rho_A}, \tag{6.5.35}$$

to eliminate the variable c_s^d, leaving twelve unknowns in our list. For the fluid components, when the solutions are dilute, we may use Henry's law:

$$\mathcal{H}_{g,A}^d = \frac{\omega_g^\gamma}{\omega_A^\gamma}, \qquad \mathcal{H}_{g,N}^\gamma = \frac{\omega_g^\gamma}{\omega_N^\gamma}, \qquad \gamma = d, w, a, \tag{6.5.36}$$

in which (here) $\mathcal{H}_{g,A}^\gamma$ and $\mathcal{H}_{A,N}^d$ are *Henry's law coefficients*, appropriately converted to ratios of mass fractions. These six relationships allow us to express six of the mass fractions in terms of the others, leaving six unknowns in our list. However, the mass fractions must also satisfy the following constraints:

$$\omega_A^d + \omega_A^w + \omega_A^a = 1, \qquad \omega_g^d + \omega_g^w + \omega_g^a = 1, \tag{6.5.37}$$

$$\omega_N^d + \omega_N^w + \omega_N^a = 1. \tag{6.5.38}$$

In order to satisfy these constraints, we select for each component a mass fraction in some particular phase, which will be its corresponding 'basis phase'. For example, suppose we select the gas phase to be the basis phase for all components (although, in general, the basis phase does not have to be the same for all components), so that the 'basis mass fractions' are ω_g^d, ω_g^w, and ω_g^a. We then use Henry's law to express the other mass fractions in terms of these, and substitute the results into (6.5.37) and (6.5.38) to obtain the following system of three linear equations:

$$\omega_g^d + \omega_g^w + \omega_g^a = 1, \tag{6.5.39}$$

$$\mathcal{H}_{g,A}^d \omega_g^d + \mathcal{H}_{g,A}^w \omega_g^w + \mathcal{H}_{g,A}^a \omega_g^a = 1, \tag{6.5.40}$$

$$\mathcal{H}_{g,N}^d \omega_g^d + \mathcal{H}_{g,N}^w \omega_g^w + \mathcal{H}_{g,N}^a \omega_g^a = 1, \tag{6.5.41}$$

which may be solved for ω_g^d, ω_g^w, and ω_g^a; these may, therefore, be considered as functions of pressure and temperature.

Thus, we are left with the following three unknowns, or *primary variables*: p_g, θ_A, θ_g, to be determined by solving the three balance equations in (6.5.31).

Obviously, a complete model also requires initial conditions and boundary conditions for the equations, in terms of the selected extensive quantities.

A. Switching Primary Variables

Sometimes, a phase may initially be absent in some portions of a considered porous medium domain. For example, a NAPL phase may initially be present in only certain portions of the domain. Then, the five variables:

$$\theta_N, p_N, \omega_N^d, \omega_N^w, \omega_N^a,$$

will not appear in the initial list of unknown variables. Also, we no longer have one of the capillary pressure constraints. This means that one of the mass fraction constraints, and the three Henry's law relationships disappear. We proceed with the elimination of unknown variables as before, except that we find that (6.5.33) gives $\theta_A + \theta_g = \phi$, so that θ_A, θ_g, which are in our final list of variables, are no longer independent of each other. We also find that the remaining linear equations, (6.5.39) and (6.5.40), includes only two equations in three unknowns. Hence, we replace one of the interdependent variables, say θ_g, by one of the mass fractions that are left, say ω_A^d, to give the three primary variables: p_g, θ_A, ω_A^d. We then move the terms with ω_A^d to the right-hand side in (6.5.39) and (6.5.40), and solve for the remaining mass fractions.

In three-phase flow, the above approach is applicable also when two phases are not present, i.e., when only one phase exists in a particular portion of the domain. For example, when only the aqueous phase exists in a certain subdomain, then a possible set of primary variables for that subdomain is p_A, ω_A^d, ω_A^a.

In another situation, a phase may disappear at some later time from a portion of a considered domain. An example is a rising water table which leaves only an entrapped air phase that eventually dissolves. Another possible case is when a phase that is not present appears later, for instance, in front of an advancing NAPL front. In such cases, we use the procedure described above to obtain a new set of primary variables. We then switch from the current set of variables to the new one.

How do we determine whether a phase appears or disappears? It is easy to tell when a phase disappears by simply checking whether its volumetric fraction θ_α goes to zero. However, the volumetric fraction cannot be used to check whether a phase that has not been present, appears, since, in such a case, that variable is not part of the set of primary variables and is not being computed (it is set to a zero value). The usual method is to compute the mass fractions (or mole fractions), using the equilibrium relationships, such as Henry's law, as if the phase is present, and see if the condition:

$$\omega_\alpha^d + \omega_\alpha^w + \omega_\alpha^a \geq 1 \tag{6.5.42}$$

is satisfied for the α-phase in question. If so, then the phase is considered to be present. The primary variables are then switched accordingly. When this algorithm is implemented in a computer code, the volumetric fraction for that phase is given some small, but nonzero, value, in order for the equations to be nonsingular at this stage of the computation. Condition (6.5.42) is equivalent to $\rho_\alpha^d + \rho_\alpha^w + \rho_\alpha^a \geq \rho_\alpha$, which states that the components are present in sufficient mass to constitute a phase.

B. Flash Calculations

The procedure whereby the mass fractions were eliminated by using Henry's law, is a simple example of what is often called *flash calculations*, whereby the composition of a system at some pressure and temperature is computed, given the known quantity (mass) of each component. The assumptions underlying Henry's law, as given by (6.5.36), may not always be satisfied. The mass fractions appearing in the balance equations remain, but are now considered as (nonlinear) functions of the mole fractions, through:

$$\omega_\alpha^\gamma(n_\alpha^d, n_\alpha^w, n_\alpha^a) = \frac{n_\alpha^\gamma M^\gamma}{n_\alpha^d M^d + n_\alpha^w M^w + n_\alpha^a M^a}, \quad \gamma = d, w, a; \ \alpha = A, N, g.$$

When the phases are not dilute solutions, then the flash calculations cannot be based on simple relationships, such as Henry's law. Instead, more general partitioning expressions must be used, such as:

$$f_g^\gamma = a_A^\gamma K_{g,A}^\gamma, \qquad f_g^\gamma = a_N^\gamma K_{g,N}^\gamma, \quad \gamma = d, w, a. \tag{6.5.43}$$

In the above equation, a_α^γ is the *activity*, f_α^γ is the *fugacity*, and $K_{\alpha,\beta}^\gamma$ is the *equilibrium constant* of the γ-component. In general, these relationships are nonlinear in the unknown concentrations, whether they be mole or mass fractions. For our system, they yield six independent equations. The three additional relationships in (6.5.37) and (6.5.38) provide a total of nine equations for the nine mass fractions in our problem.

6.5.4 Complete Model for Multiple Components

In Sect. 7.5.5, a list was given of what constitutes a complete model for a single component problem. For example, in a model with multiple species or components in a multi-phase system:

(a) A mathematical description of the boundaries of the problem domain.
(b) A list of the independent variables that describe the macroscopic state of the system. These may include (1) pressure of each fluid phase (2) volumetric fraction of each fluid and solid phase, (3) concentrations, c_α^γ, of all considered γ-components, or species, within all phases (or molar concentration, mass or molar fractions). In cases with chemical reactions, we start from a list of all involved species, before selecting the components of the problem. In the case of adsorption, F^γs for all adsorbed γ-species are also included in the list of state variables.
(c) A list of stoichiometric equations for all the reactions among the chemical species present in the system, indicating which ones are assumed to be in equilibrium.
(d) A partial differential equation that describes the mass balance of every considered γ-component within every considered α-phase. If there are equilibrium reactions, redundant component mass balance equations must be eliminated, using the procedures that have been described.
(e) A flux equation for every phase and component of a phase.
(f) Constitutive equations of phases and components. These include also thermodynamic relationships that express the partitioning of components between adjacent phases under equilibrium conditions, mass action equations, expressions for the rates of the various chemical reactions involved, and transfer functions for nonequilibrium conditions.
(g) All the constraints imposed on the independent variables. For example: the sum of mass or molar fractions must be equal to unity, the sum of volumetric fractions must equal the porosity, and the capillary pressure is equal to the difference in pressures between adjacent fluid phases.
(h) A list of primary variables whose number is equal to the number of non-redundant mass balance equations.
(i) Expressions for the various external sources and sinks.
(j) Initial conditions for each of the relevant primary variables.
(k) Boundary conditions for each of the relevant extensive quantities.

Altogether, we have 18 scalar equations for the 18 scalar variables:

$$p_\alpha, \ \mathbf{q}_\alpha, \ S_\alpha, \ \rho_\alpha, \ \alpha = w, n, g.$$

However, following the discussion in Sect. 3.9, we have only 3 (independent!) primary variables for which we have to solve the three partial differential (mass balance) equations. Obviously, each partial differential equation requires appropriate initial and boundary conditions.

References

Aitchison GD, Donald IB (1956) Effective stresses in unsaturated soils. In: Proceedings of 2nd Australian-New Zealand conference soil mechanics and foundation engineering institution of engineers, 1956

Aleman MA, Slattery JC (1985) A linear stability analysis for immiscible porous media contamination by organic compounds, 2. Numerical simulation. Water Resour. Res. 21:19–26

Arya LM, Paris JF (1981) Physicoempirical model to predict the soil moisture characteristic from particle size distribution and bulk density data. Soil Sci Soc Am J 45:1023–1030

Auriault J-L, Lebaigue O, Bonnet G (1989) Dynamics of two immiscible fluids flowing through deformable porous-media. Transp Porous Media 4:105–128

Avraam DG, Payatakes AC (1995) Flow regimes and relative permeabilities during steady-state two-phase flow in porous media. J Fluid Mech 293:207–236

Aziz K, Settari A (1979) Petroleum reservoir simulation. Applied Sciences Publishers, London, p 476

Barenblatt GI, Entov VM, Rhyzik VM (1990) Theory of fluid flows through natural rocks. Springer, Berlin, p 395

Bear J (1972) Dynamics of fluids in porous media, American Elsevier, 764 pp (also published by Dover Publications, 1988; translated into Chinese)

Bear J, Bachmat Y (1984) Transport equations in porous media-Basic equations. In: Bear J, Corapcioglu MY (eds) Fundamentals of Transport Phenomena in Porous Media. Martinud Nrijhoff, Dordrecht, pp 3–61

Bear J, Bachmat Y (1991) Introduction to modeling phenomena of transport in porous media. Kluwer Publishing Company, Dordrecht, p 553

Bear J, Cheng AH-D (2010) Modeling groundwater flow and contaminant transport. Springer, Berlin, p 834

Bear J, Verruijt A (1987) Modeling groundwater flow and pollution. D. Reidel Publishing Company, Dordrecht, p 414

Bear J, Zaslavsky D, Irmay S (1968) Physical principles of water percolation and seepage, UNESCO, 465 pp

Bentsen RG, Manai AA (1993) On the conventional cocurrent and countercurrent modeling of two-phase flow. Transp Porous Media 11:243–262

Bras RL (1990) An introduction to hydrologic science. Addison-Wesley, Reading, 643 pp

Brooks RH, Corey AT (1964) Hydraulic properties of porous media. In: Hydrology papers no. 3, Colorado State University, Fort Collins, Colorado, p 27

Brooks RH, Corey AT (1966) Properties of porous media affecting fluid flow. J Irrig Drain Div ASCE 9(2):61–87

Brutsaert W (1966) Probability laws for pore size distribution. Soil Sci 101:85–192

Buckley SE, Leverett MC (1942) Mechanism of fluid displacement in sands. Trans AIME 146:107–116

Burdine NT (1953) Relative permeability calculations from poe-size distribution data. Trans AIME 198:71–77

Campbell GS (1985) Soil physics with BASIC: transport models for soil-plant systems. Elsevier, New York, p 150

Carsel RF, Parrish RS (1988) Developing joint probability distributions of soil water retention characteristics. Water Resour Res 24(5):755–769

Childs EC, Collis-George N (1950) The permeability of porous materials. Proc R Soc Lond Ser A 2(01):392–405

Collins RE (1961) Flow of fluids through porous media. Reinhold, New York, p 270

Corey AT (1957) Measurement of water and air permeability in unsaturated soils. Proc Soil Sci Soc Am 21:7–10

Corey AT, Rathjens CH, Henderson JH, Wyllie MRJ (1956) Three-phase relative permeability. Trans AIME 207:349–351

Cueto-Felgueroso L, Juanes R (2012) Macroscopic phase-field model of partial wetting: bubbles in a capillary tube. Phys Rev Lett 108(14):144502, 5 pp
Dahle HK, Celia MA, Hassanizadeh SM (2005) Bundle-of-tubes model for calculating dynamic effects in the capillary pressure-saturation relationship. Transp Porous Media 58:5–22
Das DB, Mizrahei M (2012) Dynamic effects in capillary pressure relationships for two-phase flow in porous media: experiments and numerical analyses. AIChE J 58(12):3891–3903
Dullien FAL (1992) Porous media, 2nd edn. Academic Press, San Diego, p 574
Doughty C (2007) Modeling geologic storage of carbon dioxide: Comparison of non-hysteretic and hysteretic characteristic curves. Energy Conversion and Management, 48(6)pp. 1768–1781
Doughty C (2013) User's Guide for Hysteretic Capillary Pressure and Relative Permeability Functions in TOUGH2 Earth Sciences Division Lawrence Berkeley National Laboratory, pp 27
Dullien FAL, Dong M (1996) Experimental determination of the flow transport coefficients. Transp Porous Media 25:97–120
Edelfsen NE, Anderson ABC (1943) Thermodynamics of soil moisture. Hilgardia 15:31–298
Finsterle S, Sonnenborg TO, Faybishenko B (1998) Inverse modeling of a multi-step outflow experiments for determining hysteretic hydraulic properties. In: Proceedings of TOUGH workshop 1998, Lawrence Berkeley National Laboratory, Berkeley, California, 4–6 May 1998
Friedman SP, Seaton NA (1996) On the transport properties of anisotropic networks of capillaries. Water Resour Res 32:339–347
Gardner WR (1958) Some steady state solutions of the unsaturated moisture flow equation, with application to evaporation from a water table. Soil Sci 85:228–232
Goode PA, Ramakrishnan TS (1993) Momentum transfer across fluid-fluid interfaces in porous media: A network model. AIChE J. 39:1124–1134
Haines WB (1930) The hysteresis effect in capillary properties and the modes of moisture distribution associated therewith. J Agric Sci 20:96–105
Hassanizadeh SM, Gray WG (1990) Mechanics and thermodynamics of multiphase flow in porous media including interphase boundaries. Adv Water Res 13:169–186
Hassanizadeh M, Celia M, Dahle HK (2002) Dynamic effect in the capillary pressure-saturation relationship and its impacts on unsaturated flow. Vadose Zone J 1:38–57
Hillel D (1980) Fundamentals of soil physics. Academic Press, Dublin, p 413
Irmay S (1954) On the hydraulic conductivity of unsaturated soil. Trans Am Geophys Union 35:463–468
Joekar-Niasar V, Hassanizadeh SM, Leijnse A (2008) Insights into the relationships among capillary pressure, saturation, interfacial area and relative permeability using pore-scale network modelling. Trans Porous Media 74:201–219
Joekar-Niasar V, Doster F, Armstrong RT, Wildenschild D, Celia MA (2013) Trapping and hysteresis in two-phase flow in porous media: a pore-network study. Water Resour Res 49:4244–4256
Kak C, Slaney M (1987) Principles of computerized tomographic imaging. IEEE Press, New York
Kalaydjian F (1987) A macroscopic description of multiphase flow in porous media involving space-time evolution of fluid-fluid interface. Transp Porous Media 2:537–552
Kool JB, Parker JCP, van Genuchten MTh (1987) Parameter estimation for unsaturated flow and transport models-a review. J Hydrol 91:255–293
Land CS (1968) Calculation of imbibition relative permeability for two- and three-phase flow from rock properties. Trans Am Inst Min Metall Pet Eng 243:149–156
Lasseux D, Quintard M, Whitaker S (1996) Determination of the permeability tensors for two-phase flow. Transp Porous Media 24:107–137
Lenhard RJ, Parker JC and Kaluarachchi JJ, (1989) A model for hysteretic constitutive relations governing multiphase flow, 3. Refinement and numerical simulations. Water Resour Res. 25:1727–1736
Leverett MC (1941) Capillary behaviour in porous media. Trans AIME 142:341–358
Liang Q, Lohrenz J (1994) Dynamic method of measuring coupling coefficients of transport equations of two-phase flow in media flow. Transp Porous Media 15:771–779

Luckner L, van Genuchten MTh, Nielsen DR (1989) A consistent set of parametric models for the two-phase flow of immiscible fluids in the subsurface. Water Resour Res 25:2187–2193
Maidment DR (ed) (1993) Handbook of hydrology. McGraw-Hill, Maidenheach
Marle CM (1965) Cours de Production in Écoulements Polyphasiques, vol 4. Institut Francais du Petrole
Marle CM (1981) Multiphase flow in porous media. Editions Technip, Paris, p 267
McCord JT, Stephens DB, Wilson JL (1991) Hysteresis and state-dependent anisotropy in modeling unsaturated hillslope hydrologic processes. Water Resour Res 27:1501–1518
McCuen RH, Rawls WJ, Brakensiek DL (1981) Statistical-analysis of the Brooks-Corey and the Green-Ampt parameters across soil texture. Water Resour Res 17:1005–1013
Mishra S, Parker JC, Singhal N (1989) Estimation of soil hydraulic-properties and their uncertainty from particle-size distribution data. J Hydrol 108:1–18
Mualem Y (1973) Modified approach to capillary hysteresis based on a similarity hypothesis. Water Resour Res. 9:1324–1331
Mualem YA (1974) A conceptual model of hysteresis. Water Resour Res 10:514–520
Mualem Y (1976) A new model for predicting the hydraulic conductivity for unsaturated porous media. Water Resour Res 12(3):513–522
Mualem YA (1977) Extension of the similarity hypothesis used for modeling the soil water characteristics. Water Resour Res 13:773–780
Mualem YA (1979) Theory of universal hysteretical properties of unsaturated porous media. In: Morel-Seytoux HJ, Surface and Subsurface Hydrology, Proc. 3rd Int. Hydrology Symp. Water Resour Publ. Fort Collins, Colorado pp. 387–399 Water Resour Res 13:773–780
Mualem YA (1984) A modified dependent domain theory of hysteresis. Soil Sci 137:283–291
Muskat M (1946) The flow of homogeneous fluids through porous media. J.W. Edwards Inc, Ann Arbor (1st Ed. 1937), 763 pp
Naar J, Henderson JH (1961) An imbibition model-its application to flow behavior and the prediction of oil recovery. Trans Soc Pet Eng AIME 2(22):61–70
Nield DA, Bejan A (2013) Convection in porous media, 4th edn. Springer, Berlin, p 778
Odeh AS (1959) Effect of viscosity ratio on relative permeability. Trans AIME 2(16):346–352 and discussion by Wienaug CF, 352–353
Parker JC, Lenhard RJ (1987) A model for hysteretic constitutive relations governing multiphase flow, 1. Saturation-pressure relations. Water Resour Res 23:2187–2196
Philip JR (1957a) The theory of infiltration. 1. The infiltration equation and its solution. Soil Sci 83:345–357
Philip JR (1957b) The theory of infiltration. 2. The profile at infinity. Soil Sci 83:435–448
Philip JR (1957c) The theory of infiltration. 3. Moisture profile and relation to experiments. Soil Sci 84:163–178
Philip JR (1957d) The theory of infiltration. 4. Sorptivity and algebraic infiltration equations. Soil Sci 84:257–264
Philip JR (1957e) The theory of infiltration. 5. The influence of initial moisture content. Soil Sci 84:329–339
Philip JR (1958a) The theory of infiltration. 6. Effect of water depth over soil. Soil Sci 85:278–286
Philip JR (1958b) The theory of infiltration. 7. Soil Sci 85:333–337
Philip JR (1969) Theory of infiltration. In: Chow VT (ed) Advances in hydrosciences. Academic Press, New York, pp 215–296
Poulovassilis A (1962) The hysteresis of pore water: an application concept of independent domains. Soil Sci 97:405–412
Raats PAC (1976) Analytical solutions of a simplified flow equation. Trans ASAE 19:683–689
Rauls WJ, Brakensiek DL (1985) Prediction of soil water properties for hydrologic modeling. In: Jones EB, Ward TJ (eds) Watershed mananement in the Eiahties. Proceedings of the symposium sponsored by comm. on Wateished Managemeit, Irrigation and drainage division, ASCE. ASCE convention, Denver, Colorado 30 April–l May, pp 293–299

Reeves PC, Celia MA (1996) A functional relationship between capillary pressure, saturation, and interfacial area as revealed by. Water Resources Research 32(8):2345–2358
Rose W (1972) Some problems connected with the use of classical descriptions of fluid/fluid displacement processes. In: Proceedings of 1st international IAHR symposium on the fundamentals of transport phenomena in porous media, Haifa, Israel, pp 229–240
Rose W (1988) Measuring transport coefficients to describe coupled two-phase flows in porous media. Transp Porous Media 3:163–171
Rose W (1990) Coupling-coefficient for two-phase flow in porous space of simple geometry. Transp Porous Media 5:97–102
Rose W (1997) An upgraded viscous coupling measurement methodology. Transp Porous Media 28:221–23
Rose W, Rose D (2005) An upgraded porous medium coupled transport process algorithm. Transp Porous Media 59:357–372
Sanchez-Palencia E (1980) Non-homogeneous media and vibration theory, vol 127. Lecture notes in physics, Springer, New York
Scheidegger AE (1960a) Growth of instabilities in the displacement fronts in porous media. Physics of Fluids 1(3):94–104
Scheidegger AE (1960b) On the stability of displacement fronts in prous media. Canadian J. of Physics 38:13–162
Schneider FN, Owens WW (1970) Sandstone and carbonate two- and three-phase relative permeability characteristics. Soc Pet Eng J 10:75–84
Snell RW (1962) Three phase relative permeability and residual oil data. J Inst Pet 12:80–88
Stauffer F (1978) Time dependence of the relations between capillary pressure, water content and conductivity during drainage of porous media. In: IAHR symposium on scale effects in porous media, Thessaloniki, Greece, 29 Aug 1978
Stephens DB, Heermann S (1988) Dependence of anisotropy on saturation in a stratified sand. Water Resour Res 24:770–778
Stone HL (1970) Probability model for estimating three-phase relative permeability. Trans Soc Pet Eng AIME 49(2):214–218
Stone HL (1973) Estimation of three-phase relative permeability and residual oil data. J Can Pet Technol 12:53–61
Topp GC (1969) Soil water hysteresis measured in a sandy loam compared with the hysteretic domain model. Soil Sci Soc Am Proc 33:645–651
Topp GC (1971) Soil water hysteresis in silt loam and clay loam soils. Water Resour Res 7:914–920
van Genuchten MTh (1980) Models for describing water and solute movement through soils with large pores. American Society of Agronomy, Agronomy abstracts
Vauclin M, Haverkamp R, Vachaud G (1979) Resolution Numerique d'une Equation de Diffusion Non-Lineare. Presses Univer, Grenoble
Welge HJ (1952) A simplified method for computing oil recovery by gas or water drive. Trans AIME 195:91–98
Whitaker S (1986a) Flow in porous media. 1. A theoretical derivation of Darcy's law. Transp Porous Media 1:3–25
Wyckoff RD, Botset HG (1936) The flow of gas-liquid mixture through unconsolidated sands. Physics 7:325–345
Wylie MRJ, Gardner GHF (1958) The generalized Kozeny–Carman II. A novel approach to problems of fluid flow. World oil production sect, pp 210–228
Yuster ST (1951) Theoretical considerations of multiphase flow in idealized capillary systems. Proc 3rd World Pet Congres, II:437–445
Hassanizadeh SM, Gray WG (1993) Toward an improved description of the physics of two-phase flow. Adv Water Resour 16:53–67

Chapter 7
Modeling Transport of Chemical Species

In this chapter, we consider the transport of chemical species dissolved in one or more fluid phases that occupy the void space. We shall also consider adsorption of a chemical species on the solid phase comprising the solid matrix, dissolution of the latter and precipitation of a dissolved species. The objective is to predict how the concentrations of dissolved and adsorbed species vary with time and space within a considered porous medium domain. To describe the space and time-dependent concentration changes, we shall take into account the travel of the considered species by advection, diffusion (discussed in Sect. 7.2.2), and dispersion (introduced in Sect. 3.4.3) in the fluid phases that occupy the void space. The considered models include interphase exchanges as well as sources and sinks, including those that result from chemical reactions. As usual, we shall supplement the solute's mass balance equation by appropriate initial and boundary conditions that are required for complete, well-posed mathematical models that describe the transport of dissolved chemical species in porous medium domains, i.e., at the macroscopic level.

To facilitate the discussion, we shall review selected topics of chemistry that are required in order to understand the nature of source/sink phenomena that occur within the considered domain and should be included in the considered models. One such topic is chemical reactions that occur within the fluids that occupy the void space. They are referred to as *homogeneous reactions*. Another topic is the transfer of chemical species across fluid-fluid and fluid-solid inter-phase boundaries. These are referred to as *heterogeneous reactions*.

The solid matrix itself may be composed of portions of various chemical substances. This aspect may play an important role when chemical reactions take place across fluid-solid microscopic (e.g., adsorption/desorption and dissolution of the solid matrix) interfaces. Dissolution is an example of interface exchange, here between a the solid comprising the solid matrix and the adjacent fluid in the void space.

We shall also consider sources in the form of extraction and injection of fluids carrying the solutes through wells. As usual, once phenomena are understood at the microscopic level, where they really occur, we shall describe and incorporate them in macroscopic models.

J. Bear, *Modeling Phenomena of Flow and Transport in Porous Media*,
Theory and Applications of Transport in Porous Media 31,
https://doi.org/10.1007/978-3-319-72826-1_7

The presentation of the chemical aspects should be considered merely as very a brief introduction to some of the essentials and to the employed terminology. For digging deeper into the chemical phenomena (e.g., in connection with groundwater contamination, or CO_2 sequestration), much more knowledge and experience are required. Books on chemistry and chemical engineering should be consulted (e.g., Weber 1972; Sawyer et al. 2002; Stumm and Morgan 1995; Sposito 2004; Schwarzenbach et al. 2002; Sawyer et al. 2002).

The discussion in this chapter will include the effects of temperature, although the subject of flow and transport under nonisothermal conditions is discussed in Chap. 8.

In line with the mode of presentation in this book, the discussion in this chapter starts by considering modeling solute transport, with and without chemical reactions, at the microscopic level, i.e., at a point in a fluid continuum. Then, we continue to discuss modelling at the macroscopic level. The objective, if course, is to derive models that describe solute transport not only at laboratory scale domains, but also at large natural domains, primarily in heterogeneous geological formations. However, as is manifested by quite a large number of publications, the macroscopic level models, based on averaging the solute's mass balance equations at the microscopic level, do not correctly describe solute transport in geological formations. A brief discussion on this subject is, therefore, presented.

The material presented in this chapter, as, in fact, the material presented in the entire book should be of use for those dealing with phenomena of transport in geological formations. However, the material will also be useful to chemical engineers who design chemical reactors in the Chemical Engineering industry. Appendix A discusses *chemical reactors*, and the various phenomena of transport that occur in them.

7.1 Measures of Phase Composition

Liquid, gas, and solid phases may be comprised of many chemical species. Hydrologists and soil physicists are, usually, interested in the aqueous phase that occupies the entire void space in aquifers, or part of it (in the *vadose zone*). This phase is comprised primarily of water, with certain quantities of various *chemical species* dissolved in it. Although chemical reactions that involve solid matrix minerals, e.g., ion exchange or dissolution, do occur, and may play a significant role in changing the structure and configuration of the solid matrix, when dealing with the subsurface, it is usually assumed that the numerous minerals constituting the soil's solid matrix are represented by a single pseudo-species, referred to as 'solid', or "solid matrix". However, when we are interested in the dissolution of specific minerals that comprise part of the solid matrix, we identify them specifically. On the other hand, each fluid phase that occupies the void space, or part of it, whether a liquid or a gas, is, usually, composed of more than a single species. It is, therefore, necessary to consider the composition of each individual phase present in the void space.

As everywhere in this book, we shall use a subscript (e.g., α), to denote a *phase* and a superscript (e.g., γ) to denote a dissolved *chemical species*. When we measure the concentration at a point in a porous medium domain, or in a sample taken at such point, we shall assume that the fluid's volume is $\mathbb{V}_{o\alpha}$, i.e., the volume of the α-phase within an REV (Sect. 1.1.6) centered at the point.

The liquid phase is referred to as a 'solvent', when it is the dominant species in a phase. The dissolved chemical species is referred to as a 'solute' if it constitutes only a small portion of the phase.

The commonly used unit for expressing the mass of a chemical species is the *kilogram*, abbreviated kg. The standard unit for volume in the metric system is the cubic meter. However, we often us the *liter*, defined as the volume of one kilogram of water at 20 °C and a pressure of one atmosphere. A commonly used and very convenient SI unit for expressing the quantity of chemical species is the *mole*. Its symbol in the SI system is *mol*. It is defined as the amount of a chemical substance that contains as many elementary entities, e.g., atoms, molecules, ions as there are atoms in 12 g of carbon-12 (^{12}C). This number is the *Avogadro number*, which has a value of 6.02214×10^{23} mol^{-1}. The mole is one of the base units of the SI system, and has the unit symbol mol.

The concentration of a given chemical γ-species within a liquid α-phase (= solution) can be expressed in a number of ways.

• **Mass fraction**. The mass fraction, ω_α^γ, expresses the mass of a γ-species per unit mass of the α-fluid phase,

$$\omega_\alpha^\gamma = \frac{m_\alpha^\gamma}{m_\alpha} \equiv \frac{\rho_\alpha^\gamma}{\rho_\alpha}, \quad \sum_{(\gamma)} \omega_\alpha^\gamma = 1. \tag{7.1.1}$$

This (dimensionless) measure is applicable to a γ-species in solution in a fluid α-phase. When chemical reactions occur also in the solid phase, we may extend the values of α to include also the solid (s).

The unit ppm, 'parts per million', defines the number of grams of solute per million grams of solution.

• **Mass concentration**. The mass concentration (= mass density) of a γ-species in an α-phase, c_α^γ, expresses the mass, m_α^γ, of a γ-species, per unit volume of a fluid α-phase (usually, per liter):

$$c_\alpha^\gamma \equiv \rho_\alpha^\gamma = \frac{m_\alpha^\gamma}{\mathbb{V}_\alpha}. \tag{7.1.2}$$

The common SI units are kg/m^3 (= kg of γ per m^3 of fluid), or g/ℓ (= grams of γ per liter of fluid), or mg/ℓ (= milligrams of γ per liter of fluid). However, this is not a convenient measure when chemical reactions are involved.

• **Molarity** of a γ-species in an α-phase solution is defined as the number of γ-moles per unit volume of the α-phase solution (usually in liter):

$$\eta_\alpha^\gamma = \frac{n_\alpha^\gamma}{\mathbb{V}_\alpha}. \tag{7.1.3}$$

• **Molar phase and species density**. The molar phase density, or concentration, η_α, expresses the number of moles, n_α, of *all* species, per unit volume of an α-solution (usually liter):

$$\eta_\alpha = \frac{n_\alpha}{\mathbb{V}_\alpha}, \quad n_\alpha = \sum_{(\gamma)} n_\alpha^\gamma. \tag{7.1.4}$$

• **Molality** of a γ-species in an α-phase solution, is defined as the number of γ-moles per unit mass of the α-phase solution (usually in kg):

$$\hat{m}_\alpha^\gamma = \frac{n_\alpha^\gamma}{m_\alpha}. \tag{7.1.5}$$

Note that molality is used extensively in electrolyte solutions because, unlike molarity (see below) it is independent of temperature.

• **Mole fraction**, X_α^γ, is defined as the ratio between the number of moles of γ and the total number of moles in the α-phase:

$$X_\alpha^\gamma = \frac{n_\alpha^\gamma}{n_\alpha}, \quad n_\alpha = \sum_{(\gamma)} n_\alpha^\gamma, \quad \sum_{(\gamma)} X_\alpha^\gamma = 1, \quad \eta_\alpha X_\alpha^\gamma M^\gamma = c_\alpha^\gamma, \tag{7.1.6}$$

where M^γ is the *molecular mass* of the γ-species. It is a useful measure of concentration when modeling *reactive transport*, i.e., the flow and transport of chemical species which undergo chemical reactions. Recall that *molecular mass* is usually measured in a.m.u's (atomic molecular units; 1a.m.u = $1.66053904 \times 10^{-27}$ kg), while *molar mass* is measured in gr/mole.

• **Molar mass** of a γ-species (an atom or a combination of atoms in a molecule or an ion) is the standard atomic weight of the atom or the considered combination of atoms, multiplied by the *molar mass constant* which is equal to 1 g per mole:

$$W^\gamma = (\text{standard atomic weight of } \gamma) \times 1\,\text{kg/mol}. \tag{7.1.7}$$

For example, Molar mass of NaCl is:

$$W^{\text{NaCl}} = (22.989 + 35.453) \times 1\,\text{kg/mol} = 58.442\,\text{kg/mol},$$

$$W^{\text{H}_2\text{O}} = (2 \times 1.007 + 15.999) \times 1\text{kg/mol} = 18.013\,\text{kg/mol}.$$

• **Average molar mass of a mixture** is defined as:

$$\bar{W} = \sum_{\gamma} X^{\gamma} M^{\gamma}, \tag{7.1.8}$$

where X^{γ} and M^{γ} denote the molar fraction and the molar mass of the involved γ species in the mixture.

• **Molar concentration**. Another definition for molar concentration (in moles per unit volume), is:

$$[\gamma]_{\alpha} = \eta_{\alpha} X^{\gamma}_{\alpha} = \frac{c^{\gamma}_{\alpha}}{M^{\gamma}}. \tag{7.1.9}$$

• **Equivalent concentration**, $c^{\gamma eq}_{\alpha}$, is defined as

$$c^{\gamma eq}_{\alpha} = \frac{n^{\gamma eq}_{\alpha}}{\mathbb{V}_{o\alpha}}, \tag{7.1.10}$$

where $n^{\gamma eq}_{\alpha}$ denotes the number of *equivalents* of γ in the α-phase. It expresses the quantity of γ that reacts with, or is equal to the combined value of a specified quantity of another substance with respect to a given reaction.

Other often encountered definitions of concentration are the *equivalents per liter* ($\equiv$ eq/ℓ), defined as the number of moles of a solute, multiplied by the *valence* of the solute species, per liter of solution, and *equivalents per million, epm,* defined as the number of moles of a solute, multiplied by the *valence* of the solute species, per 10^6 gram of solution.

Electrical conductivity, EC, measures the ability of a solution to conduct electrical current. Although this is not a measure of concentration, it is included here because it is related to the quantity of ions that are present in a solution. The unit is the reciprocal of ohm-meters, or, in the SI system, siemens per meter (S/m). Often, the EC is measured in terms of the reciprocal of milli-ohms or, micro-ohms, known as milli-mhos (in mS), or micro-mhos (in μS), respectively.

In Sect. 7.3.3 we shall introduce two additional definitions: *activity*, $\{\gamma\}$, and *activity coefficient*, γ^{A}, which contribute to bridging the gap between the ideal behavior of an interacting γ-chemical species and its real one.

When any of the above measures of concentration is assigned to a point in a porous medium domain, e.g., $c = c(\mathbf{x}, t)$, it is implicitly assumed that the concentrations at points within the REV centered at the point $\mathbf{x}$ at time t do not deviate much from that average. We can refer to this condition as the 'assumption of a well mixed REV'.

7.2 Fluxes of Dissolved Species

The discussion is at the macroscopic level. As already mentioned in Sect. 3.4, the total macroscopic flux of a solute (as, in fact, of any extensive property) is made up of three fluxes: advective, diffusive and dispersive.

7.2.1 Advective Flux

We consider the transport of a γ-species dissolved in a fluid α-phase that occupies the entire void space, or part of it, at a volumetric fraction θ_α. With $\mathbf{V}_\alpha$ (dims. LT^{-1}) denoting the (intrinsic phase) average velocity of the phase, and c_α^γ (dims. M L^3) denoting the (intrinsic phase) average concentration of the solute (expressed as mass of solute per unit phase volume), the advective flux, $\mathbf{J}_{\alpha,adv}^\gamma$ (dims. $\mathrm{ML}^{-2}\mathrm{T}^{-1}$), of the considered species is given by the product:

$$\theta_\alpha \mathbf{J}_{\alpha,adv}^\gamma (\equiv \mathbf{q}_{\alpha,adv}^\gamma) = \theta_\alpha \mathbf{V}_\alpha c_\alpha^\gamma. \tag{7.2.1}$$

This flux expresses the mass of the chemical species passing through a unit area *of porous medium*, normal to $\mathbf{V}_\alpha$, per unit time. Note that we have assumed here that the areal and volumetric porosities are identical.

In the case of reactive transport, i.e., mass transport with reacting chemical species, it is more convenient to measure the mass of chemical species in *moles* (see Sect. 7.1), and determine the mass transport in terms of *averaged molar velocity*:

$$\mathbf{V}_\alpha^{mol} = \frac{1}{n_\alpha} \sum_{(\gamma)} n_\alpha^\gamma \mathbf{V}_\alpha^{\gamma,mol}, \tag{7.2.2}$$

$$\mathbf{j}_\alpha^{\gamma,mol} = n_\alpha^\gamma \mathbf{V}_\alpha^{\gamma,mol}. \tag{7.2.3}$$

We can then express the macroscopic *molar advective flux* of γ in α in terms of the number of γ-moles passing through a unit area of porous medium per unit time:

$$\theta_\alpha \mathbf{J}_{\alpha,adv}^{\gamma,moles} = \theta_\alpha \eta_\alpha^\gamma X_\alpha^\gamma \mathbf{V}_\alpha. \tag{7.2.4}$$

When a single fluid occupies the entire void space, we replace θ_α by the porosity, ϕ. When it is obvious from the text that we measure the quantity of γ in moles, the superscript "*moles*" can be omitted.

Note that the last two equations involve the α-fluid velocity, $\mathbf{V}_\alpha$, while *Darcy's law provides information on the fluid's velocity relative to the possibly moving solid*, i.e., $\mathbf{V}_r (\equiv \mathbf{V}_\alpha - \mathbf{V}_s)$.

7.2.2 *Diffusive Flux*

Diffusive flux was already introduced in Sects. 3.1.3 and 3.4.1 E. Briefly, a fluid phase is composed of a number of *chemical species* (Sect. 1.1.1), each made up of a large number of identical atoms, molecules, ions, etc., that are continuously in random motion (*Brownian motion*). At the microscopic level, i.e., within a fluid, each intensive quantity of a chemical species of a phase, e.g., its concentration, may be regarded as a *continuum*. The behavior of this continuum is obtained by averaging the relevant properties of the individual molecules that comprise it. For example, each molecule has mass, momentum, and energy. By averaging, we obtain the densities (i.e., per unit phase volume), or specific value per unit phase mass of these quantities at points within the phase. The transport of these extensive quantities at the microscopic level is obtained by averaging their transport by the individual molecules.

The diffusive flux, $\mathbf{j}^E_{dif}$, of any extensive quantity, E, was already introduced in (3.1.10). Briefly, the *(total) flux*, $\mathbf{j}^E$, of an extensive quantity, E, is expressed at the microscopic level in the form:

$$\mathbf{j}^E = e\mathbf{V}^E = e\mathbf{V} + e\left(\mathbf{V}^E - \mathbf{V}\right) = \mathbf{j}^E_{adv} + \mathbf{j}^E_{dif}, \tag{7.2.5}$$

where $\mathbf{V}$ denotes the mass averaged velocity.

A. Diffusive Mass Flux of a Chemical Species

For E denoting the mass of a γ-species, $E = m^\gamma$, $e' = c^\gamma$, the flux, denoted by $\mathbf{j}^{m^\gamma} \equiv \mathbf{j}^\gamma$, is:

$$\mathbf{j}^\gamma_{dif} = c^\gamma \mathbf{V} + c^\gamma \left(\mathbf{V}^\gamma - \mathbf{V}\right) = \mathbf{j}^\gamma_{adv} + \mathbf{j}^\gamma_{dif}, \tag{7.2.6}$$

where $\mathbf{j}^\gamma_{dif} = c^\gamma \left(\mathbf{V}^\gamma - \mathbf{V}\right)$, $\sum_{(\gamma)} \mathbf{j}^\gamma_{dif} = 0$. In words, the microscopic mass flux of a γ-species is made up of:

- an *advective mass flux*, $c^\gamma\mathbf{V}$, carried by the (mass-weighted) velocity of the phase, with respect to a fixed coordinate system, and
- a *diffusive flux*, $\mathbf{j}^\gamma_{dif}$, resulting from the random motion of the γ-molecules.

Both fluxes are in terms of the mass of chemical species per unit area *of fluid phase* within a planar cross-section.

Still at the microscopic level, we consider a fluid containing only two species: γ and δ (= *binary system*). The mass flux by molecular diffusion of the γ-species, $\mathbf{j}^\gamma_{dif}$, is expressed by *Fick's law* of molecular diffusion in the form:

$$\mathbf{j}^\gamma_{dif} = c^\gamma(\mathbf{V}^\gamma - \mathbf{V}) = -\rho\mathcal{D}^{\gamma\delta}\nabla\omega^\gamma, \qquad \mathbf{j}^\gamma + \mathbf{j}^\delta = 0, \tag{7.2.7}$$

where $\omega^\gamma = \rho^\gamma/\rho$ denotes the mass fraction of γ, and $\omega^\gamma + \omega^\delta = 1$. The scalar $\mathcal{D}^{\gamma\delta}$ is the *coefficient of molecular diffusion* (dims. L^2/T) of the γ-species in a fluid phase that contains only two species, γ and δ. The diffusive flux of the δ-species is given by $\mathbf{j}^{\delta}_{dif} = -\rho\mathcal{D}^{\delta\gamma}\nabla\omega^{\delta}$. Note that the condition $\mathbf{j}^{\gamma}_{dif} + \mathbf{j}^{\delta}_{dif} = 0$ implies that for a binary system $\mathcal{D}^{\gamma\delta} = \mathcal{D}^{\delta\gamma}$. In principle, $\mathcal{D}^{\gamma\delta} = \mathcal{D}^{\gamma\delta}(c^{\gamma})$. However, it is usually assumed, especially at low concentrations, that $\mathcal{D}^{\gamma\delta}$ is independent of c^{γ}. In general, it is a function of pressure and temperature.

When $\nabla\rho = 0$, i.e., the fluid is *homogeneous*, we may write Fick's law, (7.2.7), in terms of the concentration, c^{γ}, as:

$$\mathbf{j}^{\gamma} \equiv c^{\gamma}(\mathbf{V}^{\gamma} - \mathbf{V}) = -\mathcal{D}^{\gamma\delta}\nabla c^{\gamma}. \tag{7.2.8}$$

Fick's law, (7.2.7), also holds, as an approximation, for the diffusive flux of a γ-chemical species in a multicomponent system, as long as the δ-component is the solvent component and all components, except δ and γ, are at dilute concentrations.

In the above forms of Fick's law, as in those that will be presented throughout this chapter, we shall not take into account Onsager's coupled processes, introduced in Sect. 2.6.

The diffusive flux may also be expressed as a *molar flux*, i.e., in moles per unit time per unit fluid area. In fact, this is a more convenient way to express the diffusive flux in the case of chemical reactions as the concentrations of the involved species are expressed in moles. Fick's law for the *molar diffusive flux* (dims. moles/L^2T) of γ is:

$$\mathbf{j}^{\gamma,moles}_{dif} = \eta^{\gamma}\left(\mathbf{V}^{\gamma} - \mathbf{V}^{mol}\right) = -\eta\mathcal{D}^{\gamma\delta}\nabla X^{\gamma}. \tag{7.2.9}$$

where η is the molar phase concentration and $\eta^{\gamma}(\equiv \rho^{\gamma}/M^{\gamma})$ is the *molar γ-concentration* (= number of moles of γ per unit volume of fluid) defined by (7.1.4). Note that the flux in the diffusive flux in above equation is with respect to the molar velocity, $\mathbf{V}^{mol}$.

With X^{γ}_{α} denoting the mole-fraction of γ, Fick's law for the molar flux of a γ-species in a binary system takes the form:

$$\mathbf{j}^{\gamma,moles}_{\alpha,dif} = -\eta_{\alpha}\mathcal{D}^{\gamma\delta}_{\alpha}\nabla X^{\gamma}_{\alpha}, \quad X^{\gamma}_{\alpha} + X^{\delta}_{\alpha} = 1, \quad \mathbf{j}^{\gamma,moles}_{\alpha,dif} + \mathbf{j}^{\delta,moles}_{\alpha,dif} = 0. \tag{7.2.10}$$

The coefficient $\mathcal{D}^{\gamma\delta}_{\alpha}$ is very sensitive to the nature of the phase. It depends on the fluid's pressure and temperature. In a liquid phase it is very sensitive to the viscosity. In a gas, the value of this coefficient grows approximately as $T^{3/2}$ and is inversely proportional to the pressure.

The coefficient $\mathcal{D}^{\gamma\delta}$ in (7.2.7) and (7.2.9) are the same.

More generally, the diffusive mass flux of a γ-component is driven by spatial gradients in the *chemical potential*, $\mu^{\gamma} = \mu^{\gamma}(p, n^{\gamma}, T)$ (defined in Sect. 2.2.4). It is given by:

$$\mathbf{j}^{\gamma} = -\frac{\eta^2}{\rho RT}X^{\gamma}\mathcal{D}^{\gamma\delta}\nabla\mu^{\gamma}\big|_{p,T} \quad \left(= \frac{\eta^2}{\rho RT}X^{\delta}\mathcal{D}^{\gamma\delta}\nabla\mu^{\delta}\big|_{p,T}\right), \tag{7.2.11}$$

where $\nabla\mu^\gamma\big|_{p,T} \equiv (\partial\mu^\gamma/\partial n^\gamma)|_{p,T}\nabla n^\gamma$ denotes the gradient taken while keeping p and T fixed. For a dilute or ideal solution (Denbigh 1981), it follows from (2.2.35) that $(\partial\mu^\gamma/\partial n^\gamma)|_{p,T} = RT/n^\gamma$. Equation (7.2.11) should be used when the composition-dependent body forces, such as gravity or surface forces, affect the diffusive flux, or when the dilute or ideal solution assumption is not applicable (see Nitao and Bear 1996).

So far, we have discussed the diffusion of (electrically) neutral species. Next we shall consider electrically charged ions.

B. Diffusive Flux of Ions

The diffusion of an ion (considered as a γ-species) in an aqueous solution, away from any charged solid surface, is affected by the electrical field generated by all ions in the solution. To express this effect, an additional term is added to the flux law. In a dilute solution, with this effect, the diffusive mass flux is given by:

$$\mathbf{j}^\gamma_{dif} = -\frac{\rho\mathcal{F}}{RT}z^\gamma\mathcal{D}^\gamma\omega^\gamma\nabla\varphi_e - \rho\mathcal{D}^\gamma\nabla\omega^\gamma, \tag{7.2.12}$$

where φ_e denotes the potential of the electrical field and z^γ is the electrical charge of the γ-ion. The coefficient $\mathcal{F}$ is *Faraday's constant*, which is defined as the charge of one mole of singly-charged ions ($= 9.65 \times 10^4$ Coulombs/mole). Equation (7.2.12) is derived from the *Nernst-Planck equation* (Probstein 1994).

It is observed experimentally that in (nonorganic) electrolytic solutions, the condition of *electro-neutrality* holds: the net charge at any given point in a solution, away from charged surfaces, is essentially zero. That is,

$$\sum_{(\gamma)} z^\gamma n^\gamma = 0. \tag{7.2.13}$$

Combined with the mass balance equation, this requires that the diffusive fluxes satisfy the condition:

$$\sum_{(\gamma)} z^\gamma\mathbf{j}^\gamma/M^\gamma = 0. \tag{7.2.14}$$

The electrical field between the ions, which is proportional to the gradient, $-\nabla\varphi_e$, counteracts the tendency of molecular diffusion to disturb charge neutrality. We, therefore, substitute (7.2.12) into (7.2.14) and solve for $-\nabla\varphi_e$, obtaining:

$$-\nabla\varphi_e = \frac{RT}{\mathcal{F}}\frac{\sum_{(\gamma)} z^\gamma\mathcal{D}^\gamma\nabla\omega^\gamma/M^\gamma}{\sum_{(\gamma)}(z^\gamma)^2\mathcal{D}^\gamma\omega^\gamma/M^\gamma}. \tag{7.2.15}$$

Substituting this expression into (7.2.12) gives:

$$\mathbf{j}^{\gamma} = z^{\gamma}\rho\mathcal{D}^{\gamma}\omega^{\gamma}\frac{\sum_{(\lambda)} z^{\lambda}\mathcal{D}^{\lambda}\nabla\omega^{\lambda}/M^{\lambda}}{\sum_{(\lambda)}(z^{\lambda})^2\mathcal{D}^{\lambda}\omega^{\lambda}/M^{\lambda}} - \rho\mathcal{D}^{\gamma}\nabla\omega^{\gamma}. \tag{7.2.16}$$

In principle, this expression is the diffusive mass flux of an ionic species in an electrically neutral dilute solution; it can be used for modeling the transport of multiple ionic species (Lichtner 1995).

Calculating this kind of diffusion becomes complicated as in an electrolytic solution each species has a different diffusion coefficient (e.g., $\mathcal{D}^{H^+} = 9.31\times10^{-9}$ m^2/s, $\mathcal{D}^{OH^-} = 5.27\times10^{-9}$ m^2/s). Although a lot of data is available in the literature, data for complexes is missing. As a result, anions and cations diffuse at different rates. However, they must also maintain charge balance. Usually, the problem is resolved by *assuming* that the various aqueous diffusion coefficients are *species independent.* Accounting for species-dependent diffusion coefficients leads to much more complicated reactive transport models (Lichtner 1985).

Typical values of $\mathcal{D}^{\gamma}$ at 25 °C, for a solute in an aqueous phase, are in the range of 5–100$\times10^{-10}$ m^2/s. For example, for Ca^{2+}, $\mathcal{D}^{\gamma} = 7.9 \times 10^{-10}$ m^2/s; for K^+, $\mathcal{D}^{\gamma} = 19.6 \times 10^{-10}$ m^2/s; and for Cl^-, $\mathcal{D}^{\gamma} = 20.3 \times 10^{-10}$ m^2/s. For a dilute component in air: for water vapour, $\mathcal{D}^{\gamma} = 2.2 \times 10^{-5}$ m^2/s; and for TCE vapour, $\mathcal{D}^{\gamma} = 7.8 \times 10^{-6}$ m^2/s. The diffusivities of a broad range of compounds as a function of temperature and pressure are given by Poling et al. (2000).

C. The Macroscopic Coefficient of Molecular Diffusion

The macroscopic diffusive flux was already introduced in Sect. 3.4.2F. To obtain the macroscopic law, we follow the phenomenological approach (see (3.4.43)). We shall use the symbol $\mathbf{J}^{\gamma}_{dif}$ for the macroscopic diffusive flux, expressed in terms of mass of γ-species (or number of moles) *per unit area of fluid in the porous medium cross-section. We assume that* Fick*'s law remains valid also at the macroscopic level*, i.e., the macroscopic flux is also proportional to a driving force, which is equal to minus the gradient of the (macroscopic) concentration. Accordingly, the macroscopic diffusive flux law has the form:

$$\mathbf{J}^{\gamma}_{dif} = -\boldsymbol{\mathcal{D}}^{*\gamma}\cdot\nabla c^{\gamma}, \quad \text{or} \quad \mathbf{J}^{\gamma}_{dif} = -\rho\boldsymbol{\mathcal{D}}^{*\gamma}\cdot\nabla\omega^{\gamma}, \quad \text{or} \quad \mathbf{J}^{\gamma}_{dif} = -\eta\boldsymbol{\mathcal{D}}^{*\gamma}\cdot\nabla X^{\gamma}, \tag{7.2.17}$$

in which $\boldsymbol{\mathcal{D}}^{*\gamma}$ is the macroscopic coefficient of molecular diffusion. Note that (7.2.17) is nothing but the first term on the r.h.s. of (3.4.43).

Let us elaborate on the nature of the macroscopic coefficient of molecular diffusion.

We start from the case of saturated (i.e., single phase) flow and no solute adsorption on the microscopic solid surface within the REV. Molecular diffusion within the fluid occupying the void-space obeys Fick's law, say (7.2.8), repeated here for an isotropic porous medium in the form:

$$\mathbf{j}^{\gamma}_{dif} = -\mathcal{D}^{\gamma}\nabla c^{\gamma}. \tag{7.2.18}$$

As in the discussion on advective fluid mass flow in Sect. 4.2.5 A, we envision a parallelepiped block of porous medium with diffusion occurring within stream-tubes that extend from one face at constant concentration to the opposite face, at another constant concentration, a distance L, apart, say in the direction x. Actually, as in the discussion on the advective mass flow, the flux takes place through these tortuous stream-tubes that are longer than the distance between the opposite sides of the porous medium block, $L \rightarrow L_e > L$. Their cross-section also varies along the stream-tube. The driving force (= gradient of the solute concentration) along a stream-tube is also different from the average macroscopic driving force in the porous medium as a whole, $\Delta c^{\gamma}/L \rightarrow (\Delta c^{\gamma}/L)(L/L_e)$ and the flux $j^{\gamma}_{dif,x} \rightarrow (L/L_e)(j^{\gamma}_{dif,c})_{L_e}$. Altogether, $(j^{\gamma}_{dif,c})_{L_e} = -\mathcal{D}^{\gamma}\Delta c^{\gamma}/L_e \rightarrow (j^{\gamma}_{dif,c})_L = -\mathcal{D}^{\gamma}(\Delta c^{\gamma}/L)(L^2/L_e^2)$. When averaged over all the stream-tubes in the porous medium block, we obtain the *macroscopic Fick's law*:

$$J^{\gamma}_{i,dif} = -\mathcal{D}^{\gamma}\,(L/L_e)^2\,\frac{\partial c^{\gamma}}{\partial x_i},$$

or, generalized to 3-D and an anisotropic porous medium:

$$J^{\gamma}_{i,dif} = -\mathcal{D}^{*\gamma}_{ij}\frac{\partial c^{\gamma}}{\partial x_j}, \qquad \mathcal{D}^{*\gamma}_{ij} = \mathcal{D}^{\gamma}T^{*}_{ij}, \tag{7.2.19}$$

In this equation, the second rank symmetric tensor, T^{*}_{ij}, denotes the tortuosity of the porous medium for this case. Like other coefficients, it has to be determined experimentally for a considered porous medium.

In (7.2.19), $\boldsymbol{\mathcal{D}}^{*\gamma}$ (components $\mathcal{D}^{*\gamma}_{ij}$) is the *macroscopic coefficient of molecular diffusion*. It is a symmetric second rank tensor expressed as the product of the scalar molecular diffusivity in a fluid continuum, $\mathcal{D}^{\gamma}$, and the void space geometrical property, $\mathbf{T}^{*}$, called *tortuosity*, which is a symmetric second rank tensor. Thus, $\boldsymbol{\mathcal{D}}_{pm} = \mathbf{T}^{*}\mathcal{D}$ (see Sect. 4.2.5). For the case of variable fluid density, we use (3.4.25).

From the discussion on tortuosity in Sect. 4.2.5, it follows that the tortuosity of a fluid phase that occupies part of the void space depends on the saturation of that fluid. Thus, $\mathbf{T}^{*}_{\alpha} = \mathbf{T}^{*}_{\alpha}(\theta_{\alpha})$.

For the special case of unsaturated flow (i.e., air water flow for which we assume that the pressure in the air is constant), Millington (1959) suggested the following relationship for the tortuosity in an isotropic porous medium:

$$T^{*}(\theta) = \frac{\theta^{\frac{7}{3}}}{\phi^2}. \tag{7.2.20}$$

Recall that the flux $\mathbf{J}^{\gamma}_{\alpha,dif}$ expresses the mass of γ *per unit area of* α *in the cross section*. Because $T^{*}_{ij} \leq 1$, the value of the diffusivity in a fluid that occupies the void space of a porous medium, or part of it, is smaller than the corresponding value in a fluid body.

Sometimes a diffusing chemical species, within the wetting phase, or the non-wetting one, crosses interphase (fluid-solid, or fluid-fluid) boundaries (Sect. 7.4), e.g., due to adsorption, dissolution and volatilization. When a solute diffuses in both fluids, "solute stream-tubes" may extend from one fluid to the other. The Bear-Bachmat definition of tortuosity, expressed by (1.4.15), may then facilitate the discussion of tortuosity for such cases. For example, Case A in Sect. 1.4.2 A4, corresponds to the case with no adsorption or interphase transfer, while Case C considers interphase transfer.

In Sect. 4.2.5, we have introduced tortuosity as a phenomenon produced primarily by the difference between the tortuous shape of the microscopic stream-tubes within the void space, and the macroscopic ones. In the current subsection, we have extended this idea also to the case of molecular diffusion inside the void space (and in multiphase flow inside the phase-occupied portion of the latter), leading to the macroscopic coefficient of molecular diffusion as a product of the microscopic molecular diffusivity and the tortuosity of the porous medium, where the latter is a property (i.e., a coefficient) of the void space configuration. However, when the diffusing species can adsorb on the solid surface (Sect. 7.4.1), the idea of diffusion through fixed-in-space stream-tubes is no longer valid, and some correction must be introduced to represent the effect of solute adsorption on the solid surfaces, or crossing interphase boundaries. In fact, the very configuration of the two phases within the void space may also be affected by the mobility ratio of the two fluids.

If the phenomena of *coupled processes* are taken into account (Sect. 2.6), the diffusive flux of a solute is expressed (for example) by

$$\mathbf{J}^{\gamma}_{dif,\alpha} = -\rho_\alpha \boldsymbol{\mathcal{D}}^{*\gamma}_{\alpha} \nabla \omega^{\gamma}_{\alpha} - \Lambda^{\gamma\mathbb{H}}_{\alpha}(\theta_\alpha) \cdot \nabla T, \tag{7.2.21}$$

where $\Lambda^{\gamma\mathbb{H}}$ denotes the Soret (effect) coefficient (Sect. 2.6). This effect is usually very small and is neglected.

D. Knudsen Diffusion

As mentioned already in Sect. 3.4.1 E, Knudsen diffusion of a gas occurs when the *mean free path* of its molecules is not much larger than the characteristic dimension of the flow channel, e.g., the size of a pore or a of a capillary passage. Under such conditions, the gas molecules collide more often with the pore walls than with other gas molecules For molecular flow diffusion of such a gas, we may still use Fick's (microscopic) law in the form:

$$\mathbf{J}^{A}_{g} = -\boldsymbol{\mathcal{D}}^{A}_{K} \cdot \nabla c^{A}, \tag{7.2.22}$$

where $\boldsymbol{\mathcal{D}}^{A}_{K}$ denotes the *Knudsen diffusivity*. Cummingham and Williams (1980, p. 77) discusses and provides expressions for Knudsen diffusivity.

E. Surface Diffusion

Surface diffusion (e.g., is a phenomena which is analogous to that of molecular diffusion of molecules, atoms and ions in a bulk fluid, except that it takes place when the latter are adsorbed to a solid surface, or when they are present within the *double layer*. Similar to diffusion within the bulk fluid, this flux (here mass per unit length in the surface per unit time) is a consequence of various random phenomena at the surface, e.g., random temperature fluctuation at the surface. Both atoms that constitute part of the solid and adsorbed particles (atoms, molecules or clusters) can move by the mechanism of surface diffusion. Like in molecular diffusion, the presence of a concentration gradient will result in a net flux in the opposite direction to the concentration gradient.

Although, under certain circumstances, surface diffusion may play significant role in mass and heat transport in porous media (due to the large surface area), we shall not expand on this subject (see, for example, Cunningham and Williams 1980; Tsong 2001).

7.2.3 Dispersive Flux

The phenomenon of dispersion of any extensive property of a fluid moving within the void space has already been introduced in Sect. 3.4.3. Here, we shall focus on the dispersive flux of the mass of a dissolved chemical γ-species. The presentation considers dispersion as a macroscopic level phenomenon in a homogenous, or *slightly heterogeneous* domain. We recall, however, that real geological formations are usually *highly heterogeneous*.

To understand the meaning of dispersion, let us briefly review the work of Taylor (1953), who was one of the first to suggest a model—the Taylor model—for dispersion. He visualized a porous medium as a bundle of straight parallel circular capillary tubes. For a single capillary tube in the x direction, he studied the displacement of a liquid of solute concentration $c = 0$, by another one, miscible with the first, of concentration $c = c_o$. Both liquids have the same density and viscosity, and the laminar flow rate through the tube is $Q = const$.

We recall that for a steady flow of a single fluid through a capillary tube of constant diameter $2R$, the *parabolic velocity distribution* is given by the *Hagen-Poisseuille law* (4.1.3),

$$V(r) = 2\bar{V}\left(1 - \frac{r^2}{R^2}\right), \quad \text{with} \quad \bar{V}_r = \frac{Q}{\pi R^2}, \quad V_{max} = 2\bar{V}.$$

Fig. 7.1 shows the parabolic velocity distribution in the tube.

It is easy to show that the mass balance equation that describes the movement by advection and diffusion (Fick's law) of a solute through the tube is:

$$\frac{\partial c}{\partial t} = \mathcal{D}\left(\frac{\partial^2 c}{\partial r^2} + \frac{1}{r}\frac{\partial c}{\partial r} + \frac{\partial^2 c}{\partial x^2}\right) - 2\bar{V}\left(1 - \frac{r^2}{R^2}\right)\frac{\partial c}{\partial x}, \qquad (7.2.23)$$

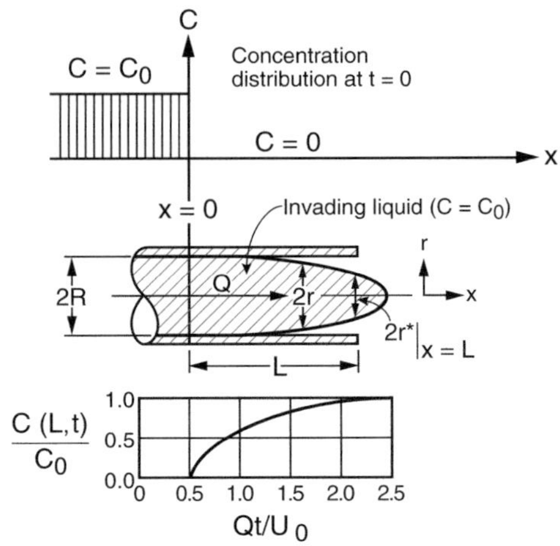

Fig. 7.1 Taylor's analysis of solute transport in a capillary tube

where $\mathcal{D}$ is the coefficient of molecular diffusion of the solute, assumed independent of the concentration, $c = c(x, r)$. Neglecting longitudinal diffusion as much smaller than the radial one, i.e., $\partial^2 c/\partial x^2 \ll \partial^2 c/\partial r^2 + (1/r)\partial c/\partial r$, the above equation reduces to:

$$\frac{\partial^2 c}{\partial r'^2} + \frac{1}{r'}\frac{\partial c}{\partial r'} = \frac{R^2}{\mathcal{D}}\frac{\partial c}{\partial t} + 2\frac{\bar{V}R^2}{\mathcal{D}}\left(\frac{1}{2} - r'^2\right)\frac{\partial c}{\partial x}, \quad r' = r/R, \tag{7.2.24}$$

with $x = 0$, $c = c_o$, and $r' = 1$, $\partial c/\partial r' = 0$. Also, $r' = 0$, c is finite. Initially, $t = 0$, $c = 0$ for $x > 0$; $c = c_o$ for $x \leq 0$.

Taylor (1953, 1954) studied and obtained two approximate solutions for the two extreme cases:

- **Case 1**: $2L/\bar{V} \ll R^2/14.4\mathcal{D}$, i.e., axial convection dominates over radial diffusion, and
- **Case 2**: $2L/\bar{V} \gg R^2/14.4\mathcal{D}$ i.e., radial diffusion dominates, or the time required for radial concentration differences to be approximately reduced by radial diffusion is short relative to the time required for longitudinal convection to cause appreciable radial concentration variations. The latter assumption is applicable to relatively low velocities.

For **Case 1**, with:

$$c(x, t) = c_o\left(1 - \frac{x^2}{4\bar{V}^2 t^2}\right), \quad c(L, t) = c_o\left(1 - \frac{U_o^2}{4Q^2 t^2}\right), \quad U_o = \pi R^2 L. \tag{7.2.25}$$

Equation (7.2.24), can be rewritten in the form:

$$\frac{\partial^2 c}{\partial r'^2} + \frac{1}{r'}\frac{\partial c}{\partial r'} = \frac{2\bar{V}R^2}{\mathcal{D}}\left(\frac{1}{2} - r'^2\right)\frac{\partial c}{\partial \xi}, \quad \xi = x - \bar{V}t, \quad r' = \frac{r}{R}, \tag{7.2.26}$$

in which $\partial c/\partial\xi$, taken as independent of r', is solved by Taylor for $\partial c/\partial\xi = 0$ at $r' = 1$. He obtained:

$$c = c^* + \frac{R^2\bar{V}}{4\mathcal{D}}\frac{\partial c}{\partial\xi}\left(r'^2 - \frac{1}{2}r'^4\right), \tag{7.2.27}$$

in which $c^* \equiv c|_{r'=0}$. With $\bar{c}$ denoting the mean concentration over a cross-section, and $\partial\bar{c}/\partial\xi \approx \partial c^*/\partial\xi$, he obtained:

$$J_c = \frac{Qc}{\pi R^2} = -\frac{R^2\bar{V}^2}{48\mathcal{D}}\frac{\partial c}{\partial\xi} = -\mathcal{D}'\frac{\partial c}{\partial\xi}. \tag{7.2.28}$$

This means that the solute is *dispersed* (i.e., it spreads out) by longitudinal advection and radial molecular diffusion, relative to a plane moving with the velocity $\bar{V}$, as if it were being diffused by a process which obeys Fick's law of molecular diffusion, but with $\mathcal{D}' = (R^2/48\mathcal{D})\bar{V}^2$. It is interesting to note the dependence of $\mathcal{D}'$ on $\bar{V}^2$. The condition that longitudinal molecular diffusion is negligible with respect to longitudinal dispersion, expressed by D', leading to the above results, is:

$$\frac{(48)^{1/2}L}{R} < \frac{L\bar{V}}{\mathcal{D}} < 4\left(\frac{L}{R}\right)^2.$$

Note that $L\bar{V}/\mathcal{D}$ is the *Peclet number*, Pe, which *defines the ratio between the rate of transport by advection and that by molecular diffusion*. Like in Fick's law, the solute flux here is proportional to the concentration gradient.

Solute mass balance during fluid displacement that involves both advection and dispersion in a capillary tube leads to the solute balance equation:

$$\frac{\partial\bar{c}}{\partial t} = \mathcal{D}'\frac{\partial^2\bar{c}}{\partial\xi^2}, \quad or \quad \frac{\partial\bar{c}}{\partial t} = \mathcal{D}'\frac{\partial^2\bar{c}}{\partial x^2} - \bar{V}\frac{\partial\bar{c}}{\partial x}. \tag{7.2.29}$$

I.C. $t = 0$ $\bar{c}(x \geq 0, 0) = 0$,
For the conditions: B.C. $t \geq 0$ $\bar{c}(x = 0, t) = \bar{c}_o$,
B.C. $t \geq 0$ $\bar{c}(x = \infty, t) = 0$,
the above equation yields (Ogata and Banks 1961):

$$\frac{\bar{c}}{\bar{c}_o} = \frac{1}{2}\left[\operatorname{erfc}\left(\frac{x - \bar{V}t}{2\sqrt{\mathcal{D}'t}}\right) + \exp\left(\frac{\bar{V}x}{D'}\right)\operatorname{erfc}\left(\frac{x + \bar{V}t}{2\sqrt{\mathcal{D}'t}}\right)\right] \tag{7.2.30}$$

in which erf x $= (2/\sqrt{\text{ß}})\int_0^x \exp(-^2)\text{d}$. The point $\bar{c}/c_o = 50\%$ moves with the fluid's average velocity, $\bar{V}$. Defining a transition zone, say, between 90 and 10% of total concentration, its width grows proportional to the square root of the velocity, and inversely proportional to the square root D' (Bear 1972, p. 585).

Other authors (e.g., Aris 1956) extended Taylor's type of analysis to a bundle of capillary tubes of different diameters, and to non-circular tubes. Using the presentation in Sect. 4.1.2, it is possible to relate capillary tube diameter to the permeability of a porous medium. Bear (1960) used a model consisting of an array of small cells with interconnecting short tubes. He assumed that a liquid carrying a dissolved solute enters a cell occupied by a liquid at a different solute concentration, displaces part of it, and then mixes (perfect mixing) with the remaining part to form a new concentration. The liquid is transferred from one cell to the next by the fluid's discharge rate, or average velocity. He solved for an array of N such cells, and, for a sufficiently large number of cells, obtained the final concentration in the form of:

$$\frac{c(x,t)}{c_o} = \frac{1}{\sqrt{4\pi D' t}} \exp\left(-\frac{(x-\bar{x})^2}{4D't}\right), \tag{7.2.31}$$

where $\bar{x} = \bar{V}t$, and $D' = a_L\bar{V}$, and $a_L = \frac{1}{2}(\Delta\ell')^2/\Delta\ell$ is a medium property that he called *longitudinal dispersivity*. Thus, in both the mixing cell model and in the capillary tube model, the coefficient of dispersion, D', depends on the velocity. However, in a mixing cell model, the coefficient of dispersion is proportional to the *first power* of the velocity, while in a capillary tube model, it is proportional to $\bar{V}^2$. To understand the reason for this difference, one should recall that in a capillary tube model, dispersion results from solute transport at different velocities along different microscopic streamlines within the void space (say, as a collection of interconnected network of cells). Solute is exchanged by diffusion between adjacent stream-tubes. The velocity itself remains unchanged. In fact, the velocity appears in the form of the path length $\bar{V}t$, independent of the velocity. In the array-of-cells model, there is no velocity distribution and a perfect mixing is caused only by diffusion. The velocity appears as a parameter. In a real porous medium, especially at field scale, with significant heterogeneity, the appropriate model may be some combination of the two models. Before leaving this subject, it may be interesting to mention one more effort to construct a theoretical model for dispersion, especially to find how the dispersive flux is related to the fluid's velocity.

One more model–the *random walk model*–will be discussed in Sect. 7.6.2 A.

Altogether, we have introduced a few models to explain the spreading of a solute relative to mean flow. One model is based on the fact that we have decided to express the advective flux at a point in a porous medium domain as the product of the *average* velocity and the *average* concentration at that point (see (3.4.35)). The other models are combination of flow at different velocities within the void space, with exchange by molecular diffusion between adjacent stream-tubes and the averaging over elementary such tubes. The third mechanism is some kind of mixing, in addition to the translation. the last models, and various combinations of them seems more appropriate for solute transport in highly heterogeneous domains. In all models, the driving force is the concentration gradient. As a first simplification we assume the first power of ∇c, i.e., the assumption of Fickian type transport by dispersion.

With the above discussion in mind, our next task is to express the dispersive flux, introduced above, in terms of averaged (and *measurable*) quantities, such as averaged velocity and averaged concentration. Investigations starting around the mid-50's (e.g., Josselin and de Jong 1958; Saffman 1959; Bear 1961a, b; Scheidegger 1961; Bear 1972; and Bear and Bachmat 1991, p. 401), have led to the conclusion that the dispersive flux of a chemical species (per unit area of fluid) in a porous medium can be expressed as a *Fickian-type* law (i.e., a law that resembles Fick's (linear) law of molecular diffusion) in the form:

$$\mathbf{J}_{dis} \equiv \overline{\mathring{c}\mathring{\mathbf{V}}}^f = -\mathbf{D}\cdot\nabla\overline{c}^f, \tag{7.2.32}$$

or, in indicial notation, making use of Einstein's summation convention::

$$J_{dis,i} \equiv \overline{\mathring{c}\mathring{V}_i}^f = -D_{ij}\frac{\partial\overline{c}^f}{\partial x_j}, \tag{7.2.33}$$

where the D_{ij}'s (dims. L^2/T) are components of a coefficient $\mathbf{D}$, called the *coefficient of (mechanical, or advective) dispersion*, or the *dispersion coefficient*. This coefficient is a second rank symmetric tensor that relates the flux vector $\mathbf{J}_{dis}$ to the driving force vector $-\nabla\overline{c}^f$. Equation (7.2.33) is valid for the general case of an anisotropic porous medium. The dispersion coefficient is characterized by:

- The D_{ij}–matrix is *non-negative definite* (or positive definite). This is a consequence of thermodynamics: the *rate of entropy production*, $\dot{\mathcal{S}}$, is related to the thermodynamic driving force, $\mathbf{X}$, and the thermodynamic flux, $\mathbf{Y}$, (referred to by De Groot and Mazur (1962) as *conjugated flux and force*, respectively) by $\dot{\mathcal{S}} = Y_i X_i$. Here, the driving force $\mathbf{X}$ is proportional to the negative concentration gradient, $-\nabla\overline{c}^f$. In this case, the rate of entropy production can be expressed by:

$$\dot{\mathcal{S}} = \chi\left(-D_{ij}\frac{\partial\overline{c}^f}{\partial x_j}\right)\times\chi\left(-\frac{\partial\overline{c}^f}{\partial x_i}\right) \geq 0, \quad \text{or} \quad \chi^2 D_{ij}\frac{\partial\overline{c}^f}{\partial x_j}\frac{\partial\overline{c}^f}{\partial x_i} \geq 0, \tag{7.2.34}$$

  in which, $\mathbf{Y} = \chi\mathbf{J}_{dis} = -\chi\mathbf{D}\cdot\nabla\overline{c}^f$ and $\mathbf{X} = -\chi\nabla\overline{c}^f$. In the above, χ is a parameter that depends on the considered extensive quantity; for each such quantity, it transforms the flux and the driving force, in the form of a gradient of an appropriately considered scalar (here $\nabla\overline{c}^f$), into conjugated thermodynamic flux and force (De Groot and Mazur 1962).
- The D_{ij}–matrix is *symmetric*, i.e.,

$$D_{ij} = D_{ji}. \tag{7.2.35}$$

  This is a consequence of the conjugated force and flux relation (De Groot and Mazur 1962), i.e., they satisfy:

$$\frac{\partial Y_i}{\partial X_j} = \frac{\partial Y_j}{\partial X_i}. \tag{7.2.36}$$

Because we have circumvented the need to know the details (of velocity and concentration) at the microscopic level by 'escaping' to the macroscopic level, we are required to introduce a set of coefficients, in this case, D_{ij}. This situation occurs whenever we try to overcome the lack of information about details of interphase surfaces at the microscopic level by switching to the macroscopic level.

It is interesting to note that although Darcy's law, and the Fick's type law that governs the dispersive flux, (7.2.33), look similar, there is a basic difference between the coefficients K_{ij} and D_{ij}: the former is a function only of the microscopic geometry of the void space (and of fluid properties), while the latter depends also on the macroscopic velocity field. Another difference is that there is no microscopic Darcy law, as the latter is an approximation of a momentum balance equation and is not a phenomenological law.

Several authors (e.g., Nikolaevski 1959; Bear 1961a; Scheidegger 1961) have derived the following expression for the components D_{ij}:

$$D_{ij} = \mathrm{a}_{ijk\ell}\frac{V_k V_\ell}{V}, \tag{7.2.37}$$

in which $\mathrm{a}_{ijk\ell}$ is a coefficient called *dispersivity*, and $V_k \equiv \overline{V_k}^f$. Henceforth, for simplicity, we shall continue to drop the notation for intrinsic phase averaging. Bear and Bachmat (1967, 1991) suggested:

$$D_{ij} = \mathrm{a}_{ijk\ell}\frac{V_k V_\ell}{V} f(\mathrm{Pe}, r), \quad f(\mathrm{Pe}, r) = \frac{\mathrm{Pe}}{\mathrm{Pe} + 1 + r}, \tag{7.2.38}$$

where V ($\equiv |\mathbf{V}|$) is the magnitude of the average velocity, r represents the ratio between characteristic lengths, in the direction of the flow and normal to it, within a pore, and Pe is a *Peclet number* defined by:

$$\mathrm{Pe} = \frac{V\Delta_f}{\mathcal{D}_f}, \tag{7.2.39}$$

which *expresses the ratio between the rates of transport of the considered mass of chemical species, respectively, by advection and by diffusion*, both at the macroscopic level. In this definition, Δ_f is the *hydraulic radius* of the fluid occupied portion of the void space, serving as a characteristic length of the void space, and $\mathcal{D}_f$ denotes the coefficient of molecular diffusion in the fluid phase. Since $r = O(1)$, $f(\mathrm{Pe}^\gamma, r)$ is an increasing function of Pe^γ, but at a decreasing rate. For $\mathrm{Pe}^\gamma \ll 1$, $f(\mathrm{Pe}^\gamma, r) = O(\mathrm{Pe}^\gamma)$. For $\mathrm{Pe}^\gamma \gg 1$, $f(\mathrm{Pe}^\gamma, r) \approx 1$. Henceforth, as is common in practice, we shall assume $f \approx 1$.

As is common in practice, we shall assume $f(\mathrm{Pe}, r) \approx 1$, so that the coefficient of dispersion is expressed in the form (7.2.37).

Note that in the latter equation, the coefficient of dispersion is proportional to the first power of the velocity, contrary to the results of the analysis presented in Sect. 3.4.4. Currently, there is no experimental validation of the proportionality to V^2, and, therefore, we shall focus on the commonly accepted relationship (7.2.37).

The coefficients $a_{ijk\ell}$ (dims. L) appearing in (7.2.37) are components of a fourth rank tensor, **a**, called the *dispersivity* of the porous medium. It expresses the effect, on the flow, of the microscopic configuration of the interface between the considered fluid phase and all other phases within the REV. In a saturated system, this interface is that between the fluid and the solid. When, in multi-phase flow, a fluid occupies only part of the void space, each of the dispersivity components, $a_{ijk\ell}$, is a function of the volumetric fraction of the fluid.

In a three-dimensional space, the dispersivity tensor, $\mathrm{a}_{ijk\ell}$, has $3^4 = 81$ components. However, because of various symmetry considerations, the number of independent coefficients is smaller. Specifically:

(a) From the expression for the rate of entropy production, $\dot{\mathcal{S}}$, and following the discussion leading to (7.2.34), we have:

$$\dot{\mathcal{S}} = \chi\left(-D_{ij}\frac{\partial \overline{c}^f}{\partial x_i}\right) \times \chi\left(-\frac{\partial \overline{c}^f}{\partial x_j}\right) = \chi^2 \mathrm{a}_{ijk\ell}\frac{\partial \overline{c}^f}{\partial x_i}\frac{\partial \overline{c}^f}{\partial x_j}\frac{V_k V_l}{V} \geq 0. \quad (7.2.40)$$

It follows that $a_{ijk\ell}$ is positive definite. This means that all *principal minors* of $a_{ijk\ell}$ are positive.

(b) The values of the $a_{ijk\ell}$ are invariant under the permutation of indices:

$$a_{ijk\ell} = a_{ij\ell k}, \quad a_{ijk\ell} = \mathrm{a}_{jik\ell}. \quad (7.2.41)$$

Hence, only 36 of the 81 components are *independent* of each other. It is interesting to note that the 36 components are constrained by $2^6 - 1 = 63$ constraints. As the material has more symmetry properties, the number of independent coefficients decreases, until, when the material is isotropic, this number is reduced to two (Bear et al. 2009).

A. Isotropic Porous Medium

In an *isotropic porous medium*, it has been demonstrated (Bear and Bachmat 1991; see also Sirotine and Chaskolskaya 1984, p. 651–2) that the 36 independent components reduce to *two*. This can be shown by considering fourth rank tensors that satisfy the relationships (7.2.41) and are invariant under the action of full rotational (orthogonal) symmetry. The two coefficients are designated as a_L and a_T, and are called the *longitudinal* and the *transverse dispersivities* of the porous medium, respectively. The parameter a_L is a length that characterizes the microscopic level heterogeneity within the REV. Furthermore, by the positive definiteness of $\mathrm{a}_{ijk\ell}$, it follows that:

$$\mathrm{a}_L \geq 0, \quad \mathrm{a}_T \geq 0. \quad (7.2.42)$$

De Josselin de Jong (1958) and laboratory column experiments (e.g., Bear 1961b) have shown that a_T is 8–24 times smaller than a_L.

In terms of a_L and a_T, the components of the dispersivity tensor for an *isotropic porous medium* are expressed in the form:

$$a_{ikj\ell} = a_T\delta_{ij}\delta_{k\ell} + \frac{a_L - a_T}{2}(\delta_{ik}\delta_{j\ell} + \delta_{i\ell}\delta_{jk}), \tag{7.2.43}$$

where δ_{ij} is the Kronecker delta. The coefficient of dispersion can then be expressed as

$$D_{ij} = \left[a_T\delta_{ij} + (a_L - a_T)\frac{V_iV_j}{V^2}\right]V, \quad V = |\mathbf{V}|, \tag{7.2.44}$$

in which V_i denotes the ith component of the average velocity vector $\mathbf{V}$.

In Cartesian coordinates, with V_x, V_y, and V_z denoting average velocity components in the x, y, and z directions, respectively, we obtain from (7.2.44):

$$\begin{aligned}
D_{xx} &= \left[a_T + (a_L - a_T)\frac{V_x^2}{V^2}\right]V = \frac{1}{V}\left(a_L V_x^2 + a_T V_y^2 + a_T V_z^2\right),\\
D_{yy} &= \left[a_T + (a_L - a_T)\frac{V_y^2}{V^2}\right]V = \frac{1}{V}\left(a_T V_x^2 + a_L V_y^2 + a_T V_z^2\right),\\
D_{zz} &= \left[a_T + (a_L - a_T)\frac{V_z^2}{V^2}\right]V = \frac{1}{V}\left(a_T V_x^2 + a_T V_y^2 + a_L V_z^2\right),\\
D_{xy} &= \left[(a_L - a_T)\frac{V_xV_y}{V^2}\right]V = D_{yx},\\
D_{xz} &= \left[(a_L - a_T)\frac{V_xV_z}{V^2}\right]V = D_{zx},\\
D_{yz} &= \left[(a_L - a_T)\frac{V_yV_z}{V^2}\right]V = D_{zy}.
\end{aligned} \tag{7.2.45}$$

Like any second rank tensor, $\mathbf{D}$ also has *three principal directions*. Using these principal directions as Cartesian coordinate axes, x_1, x_2, x_3, we may write $\mathbf{D}$, in the matrix form:

$$\mathbf{D} = \begin{bmatrix} D_{x_1x_1} & 0 & 0 \\ 0 & D_{x_2x_2} & 0 \\ 0 & 0 & D_{x_3x_3} \end{bmatrix}. \tag{7.2.46}$$

In the special case of uniform flow, say $V_x = V$, $V_y = V_z = 0$, Eq. (7.2.45) reduces to $D_{xx} = a_L V$, $D_{yy} = a_T V$, $D_{zz} = a_T V$, $D_{xy} = \mathrm{D}_{xz} = D_{yz} = 0$; or, in matrix form:

$$\mathbf{D} = \begin{bmatrix} a_L & 0 & 0 \\ 0 & a_T & 0 \\ 0 & 0 & a_T \end{bmatrix} V = \begin{bmatrix} D_L & 0 & 0 \\ 0 & D_T & 0 \\ 0 & 0 & D_T \end{bmatrix}, \tag{7.2.47}$$

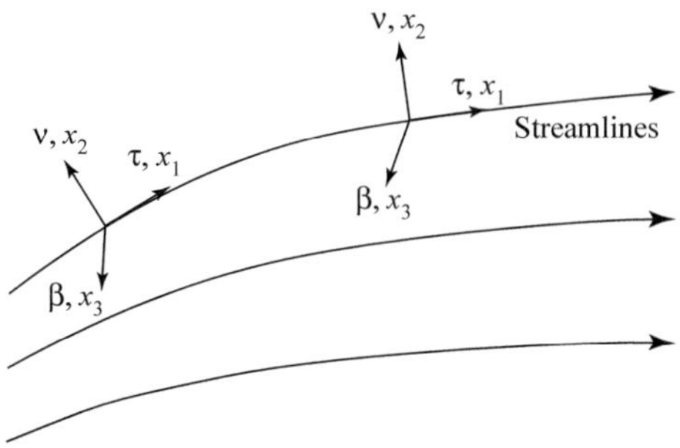

Fig. 7.2 Principal directions of dispersion coefficient in an isotropic porous medium

where D_L and D_T are, respectively, the longitudinal and transversal dispersion coefficients of an isotropic porous medium.

We have already mentioned that the tensor $\mathbf{D}$ (and its principal directions) depends also on the (macroscopic) velocity field. Specifically, if we consider a point on a macroscopic (instantaneous) streamline in a flow domain, we may construct at that point:

- a unit vector, $\boldsymbol{\tau}$, in the direction of the *tangent* to the streamline, (i.e., in the direction of the flow),
- a unit vector, $\boldsymbol{\nu}$, called the *principal normal* to the streamline (defined by $\kappa\boldsymbol{\nu} = d\boldsymbol{\tau}/ds$, where s is the distance measured along the streamline, and κ is the curvature of the streamline at the point), and
- a unit vector, $\boldsymbol{\beta}$ $(= \boldsymbol{\tau} \times \boldsymbol{\nu})$, normal to both $\boldsymbol{\tau}$ and $\boldsymbol{\nu}$ (Fig. 7.2).

In an *isotropic porous medium*, the principal directions of the tensor $\mathbf{D}$ *coincide with the directions of these three unit vectors*. As such, as the velocity varies, these directions may vary from point to point and in time.

If, *locally*, we select $\boldsymbol{\tau}$, $\boldsymbol{\nu}$ and $\boldsymbol{\beta}$, as *basis vectors* of the coordinate system, x_1, x_2, x_3, then $\mathbf{D}$ takes the form (7.2.46). In such a case, $D_{x_1x_1}$ is called *coefficient of longitudinal dispersion*, while $D_{x_2x_2}$ and $D_{x_3x_3}$ are called *coefficients of transverse dispersion*.

B. Anisotropy with Axial Symmetry

In an *anisotropic porous medium*, the number of independent dispersivity coefficients is larger, depending on the kind of symmetry exhibited by the anisotropic medium. As an example, consider an axially symmetrical porous medium (= *transverse isotropy*), i.e., a porous medium with one axis of rotational symmetry, with the vector $\mathbf{e}$ (components e_i) indicating the axis of symmetry. For such medium, there exist *six* independent $\mathsf{a}_{ijk\ell}$-coefficients. The dispersivity components can then be expressed by (Bear et al. 2009):

$$a_{ijk\ell} = a_1\delta_{ij}\delta_{k\ell} + \frac{a_2}{2}(\delta_{ik}\delta_{j\ell} + \delta_{i\ell}\delta_{jk}) + a_3 e_i e_j \delta_{k\ell} + a_4 e_k e_\ell \delta_{ij}$$
$$+ \frac{a_5}{2}(e_i e_k \delta_{j\ell} + e_j e_k \delta_{i\ell} + e_i e_\ell \delta_{jk} + e_j e_\ell \delta_{ik}) + a_6 e_i e_j e_k e_\ell, \quad (7.2.48)$$

with a_1 through a_6 indicating the six independent dispersivity coefficients. Fel and Bear (2010) determined the constraints that the six a_i's have to satisfy as a consequence of the positive definiteness of the entropy production, expressed by (7.2.34). We note that by dropping terms associated with e_i in (7.2.48), we obtain the isotropic case, described by (7.2.43), i.e., with $a_1 = a_T$, and $a_2 = a_L - a_T$, and $a_3 = a_4 = a_5 = a_6 = 0$. These six dispersivity coefficients, a_1–a_6, *are properties of the porous medium only*, meaning that they are independent of the flow taking place in the porous medium, and the chosen coordinate system.

Based on (7.2.37), the corresponding expression for the dispersion coefficients D_{ij} is:

$$D_{ij} = \left[a_1 \delta_{ij} + a_2 \frac{V_i V_j}{V^2} + a_3 e_i e_j + a_4 \delta_{ij} \frac{(V_k e_k)^2}{V^2} \right.$$
$$\left. + a_5 \frac{V_k e_k}{V} \frac{V_i e_j + V_j e_i}{V^2} + a_6 e_i e_j \frac{(V_k e_k)^2}{V^2} \right] V. \qquad (7.2.49)$$

We note that the dispersion coefficient, which is used for determining the dispersive flux by means of (7.2.32), depends not only on the porous medium (through the dispersivity coefficients), but also on the velocity vector. As velocity may vary in space and time, so does the dispersion coefficients.

In order to model solute transport in a transversely isotropic porous medium under general flow conditions, we need to determine the *six* independent dispersivity coefficients. These a_i coefficients can be determined by conducting tracer tests in the field, and comparing tracer concentrations within a plume with available analytical or numerical solutions.

As the expression for dispersive flux, say, (7.2.32), involve six dispersion coefficients D_{ij} (six, instead of nine, because of the symmetry $D_{ij} = D_{ji}$), the first step in a parameter estimation procedure is to determine these six components of the dispersion coefficient. Given a transversely isotropic aquifer with known axis of symmetry (i.e., known vector components e_i), and known flow (i.e., known V_i), in principle, it is possible to determine the dispersion coefficients from information on observed concentrations during a controlled experiment.

Let us examine some special cases. Consider a horizontally layered material, with the z-axis coinciding with the axis of material symmetry, i.e., $e_3 = 1$ and $e_1 = e_2 = 0$. We express (7.2.49) as:

$$D_{xx} = \left(a_1 + a_2 \frac{V_x^2}{V^2} + a_4 \frac{V_z^2}{V^2} \right) V,$$

$$D_{yy} = \left(a_1 + a_2 \frac{V_y^2}{V^2} + a_4 \frac{V_z^2}{V^2} \right) V,$$

$$D_{zz} = \left[a_1 + a_3 + (a_2 + a_4 + 2a_5 + a_6) \frac{V_z^2}{V^2} \right] V,$$

$$D_{xy} = D_{yx} = a_2 \frac{V_x V_y}{V^2} V,$$

$$D_{xz} = D_{zx} = (a_2 + a_5) \frac{V_x V_z}{V^2} V,$$

$$D_{yz} = D_{zy} = (a_2 + a_5) \frac{V_y V_z}{V^2} V. \tag{7.2.50}$$

By eliminating the factor $(a_2 + a_5)$ between the fifth and sixth equations in (7.2.50), we can clearly see that D_{xz} and D_{yz} are related to each other. Further analysis shows that the matrix is of rank four, and the following constraints must be satisfied in order for the system of equations to have a solution:

$$D_{yz} = \frac{V_y}{V_x} D_{xz}, \qquad D_{xx} = D_{yy} + \frac{V_x^2 - V_y^2}{V_x V_y} D_{xy}. \tag{7.2.51}$$

Hence, *only four of the six dispersion coefficients are independent.*

Often, it is convenient to use a *local coordinate system* that coincides with the flow direction at the considered location. For example, in the case of uniform flow in the direction of the x_1-axis (see Fig. 7.2), such that $V_1 = V$ and $V_2 = V_3 = 0$, we can show that

$$D_{xy} = \frac{e_y}{e_z} D_{xz}, \qquad D_{yy} = D_{zz} + \frac{e_y^2 - e_z^2}{e_y e_z} D_{yz}, \tag{7.2.52}$$

i.e., again, only four of the dispersion coefficients are independent.

The main conclusions of the above analysis are:

(a) In a single field experiment, in which the flow conditions remain unchanged, it is possible to determine only four dispersion coefficients at any one location, due to the required interdependency given either by (7.2.51) or by (7.2.52).
(b) Given these four independent dispersion coefficients, it is not possible to resolve the six dispersivity coefficients, a_1–a_6.
(c) However, as demonstrated below (see also Fel and Bear 2010), it is possible to determine the six dispersivity coefficients if two experiments are conducted.
(d) In a forward modeling problem, in which values of six dispersion coefficients are required as input, one needs to check the consistency of the assigned dispersion values. These values need to be either determined from (7.2.49), based on the six dispersivity coefficients, or satisfy the relations as shown in (7.2.51) or (7.2.52).

Next, let us consider two special flow cases in the layered medium considered above. In the following discussion, we shall choose the z-axis to coincide with the material axis of symmetry, i.e., $e_3 = 1$ and $e_1 = e_2 = 0$.

In the first case, we consider uniform flow *normal* to the layers, that is, in the z-direction, such that $V_3 = V$ and $V_1 = V_2 = 0$. Using this condition in (7.2.49), we obtain:

$$\mathbf{D}^V = \begin{bmatrix} a^V_{TH} & 0 & 0 \\ 0 & a^V_{TH} & 0 \\ 0 & 0 & a^V_{LV} \end{bmatrix} V, \qquad \begin{aligned} a^V_{TH} &= a_1 + a_4, \\ a^V_{LV} &= a_1 + a_2 + a_3 + a_4 + 2a_5 + a_6, \end{aligned} \tag{7.2.53}$$

where the superscript $(.)^V$ is used to emphasize that the flow direction is vertical, a^V_{TH} is the transverse dispersivity in the horizontal direction (only one value because of the isotropy in the horizontal plane), and a^V_{LV} is the longitudinal dispersivity in the vertical direction. Altogether, to describe dispersion in a layered horizontal porous medium, when the flow is uniform and normal to the layers, we need only one longitudinal and one transversal dispersivities.

As a second case, we consider uniform flow *parallel* to the layers, say, in the $+x$-axis direction, such that $V_1 = V$ and $V_2 = V_3 = 0$. Equation (7.2.49) becomes:

$$\mathbf{D}^H = \begin{bmatrix} a^H_{LH} & 0 & 0 \\ 0 & a^H_{TH} & 0 \\ 0 & 0 & a^H_{TV} \end{bmatrix} V, \qquad \begin{aligned} a^H_{LH} &= a_1 + a_2, \\ a^H_{TH} &= a_1, \\ a^H_{TV} &= a_1 + a_3, \end{aligned} \tag{7.2.54}$$

where a^H_{TH} and a^H_{TV} are, respectively, the transverse dispersivities in the horizontal and in the vertical directions, and a^H_{LH} is the longitudinal dispersivity in the horizontal direction. Thus, to describe dispersion in a layered horizontal porous medium, when flow is uniform and parallel to the layers, we need one longitudinal and two transversal dispersivities.

As observed in the cases discussed above, under uniform flow conditions, we can only determine two, three, or four independent dispersion coefficients in a single experiment, depending on whether the flow is perpendicular, parallel, or at an angle, to the material symmetry axis. This implies that *at least two flow tests in different flow directions are needed, and one of the two directions must be inclined with respect to the direction of the material symmetry axis*. For example, if we conduct a horizontal flow test, and obtain result as in (7.2.54), we can determine three dispersivity coefficients, a_1, a_2 and a_3. For the second test, the flow should be neither in the vertical, nor in the horizontal, direction, as there will not be sufficient information to determine the remaining three coefficients. The flow of the second test must be in an inclined direction with the horizontal plane and the vertical axis, which will provide four additional equations. The remaining three coefficients can then be determined under over-determinacy condition. Similar statement was presented by Fel and Bear (2010) for the special case of flow in the horizontal direction, and making a 45° angle with the axis of symmetry.

In the above, we have assumed that the direction of the axis of symmetry is known *a priori*, i.e., we know the three values: e_1, e_2 and e_3 that appear in (7.2.49). If this direction is not known, we have to use the experimental data to solve the inverse problem also for two of these three components of $\mathbf{e}$ (because $e_1^2 + e_2^2 + e_3^2 = 1$), for a total of 8 unknown values. In this case, two flow tests in two different inclined directions (with respect to the materials axis of symmetry) are sufficient for the determination of these 8 unknowns.

C. Anisotropy with Tetragonal Symmetry

As an example of such porous medium material, we may consider one that is made up of orderly packed solid boxes $a \times b \times c$, with equal spacing between the boxes in all directions (or cubes with 3 different spaces). For this case, the 36 independent $a_{ijk\ell}$-components can be expressed by 7 independent, parameters, which are subject to certain constraints (Bear et al. 2009). It is interesting to note that this case is not identical to the case with axial symmetry (such as a stratified aquifer), considered above. Here, we also need information on the directions in which the boxes, $a \times b \times c$, are positioned in space, e.g., in the form of two of the three e_i's. This case is analyzed in detail by Bear et al. (2009).

D. Anisotropy with Orthorhombic Symmetry

An example is a porous medium material made up of orderly packed solid boxes $a \times b \times c$ with equal spacing between the boxes in all directions (or cubes with three different spaces). For this case, the 36 independent dispersivity components can be expressed by *twelve* independent parameters. We also need information on the directions in which the boxes, $a \times b \times c$, are oriented in space (and this, as indicated earlier, requires information on two e_i's).

It is possible to analyze three special cases of flow, each one with uniform flow parallel to one of the three axes. To describe dispersion in each of these three cases we need only three coefficients: a longitudinal dispersivity and two transversal ones.

In each of the material symmetry cases discussed above, the number of independent coefficients is accompanied by a number of constraints that these coefficients have to satisfy. The information concerning the number of independent coefficients and the constraints among them (Bear et al. 2009) is important when experiments are conducted aimed at determining the values of these coefficients for a specific porous medium, by using an inverse method.

Similar to the discussion presented with respect to the experimental procedure for determining the dispersivity coefficients in the case of transverse isotropy, here also, a number of independent experiments will be required.

E. Another Model for Dispersion in Anisotropic Domains

Some authors, on the basis of field observations, have suggested that for flow parallel to the horizontal stratification in a stratified (= layered) aquifer, transverse dispersion is much smaller in the vertical direction than in the horizontal one, i.e., $a^H_{TH} \gg a^H_{TV}$ in (7.2.54) (Robson 1974, 1978; Garabedian et al. 1991; Gelhar et al. 1992). Based on the above observation, Burnett and Frind (1987) (see also Jensen et al. 1993; Zheng and Bennett 1995) suggested a 'working model' for transversely isotropic porous medium, in which the dispersion tensor is defined by *three* dispersivities only (rather than six, see Sect. 7.2.3 B): a longitudinal dispersivity, a_L, and two transversal dispersivities: a horizontal one, a_{TH}, and a vertical one, a_{TV}. The components of the dispersion tensor in three dimensions, with the z-axis as the axis of material symmetry, are presented as:

$$
\begin{aligned}
D_{xx} &= \frac{1}{V}\left(a_L V_x^2 + a_{TH} V_y^2 + a_{TV} V_z^2\right), \\
D_{yy} &= \frac{1}{V}\left(a_{TH} V_x^2 + a_L V_y^2 + a_{TV} V_z^2\right), \\
D_{zz} &= \frac{1}{V}\left(a_{TV} V_x^2 + a_{TV} V_y^2 + a_L V_z^2\right), \\
D_{xy} = D_{yx} &= \frac{1}{V}\left(a_L - a_{TH}\right) V_x V_y, \\
D_{xz} = D_{zx} &= \frac{1}{V}\left(a_L - a_{TV}\right) V_x V_z, \\
D_{yz} = D_{zy} &= \frac{1}{V}\left(a_L - \mathrm{a}_{TV}\right) V_y V_z.
\end{aligned}
\tag{7.2.55}
$$

These expressions can be compared with those for the isotropic case, (7.2.45). Burnett and Frind (1987) further *assumed* that $\mathrm{a}_{TH} \gg a_{TV}$. The relations presented in (7.2.55), however, are not consistent with (7.2.49). In fact, Lichtner et al. (2002) have demonstrated that (7.2.55) does not conform with tensor transformation rules, suggesting that it is not an acceptable model.

Based on a turbulence model investigated by Batchelor (1959), using a method introduced by Robertson (1940), Poreh (1965) suggested a model that is based on four dispersivity coefficients:

$$
D_{ij} = \left[\alpha_1 \delta_{ij} + \alpha_2 \frac{V_i V_j}{V^2} + \alpha_3 e_i e_j + \frac{\alpha_4}{2} \frac{e_i V_j + e_j V_i}{V}\right] V. \tag{7.2.56}
$$

Lichtner et al. (2002) examined the Poreh (1965) model and discussed the need for introducing $\cos\theta$ as a factor in the constitutive model. As a result, they proposed a four parameter model when the principal axes are not aligned with the flow; they suggested three parameters.

F. The Role of Diffusion in Dispersion

Although, in this section, we have discussed the phenomena of diffusion and dispersion separately, and presented expressions for their fluxes, we have to understand that they are inseparable. The difference between them stems from the very definition of dispersion: molecular diffusion always exists in a fluid, whether stationary or in motion, as long as a concentration gradient of some dissolved chemical species exists. For dispersion, we need *two* conditions: (1) fluid motion *and* a (2) concentration gradient. Consider an imaginary case of steady flow of a fluid in the void space of a porous medium containing a dissolved tracer, but without diffusion. Spreading of an initial quantity of tracer, say within a fluid that occupies the void space of an REV, will still occur, as the tracer will be carried within the tortuous stream-tubes at variable velocity. The spreading within a stream-tube can be by advection. However, each fluid particle will stay in its original stream-tube. Theoretically, reversing the flow, will bring all tracer particles back to the original position in the REV.

Only diffusion, driven by lateral concentration gradient can cause tracer particles to move *laterally*, from one stream-tube to the next. An important feature of *spreading by diffusion is that it is irreversible*. Altogether, because of diffusion, there is no way to bring the tracer particles back to their original position. Thus, dispersion is irreversible.

Dispersion is velocity dependent, while diffusion is not. Thus, at high velocity, dispersion dominates, while diffusion dominates at low velocity. The dimensionless Peclet number, Pe, defined in (7.2.39), provides a measure of the ratio between the two.

An interesting issue arises in the following case. We consider an aquifer with uniform flow, say, specific discharge q_o in the $+x$-direction. Suppose tracer labelled water is injected during a relatively short period through a well at (0, 0) into the aquifer, creating a "stain" of labelled water. Along the x-axis, we have $\partial c/\partial x < 0$ in the downstream direction, i.e., $x > 0$, and $\partial c/\partial x > 0$ in the upstream direction, i.e., $x < 0$. The tracer advective + dispersive flux along the x-axis is expressed by:

$$q_c = q_o c - q a_L \frac{\partial c}{\partial x}, \quad q = |q_o|.$$

For $x > 0$, $\partial c/\partial x < 0$, and the solute will continue to be displaced downstream. However, for $x < 0$, the unlabeled water pushes the labelled water, but it is possible that the above expression for q_c will lead to $q_c < 0$, i.e., tracer transport against the flow in the aquifer, a result that does not make sense. This seems to be an inconsistency in the dispersion theory presented here (see, for example, Simpson 1978).

G. Some Comments on Dispersion in Single and Multiphase Flow

- Dispersion is a consequence of using averaged fluid velocity instead of the actual velocity within the void space. Hence, in the case of multiphase flow, a dispersivity is defined for each of the phases present in the void space and that dispersivity depends on the saturation, $a_{ijkl,\alpha} = a_{ijkl,\alpha}(\theta_\alpha)$. Very little information is available on this subject.
- A special interesting case is that of unsaturated flow, i.e., air and water that occupy the void space. In Sect. 6.1 we have introduced the concepts of *irreducible water saturation* and that of *immobile water*. The presence of such domains certainly affect the water's velocity distribution within the void space, thus affecting the dispersivity of the water occupying part of the void space. Furthermore, a solute may be exchanged between the mobile and the immobile potions of the void space. This situation will also affect the dispersivity of the water that occupies part of the void space.
- **Ion exclusion**. In the case of an ionic species–an anion or a cation–in water. Because of the electrical charge on certain solid surfaces, the considered species may be repelled from the solid wall, where water velocity is small (recalling that we assume that water adsorbs to the solid wall). Thus, the considered species moves mainly in the regions of higher velocity within the void space. The average velocity of the water that carries and disperses the solute is, thus, higher than for a

non-ionic solute. As a consequence, the advective flux will be higher, and so will the coefficient of dispersion, which is proportional to the average velocity. This phenomenon has also been called *charge exclusion* (e.g., Gvirtzman et al. 1989; Gvirtzman and Gorelick 1991).

- **Size exclusion** Some molecules or ions are so large that their travel is restricted to the larger pores. As a consequence, they are carried (by advection) at a higher average fluid velocity. The higher average velocity also results in a higher coefficient of dispersion.

Although we have considered here dispersion in connection with the spreading of the mass of a dissolved species, the same ideas are valid also for thermal energy (heat) by a diffusive process (= conduction) and by thermal dispersion.

7.2.4 Field Scale Solute Dispersion

In Sect. 3.4.3, the phenomenon of solute dispersion was shown to be a consequence of *porous medium heterogeneity at the microscopic scale*, i.e., due to the presence of both a solid matrix and a void space within the REV. A grain, or pore diameter, or the hydraulic radius of the void space, was suggested as the *scale* of this heterogeneity. The latter produces velocity variations *within* the void space. The dispersive flux, a macroscopic level concept, obtained by averaging over an REV, was introduced as a means for circumventing the need to know the details of the velocity distribution at the microscopic level. We have also introduced the dispersive flux as a Fickian-type model, with the concentration gradient as the driving force and a dispersion coefficient that is proportional to the average velocity according to some models, and to the square of that velocity according to some other.

Actually, efforts to verify the Fickian-type solute transport model introduced so far in this chapter, with a coefficient of dispersion proportional to the first power of the average velocity, by comparing model predictions with field observations, showed that the model is valid for rather small and very homogeneous domains. Often, the term "laboratory scale" is used to describe such domains. In fact, some authors (e.g., Levy and Berkowitz 2003) claim that the Fickian model is not valid even for very short (assumed homogeneous) sand columns (i.e., at 'laboratory scale').

As emphasized throughout this book, a characteristic feature of all subsurface (or geological) domains, and here we focus only on such domains, is that they are *highly heterogeneous* with respect to their macroscopic coefficients, e.g., porosity and (mainly) permeability. We have suggested the term 'megascopic level', obtained by smoothing out variations at the macroscopic level, and introduced the concept of 'scale of heterogeneity', indicating that at the macroscopic level, variations, say in permeability, may occur at different scales. In fact, this *multiple scale heterogeneity is a dominant factor in geological formations*. Because pressure propagates very fast, the effect of this inherent heterogeneity is less noticeable when considering fluid flow. However, its effect is significant on the mass transport of a dissolved chemical species. There exist abundant evidence in the literature that shows that the dispersion model discussed so far in this section does not describe what really happens in geological formations, or under "field scale" conditions.

Megascopic Scale Averaging

It should be possible to solve a transport problem at the macroscopic level in any heterogeneous domain in which the spatial variations of the permeability and of the other relevant coefficients are known. Indeed, in small scale field problems, e.g., in the vicinity of an injection well, or within a small distance downstream of a pollution source, the formation properties (porosity, permeability, dispersivity) may be known (or estimated), and the problem of predicting the concentration distribution of an injected solute can be solved by making use of the (macroscopic level) model described earlier in this chapter. However, when considering a solute, or pollution plume that advances a large distance, sometimes kilometers, we face a situation similar to that which is encountered at the microscopic level, viz., that the detailed information about the spatial variation of the relevant parameters is unknown, due to the heterogeneity inherent in such domains. The way we overcome the lack of information about the heterogeneity at the microscopic level (resulting from pore scale heterogeneity) is to use homogenization, or averaging over an REV, as discussed in Sect. 1.1. The same averaging, or smoothing approach, may also be applied to heterogeneities that are encountered at the macroscopic level, to obtain a continuum at the *megascopic level*. In Sect. 1.1.6, this kind of averaging volume was referred to as the *Representative Macroscopic Volume* (RMV). The characteristic size of this volume, ℓ^*, is constrained by:

$$d^* \ll \ell^* \ll L,$$

where L is a length characterizing the porous medium domain; d^* is the length that characterizes the macroscopic heterogeneity that we wish to smooth out. Similar to what happens during microscopic-to-macroscopic smoothing, here, the information about the heterogeneity at the macroscopic level appears at the megascopic level in the form of various coefficients.

Denoting the volume of an RMV by $\overline{\mathbb{V}}_o$, and the *macroscopic value* of e by $\overline{e}$, we define the *megascopic value* of e by:

$$\overline{\overline{e}}(\mathbf{x}, t) = \frac{1}{\overline{\mathbb{V}}_o} \int_{\overline{\mathbb{V}}_o} \overline{e}(\mathbf{x}', t; \mathbf{x})\, d\overline{\mathbb{V}}_o(\mathbf{x}'), \qquad (7.2.57)$$

where $\mathbf{x}$ and $\mathbf{x}'$ denote the centroid of the RMV and a point (of the porous medium regarded as a continuum) inside it, respectively. With this definition, we may now derive the total flux of a γ-component at the megascopic level, by averaging (3.4.31) over an RMV. For saturated flow, we obtain:

$$\begin{aligned}\overline{\overline{\phi \mathbf{J}^\gamma}} &= \overline{\overline{\overline{c}^f \mathbf{q}}} + \overline{\overline{\phi\left(\mathbf{J}^\gamma + \mathbf{J}^{*\gamma}\right)}} \\ &= \overline{\overline{\overline{c}^f}}\,\overline{\overline{\mathbf{q}}} + \overline{\overline{\hat{\overline{c}}^f\,\hat{\mathbf{q}}}} + \overline{\overline{\phi\left(\mathbf{J}^\gamma + \mathbf{J}^{*\gamma}\right)}} \\ &\approx \overline{\overline{\overline{c}^f}}\,\overline{\overline{\mathbf{q}}} + \overline{\overline{\hat{\overline{c}}^f\,\hat{\mathbf{q}}}},\end{aligned} \tag{7.2.58}$$

where a double bar over a macroscopic value indicates a megascopic value obtained by averaging over an RMV, with $\overline{c} = \phi\overline{c}^f$, and $\hat{(..)}$, defined by:

$$\hat{\overline{(..)}}^f = \overline{(..)}^f - \overline{\overline{\overline{(..)}^f}},$$

is the deviation of a macroscopic value at any point within an RMV from its average over the RMV. We note that the flux on the left-hand side of (7.2.58) (and hence all other terms) is per unit area of porous medium.

As could have been expected, the megascopic total flux contains two *new* additional dispersive fluxes, which result from the variability of the relevant macroscopic quantities. One is $\overline{\overline{\hat{\overline{c}}^f\,\hat{\mathbf{q}}}}$, which will be referred to as the *macrodispersive flux* of the chemical species. The other is the average over the RMV of the sum of the dispersive and diffusive fluxes at the macroscopic level. Note that on the last line of (7.2.58), we have neglected the second dispersive flux as being much smaller than the first.

Altogether, the total flux is again the sum of an advective mass flux and a dispersive one. There is no reference here to the diffusive flux, as we have neglected it. At very low velocities, we may not neglect the average of the macroscopic diffusive flux.

We have to express the dispersive flux at the megascopic level in terms of megascopic quantities, in the same manner as is done in the description of transport at the macroscopic level. We usually *assume* that a Fickian-type dispersion law, e.g., (7.2.32), is also valid for describing the macrodispersive flux. A *macrodispersivity*, A_{ijkm}, can be defined in the same way as the dispersivity was defined earlier in (7.2.37). Bear (1979), while developing the vertically integrated mass balance equation for a component of a phase, suggested for the general case of an anisotropic porous medium, an expression for macrodispersivity in the form:

$$A_{ijkm} = \frac{\overline{\hat{K}_{in}\hat{K}_{j\ell}}}{\overline{K_{kn}}\;\overline{K_{m\ell}}}\,\widetilde{L}, \tag{7.2.59}$$

where K_{ij} denotes the ij-th component of the hydraulic conductivity tensor, and $\widetilde{L}$ is a length that characterizes the inhomogeneity of the aquifer, resulting from stratification. It is a fourth rank tensor, which is analogous to the dispersivity at the macroscopic level (i.e., with A_L and A_T, etc.). In an isotropic porous medium, the macrodispersivity reduces to a scalar. Gelhar (1976) and Gelhar et al. (1979) analyzed the dependence of macrodispersion on permeability variations. For horizontal flow in an isotropic confined aquifer, they suggested that:

$$A_L = \frac{1}{3}\frac{L_1^2\sigma_{\ln k}^2}{a_T}, \tag{7.2.60}$$

in which L_1 is a correlation distance (= distance along which permeabilities are still correlated), $\sigma_{\ln k}$ is the standard deviation of $\ln k$, and a_T is the transverse dispersivity.

Altogether, we may summarize this approach by suggesting that dispersion and macrodispersion are analogous phenomena, in that both are consequences of velocity variations that are due to heterogeneity, but at different scales. Dispersion arises from velocity variations *within* the void space (i.e., at the microscopic level), caused by the presence of the solid surfaces. Macrodispersion is produced by macroscopic velocity variations, caused by variations (i.e., heterogeneity) in the permeability and porosity. In both cases, the flux is the sum of an advective flux and a (hydrodynamic) dispersive one, written at the respective levels. The structure of the coefficient of dispersion is the same in both cases, and so is the relationship between the coefficient of dispersion, the dispersivity, and the average velocity. In practice, we use exactly the same mathematical model (except that in the case of field scale, we usually neglect the flux due to molecular diffusion), but select the magnitude of the dispersivity according to the scale of heterogeneity.

In laboratory column experiments, the porous medium is more or less homogeneous, say with respect to permeability and porosity. The scale of heterogeneity is that of the size of a grain or a pore. Indeed, the magnitude of longitudinal dispersivity found in many column experiments is approximately equal to a pore- or grain-size. However, under field conditions, the scale of heterogeneity, due to variability in permeability and porosity, is much larger. In fact *this scale grows with the size of the domain*, i.e., the distance between the source of concentration and the point of observation. Gelhar et al. (1992) compiled a large number of field experiments and presented the observed longitudinal dispersivity, A_L, as a function of the travel distance, L_s, as shown in Fig. 7.3. It is clear that macrodispersivity is proportional to the size of the field, although the data shows a wide range of scatter. Lallemand-Barrés and Peaudecerf (1978) analyzed published values of plume concentration and showed that, on the average, the dispersivity increases with the distance (between a few meters and 10 km), between the source and the point of observation. As a 'rule of thumb', they concluded that the dispersivity can be approximated as 1/10 of the distance traveled by the plume. This is often referred to as a 'scale effect'.

Based on the argument of self-similar (fractal) hierarchy of logarithmic hydraulic conductivity, Neumann (1990) suggested a universal scaling law and presented the following equations based on the least square fit of the data:

$$A_L = 0.017\, L_s^{1.5}; \quad L_s \leq 100 \text{ m}; \tag{7.2.61}$$

$$A_L = 0.32\, L_s^{0.83}; \quad L_s > 100 \text{ m}. \tag{7.2.62}$$

These two empirical formulas are plotted in Fig. 7.3. Gelhar et al. (1992, 1993), however, cautioned the use of these power laws by pointing out the large scatter in data (2–3 orders of magnitude) in Fig. 7.3.

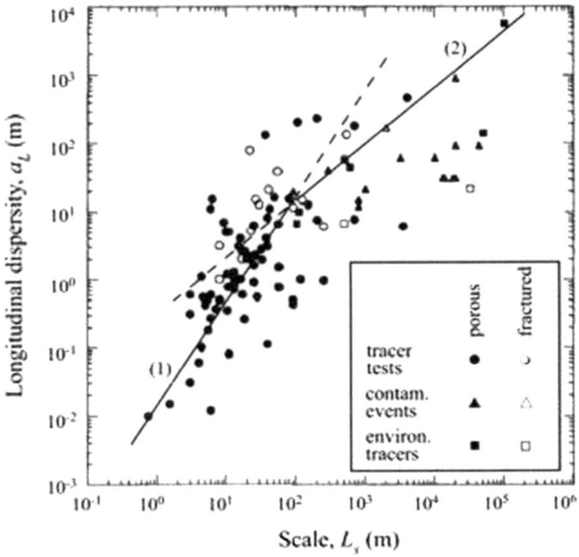

Fig. 7.3 Longitudinal dispersivity versus plume travel distance for various types of observations and media (Gelhar et al. 1992). Line marked as (1): Eq. (7.2.61); and (2): Eq. (7.2.62)

Gelhar and Axnes (1983) and Dagan (1984) (see also (7.2.60)) showed that the longitudinal dispersivity is also proportional to the product of the variance of the logarithm of the hydraulic conductivity, and the correlation length scale, i.e.,

$$A_L \sim L_1 \sigma^2_{\ln k}. \tag{7.2.63}$$

This can explain the range of scatter observed in Fig. 7.3.

Another issue that may be raised in connection with the above averaging (over an RMV) approach is that of *monitoring*. In the averaged model, the megascopic concentration, $\overline{\overline{c^f}}$, should be measured over the volume of an RMV. Can this be done in practice?

At the end of Sect. 7.1, we introduced the notion of the "well-mixed" REV. However, it questionable whether we can assume that a considered solute is "well mixed" also within an RMV, in the case of highly heterogeneous domains, especially geological ones. Two issues will be raised. One is the possibility that the pore space of an RMV may contain a significant sub-domain of dead-end pores (Sect. 1.1.7), in which flow is very slow, so that the notion of a *well-mixed* RMV at (i.e., in the vicinity of) a point may be questionable. A second issue is that raised already in Sect. 1.1.4 E in connection with REV averaging of solute concentration and of heat. For an average over an RMV to represent what happens in its interior, it has to be "well mixed", and this requires a sufficiently small Peclet number within it. This ensures good mixing.

To summarize the discussion so far, under field conditions, we observe the spreading, or dispersion of an advancing plume as it advances in a geological formation. The reason is the heterogeneity inherent in geological formations.

7.3 Mass Balance Equation for Reacting Species

Often, phenomena of mass transport of single and multiple fluid phases in porous media involve reactions among chemical species that are present in a considered fluid. In many types of chemical reactors (see App. A), the objective is to provide the appropriate environment and conditions in which desired chemical reactions take place. Reactions may also take place among chemical species in the fluid(s) and those comprising the solid matrix. Many aquifer clean-up techniques rely on chemical reactions. Some methods utilize chemical or biological reactions to degrade a contaminant into harmless products by introducing appropriate reactants or nutrients into the subsurface, either through an injection well, or by placement into a permeable trench, also known as a *Permeable Reactive Barrier* (PRB) (e.g., Czurda and Haus 2002). Chemical reactions occur also in a Carbon Capture and Storage (CCS) project, in which CO_2 is injected into deep usually brine containing formations (e.g., Al-Khoury and Bundschuh 2014). The CO_2 that dissolves in the brine participates in a variety of chemical reactions with the various species that comprise the brine.

7.3.1 *Species Balance Equations*

As in the case of mass of a fluid phase, the expression for the mass flux of a chemical species contains two variables: the flux of the species and its concentration. An additional equation is called for–the mass balance for the species.

The mass balance equation for a γ-species in a fluid (liquid, or gas) α-phase that occupies the entire void space was already presented as (3.6.2). Repeated here for convenience, recalling that $c^\gamma_\alpha \equiv \rho_\alpha \omega^\gamma_\alpha$, $\mathbf{J}^{m^\gamma}_{\alpha,adv} \equiv \mathbf{J}^\gamma_{\alpha,adv} = \rho_\alpha \omega^\gamma_\alpha \mathbf{V}_\alpha$, we have:

$$\frac{\partial \phi \rho_\alpha \omega^\gamma_\alpha}{\partial t} = -\nabla \cdot \phi \left(\rho_\alpha \omega^\gamma_\alpha \mathbf{V}_\alpha + \mathbf{J}^{m^\gamma}_{\alpha,dif} + \mathbf{J}^{m^\gamma}_{\alpha,dis} \right) + f^{m^\gamma}_{s\to\alpha} + \phi \rho_\alpha \Gamma^{m^\gamma}_\alpha. \qquad (7.3.1)$$

In the above equation, the symbol $f^{m^\gamma}_{s\to\alpha} (\equiv f^\gamma_{s\to\alpha})$ denotes the transport of γ-mass from the solid (s) phase to the α-phase across their common interphase boundaries, with $f^{m^\gamma}_{s\to\alpha} = -f^{m^\gamma}_{\alpha\to s}$. Obviously, solid-fluid transfer, e.g., due to adsorption on the solid, to solid dissolution and to precipitation on the solid, call for the introduction of the mass balance equation for the relevant species in the thin layer next to the solid's surface (e.g., (7.4.9)).

The term $\phi \rho_\alpha \Gamma^{m^\gamma}_\alpha$ denotes the source (= negative of sink) of γ-mass in α per unit volume and unit time, say by chemical reactions (see Sect. 7.3.3). Often, in dealing with ground water flow in aquifers, this term is used for expressing added mass of γ by injection wells (see Sect. 7.3.2) per unit volume of porous medium. In what follows, we shall focus on sources of γ that are due only to chemical reactions.

Still considering single phase (α) flow, expressing the sum of diffusive + dispersive m^γ-fluxes, $\mathbf{J}^{m^\gamma}_{\alpha,dif} + \mathbf{J}^{m^\gamma}_{\alpha,dis}$, by $-\mathbf{D}'_\alpha \cdot \nabla c^\gamma_\alpha$, and assuming that the fluid's mass balance equation can be approximated by $\nabla\cdot\phi\rho_\alpha\mathbf{V}_\alpha(\equiv\nabla\cdot\rho_\alpha\mathbf{q}_\alpha)=0$, the solute mass balance equation for $c^\gamma_\alpha=c^\gamma_\alpha(\mathbf{x},t)$, reduces to:

$$\phi\frac{\partial c^\gamma_\alpha}{\partial t}=-\phi\mathbf{V}_\alpha\cdot\nabla c^\gamma_\alpha-\nabla\cdot(\mathbf{D}'_\alpha\cdot\nabla c^\gamma_\alpha)+f^{m^\gamma}_{s\to\alpha}+\phi\rho_\alpha\Gamma^{m^\gamma}_\alpha. \tag{7.3.2}$$

often called *advection-dispersion-reaction equation*, abbreviated ADE, although this is nothing but the mass balance equation for the considered γ-species.

Into (7.3.1), as into the other γ-balance equations, we can now insert appropriate expressions for the diffusive and dispersive fluxes of a considered γ-species, for the transfer of γ from the solid to the fluid, as well as source terms that express the source/sink term by injection and pumping and the production of γ in the fluid by chemical reactions.

For an α-phase in two phase (α,β) flow, (7.3.1) takes the form:

$$\frac{\partial\phi S_\alpha\rho_\alpha\omega^\gamma_\alpha}{\partial t}=-\nabla\cdot\phi S_\alpha\left(\rho_\alpha\omega^\gamma_\alpha\mathbf{V}_\alpha+\mathbf{J}^\gamma_{\alpha,dif}+\mathbf{J}^\gamma_{\alpha,dis}\right)+\sum_{\delta=\beta,s}f^{m^\gamma}_{\delta\to\alpha}+\phi S_\alpha\rho_\alpha\Gamma^\gamma_\alpha, \tag{7.3.3}$$

in which S_α denotes α-saturation, $c^\gamma_\alpha=\rho_\alpha\omega^\gamma_\alpha$, and the sum is over all processes of mass transfer from the solid (noting that the solid may be composed of several minerals/phases) and from all non-α fluid phases.

The above mass balance equation for a γ-species in an α-phase, can also be written in terms of the mole fraction, X^γ_α:

$$\begin{aligned}\frac{\partial}{\partial t}\phi S_\alpha\eta_\alpha X^\gamma_\alpha=&-\nabla\cdot\phi S_\alpha\eta_\alpha\left(X^\gamma_\alpha\mathbf{V}_\alpha-\mathcal{D}_{\alpha,dif}\cdot\nabla X^\gamma_\alpha-\mathbf{D}_{\alpha,dis}\cdot\nabla X^\gamma_\alpha\right)\\&+\sum_{\delta=\beta,s}f^\gamma_{\delta\to\alpha}+\phi S_\alpha\rho_\alpha\Gamma^\gamma_\alpha,\end{aligned} \tag{7.3.4}$$

in which X^γ_α denotes the mole-fraction of the γ-species in the α-phase, η_α is the molar density of the α-phase $(=n_\alpha/\mathbb{V}_\alpha)$, $\eta_\alpha X^\gamma_\alpha=[c^\gamma_\alpha]$ denotes the molar concentration of a γ-species in an α-phase, $\phi S_\alpha\mathbf{V}_\alpha\equiv\mathbf{q}_\alpha$ denotes the specific discharge of the α-phase, and $\mathcal{D}_{\alpha,dif}$ and $\mathbf{D}_{\alpha,dis}$ are, respectively, the (second rank tensor) coefficients of molecular diffusion and of dispersion in the α-phase within a porous medium, $\sum_{\delta=\beta,s}f^\gamma_{\delta\to\alpha}$ denotes the rate of γ-species transfer per unit volume of the porous medium, from all non-α-phases to the α-phase, and Γ^γ_α is the rate of production of the mass of γ per unit mass of the α-phase). Each term in the above equation expresses the number of moles of γ per unit volume of porous medium per unit time.

So far, we have written *species mass balance* equations for a γ-species within an α-phase present in the void space, noting that the same species may be present in more than one phase (actually, also adsorbed on the solid, or present in a mineral). Under thermodynamic equilibrium, the concentrations of γ in adjacent phases are related to each other. With this in mind, another option is to make use of the notion of *component*, defined in Sect. 1.1.1, and write *component mass balance equations*. In such equations, we write a balance equation for a considered species which is a component (indicated as a superscript) present in all phases comprising the porous medium domain. If necessary, we can include the solid also as a participating phase.

When the same γ-species is present in all fluid phases, and possibly (as adsorbed) also on the solid matrix, especially when chemical equilibrium is assumed, it is convenient to write a single mass balance equation for every γ-species in/on all phases.

Consider the case of two fluid phases, a liquid and a gas denoted by subscripts ℓ and g that, together, occupy the entire void space ($\theta_\ell + \theta_g = \phi$), and solid phase ($\phi_s = 1 - \phi$), and 3 chemical species, to which we'll now refer as *components*, which can be present in both the liquid and the gas. An example is the case of: *carbon dioxide* (CO_2), *methane* (NH_4), and *water* (H_2O). From (7.3.3), with $f^\gamma_{\alpha\to\beta} = f^\gamma_{\beta\to\alpha}$, we obtain for every γ-species, regarded as a γ-component:

$$\frac{\partial}{\partial t}\phi\sum_{\delta=\alpha,\beta} S_\delta \rho_\delta \omega^\gamma_\delta = -\nabla\cdot\phi\sum_{\delta=\alpha,\beta} S_\delta\left(\rho_\delta\omega^\gamma_\delta \mathbf{V}_\delta + \mathbf{J}^\gamma_{\delta,dif} + \mathbf{J}^\gamma_{\delta,dis}\right) + \phi\sum_{\delta=\alpha,\beta} S_\delta\rho_\delta\Gamma^\gamma_\delta, \tag{7.3.5}$$

Or, in terms of the mole fraction, X^γ_δ:

$$\frac{\partial}{\partial t}\phi\sum_{\delta=\alpha,\beta} S_\delta \eta_\delta X^\gamma_\delta = -\nabla\cdot\phi\sum_{\delta=\alpha,\beta} S_\delta\left(\eta_\delta X^\gamma_\delta \mathbf{V}_\delta + \mathbf{J}^\gamma_{\delta,dif} + \mathbf{J}^\gamma_{\delta,dis}\right) + \phi\sum_{\delta=\alpha,\beta} S_\delta\rho_\delta\Gamma^\gamma_\delta. \tag{7.3.6}$$

As a second example, consider the case of two phases: a liquid (ℓ) and a gas (g). The ℓ-phase is composed of two species: water (w) and dissolved air (a). The gas (g) is composed of air (a) and water vapour (w). Here, we write two balance equations: one for the water component (in both ℓ, g phases):

$$\frac{\partial}{\partial t}\phi\sum_{\delta=l,g} S_\delta \rho_\delta \omega^w_\delta = -\nabla\cdot\phi\sum_{\delta=l,g} S_\delta\left(\rho_\delta\omega^w_\delta \mathbf{V}_\delta + \mathbf{J}^w_{\delta,dis+dif}\right) + \phi\sum_{\delta=l,g} S_\delta\rho_\delta\Gamma^w_\delta, \tag{7.3.7}$$

and one for the air component (in the ℓ and g phases):

$$\frac{\partial}{\partial t}\phi\sum_{\delta=l,g} S_\delta \rho_\delta \omega^a_\delta = -\nabla\cdot\phi\sum_{\delta=l,g} S_\delta\left(\rho_\delta\omega^a_\delta \mathbf{V}_\delta + \mathbf{J}^a_{\delta,dis+dif}\right) + \phi\sum_{\delta=l,g} S_\delta\rho_\delta\Gamma^a_\delta. \tag{7.3.8}$$

Further to the above presentation, let us consider the case of a salt (say, NaCl). We focus on the sodium (denoted by Na) which is dissolved in the liquid, and is also adsorbed (from the liquid) on the solid (s). We then add a balance equation for the sodium (as a third component), again, for the porous medium as a whole:

$$\frac{\partial}{\partial t}\left[\phi S_l \rho_l \omega_l^{Na} + (1-\phi)\rho_s \omega_s^{Na}\right] = -\nabla\cdot\phi S_l \mathbf{J}^{Na}_{total,l} + \phi S_l \rho_l \Gamma_l^{Na} + (1-\phi)\rho_s \Gamma_s^{Na}, \tag{7.3.9}$$

where $f^{Na}_{l\to s} = -f^{Na}_{s\to l}$, and we have allowed for sources, say by chemical reactions, both in the liquid and on the solid.

Finally, consider the case of three components: water (H_2O), methane (CH_4), carbon dioxide (CO_2) indicated by superscripts by w, m, c respectively, that, together, occupy the void space in the form of two phases: a liquid (ℓ) and a gas (g). We write three component mass balance equations: one for the water component (in both phases):

$$\frac{\partial}{\partial t}\phi \sum_{\delta=l,g} S_\delta \rho_\delta \omega_\delta^w = -\nabla\cdot\phi \sum_{\delta=l,g} S_\delta \left(\rho_\delta \omega_\delta^w \mathbf{V}_\delta + \mathbf{J}^w_{\delta,dis+dif}\right) + \phi \sum_{\delta=l,g} S_\delta \rho_\delta \Gamma_\delta^w, \tag{7.3.10}$$

one for the (dissolved and gaseous) methane component:

$$\frac{\partial}{\partial t}\phi \sum_{\delta=l,g} S_\delta \rho_\delta \omega_\delta^m = -\nabla\cdot\phi \sum_{\delta=l,g} S_\delta \left(\rho_\delta \omega_\delta^m \mathbf{V}_\delta + \mathbf{J}^m_{\delta,dis+dif}\right) + \phi \sum_{\delta=l,g} S_\delta \rho_\delta \Gamma_\delta^m, \tag{7.3.11}$$

and one for the (dissolved and gaseous) carbon dioxide component:

$$\frac{\partial}{\partial t}\phi \sum_{\delta=l,g} S_\delta \rho_\delta \omega_\delta^c = -\nabla\cdot\phi \sum_{\delta=l,g} S_\delta \left(\rho_\delta \omega_\delta^c \mathbf{V}_\delta + \mathbf{J}^c_{\delta,dis+dif}\right) + \phi \sum_{\delta=l,g} S_\delta \rho_\delta \Gamma_\delta^c. \tag{7.3.12}$$

Note that the γ-mass balances presented above are written in terms of the phase velocities $\mathbf{V}_\alpha$, while Darcy's law provides information on the relative velocity, $\mathbf{V}_\alpha - \mathbf{V}_s$. When, albeit very seldom, we cannot assume that $\mathbf{V}_s \ll \mathbf{V}_\alpha$, we have to regard $\mathbf{V}_s$ as an additional variable and to solve the case of flow and transport in a deformable porous medium (see Chap. 9).

7.3.2 Injection and Pumping of a γ-Species Through Wells

Injection and pumping through wells, regarded as point sources and sinks of fluid mass, in a porous medium domain, have been presented in Sect. 5.1.1. There, in (5.1.5), the term $\rho_m Q_m(\mathbf{x}_m, t)\delta(\mathbf{x} - \mathbf{x}_m)$, expresses the mass of the considered fluid added, per unit time, through a point of injection (m) to a porous medium domain, per unit volume of porous medium. Extending the same approach to the case of a γ-solute, at mass-fraction ω^γ, the source term, $\phi S_\alpha \rho_\alpha \Gamma_\alpha^\gamma$, appearing in the solute balance equation (7.3.4), takes the form

$$\phi S_\alpha \rho_\alpha \Gamma_\alpha^\gamma = \sum_{(M)} \rho_m \omega_m^\gamma Q_m(\mathbf{x}_m, t)\delta(\mathbf{x} - \mathbf{x}_m), \tag{7.3.13}$$

where M indicates the number of m-wells, and δ denotes the *Dirac delta function.*

7.3.3 Chemical Reactions

In the previous subsection, the product $\phi S_\alpha \rho_\alpha \Gamma_\alpha^\gamma$, appearing in the mass balance equation (7.3.3) expresses the rate at which the mass of a γ-species is produced at a point in a fluid phase, per unit time, per unit volume of porous medium, *due to chemical reactions, or decay phenomena* in the α-phase. The objective of this subsection is to discuss such sources (= negative sinks). These kinds of reactions are often referred to as '*homogeneous reactions*', because they occur within a single phase.

A. Conditions for Chemical Equilibrium

We are dealing with chemical reactions that occur among dissolved chemical species that are present in a fluid that occupies the void-space or part of it. The phase may be stationary or moving.

In order to interact, chemical species, e.g., in the form of ions, must collide at the proper 'orientation', and have the required amount of energy to perform the interaction, that is, to break and make chemical bonds. Another factor is the chemical structure of the interacting species. In order to interact, a chemical species may have to follow a rather intricate path before it can interact with another species. This is especially true in the case of large organic molecules. Obviously, the net rate of production (or consumption) of chemical species participating in a reaction, as indicated by the number of molecules (and ions) reacting per unit time, is a statistical quantity. In order to describe a reaction at the microscopic level, its rate must be statistically averaged over the μREV. Therefore, we require that the size of the μREV be much greater than the mean distance traveled by a molecule before reacting with another molecule.

The macroscopic rate of reaction of a species at a point in a porous medium domain is defined as the volume-averaged microscopic rate over all points in the REV that is centered at the considered point. *For homogeneous reactions, it is usually assumed that the deviations in the thermodynamic state within the REV are sufficiently small such that the same form of the microscopic rate law can be used at the macroscopic level.*

When there is no net transfer of mass into a μREV, each reaction will eventually reach equilibrium. This means that the rates of consumption and production are equal, and the net rate, which is equal to the difference between forward and backward rates vanishes. Each reaction has a characteristic time constant that indicates how fast it reaches equilibrium. A reaction may actually be the result of several simpler ones, often referred to as *elementary reactions*. Some such reactions may be considerably slower than others. The slowest reaction is the *rate-limiting reaction*; it will determine the speed of the overall reaction.

The reaction in a given system may be characterized as falling into one of two groups: *equilibrium* and *kinetic reactions*. This classification depends on the time scales inherent in the aspects of the considered problem. The importance of this classification is that it facilitates the identification of the type of a considered reaction. The Damköehler number (see below) determines whether a reaction satisfies local equilibrium conditions, or is a kinetic one.

When all the reactions in a given system are equilibrium ones, then the μREV is said to be in a state of *local chemical equilibrium*. If every μREV in a domain satisfies this condition, then the domain is said to satisfy the condition of *local chemical equilibrium* at the microscopic level.

Obviously, in the case of multiple reactions that occur simultaneously, we may be in a situation of *partial chemical equilibrium*. This occurs when some reactions are rather quick to occur, while others take a much longer time.

Any experiment intended to determine the time of a reaction will involve some time interval required for 'mixing' by flow and diffusion. This is the time required for bringing species sufficiently close to each other, thus enabling them to interact. A well-mixed batch reactor experiment provides information about the behavior of a closed system at a microscopic point, i.e., within a μREV.

In order to move now to macroscopic equilibria and kinetics, let us consider an REV as a domain at the macroscopic level surrounded by constant pressure and temperature conditions. First, suppose there is no mass transfer across the outer boundaries of the domain. With the aid of diffusion, unless a reactant completely dissolves. a reaction will eventually reach a state of equilibrium; this will happen when the rate of disappearance of all reactants equals that of appearance of all products. Also, at equilibrium, any initial (microscopic) concentration gradient within the REV will eventually disappear due to diffusion, so that concentration becomes uniform over the REV. Each reaction has a characteristic time constant that defines how fast it reaches equilibrium. If the concentrations are initially uniform, the time constant of a reaction in an REV is equal to the time constant of the same reaction in a μREV in which the values of the thermodynamic state variables are equal to those within the REV. Otherwise, the two time constants can be different, because the reaction rates will vary over the REV in accordance with the spatially varying concentration. This observation motivates the requirement that concentration gradients within an REV be sufficiently small so that the time constant for a reaction be approximately equal to those inside a μREV under similar conditions.

In a porous medium system, an REV is never a closed system. Mass of fluids and chemical species are continuously transported by diffusion and advection within the REV, in addition to chemical reactions which may occur. Let us consider the case in which mass transfer occurs at the boundary of the REV. There exist time constants associated with changes in species concentrations produced by advective and diffusive fluxes across the outer boundaries of the REV and inside the REV itself. If the time constant for a reaction is much smaller than these time constants, then the reaction is an *equilibrium reaction*; otherwise, it is a *kinetic reaction*.

If all of the reactions in an REV are equilibrium ones, then the REV is said to be in *local chemical equilibrium*. If every REV in a porous medium domain is in chemical equilibrium, then the domain is said to obey the assumption of *local chemical equilibrium at the macroscopic level*.

Given the time required for a reaction to reach equilibrium, we wish to discuss the conditions under which the assumption of equilibrium within an REV—usually referred to as the *Assumption of Local Equilibrium* is justified. Here, we focus on homogeneous reactions.

The use of dimensionless numbers as a tool for comparing terms in a balance equation, which, actually, means comparing processes in a transport problem, is presented in detail in Sect. 3.10. There, in the example of modeling contaminant transport, we introduced dimensionless numbers that are relevant to the discussion here: the Strouhal (St), the Peclet (Pe) and the Damköhler (Dm^I), Dm^{II}) numbers. They are defined for an REV as the domain of interest:

$$\begin{aligned}
\mathrm{St} &\equiv \frac{L_c}{V_c t_c} = \frac{t_{c,\mathrm{adv}}}{t_c},\\
\mathrm{Pe} &\equiv \frac{L_c V_c}{\mathcal{D}} = \frac{t_{c,\mathrm{dif}}}{t_{c,\mathrm{adv}}},\\
\mathrm{Dm}^I &\equiv \frac{L_c/V_c}{t_{c,\mathrm{react}}} = \frac{t_{c,\mathrm{adv}}}{t_{c,\mathrm{react}}},\\
\mathrm{Dm}^{II} &\equiv \frac{L_c^2/\mathcal{D}}{t_{c,\mathrm{react}}} = \frac{t_{c,\mathrm{dif}}}{t_{c,\mathrm{react}}}.
\end{aligned} \tag{7.3.14}$$

where t_c is a characteristic time, e.g., the duration of the period over which the problem is being modeled, the characteristic length, L_c, is the size of the REV, the characteristic fluid velocity, V_c, is the maximum fluid velocity within the REV, and $\mathcal{D}$ is the coefficient of diffusion at the microscopic level.

The *characteristic time of reaction*, $t_{c,\mathrm{react}}$, is determined by a batch reactor experiment. It is defined as the time at which the concentration of an important species in the reaction, $c(t)$, decays to the extent that a characteristic concentration change satisfies:

$$\Delta c_c \approx \left|c(t_{c,\mathrm{react}}) - c(0)\right|. \tag{7.3.15}$$

For a first order reaction, (see (7.3.61)), the characteristic time is $1/k$. This is the time it takes to reach steady state.

We also need the dimensionless diffusion time, $t^\star_{\mathrm{dif}}$, and the dimensionless reaction time, $t^\star_{react}$, defined by:

$$t^\star_{dif} \equiv \frac{L_c^2/\mathcal{D}}{t_c} = \frac{t_{c,dif}}{t_c} = \mathrm{St}\cdot\mathrm{Pe},$$
$$t^\star_{react} \equiv \frac{t_{c,\mathrm{react}}}{t_c} = \frac{\mathrm{St}}{\mathrm{Dm}^I}. \tag{7.3.16}$$

In order to ensure chemical equilibrium, we require that:

- The characteristic time for diffusion must be much smaller than the characteristic time of the considered problem, so that

$$t^\star_{\mathrm{dif}} \ll 1, \quad \text{or, equivalently,} \quad \mathrm{St} \ll \mathrm{Pe}. \tag{7.3.17}$$

- The characteristic time for the reaction be much smaller than that for advection and diffusion, i.e.,

$$\mathrm{Dm}^I \gg 1, \quad \mathrm{Dm}^{II} \gg 1. \tag{7.3.18}$$

- The characteristic time for the reaction be much smaller than the characteristic time of the problem:

$$t^\star_{\mathrm{react}} \ll 1, \quad \text{or, equivalently,} \quad \mathrm{St} \ll \mathrm{Dm}^I. \tag{7.3.19}$$

Altogether, conditions (7.3.17) through (7.3.19) ensure local chemical equilibrium of the reaction within an REV.

The discussion presented above on conditions for chemical equilibrium addressed, primarily, solute transport that occurs at the macroscopic scale. Under field conditions, to which we refer as 'formation scale', the assumption of chemical equilibrium is more complicated.

B. Chemical Reactions

We start by considering reactions at a point in a porous medium domain. Subsequently, we shall combine these reactions with solute transport in a porous medium domain. Examples of chemical reactions in reactors in the chemical industry are presented in App. A.

Chemical reactions were already briefly introduced in Sect. 2.2.6 in connection with Gibbs free energy.

An irreversible chemical reaction at a point in a liquid continuum is described by expressions like:

$$\mathrm{A} \rightarrow \mathrm{B} + \mathrm{C}, \qquad \mathrm{A} + \mathrm{B} \rightarrow \mathrm{AB},$$
$$\mathrm{A} + \mathrm{A} \rightarrow \mathrm{A}_2, \qquad a\mathrm{A} + b\mathrm{B} \rightarrow c\mathrm{C} + d\mathrm{D}, \tag{7.3.20}$$

in which A, B, AB, C, and D are species, or compounds, present in the liquid, and a, b, c, d are positive numbers. Any of the equations in (7.3.20) is referred to as a *stoichiometric equation*. For a reaction proceeding from left to right, the species on the left-hand side are referred to as *reactants*, while those on the right-hand side are called *products*. The last equation describes two reactants and two products. When a reaction is also *reversible*, the symbol $\rightarrow$ is replace by $\rightleftharpoons$.

The general form of a stoichiometric equation is:

$$\nu^{\text{A}}\text{A} + \nu^{\text{B}}\text{B} + \ldots \rightarrow \nu^{\text{P}}\text{P} + \nu^{\text{Q}}\text{Q} + \ldots. \tag{7.3.21}$$

where $\nu^{\text{A}}, \ldots, \nu^{\text{P}}, \ldots$, with $\nu^{\gamma} > 0$, are numbers to be discussed below.

In all reactions, the involved species may be atoms, molecules, free radicals, or ions. In each stoichiometric equation, the same total number of each kind of atom appears on both sides. In a closed batch reactor, products or reactants may disappear as a consequence of processes that occur. This balance enables the calculation of the amounts of every kind of involved atom or compound.

The *reaction rate* expresses the decrease in the concentration of a reactant, or the increase in that of a product, per unit time.

Consider a homogeneous reaction described by:

$$\text{A} + \text{B} \rightarrow \text{C}. \tag{7.3.22}$$

The *reaction rate*, R_r (dims: (moles/unit vol.)/unit time), is defined as:

$$R_r = -\frac{d[\text{A}]}{dt} = -\frac{d[\text{B}]}{dt} = \frac{d[\text{C}]}{dt}, \tag{7.3.23}$$

in which $[\text{A}] \equiv c^{\text{A}}/M^{\text{A}}$ represents the *molar concentration* of A, with M^{A} denoting the molar mass of A. Thus, the *reaction rate expresses the number of moles (produced or disappearing) per unit volume of solution per unit time*. We note that we have defined a *single* rate for the entire reaction.

The derivatives in the above equation have their usual meaning, i.e.:

$$\frac{dc^{\text{A}}}{dt} = \lim_{\Delta t \to 0} \frac{c^{\text{A}}|_{t+\Delta t} - c^{\text{A}}|_t}{\Delta t}. \tag{7.3.24}$$

For a more complicated chemical reaction, e.g., expressed by the stoichiometric equation:

$$\text{A} + 2\text{B} \rightarrow 3\text{C} + \text{D}, \tag{7.3.25}$$

the *reaction rate*, R_r, is given by:

$$R_r = -\frac{d[\text{A}]}{dt} = -\frac{1}{2}\frac{d[\text{B}]}{dt} = \frac{1}{3}\frac{d[\text{C}]}{dt} = \frac{d[\text{D}]}{dt}. \tag{7.3.26}$$

For the general form of the stoichiometric equation (7.3.21), the reaction rate can be expressed as:

$$R_r = -\frac{1}{\nu^{\mathrm{A}}}\frac{d[\mathrm{A}]}{dt} = -\frac{1}{\nu^{\mathrm{B}}}\frac{d[\mathrm{B}]}{dt} = \ldots = \frac{1}{\nu^{\mathrm{P}}}\frac{d[\mathrm{P}]}{dt} = \frac{1}{\nu^{\mathrm{Q}}}\frac{d[\mathrm{Q}]}{dt} = \ldots, \tag{7.3.27}$$

in which the ν^γ, for $\gamma = \mathrm{A},\ \mathrm{B}, \ldots, \mathrm{P},\ \mathrm{Q}, \ldots$, are the *stoichiometric coefficients*. They describe the relative number of moles of each reactant and those of each product that participate in the considered reaction.

The rate of reaction depends on the concentration of the participating species and on the pressure and on the temperature, i.e., $R_r = R_r(p, T, [A], [B], \ldots)$. It can be obtained only experimentally.

Let us introduce here some remarks concerning the source term, $\rho_\alpha \Gamma^\gamma_\alpha$, that appears in the γ-balance equations presented above, e.g., (7.3.4). In general, this source term expresses sources like injection through wells, and added mass of a chemical species by chemical reactions. In what follows, we shall focus on the latter.

When a number of γ-species participate in a reversible chemical reaction, the stoichiometric equation can be written in the compact form:

$$\sum_{(\gamma)} \nu^\gamma \mathcal{M}^\gamma \rightleftharpoons 0, \tag{7.3.28}$$

in which $\mathcal{M}^\gamma$ denotes the chemical symbol for the respective γ-species and ν^γ denotes the corresponding *stoichiometric coefficient*. Following standard convention, $\nu^\gamma < 0$ for a reactant, and $\nu^\gamma > 0$ for a product.

The corresponding *rate of reaction*, R_r, measured as moles per liter per second, is given by:

$$R_r = \frac{1}{\nu^\gamma}\frac{d[\gamma]}{dt}, \tag{7.3.29}$$

in which $[\gamma] \equiv \rho^\gamma / M^\gamma$ represents the molar concentration of the γ-species, with M^γ denoting the molar mass of γ. Thus, R_r expresses the number of moles that are produced, or that disappear, per unit volume of solution per unit time.

In many cases, the considered γ-species participates in a number (j) of chemical reactions that occur simultaneously. In such cases, the rate of production of the mass of γ, per unit mass of the phase, denoted by Γ^γ, is expressed in the form:

$$\rho\Gamma^\gamma = M^\gamma \sum_{(j)} \nu^\gamma_j R_{r,j} \tag{7.3.30}$$

where M^γ denotes the *molar mass* of the γ-component, and ν^γ_k / M^γ is proportional to the *stoichiometric coefficient* appearing with the γ-component in the equation that describes the kth chemical reaction. The above expression for $\rho\Gamma^\gamma$ can now be inserted in the mass balance equation (7.3.4) for the γ-species.

For the *irreversible reaction* (7.3.21), the reaction rate often takes the form:

$$R_r = k[\mathrm{A}]^{\lambda^{\mathrm{A}}}[\mathrm{B}]^{\lambda^{\mathrm{B}}}\ldots, \tag{7.3.31}$$

in which the product is only over the reactant species, and the λ^γ's are powers, which, in general, are not necessarily equal to the ν^γ's. Equation (7.3.31) is an example of a *rate law*. The coefficient k is called the *rate constant* of the reaction. The ratelaw expresses the reaction rate, R_r, determined *experimentally*, as a function of the concentrations of all reactants present in the solution. The reaction expressed by (7.3.31) is said to be λ^{A}-order in A, λ^{B}-order in B, etc. The total order of the rate law is the sum of these exponents. For example, for the reaction:

$$[\mathrm{A}] + [\mathrm{B}] \rightarrow [\mathrm{C}], \tag{7.3.32}$$

we have:

$$R_r \equiv -\frac{d[\mathrm{A}]}{dt} = k[\mathrm{A}][\mathrm{B}]. \tag{7.3.33}$$

A *reversible reaction* like:

$$\mathrm{A} + 2\mathrm{B} \rightleftharpoons \mathrm{C}, \tag{7.3.34}$$

consists of a forward reaction and a reverse one:

$$\mathrm{A} + 2\mathrm{B} \xrightarrow{k_{\mathrm{for}}} \mathrm{C}, \quad \text{and} \quad \mathrm{A} + 2\mathrm{B} \xleftarrow{k_{\mathrm{rev}}} \mathrm{C}. \tag{7.3.35}$$

The rate of a reversible reaction is the difference between the reaction rates of the forward and reverse reactions. The forward reaction rate may be written as:

$$R_{r,\mathrm{for}} = k_{\mathrm{for}}[\mathrm{A}][\mathrm{B}]^2, \tag{7.3.36}$$

and the reverse reaction rate as:

$$R_{r,\mathrm{rev}}\left(= \frac{d[\mathrm{C}]}{dt}\right) = k_{\mathrm{rev}}[\mathrm{C}], \tag{7.3.37}$$

with

$$\frac{d[A]}{dt} = -\frac{1}{2}\frac{d[B]}{dt} = \frac{d[C]}{dt} = R_r.$$

The resulting reaction rate, R_r, is:

$$R_r = R_{r,\mathrm{for}} - R_{r,\mathrm{rev}} = k_{\mathrm{for}}[\mathrm{A}][\mathrm{B}]^2 - k_{\mathrm{rev}}[\mathrm{C}], \tag{7.3.38}$$

i.e., the net rate is the difference between the forward and backward rates. At equilibrium the net rate is zero and the forward and backward rates are non-zero, but equal, i.e., $R_{r.for} = R_{r,rev}$, $R_r = 0$. This leads to:

$$\frac{R_{r,for}}{R_{r,rev}} = \frac{[\mathrm{C}]}{[\mathrm{A}][\mathrm{B}]^2} = K_{eq} = \frac{k_{for}}{k_{rev}}, \tag{7.3.39}$$

where K_{eq} is the *equilibrium constant*, to be further discussed below.

For ideal, or dilute, solutions, where the solvent activity of the solvent can be assumed to be unity, the rate law for an elementary reversible reaction can be shown to be (De Groot and Mazur 1962):

$$R_r = k_{\text{for}} \prod_{(\gamma,\ \nu^\gamma<0)} [\mathrm{X}^\gamma_{\text{react}}]^{-\nu^\gamma} - k_{\text{rev}} \prod_{(\gamma,\ \nu^\gamma>0)} [X^\gamma_{prod}]^{\nu^\gamma}, \tag{7.3.40}$$

where $[X^\gamma]$ denotes the concentration (in this case, in terms of mole fraction) of the (reactant or product) γ-species, and we have followed the standard sign convention for ν mentioned earlier.

Consider the reversible reaction:

$$[\mathrm{A}] \rightleftharpoons [\mathrm{B}],$$

in which both forward and reverse reactions are first-order (see below), but with different constants:

$$\mathrm{A} \xrightarrow{k_f} \mathrm{B}; \qquad R_{rf} = -\left.\frac{d[\mathrm{A}]}{dt}\right|_1 = k_f[\mathrm{A}] = \left.\frac{d[\mathrm{B}]}{dt}\right|_1, \tag{7.3.41}$$

$$\mathrm{A} \xleftarrow{k_r} \mathrm{B}; \qquad R_r = -\left.\frac{d[\mathrm{B}]}{dt}\right|_2 = k_r[\mathrm{B}] = \left.\frac{d[\mathrm{A}]}{dt}\right|_2, \tag{7.3.42}$$

where subscripts $f \equiv for, r \equiv rev$. In this equation, we note how, simultaneously, A is depleted by the forward reaction, at a rate $k_f[\mathrm{A}]$, and produced by the reverse reaction, at a rate $k_r[\mathrm{B}]$. The net rate of production of A is:

$$\frac{d[\mathrm{A}]}{dt} \equiv \left.\frac{d[\mathrm{A}]}{dt}\right|_1 + \left.\frac{d[\mathrm{A}]}{dt}\right|_2 = -k_f[\mathrm{A}] + k_r[\mathrm{B}]. \tag{7.3.43}$$

From:

$$\frac{d[\mathrm{B}]}{dt} \equiv \left.\frac{d[\mathrm{B}]}{dt}\right|_1 + \left.\frac{d[\mathrm{B}]}{dt}\right|_2 = -\frac{d[\mathrm{A}]}{dt}, \tag{7.3.44}$$

it follows that:

$$\frac{\mathrm{d}}{\mathrm{d}t}\Big([\mathrm{A}] + [\mathrm{B}]\Big) = 0, \quad \text{or,} \quad [\mathrm{A}] + [\mathrm{B}] = [\mathrm{A}]\big|_{t=0} + [\mathrm{B}]\big|_{t=0}. \tag{7.3.45}$$

Thus, if $[\mathrm{B}]\big|_{t=0} = 0$, we may integrate (7.3.43) to obtain:

$$[\mathrm{A}] = \left(\frac{k_r + k_f\, e^{-(k_f + k_r)t}}{k_f + k_r} \right) [\mathrm{A}]\big|_{t=0}. \tag{7.3.46}$$

As $t \to \infty$, the concentrations reach their equilibrium values:

$$[\mathrm{A}] = \frac{k_r}{k_f + k_r} [\mathrm{A}]\big|_{t=0}, \qquad [\mathrm{B}] = \frac{k_f}{k_f + k_r} [\mathrm{A}]\big|_{t=0}. \tag{7.3.47}$$

Thus, at equilibrium:

$$k_f [\mathrm{A}] = k_r [\mathrm{B}], \quad \text{and} \quad K_{eq} = \frac{k_f}{k_r}, \tag{7.3.48}$$

where K_{eq} is referred to as the *(thermodynamic) equilibrium constant* of the considered reaction. If K_{eq} is known for a reaction, and one of the rate constants is also known, the other one can be determined by this relationship.

For the reaction:

$$\mathrm{A} + \mathrm{B} \underset{kr}{\overset{k_f}{\rightleftharpoons}} \mathrm{C} + \mathrm{D}, \tag{7.3.49}$$

which is second order in both directions, suppose:

$$\text{Forward reaction}: \quad \left.\frac{d[\mathrm{A}]}{dt}\right|_f = -k_f [\mathrm{A}][\mathrm{B}], \tag{7.3.50}$$

$$\text{Reverse reaction}: \quad \left.\frac{d[\mathrm{A}]}{dt}\right|_r = k_r [\mathrm{C}][\mathrm{D}], \tag{7.3.51}$$

$$\text{Net gain in A}: \quad \frac{d[\mathrm{A}]}{dt} = -k_f [\mathrm{A}][\mathrm{B}] + k_r [\mathrm{C}][\mathrm{D}]. \tag{7.3.52}$$

At equilibrium, the net gain in A vanishes, so that:

$$K_{eq} = \frac{[\mathrm{C}][\mathrm{D}]}{[\mathrm{A}][\mathrm{B}]} = \frac{k_f}{k_r}. \tag{7.3.53}$$

This equation is a special case of the *law of mass action* to be discussed later. The coefficient K_{eq} $(= k_f / k_r)$ is the *equilibrium constant* defined above. The law of mass action is valid for any type of reaction, e.g., the dissolution of minerals, the formation of complexes between dissolved species, and the dissolution of gases in water.

In reality, however, experiments with dissolved species lead to the conclusion that the mass action law is valid only when written in terms of modified values of species concentrations called *activities*. Thus, for the reaction described by the general stoichiometric equation:

$$a\mathrm{A} + b\mathrm{B} + \cdots \rightleftharpoons p\mathrm{P} + q\mathrm{Q} + \cdots, \tag{7.3.54}$$

the *law of mass action* takes the form:

$$K_{eq} = \frac{\{\mathrm{P}\}^p \{\mathrm{Q}\}^q \ldots}{\{\mathrm{A}\}^a \{\mathrm{B}\}^b \ldots}, \tag{7.3.55}$$

in which $\{\gamma\}$, $\gamma = P, Q, A, B$. Denotes the *activity* of the indicated species

In Sect. 2.2.4 we have introduced the chemical potential, of an A-species:

$$\mu^{A} = \mu^{*A} + RT \ln X^{A}, \tag{7.3.56}$$

where μ^{*A} is a function only of p and T. However, the above relationship is valid only for ideal solution (Denbigh 1981, p. 270).

The *activity* $\{A\}$ of a species A, is related to its *molal concentration*, or *molality* $\hat{m}^{\mathrm{A}}$, which is the number of moles of the latter per unit mass (e.g., kg) of solvent (usually, in the context of aqueous solutions), by:

$$\{\mathrm{A}\} = \gamma^{\mathrm{A}} \hat{m}^{\mathrm{A}} \quad \rightarrow \quad \gamma^{\mathrm{A}} = \frac{\{\mathrm{A}\}}{\hat{m}^{\mathrm{A}}}, \tag{7.3.57}$$

where γ^{A} is the dimensionless *activity coefficient* of A. For dilute solutions, $\gamma^{\mathrm{A}} \approx 1$ and $\{\mathrm{A}\} \approx \hat{m}^{\mathrm{A}}$. This definition of the activity coefficient is the standard one used in geochemistry. Note that γ^{A} depends on the standard state selected for the species, and that $\gamma^{\mathrm{A}} \rightarrow 1$ as $\hat{m}^{\mathrm{A}} \rightarrow 0$.

For an ionic aqueous species, the activity coefficient, as defined by (7.3.57), is given by various empirical formulas, e.g., by the *Debye-Hueckel expression* for the activity coefficient, as modified by Helgeson (1969):

$$\log \gamma^{\mathrm{A}} = -\frac{\mathcal{A}\,(z^{\mathrm{A}})^2 \sqrt{I}}{1 + \mathcal{B} r^{\gamma} \sqrt{I}} + \mathcal{C} I, \tag{7.3.58}$$

where z^{A} is the charge on the A-species and r^{γ} denotes the effective diameter of the hydrated ion (in cm). The coefficients $\mathcal{A}$, $\mathcal{B}$, and $\mathcal{C}$ are temperature-dependent constants that are independent of γ. The symbol I denotes the *ionic strength* of the solution, defined by

$$I \equiv \frac{1}{2} \sum_{(\gamma)} \hat{m}^{\gamma} \, (z^{\gamma})^2, \tag{7.3.59}$$

where superscript γ denotes the γth ionic species. This equation gives good agreement with experimental data for ionic strengths up to around 1 molal solution. At higher ionic strengths, more complicated expressions are required (e.g., Pitzer 1979).

When a species participates in several chemical reactions that cause its concentration within a fluid phase to increase (or decrease), we express the strength of the source (= rate of production) of that species in the macroscopic balance equation of the latter, say, (7.3.3):

$$\phi S_\alpha \rho_\alpha \Gamma_\alpha^\gamma = \phi S_\alpha M^\gamma \sum_{(j)} \nu_j^\gamma R_{r,j}, \tag{7.3.60}$$

in which $R_{r,j}$ is the reaction rate of the jth homogeneous chemical reaction in the fluid α-phase and ν_j^γ is the stoichiometric coefficient of the γ-species in the jth reaction. This rate of production is in addition to the rates of production resulting from other sources.

Although we have referred above to the characteristic time of the reaction described by a given stoichiometric equation, often, the actual reaction goes through a number of intermediate steps that do not appear explicitly in the stoichiometric equation and in the corresponding rate law. However, when such an intermediate step is much slower than the one explicitly referred to, it dictates the *rate-determining*, or *rate-limiting step* of the overall reaction.

C. First and Higher Order Reactions

Consider the case:

$$\mathrm{A} \rightarrow \mathrm{B}.$$

The *first-order rate law* for the consumption of a reactant A, is expressed in the form:

$$R_r \equiv -\frac{d[\mathrm{A}]}{dt} = k[\mathrm{A}], \tag{7.3.61}$$

in which k is referred to as the *first-order rate constant* (dims. T^{-1}). An example of a first-order reaction is the radioactive decay of tritiated water, HTO, where T stands for tritium ($\equiv H^3$), to ordinary water, H_2O: HTO $\rightarrow$ H_2O, for which the rate constant is $k = 1.78 \times 10^{-9}\,\mathrm{s}^{-1}$.

By integrating (7.3.61) from $[\mathrm{A}] = [\mathrm{A}_o]$ at $t = 0$, to any time, t, we obtain:

$$[\mathrm{A}](t) = [\mathrm{A}_o]e^{-kt}, \tag{7.3.62}$$

often referred to as the *integrated rate law*. A plot of $\ln([\mathrm{A}](\mathrm{t})/[\mathrm{A}_o])$ versus time, will yield a straight line with a slope $-k$. A larger k indicates a faster rate of disappearance, or decay, of the A-species. Using (7.3.61), we may define the *half-life*, $t_{1/2}$, of the A-species in the considered reaction, i.e., the time in which its concentration will be reduced by a factor 2 is:

$$t_{1/2} = \frac{\ln 2}{k} = \frac{0.693}{k}. \tag{7.3.63}$$

In a first-order reaction, e.g., (7.3.61), the half-life of the reactant is independent of the concentration.

Radioactive and certain other decay phenomena, A $\rightarrow$ Products, may be expressed as the first-order rate law:

$$\frac{d\mathrm{N}}{dt} = -\lambda \mathrm{N}, \tag{7.3.64}$$

in which λ is the *first order rate constant* for the radioactive decay, and N is the number of atoms of the radioactive material. We can also use molar concentrations instead of N. Integrating the above expression from $N = N_o$ at $t = 0$, to any t, gives:

$$N(t) = N_o e^{-\lambda t}. \tag{7.3.65}$$

The half-life is defined by (7.3.63), in which k is replaced by λ. In principle, no equilibrium can be reached until the radioactive material has completely disappeared.

Let us add a few words on higher order reactions. Consider, for example, the case:

$$\mathrm{A} + \mathrm{A} \rightarrow \mathrm{A}_2,$$

with the second order rate of production (of A) law given by:

$$\frac{d[\mathrm{A}]}{dt} = -k'[\mathrm{A}]^2, \tag{7.3.66}$$

in which k' (positive for a product) is referred to as a *second order rate constant* (dims. $\mathrm{M}^{-1}\mathrm{T}^{-1}$). By integration, as above, we obtain:

$$[\mathrm{A}](t) = \frac{[\mathrm{A}_o]}{1 + k'[\mathrm{A}_o]t}. \tag{7.3.67}$$

In this case, the half-life of A is given by:

$$t_{1/2} = \frac{1}{k'[\mathrm{A}_o]}.$$

As a second example, consider the case:

$$\mathrm{A} + \mathrm{B} \rightarrow \mathrm{C},$$

with the *second order rate law* (but first order in the reactants A and B):

$$\frac{d[\mathrm{A}]}{dt} = k''[\mathrm{A}][\mathrm{B}], \tag{7.3.68}$$

in which k'' (dims. $\mathrm{M}^{-1}\mathrm{T}^{-1}$) is a *second-order rate constant*. In this case, integration makes use of the stoichiometry of the reaction. Formally, we may write:

$$t_{1/2} = \frac{\ln 2}{k''[\mathrm{B}]},$$

with $[\mathrm{B}] = [\mathrm{B}]_o$, and $k''[\mathrm{B}]$ referred to as a *pseudo-first-order rate constant*. However, since [B] is a function of time, we cannot determine the half-life, $t_{1/2}$, unless $[\mathrm{B}] \approx$ constant.

When we consider a γ-species in a fluid α-phase (concentration c^γ_α) that undergoes radioactive decay *within a porous medium domain*, the sink (= negative source) term, expressing the rate of disappearance of the species in a macroscopic mass balance equation, say, (7.3.3), is given by:

$$\phi S_\alpha \rho_\alpha \Gamma^\gamma_\alpha = -\phi S_\alpha \lambda c^\gamma_\alpha. \tag{7.3.69}$$

With Γ_s representing the rate of radioactive decay of an adsorbed radioactive species, the source term may be expressed as:

$$\Gamma_s = -\lambda F^\gamma, \tag{7.3.70}$$

where F^γ expresses the mass of the adsorbate per unit mass of solid.

For decay, or degradation of a γ-species in a fluid α-phase, or of an adsorbate, the sink term takes in the form:

$$\phi S_\alpha \rho_\alpha \Gamma^\gamma_\alpha = -\theta_\alpha k^\gamma_\alpha c^\gamma_\alpha, \qquad \rho_b \Gamma^\gamma_s = -\rho_b k^\gamma_s F^\gamma, \tag{7.3.71}$$

where k^γ_α is a *degradation rate constant* for the species in the fluid phase.

D. Catalytic Reactions

A *catalyst* is a substance that enhances the rate of a chemical reaction, i.e., increasing the reaction rate, R_r, although it remains unchanged by the time the reaction has ended. A chemical reaction in which a *catalyst* in involved is called a *catalytic reaction*. Often, a very small quantity of a catalyst is required. The reason for this phenomenon is that with a catalyst, the reaction requires less *activation energy*. This is the kind of reaction that occurs, for example, in catalytic reactors discussed in App. A.

An example of a catalytic process, consider the overall reaction: $P + Q \rightleftharpoons R$. With a catalyst C, it takes the form:

$$\begin{array}{rcl} P + C & \leftrightarrows & PC \\ PC + Q & \leftrightarrows & PQC \\ PQC & \leftrightarrows & RC \\ RC & \leftrightarrows & R + C. \end{array} \tag{7.3.72}$$

An example is the catalytic reaction that produces *ammonia* ($NH_{3(g)}$) from N_2 and $3H_2$:

$$\begin{array}{rcl} N_{2(g)} + \quad 2C & \leftrightarrows & 2NC \\ 3H_{2(g)} & \leftrightarrows & 6H_{(g)} \\ 2NC + \quad 6H_{(g)} & \leftrightarrows & 2NH_3C \\ 2NH_3C & \leftrightarrows & 2NH_{3(g)} + 2C. \end{array} \tag{7.3.73}$$

In both cases, we note the role of the catalyst C. If we add all left hand side and all right hand side, the catalyst C, as well as NH_3C disappear. In App. A, we discuss the catalysis in a catalytic converter.

E. Matrix Representation of Reactions

Consider the case in which NS γ-species participate in NR *independent* (homogeneous) chemical reactions represented in the compact form:

$$\sum_{\gamma=1}^{\mathrm{NS}} \nu_i^\gamma \mathcal{M}^\gamma = 0, \quad i = 1, \ldots, \mathrm{NR}, \tag{7.3.74}$$

or, replacing superscript γ by subscript j:

$$\sum_{j=1}^{\mathrm{NS}} \nu_{ij} \mathcal{M}_j = 0, \quad \begin{cases} i = 1, \ldots, \mathrm{NR}, \\ j = 1, \ldots, \mathrm{NS}, \end{cases} \quad \begin{cases} \nu_{ij} > 0 \text{ for products,} \\ \nu_{ij} < 0 \text{ for reactants,} \\ \nu_{ij} = 0 \text{ for species that} \\ \quad \text{do not participate,} \end{cases} \tag{7.3.75}$$

recalling that $\mathcal{M}_j$ represents the chemical formula of the j's species and ν_{ij} represents the stoichiometric coefficient of the jth species in the ith reaction. Each reaction has its own rate R_r. In the above equation, we may invoke *Einstein's summation convention* and omit the sum symbol. Thus, the left hand side of the above equation is a product of the matrix ν_{ij} by the vector $\mathcal{M}^j$, or $\boldsymbol{\nu} \cdot \boldsymbol{\mathcal{M}} = 0$:

$$\begin{bmatrix} \nu_1^1 & \nu_1^2 & \nu_1^3 & \ldots & \nu_1^{NS} \\ \nu_2^1 & \nu_2^2 & \nu_2^3 & \ldots & \nu_2^{NS} \\ \vdots & \vdots & \vdots & \ddots & \vdots \\ \nu_{NR}^1 & \nu_{NR}^2 & \nu_{NR}^3 & \ldots & \nu_{NR}^{NS} \end{bmatrix} \begin{bmatrix} \mathcal{M}_1 \\ \mathcal{M}_2 \\ \mathcal{M}_3 \\ \vdots \\ \mathcal{M}_{NS} \end{bmatrix} \leftrightharpoons 0, \tag{7.3.76}$$

Note that if a species doesn't take part in a reaction, we just set $\nu_r^\gamma = 0$. We assume that the reactions are linearly independent, i.e., no reaction can be written in terms of a linear combination of other reactions. This condition is equivalent to saying that the rows of the matrix of stoichiometric coefficients: are, in the terminology of linear algebra, *linearly independent*. It is, then, always possible to rewrite the transformed set of reactions, $\mathbf{Q} = \boldsymbol{\lambda} \cdot \mathbf{P}$, in the form:

$$\mathrm{Q}^r \rightleftharpoons \sum_{\gamma=1}^{NC} \lambda_r^\gamma \mathrm{P}^\gamma, \quad r = 1, \ldots, \mathrm{NR}, \quad \gamma = 1, \ldots, \mathrm{NC}, \quad \lambda_r^\gamma = -\sum_{r=1}^{NR} \nu_r^\gamma [\nu_r]^{-1}, \tag{7.3.77}$$

where the P^γ ($\gamma = 1, \ldots, \mathrm{NC}$)'s are the set of NC *primary, or basis, species*, also referred to as *components* (see below), the Q^r's ($r = 1, \ldots, \mathrm{NR}$) are the set of *secondary species*, and (the 2nd rank matrix) λ_r^γ is another kind of stoichiometric coefficients (see the references given by Lichtner (1996)). Note that a distinct secondary species is associated with each reaction, while all other species in the reaction are primary. This equivalent representation of the set of reactions is called a *canonical form* Lichtner 1985). For a given set of reactions, this form is not necessarily unique. It is constructed by first identifying a (non-unique) set of NC = NS − NR primary species from which all other NR species can be expressed through the appropriate reactions.

The canonical form is convenient for cataloging and storing properties of reactions in a database, with the common ionic forms of each element often used as the primary species (e.g., the common ionic forms of iron are Fe^{2+} and Fe^{3+}). All species participating in a reaction can be written in terms of their primary forms; their properties can be stored together with the stoichiometric coefficients for the canonical reaction associated with each of the secondary species. The canonical form is also advantageous in certain numerical models.

An important advantage of the canonical form is that the law of mass action for each reaction takes the form:

$$\{\mathrm{Q}^r\} = \frac{1}{K_r} \prod_{\gamma=1}^{NC} \{\mathrm{P}^\gamma\}^{\lambda_r^\gamma}, \tag{7.3.78}$$

so that the activity (denoted as {..}) of each secondary species is expressed directly in terms of those of the primary species.

F. Temperature Dependence of Reaction Rate

It is found empirically that the rate of many chemical reactions increases with a rise in temperature, following the relationship:

$$k = Ae^{-E_a/RT}, \tag{7.3.79}$$

known as the *Arrhenius equation* (e.g., Lasaga and Kirpatrick 1981). Here, A is a *pre-exponential factor*, and E_a is like an *activation energy* that expresses the minimum energy required for reactants to form products.

G. Heat of Reaction

In Sect. 2.2.6, we have already mentioned that like phase change, a chemical reaction is also associated with energy (expressed as heat or enthalpy) change: heat is absorbed (i.e., taken from the environment), or released (to the environment) during the reaction. Pressure and temperature remain unchanged. The reason is due to the difference in the sum of internal energies of the products and the sum of the internal energies of the reactants:

$$\Delta\mathbb{E}_r = \sum \mathbb{E}_{\text{products}} - \sum \mathbb{E}_{\text{reactants}},$$

in which subscript r indicates the reaction. For example, in terms of molar *enthalpy*, h, we can write:

$$\Delta h_{\text{reaction}} (= \Delta h_r) = h|_{\text{products}} - h|_{\text{reactants}}.$$

where h is measured, for example, in terms of Joules per mole. Heat is released when the energy of the reactants is larger than that of the products. Heat is absorbed when the energy of the products is larger than that of the reactants. For example, for the reaction:

$$H_{2(g)} + \tfrac{1}{2}O_{2(g)} \rightarrow H_2O_{(liq)},$$

with the pressure (= say, 1 atmosphere) remaining constant, and at unchanged 25 °C, heat equivalent to 58 kcal/mole is released. This is the *heat of formation*; it is an *exothermic reaction*. In an *endothermic reaction*, ($\Delta h > 0$) heat is absorbed.

Another example is:

$$\begin{aligned}
CaO_{(s)} &\rightarrow Ca_{(s)} + \tfrac{1}{2}O_{2(g)}, && \Delta H = 151.9 \text{ kcal},\\
CO_{2(g)} &\rightarrow C_{(s)} + O_{2(g)} && \Delta H = 94.1 \text{ kcal}\\
Ca_{(s)} + C_{(s)} + \tfrac{3}{2}O_{2(g)} &\rightarrow CaCO_{2(s)} && \Delta H = -288.5 \text{ kcal},
\end{aligned}$$

\- -

$$CaO_{(s)} + CO_{2(g)} \rightarrow CaCO_{3(s)} \qquad \Delta H = -42.5 \text{ kcal},$$

7.3.4 Transport of Chemically Reacting Species

Let us elaborate on how we handle multiple chemically interacting species.

For convenience, let us introduce the *balance operator* symbol $\mathcal{B}(c)$, defined as:

$$\mathcal{B}(c^\gamma) \equiv \frac{\partial}{\partial t}\theta c^\gamma + \nabla \cdot \theta \left(c^\gamma \mathbf{V} - \mathbf{D}_h \cdot \nabla c^\gamma\right). \tag{7.3.80}$$

It expresses the net rate of accumulation of the γ-species, per unit volume of porous medium, due to interphase transfers and to sources, including those due to chemical reactions. When we consider a model that involves a number of fluid phases, we shall use the mass balance operator symbol, $\mathcal{B}^\gamma_\alpha(c^\gamma_\alpha)$, defined in (7.3.80), in which θ_α and $\mathbf{V}_\alpha$ are the volumetric fraction and the velocity of the α-phase in the mass balance equation of the γ-species in the α-phase.

We note that from the linearity of the $\mathcal{B}$-operator, it follows that:

$$\mathcal{B}([\gamma]) = \frac{1}{M^\gamma}\mathcal{B}(c^\gamma), \tag{7.3.81}$$

where M^γ is the molecular mass of the species. Therefore, $\mathcal{B}([\gamma])$ is, formally, equal to the net rate of accumulation of the γ-species, in number of moles per unit volume of porous medium per unit time.

In general, because $\mathbf{D}_h$ depends on the coefficient of molecular diffusion, the operator $\mathcal{B}$ depends on the considered component. However, in what follows, we shall first assume that $\mathcal{B}$ is independent of γ. After that, we shall consider the more general case of a γ-dependent $\mathcal{B}$, denoting it as $\mathcal{B}^\gamma$.

For the NS interacting species that are present in the fluid phase, we have NS mass balance equations of the form of (3.6.2), with the source term given by (7.3.1). In terms of the $\mathcal{B}(c^\gamma)$, we can express the balance equations for all γ-species present in the pore space in the compact form:

$$\mathcal{B}(c^\gamma) = M^\gamma \sum_{j=1}^{NR'} \nu_j^\gamma R_{\text{pm},\,j}, \quad \gamma = 1, 2, \ldots, \text{NS}, \tag{7.3.82}$$

where $R_{\text{pm},\,j} (\equiv \theta R_{r,j})$ denotes the rate of the jth reaction, expressed in terms of the number of reacting moles per unit volume of porous medium per unit time, with

$$\theta \rho \Gamma^\gamma = M^\gamma \sum_{j=1}^{NR'} \nu_j^\gamma R_{\text{pm},\,j}, \tag{7.3.83}$$

and NR$'$ refers to the total number of homogeneous and heterogeneous reactions.

Or, in terms of the molar concentration, in the form:

$$\mathcal{B}([\gamma]) = \sum_{j=1}^{NR'} \nu_j^\gamma R_{\text{pm},\,j}, \quad \gamma = 1, 2, \ldots, \text{NS}. \tag{7.3.84}$$

In each of the above sets of NS mass balance equations, we have NS molar concentration variables, say $[\gamma]$. If all reaction rates, $R_{\text{pm},\,j}$, are assumed to be known functions of the concentrations, then we have NS equations in NS unknowns. However, often, some or all of the reaction rates are not known. On the other hand, because of the mass action law, not all species' concentrations are independent of each other, and some of the balance equations become redundant.

A. Primary and Secondary Species

In Sect. 1.1.1, we defined *components* as the smallest set of species required to completely define the chemical composition of a phase under equilibrium conditions.

Thus, the primary species may be considered as the components of such a system. Note that the set of components is not unique.

The chemical analysis of an aqueous solution is often reported in terms of the number of moles of the various elements in the system. It is possible to translate the number of moles of each element into the corresponding number of moles of each primary species. As illustrated in the following example, this translation can be easily done by associating a unique primary species with each element.

Example of the Canonical Form:

Consider the set of reactions:

$$\begin{aligned} OH^- + H^+ &\rightleftharpoons H_2O, \\ CO_3^{2-} + H^+ &\rightleftharpoons HCO_3^-, \\ CO_{2(aq)} + H_2O &\rightleftharpoons HCO_3^- + H^+, \\ HSiO_3^- + H^+ &\rightleftharpoons SiO_{2(aq)} + H_2O. \end{aligned}$$

Selecting H^+, H_2O, HCO_3^-, $SiO_{2(aq)}$ as the set of primary species, the canonical form of the reactions is:

$$\begin{aligned} OH^- &\rightleftharpoons -H^+ + H_2O, \\ CO_3^{2-} &\rightleftharpoons -H^+ + HCO_3^-, \\ CO_{2(aq)} &\rightleftharpoons H^+ - H_2O + HCO_3^-, \\ HSiO_3^- &\rightleftharpoons -H^+ + H_2O + SiO_{2(aq)}. \end{aligned}$$

Suppose an experimental analysis of a solution yields the number of moles of the elements: H, O, C, and Si. To convert to the amount of each primary species, we may associate each mole of H^+ with one mole of H, each mole of H_2O with one mole of O, each mole of HCO_3^- with a mole of C, and each mole of $SiO_2(aq)$ with one mole of Si.

An alternative canonical form is obtained by choosing OH^-, H_2O, CO_3^{2-}, $HSiO_3^-$ as the set of primary species. We then obtain:

$$\begin{aligned} H^+ &\rightleftharpoons -OH^- + H_2O, \\ HCO_3^- &\rightleftharpoons -OH^- + H_2O + CO_3^{2-}, \\ CO_{2(aq)} &\rightleftharpoons -2OH^- + H_2O + CO_3^{2-}, \\ SiO_{2(aq)} &\rightleftharpoons HSiO_3^- - OH^-. \end{aligned}$$

We may associate each mole of OH^- with a mole of O, each mole of H_2O with two moles of H, each mole of CO_3^{2-} with a mole of C, and each mole of $HSiO_3^-$ with a mole of Si.

B. Speciation

The procedure for determining the concentrations of all *species* in a given chemical system under equilibrium, given the total amount of all relevant *components* is called *speciation*. Here, we consider speciation within a single-phase solution.

Let $n^{\gamma}_{\text{total}}$ denote the mole fraction of the γ component (= basis species) contained within all species in a solution (i.e., number of γ-moles per mole of solution). Then, the molar balance equations for the components are:

$$n^{\gamma}_{\text{total}} = n^{\gamma} + \sum_{(r)} \lambda^{\gamma}_{r} n^{\prime r}, \quad \gamma = 1, \ldots, \text{NC}, \tag{7.3.85}$$

where n^{γ} is the mole fraction of the γ-component, subscript r runs over all reactions, the $n^{\prime r}$'s are the mole fractions of the secondary species associated with the rth reaction, and the λ^{γ}_{r}'s are the stoichiometric coefficients for the γ-components in the rth reaction. Note that this yields a system of NC equations, where NC is the number of components.

The mole fractions can be written in terms of molar concentrations by the relationships:

$$n^{\sigma} = \hat{m}^{\sigma} / \sum_{(\delta)} \hat{m}^{\delta}, \tag{7.3.86}$$

where the sum is taken over all species. Therefore, (7.3.85) can be viewed as written in terms of molar concentrations, $\hat{m}^{\delta}$, thus constituting the set of unknown variables that needs to be determined.

We also need the system of NR equations given by (7.3.78). It is assumed that the activity of each species is a known function of the $\hat{m}^{\delta}$'s, using, for example, (7.3.57) and (7.3.58), so that (7.3.78) is a system of equations with $\hat{m}^{\delta}$ as the unknown variables. Therefore, (7.3.78) and (7.3.85) together form a system of NC + NR = NS equations in the NS unknowns, $\hat{m}^{\delta}$. This system is nonlinear, and, therefore, it must, usually, be solved by numerical means.

Because, in many cases, the concentrations of certain species can vary by many orders of magnitude, it is, usually, preferable to use the logarithm of concentrations as the unknown variables, in order to avoid an ill-conditioned system of equations (Wolery 1983). In some cases, the set of primary species may have to be changed in order to obtain a suitable set.

PHREEQC (Appelo and Postma 2005; Parkhurst and Appelo 1999) is probably the most commonly used public-domain computer program designed to perform a wide variety of low temperature aqueous geochemical speciation calculations for natural waters of the kind described above. It is based on the ion-association aqueous model described above. It can: (1) perform speciation and saturation-index calculations in a batch-reactor. It can also solve a one-dimensional flow and transport model involving reversible reactions, which include aqueous, mineral, gas, solid-solution, surface-complexation, and ion-exchange equilibria, and irreversible reactions, such as specified mole transfers of reactants, kinetically controlled reactions, mixing of solutions, and temperature changes. In fact, PHREEQC can also perform *inverse modeling*, which finds sets of mineral and gas mole transfers that account for differences in composition between waters, within specified compositional uncertainty limits. PHREEQC version 2, can also perform: kinetically controlled reactions, solid-solution equilibria, fixed-volume gas-phase equilibria, variation of the number of exchange or surface sites in proportion to a mineral or kinetic reactant, diffusion or dispersion in 1-D transport, 1-D transport coupled with diffusion into stagnant zones, and isotope mole balance in inverse modeling.

C. Equilibrium Reactions

We start from the case in which all reactions are in equilibrium, i.e., $\mathrm{NR} = \mathrm{NR}_{\mathrm{eq}}$. In Sect. 7.3.4 A, we saw that any set of independent equilibrium reactions can be transformed into the *canonical* form:

$$\mathrm{Q}^i \rightleftharpoons \sum_{j=1}^{NC} \nu_i^{\mathrm{P}^j} \mathrm{P}^j, \qquad i = 1, \ldots, \mathrm{NR}\,(= \mathrm{NR}_{\mathrm{eq}}), \tag{7.3.87}$$

in which P^j $(j = 1, \ldots, \mathrm{NC})$ is the set of *primary species* (or *components*), Q^i $(i = 1, \ldots, \mathrm{NR}_{\mathrm{eq}})$ is the set of *secondary species*, and $\nu_i^{\mathrm{P}^j}$ denotes the stoichiometric coefficient of the ith canonical reaction associated with the primary species P^j. Here, $\mathrm{NC} = \mathrm{NS} - \mathrm{NR}_{\mathrm{eq}}$ is the number of *components*. When the reactions are written in a canonical form, then the system of balance equations (7.3.84), becomes:

$$\mathcal{B}([Q^i]) = -R_{\mathrm{pm}\,i}, \qquad i = 1, 2, \ldots, \mathrm{NR}_{\mathrm{eq}}, \tag{7.3.88}$$

$$\mathcal{B}([P^j]) = \sum_{i=1}^{NR_{\mathrm{eq}}} \nu_i^{\mathrm{P}^j} R_{\mathrm{pm}\,i}, \quad j = 1, 2, \ldots, \mathrm{NC}. \tag{7.3.89}$$

Substituting the reaction rates, $R_{\mathrm{pm}\,i}$, from (7.3.88) into (7.3.89), gives:

$$\mathcal{B}([P^j]) = -\sum_{i=1}^{NR_{\mathrm{eq}}} \nu_i^{\mathrm{P}^j} \mathcal{B}([Q^i]), \qquad j = 1, 2, \ldots, \mathrm{NC}. \tag{7.3.90}$$

From the linearity of the balance operator, we then obtain:

$$\mathcal{B}([*P^j]) = 0, \qquad j = 1, 2, \ldots, \mathrm{NC}, \tag{7.3.91}$$

where the total concentration (in units of molar concentration) of a primary species, P^j, is defined as:

$$[*P^j] \equiv [P^j] + \sum_{i=1}^{NR_{\mathrm{eq}}} \nu_i^{\mathrm{P}^j} [Q^i], \qquad j = 1, 2, \ldots, \mathrm{NC}. \tag{7.3.92}$$

Note that the total molar concentration $[*P^j]$ can be positive or negative, depending on the sign of the stoichiometric coefficients $\nu_i^{\mathrm{P}j}$, and the relative magnitudes of the primary and secondary species concentrations (Lichtner 1985).

Any given set of reactions written in canonical form can lead to a unique decomposition of every species in terms of the primary species. For example, if CO_3^{2-}, and H^+ are primary species, then one mole of the secondary species H_2CO_3 contains one mole of CO_3^{2-}, and two moles of H^+. In this way, it is possible to consider the total number of moles of a primary species as it exists within all species present in a system. This number, per unit volume of the phase, is defined as the *total concentration of the primary species*. It is common to associate a *chemical element* with a primary species, where the *element* is a constituent of the primary species. If there are N moles of an element per mole of primary species, and the element does not occur in any other primary species, then the total concentration of that element is equal to N times the total concentration of the primary species.

The set of NC balance equations (7.3.91) must be solved in terms of the total concentrations, $[*P^j]$. Note that the equations are decoupled, so that they may be solved individually. Boundary conditions and initial conditions involving the primary species must be expressed in terms of the total concentration, using the mass action law (7.3.78).

For a dilute solution, the mass action law can be written as:

$$[Q^i] = \frac{1}{K'_{\mathrm{eq}\,i}} \prod_{j=1}^{NC} ([P^j]^{\nu^{\mathrm{P}j}}). \tag{7.3.93}$$

Then, substituting this expression into (7.3.92), gives:

$$[*P^j] \equiv [P^j] + \sum_{i=1}^{NR_{\mathrm{eq}}} \left(\frac{\nu_i^{\mathrm{P}j}}{K'_{\mathrm{eq}\,i}} \prod_{k=1}^{NR_{\mathrm{eq}}} ([P^k])^{\nu^{\mathrm{P}k}} \right), \quad j = 1, 2, \ldots, \mathrm{NC}, \tag{7.3.94}$$

which is a function solely of the primary species concentrations. Note that $[* \ldots]$ can be positive, zero, or negative. Thus, in this case, the total concentrations are easily computed from the primary species concentrations. However, in order to obtain the primary species concentrations from the total concentrations, a nonlinear equation must, in general, be solved numerically. Once the primary species are found, then (7.3.93) can be used to obtain the concentrations of the secondary species.

When the dilute solution assumption is not valid, the law of mass action to be used is (7.3.78), which is rewritten here as:

$$\{\mathrm{Q}^i\} = \frac{1}{K_{\mathrm{eq}\,i}} \prod_{j=1}^{NC} \{\mathrm{P}^j\}^{\nu^{\mathrm{P}j}}, \tag{7.3.95}$$

where the {..}'s denote the activities of the species, each of which is usually some nonlinear function of the other species concentrations. Thus, to convert from primary concentrations to total concentrations, as required for the initial and boundary conditions, we need to first find the concentrations of the secondary species in terms the primary species by solving the nonlinear system of equations presented in (7.3.95), and then substitute these concentrations into (7.3.91).

D. Nonequilibrium Reactions

We now consider the case where some of the reactions are *kinetic*, i.e., not in equilibrium (say, $\mathrm{NR_{eq}} < \mathrm{NR}$). Let $\mathrm{NR_{ne}}$ $(= \mathrm{NR} - \mathrm{NR_{eq}})$ represent the number of nonequilibrium reactions. As in the case of equilibrium, we select the primary and secondary species on the basis of the canonical form of the equilibrium reactions. As before, the stoichiometric coefficients are denoted as $\lambda_i^{\mathrm{P}^j}$. The nonequilibrium reactions do not have to be expressed in any special form; they include the usual stoichiometric coefficients, ν_k^γ, for the γ-species participating in the kth nonequilibrium reaction $(k = 1, 2, \ldots, \mathrm{NR_{ne}})$. However, it is often convenient to write them also in canonical forms. If there are no equilibrium reactions, then there are no secondary species, and all species are primary. The resulting balance equations are:

$$\mathcal{B}^{\mathrm{Q}^i}([Q^i]) = -R_{\mathrm{pm}\,i}^{\mathrm{eq}} + \sum_{k=1}^{NR_{\mathrm{ne}}} \nu_k^{\mathrm{Q}^i} R_{\mathrm{pm}\,k}^{\mathrm{ne}}, \qquad i = 1, 2, \ldots, \mathrm{NR_{eq}}, \tag{7.3.96}$$

$$\mathcal{B}^{\mathrm{P}^j}([P^j]) = \sum_{i=1}^{NR_{\mathrm{eq}}} \nu_i^{\mathrm{P}^j} R_{\mathrm{pm}\,i}^{\mathrm{eq}} + \sum_{k=1}^{NR_{\mathrm{ne}}} \nu_k^{\mathrm{P}^j} R_{\mathrm{pm}\,k}^{\mathrm{ne}}, \quad j = 1, 2, \ldots, \mathrm{NC}, \tag{7.3.97}$$

where the number of *components* is given by $\mathrm{NC} \equiv \mathrm{NS} - \mathrm{NR_{eq}}$. Note that now we have allowed the balance operator to be different for different γ-species, so that the coefficient $\mathbf{D}_h$ can depend on the relevant species, say, because of molecular diffusion. Also, some of the species may now be immobile on the solid phase, or they may have different advective velocities, such as in the case of colloids, or in the presence of ion exclusion phenomena. We have also made a distinction between the equilibrium reaction rates, $R_{\mathrm{pm}\,i}^{\mathrm{eq}}$, and the nonequilibrium reaction rates, R_{ki}^{ne}. We assume that the R_{ki}^{ne}'s are known functions of the species concentrations.

By substituting the equilibrium reaction rates appearing in (7.3.96) into (7.3.97), we obtain the system of balance equations:

$$\mathcal{B}^{\mathrm{P}^j}([P^j]) + \sum_{i=1}^{NR_{\mathrm{eq}}} \nu_i^{\mathrm{P}^j} \mathcal{B}^{\mathrm{Q}^i}([Q^i]) = \sum_{k=1}^{NR_{\mathrm{ne}}} a_k^{\mathrm{P}^j} R_{\mathrm{pm}\,k}^{\mathrm{ne}}, \quad j = 1, 2, \ldots, \mathrm{NC}, \tag{7.3.98}$$

where:

$$a_k^{\mathrm{P}^j} \equiv \nu_k^{\mathrm{P}^j} + \sum_{i=1}^{NR_{\mathrm{eq}}} \lambda_i^{\mathrm{P}^j} \nu_k^{\mathrm{Q}^i}. \tag{7.3.99}$$

This set of NC partial differential equations, combined with the NR_{eq} algebraic equations (7.3.95), gives a total of NS equations in the NS unknown species concentrations, $[\gamma]$. This type of coupled equations is called a system of *algebraic-differential equations* (ADE) (although the term is most often used in the context of ordinary, not partial, differential equations).

We may express (7.3.98) by using the total concentrations as the unknowns, as long as the $\mathcal{B}^{\gamma}$-operators are independent of γ. However, this may not be advantageous whenever the nonequilibrium rates, R_{ki}^{ne}, are functions of the species concentrations, instead of the total concentrations. In some cases, it may be better to simply regard the species concentrations as the unknown variables. If the solution is dilute, the mass action law, (7.3.93), can be used to eliminate the secondary species concentrations, reducing (7.3.98) to a smaller set of NC balance equations, with NC primary species concentrations, $\hat{c}^{\text{P}^j}$,

The system of equations described above can be quite nonlinear and the involved concentrations may vary over many orders of magnitude, causing the system to be numerically ill-conditioned. To overcome this difficulty, the logarithm of the concentrations are often used as the unknown variables instead of the concentrations themselves, similar to what is done in dealing with speciation (Sect. 7.3.4 B). Also, whenever the concentration of a primary species becomes very small, the set of primary variables may have to be changed.

We emphasize that (7.3.98) may include equilibrium solid-fluid reactions through the corresponding mass action law (7.3.95), and nonequilibrium reactions through appropriate mass balance equations and reaction rates. We also recall that we have referred to the exchange of a species between two adjacent phases as a 'heterogenous reaction'; it may occur under equilibrium or nonequilibrium conditions. It is, thus, included in this analysis. In fact, the method of summing component balance equations to eliminate exchange terms, which we have presented earlier for heterogeneous reactions, is a special case of the more general procedure presented here.

Let us demonstrate the above procedure, consider the following examples.

Example 1 Consider a case of saturated flow, with two chemical species, A and B (NS = 2), which participate in a single fast chemical reversible reaction $\text{A} \rightleftharpoons \text{B}$ (NR = 1). We assume that the system is continuously in equilibrium ($\text{NR}_{\text{eq}} = 1$). The reaction is already written in its canonical form:

$$\text{A} \rightarrow \text{B}, \qquad \lambda^{\text{B}} = 1, \tag{7.3.100}$$

where A is a secondary species and B is the primary species. The two mass balance equations are:

$$\begin{aligned} \mathcal{B}([A]) &= -R_{\text{pm}}, \\ \mathcal{B}([B]) &= \lambda^{\text{B}} R_{\text{pm}} = R_{\text{pm}}, \quad \text{as} \quad \lambda^{\text{B}} = 1, \end{aligned} \tag{7.3.101}$$

leading to the balance equation:

$$\mathcal{B}([A]) + [B]) \equiv \mathcal{B}([B]) = 0. \tag{7.3.102}$$

The component with total concentration $[*B]$ is a conservative one. No sources appear in its balance equation.

We now solve the last equation for $[*B] = [*B](\mathbf{x}, t)$ within the considered domain, subject to appropriate initial and boundary conditions on $[*B]$. Since we have assumed equilibrium conditions, we use the mass action law for a dilute solution:

$$K'_{eq} = \frac{[B]}{[A]}, \tag{7.3.103}$$

to solve *algebraically* for $[A]$ and $[B]$:

$$[A] = \frac{1}{1 + K'_{eq}}[*B], \qquad [B] = \frac{K'_{eq}}{1 + K'_{eq}}[*B]. \tag{7.3.104}$$

The number of degrees of freedom is presented in Sect. 3.9. According to the phase rule, we note that in this problem, we have two chemical species, but only one chemical degree of freedom (for which a PDE has to be solved).

Example 2 We return to the first example of $A \rightleftharpoons B$, but now the chemical reaction is slow, i.e., the system is under nonequilibrium conditions. For this case, $R_{pm} = \theta f'([A], [B])$. For example, $f'([A], [B]) = k[A][B]$.

Again, with $\nu^A = -1$, $\nu^B = 1$, the two balance equations are:

$$\begin{aligned} \mathcal{B}([A]) &= \nu^A R_{pm} = -\theta f'([A], [B]), \\ \mathcal{B}([B]) &= \nu^B R_{pm} = \theta f'([A], [B]). \end{aligned} \tag{7.3.105}$$

By eliminating $f'([A], [B])$, we obtain the PDE:

$$\mathcal{B}\left([A] + [B]\right) \equiv \mathcal{B}([*B]) = 0. \tag{7.3.106}$$

We solve this *homogeneous* PDE, subject to appropriate initial and boundary conditions. We obtain the solution in the form of $[*..] = [*B](\mathbf{x}, t)$. We then have to solve another, this time *inhomogeneous* PDE:

$$\mathcal{B}\left([B]\right) = \theta f'\left([*B] - [B], [B]\right), \tag{7.3.107}$$

for $[B]$. Finally, we compute $[A](\mathbf{x}, t)$ by using $[A] = [*B] - [B]$.

In Example 1, the number of chemical degrees of freedom was one. Here it is two, since we cannot use the *law of mass action*. Therefore, we need to solve for two unknowns, $[*B]$ and $[B]$.

Example 3 Another example of equilibrium reactions is the case of the *carbonate* system. This system contains NS = 7 chemical species:

$$H_2O,\ OH^-,\ H^+,\ H_2CO_3,\ HCO_3^-,\ CO_3^{2-},\ CO_{2(aq)},$$

which participate in the following $NR_{eq} = 4$ equilibrium chemical reactions:

$$\begin{array}{lll} H_2O & \rightleftharpoons H^+ & + OH^-, \\ H_2CO_3 & \rightleftharpoons HCO_3^- & + H^+, \\ HCO_3^- & \rightleftharpoons CO_3^{2-} & + H^+, \\ CO_{2(aq)} + H_2O & \rightleftharpoons H_2CO_3. & \end{array} \tag{7.3.108}$$

Thus, the number of components is $NC = NS - NR_{eq} = 3$.

The carbonate system is important, because, in natural systems, it includes some of the important reactions affecting the pH ($\equiv -\log a^{H^+}$), which has a major effect on other aqueous and mineral reactions. Note that in the above example, we have ignored carbonate complexes involving cations such as $MgCO_3$, $CaCO_3$, $CaHCO_3^+$, and $MgHCO_3^+$. Also note that $CO_2(aq)$ and H_2CO_3 are usually treated as equivalent species.

For the sake of illustration, let us select the species:

$$P^1{=}H^+,\ P^2{=}H_2CO_3,\ P^3{=}H_2O,$$

as the primary species, and:

$$Q^1{=}OH^-,\ Q^2{=}HCO_3^-,\ Q^3{=}CO_3^{2-},\ Q^4{=}CO_{2(aq)},$$

as secondary species.

The corresponding reactions, written in canonical form, are given by:

$$\begin{array}{lll} R_1: & OH^- & \rightleftharpoons H_2O - H^+, \\ R_2: & HCO_3^- & \rightleftharpoons H_2CO_3 - H^+, \\ R_3: & CO_3^{2-} & \rightleftharpoons HCO_3^- - H^+ = H_2CO_3 - 2H^+, \\ R_4: & CO_{2(aq)} & \rightleftharpoons H_2CO_3 - H_2O, \end{array} \tag{7.3.109}$$

with the corresponding stoichiometric coefficients given by:

$$\left[\lambda_i^{Q^j}\right] = \begin{bmatrix} -1 & 0 & 1 \\ -1 & 1 & 0 \\ -2 & 1 & 0 \\ 0 & 1 & -1 \end{bmatrix}. \tag{7.3.110}$$

Note how in the above equation, every primary species reaction has coefficients that multiply every secondary species, Q_j; the coefficients λ^P of the secondary species are all equal to 1 by the definition of the canonical formulation.

The species H_2CO_3 differs from the species CO_2 by a single water (H_2O) molecule and, therefore, these species are considered equivalent since they only differ by hydration and one of them can be omitted (usually H_2CO_3). In this context, it would be useful to point out that all species present in an aqueous solution are hydrated, i.e., surrounded by some number of loosely bound water molecules.

The mass balance equations for the four secondary species are:

$$\mathcal{B}([Q^1]) = \mathcal{B}([OH^-]) \quad = -R_{\text{pm}\,1}, \tag{7.3.111}$$

$$\mathcal{B}([Q^2]) = \mathcal{B}([HCO_3^-]) \quad = -R_{\text{pm}\,2}, \tag{7.3.112}$$

$$\mathcal{B}([Q^3]) = \mathcal{B}([CO_3^{2-}]) \quad = -R_{\text{pm}\,3}, \tag{7.3.113}$$

$$\mathcal{B}([Q^4]) = \mathcal{B}([CO_{2(aq)|}]) = -R_{\text{pm}\,4}. \tag{7.3.114}$$

The mass balance equations for the three primary species are given by:

$$\begin{bmatrix} \mathcal{B}([P^1]) \\ \mathcal{B}([P^2]) \\ \mathcal{B}([P^3]) \end{bmatrix} = \begin{bmatrix} -1 & -1 & -2 & 0 \\ 0 & 1 & 1 & 1 \\ 1 & 0 & 0 & -1 \end{bmatrix} \begin{bmatrix} R_{\text{pm}\,1} \\ R_{\text{pm}\,2} \\ R_{\text{pm}\,3} \\ R_{\text{pm}\,4} \end{bmatrix}, \tag{7.3.115}$$

where the middle matrix in the above equation is the transpose of the one in (7.3.110). Thus, rewriting (7.3.115) as individual equations, we obtain:

$$\mathcal{B}([P^1]) \equiv \mathcal{B}([H^+]) \quad = -R_{\text{pm}\,1} - R_{\text{pm}\,2} - 2R_{\text{pm}\,3}, \tag{7.3.116}$$

$$\mathcal{B}([P^2]) \equiv \mathcal{B}([H_2CO_3]) = R_{\text{pm}\,2} + R_{\text{pm}\,3} + R_{\text{pm}\,4}, \tag{7.3.117}$$

$$\mathcal{B}([P^3]) \equiv \mathcal{B}([H_2O]) \quad = R_{\text{pm}\,1} - R_{\text{pm}\,4}. \tag{7.3.118}$$

Substituting (7.3.111) through (7.3.114) into the right-hand sides of the above equations, and from the definition of total concentrations, we obtain:

$$\begin{aligned} \mathcal{B}([*H^+]) = 0, \quad & [*H^+] \equiv [H^+] - [Q^1] - [Q^2] - 2[Q^3], \\ \mathcal{B}([*H_2CO_3]) = 0, \quad & [*H_2CO_3] \equiv [H_2CO_3] + [Q^2] + [Q^3] + [Q^4], \\ \mathcal{B}([*H_2O]) = 0, \quad & [*H_2O] \equiv [H_2O] + [Q^1] - [Q^4], \end{aligned} \tag{7.3.119}$$

in which the meaning of the $*$-symbol is explained in (7.3.94). There is a one-to-one correspondence between the number of moles of carbon and the number of moles of the primary species H_2CO_3. Thus:

$$[*H_2CO_3] \equiv [H_2CO_3] + [HCO_3^-] + [CO_3^{2-}] + [CO_{2(aq)}] \tag{7.3.120}$$

is the total number of moles of carbon per unit volume of aqueous phase. The negative of the total concentration $[H^+]$ is the *total alkalinity* of the system. It is defined as the equivalent amount of a base that is titratable with a strong acid.

The results of laboratory analyses of groundwater samples are often reported in terms of the total concentrations of chemical elements. Hence, in the model that represents the groundwater transport problem, the initial conditions will likely be presented also in terms of these units.

When the initial conditions are given in terms of the concentrations of the primary species, instead of total concentrations, we need to determine also the secondary species concentrations, in order to determine simultaneously the total concentrations. When the solution is diluted, we may use the following *law of mass action* for the considered reactions:

$$[OH^-] = \frac{[H_2O]}{[H^+]K'_{eq1}}, \qquad [HCO_3^-] = \frac{[H_2CO_3]}{[H^+]K'_{eq2}}, \tag{7.3.121}$$

$$[CO_3^{2-}] = \frac{[H_2CO_3]}{([H^+])^2 K'_{eq3}}, \quad [CO_{2(\mathrm{aq})}] = \frac{[H_2CO_3]}{[H_2O]K'_{eq4}}, \tag{7.3.122}$$

in which $[H_2O] \approx 1$, and the K'_{eq}s are known equilibrium coefficients.

Often, the quantity of water (H_2O) involved in the reactions is in excess, so that the balance equation for water is not needed. Then, the number of components is reduced from three to two, NC = 2, and the remaining two balance equations are:

$$\mathcal{B}([H^+]) = 0, \quad \text{and} \quad \mathcal{B}([H_2CO_3]) = 0. \tag{7.3.123}$$

Note that in the mass action equation the activity of H_2O is used, with the activity of H_2O approximately 1.

Example 4 Here, we wish to consider an example with cation exchange (Sect. 7.4.2) and two phases: an aqueous solution and a solid (Kinzelbach 1992). The example is the same as Example 3 above, except for the *cation exchange*. The species in the aqueous phase are mobile, while those on the solid are immobile. The species in solution participate in the same chemical reactions as in Example 3, where we have assumed that the system is continuously under conditions of equilibrium. In addition to the species in Example 3, we have the following cations in the aqueous solution:

$$\mathrm{Ca}^{2+}, \mathrm{Mg}^{2+}, \mathrm{Na}^+, \mathrm{K}^+,$$

and their counterparts adsorbed on the solid surface:

$$\mathrm{Ca}^{2+}_{(s)}, \mathrm{Mg}^{2+}_{(s)}, \mathrm{Na}^+_{(s)}, \mathrm{K}^+_{(s)}.$$

Altogether, we have NS = 15 chemical species.

The ion exchange reactions are:

$$\begin{aligned}
\mathrm{Mg}^{2+}_{(s)} + \mathrm{Ca}^{2+} &\rightleftharpoons \mathrm{Mg}^{2+} + \mathrm{Ca}^{2+}_{(s)}, \\
2\mathrm{Na}^+_{(s)} + \mathrm{Ca}^{2+} &\rightleftharpoons 2\mathrm{Na}^+ + \mathrm{Ca}^{2+}_{(s)}, \\
2\mathrm{K}^+_{(s)} + \mathrm{Ca}^{2+} &\rightleftharpoons 2\mathrm{K}^+ + \mathrm{Ca}^{2+}_{(s)},
\end{aligned} \tag{7.3.124}$$

The entire set of reactions, written in canonical form, is:

$$
\begin{array}{llcl}
R_1: & OH^- & \rightleftharpoons & H_2O - H^+, \\
R_2: & HCO_3^- & \rightleftharpoons & H_2CO_3 - H^+, \\
R_3: & CO_3^{2-} & \rightleftharpoons & H_2CO_3 - 2H^+, \\
R_4: & CO_{2(aq)} & \rightleftharpoons & H_2CO_3 - H_2O, \\
R_5: & Mg^{2+}_{(s)} & \rightleftharpoons & Mg^{2+} + Ca^{2+}_{(s)} - Ca^{2+}, \\
R_6: & Na^{+}_{(s)} & \rightleftharpoons & Na^{+} + \frac{1}{2}Ca^{2+}_{(s)} - \frac{1}{2}Ca^{2+}, \\
R_7: & K^{+}_{(s)} & \rightleftharpoons & K^{+} + \frac{1}{2}Ca^{2+}_{(s)} - \frac{1}{2}Ca^{2+},
\end{array}
\tag{7.3.125}
$$

where we selected the following primary species:

$$
\begin{gathered}
P^1 = H^+, P^2 = H_2CO_3, P^3 = H_2O, \\
P^4 = Mg^{2+}, P^5 = Na^+, P^6 = K^+, P^7 = Ca^{2+}, P^8 = Ca^{2+}_{(s)}.
\end{gathered}
$$

The secondary species are:

$$
\begin{gathered}
Q^1 = OH^-, Q^2 = HCO_3^-, Q^3 = CO_3^{2-}, Q^4 = CO_{2(aq)}, \\
Q^5 = Mg^{2+}_{(s)}, Q^6 = Na^+_{(s)}, Q^7 = K^+_{(s)}.
\end{gathered}
$$

The resulting stoichiometric matrix for the primary species, obtained from the canonical form of the reactions, is:

$$
\left[\lambda_i^{Q^j}\right] = \begin{bmatrix}
-1 & 0 & 1 & 0 & 0 & 0 & 0 & 0 \\
-1 & 1 & 0 & 0 & 0 & 0 & 0 & 0 \\
-2 & 1 & 0 & 0 & 0 & 0 & v0 & 0 \\
0 & 1 & -1 & 0 & 0 & 0 & 0 & v0 \\
0 & 0 & 0 & 1 & 0 & 0 & -1 & 1 \\
0 & 0 & 0 & 0 & 1 & 0 & -\frac{1}{2} & \frac{1}{2} \\
0 & 0 & 0 & 0 & 0 & 1 & -\frac{1}{2} & \frac{1}{2}
\end{bmatrix}.
\tag{7.3.126}
$$

The balance equations for the secondary species, which have not already been given in Example 3, are:

$$
\begin{aligned}
\mathcal{B}_{\text{ads}}([Q^5]) \equiv \mathcal{B}_{\text{ads}}([Mg^{2+}_{(s)}]) &= -R_{\text{pm}\,5} = f^{\text{Mg}^{2+}}_{\ell\to s}, \\
\mathcal{B}_{\text{ads}}([Q^6]) \equiv \mathcal{B}_{\text{ads}}([Na^{+}_{(s)}]) &= -R_{\text{pm}\,6} = f^{\text{Na}^{+}}_{\ell\to s}, \\
\mathcal{B}_{\text{ads}}([Q^7]) \equiv \mathcal{B}_{\text{ads}}([K^{+}_{(s)}]) &= -R_{\text{pm}\,7} = f^{\text{K}^{+}}_{\ell\to s}.
\end{aligned}
\tag{7.3.127}
$$

In the above equations, the molar concentration, [...], of an adsorbed cation species is defined as the number of moles attached to the solid per unit surface area of the solid; the mass balance equation operator is defined by (7.3.80).

By taking the transpose of the stoichiometric matrix, we obtain the balance equations for the primary species, in the form:

$$
\begin{aligned}
\mathcal{B}_\ell([P^4]) \equiv \quad \mathcal{B}_\ell([Mg^{2+}]) &= R_{\mathrm{pm}\,5} = -f_{\ell\to s}^{\mathrm{Mg}^{+2}}\,. \\
\mathcal{B}_\ell([P^5]) \equiv \quad \mathcal{B}_\ell([Na^{+}]) &= R_{\mathrm{pm}\,6} = -f_{\ell\to s}^{\mathrm{Na}^{+}}, \\
\mathcal{B}_\ell([P^6]) \equiv \quad \mathcal{B}_\ell([K^{+}]) &= R_{\mathrm{pm}\,7} = -f_{\ell\to s}^{\mathrm{K}^{+}}, \\
\mathcal{B}_\ell([P^7]) \equiv \quad \mathcal{B}_\ell([Ca^{2+}]) &= R_{\mathrm{pm}\,5} + \tfrac{1}{2}R_{\mathrm{pm}\,6} + \tfrac{1}{2}R_{\mathrm{pm}\,7}, \\
&= -f_{\ell\to s}^{\mathrm{Mg}^{2+}} - \tfrac{1}{2}f_{\ell\to s}^{\mathrm{Na}^{+}} - \tfrac{1}{2}f_{\ell\to s}^{\mathrm{K}^{+}}, \\
\mathcal{B}_{\mathrm{ads}}([P^8]) \equiv \mathcal{B}_{\mathrm{ads}}([Ca_{(s)}^{2+}]) &= -R_{\mathrm{pm}\,5} - \tfrac{1}{2}R_{\mathrm{pm}\,6} - \tfrac{1}{2}R_{\mathrm{pm}\,7}, \\
&= +f_{\ell\to s}^{\mathrm{Mg}^{2+}} + \tfrac{1}{2}f_{\ell\to s}^{\mathrm{Na}^{+}} + \tfrac{1}{2}f_{\ell\to s}^{\mathrm{K}^{+}}.
\end{aligned} \tag{7.3.128}
$$

As before, we eliminate the balance equations for the secondary species to obtain the balance equations for the primary species:

$$
\begin{aligned}
\mathcal{B}_\ell([Mg^{2+}]) + \mathcal{B}_{\mathrm{ads}}([Mg_{(s)}^{2+}]) &= 0, \\
\mathcal{B}_\ell([Na^{+}]) + \mathcal{B}_{\mathrm{ads}}([Na_{(s)}^{+}]) &= 0, \\
\mathcal{B}_\ell([K^{+}]) + \mathcal{B}_{\mathrm{ads}}([K_{(s)}^{+}]) &= 0, \\
\mathcal{B}_\ell([Ca^{2+}]) + \mathcal{B}_{\mathrm{ads}}([Ca_{(s)}^{2+}]) &= 0, \\
\mathcal{B}_{\mathrm{ads}}([*Ca_{(s)}^{2+}]) &= 0.
\end{aligned} \tag{7.3.129}
$$

In the above equations, the concentration:

$$
[Ca_{(s)}^{2+}] \equiv [Ca_{(s)}^{2+}] + [Mg_{(s)}^{+}] + \frac{1}{2}[Na_{(s)}^{+}] + \frac{1}{2}[K_{(s)}^{+}] \tag{7.3.130}
$$

may be interpreted as the total amount of cations on the solid surface, in equivalent moles of a *doubly-charged cation* per unit volume of porous medium. This concentration is equal to twice the *cation exchange capacity* measured in equivalents, using a solution with a *singly-charged cation*.

We have four balance equations, in addition to the ones in Example 3, for a total of eight balance equations. There are eleven unknowns:

$$
\begin{aligned}
&[*H^{+}], [*H_2CO_3], [*H_2O], \\
&[Ca^{2+}], [Mg^{2+}], [Na^{+}], [K^{+}], [Ca_{(s)}^{2+}], [Mg_{(s)}^{2+}], [Na_{(s)}^{+}], [K_{(s)}^{+}].
\end{aligned} \tag{7.3.131}
$$

The last three unknown concentrations can be expressed in terms of the other ones, using the three equilibrium conditions for the cation exchange reactions:

$$
[Mg_{(s)}^{+2}] = \frac{[Mg^{+2}][Ca_{(s)}^{+2}]}{K'_{\mathrm{Ca/Mg}}[Ca^{+2}]}, \qquad \left([Na_{(s)}^{+}]\right)^2 = \frac{\left([Na^{+}]\right)^2[Ca_{(s)}^{+2}]}{K'_{\mathrm{Ca/Na}}[Ca^{+2}]}, \tag{7.3.132}
$$

$$
\left([K_{(s)}^{+}]\right)^2 = \frac{\left([K^{+}]\right)^2[Ca_{(s)}^{+2}]}{K'_{\mathrm{Ca/K}}[Ca^{+2}]}, \tag{7.3.133}
$$

where $K'_{\text{Ca/Mg}}$, $K'_{\text{Ca/Na}}$, and $K'_{\text{Ca/K}}$ are known selectivity coefficients (see any chemistry textbook, e.g., Schwarzenbach et al. 2002, and Sparks 2003) written here for molar concentrations.

By inserting these expressions into the balance equations, we obtain a set of eight equations in eight unknowns. The above equilibrium relationships assume a dilute solution. The concentrations need to be replaced by their respective activities, and a system of nonlinear equations for the mass action laws must be solved, together with the balance equations.

Example 5 This example is similar to Example 3, except that we introduce an additional species, CO_2(g) that can participate in a non-equilibrium heterogeneous reaction— its dissolution in the aqueous phase. We also have a mineral called *calcite*, $CaCO_3$(s), which can dissolve in the aqueous phase as a non-equilibrium reaction. Precipitation of the mineral is also possible. This chemical system is called the *calcium carbonate* system.

So far, we have considered the solid matrix as consisting of a single inert solid phase. In fact, for reacting minerals, the solid matrix may be regarded as comprised of several phases. In this example, the solid matrix consists partly of the calcite mineral (subscript $calcite$) and partly of a non-reactive solid (subscript $inert$). The volumetric fractions of these phases will be denoted by θ_{calcite} and θ_{inert}, respectively. It is assumed that θ_{inert} is known. Note that the porosity in this model varies with time as it is related to θ_{calcite} through:

$$1 - \phi = \theta_{\text{calcite}} + \theta_{\text{inert}}, \qquad \Delta\phi = -\Delta\theta_{\text{calcite}}. \tag{7.3.134}$$

We select the following species as primary:

$$\text{P}^1 = \text{H}^+, \text{P}^2 = \text{H}_2\text{CO}_3, \text{P}^3 = \text{H}_2\text{O}, \text{P}^4 = \text{CO}_{2(g)}, \text{P}^5 = \text{CaCO}_{3(s)},$$

and:

$$\text{Q}^1 = \text{OH}^-, \text{Q}^2 = \text{HCO}_3^-, \text{Q}^3 = \text{CO}_3^{2-}, \text{Q}^4 = \text{CO}_{2(aq)},$$

as secondary. The corresponding reactions, with the equilibrium reactions presented in canonical form, are:

$$\begin{array}{rrcll}
R_1^{\text{eq}}: & \text{OH}^- & \rightleftharpoons & \text{H}_2\text{O} & - \text{H}^+, \\
R_2^{\text{eq}}: & \text{HCO}_3^- & \rightleftharpoons & \text{H}_2\text{CO}_3 & - \text{H}^+, \\
R_3^{\text{eq}}: & \text{CO}_3^{2-} & \rightleftharpoons & \text{H}_2\text{CO}_3 & - 2\text{H}^+, \\
R_4^{\text{eq}}: & \text{CO}_{2(aq)} & \rightleftharpoons & \text{H}_2\text{CO}_3 & - \text{H}_2\text{O}, \\
R_1^{\text{ne}}: & \text{CO}_{2(g)} & \rightleftharpoons & \text{CO}_{2(aq)}, & \\
R_2^{\text{ne}}: & \text{CaCO}_{3(s)} & \rightleftharpoons & \text{Ca}^{2+} & + \text{CO}_3^{2-}.
\end{array} \tag{7.3.135}$$

The non-equilibrium reactions do not have to be presented in canonical form, but they must be consistent with whatever rate law is used.

For non-equilibrium reaction rates, we use the rate law:

$$R^{\text{ne}}_{\text{pm}\,1} = \Sigma_{\ell,g}\alpha^{\star\text{CO}_2}\left\{[CO_{2(g)}] - [CO_{2(\text{aq})}]/K^{'\text{CO}_2}_{\text{eq}}\right\}. \tag{7.3.136}$$

Using the rate law proposed by Aagaard and Helgeson (1983) on the basis of the transition state theory, we obtain:

$$R^{\text{ne}}_{\text{pm}\,2} = \Sigma_{\ell,\text{calcite}}k_{\text{calcite}}\left\{1 - \frac{[Ca^{+2}][CO_3^{2-}]}{K^{'\text{calcite}}_{\text{eq}}}\right\}, \tag{7.3.137}$$

where $K^{'\text{CO}_2}_{\text{eq}}$ and $K^{'\text{calcite}}_{\text{eq}}$ are the equilibrium coefficients for the reactions, $\alpha^{\star\text{CO}_2}$ is a transfer coefficient, and k_{calcite} is a rate constant. There also exist calcite rate laws based on experiment. The specific interfacial areas, $\Sigma_{\ell,g}$ and $\Sigma_{\ell,\text{calcite}}$, are functions of the fluid and solid volumetric fractions.

The balance equations for the secondary species are:

$$\begin{aligned}
\mathcal{B}_\ell([Q^1]) \equiv \;\; \mathcal{B}_\ell([OH^-]) &= -R^{\text{eq}}_{\text{pm}\,1},\\
\mathcal{B}_\ell([Q^2]) \equiv \mathcal{B}_\ell([HCO_3^-]) &= -R^{\text{eq}}_{\text{pm}\,2},\\
\mathcal{B}_\ell([c^{Q^3}]) \equiv \mathcal{B}_\ell([cCO_3^{2-}]) &= -R^{\text{eq}}_{\text{pm}\,3} + R^{\text{ne}}_{\text{pm}\,2},\\
\mathcal{B}_\ell([Q^4]) \equiv \mathcal{B}_\ell([CO_{2(\text{aq})}]) &= -R^{\text{eq}}_{\text{pm}\,4} + R^{\text{ne}}_{\text{pm}\,1}.
\end{aligned} \tag{7.3.138}$$

The balance equations for the primary species are:

$$\begin{aligned}
\mathcal{B}_\ell([P^1]) \equiv \qquad \mathcal{B}_\ell([c^{\text{H}^+}]) &= -R^{\text{eq}}_{\text{pm}\,1} - R^{\text{eq}}_{\text{pm}\,2} - 2R^{\text{eq}}_{\text{pm}\,3},\\
\mathcal{B}_\ell([P^2]) \equiv \quad \mathcal{B}_\ell([H_2CO_3]) &= R^{\text{eq}}_{\text{pm}\,2} + R^{\text{eq}}_{\text{pm}\,3} + R^{\text{eq}}_{\text{pm}\,4},\\
\mathcal{B}_\ell([P^3]) \equiv \qquad \mathcal{B}_\ell([H_2O]) &= R^{\text{eq}}_{\text{pm}\,1} - R^{\text{eq}}_{\text{pm}\,4},\\
\mathcal{B}_g([P^4]) \equiv \quad \mathcal{B}_g([CO_{2(g)}]) &= -R^{\text{ne}}_{\text{pm}\,1},\\
\mathcal{B}_{\text{calcite}}([P^5]) \equiv \mathcal{B}_{\text{calcite}}([CaCO_{3(s)}]) &= -R^{\text{ne}}_{\text{pm}\,2}.
\end{aligned} \tag{7.3.139}$$

The molar concentration of the precipitating species (in this case, calcite) is defined as the number of moles per unit volume of porous medium, with the balance operator for calcite taking the form:

$$\mathcal{B}_{\text{calcite}}([CaCO_{3(s)}]) \equiv \frac{\partial[CaCO_{3(s)}]}{\partial t}. \tag{7.3.140}$$

We solve for the equilibrium reaction rates of the secondary species in the balance equations, and then substitute these rates into the balance equations of the primary species to obtain the following final set of balance equations for each component:

$$\begin{aligned}
\mathcal{B}_\ell([*P^1]) \equiv \quad & \mathcal{B}_\ell([*H^+]) &= -2R^{\text{ne}}_{\text{pm}\,2}, \\
\mathcal{B}_\ell([P^2]) \equiv \quad & \mathcal{B}_\ell([*H_2CO_3]) &= R^{\text{ne}}_{\text{pm}\,1} + R^{\text{ne}}_{\text{pm}\,2}, \\
\mathcal{B}_\ell([*P^3]) \equiv \quad & \mathcal{B}_\ell([*H_2O]) &= -R^{\text{ne}}_{\text{pm}\,1}, \\
\mathcal{B}_g([P^4]) \equiv \quad & \mathcal{B}_g([CO_{2(g)}]) &= -R^{\text{ne}}_{\text{pm}\,1}, \\
\mathcal{B}_s([P^5]) \equiv \quad & \mathcal{B}_s([CaCO_{3(s)}]) &= -R^{\text{ne}}_{\text{pm}\,2}.
\end{aligned} \tag{7.3.141}$$

Altogether, we have here NS $= 9$ species concentrations, NC $= 5$ balance equations, and $\text{NR}_{\text{eq}} = 4$ mass action laws for the equilibrium reactions, i.e., a total of NC + NR_{eq} equations. The mass action laws can be used to eliminate the secondary species concentrations from the balance equations, and we have NC $= 5$ balance equations for the NC $= 5$ primary variables. In some formulations, however, the *total concentration* is solved for as an independent variable. This has the advantage that the accumulation and flux terms in the mass balance equations are linear in the total concentrations, which is useful if no heterogeneous reactions are considered. However, the above is limited to cases where diffusion is independent of the concentration of the ionic species.

From (7.3.134), we obtain the change in porosity in the form $-\Delta\theta_{\text{calcite}}$ $(= -\Delta([CaCO_{3(s)}]M^{\text{CaCO}_{3(s)}}/\rho_{\text{calcite}}))$. When changes in porosity and tortuosity, relative to their initial values are significant, the flow field based on the unaltered values cannot be used in the transport and flow equations must be solved simultaneously as a coupled set of equations.

7.4 Interphase Mass Transfers

Consider two fluid phases in a porous medium domain: two liquids or a liquid and a gas that, together occupy the void-space. In the *macroscopic* mass balance equation for a γ-species in an α-phase, e.g., (7.3.3), the term $f^{\gamma}_{\beta\to\alpha}(= -f^{\gamma}_{\alpha\to\beta})$ expresses the rate of γ-mass transferred from a β-fluid phase to a considered α-phase (or α to β), per unit volume of porous medium, per unit time. We may also encounter $f^{\gamma}_{\alpha\to s}$, expressing the transfer from the α-phase to the solid (s), or $f^{\gamma}_{s\to\alpha}$ for the transfer from solid to fluid. All f-terms express the mass of γ *per unit volume of porous medium* per unit time transferred through the entire area of the interphase boundary surface within the REV around the considered (macroscopic) point. In fact, interphase transfers, of all extensive quantities, across interphase (microscopic) surfaces, are included also in (1.4.71) developed within the framework of the Hassanizadeh and Gray averaging approach described in Sect. 1.4.2C. Whitaker (1999) also presents a species mass balance equation for the interface between two phases.

To obtain an expression for the (macroscopic) $f^{\gamma}_{\alpha\to\beta}$-term, we have first to understand and express what happens at a point on an $\alpha-\beta$-interface at the microscopic level. Following the phenomenological approach, once we understand and express what happens at the microscopic level, say per unit surface area of an interphase boundary, we can easily transform this information to obtain the behavior per unit volume of a porous medium. Thus, by taking into account the specific surface area, i.e., interphase surface area per unit volume of porous medium, we'll obtain the sought value of $f^{\gamma}_{\alpha\rightarrow\beta}$.

Interphase transfer processes are often referred to as *heterogeneous reactions*.

In Sect. 2.2.5 we have considered what happens at the interface between two fluid phases that occupy the void-space. For example, we noted the laws of *Henry* and *Raoult* that govern what happens at a point on such an interface. However, there, we have not discussed the transfer of extensive quantities of the relevant phases across interphase (microscopic) boundaries. This subject is discussed here. We shall start by considering the transport of mass of a chemical species at a (microscopic) point on an interphase surface, but eventually lead to the macroscopic exchange rate, $f^{\gamma}_{\alpha\rightarrow\beta}$.

Several processes γ-mass transfer can occur in a number of forms:

- Transfer of a γ-species from a liquid to a gas and from a liquid to a liquid.
- Adsorption, i.e., the transfer γ from a liquid or a gas to a solid (and desorption).
- Ion exchange between a solvent and the solid, or between a liquid and an adjacent liquid.
- Volatilisation (from a liquid to a gas) and condensation (from a gas to a liquid).
- Dissolution (from a solid to a liquid) and precipitation (from a liquid to a solid). Dissolution may occur also when certain solids are added to a liquid. e.g., sodium chloride added to water:

$$\mathrm{NaCl}_{(s)} = \mathrm{Na}^{+}_{(aq)} + \mathrm{Cl}^{-}_{(aq)}.$$

In what follows, we shall discuss these processes in details, with the objective of leading to an appropriate expression for the f^{γ}-term. Mass transfer is demonstrated in many of the examples presented in Appendix A.

Let $\mathcal{S}_{\alpha,\beta}$ denote the microscopic interface between two phases: α and β. We may use (7.3.15) to define a characteristic reaction time, $t_{c,react}$. However, in this case, the characteristic time also depends on the involved volumes of the phases relative to the interface area. In fact, the characteristic time is proportional to the volume-to-surface ratio $\mathbb{V}_\alpha/S_{\alpha,\beta}\ (=\ell_\alpha)$, where ℓ_α is a characteristic length of the phase-occupied domain. Thus, $\ell_\alpha/t_{c,react}(\equiv\kappa_\alpha)$ is a quantity that is independent of the size of the experimental system and is, therefore, a true characteristic of the reaction ($\equiv$ transfer) at the interface.

Next we consider a porous medium containing the two phases. The (microscopic) characteristic length L_c for the α-phase is equal to ℓ_α. The relevant dimensionless numbers: the Peclet number for the α-phase, Damköhler and Strouhal numbers are presented in Sect. 3.10. They are:

$$(\mathrm{Pe})_\alpha = \frac{(V_\alpha)_c\,\ell_\alpha}{\mathcal{D}_\alpha},$$

$$(\mathrm{Dm}^I)_\alpha = \frac{\ell_\alpha/(V_\alpha)_c}{t_{c,\mathrm{react}}} = \frac{1/(V_\alpha)_c}{1/\kappa_\alpha},$$

$$(\text{St})_\alpha = \frac{\ell_\alpha}{(V_c)_\alpha t_c},$$

$$(\text{Dm}^{II})_\alpha = \frac{\ell_\alpha^2/\mathcal{D}_\alpha}{t_{c,\text{react}}} = \frac{\ell_\alpha/\mathcal{D}_\alpha}{1/\kappa_\alpha}. \tag{7.4.1}$$

Similar expressions apply also to the dimensionless numbers of the β-phase. Therefore, the conditions for equilibrium of a heterogeneous reaction must hold for both phases.

In what follows, we shall discuss these processes and how to express each of them in terms of relevant state variables.

As we shall see, each of the f^γ-transfer expressions involves a *geometrical coefficient* that is associated with the actual microscopic configuration of the relevant surface within the void space, e.g., the solid-fluid surface, or the characteristic distance between the fluid and solid surfaces within the void space. Although they can be estimated, like all macroscopic coefficients related to the geometry of the phases within an REV, these coefficients have to be determined *experimentally*. For manufactured porous media, especially those that have a repetitive microscopic structure, it is possible to *estimate* the values of these coefficients. In two-phase flow, the specific interfacial area will depend on the saturation of the phases involved.

7.4.1 Adsorption

Adsorption and *desorption* are terms used for fluid-to-solid and solid-to-fluid mass transfers, with 'fluid' referring here to both a liquid and a gas. An example for the latter case is the removal of toxic gases, e.g., stack gases like SO_2 into a solvent. This creates a thin film of the *adsorbate* on the surface of the *adsorbent*. We distinguish:

- *Chemisorption* when the adsorbed ions, atoms or molecules, form a chemical bond with atoms or molecules of the adsorbent.
- *Physical adsorption* is when a species is attached to the solid by *weak* physical bonds, like *van der Walas (intermolecular) forces* and hydrogen bonds.

The macroscopic balance equation of a chemical species, say (7.3.1), includes a term ($f^{m^\gamma}_{s\to\alpha}$) that expresses the mass of a γ-species passing from the solid to the fluid, across their common (microscopic) interface, per unit volume of porous medium. This term expresses solid desorption. The opposite, i.e., from the fluid to the solid ($s \to \alpha$) expresses *adsorption*. Thus, essentially, this is a microscopic level that takes place on the internal solid surface of the solid matrix and, when the latter is porous, also on the solid-fluid interface within the solid matrix.

In the discussion below, we shall assume that the solid matrix is composed of a single substance, although, in principle, it may be composed of a mixture of two or more substances. Adsorption occurs at a point on the fluid-solid interface. However, as everywhere else in this book, eventually, we are interested in adsorption at the macroscopic level, i.e., per unit volume of porous medium, assigned to the macroscopic point in the considered domain.

A. The Concept

Adsorption, or *sorption*, the opposite of *desorption*, is the process in which mass of a chemical species dissolved in the liquid, or molecules of a gas (= *adsorbate*) that occupies the void space, or part of it. Adheres to and accumulates on the surface of a solid (= *adsorbent*). This phenomenon can take on two forms:

- *Physical adsorption* in which molecules, say of a gas, or of a liquid, adsorb to the solid surface by *van der Waals forces,* hydrogen bonds, or electrical forces.
- *Chemi-sorption*, where chemical species present in the solid's surface actually interact *chemically* with dissolved chemical species present in the liquid.

An interesting observation here is that the valance of the solid's surface is usually not electrically neutral, as the atoms are built to bond in three dimensions. In principle, we are considering what happens at a solid-fluid interface. Actually, under a powerful microscope, that allows observation at the molecular level, this 'surface' is much more complicated; it allows various types of bonds between ions and molecules on both sides of the assumed smooth solid-fluid interface. In principle, chemi-sorption may also take place between a gaseous phase and a solid.

Switching to the microscopic level, one option is to assume that equilibrium exists at every point on the fluid-solid interface. Primarily, this means equality of the chemical potentials on both sides of the liquid-solid interface. Or, we can express the same relationship by a law which is analogous to Henry's law. Eventually, for (macroscopic level) modeling purposes, we have to translate our understanding of what happens at a point on a (microscopic) solid-fluid interface to a description of the solid-fluid interaction per unit volume of porous medium, i.e., taking into account the specific area of the solid-fluid interface. In earlier subsections, we have been using the symbol $f^{\gamma}_{f\to s}$ to denote the mass of dissolved γ-species that moves from the fluid to be adsorbed on the solid matrix, per unit volume of porous medium.

With the above comments on adsorption, a simple (and common) treatment of *adsorption under equilibrium conditions* makes use of a tool referred to as *adsorption isotherm* introduced in Sect. 2.1. Under such conditions, the amount of adsorbed species on a solid matrix at a point in a porous medium domain, i.e., within an REV, depends on the solid and is solely a function of the concentration of the species in the liquid. This assumption is valid as long as the concentration of all other dissolved species affecting adsorption do not change appreciably in time. In general, however, this condition does not hold and a more complicated analysis is required, involving also the reactions on the solid surface. The term sorption also includes ion exchange and surface complexation. Examples of such reactions are introduced below.

In some adsorption theories (e.g., Weber 1972), the solid is always assumed to be covered by a (wetting) fluid *boundary layer*, or *film*, that has properties and composition different from those of the bulk fluid. Then, the term 'equilibrium' mentioned above means equilibrium between the adsorbed species and the concentration of the species in that film. To obtain a macroscopic description of adsorption, we make certain assumptions, e.g., that because of diffusion and the short distances involved, the average concentration, say in an REV, is the same as that close to the solid, so that we can express the isotherm in terms of average concentration.

To understand what happens at a liquid-solid interface we shall make use of the concept of a'film', or 'boundary layer', discussed in more detail in Sect. 7.4.3. Because advective liquid flow within this boundary layer is negligible, to reach the solid surface, the adsorbate has first to pass from the bulk solution through this layer by *molecular diffusion*. Then, after passing through the boundary layer, the adsorbate can interact with the solid. The desorbed species can return to the bulk solution in a similar way.

We should make a distinction between *adsorption*, as defined above, and *absorption*. The latter term is used when the *solid matrix itself is porous*, albeit with tiny pores, which means a *huge* void-solid internal surface area per unit volume of solid matrix. The terms *micropores* and *macropores* are often used. When permeability inside the saturated solid matrix, or grains, is very low, solute advection with the liquid is not possible. However, species diffusion may still occur. A dissolved species can *diffuse* into and within the (saturated) porous solid matrix, and adsorb on the internal surface within it. We use the term *absorption* to indicate the solute that enters (= absorbed by) the porous solid. A porous medium as described here is often referred to as *a double porosity porous medium*, (Sect. 1.1.7 A). Charcoal, often referred to as 'active carbon' is an example of a solid matrix of this kind. It has a network of interconnected tiny pores, providing a huge surface area for adsorption. A very large mass of a chemical species can adsorb on such surface, per unit volume of porous medium.

Note that the term 'double porosity medium' is used also for fractured rock domains, in which the blocks are porous (Sect. 1.3.6 B). However, the pores in the blocks may be larger, allowing (single or multi-phase) flow in the porous blocks.

In many cases, when considering the rate of adsorption, or the characteristic time involved, the rate determining (or rate limiting) step is not the chemical interaction with the solid, but the diffusion through the film and (in the case of a porous matrix) through the tiny pores within the solid matrix.

When adsorption of a dissolved chemical species takes place in saturated (i.e., single phase) flow, the total mass of a considered species, say within every REV of the porous medium, is *partitioned* between the solution and the (surface of the) solid matrix. Any increase in the quantity of a considered species in the liquid is associated with an appropriate increase in its quantity on the solid, and vice versa. Obviously, there is a limit to the quantity of a chemical species that can adsorb on the solid. In *desorption*, the quantity of the species on the solid decreases; this is associated with an appropriate increase in the species' quantity in solution.

B. Adsorption Isotherm

An *adsorption isotherm* is the expression that relates the quantity of a species adsorbed on the solid to its quantity in the liquid phase that occupies the void space (or part of it), *at a fixed temperature*, under conditions of (chemical) equilibrium between the two quantities.

Let the symbol F^A denote the mass of the A-species (= adsorbate) adsorbed on the solid (= adsorbent), per unit mass of the latter. Note that the concentration F^A may be measured in kg/kg, or in moles/kg, while the concentration in the liquid, c^A, is measured in kg/ℓ, or in moles/ℓ. Although it would seem more natural to define the mass of the species on the solid per *unit surface area of the solid*, the reference to 'unit mass of solid', stems from the way this quantity is measured in the laboratory.

Different adsorbate-adsorbent pairs have different isotherms, stemming from the different mechanisms involved. The isotherm for a given adsorbate-adsorbent pair can be obtained by performing a *batch adsorption experiment*. A fixed amount of porous medium (e.g., soil) is mixed in separate containers with an aqueous solution at different concentrations, and the change in the latter, resulting from adsorption, is recorded as time elapses, until the systems reach equilibrium. By performing a mass balance within each container, the adsorbed quantity is computed, yielding a point on the isotherm. If the time for reaching equilibrium is too long, equilibrium may not be assumed. A review of a large number of adsorption isotherms is provided by Foo and Hameed (2010).

Following are examples of some more commonly used isotherms for a specified γ-species:

- Freundlich (1907) suggested the *nonlinear isothermn*

$$F^{\gamma} = b(c^{\gamma})^n, \tag{7.4.2}$$

where b and the power n are constant coefficients (functions of temperature), and c^{γ} denotes the concentration of the γ-adsorbate in the solution. The case $n < 1$ means that as F^{γ} increases, it becomes more difficult for additional quantities of the adsorbate to be adsorbed. The opposite situation is described by $n > 1$.

- For $n = 1$, and replacing the symbol b by the more commonly used symbol K_d, the relationship (7.4.2) reduces to the *linear adsorption isotherm*:

$$F^{\gamma} = K_d^{\gamma} c^{\gamma}. \tag{7.4.3}$$

The coefficient K_d^{γ}, which expresses the affinity of the γ-species for the solid, relative to that for the liquid (usually for an aqueous phase), is called the *distribution coefficient*, or *partitioning coefficient* of γ in the considered fluid. From (7.4.3), it follows that K_d^{γ} $(\equiv F^{\gamma}/c^{\gamma})$, with $K_d^{\gamma} = K_d(c^{\gamma}, T)$, gives, *at every instant*, the mass of the adsorbed γ-species on the solid, per unit mass of the latter, per unit concentration of that species in the liquid phase. It describes the partitioning of the total amount of the species between the solid surface and the liquid phase, say, in a unit volume of porous medium. We note that K_d^{γ} has the dimensions of γ-mass per unit volume, and should be described by the corresponding units (e.g., kg/ℓ).

It is always possible to rewrite the isotherm in terms of other measures of solute concentration, (e.g., moles/ℓ).

Sometimes, K_d^{γ} for an adsorption process differs from that for the desorption one. This implies that the process is not completely reversible. Another observation is that, often, especially in chemisorption, there exists a limit to the adsorptive capacity of a solid surface. This requires a modification of the isotherm (7.4.3).

In unsaturated (air-water) flow, as the larger pores are occupied by air, part of the solid's surface is less readily accessible to pore water. This may make K_d^γ in (7.4.3) a function of the saturation. On the other hand, we recall that water is usually the *wetting liquid*, and, as such, it is everywhere adjacent to the solid surface, albeit as a very thin film, with diffusion of chemical species through it. Hence, we may conclude that (7.4.3) is valid also in unsaturated flow, unless the moisture content is very low, a situation that, under certain conditions, may occur, for example, close to ground surface.

• Langmuir (1915, 1918) suggested the nonlinear equilibrium isotherm:

$$F^\gamma = \frac{k_3 c^\gamma}{1 + k_4 c^\gamma}, \quad k_3, k_4 = \text{constant coefficients.} \tag{7.4.4}$$

Note that $F^\gamma \to \text{const.}$ when $k_4 c^\gamma \gg 1$.

• Lindstrom et al. (1971) and Van Genuchten (1974), present the nonlinear isotherm:

$$F^\gamma = k_5\, c^\gamma \exp(-2k_6 F^\gamma), \quad k_5, k_6 = \text{constant coefficients.} \tag{7.4.5}$$

For $2k_6 F^\gamma \ll 1$, we have $F^\gamma \simeq k_5 c^\gamma/(1 + 2k_5 k_6 c^\gamma)$ and the isotherm reduces to Langmuir one.

Some of the above isotherms can be expressed as $F^\gamma = K_d(c^\gamma)c^\gamma$. Thus, for any γ-species, we have:

$$\frac{m^\gamma\big|_{\text{in the fluid(w)}}}{m^\gamma\big|_{\text{on the solid(s)}}} = \frac{c^\gamma\, \mathbb{V}_w}{F^\gamma\, m_s} = \frac{1}{\frac{\rho_b K_d^\gamma(c^\gamma)}{\phi}}, \tag{7.4.6}$$

$$\frac{m^\gamma\big|_{\text{in the fluid(w)}}}{m^\gamma\big|_{\text{in the porous medium}}} = \frac{c^\gamma\, \mathbb{V}_w}{c^\gamma\, \mathbb{V}_w + F^\gamma\, m_s} = \frac{1}{1 + \frac{\rho_b K_d^\gamma(c^\gamma)}{\phi}}, \tag{7.4.7}$$

in which ϕ is the porosity, and ρ_b denotes the bulk density of the solid matrix.

Altogether, an adsorption isotherm provides information on the quantity of an adsorbed species *when a solid and an aqueous phase are in equilibrium*, i.e., when the net rate of mass transfer of the species between the aqueous liquid and the solid is zero.

In most cases, the time characterizing the adsorption reaction may be sufficiently small, relative to the times characterizing advection and diffusion in the liquid phase within the void space, so that equilibrium may be assumed to prevail. However, we may encounter cases where equilibrium is not a valid assumption, and the kinetic approach and rate of reaction, discussed earlier in this subsection has to be taken into account.

In the case of adsorption in two-phase (α and β) flow, the rates of interphase transfer from α to s and from β to s are expressed by $f^\gamma_{\alpha\to s}$ and $f^\gamma_{\beta\to s}$. We recall that only one of the two fluids is wetting the solid. Also, a chemical species that can adorb on the solid may be present only in one of the fluids.

C. Retardation

Let us introduce the concept of retardation through the case of single phase flow, i.e., the void space is fully occupied by a liquid α-phase. We can eliminate the term expressing the rate of interphase γ-mass transfer by summing up the balance equations for the considered γ-species: (7.3.3) for the liquid, i.e., $\delta = s$ only:

$$\frac{\partial \phi \rho_\alpha \omega_\alpha^\gamma}{\partial t} = -\nabla \cdot \phi \left(\rho_\alpha \omega_\alpha^\gamma \mathbf{V}_\alpha + \mathbf{J}_{\alpha,dif}^\gamma + \mathbf{J}_{\alpha,dis}^\gamma \right) - f_{\alpha \to s}^\gamma + \phi \rho_\alpha \Gamma_\alpha^\gamma, \tag{7.4.8}$$

and for the solid phase:

$$\frac{\partial (\rho_b F^\gamma)}{\partial t} = f_{\alpha \to s}^\gamma + \rho_b \Gamma_s^\gamma. \tag{7.4.9}$$

By summing up the above two equations, we obtain:

$$\frac{\partial (\phi \rho_\alpha \omega_\alpha^\gamma + \rho_b F^\gamma)}{\partial t} = -\nabla \cdot \phi \left(\rho_\alpha \omega_\alpha^\gamma \mathbf{V}_\alpha + \mathbf{J}_{\alpha,dif}^\gamma + \mathbf{J}_{\alpha,dis}^\gamma \right) + \rho_b \Gamma_s^\gamma + \phi \rho_\alpha \Gamma_\alpha^\gamma, \tag{7.4.10}$$

When we use the linear isotherm:

$$F_\alpha^\gamma = (K_d)_\alpha^\gamma \rho_\alpha \omega_\alpha^\gamma, \tag{7.4.11}$$

we can rewrite (7.4.10) in the form:

$$\frac{\partial (\phi \rho_\alpha \omega_\alpha^\gamma + \rho_b F^\gamma)}{\partial t} = -\nabla \cdot \phi \left(\rho_\alpha \omega_\alpha^\gamma \mathbf{V}_\alpha + \mathbf{J}_{\alpha,dif}^\gamma + \mathbf{J}_{\alpha,dis}^\gamma \right) + \rho_b \Gamma_s^\gamma + \phi \rho_\alpha \Gamma_\alpha^\gamma, \tag{7.4.12}$$

or, with $R_d^\gamma = const.$, and $\partial \phi / \partial t = \partial \rho_b / \partial t = 0$,

$$R_d^\gamma \phi \frac{\partial \rho_\alpha \omega_\alpha^\gamma}{\partial t} = -\nabla \cdot \phi \left(\rho_\alpha \omega_\alpha^\gamma \mathbf{V}_\alpha + \mathbf{J}_{\alpha,dif}^\gamma + \mathbf{J}_{\alpha,dis}^\gamma \right) + \rho_b \Gamma_s^\gamma + \phi \rho_\alpha \Gamma_\alpha^\gamma, \tag{7.4.13}$$

in which:

$$R_d^\gamma \equiv 1 + \frac{\rho_b K_d^\gamma}{\phi} \quad (> 1) \tag{7.4.14}$$

is called the *retardation factor* of the γ-species.

To understand the reason for this name, consider the case of a liquid that occupies the entire void space, and (a) no external sources or sinks exist, (b) $\rho =$ const., $\rho_s =$ const., and $\partial \phi / \partial t = 0$, (c) no degradation or decay phenomena take place, (d) the considered γ-species adsorbs on the solid under conditions of equilibrium, following a linear isotherm, with $K_d^\gamma > 0$ and $\partial K_d^\gamma / \partial t = 0$, and (e) diffusion is negligible with respect to dispersion. Then, (7.4.13) reduces to the form:

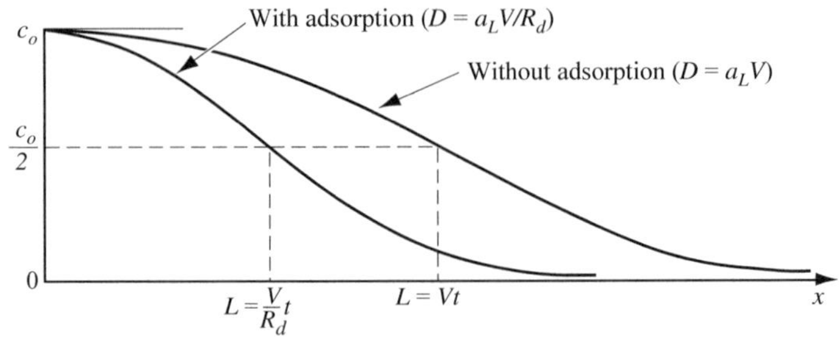

Fig. 7.4 Effect of retardation

$$\phi \frac{\partial \omega^\gamma}{\partial t} = -\nabla \cdot \phi \left(\omega^\gamma \frac{\mathbf{V}}{R_d^\gamma} - \frac{\mathsf{D}}{R_d^\gamma} \cdot \nabla \omega^\gamma \right). \tag{7.4.15}$$

Under identical assumptions, except (d), i.e., in the absence of adsorption, we obtain the same equation, but with $R_d^\gamma = 1$.

Altogether, the cases with adsorption and without it are similar, except that in the former equation, the average fluid velocity is replaced by $\mathbf{V}/R_d^\gamma$, and the coefficient of hydrodynamic dispersion is replaced by D_h/R_d^γ. Thus, under the assumption of equilibrium adsorption, described by a linear isotherm, the effect of adsorption is to *retard* the advance of the component (as part of it is adsorbed onto the solid). Instead of advancing with the fluid, moving at a velocity $\mathbf{V}$, the mean movement of the contaminant is at the reduced, or retarded velocity, $\mathbf{V}/R_d^\gamma$. At the same time, spreading by dispersion occurs *as if* the coefficient of mechanical dispersion, D, is also reduced by the factor R_d^γ. For a constant R_d, we may rewrite the l.h.s. of (7.4.13) in the form $\phi \partial \rho_\alpha \omega_\alpha^\gamma / \partial(t/R_d^\gamma)$, i.e., R_d^γ expresses a change in time scale.

Figure 7.4 shows the effect of retardation in the example of a semi-infinite porous medium column, with $c^\gamma = c_o^\gamma$ at $t = 0$, and with $c = 1.0$ at $x = 0$ for $t \geq 0$. The curves were obtained by an analytical solution. We note that in the case with adsorption, the point $c = 0.5$ advances at a speed V/R_d^γ, and that the curve is steeper, indicating, apparently, a smaller coefficient of hydrodynamic dispersion.

D. Effect of Surface Diffusion

Surface diffusion of an adsorbed species, briefly introduced in Sect. 7.2.2 E, may affect the adsorption-desorption phenomena, we shall not discuss this topic here.

7.4.2 Ion Exchange

In ion exchange reactions, already mentioned in Example 4 (Sect. 7.3.4), ions that are held by electrostatic forces to a charged functional group on the surface of the solid matrix are exchanged by ions of a similar charge present in the aqueous solution occupying the void space. Multivalent exchange is also possible, e,g., $Ca^{++} \leftrightarrow 2Na^+$.

The exchange continues until equilibrium is reached for all ions present in the porous medium domain. A process of this kind occurs, for example, in synthetic ion exchange resins. For example, the cation exchange:

$$HR_{(s)} + Na^{+}_{(aq)} \rightleftarrows NaR_{(s)} + H^{+}_{(aq)},$$

where R denotes a resin, or the anion exchange:

$$ROH_{(s)} + (Cl)^{-}_{(aq)} \rightleftarrows RCl_{(s)} + OH^{-}_{(aq)},$$

were R represents the organic part of a resin.

In the soil, ion exchange processes occur primarily on clay minerals and on oxides/oxihydroxides (in connection with both cation and anion exchange). The explanation is based on the observation that oxides/oxihydroxides can be positively charged (anion exchange), or negatively charged (cation exchange), depending on their point of zero charge and the pH of the water. When not charged, they are not available for ion-exchange. It is of interest to note that, as charge forces act over larger distances, compared to hydrophobic adsorption, ion-exchange is an extremely fast process.

Ion exchange is also an important water treatment process commonly used for water softening or demineralization. It is also used to remove other certain substances from the water in processes such as de-alkalization, de-ionization, and disinfection. In all these examples, unwanted dissolved ions are exchanged for other ions with a similar charge that are present on the solid's surface.

Some ion exchange theories envision the presence of a thin liquid boundary layer (= film) that covers the solid. The time characterizing ion exchange depends on the relative times of (1) transport of the ions from the bulk solution to the boundary layer and (2) diffusion through the layer. In the case of a porous solid matrix, ions diffuse into the porous solid to adhere on the internal surface of the porous solid. Ion exchange involves also the transport of the released ions back to the bulk solution. The limiting rate is often dictated by the various diffusive steps, rather than by the actual exchange process (Weber 1972).

Without going into details, which can be found in the literature, especially that dealing with clay minerals (e.g., Grim 1968), the structure of most clay minerals can be described as composed of layers of aluminum silicates, each layer being made up of sheets of SiO_4 tetrahedral units and $Al(OH)_xO_{6-x}$ octahedral ones, where the two kinds share some oxides with each other. For example, *kaolinite*, $Al_2Si_2O_5(OH)_4$, is composed of tetrahedral SiO_4 and octahedral $Al(OH)_4O_2$ sheets. The Al^{3+} ion in the octahedral unit may be substituted by such ions as Mg^{2+}, Fe^{2+}, Mg^{2+}, and Mn^{2+}. This isomorphic substitution creates an excess negative charge on the sheets, such that the clay surface can attract other positive ions from the solution.

Reversible ion exchange for a univalent component may take the form:

$$X^{+} + AS \rightleftharpoons A^{+} + XS, \tag{7.4.16}$$

in which X^+ is a dissolved cationic species, XS is the species in the adsorbed state on the solid, denoted by S, and A^+ is the cation, initially on the solid, remaining in the liquid phase. Upon reaching equilibrium, we obtain, according to the *law of mass action* (Sect. 7.3.3 A; and, e.g., Chang and Cruickshank 2003),

$$K^{X}_{eqA} = \frac{\{XS\}\left\{A^{+}\right\}}{\{AS\}\left\{X^{+}\right\}}, \tag{7.4.17}$$

where K^{X}_{eqA} is the *equilibrium coefficient* for the exchange reaction, and the bracketed terms represent activities (Sect. 7.3.3 A).

Ion exchange is further discussed in Appendix A.

7.4.3 Gas to Liquid γ-Mass Transfer

We consider a γ chemical species dissolved in a liquid that maintains an interface with an adjacent gas domain. A stream of γ-atoms or molecules crosses the interface–from the liquid to the gas and back. The transferred substance is moving in and out through the interface. Volatilization takes over when more vapour exits the interface than vapour condenses on it.

The term *dissolution* refers here to the case in which a chemical species in a gas dissolves in an adjacent liquid body, in excess of the stream of these molecules leaving the liquid. These phenomenon are encountered, for example, in groundwater contamination. Water (fresh or saline) and air that together occupy the void space may serve as an example. Hydrocarbon gas and saline water in a gas reservoir, may serve as another example. Appelo and Postma (2005, p. 490) present the example of benzene, $C_6H_{6(g)}$, and liquid water:

$$C_6H_{6(aq)} \leftrightarrows C_6H_{6(g)}, \quad \text{for which} \quad K_H = \frac{p_{C_6H_6(g)}}{M_{C_6H_6(aq)}}, \tag{7.4.18}$$

where M denotes molar density, and discuss the γ-mass exchange between them.

Our objective here is to determine $f^{\gamma}_{\alpha\to\beta}$, i.e., the rate at which the γ-species is transferred across a liquid-gas interface.

Let us use this case to introduce the *two-film model* which is a convenient tool to envision and evaluate interphase mass transfer, say of a γ-species present in two adjacent phases separated by a common interface. The *film* is sometimes referred to as *boundary layer*. Each of the surfaces that bounds the film is located such that up to it the presence of the phase on the other side of the film is hardly felt.

Here we are considering gas dissolution, i.e., when a γ-species is transferred from a gaseous phase (g) to a liquid (ℓ) one. Figure 7.5 shows a (microscopic) interface segment of an (assumed sharp) interface between a gas, and a liquid. On each side of this interface we envisage a *thin stagnant layer*, or '*film*', of the relevant fluid. The idea, as suggested by Nernst (1904) for a single film and Whitmanm (1923) for a

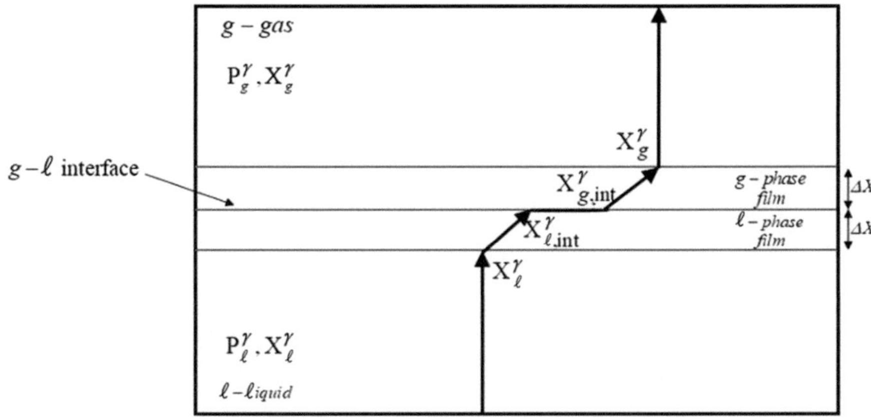

Fig. 7.5 Two-film model for interphase $\ell \rightarrow g$-transfer

double film, is that the resistance to the mass transfer *by diffusion* across the interface occurs across these *two stagnant thin films*, or *boundary layers*. While in the bulk fluid zones, we may encounter both γ-advection and diffusion, across the two films we have *only diffusion*, at a rate proportional to the *concentration gradient* across each of the two films. These gradients act as *driving forces. The interface itself is assumed to offer no resistance to the mass transfer*. It is assumed that *equilibrium between the two phases exists at the interface itself.* Thus, there is no change in the concentration relationship between the species concentration at/across the interface; the values on both sides of the interface are determined by the thermodynamic relationship between them, say Henry's law (Sect. 2.2.5), or similar laws of equilibrium. Altogether, a gas molecule starting at a point in the interior of the gas domain has to travel through the gas film, cross the interface, cross the liquid film, and end up in the liquid domain.

With the nomenclature shown on Fig. 7.5, and replacing the gradient in Fick's law by a difference over distance, we us (7.1.4) and (7.2.9) to write:

$$\text{From bulk liquid to interface}: \quad j_{\ell,film}^{\gamma} = \eta_\ell \mathcal{D}_\ell^\gamma \frac{X_\ell^\gamma - X_{\ell,int}^\gamma}{\Delta x_\ell}. \tag{7.4.19}$$

$$\text{From interface to bulk gas}: \quad j_{g,film}^{\gamma} = \eta_g \mathcal{D}_g^\gamma \frac{X_{g,int}^\gamma - X_g^\gamma}{\Delta x_g}, \tag{7.4.20}$$

where η_α denotes the molar α-phase density, and the fluxes are in moles per unit area per unit time.

Note that, the quotients $\Delta x_g/\mathcal{D}_g^\gamma$ and $\Delta x_\ell/\mathcal{D}_\ell^\gamma$ act as two 'resistances' to the diffusive fluxes. The model presented here may therefore be also called the '*two-resistance model*'.

On the interface, the two phases are in equilibrium, so that any Henry-like law takes the form:

$$\frac{X^{\gamma}_{g,int}}{X^{\gamma}_{\ell,int}} = \mathcal{H}^*_{g,\ell}. \tag{7.4.21}$$

Since $j^{\gamma}_{\ell,film} = j^{\gamma}_{g,film} = j^{\gamma}_{g\to\ell}$, with (7.4.21), we obtain:

$$j^{\gamma}_{g\to\ell} = \frac{X^{\gamma}_g/\mathcal{H}^*_{g,\ell} - X^{\gamma}_\ell}{\frac{\Delta x_g}{\eta_g \mathcal{D}^{\gamma}_g}\frac{1}{\mathcal{H}^*_{g,\ell}} + \frac{\Delta x_\ell}{\eta_\ell \mathcal{D}^{\gamma}_\ell}}, \tag{7.4.22}$$

Note that the denominator in the above equation may be regarded as the *resistance* of the interface to the transfer of mass. It depends on the transported species and the nature of the two phases, but it also involves some characteristic distances of the interface.

Actually, the above two fluxes are not necessarily equal if we take into account the possibility of accumulation of mass in the surface itself, as suggested in (1.4.71). However, here we shall assume that these two fluxes are equal, as γ-mass does not accumulate in/on the surface. Surfactants were introduced in Sect. 2.4.1 B. A γ-mass balance, say per unit area of interface, with or without the spreading of the surfactant within the interface, will lead to a model for a surfactant.

Finally, as our objective is to find an expression for the mass transfer term, $f^{\gamma}_{\alpha\to\beta}$ that appears in the *macroscopic* γ-mass balance equation, we have to relate the flux $j^{\gamma}_{\alpha\to\beta}$ to $f^{\gamma}_{\alpha\to\beta}$. Since the former is the flux through an interface of unit area, it is obvious that we have to multiply $j^{\gamma}_{\alpha\to\beta}$ by the *surface area of the* $\alpha - \beta$*-interface, per unit volume of porous medium*, $\Sigma_{\alpha\beta}$. This area of the α-β surface depends on the saturation (S_α) of the α-phase. The Δx values representing the thickness of the films are also geometrical parameters. Thus,

$$f^{\gamma}_{g\to\ell} = j^{\gamma}_{g\to\ell}\Sigma_{g,\ell}(S_\ell). \tag{7.4.23}$$

Altogether, based on the above development, we may now express the term $f^{\gamma}_{g\to\ell}$ in the form:

$$f^{\gamma}_{g\to\ell} = K^{\gamma}_{g,\ell}(X^{\gamma}_g/\mathcal{H}^* - X^{\gamma}_\ell), \tag{7.4.24}$$

where

$$K^{\gamma}_{g,\ell} = K^{\gamma}_{g,\ell}(g, \ell, S_g) = \frac{\Sigma_{g,\ell}(S_\ell)}{\frac{\Delta x_g}{\eta_g \mathcal{D}^{\gamma}_g}\frac{1}{\mathcal{H}^*_{g,\ell}} + \frac{\Delta x_\ell}{\eta_\ell \mathcal{D}^{\gamma}_\ell}} \tag{7.4.25}$$

is the coefficient of gas to liquid mass transfer to be determined experimentally for any gas, liquid and solid matrix.

It is interesting to note that the mass transfer considered here is due only to diffusion; it is assumed to be independent of the advective flow of the two phases–a wetting fluid and a non-wetting one—that simultaneously occupy and move through the void space.

7.4.4 Liquid to Liquid γ-Mass Transfer

We consider the case in which the void-space is occupied by two *immiscible liquids*– a wetting liquid and a non-wetting one. Water and oil in a petroleum reservoir may serve as an example. The case of two liquids in a chemical reactor is presented in Appendix A.

Actually, the two-film model underlying the presentation in the previous subsection, where we have considered mass transfer between a gas and a liquid, and vice-versa, is applicable also to the case of mass transfer between two immiscible liquids that together occupy the void space of a porous medium domain.

We consider mass transfer of a γ-species that dissolves in two immiscible liquid phases that together occupy the void space of a porous medium domain. The dissolved species can cross the (microscopic) interface between the two liquids. In a petroleum reservoir, the dissolution of a hydrocarbon species in water, may serve as examples. Partitioning *at the interface* is assumed to occur almost instantaneously; this means that chemical equilibrium is always assumed there. For the sake of simplicity, let us assume that no chemical reactions occur *within* the fluid phases. However, non-equilibrium conditions in the phases may exist in the form of concentration gradients and movement of species. Interphase mass transfer rates are, therefore, controlled by diffusive and advective transport of species as occurring within each fluid phase close to the interface.

As in the cases of liquid to gas mass transfer, our objective to find an expression for the case of liquid to liquid, say non-wetting (nw) to wetting (w), mass transfer of a γ-chemical species. Figure 7.6 shows a point on the interface between a wetting and a non-wetting liquid. We note how the wetting liquid covers the solid surface, also in non-wetting liquid domain. Following the two-film model presented earlier, the concentrations on the interface are c_w^γ on the w-side and c_{nw}^γ on the nw-side.

Thus, from bulk nw-liquid to interface:

$$j_{nw,film}^\gamma = \rho_{nw}\mathcal{D}_{nw}^\gamma \frac{c_{nw}^\gamma - c_{nw,int}^\gamma}{\Delta x_{nw}}. \tag{7.4.26}$$

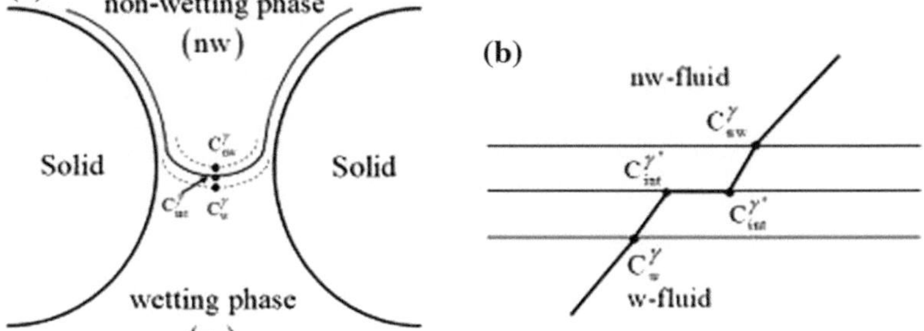

Fig. 7.6 Mass transfer across liquid–liquid interface

From interface to bulk w-liquid:

$$j^{\gamma}_{w,film} = \rho_w \mathcal{D}^{\gamma}_w \frac{c^{\gamma}_{w,int} - c^{\gamma}_w}{\Delta x_w}, \tag{7.4.27}$$

where the fluxes are in mass, say per unit interface area per unit time.

On the interface, the two phases are in equilibrium, so that Henry's law is:

$$\frac{c^{\gamma}_{nw,int}}{c^{\gamma}_{w,int}} = \tilde{\mathcal{H}}^{\gamma}_{wn,w}, \tag{7.4.28}$$

where $\tilde{\mathcal{H}}^{\gamma}_{wn,w}$ is an appropriate Henry coefficient.

Since $j^{\gamma}_{w,film} = j^{\gamma}_{nw,film} = j^{\gamma}_{nw\to w}$, with (7.4.28), we obtain:

$$j^{\gamma}_{nw\to w} = \frac{c^{\gamma}_{nw}/\tilde{\mathcal{H}} - c^{\gamma}_w}{\frac{\Delta x_{nw}}{\rho_{nw}\mathcal{D}^{\gamma}_{nw}}\frac{1}{\mathcal{H}^*_{g,\ell}} + \frac{\Delta x_w}{\rho_w \mathcal{D}^{\gamma}_w}}, \tag{7.4.29}$$

With $\Sigma_{w,nw} = \Sigma_{w,nw}(S_w)$ denoting the specific surface area of the $w-nw$-interface, the γ-mass transferred from the non-wetting fluid to the wetting one, per unit volume of porous medium is $f^{\gamma}_{nw\to w} = j^{\gamma}_{nw\to w}\Sigma_{w,wn}$. Altogether, we can write:

$$f^{\gamma}_{nw\to w} = K^{\gamma}_{nw\to w}\left(c^{\gamma}_{nw}/\tilde{\mathcal{H}} - c^{\gamma}_w\right), \quad K^{\gamma}_{nw\to w} = \frac{\Sigma_{nw,w}(S_{nw})}{\frac{\Delta x_{nw}}{\rho_{nw}\mathcal{D}^{\gamma}_{nw}}\frac{1}{\mathcal{H}^*_{g,\ell}} + \frac{\Delta x_w}{\rho_w \mathcal{D}^{\gamma}_w}}, \tag{7.4.30}$$

where $K^{\gamma}_{nw\to w}$ is an experimentally determined liquid-to-liquid mass transfer coefficient, to be determined experimentally.

As γ-particles move across the film-interface-film domain representing the fluid-fluid interface, a characteristic time is associated with each step. The characteristic time of the entire process is determined by the slowest process. This will be the 'rate limiting step' for the entire transfer process. Usually, diffusion through the liquid phase is the 'rate limiting step'.

In Sect. 7.3.3, we have already discussed the conditions that justify the assumption of equilibrium for a heterogeneous reaction in an REV. The conditions were expressed in terms of the Strouhal, Peclet, and Damköhler numbers, using a characteristic length ℓ_α that is equal to the local volume-to-surface area ratio of the phase. If condition (7.3.19), based on these numbers, is satisfied, then the reaction at the interface is fast enough so that it may be considered to be in equilibrium. However, it is still possible that condition (7.3.17) does not hold, because advection might dominate over diffusion, or diffusion could take place slowly. In such a case, we need expressions for the rate of interphase mass transfer. Usually such expressions are derived empirically.

Essentially, the two film model *assumes*, as a good approximation, that the concentration within each film varies linearly with the distance from the interface. Such a situation may be expected to hold for a *diffusion-dominated system* under quasi-steady conditions, with no chemical reactions. However, if the system is dominated by advection, then the concentration profile, and, hence, the resulting mass transfer coefficients, will depend also on the magnitude of the mean fluid velocity.

Like in the case of gas-to liquid mass transfer discussed above, the mass exchange is not affected by the advective fluxes in the liquids, which may be co-current or counter-current.

Although we have used the concentration, c^γ, in the above formal derivations, we could have used the mass fraction $\omega^\gamma (= c^\gamma/\rho_\alpha)$ instead, as it is convenient to use ω^γ when ρ is (almost) unchanged.

7.4.5 Solubility and Precipitation

These phenomena often occur during reactive transport through the void-space of a porous medium domain. Our objective here is to present the conditions under which mass transfer resulting from dissolution and precipitation occur in a porous medium domain in which a liquid with a dissolved chemical species is transported through the void space.

• A. Saturation index

Under certain (c, p, T)-conditions, a solid which is in contact with a liquid that occupies the void-space, will dissolve in the liquid. We refer to this process as *solid dissolution*. In general, it is not necessarily the 'solid' as such that dissolves, but only certain chemical species that constitute part of the solid's (very thin) layer that is in contact with the liquid.

Precipitation is the opposite of dissolution. Under certain (c, p, T)-conditions, when two solutions that contain the ions of a salt are mixed, some salt will precipitate, i.e., it will emerge from the solution as a solid. Under any prevailing $p.T$ conditions, there exists a limit at which any additional solid salt added to the solution cannot be dissolved. Instead, the additional salt will remain undissolved. The solution is said to be 'saturated' with respect to that solid under the prevailing conditions. The solute concentration at that point is referred to as *solubility* of the considered solute in the solvent, under the prevailing p, T conditions.

Whether, under prevailing p, T conditions, saturation or precipitation will occur or not depends on a *saturation index*. (see (7.4.32)).

There is a limit to the amount of ions (at a given p and T) that can be present in a solution. Any addition of ions beyond that limit cannot dissolve; instead ions will merge into a solid that will precipitate.

A simple starting point is the *law of mass action* presented as Sect. 7.3.3 B.

Following (7.3.28), a precipitation/dissolution reaction can be written in the canonical form of the stoichiometric equation:

$$\sum_{(\gamma)} \nu_m^\gamma A^\gamma \leftrightharpoons M_m, \tag{7.4.31}$$

in which γ as a superscript denotes a γ-species, M_m denotes a considered mineral, A^γ denotes a primary γ-species, and ν_m^γ denotes the corresponding *stoichiometric coefficient* for the considered m-mineral. The *saturation index*, SI_m, for the m-mineral is defined as:

$$SI_m = K_{eq,m} Q_m. \tag{7.4.32}$$

where $K_{eq,m}$, defined in (7.3.39), (7.3.48), and (7.3.55), denotes the equilibrium constant for the considered m-mineral, and Q_m denotes the *activity product* of the m-mineral, defined by;

$$Q_m = \prod_{(\gamma)} \left(\{\gamma\}\hat{m}^\gamma\right)^{\nu_m^\gamma}, \tag{7.4.33}$$

where $\hat{m}^\gamma$ denotes the *molality* of the γ primary species, and $\{\gamma\}$ denotes the *activity coefficient* of the γ-species.

With the above definitions, the criterion for dissolution/precipitation is:

$$SI_m = \begin{cases} > 1, \text{ precipitation}, \\ = 1, \text{ equilibrium}, \\ < 1, \text{ dissolution}. \end{cases} \tag{7.4.34}$$

• **B. Dissolution**

An *ionic solid* is a solid built of negative and positive ions that attract each other. When placed in a liquid, e.g., water, the solid's ions are attracted to the liquid's ones and the solid dissolves. An equilibrium is established between the ions in the saturated solution and the ions remaining in the solid at their contact domain. For example (Sienko and Plane 1966, p. 272), in the case of excess silver chloride in contact with a saturated solution of silver chloride, we have:

$$AgCl_s \rightleftarrows Ag^+ + Cl^-,$$

with equilibrium established between the ions in solution and the excess AgCl. Making use of the equilibrium relationship (7.3.55), we write:

$$\frac{\{Ag^+\}\{Cl^-\}}{\{AgCl_{(s)}\}} = K_{eq}, \tag{7.4.35}$$

where we recall that $\{(..)\}$ denotes the activity of (..), and that the activity of a pure solid is equal to one, $\{AgCl_{(s)}\} = 1$, independent of the solid-solution area of contact. It follows that to determine whether a solid will dissolve or not, we have to examine:

$$\{Ag^+\}\{Cl^-\} \lesseqgtr \mathrm{K_{eq}}, \tag{7.4.36}$$

The actual value of K_{eq} is determined experimentally.

The product on the l.h.s. of the above equation is called the *ion product*. In other words, the ion product must equal K_{sp} when the ions of a saturated solution are in equilibrium with excess solid.

The case of CO_2 disposal in deep brine containing formations may serve as another example. As the brine becomes acidic, it tends dissolves minerals that comprise the solid matrix, increasing the porosity and permeability of the void space.

As a third example, consider the case of *barium sulfate* ($BaSO_4$). In solution, we have:

$$BaSO_4 \leftrightarrows Ba^{2+} + SO_4^{\cdot} \tag{7.4.37}$$

Since $SI_{BaSO_4} = 1$, we have:

$$K_{eq} = \{Ba^{2+}\}\{SO_4^{2-}\} = (3.9 \times 10^{-5})(3.9 \times 10^{-5}) = 1.5 \times 10^{-9}. \tag{7.4.38}$$

to be compared with the appropriate value of $K_{eq}(p, T)$ defined above.

Thus, for any solution containing Ba^{2+} and SO_4^{2-} in equilibrium with solid $BaSO_4$, the product of the concentrations of Ba^{2+} and SO_4^{2-} is equal to 1.5×10^{-9}. Since K_{eq} is a very small number, $BaSO_4$ is practically an insoluble salt. When $\{Ba^{2+}\}\{SO_4^{2-}\} < 1.5 \times 10^{-9}$, the solution is unsaturated and $BaSO_4$ must dissolve to increase the concentrations of both Ba^{2+} and SO_4^{-}. However, when $\{Ba^{2+} \times SO_4^{2-}\} > 1.5 \times 10^{-9}$, the solution is supersaturated and precipitation will take place (in order to reduce the concentrations).

The solid-to-solution mass transfer flux associated with solid dissolution can be expressed by using the *two-thin-films model*, presented in the previous subsections. We assume that next to the solid surface, we have a thin film (film1) with solid ions at saturation concentration at the prevailing p, T conditions, $c^A_{s,max}$. Next to it, we have a second thin film (film2), of thickness Δx_{film2}, across which the ion concentration drops to that of the solution in the void space, c^A_ℓ. We may then express the flux from the solid to the liquid by:

$$j^A_{s\to\ell} = \mathcal{D}^A \frac{\left(c^A_{s,max} - c^A_\ell\right)}{\Delta x_{film2}}, \tag{7.4.39}$$

in which c^A_ℓ denotes the concentration in the liquid, say, average concentration in the liquid, and $\mathcal{D}^A$ is the coefficient of molecular diffusion in the liquid ℓ.

To obtain the flux due to dissolution, per unit volume of porous medium, we have to multiply $j^A_{s\to\ell}$ by the specific $s - \ell$ surface, $\Sigma_{s,\ell}$, leading to:

$$f^A_{s\to\ell} = j^A_{s\to\ell}\Sigma_{s,\ell}, = K_{s\to\ell}\left(c^A_{s,max} - c^A_\ell\right), \quad K_{s\to\ell} = \frac{\mathcal{D}^A \Sigma_{s,\ell}}{\Delta x_{film2}} \tag{7.4.40}$$

where $K_{s\to\ell}$ is the *solid's dissolution mass transfer coefficient*.

The two concentrations in the above equation are average values at the considered point in the porous medium domain. It may be interesting to comment that actually the value of c_ℓ^A should be the average over an REV of the concentration close to the solid surface, rather than the average over the REV. The latter concentration is lower than the average. This comment can be understood if we recall the development by Taylor model presented in Sect. 7.2.3. Taylor's solution for solute transport through a capillary tube, makes a clear distinction between the average concentration and the distribution of solute concentration at the microscopic level.

• **C. Precipitation**

An example may be useful (Sienko and Plane 1966, p. 276). We mix 50 ml of 5.0×10^{-4} moles of $Ca(NO_3)_2$ with 50ml of 2.0×10^{-4} moles of NaF. Together we have 100ml of solution. The K_{eq} of CaF_2 is 1.7×10^{-10}. In general:

$$Ca(NO_3)_{2(s)} + 2NaF \leftrightarrows CaF_2 + 2NaNO_3.$$

We have to determine whether precipitation will occur or not.

In the considered mixture, because of the 2-fold dilution of the $Ca(NO_3)_2$, we have 2.5×10^{-4} moles of Ca^{2+}. The F is also diluted in the mixture to 1.0×10^{-4} moles.

To determine whether or not we'll have precipitation, we have to determine the value of K_{eq}:

$$[Ca^{2+}][F^{2-}]^2 = (2.5 \times 10^{-4})(1.0 \times 10^{-4})^2 = 2.5 \times 10^{-12}.$$

Since the above value is smaller than the K_{eq} of CaF_2, $(= 1.7 \times 10^{-10})$, *precipitation will not occur*. The resulting mixture will be an unsaturated solution with respect to the mixture of solutions.

7.5 Complete Solute Transport Model

When we consider a single, non-reacting and non-adsorbing dissolved chemical species, the single dependent variable for which a solution is sought is the concentration of that species in the fluid, say $c^\gamma(\mathbf{x}, t)$. Obviously, we may use ω^γ, or X^γ, as variables. In multi-phase flow, say, α and β, the same species may be present in more than one phase within an REV. Phase transfer may occur and we have to consider ω_α^γ and ω_β^γ, as additional variables.

In what follows, as long as we consider a single phase, which may occupy only part of the void space, no special subscript will be used to denote that phase. Similarly, no superscript will be used to denote the considered chemical species, as long as we are considering only a single species.

Because the velocity distribution within a considered domain is required as input information to the problem of transport of a chemical species, it is always necessary to solve simultaneously also the problem of transport of the (total) mass of the fluid phases that carry the considered chemical species, before solving the component's transport problem. Obviously, when the component's concentration affects the fluid's density (and possibly also its viscosity), the two problems are coupled and must be solved simultaneously.

To obtain a unique solution of a balance equation in a given porous medium domain, we have to specify initial and boundary conditions. In the case considered here, that balance equation is of the mass of the considered species. The basic ideas underlying the concept of boundary conditions have been presented in Sect. 5.2, where we have discussed boundary conditions for flow, i,e., fluid mass transport. As we have done there, we shall continue to make the assumption that the boundary, which is material with respect to the solid matrix, is an abrupt surface, defined by the equation $F = F(\mathbf{x}, t)$.

As in the case of flow, we start by discussing the general boundary condition, which expresses the continuity of the flux of the considered chemical species across a boundary. We shall then present a number of the more commonly encountered particular cases of practical interest.

7.5.1 General Boundary Condition

We wish to solve an equation that expresses the mass balance of a chemical species transported in a fluid phase. As stated in Sect. 5.2.3, in the absence of sources or sinks *on a boundary*, the general boundary condition for any extensive quantity, is that *there is no-jump in the total flux of that quantity across the boundary*. This is clearly expressed by the no-jump condition (5.2.6). If such sources are present, the jump is equal to the strength of the sources. For a solute transported by a fluid phase, this condition states that in the absence of sources and sinks of the considered solute on the boundary, which is usually the case, the component normal to the boundary of the total flux of that solute, with respect to the (possibly moving) boundary, undergoes no-jump as the latter is crossed. Thus, with θ denoting the volumetric fraction of the phase, $\boldsymbol{\nu}$ denoting the unit vector normal to the boundary, and $\mathbf{u}$ denoting the velocity of the latter, this statement takes the form:

$$[\![\ \theta\,[c(\mathbf{V}-\mathbf{u})-\mathbf{D}_h\cdot\nabla c]\ ]\!]_{1,2}\cdot\boldsymbol{\nu}=0, \tag{7.5.1}$$

where $[\![\ (..)\ ]\!]_{1,2} \equiv (..)\big|_1 - (..)\big|_2$ denotes the jump in (..) from side 1 to side 2 of the boundary. For example, sides 1 and 2 may represent the internal side and the external one, respectively. Obviously, we may express the above equation in terms of ω_α^γ, or X_α^γ.

Since we have assumed that the boundary is material with respect to the solid matrix, viz.,

$$(\mathbf{V}_s-\mathbf{u})\big|_1=(\mathbf{V}_s-\mathbf{u})\big|_2=0,$$

we may rewrite (7.5.1) in the form:

$$[\![\ (c\mathbf{q}_r - \theta \mathbf{D}_h \cdot \nabla c) \ ]\!]_{1,2} \cdot \boldsymbol{\nu} = 0, \tag{7.5.2}$$

where $\mathbf{q}_r$ ($\equiv \theta(\mathbf{V}_f - \mathbf{V}_s)$) (= specific discharge relative to the solid) is expressed by Darcy's law, or by any other motion equation.

For (7.5.2) to become a condition for c on a boundary, information on what happens on the external side of the latter (in this case, the total flux relative to the boundary) *must be known* as a function of space and time.

In addition to boundary conditions that stem from the continuity of flux, as presented above, a second fundamental assumption exists, namely, that of continuity of the value of the intensive quantity of the considered phase, or of the considered chemical species, as the boundary is approached from both sides. We usually know the value on the external side, and assume that *at every point on a boundary, there exists no discontinuity in the (intrinsic phase average of) scalar intensive quantities*, such as density, temperature and (here) concentration. The motivation for making this assumption is that otherwise, the infinite gradient associated with a discontinuity will instantaneously equalize the values of such quantities on both sides of the boundary by the process of diffusion (= conduction, for heat).

In what follows, we shall apply the general condition (7.5.2) to a number of particular cases. We shall assume that all boundaries are stationary, except for the phreatic surface.

When necessary, subscripts 1 and 2 will be used to denote the internal and external sides of a boundary surface, respectively. The latter is described by $F(\mathbf{x}, t) = 0$, with a normal unit vector given by $\mathbf{n} = \nabla F / |\nabla F|$. We shall use c to denote concentration, although other measures may also be used. We shall assume that the considered phase occupies only part of the void space, at volume fraction θ, with the possibility that $\theta = \phi$.

7.5.2 *Particular Cases*

A. Boundary of Prescribed Concentration

When the value of $c(\mathbf{x}, t)$ are imposed as a known function, $f^{(1)}(\mathbf{x}, t)$, at all points of a boundary segment, $\mathcal{B}$, due to phenomena that take place on the external side of the considered domain, independent of what happens within the latter, we employ the boundary condition

$$c(\mathbf{x}, t) = f^{(1)}(\mathbf{x}, t) \qquad \text{on} \quad \mathcal{B}, \tag{7.5.3}$$

where $c(\mathbf{x}, t)$ denotes the concentration on the internal side of the boundary and $f^{(1)}(\mathbf{x}, t)$ is a known function that represents c on the external side. This is a first type, or *Dirichlet* boundary condition.

Although (7.5.3) is used very often, probably because of its simplicity, it is not a direct consequence of the general condition of no-jump in the normal component of the total solute flux across a boundary. Instead, as explained in the previous subsection, it is based on the assumption of no jump in the values of scalar intensive quantities. Whenever possible, we should avoid using this condition as, usually (except, for example, at a boundary of inflow from a reservoir of known concentration), we do not have information concerning solute concentration on the external side of a boundary.

B. Boundary of Prescribed Solute Flux

When phenomena occurring in the external domain impose a known total flux, $f^{(2)}(\mathbf{x}, t)$, of the considered component normal to a boundary segment, $\mathcal{B}$, at all points of the latter, *regardless* of what happens within the considered domain itself, the condition obtained from (7.5.2) is

$$(c\mathbf{q}_r - \theta\mathbf{D}_h\cdot\nabla c)\cdot\boldsymbol{\nu} = f^{(2)}(\mathbf{x}, t) \qquad \text{on} \quad \mathcal{B}. \tag{7.5.4}$$

Since both c and ∇c are involved in (7.5.4), this is a *Cauchy*, or *third type boundary condition*. When $\mathbf{q}_r = 0$, Eq. (7.5.4) reduces to a *Neumann*, or *second type boundary condition*.

A boundary of special interest is the *impervious boundary*. For such a boundary, with $f^{(2)}(\mathbf{x}, t) = 0$, and $\mathbf{q}_r\cdot\mathbf{n} = 0$, Eq. (7.5.4) reduces to

$$(\mathbf{D}_h\cdot\nabla c)\cdot\boldsymbol{\nu} = 0 \qquad \text{on} \quad \mathcal{B}. \tag{7.5.5}$$

This is a particular case of a *Neumann* boundary condition.

For an impervious boundary surface that coincides with the vertical xz-plane, with $V_y = 0$, $V_x, V_z \neq 0$, $\nu_y = 1$, $\nu_x, \nu_y = 0$, $D_{hyx} = D_{hyz} = 0$, $D_{hyy} = a_T V + \mathcal{D}^*$, the condition (7.5.5), of zero total flux normal to an impervious boundary, reduces to

$$\left(a_T V + \mathcal{D}^*\right)\frac{\partial c}{\partial y} = 0, \quad \text{or} \quad \frac{\partial c}{\partial y} = 0.$$

C. Boundary Between Two Porous Media

Along such a boundary, we allow for the possible existence of discontinuities in all solid matrix characteristics, e.g., ϕ, k_{ij}, $a_{ijk\ell}$, etc. Neither the concentration nor the flux are a-priori known on the boundary. Actually, each side serves as an external side to the other one.

Two conditions must be satisfied on such a boundary. The first is that of no-jump in component concentration, expressed in the form:

$$c\big|_1(\mathbf{x}, t) = c\big|_2(\mathbf{x}, t) \qquad \text{on} \quad \mathcal{B}. \tag{7.5.6}$$

The second condition is that of continuity in the normal component of the total flux of the considered component,

$$(c\mathbf{q}_r - \theta\mathbf{D}_h\cdot\nabla c)\big|_1\cdot\boldsymbol{\nu} = (c\mathbf{q}_r - \theta\mathbf{D}_h\cdot\nabla c)\big|_2\cdot\boldsymbol{\nu}. \tag{7.5.7}$$

Because $\mathbf{q}_r|_1\cdot\boldsymbol{\nu} = \mathbf{q}_r|_2\cdot\boldsymbol{\nu}$, and $c|_1 = c|_2$, the last equation reduces to

$$(\theta\mathbf{D}_h\cdot\nabla c)\big|_1\cdot\boldsymbol{\nu} = (\theta\mathbf{D}_h\cdot\nabla c)\big|_2\cdot\boldsymbol{\nu}. \tag{7.5.8}$$

The reason for requiring *two* conditions on such a boundary, rather than one, stems from the observation that the partial differential (balance) equations cannot be solved for domains with discontinuities in the coefficients. To overcome this difficulty, we divide the problem domain along the surfaces of discontinuity into sub-domains in each of which no such discontinuity exists. We write a complete model for each of these sub-domain. Such a model requires that conditions be specified also along the surface of discontinuity (which now serves as a boundary to both sub-domains). We need one condition for each side (= sub-domain), for a total of two conditions. Because each of these conditions involves the variables for the two adjacent sub-domains, the two models are coupled, and have, therefore, to be solved simultaneously.

D. Boundary with a Body of Liquid

We consider the boundary between a porous medium domain (pm) and a body of fluid (fb), assumed to be a 'well-mixed' domain, that is, at a *known* uniform concentration of the considered component. A large lake and a river may serve as examples (Fig. 7.7). For the gaseous phase in the soil, the atmospheric air above ground surface may serve as another example.

To simplify the discussion, we consider a saturated porous medium domain in hydraulic contact with a 'well-mixed' body of fluid which contains the considered component at a known uniform concentration, c^H. The boundary is assumed stationary.

The condition of no-jump in the normal component of the total flux of the considered component, takes the form of:

$$(c^H\mathbf{V})\big|_{fb}\cdot\boldsymbol{\nu} - \left(c\mathbf{q}_r + \phi\mathbf{J}^\gamma + \phi\mathbf{J}^{*\gamma}\right)\big|_{pm}\cdot\boldsymbol{\nu} = 0, \tag{7.5.9}$$

where $\phi\big|_{fb} = 1$. Equation (7.5.9) expresses the continuity in the mass flux across the boundary, of the component in the water, by advection, diffusion and dispersion. Since the fluid body is assumed to be at a uniform concentration, only advection takes place in it. The use of $\mathbf{q}_r$ stems from the assumption that the boundary is material with respect to the solid matrix. When $\mathbf{V}_s = 0$, we have $\mathbf{q}_r \equiv \mathbf{q}$.

Expressing the dispersive and diffusive fluxes in terms of ∇c, we may rewrite (7.5.9) in the form:

$$(c^H\mathbf{V})\big|_{fb}\cdot\boldsymbol{\nu} - (c\mathbf{q}_r - \phi\mathbf{D}_h^\gamma\cdot\nabla c)\big|_{pm}\cdot\boldsymbol{\nu} = 0. \tag{7.5.10}$$

Fig. 7.7 A well-mixed zone

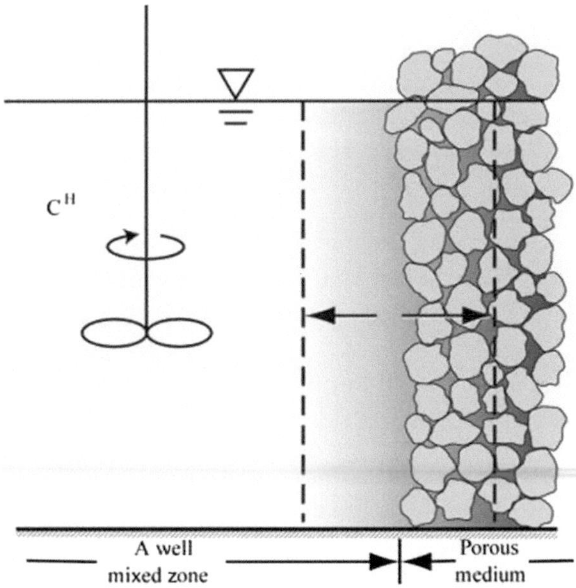

Consequently, when $\mathbf{V}\big|_{fb}\cdot\boldsymbol{\nu} = \mathbf{q}_r\cdot\boldsymbol{\nu} = 0$, i.e., no advection takes place across the boundary, the dispersive flux, $\mathbf{J}^{*\gamma}$, vanishes, and (7.5.10) reduces to

$$(\mathcal{D}^{*\gamma}\cdot\nabla c)\big|_{pm}\cdot\boldsymbol{\nu} = 0. \tag{7.5.11}$$

This implies that no transport of mass by molecular diffusion takes place across such a boundary, even when $c\big|_{pm} \neq c^H$. This conclusion is unacceptable. Under the physical conditions of this example, we should expect transport by molecular diffusion of the component to take place between the porous medium domain and the adjacent fluid body, as this remains the only possible mode of transport.

The error in the conclusion expressed by (7.5.11) stems from the assumption that a 'well-mixed' zone exists on the external side of the boundary. This assumption, in the absence of advection, combined with the sharp boundary approximation, must yield no mass flux by diffusion across it. In order to reinstate the diffusive-dispersive flux, which takes place in reality, we introduce the concept of a *transition zone*, *boundary layer*, or *buffer zone*, at the boundary (Fig. 7.7). We may associate the width of this transition zone, Δ, with the magnitude of an REV, assuming that the abrupt boundary passes through its midpoint. Instead of the boundary between the body of fluid and the porous medium, we now consider the boundary between the latter and the transition zone. Assuming that the sum of dispersive and diffusive fluxes through the transition zone is proportional to the average concentration gradient, and that the latter is proportional to the concentration difference $c^H - c$, we express the condition of continuity of flux at the boundary by

$$c^H \mathbf{V}\big|_{fb} \cdot \boldsymbol{\nu} + \alpha^\star (c^H - c) = (c\mathbf{q}_r - \phi \mathbf{D}_h^\gamma \cdot \nabla c)\big|_{pm} \cdot \boldsymbol{\nu}, \tag{7.5.12}$$

where $\alpha^\star$ is a *transfer coefficient*, such that $\alpha^\star(c^H - c)$ represents the sum of diffusive and dispersive fluxes through the transition zone.

Since, $\mathbf{V}\big|_{fb} \cdot \boldsymbol{\nu} = \mathbf{q}_r \cdot \boldsymbol{\nu}$, Eq. (7.5.12) reduces to

$$(c^H - c\big|_{pm})(\mathbf{q}_r \cdot \boldsymbol{\nu} + \alpha^\star) = -\phi \mathbf{D}_h^\gamma \cdot \nabla c\big|_{pm} \cdot \boldsymbol{\nu}, \tag{7.5.13}$$

which now serves as the boundary condition.

In the absence of advection, or when $|\mathbf{q}_r \cdot \boldsymbol{\nu}| \ll \alpha^\star$, Eq. (7.5.13) reduces to

$$\alpha^\star (c^H - c\big|_{pm}) = -\phi \boldsymbol{D}^{*\gamma} \cdot \nabla c\big|_{pm} \cdot \boldsymbol{\nu}. \tag{7.5.14}$$

We note that if we accept (7.5.13), then $c^H\big|_{fb} \neq c\big|_{pm}$ on the boundary, i.e., a jump in concentration takes place on the boundary. This is a consequence of introducing the transition zone and the 'well-mixed zone' approximation.

When $|\mathbf{q}_r \cdot \boldsymbol{\nu}| \gg \alpha^\star$, Eq. (7.5.13) reduces to

$$(c^H - c\big|_{pm})\mathbf{q}_r \cdot \boldsymbol{\nu} = -\phi \mathbf{D}_h^\gamma \cdot \nabla c\big|_{pm} \cdot \boldsymbol{\nu}, \tag{7.5.15}$$

which is a boundary condition of the third type, identical to (7.5.10), yet is based on different reasoning.

It is interesting to note here the comment concerning *discharge averaged concentration* presented at the end of Sect. 7.5.2 A.

E. Phreatic Surface

We recall that the phreatic surface is defined as the surface at every point of which the water pressure is atmospheric. It, thus, serves as the upper boundary of the saturated domain. At the same time, it serves also as the lower boundary of the unsaturated one.

The condition for fluid mass transport at a phreatic surface is presented in Sect. 5.2.4 E. Here we consider the In all cases, the boundary conditions are derived from the requirement of no-jump in the solute's's flux normal to the phreatic surface.

As has already been stated in Sect. 5.2.4 E, the main difficulty associated with the phreatic surface as a boundary is that its shape and position are not known *a-priori*. In fact, in flow problems, they are the objective of modeling. The shape of the phreatic surface can be described by the equation $F(\mathbf{x}, t) = 0$. However, since the pressure is atmospheric everywhere on this surface, or $h(\mathbf{x}, t) = z$, i.e., the piezometric head is equal to the elevation, the equation that describes the shape of the phreatic surface may be written also (in cartesian coordinates) in the form:

$$F = F(\mathbf{x}, t) = h(x, y, z, t) - z = 0. \tag{7.5.16}$$

When this surface moves at a velocity $\mathbf{u}$, we have

$$\boldsymbol{\nu} = \frac{\nabla F}{|\nabla F|}, \quad \nabla F = \nabla(h - z), \quad \frac{dF}{dt} \equiv \frac{\partial F}{\partial t} + \mathbf{u}\cdot\nabla F = 0,$$
$$\frac{\partial F}{\partial t} = \frac{\partial h}{\partial t} = -\mathbf{u}\cdot\boldsymbol{\nu}|\nabla F|, \tag{7.5.17}$$

where $\boldsymbol{\nu}$ denotes the unit vector normal to the phreatic surface.

CASE A: The phreatic surface is an upper boundary of the saturated zone. As is usually done in groundwater hydrology, we neglect the details of the movement of water through the unsaturated zone, and consider only some mean value of natural replenishment, $\mathbf{N}$, infiltrating at ground surface, percolating through the unsaturated zone, and reaching the phreatic surface. Let the concentration of the considered component in the infiltrating water, as it approaches the phreatic surface, be denoted by c_N. We assume that (here and elsewhere in this section), in spite of concentration differences, the mass density of the water remains constant.

In the unsaturated zone, just above the phreatic surface, the moisture content is assumed to be equal to $\theta = \theta_{wr} =$ constant. The component's total flux, relative to the moving phreatic surface, is given by $c_N\,(\mathbf{N} - \theta_{wr}\mathbf{u})\cdot\boldsymbol{\nu}$. The total flux in the saturated zone, relative to the moving phreatic surface, is expressed by $\phi\,[c(\mathbf{V} - \mathbf{u}) - \mathbf{D}_h\cdot\nabla c]$. Thus, the no-jump condition can be expressed as:

$$\{\phi\,[c(\mathbf{V} - \mathbf{u}) - \mathbf{D}_h\cdot\nabla c]\}\cdot\boldsymbol{\nu} = c_N(\mathbf{N} - \theta_{wr}\mathbf{u})\cdot\boldsymbol{\nu}.$$

When combined with the flow boundary condition (5.2.26), we obtain

$$(c - c_N)\left(\mathbf{N}\cdot\nabla F + \theta_{wr}\frac{\partial F}{\partial t}\right) - \phi(\mathbf{D}_h\cdot\nabla c)\big|_{sat}\cdot\nabla F = 0. \tag{7.5.18}$$

We may insert $F = h(x, y, z, t) - z$ in the last equation. This is a third type boundary condition for c. We note that $[\![\ c\ ]\!]_{sat,unsat} \equiv c - c_N \neq 0$. Thus, the unsaturated zone just above the phreatic surface acts as a 'well-mixed zone' in the sense discussed earlier. However, we have simplified the expression for the flow through the transition zone by neglecting the dispersive-diffusive flux through it.

CASE B: The phreatic surface is the lower boundary of the unsaturated zone. We shall now assume that the saturated zone is a 'well-mixed zone' at concentration $c\big|_{sat}$. The no-jump condition takes the form:

$$\phi\,[(\mathbf{V} - \mathbf{u})c]\,\big|_{sat}\cdot\boldsymbol{\nu} + \alpha^\star(c\big|_{sat} - c) = \theta\,[(\mathbf{V} - \mathbf{u})c - \mathbf{D}_h\cdot\nabla c]\cdot\boldsymbol{\nu}, \tag{7.5.19}$$

or, with $\phi \approx \theta$ at the phreatic surface,

$$\left[(\mathbf{V} - \mathbf{u}) + \alpha^\star\right]\left(c\big|_{sat} - c\right)\cdot\boldsymbol{\nu} = -\,(\mathbf{D}_h\cdot\nabla c)\cdot\boldsymbol{\nu}. \tag{7.5.20}$$

In this expression, we have to know, or assume, $c\big|_{sat}$. We can always use (7.5.17) to express $\boldsymbol{\nu}$ and $\mathbf{u}\cdot\boldsymbol{\nu}$ in terms of h.

F. Seepage Face

The seepage face is discussed in Sect. 5.2.4F. Here, the water leaving the porous medium domain through a seepage face carries the dissolved component.

Because there is no porous medium on the external side of this boundary, the condition of continuity of flux of a component takes the form:

$$(\phi c\mathbf{V} - \phi\mathbf{D}_h\cdot\nabla c)\big|_{pm}\cdot\mathbf{n} = (c\mathbf{V})\big|_{env}\cdot\boldsymbol{\nu}, \tag{7.5.21}$$

where symbols 'pm' and 'env' denote the porous medium domain and its external environment, respectively, and we have assumed a stationary seepage face, $\mathbf{u} \equiv 0$. With

$$c\big|_{pm} = c\big|_{env}, \qquad (\phi\mathbf{V})\big|_{pm}\cdot\boldsymbol{\nu} = \mathbf{V}\big|_{env}\cdot\boldsymbol{\nu},$$

i.e., assuming neither volatilization, nor precipitation, (7.5.21) reduces to the boundary condition

$$(\mathbf{D}_h\cdot\nabla c)\big|_{pm}\cdot\boldsymbol{\nu} = 0. \tag{7.5.22}$$

This is a boundary condition of the second type.

G. Discharge Dependent Boundary Condition

Sometimes, in a case of solute transport through a geological formation, a boundary is chosen *arbitrarily*, to delineate the investigated domain of interest. An outflow boundary is another case in which a solute leaves the domain, but we do not know, a-prior, at what concentration or flux. Especially, we do not know the concentration gradient that controls the dispersive and the diffusive fluxes. On such boundary, neither the concentration nor its flux are a-priori known. What condition should be specified on such boundary?

Information for such boundary can be obtained by collecting the fluid emerging through a boundary segment, determining the time and space dependent solute concentration as the condition along such boundary. This concentration is often referred to as 'discharge dependent concentration boundary'. However, we really cannot assign a boundary condition to such domain. One possible approximation is to assume that this domain is sufficiently far away from where the real 'action' is so that the initial condition remains unchanged (we often use the term "clamped") along such boundary. It is then a boundary of the first kind.

There exist certain numerical approximate (mostly, iterative) techniques for this kind of boundary, but these are beyond the scope of this book.

7.5.3 Initial Condition

Initial conditions state the spatial distribution of the considered state variable, here the concentration of the considered component, at some initial time, usually taken as $t = 0$. We require that the solution $c(\mathbf{x}, t)$ satisfies

$$c(\mathbf{x}, 0) = f^{(3)}(\mathbf{x}), \tag{7.5.23}$$

where $f^{(3)}$ is a known function.

7.5.4 A Comment on Primary Variables Switching

The subject of degrees of freedom was discussed in Sect. 3.9. We have shown there that *Gibbs phase rule* (3.9.2), repeated here for convenience:

$$\text{NF} = \text{NC} - \text{NP} + 2, \tag{7.5.24}$$

is used for determining the number of thermodynamic degrees of freedom (NF) for a problem with NP phases and NC components.

In Sect. 6.4.1 we have mentioned the need for primary variables switching. In Sect. 2.3.1 we have discussed the number of degrees of freedom for NP fluid phases and NC components. Let us now consider the case of 2 phases and two components.

7.5.5 Complete Model for a Single Solute

The mathematical model of a multi-phase flow problem, combined with the transport of a single chemical component, possibly with first order decay, adsorption, and volatilization, consists of the following parts:

- A mathematical description of the configuration of the surface that bounds the porous medium problem domain.
- A list of the dependent variables. These are the concentrations, say, c_α^γ, or the mole fractions, X_α^γ, of the considered γ-component within all α-fluid phases present in the system. In the case of adsorption, $F_{\alpha\to s}^\gamma$ is included in the list of state variables. For the flow model, depending on the number of fluid phases that are in motion, we add such variables as piezometric heads, pressures, saturations, etc.
- Flux equations for the mass of the considered fluid phases. Darcy's law is usually employed.
- Partial differential ($\equiv$ mass balance) equations for the relevant fluid phases.
- Mass and momentum balance equations for the solid, when the latter is deformable.

- Partial differential equations that describe mass balances of the considered chemical species within all fluid phases present in the system and (in the case of adsorption) on the solid. These balance equations may contain source terms that correspond to decay, and chemical reactions, and terms that express phase transfers, e.g., due to adsorption, and volatilization of the solutes.
- Dispersive, and diffusive flux equations for the mass of the considered species.
- Constitutive equations for the fluid phases, and for the solid (in the case of a deformable solid). These include also thermodynamic relationships that describe the partitioning of the species between adjacent phases under equilibrium conditions, or transfer functions for non-equilibrium transfers.
- Expressions for the various external sources and sinks for the mass of the considered fluid phases and the considered species.
- In the case of reacting species, we'll need all the relevant chemical reactions and various related coefficients.
- Initial conditions for each of the relevant state variables.
- Boundary conditions for each of the relevant extensive quantities—mass of fluid phases, and of the considered species.
- Numerical values, or functional relationships for all the coefficients that appear in the various balance equations and constitutive relations included in the model.

The content of a problem statement, especially the separation into a flow and contaminant transport sub-problems, has been introduced in a rather simplified form in order to emphasize the structure of a mathematical model of contaminant transport, and the roles played by the various types of equations (balance, flux, constitutive, and definitions). Based on the standard content of a model, as presented above, we usually end up with a large number of variables that describe the state of the system. To obtain a closed set of equations, within the framework of a *well posed problem*, we need an equal number of equations. However, following the discussion on *primary variables* (or *degrees of freedom*), presented in Sect. 3.9, the next step is to determine the number of *primary variables* (or *degrees of freedom*) of the problem. The number of partial differential equations of balance that has to be solved is then equal to the number of the primary variables. All other variables are obtained from the known values of the selected primary variables, using the remaining equations.

Actually, when we assume that interphase mass transfers take place *under equilibrium conditions*, a single mass balance equation can be written for the component in the porous medium as a whole. The single variable is, then, the concentration of the component within some fluid phase in the porous medium. The boundary condition for such an equation is on the concentration or total mass flux of the component in the porous medium as a whole. Similarly, initial conditions are written for the concentration or total mass of the component in all the phases present in the system.

The statement of the transport model for a particular site must also include information on all relevant porous medium parameters, such as porosity, permeability and dispersivity. Information is also required concerning the coefficients that appear in the constitutive relations, e.g., decay and growth coefficients, partitioning coefficients and reaction rate coefficients.

7.5.6 *Multiple Reacting-Species in Multiple Phases*

Consider the case of two or three fluid phases that together occupy the void space. Phase changes may occur. The phases involve a number of chemical species with the possibility of chemical reactions within the phases and interphase transfers. The flow and transport may be under non-isothermal conditions with T denoting the temperature which is common to all phases *at a point*. All these cases have been presented and discussed throughout the book.

With X_α^γ denoting the molar concentrations of the γ-species in the α-phase, we have the following constraints on the problem variables:

$$\sum_{\gamma=1}^{\rm NC} X_\alpha^\gamma = 1, \quad {\rm NP} = 2 \ \text{ or } \ 3. \tag{7.5.25}$$

For the saturations, we write the constraint

$$\sum_{\alpha=1}^{\rm NP} S_\alpha = 1, \quad {\rm NP} = 2 \ \text{ or } \ 3. \tag{7.5.26}$$

Under the assumption of chemical equilibrium (Sect. 2.1) among the phases,

When non-isothermal conditions are stipulated, we need to include in the model also the energy balance equation, say (8.4.15), and, obviously, provide appropriate initial and boundary conditions for that equation.

Actually, we have introduced this subsection only for the sake of completeness, as all the information (balance equations, constitutive equations, etc.) required for the construction of well posed models of "reactive transport", the term often used for the transport of multiple reacting chemical species in multiple fluid phases, has been presented and discussed throughout the book:

- Multi-phase flow is presented and discussed in Chap. 6.
- Multiple interacting chemical species, including chemical reactions and interphase mass transfer are discussed in the current chapter.
- Non-isothermal conditions are considered in Chap. 8.
- Flow and transport in the void-space of deformable porous media are considered in Chap. 9.
- In Appendix A, we have presented models of chemical reactors encountered in the Chemical Industry, in which phenomena of interphase mass transfers and chemical reactions occur in multi-phase flow under non-isothermal conditions.

In view of all these presentation, there is no need at this point to elaborate here on modeling chemical reactions in multi-phase flow.

7.6 Stochastic Modeling and CTRW

7.6.1 Comments on the Stochastic Approach

In Sect. 1.1.6, we discussed the conditions that have to be satisfied in order to facilitate the treatment of a geological formation as a continuum. Here, 'treatment' means the setting up of a well-posed mathematical (continuum) model of flow and transport in a specified porous medium domain and solving it in order to predict the domain's future behavior in response to imposed excitations in the form of external sources and boundary conditions. However, we have emphasized that most geological formations, e.g., petroleum reservoirs, groundwater aquifers and geothermal reservoirs, are *highly heterogeneous* to the extent that an REV, or an RMV cannot be found. The use of the continuum approach for such formations if often questioned.

In certain cases, heterogeneity of a geological domain manifests itself in the form of subdomains, or layers, each of which may be homogeneous, or slightly heterogeneous, so that a solution to the difficulties of high heterogeneity can be found by dividing the considered domain into sub-domains, each of which is homogeneous, or only slightly heterogeneous.

In fact, even if an RMV can be found, or *assumed to exist*, strong heterogeneity of a geological formation may produce phenomena which do not exist, or have been neglected as insignificant, in the case of homogeneous, or slightly heterogeneous formations. In addition, usually, there is not enough information concerning the spatial variability of the relevant coefficients, because of the high cost of drilling and data acquisition. Interpolation does not solve the problem, as it leaves a high level of uncertainty as to the values of the considered coefficients between points at which actual data are available.

In spite of the lack of sufficient information about coefficients, in the real world, *decisions (based on predicting excitation-response relations) have to be made* with respect to flow and transport in such (highly heterogenous) reservoirs. This led to efforts to find techniques for coping with the lack of sufficient information concerning model coefficients, obviously at the cost of a certain degree of uncertainty in the predicted values.

We consider phenomena of transport of mass of fluid phases and of dissolved substances in large, often deep, geological formations. It is obvious that no model (e.g., of the continuum-type discussed throughout this book) can provide predictions of flow and transport regimes unless information is available on the spatial distributions of model coefficients. Unfortunately, in many situations, sufficiently accurate information on formation coefficients (i.e., ones that appear in these models) cannot be obtained, due to the lack of sufficiently good monitored information (in terms of quality and quantity). Under such conditions, most planners and decision-makers will be satisfied with information on the future transport of an extensive quantity within an investigated formation provided in the form of a *statistical description of the transport of that quantity*. Forecasts may take the form not of the future water levels, or concentrations, at a considered point, but of the *statistics* of possible values of such variables at that point. Values of coefficients at a point still should be interpreted as representing the behavior of the porous medium in the vicinity of the point, but there is no need to specify the porous medium domain represented by the point value. For many decision making purposes, this is sufficient and valuable information. The trigger for using such techniques is always insufficient information on the spatial distribution of formation coefficients, and the usefulness of a statistical description of model predictions.

Stochastic modeling is an important branch of modeling transport phenomena, especially in heterogeneous geological formations, in which statistical description of model coefficients leads to statistics of predicted values of state variables. There is no doubt that this is a very important tool for dealing with problems of flow and transport in highly heterogenous domains, and this means practically *in all large and deep geological formations*. One such technique, the Monte Carlo simulation technique, has been presented in Sect. 5.4. Stochastic modeling is beyond the scope of this book. The reader should consult the literature, e.g., Gelhar (1986), the excellent text by Dagan (1989), and Dagan and Neuman, eds. (1997).

7.6.2 Statistical Approaches and the CTRW Method

The usual dispersion-based solute transport models fail to describe what happens in a geological formation, especially, but not only, in large scale cases. Significant deviations of field observations of solute concentration from model predicted values are observed in most cases. This is usually attributed to the *high heterogeneity* of geological formations. The commonly accepted conclusion is that the usually employed dispersion model, in which the dispersive flux is Fickian-type law and a dispersivity that depends on the size of the considered field, is inappropriate, especially in the case of solute plumes. The CTRW (= Continuous Time Random Walk) method presented below is an attempt to overcome some of the above difficulties.

In Sect. 7.2.4, we have *upscaled* the solute transport model, from the macroscopic level the megascopic one, by the same averaging approach as used for the upscaling from the microscopic to the macroscopic levels. We have assumed that the dispersive flux, introduced by the averaging procedure, is *Fickian*, i.e., proportional to the concentration gradient. In recent years, some researchers of dispersive flux, on the basis of field observations of solute plume propagation, reached the conclusion that under field conditions (and some concluded that even under laboratory ones) the dispersive flux is often *non-Fickian*. They related this conclusion mainly to the *heterogeneity of the porous medium domain*, which, as emphasized earlier, is an *inherent characteristic of natural domains, especially when the latter are large*. The heterogenous nature of a domain introduces phenomena that have not been represented in the discussion on macroscopic level modeling. Furthermore, they show that for the Fickian dispersive flux to be valid, the solute migration time must be sufficiently long before the solute mass balance equation (or the *advection-dispersion equation*, as it is sometimes referred to) becomes applicable. According to Levy and Berkowitz

(2003) and Cortis et al. (2004), the above observation is valid even in relatively homogeneous geological formations, say, with respect to porosity and permeability. Actually these authors concluded that non-Fickian behavior is exhibited even in rather short laboratory columns (e.g., demonstrated in the form of long time tails of solute concentration). Their analysis showed that the motion and spreading of solute plumes are characterized by distinct temporal scaling, i.e., the time dependence of the spatial moments does not correspond to the normal (or Gaussian) distribution that characterizes Fickian-type transport.

We recall that porous medium coefficients, like porosity and permeability, are introduced in the microscopic-to-macroscopic averaging process, to replace the (unavailable) information concerning the geometry of (microscopic) interphase boundaries. In the passage to higher levels of continuum, the same kind of coefficients were used, e.g., permeability and porosity, adjusted to the higher scale, presumably accounting also for the field-scale heterogeneity. Is that the right way to account for real formation heterogeneity?

Based on the above observations,

- that the dispersive flux cannot be described by a Fickian type law,
- that geological domains are heterogeneous, often highly so, characterized by a multi-scale heterogeneity, i.e., depending on the size of the advancing plume, and
- that the parameters that affect solute spreading are time dependent,

some researchers proposed approaches other than stating and solving the solute's mass balance PDE also in highly heterogeneous porous media with scale affected coefficients. These approaches include the stochastic modeling introduced above.

The basic idea underlying the statistical approaches to solute transport in porous media is that it is impossible to provide an exact mathematical description of the motion of a single solute particle of a species dissolved in a moving fluid, a description that is required for any forecast of spreading, or dispersion, of a cloud of such particles. The path of a single species particle, say an ion or a molecule, should not be identified with that of a liquid particle, which is an instantaneous *ensemble* of a large number of molecules the identity of which changes as a result of molecular diffusion. The path of a species particle may be visualized as the vectorial result of two motions: one along pathlines (or streamlines in steady state) of liquid particles, and another one which involves the passage from one pathline to an adjacent one by molecular diffusion. The nature of both motions, the first determined by the intricate internal geometry of the void space, and the second by the random character of molecular diffusion, prevents any deterministic prediction of paths of solute particles.

A. Some Early Statistical Approaches to Dispersion

state. Instead of considering a *single* state of the system, we consider *many* such states, each of which is likely to occur. We refer to a all of them together as an *ensemble* of states. This ensemble contains all the states that cannot be distinguished macroscopically because we are ignorant as to the microscopic details of the considered system. Then, the expected value of an observable quantity is the average of that quantity over all states of the ensemble. For these expectation values, we try to deduce predictions from what possible at any single given time. Then ensemble average and time average variables may be interchanged. This is the *ergodic hypothesis*.

Around 1950, at the early stages of research on 'miscible displacement' (later called 'solute dispersion'), the movement of a solute was envisaged as the *random movement* of solute mass particles. The randomness was envisaged as a consequence of the random structure of the void space. Bear (1972, pp. 587–603) summarizes some of these statistical models. The underlying postulate of the statistical approach is that although it is impossible to predict the exact path of an *individual* solute particle, one may employ the rules of probability theory to predict the spatial distribution, at any later time, of a *cloud of many* solute particles that (1) were, initially, at a close proximity, and (2) move under the same average conditions. Obviously, the flow itself obeys the laws of physics, and our reference to 'random motion', or 'random walk', is a way to express our ignorance as to the motion's details. The random motion of an anonymous particle is assumed to represent the motion of an *ensemble* of many such particles (see any book on stochastic processes, e.g., Todorovic 1992). The probability distribution of the location of a single particle is interpreted as representing the spatial distribution of the concentration of a cloud of tracer particles originating from the neighborhood of a certain point at a certain time, and moving under the same average flow conditions. Rather than investigate a specific porous medium, Scheidegger (1958, and a review in Bear 1972, pp. 590–592) suggested to investigate an *ensemble* of porous media, assumed to be macroscopically identical, i.e., having identical macroscopic properties. This analysis led to the description of particle spreading as a *random walk*. Scheidegger's work was based on the following assumptions:

(a) The homogeneous isotropic porous medium constitutes an ensemble of a large number of systems, or samples, having identical macroscopic characteristics.
(b) Laminar flow of a homogeneous fluid, regarded as a continuum. Each of the ensemble particles has its own path,
(c) The motion of a particle through a specific porous medium sample is made up of a sequence of straight elementary displacements of equal duration.
(d) The direction and length of each displacement take on *random* values, with no correlation among displacements.

The above are actually the characteristics of 'random walk'.

Under these conditions, a sufficiently long path of a particle through a specific system incorporates all the conditions encountered in the ensemble of systems. Hence, we may apply the *ergodic property*, which enables us to interpret a temporal average along a single path as a spatial average over a large ensemble of paths.

The *one-dimensional random walk* model is, probably, the simplest statistical model that describes dispersion. In this model, a solute particle performs *displacements* along a straight line in the form of a series of steps taken either forward, say to the right, or backward, i.e., to the left, with equal probability of 50%. After N such steps, the particle may be at any of the points:

$$-N, -N+1, \ldots -1, 0, +1, \ldots, N-1, N$$

Assuming that each step is *equally likely* to be taken in either direction, independent of the history of all proceeding displacements, the probability of any sequence of N steps is $\left(\frac{1}{2}\right)^N$. Hence, the probability $P(M, N)$ that a particle will arrive at point M after N displacements is:

$$\left(\tfrac{1}{2}\right)^N \times \begin{pmatrix}\text{the number of distinct sequences which} \\ \text{will lead to a point } M \text{ after } N \text{ steps.}\end{pmatrix}.$$

In order to arrive at M after N steps, $(N + M)/2$ steps must be taken in the positive direction, while (N – M)/2 steps must be taken in the negative one. Hence, the total number of distinct sequences is:

$$P(M, N) = \frac{N!}{[(M + N)/2]![(N - M)/2]!}\frac{1}{2}^N,$$

which is *Bernoulli's distribution*. The *root mean square displacement* is $N^{1/2}$. As $N \to \infty$, and $M \ll N$, we obtain:

$$P(M, N) = (2\pi N)^{1/2} \exp(-M^2/2N).$$

With ℓ denoting the length of a step, and $x = M\ell$. The probability element $P(x)\Delta x$ that the particle is likely to be in the interval $(x, x + \Delta x)$ after N displacements, with $\Delta x > \ell$ is:

$$P(x, N)\Delta x = P(M, N)\cdot(\Delta x/2\ell),$$

or:

$$P(x, N) = \frac{1}{2\pi N\ell^2} \exp\left(\frac{-x^2}{2N\ell^2}\right). \tag{7.6.1}$$

If a particle undergoes n displacements of length l per unit time, i.e., $u = n\ell$, the probability that it will find itself between x and $x + \Delta x$ at time t is:

$$P(x, t)\Delta x = \frac{1}{2\sqrt{\pi Dt}} \exp\left(\frac{-x^2}{4Dt}\right)\Delta x, \quad D = n\ell^2 = \frac{l}{2}u. \tag{7.6.2}$$

When, in the random walk described above, the duration of a time step of length ℓ is Δt, and the probability of the particle taking a step in the $+x$-direction is p, with $q = 1 - p$ denoting the probability that the particle will take a step in the $-x$ direction, we obtain:

$$P(x, t + \Delta t) = pP|_{(x-\Delta x,t)} + q\ell^2|_{(x+\Delta x,t)}.$$

Developing in a Taylor series and (1) neglecting terms beyond the second order, (2) passing to the limit with $\Delta x \to 0$, $\Delta t \to 0$, but such that the mean and the variance of the displacement will remain finite for all values of t, and (3) taking p and q of $O(\Delta x)$, we obtain for P:

$$\frac{\partial P}{\partial t} = D\frac{\partial^2 P}{\partial x^2} - V\frac{\partial P}{\partial x}, \tag{7.6.3}$$

in which $D = (\Delta x)^2/2\Delta t$, $V = 2(p-q)D/\Delta x$. This is the partial differential equation that describes random walk after a large number of elementary displacements. Its elementary solution is:

$$P(x,t) = \frac{1}{\sqrt{4\pi Dt}}\exp\frac{-(x-Vt)^2}{4Dt}, \tag{7.6.4}$$

which is the same as (7.6.2).

Scheidegger (1954, 1958) suggested that the application of the statistical approach requires:

(a) An assumption about the statistical (average) properties of the porous medium (= the ensemble).
(b) An assumption on the micro-dynamics of the flow, i.e., on the relationships between the forces, the liquid properties, and the resulting velocity during each small time step; in general, the flow is assumed laminar.
(c) A choice of a type of statistics to be employed, i.e., the probability of occurrence of events during small time intervals within the chosen ensemble. This may take the form of correlation functions between velocities at different points or different times, or joint-probability densities of the local velocity components of the particle as functions of time and space, or a probability of an elementary particle displacement. In the case of dispersion considered here, the selected correlation function determines the type of dispersion equation derived.

The statistical approaches are based on the movement of (here solute) particles. As the total particle travel time becomes much larger than the time interval during which its successive (local) velocities are still correlated, the total displacement may be considered as a sum of a large number of elementary displacements which are *statistically independent*. Then, the *probability distribution* of the particle's total displacement tends to the *normal (Gauss) distribution*. This observation is based on the *central limit theorem* of probability, which states that no matter what the probability distribution of each elementary displacement (or of the velocity during each elementary displacement) is, as the number of steps increases, the probability distribution of the total displacement tends to normal. In view of the *ergodic principle*, this distribution also represents the spatial distribution of displacements of a cloud of initially close particles.

In order to express these concepts in a mathematical form, let $\bar{\mathbf{V}}$ denote the average velocity vector of uniform flow of a homogeneous fluid through a homogeneous porous medium of infinite areal extent. At any instant t, the velocity component $\bar{V}_i$ of a marked fluid (i.e., solute) particle is expressed by $V_i(t) = \bar{V}_i + \hat{V}_i(t)$, with

$\hat{V}_i(t)$, $1 = 1, 2, 3,,$ denoting the (random) deviation from the average. Scheidegger suggests the following development:

The location of a particle that starts at the origin at $t = 0$, reaching $\hat{x}(t)$ at time t:

$$\hat{x}_i(t) = x_i(t) - \bar{x}_i(t), \quad i = 1, 2, 3. \tag{7.6.5}$$

A particle starting from the origin at $t = 0$, will reach a point $\hat{x}_i(t)$ at time t, such that:

$$d\hat{x}_i(t) = \hat{V}_i(t)dt, \quad \hat{x}_i(t) = \int_0^t \hat{V}_i(\theta)d\theta. \tag{7.6.6}$$

The spread of the particles' possible displacements around its average displacement, $\bar{x}_i(t)$, is described by the *matrix of covariances* (= correlation moments) $\overline{\hat{x}_i\hat{x}_j}$. Thus, with $f(x_i, x_j)$ denoting the *joint probability function* of x_i and x_j, we have:

$$\begin{aligned} Cov(x_i, x_j) = \overline{\hat{x}_i\hat{x}_j} &= \int_{-\infty}^{+\infty}\int_{-\infty}^{+\infty} (x_i - \bar{x}_i)(x_j - \bar{x}_j) f(x_i, x_j) dx_i dx_j \\ &= \overline{x_i x_j} - \bar{x}_i\bar{x}_j, \quad f(x_i, x_j) = \text{correlation function}. \end{aligned} \tag{7.6.7}$$

The covariances become the variances σ^2 of x_i:

$$\mu_{ii} = \sigma^2(x_i) = \text{Var}(x_i) = \overline{(\hat{x}_i)^2}, \tag{7.6.8}$$

with σ denoting the standard deviation of x_i.

The two components $V_i(t)$ and $V_j(t)$ of $\mathbf{V}(t)$ in the i and j directions, are two stationary random functions of time. This means that the correlation coefficient, $r(t, t')$ is independent of t, t', but depends only on the interval $(t - t')$ between them:

$$r_{ij}(t, t') = \frac{\overline{\hat{V}_i(t)V_j(t')}}{\sigma[V_i(t)]\sigma[V_j(t')]}, \quad \text{(no summation)}, \tag{7.6.9}$$

and we require that the average, the variance and the other moments remain constant in time, i.e., $r_{ij}(t, t') = r_{ij}(t, t + \tau) = r_{ij}(\tau)$, $\tau = t' - t$. We consider the derivative:

$$\begin{aligned} \overline{\frac{d}{dt}[\hat{x}_i(t)\hat{x}_j(t)]} &= \frac{d\hat{x}_i(t)}{dt}\hat{x}_j(t) + \hat{x}_i(t)\frac{d\hat{x}_j(t)}{dt} \\ &= \int_0^t \hat{V}_i(t)\hat{V}_j(\theta)d\theta + \int_0^t \hat{V}_i(\theta)\hat{V}_j(t)d\theta. \end{aligned} \tag{7.6.10}$$

By averaging, we obtain:

$$\overline{\frac{d}{dt}[\hat{x}_i(t)\hat{x}_j(t)]} = 2r_{ij}(\theta)\overline{\hat{V}_i(t)\hat{V}_j(t)} \int_0^t r_{ij}(\tau)d\tau, \tag{7.6.11}$$

in which $r_{ij}(\tau)$ is the *coefficient of correlation* between the velocity component $V_i(t)$ and $V_i(t+\tau)$. From the above equation, we obtain:

$$\begin{aligned}\overline{\hat{x}_i(t)\hat{x}_j(t)} &= 2r_{ij}(0)\overline{\hat{V}_i(t)\hat{V}_j(t)}\int_0^t\int_0^\theta r_{ij}(\tau)d\tau d\theta\\ &= 2r_{ij}(0)\overline{\hat{V}_i(t)\hat{V}_j(t)}\int_0^t(t-\tau)r_{ij}(\tau)d\tau.\end{aligned} \tag{7.6.12}$$

Because $V_i(t)$ and $V_j(t)$ are stationary random functions, $\overline{\hat{V}_i(t)\hat{V}_j(t)}$ is a constant, independent of t. If t_o is the time during which velocities are still effectively correlated (i.e., *correlation time*), then:

$$t_o = \int_0^\infty r_{ij}(\tau)d\tau. \tag{7.6.13}$$

Scheidegger (1954, 1958) investigated two cases:

- $t \gg t_o$. then (7.6.12) reduces to

$$\overline{\hat{x}_i(t)\hat{x}_j(t)} \approx 2r_{ij}(0)\overline{\hat{V}_i(t)\hat{V}_j(t)}t_o t, \tag{7.6.14}$$

which can be written as:

$$D_{ij} = \frac{\overline{\hat{x}_i\hat{x}_j}}{2t} = r_{ij}(0)\overline{\hat{V}_i\hat{V}_j}t_o. \tag{7.6.15}$$

For $i=j$, $r_{ij}(0)=1$, and:

$$D_{ii} = \frac{\overline{\hat{x}_i^2}}{2t} = \overline{\hat{V}_i^2}t_o;\quad \sigma^2(x_i) = 2D_{ii}t. \tag{7.6.16}$$

- $t \ll t_o$. Scheidegger (1960) obtains:

$$\begin{aligned}&\overline{\hat{x}_i\hat{x}_j} = r_{ij}(0)\overline{\hat{V}_i\hat{V}_j}t^2,\quad D_{ij} = \frac{\overline{\hat{x}_i\hat{x}_j}}{2t} = \frac{1}{2}r_{ij}(0)\overline{\hat{V}_i\hat{V}_j}t,\\ &D_{ii} = \frac{\overline{\hat{V}_i^2}}{2t} = \frac{\overline{\hat{V}_i^2}}{2}t,\qquad \sigma^2(x_i) = \overline{\hat{V}_i^2}t^2,\end{aligned} \tag{7.6.17}$$

which expresses the fact that for very short time intervals, there is no random process and every particle progresses at its velocity. If each elementary displacement, $d\hat{x}_i(t)$, is a random variable, the total deviation of the particle's position from its average position tends to a normal distribution only if $t \gg t_o$. This is a consequence of the *central limit theorem* (Chandrasekhar 1943, see Bear 1972, p. 594):

$$\Phi(x_1, x_2, x_3, t) = \frac{1}{(2\pi)^{3/2}|\mu|^{1/2}} \exp\left[-\frac{1}{2}\frac{\mu'_{ij}}{|\mu|}(x_i - \overline{x}_i)(x_j - \overline{x}_j)\right], \qquad (7.6.18)$$

in which $|\mu|$ is the determinant of the correlation matrix $[\mu_{ij}]$, $i, j = 1, \ldots, n$, and the summation convention is employed. The symbol Φ denotes the relative solute mass concentration in percent of total mass per unit volume of porous medium at the considered point. The solute concentration, c is related to Φ by $c = \Phi m/\phi$. Note that $\Phi(x_1, x_2, x_3, t) = \Phi_1(x_1, t)\Phi_2(x_2, t)\Phi_3(x_3, t)$, where $\Phi_i(x_i, t), i = 1, 2, 3$, is a normal density function of a single random variable x_i at time t_3. Thus, in the one-dimensional case,

$$\Phi_x(x, t) = \frac{1}{(2\pi)^{1/2}\sigma} \exp\left[-\frac{(x - \overline{x})^2}{2\sigma^2}\right], \quad \sigma = (2Dt)^{1/2}, \quad D \equiv D_{ii}. \qquad (7.6.19)$$

The above results do not show the relationships between σ_i, or μ_{ij}, and statistically averaged medium and flow parameters. This information may be obtained by choosing the microdynamics of the flow within each elementary time interval. However, the above discussion explains why the solute distribution predicted by solving the macroscopic solute mass balance equation for the spreading of particles from a point source takes the form of a normal (Gaussian) distribution, with $\sigma^2 = \sqrt{2Dt}$.

Under these conditions, Scheidegger's analysis for uniform flow lead to the normal (Gaussian) distribution after a sufficiently large number of steps:

$$P(x, y, z, t) = \frac{1}{4\pi Dt} \exp\left(-\frac{(x - \bar{x})^2 + (y - \bar{y})^2 + (z - \bar{z})^2}{4Dt}\right),$$

in which $D = \sigma^2/2t = \sigma^2/2\Delta t$. $\bar{x} = V_x t, \bar{y} = V_y t$, $\bar{z} = V_z t$, with V denoting the uniform velocity. In view of the ergodic property, the above equation describes the spatial distribution at time t of a large number of particles that started ($t = 0$) from the vicinity of the same origin, and travelled along independent paths under similar statistical conditions. Such an ensemble of particles is normally distributed around its center, which travels at the average flow velocity.

Chandrasekhar (1943; see Bear 1972, p. 594) presented a discussion on the general, three-dimensional *random walk*, using *Markov's method*. His analysis of random displacements in which the particles undergo, on the average, $\bar{n}$ displacements per unit time, $n = \bar{n}t$ $(r_1, r_2, \ldots)$, led to the probability distribution of the particles' locations, after n displacements, $n = \bar{n}t$, we define the average velocity of the particle by $V_x = \bar{n}\bar{x} = n\bar{x}/t = \bar{X}/t$ and similar expressions for y and z. Then, in the form of the *normal distribution*:

$$W(R)dR = \frac{1}{(2\pi)^{3/2}\sigma_X\sigma_Y\sigma_Z} \exp\left\{-\frac{(X - \bar{X})^2}{2\sigma_X^2} - \frac{(Y - \bar{Y})^2}{2\sigma_Y^2} - \frac{(Z - \bar{Z})^2}{2\sigma_Z^2}\right\}, \qquad (7.6.20)$$

in which $X = \sum_1^n x_i$, $\bar{X} = (1/n) \sum_1^n x_i$, etc. Eventually, his analysis led to:

$$W(R,t)dR = \frac{dXdYdZ}{(4\pi t)^{3/2}(D_{11}D_{22}D_{33})^{1/2}} \exp\left\{-\frac{(X-V_x t)^2}{4D_{11}t} - \frac{(Y-V_y t)^2}{4D_{22}t} - \frac{(Z-V_z t)^2}{4D_{33}t}\right\}, \tag{7.6.21}$$

in which $D_{11} = n\bar{X^2}/2t = \sigma_x^2/2t$, etc. The above equation is an elementary solution of the PDE:

$$\frac{\partial W}{\partial t} = D_{ik}\frac{\partial^2 W}{\partial X_i \partial X_k} - V_i\frac{\partial W}{\partial X_i}, \tag{7.6.22}$$

which is nothing but a simplified form of the dispersion equation (called ADE):

$$\frac{\partial c}{\partial t} = D\frac{\partial^2 c}{\partial x^2} - V\frac{\partial c}{\partial x}, \tag{7.6.23}$$

with D denoting the coefficient of dispersion.

Dankwerts (1953) also made use of the random walk approach. However, he added the assumption of *residence time of an elementary step*, i.e., the time during which a fluid particle passes through an elementary channel, or through an elementary length of the medium. In his *random residence time* model, the flow takes place through a sequence of cells in which complete mixing takes place. Josselin and de Jong (1958) was probably the first to express, in an analytical way, the fact that the longitudinal dispersion is larger than the transversal one. In earlier random walk models, each unit step is characterized by the lack of relationship among displacements and with respect to the main flow direction. Also, all steps had the same duration. In de Jong's work, the residence time of a particle in a channel depends on the direction of the channel with respect to that of the uniform (averaged) fluid velocity. All channels have the same length, but the velocity in them varies with their direction. This leads to a random walk in which each particle undergoes a different number of elementary displacements in a given time interval. A similar analysis was presented also by Saffman (1959, 1960).

It is interesting to note that as the total travel time of a solute particle becomes much larger than the time interval during which its successive (local) velocities are still correlated, the total displacement may be considered as a sum of a large number of elementary displacements which are statistically independent of each other. Then, the probability distribution of the particle's total displacement tends to the *normal (Gauss) distribution*. This is based on the *central limit theorem* of probability, which states that no matter what the probability distribution is for the elementary displacements (or for the velocity during these displacements), as the number of steps increases, the probability distribution of the total displacement tends to normal. In view of the ergodic principle, this distribution also represents the spatial distribution of displacements of a cloud of initially close particles.

Todorovic (1992, p. 17) suggests that random walk representing the movement of particles in the void space of a porous medium should be with time intervals, or delays, between successive particle displacements. These can be attributed to the effect of the very slow motion in dead-end pores. For sufficiently long times, he found that the coefficient of dispersion D is proportional to V^n, $n > 1$. This should be compared with a random walk development by Scheidegger (1960, p. 193) that resulted in $D = \text{const.} \times V^n$, with $n = 2$ in a dynamic model, and $n = 1$ in a geometrical model.

B. The Continuous Time Random Walk (CTRW) Method

The *Continuous Time Random Walk* (abbrev. CTRW) method is a generalized statistics-based *random walk* method, applicable to a variety of subjects in many disciplines. It is a random walk model in which particle transitions between different states are controlled by a *probabilistic waiting-time distribution function*. This waiting time is a feature that does not exist in the usual random walk models. Here we focus on its application to solute transport, and the relevant particles are solute particles carried by the fluid moving in the void space.

The random walk approach to solute transport was introduced above as a background to the CTRW method considered here. We are presenting this method here because of its ability to handle non-Fickian dispersion and certain effects of heterogeneity (at all levels) of geological formations (e.g., Berkowitz et al. 2008, p. 226). The method was initially introduced by Montroll and Weiss (1965) as a purely mathematical generalization of the regular random walk. It was first applied to transport problems by Scher and Lax (1973a, b). Further developments and applications were published, among others, by Scher and Montroll (1975), Metzler and Klafter (2000), Berkowitz et al. (2002), Dentz and Berkowitz (2003), Cortis and Berkowitz (2004), Berkowitz et al. (2006), and Rhodes et al. (2008).

Like in a random walk, also in the CTRW method, we consider the random movement and spreading of a cloud of labeled (e.g., solute) particles that execute a series of movements at a variable velocity. However, in the CTRW method, these random elementary displacements are:

- Strongly affected by properties of the void space configuration, as manifested by the relevant porous medium properties, especially by the heterogeneity inherent in geological formations, and
- Characterized by a time interval separating successive particle displacements, for example, due to delays associated with *dead end pores*, or, in general, zones of slow fluid movement. These may require the introduction of a distribution of delay intervals between random walk steps.

The CTRW is a *random walk model* (see discussion in Sect. 7.6.2 A) that is not based on solving the solute's mass balance equation referred to as '*Advection-Dispersion Equation (ADE)*'. Instead, it is based on solving a different mass balance formulation. Berkowitz et al. (2000) and Cortis et al. (2004) show that the ADE can be obtained as the limiting case in a very homogeneous domain.

An important feature of the CTRW method is that it can incorporate critical effects of porous medium heterogeneity. For example, CTRW can distinguish between *porosity* and *effective porosity*. The latter is equal to the difference between the porosity and the *dead end porosity* (Sect. 1.1.7 A). The CTRW also takes into account residence times and delays of particles between consecutive steps in the random walk model, as a consequence of various delay processes, dead-end pores, slow transport zones, and trapping zones. It also takes into account the notion of 'preferential pathways', as well as 'mobile-immobile' portions of the porous medium domain. These are typical characteristics of heterogeneous geological domains. The method can also take into account the inherent multi-scale heterogeneity of geological domains (Berkowitz et al. 2006).

In the CTRW approach, the transport of a solute is envisioned as the movement of a series of particles that move between nodal points at a specified probability. This is similar to the basic idea of random walk presented earlier in this section, except that the random walk is generalized here by considering a more general probability function that governs the passage of a particle from one stage to the next. A major building block of the CTRW method is the probability, $\psi(t; i, j)$, of a particle arriving at time t at node i 'hopping' to another node (j) at time $t + \Delta t$.

Berkowitz et al. (2006) and Berkowitz et al. (2008) show that a wide range of particle spreading patterns can be described by the *continuum-level equation*:

$$u\tilde{c}(s, u) - c_o(s) = -\tilde{M}(u)\left[\mathbf{V}_\psi \cdot \nabla \tilde{c}(s, u) - D_\psi \cdot \nabla\nabla \tilde{c}(s, u)\right], \qquad (7.6.24)$$

where the $\tilde{(.)}$ denotes the *Laplace transform* of (.), $c_o(s)$ represents the initial value of c, $\mathbf{V}_\psi$ is the 'transport velocity', which is different from the fluid's flow velocity $\mathbf{V}$, and $\tilde{M}$ represents a '*memory function*', defined by:

$$\tilde{M}(u) = \bar{t}u\frac{\tilde{\psi}(u)}{1 - \tilde{\psi}(u)}, \qquad (7.6.25)$$

in which $\bar{t}$ denotes a characteristic time. Berkowitz et al. (2000) emphasize that the 'dispersion coefficient' D_ψ in (7.6.24) has a different physical interpretation than the coefficient of dispersion appearing in the classical ADE.

Under relatively homogeneous conditions, $\tilde{M}(u) = 1$ and (7.6.24) reduces to the usual ADE. With other choices of $\tilde{M}(u)$, other common equations, such as *multirate* and *mobile-immobile* equations can be recovered. Berkowitz et al. (2000) show that the CTRW method can also be used to handle solute transport that takes into account delays as a result of the presence of trapping, low permeability lenses, adsorption/desorption and various delay mechanisms. Altogether, a key result of the CTRW approach is its solute transport equation, say (7.6.24), represents a significant generalization of the ADE.

A variety of specific mathematical formulations of the CTRW approach have been considered to date. Obviously the actual implementation of CTRW requires appropriate numerical solutions. The CTRW method is outlined below.

Let $P_n(\ell)$ denote the probability of a particle arriving at the position ℓ after n steps, and let $p(\ell, \ell')$ denote the probability that the particle will move from ℓ to ℓ' under the constraint that $\sum_{\ell'} p(\ell, \ell') = 1$. Then, the considered kind of random walk obeys

$$P_{n+1}(\ell) = p(\ell, \ell')P_n(\ell'). \tag{7.6.26}$$

Next, the time variable n in the above equation is converted into a continuous time, t, while keeping the spatial distribution discrete. With this, the probability per unit time, $R(s, t)$, of a particle just arriving at a site s at time t, can be written as a function of the probability of moving between two discrete locations separated by a length s, with a difference in arrival times of t, expressed in the form (Scher and Lax 1973b):

$$R(s, t) = \sum_{s'} \int_0^t \psi(s - s', t - \tau)R(s', \tau)d\tau. \tag{7.6.27}$$

where ψ denotes transit time probability. The next step is to impose a periodic boundary conditions on the lattice (Scher and Lax 1973b). This takes the form

$$\mathbf{s} = \sum_i s_i \mathbf{a}_i, \quad , |\mathbf{a}_i| = a; \tag{7.6.28}$$

$$s_i + j_i N \to s_i, \quad i, j \in I, \tag{7.6.29}$$

where a is the lattice constant and s are integers given by condition (7.6.29). If Na is the length of the lattice condition, (7.6.29) can be rewritten as:

$$s_i = -\frac{N-1}{2}, \ldots, \frac{N-1}{2}, \tag{7.6.30}$$

The next step is to make a distinction between the probability of a particle just arriving at a site and the probability that it will remain at that site for a given time before taking the next jump. This requires an analogy to the P_n in (7.6.26). Using the concept of the Master Equation (Shlesinger 1996) and an ensemble average approach (Klafter and Silbey 1980) leads to:

$$\frac{\partial P(s,t)}{\partial t} = \sum_{s'} \int_0^t \phi(s - s'), (t - \tau)P(s', \tau)d\tau - \int_0^t \phi(s' - s), (t - \tau)P(s, \tau)d\tau, \tag{7.6.31}$$

where $P(s, t)$ is the probability of a walker being at s at a time t. Equation (7.6.27) is related to Eq. (7.6.31) through the following integral (Berkowitz et al. 2001)

$$P(s, t) = \int_0^t \Pi(t - \tau)R(s, \tau)d\tau, \quad \Pi(t) = 1 - \int_0^t \psi(\tau)d\tau, \tag{7.6.32}$$

where $\Pi(t)$ denotes the probability of remaining at s, and

$$\psi(\tau) = \sum_{s} \psi(s, \tau), \quad \psi(s, \tau) = \mathfrak{L}^{-1}\left(\frac{u\tilde{\psi}(s, u)}{1 - \tilde{\psi}(u)}\right), \tag{7.6.33}$$

with $\tilde{\psi}$ denoting the *Laplace transform* of ψ.

The CTRW approach accounts naturally for transport in preferential pathways, with mass transfer to stagnant and slow flow regions; CTRW can account for these physical transport mechanisms, as well as other factors that influence transport of reactive contaminants, such as sorption. Of specific interest here are the analyses by Cortis and Berkowitz (2004) of transport in partially saturated, laboratory columns. Three typical breakthrough curves from a series of miscible displacement experiments in partially saturated soils, presented by Nielsen and Biggar (1961, 1962) and Jardine et al. (1993) were re-analysed using the CTRW approach. They measured breakthrough curves on undisturbed cylindrical soil columns (8.5 cm diameter, 24 cm length). The soil columns were saturated with 0.05 M $CaCl_2$ from the bottom, and were then allowed to drain. Bromide was used as the non-reactive, passive tracer. Cortis and Berkowitz (2004) re-examined three breakthrough curves obtained from three different degrees of saturation. The CTRW solutions were found to reproduce the breakthrough behavior far more effectively than the advection-dispersion equation solution. Nielsen and Biggar (1961, 1962) reported systematic deviations in the calculated parameter values using the advection-dispersion equation from the experimental data, which displayed non-Fickian transport behavior. Again, the CTRW solutions were found to reproduce the breakthrough behavior far more effectively.

7.7 Colloidal and Nanoparticle Transport

Colloids in the soil were introduced in Sect. 1.3.2. here we shall discuss colloids as possible carriers of contaminants and the models that describe the transport of such colloids in a porous medium domain.

7.7.1 Mass Balance Equations for a Contaminant

We assume that the contaminant does not, in any way, affect the transport of the colloids, so that the colloid balance equations, (7.7.19)–(7.7.21), do not depend on the contaminant concentration. Thus, the colloid balance equations may be solved independently of the contaminant transport problem. The resulting colloidal concentrations and colloid transfer rates will then be used in the contaminant balance equation to determine the contaminant concentrations. This subject will be discussed as the next step.

The mass balance equation for the contaminants in the aqueous phase (volumetric fraction $\theta \equiv \theta_\ell$) is

$$\mathcal{B}_\ell(c_\ell) = f_{ads\to\ell} + f_{m\to\ell} + f_{im\to\ell} + f_{int\to\ell}. \tag{7.7.1}$$

For the adsorbed contaminant, the mass balance equation is

$$\mathcal{B}_{ads}(c_{ads}) = -f_{ads\to\ell}. \tag{7.7.2}$$

For the contaminant on the mobile colloids, the mass balance equation is

$$\mathcal{B}_m(c_m) = -f_{m\to\ell} + f_{im\to m} + f_{int\to m}, \tag{7.7.3}$$

For the contaminant on the immobile colloids, the mass balance equation is

$$\mathcal{B}_{im}(c_{im}) = -f_{im\to\ell} - f_{im\to m}, \tag{7.7.4}$$

and for the contaminant attached to the air-water interface, the mass balance equation is

$$\mathcal{B}_{int}(c_{int}) = -f_{int\to\ell} - f_{int\to m}. \tag{7.7.5}$$

In the above equations, the mass balance operators are defined by:

$$\mathcal{B}_\ell(c_\ell) \equiv \frac{\partial(c_\ell\theta)}{\partial t} + \nabla\cdot\theta\left(c_\ell\mathbf{V}_\ell - \mathbf{D}_{h\ell}\cdot\nabla c_\ell\right), \tag{7.7.6}$$

$$\mathcal{B}_m(c_m) \equiv \frac{\partial(c_m c_{col,m}\sigma_{col}\theta)}{\partial t} + \nabla\cdot\theta c_m\sigma_{col,m}\left(c_{col,m}\mathbf{V}_{col,m} - \mathbf{D}_{col}\cdot\nabla c_{col,m}\right), \tag{7.7.7}$$

$$\mathcal{B}_{ads}(c_{ads}) \equiv \frac{\partial(c_{ads}\Sigma_{s,\ell})}{\partial t}, \qquad \mathcal{B}_{im}(c_{im}) \equiv \frac{\partial(c_{im}c_{col,im}\sigma_{col})}{\partial t}, \tag{7.7.8}$$

$$\mathcal{B}_{int}(c_{int}) \equiv \frac{\partial(c_{int}c_{col,int}\sigma_{col})}{\partial t}, \tag{7.7.9}$$

where σ_{col} is the surface area on the colloid particles per unit mass of the colloids. Note that c_m, c_{im}, and c_{int}, are defined as the mass of contaminant per unit surface area of their respective colloidal particle type, and c_{ads} is the mass of contaminant per unit surface of the solid phase.

The exchange terms for the contaminants on the colloidal particles are defined by:

$$f_{im\to m} \equiv c_{im}f^*_{col,im\to m} - c_m f^*_{col,m\to im} \tag{7.7.10}$$

and

$$f_{int\to m} \equiv -c_{int}f^*_{col,int\to m} + c_m f^*_{col,int\to m}, \tag{7.7.11}$$

where

$$f^*_{col,im\to m} \equiv \max(f_{col,im\to m}, 0), \quad f^*_{col,m\to im} \equiv \max(-f_{col,im\to m}, 0),$$

$$f^*_{col,int\to m} \equiv \max(f_{col,int\to m}, 0), \quad f^*_{col,int\to m} \equiv \max(-f_{col,int\to m}, 0),$$

and the terms $f_{col,im\to m}$ and $f_{col,int\to m}$ are computed from (7.7.24) and (7.7.25).

The terms that express the transfer from the colloids to the water phase are modeled as:

$$f_{ads\to\ell} = (\Sigma_{s,\ell})\alpha^\star_{ads\to\ell}(c_{ads} - \rho_b K_d c_\ell/\Sigma_{s,\ell}), \tag{7.7.12}$$

$$f_{m\to\ell} = \sigma_{col} c_{col,m} \alpha^\star_{m\to\ell}(c_m - K_{eq,col} c_\ell), \tag{7.7.13}$$

$$f_{im\to\ell} = \sigma_{col} c_{col,im} \alpha^\star_{im\to\ell}(c_{im} - K_{eq,col} c_\ell), \tag{7.7.14}$$

$$f_{int\to\ell} = \sigma_{col} c_{col,int} \alpha^\star_{int\to\ell}(c_{int} - K_{eq,col} c_\ell). \tag{7.7.15}$$

The coefficients K_d and $K_{eq,col}$ are defined such that under equilibrium conditions, we have:

$$c_{ads}\Sigma_{s,\ell} = \rho_b K_d c_\ell, \tag{7.7.16}$$

$$c_m = K_{eq,col} c_\ell, \quad c_{im} = K_{eq,col} c_\ell, \quad c_{int} = K_{eq,col} c_\ell. \tag{7.7.17}$$

Altogether, we have five mass balance equations in the five concentrations: c_ℓ, c_{ads}, c_m, c_{im}, and c_{int}.

If equilibrium adsorption of the contaminant on the solid and on colloidal particles is assumed, then all the mass balance equations may be combined, thus eliminating all of the exchange terms. We then obtain the following *single* mass balance equation, in terms of c_ℓ:

$$\begin{aligned}\frac{\partial}{\partial t}&\left\{c_\ell\left[(\theta + \rho_b K_d) + \sigma_{col} K_{eq,col}\left(c_{col,m}\theta + c_{col,im} + c_{col,int}\right)\right]\right\}\\ &+\nabla\cdot\theta\Big[(1 + \sigma_{col} c_{col,m} K_{eq,col}\alpha_{col}) c_\ell \mathbf{V}_\ell\\ &\qquad -\mathbf{D}_{h\ell}\cdot\nabla c_\ell - c_\ell \sigma_{col} K_{eq,col}\mathbf{D}_{col}\cdot\nabla c_{col,m}\Big] = 0.\end{aligned} \tag{7.7.18}$$

The general flux boundary condition for the above balance equation is

$$\begin{aligned}[\![\ &\theta(1 + \sigma_{col} c_{col,m} K_{eq,col}\alpha_{col}) c_\ell \mathbf{V}_\ell - \theta \mathbf{D}_{h\ell}\cdot\nabla c_\ell\\ &\qquad -\theta c_\ell \sigma_{col} K_{eq,col}\mathbf{D}_{col}\cdot\nabla c_{col,m}\ ]\!]_{1,2}\cdot\mathbf{n} = 0,\end{aligned}$$

in which the mobile colloid concentration, $c_{col,m}$, must first be obtained by solving the colloid mass balance equations.

area. . What is their role in contaminant transport (e.g., the transport of adsorbed contaminants). However, evidence gathered over the last few decades, indicate that they are capable of traveling through the void space. Because of their size, their surface area per unit volume is huge, thus they can adsorb large quantities of various chemicals, including contaminants.

7.7.2 Colloids as Carriers of Contaminants

So far in this chapter, we have considered only contaminants that are transported within a fluid in the form of dissolved species or components. Contaminants that are adsorbed on or are part of a solid phase have been assumed to be immobile. However, there exists enough evidence (e.g., McCarty and Zachara 1989; McCarty and Degueldre 1993) that contaminants may travel also *on* colloidal size solid particles (viz., particles with diameters less than $10\,\mu$m), e.g., clay-like particles, or microorganisms, that are transported by groundwater. Organic colloidal particles, or *humic colloids*, present close to ground surface, are of special interest, because of their ability to adsorb and absorb non-polar organic contaminants. These colloids include humic substances, viruses, and the organic coating on very tiny inorganic particles. In Sect. 1.3.2, we have mentioned the presence of colloidal size matter in the soil, and the possibility that such matter may be a significant carrier of contaminants. This occurs when contaminants adsorb on the (external and internal) surfaces of colloidal particles. Thus, colloidal matter may serve as carrier of contaminants. In some cases, a contaminant, such as a radionuclide, participates in the formation of a colloid and is transported as a constituent, rather than as an adsorbate on the particles. Altogether, a contaminant may be transported by colloidal particles. An introduction to soil colloidal particles and is presented in Sect. 1.3.2.

Colloidal particles that are repelled from solid walls will have a mean velocity larger than that of the fluid. Because of their size, colloidal particles may also be affected by the phenomenon of *size-exclusion* (Sect. 7.2.3). Such particles will mix only with a portion of the fluid rather than with the entire fluid volume in the pore space of an REV; therefore, their mean velocity will be higher than that of a conservative contaminant.

Following Corapcioglu and Choi (1996), we shall present below a model for colloidal transport in a porous medium filled with two fluid phases: water and air. A single non-reacting and non-volatile contaminant will be considered that can adsorb onto an immobile solid phase (ads), onto an immobile colloidal phase (im), and onto a mobile colloidal phase (m) attached to the solid. Because the contaminant is non-volatile, we only consider transport in the water phase (ℓ). The colloids are assumed to be *hydrophobic* so that they have a tendency to attach themselves to the air-water interface. We shall, therefore, represent this interface as a pseudo-phase (int). This interface is assumed to be immobile, and its surface area to be a function of water saturation.

7.7.3 Mass Balance Equations for Colloids

Using the balance operator notation, $\mathcal{B}_\alpha$, introduced earlier, the mass balance equation for mobile non-aggregated colloidal particles is

$$\mathcal{B}_{col,m}(c_{col,m}) = f_{col,im\rightarrow m} + f_{col,int\rightarrow m}. \tag{7.7.19}$$

The mass balance equation for immobile aggregated colloid particles is

$$\mathcal{B}_{col,im}(c_{col,im}) = -f_{col,im\rightarrow m}. \tag{7.7.20}$$

The mass balance equation for colloidal particles attached to the air-water interface is

$$\mathcal{B}_{col,int}(c_{col,int}) = -f_{col,int\rightarrow m}. \tag{7.7.21}$$

The mass balance operators appearing in these equations are defined as:

$$\mathcal{B}_{col,m}(c_{col,m}) \equiv \frac{\partial(c_{col,m}\theta)}{\partial t} + \nabla\cdot\theta\left(c_{col,m}\mathbf{V}_{col,m} - \mathbf{D}_{col,m}\cdot\nabla c_{col,m}\right), \tag{7.7.22}$$

and

$$\mathcal{B}_{col,im}(c_{col,im}) \equiv \frac{\partial(c_{col,im}\Sigma_{s,\ell})}{\partial t}, \quad \mathcal{B}_{col,int}(c_{col,int}) \equiv \frac{\partial(c_{col,im}\Sigma_{\ell,g})}{\partial t}. \tag{7.7.23}$$

The concentration $c_{col,m}$ expresses the mass of mobile colloids per unit volume of water phase. The concentrations $c_{col,im}$ and $c_{col,int}$ express the mass of colloids per unit surface area of their respective (immobile) interfaces.

It is assumed that the specific interfacial surface areas, $\Sigma_{s,\ell}$ and $\Sigma_{\ell,g}$, are known functions of the saturation or of the volumetric fraction of the water phase.

The following forms will be used for the terms that represent the transfer from the mobile colloidal particles and the immobile particles

$$f_{col,im\rightarrow m} = \Sigma_{\ell,s}(k_{im\rightarrow m}c_{col,im} - k_{m\rightarrow im}c_{col,m}), \tag{7.7.24}$$

and the transfer of immobile particles to the air-water interface:

$$f_{col,int\rightarrow m} = \Sigma_{g,\ell}(k_{int\rightarrow m}c_{col,int} - k_{m\rightarrow int}c_{col,m}). \tag{7.7.25}$$

We shall express the velocity of the mobile colloidal particles as:

$$\mathbf{V}_{col} = \alpha_{col}\mathbf{V}_\ell, \tag{7.7.26}$$

where α_{col} $(\equiv |\mathbf{V}_{col}|/|\mathbf{V}_\ell|)$ denotes the ratio between the mean velocities of the mobile colloidal particles and of the water phase. This coefficient depends on the size of the colloidal particles and on ion exclusion effects; in principle, $\alpha_{col} > 1$. It will also depend on the prevailing chemical conditions, and, perhaps, on the velocity of the fluid. In determining the value of this coefficient experimentally, it must be remembered that $\mathbf{V}_{col}$ is the mean velocity of the non-aggregated particles that does not include the apparent loss in velocity due to particle aggregation. That effect is included, elsewhere, through the exchange terms in the mass balance equations.

7.8 Electromigration and Electrokinetics

Electric charge is also an extensive quantity that can accumulate in and be transported through a porous medium domain, provided the liquid occupying the void space and/or the solid matrix are electrically conductive. This occurs, for example, when the interconnected void space of a porous geological formation is occupied by an electrically conducting liquid, e.g., water with dissolved salts. At the microscopic level, Ohm's law, e.g., $\mathbf{j}_\alpha^{EL} = -\kappa_\alpha^{EL}\nabla\Phi_\alpha^{EL}$ expresses the flux in an electrically EL-conducting substance, and Φ^{EL} denotes the voltage. The microscopic charge balance equation, $\partial\rho^{EL}/\partial t = -\nabla\cdot\mathbf{j}^{EL}$, where ρ^{EL} denotes the charge density, can also be written. The modeling routine discussed in this book is applicable.

However, there are some special electrical phenomena that are associated primarily with soils and with geological formations. One such phenomenon is the *double layer* introduced in Sect. 1.3.1. The conduction of electric currents through (or, equivalently, the corresponding electrical resistivity of) a porous geologic domain is the basis of an array of powerful geophysical (e.g., electromagnetic) methods used for determining the occurrence, saturation and eventually the volume of subsurface fluids of economic importance such as oil and gas. Some additional special features are introduced below. The objective is merely to introduce certain definitions that may be useful to the reader. Addition information can be found in the literature (e.g., Kirby 2010).

A. Electroosmosis

When an electrical field is applied to a porous medium (fully or partly) saturated saturated by an electrically conducting (= ionic) fluid, the movement of the fluid is enhanced. The added motion of the fluid is called *electroosmosis*. It is the result of the force that acts on ions in the double layer (Sect. 1.3.1).

B. Electrokinetic and Streaming Potential

These phenomena occur in the subsurface when a liquid that has certain electrical properties passes through a porous medium with different electrical properties. *Electro-osmotic flow* is caused by the *Coulomb force* induced by an electric field on the net mobile electric charge in a solution. When the pore space is fully, or partially, saturated by an electrolytic solution, such as water with dissolved salts, a layer of mobile ions, known as an *electrical double layer* (Sect. 1.3.1), or *Debye layer*, forms at the fluid-solid interface.

When an electric field is applied to a fluid that occupies the void space, especially in a very fine-grained medium, the net charge in the *electrical double layer* may force liquids to move by the resulting Coulomb force. This resulting flow is called *electro-osmotic flow* (Barbour and Fredlund 1989).

A number of additional examples associated with electrical charge in porous media are briefly introduced in Sect. 7.8. The objective of the presentation in this section is just to bring the subjects to the attention of the reader, as it is often mentioned in the literature.

C. Electromigration

Electromigration is the movement of ions in an electrolytic solution that occupies the void space in response to an applied electric field. When two electrodes are placed some distance apart in a saturated porous medium domain, and an electric voltage is applied, the anions (= negative ions) in the fluid will travel towards the anode (= positively charged electrode), while the cations (= positive ions) will travel towards the cathode (= negatively charged electrode). This phenomenon is often mentioned as a *remediation technique*, i.e., removal of contaminants, in low permeability formations. If the electrodes are placed in an electrically conductive domain, such as an aquifer, and a DC voltage is applied, then cations, such as those of heavy metals or uranium, will be drawn to the cathode, where they can be removed. Anions, such as those of cyanide or nitrate, can also be removed, since they will be drawn to the anode. An advantage of electro-migration is that it may be applied to low permeability sediments, such as clays.

However, chemical speciation in a fluid, and adsorptive properties of mineral surfaces, may be strongly affected by the fluid's pH with consequences that may run counter to the ultimate remediation goals. At the cathode, water dissociates into hydrogen gas and hydroxyl ions, through the reaction $2H_2O + 2e^- \rightarrow 2OH^- + H_2(g)$. At the anode, water dissociates into oxygen gas and hydrogen ions by the reaction $2H_2O \rightarrow 4H^+ + O_2(g) + 4e^-$. The addition of hydroxyl ions increases the pH in the region around the cathode, while the added hydrogen ions decrease the pH in the region near the anode. In the region of low pH, metallic ions will go into solution by desorption from the solid, or by other reactions, such as dissolution of metal solids. These ions will be drawn to the cathode, where they can be removed. However, the pH in the solution increases as the ions move towards the cathode, causing immobilization of some or most of the ions by precipitation, formation of complexes, or solid adsorption, before the cathode is reached (Probstein and Hicks 1993).

D. Electrokinetic Phenomena

Electrokinetic phenomena are the phenomena resulting from the coupling between an electric field and the mechanical motion of a fluid, or of a chemical species within a fluid. One such phenomenon is the *double layer* mentioned in.

At the solid-liquid interface that bounds the pore space in a porous medium, an electric potential difference is created. The charge on the liquid side will be opposite to that on the solid surface.

Three types of electrokinetic phenomena may be mentioned:

- **Electroosmosis**. This is the generation of an advective flow field, usually that of an aqueous phase, by the application of an electric field. The fluid is pulled along by the viscous drag from the moving positive ions that are concentrated in the double layer, because of their attraction to the cathode. Thus, the application of a DC voltage to a pair of electrodes placed in an aquifer will produce water at the cathode. Note that electro-migration of ions in the water will also take place. This method could be used to remove contaminants from clay formations, which, because of their low permeability, are difficult to clean otherwise (Shapiro and Probstein 1993).

- **Electrophoresis**. Here we have the movement of charged particles, or a charged surface (and any attached material), in response to an electrical field. This phenomenon may find application in the removal of contaminated colloids or polar organic phases.

- **Zeta and streaming potentials**. When a solid comes into contact with liquid, an electric potential difference between the two comes arises at the interface. Consequently, the liquid will be charged oppositely to the solid (wall). Ions of the liquid accumulate near the solid surface and produce an *electrical double layer* at the solid-liquid interface. Thus, when a liquid is forced to flow through a capillary tube, the liquid stream carries with it part of the mobile part of that electrical double layer near the walls of the capillary. The convection at a local velocity V of an electrical charge of density e at a local velocity V_e results in a *streaming current* eV_e, where e denotes the current's density. As a consequence of the streaming current, a potential difference will be established between the ends of the capillary which, in turn, generates a conductive current opposite in sense to the streaming current. At steady state, the two currents are equal, and the potential difference induced will be the *streaming potential*. This streaming current and potential have been shown to be related to the pressure gradient that forces the liquid's flow. This phenomenon has been shown also to exist also when flow occurs through porous medium domains (e.g., Adamson 1982; Davies and Rideal 1963). The flow induced by the streaming potential is superimposed on any flow caused by a pressure gradient, as described by Darcy's law.

References

Aagaard P, Helgeson HC (1983) Activity/composition relations among silicates and aqueous solutions: II. Chemical and thermodynamic consequences of ideal mixing of atoms on homological sites in montmorillonites, illites and mixed layer clays. Clays Clay Min 31:207–217

Adamson AW (1982) Physical chemistry of surfaces, 4th edn. Wiley, New York, p 664

Al-Khoury R, Bundschuh J (eds) (2014) Computational models for CO_2 geo-sequestration, and compressed air energy storage. CRC Press, Boca Raton, 531 p

Appelo CAJ, Postma D (2005) Geochemistry, groundwater and pollution, 2nd edn. CRC Press, Boca Raton, p 652

Aris R (1956) On the dispersion of a solute in a fluid flwing through a tube. Proc R Soc A 235:67–77

Barbour SL, Fredlund DG (1989) Mechanisms of osmotic flow and volume change in clay soils. Can Geotech J 26:551–562
Batchelor GK (1959) The theory of homogeneous turbulence. Cambridge Univ. Press, Cambridge
Bear J (1960) The transition zone between fresh and salt waters in coastal aquifers, PhD thesis, University of California, Berkeley
Bear J (1961a) On the tensor form of dispersion. J Geophys Res 66:1185–1197
Bear J (1961b) Some experiments on dispersion. J Geophys Res 66:2455–2467
Bear J (1972) Dynamics of fluids in porous media. American Elsevier, 764 p. (Also published by Dover Publications, 1988; translated into Chinese)
Bear J (1979) Hydraulics of groundwater. McGraw-Hill, New York, 569 p. (Also published by Dover Publications, 2007; translated into Chinese)
Bear J, Fel L, Zimmels Y (2009) Effects of material symmetry on coefficients of transport in anisotropic porous media. Trans Porous Media 62:347–361
Bear J, Bachmat Y (1991) Introduction to modeling phenomena of transport in porous media. Kluwer Publishing Co, Dordrecht, 553 p
Berkowitz B, Scher H, Silliman SE (2000) Anomalous transport in laboratory-scale, heterogeneous porous media. Water Resour Res 36(1):149–158
Berkowitz B, Klafter J, Metzler R, Scher H (2002) Physical pictures of transport in heterogeneous media: advection-dispersion, random walk and fractional derivative formulations. Water Res Res 38(10)
Berkowitz B, Cortis A, Dentz M, Scher H (2006) Modeling non-Fickian transport in geological formations as a continuous time random walk. Rev Geophys 44(2)
Berkowitz B, Dror I, Yaron B (2008) Contaminant geochemistry: interactions and transport in the subsurface environment. Springer, Heidelberg, 412 p
Burnett RD, Frind EO (1987) Simulation of contaminant transport in 3 dimensions. 2 Dimensionality effects. Water Resour Res 23:695–705
Chandrasekhar S (1943) Stochastic problems in physics and astronomy. Rev Mod Phys 15:1–89
Chang R, Cruickshank B (2003) Chemistry, 8th ed. McGraw-Hill, New York, 1120 p
Corapcioglu MY, Choi H (1996) Modeling colloid transport in unsaturated porous media and validation with laboratory column data. Water Resour Res 32(12):3437–3449
Cortis A, Berkowitz B (2004) Anomalous transport in "classical" soil and sand columns. Soil Sci Soc Am J 68:1539–1548
Cortis A, Gallo C, Scher H, Berkowitz B (2004) Numerical simulation of non-Fickian transport in geological formations with multiple-scale heterogeneities. Water Resour Res 40(4):1–16
Cunningham RE, Williams RJJ (1980) Diffusion in gases and porous media. Springer Science + Business Media, LLC, 275 p
Czurda KA, Haus R (2002) Reactive barriers with fly ash zeolites for in situ groundwater remediation. Appl Clay Sci 21:13–20
Dagan G (1989) Flow and transport in porous formations. Springer, New York, 465 p
Dagan G (1984) Solute transport in heterogeneous porous formations. J Fluid Mech 145:151–177
Dagan G, Neuman SP (eds) (1997) Subsurface flow and transport: a stochastic approach. Cambridge University Press, Cambridge UK, p 241
Dankwerts PV (1953) Continuous flow systems (distribution of residence times). Chem Eng Dvi 1(2):1–13
Davies JT, Rideal EK (1963) Interfacial Phenomena. Academic Press, New York, p 480
De Groot SR, Mazur P (1962) Non-equilibrium thermodynamics. North-Holland Publishing Co, Amsterdam, 510 p
De Josselin de Jong G (1958) Longitudinal and transverse diffusion in granular deposits. Trans Am Geophys Union 39:67–74
Denbigh KG (1981) The principles of chemical equilibrium, 4th edn. Cambridge University Press, Cambridge, 494 p
Dentz M, Berkowitz B (2003) Transport behavior of a passive solute in continuous time random walks and multi-rate mass transfer. Water Resour Res 39(5)

Foo KY, Hameed BH (2010) Insights into the modeling of adsorption isotherm systems. Chem Eng J 156(1):2–10
Freundlich H (1907) Uber die Adsorption in Losungen. Z Phys Chem 57:385
Garabedian SP, LeBlanc DR, Gelhar LW, Celia MA (1991) Large-scale natural gradient tracer test in sand and gravel, Cape Code, Massachusetts, 2. Analysis of spatial moments for a nonreactive tracer. Water Resour Res 27:911–924
Gelhar LW (1976) Stochastic analysis of flow in aquifers. In AWRA Symposium on Advances in Groundwater Hydrology, Chicago, Ill
Gelhar LW (1986) Stochastic subsurface hydrology from theory to applications. Water Resour Res 22:135S–145S
Gelhar LW (1993) Stochastic subsurface hydrology. Prentice-Hall, Englewood Cliffs, NJ
Gelhar LW, Axnes CL (1983) Three dimensional stochastic analysis of macrodispersion in aquifers. Water Resour Res 19:161–180
Gelhar LW, Gutjhar AL, Naff RL (1979) Stochastic analysis of macrodispersion in a stratified aquifer. Water Resour Res 15:1387–1397
Gelhar LW, Welty C, Rehfekdt KR (1992) A critical-review of data on field-scale dispersion in aquifers. Water Resour Res 28:1955–1974
Grim RE (1968) Clay mineralogy, 2nd edn. McGraw-Hill, New York, p 596
Gvirtzman H, Gorelick SM (1991) Dispersion and advection in unsaturated porous-media enhanced by anion exclusion. Nature 352:793–795
Gvirtzman H, Magaritz M (1989) Water and anion transport in the unsaturated zone traced by environmental tritium. In: Bar-Yosef B, Barrow NJ, Goldshmidt J (eds) Inorganic contaminants in the Vadose zone. Ecological studies, vol. 74. Springer, Berlin, pp 190–198
Helgeson HC (1969) Thermodynamics of hydrothermal systems at elevated temperatures and pressures. Am J Sci 267:729–804
Jardine PM, Jacobs GK, Wilson GV (1993) Unsaturated transport processes in undisturbed heterogeneous porous media: I. Inorganic contaminants. Soil Sci Soc Am Proc 57:945–953
Kinzelbach W (1992) Numerische Methoden zur Modellierung des Transports von Schadstoffen im Grundwasser, 2nd edn. Oldenbourg Verlag, München, p 313
Kirby BJ (2010) Micro- and nanoscale fluid mechanics: transport in microfluidic devices. Technology and engineering. Cambridge University Press, Cambridge, 536 p
Klafter J, Silbey R (1980) Derivation of continuous-time random walk equations. Phys Rev Let 44(2):55–58
Lallemand-Barrés A, Peaudecerf P (1978) Recherche de relationsentre les valeurs measurées de la dispersivité macroscopique d'un milieu aqiufère, ses autres caractéritsique et les conditions de mesure. Bull Bur Rech Géol Min Sêr 2(Sec. III):277–284
Langmuir I (1915) Chemical reactions at low temperatures. J Am Chem Soc 37:1139
Langmuir I (1918) The adsorption of gases on plane surfaces of glass, mica and platinum. J Am Chem Soc 40:1361–1403
Lasaga AC, Kirpatrick RJ (eds) (1981) Kinetics of geochemical processes. Reviews in Minerology, Minerological Society of America, Book Crafters Inc, Chelsea, Michigen
Levy M, Berkowitz B (2003) Measurement and analysis of non-Fickian dispersion in heterogeneous porous media. J Contamin Hydrol 64:203–226
Lichtner PC (1995) Principles and practice of reactive transport modeling. In: Murakami T, Ewing RC (eds) Scientific basis for nuclear waste management, XVIII. Materials Research Society Proceedings, vol 353. Pitsburg, PA, pp 117–130
Lichtner PC (1996) Continuum formulation of multicomponent-multiphase reactive transport. In: Lichtner PC, Steefel CI, Oelkers EH (eds) Reactive transport in porous media. Reviews in mineralogy, vol 34. Mineralogical Society of America, Washington, D.C
Lichtner PC (1985) Continuum model for simultaneous chemical reactions and mass transport in hydrothermal systems. Geochimica and Cosmochimica Acta 49:779–800
Lichtner PC, Kelkar S, Bruce Robinson B (2002) New form of dispersion tensor for axisymmetric porous media with implementation in particle tracking. Water Resour Res 38(8):1146–1168

Lindstrom FT, Boersma L, Stockard D (1971) A theory on the mass transport of previously distributed chemicals in a water saturated sorbing porous medium. Soil Sci 112:291–300
McCarty PL, Degueldre C (1993) Sampling and characterization of colloids and particles in ground water for studying their role in contaminant transport. In: van Leeuwen HP, Buffle J (eds) Environmental particles. IUPAC environmental analytical and physical chemistry series, vol II. Lewis Publishing, Chelsea, MI
McCarty PL, Zachara JM (1989) Subsurface transport of contaminants. Environ Sci Technol 23(5):496–502
Metzler R, Klafter J (2000) The random walk's guide to anomalous diffusion: a fractional dynamics approach. Phys Rep 339(1):1–77
Millington RJ (1959) Gas diffusion in porous media. Science 130:100–102
Montroll EW, Weiss GH (1965) Random walks on lattices. II. J Math Phys 6(2):167–181
Nernst W (1904) Theorie der Reaktionsgeschwindigkeit in heterogenen Systemen. Z Phys Chem 47:52–55
Neumann SP (1990) Universal scaling of hydraulic conductivities and dispersivities in geologic media. Water Rsourc Res 268:1749–1758
Nielsen DR, Biggar JW (1961) Miscible displacement in soils: I. Experimental information. Soil Sci Soc Am Proc 25:1–5
Nielsen DR, Biggar JW (1962) Miscible displacement in soils: III. Theoretical considerations. Soil Sci Soc Am Proc 26:216–221
Nikolaevski VN (1959) Convective diffusion in porous media. J Appl Math Mech (PMM) 23:1042–1050
Nitao JJ, Bear J (1996) Potentials and their role in transport in porous media. Water Resour Res 32:225–250
Parkhurst DL, Appelo CAJ (1999) User's Guide in PHREEQC (version 2) A computer program for speciation, barch reaction, pne-dimensional transport, and inverse geochemical calculations. US Geological Services Water-Resources Investigations Report 99–4259, 312 p
Pitzer KS (1979) Theory: ion interaction approach. In: Pytokowicz RM (ed) Activity coefficients in electrolyte solutions, vol I. CRC Press, Boca Raton, pp 157–208
Poling BE, Prausnitz JM, O'Connell JP (2000) Properties of Gases and Liquids, 5th edn. McGraw-Hill, New York
Poreh M (1965) Dispersivity tensor in isotropic and axisymmetric mediums. J Geophys Res 70:3909–3914
Probstein RL (1994) Physicochemical hydrodynamics: an introduction, 2nd edn. Wiley Inc, New York, 400 p
Probstein RL, Hicks RE (1993) Removal of contaminants from soil by electric fields. Science 260:498–503
Rhodes ME, Bijeljic B, Blunt MJ (2008) Pore-to-field simulation of single-phase transport using continuous time random walks. Adv Water Resour 31:1527–1539
Robertson HP (1940) The invariant theory of isotropic turbulence. Proc Cambridge Phil Soc 36:209–223
Robson SG (1974) Feasibility of digital water-quality modeling illustrated by application at Barstow, California. US Geological Survey Water-Resources Investigations Report 46–73, 66 p
Robson SG (1978) Application of digital profile modeling techniques to groundwater solute transport at Barstow, California. US Geological Survey Water Supply Paper 2050, 28 p
Saffman PG (1959) A theory of dispersion in a porous medium. J Fluid Mech 6:321–349
Saffman PG (1960) Dispersion due to molecular diffusion and macroscopic mixing in flow through a network of capillaries. J Fluid Mech, GB 7(2):194–208
Sawyer CN, McCarty PL, Parkin GF (2002) Chemistry for environmental engineering and science, 5th edn. McGraw-Hill, New York, 768 p
Scheidegger AE (1958) The physics of flow through porous media. University of Toronto Press, Toronto

Scheidegger AE (1960) The physics of flow through porous media, 2nd edn. University of Toronto Press, Toronto, 313p
Scheidegger AE (1961) General theory of dispersion in porous media. J Geophys Res 66:3273–3278
Scher H, Lax M (1973a) Stochastic transport in a disordered solid. I Theory. Phys Rev B 7(10):4491–4502
Scher H, Lax M (1973b) Stochastic transport in a disordered solid. II. Impurity conduction. Phys Rev B 7(10):4502–4519
Scher H, Montroll EW (1975) TAnomalous transit time dispersion in amorphous solids. Phys Rev B 12(6):2455–2477
Schwarzenbach RP, Gschwend PM, Imboden DM (2002) Environmental organic chemistry, 2nd edn. Wiley Interscience, New York, 1328 p
Shapiro AP, Probstein RF (1993) Removal of contaminants from saturated clay by electroosmosis. Environ Sci Technol 27(2):283–291
Shlesinger MF (1996) Random processes. Encyclopedia of applied physics, vol 16. VCH Publishers Inc, Weinheim
Sienko MJ, Plane RA (1966) Chemistry: principles and properties. McGraw Hill, New York, 623 p
Simpson ES (1978) A note on the structure of the dispersion coefficient. Geol Soc Am Abstr Programs 393
Sirotine Y, Chaskolskaya M (1984) Fondaments de la physique des crystaux. Edition Mir, 680 p. (Russian edn. 1975)
Sposito G (2004) The surface chemistry of natural particles. Oxford University Press, New York, 256 p
Taylor GI (1953) Dispersion of soluble matter in solvent flowing slowly through a tube. In: Proceedings of the royal society A, No. 1137, vol 219, pp 186–203
Taylor GI (1954) The dispersion of matter in turbulent flow through a pipe. In: Proceedings of the royal society, London, series A, vol 223, pp 446–468
Todorovic P (1992) An introduction to stochastic processes and their applications. Springer, New York
Tsong TT (2001) Mechanisms of surface diffusion. Progr Surf Sci 67235–248:2001
Van Genuchten MTh (1974) Mass transfer studies of sorbing porous media. PhD Thesis, New Mexico State University, La Cruz, NM
Weber WJ (1972) Physicochemical processes for water quality. Wiley, New York, 640 p
Whitaker S (1999) The method of averaging. Kluwer Academic Publishing, Dordrecht, 219 p
Whitmanm WG (1923) The two-film theory of gas absorption. Chem Math Eng 29(4):146–148
Wolery TJ (1983) EQ3NR - a computer program for geochemical aqueous speciation-solubility calculations: user's guide and documentation, Lawrence Livermore National Laboratory, report no. UCRL-53414, Livermore, California
Zheng C, Bennett CD (1995) Applied contaminant transport modeling, theory and practice. Van Nostrand Reinhold, 440 p
Berkowitz B, Kosakowski G, Margolin G, Scher H (2001) Application of continuous time random walk theory to tracer test measurements in fractured and heterogeneous porous media. Ground water 39(4):593–604
Fel LG, and Bear J (2010) Dispersion and dispersivity tensors in saturated porous media with uniaxial symmetry. Trans in Porous Media 85(1):259–268
Jensen KH, Bitsch K, and Bjerg PL (1993) Large-scale dispersion experiments in a sandy aquifer in Denmark—Observed tracer movements and numerical-analyses. Water Resour Res 29:673–696
Ogata A, and Banks RB (1961) A solution of the differential equation of longitudinal dispersion in porous media, U.S. Geological Survey, Professional Paper, 411-A
Scheidegger AE (1954) Statistical hydrodynamica in porous media. J Appl Phys 25:994–1001
Sparks DL (2003) Environmental Soil Chemistry. Academic Press, pp 352
Stumm W, Morgan JJ (1995) Aquatic chemistry, chemical equilibria and rates in natural waters, 3rd ed. Wiley Interscience, pp 1040

Chapter 8
Modeling Energy and Mass Transport

In Chaps. 5 and 6, we considered mass transport of a single fluid phase, and of multiple fluid phases. Most of the presentation focused on flow under isothermal conditions, although we did mention that when non-isothermal conditions prevail, fluid density and viscosity depend also on temperature. In Chap. 7, we discussed the transport of mass of chemical species in single phase flow, and in the flow of multiple multi-species phases. Again, we did mention the possible effects of temperature variations, and the possibility of heat sources and sinks as a result of endogenic and exogenic reactions, but, in general, the underlying assumption was that the flow and transport take place under *isothermal conditions*. In this chapter, we remove this constraint and discuss energy transport under non-isothermal conditions, i.e., simultaneous mass and energy transport. The effects of natural thermal gradients, produced by solar radiation at ground surface, on water and water vapour movement, as well as on the chemical and biological behavior in the subsurface, may serve as examples of interest to soil scientists. In dealing with contaminated groundwater, a number of remediation techniques are associated with heating the soil, and injection of steam. Other interesting cases that require knowledge of heat and mass transport in porous media are the storage of energy in aquifers, or in the unsaturated zone, the production of geothermal energy, the disposal of CO_2 in deep geological formations, the geological storage of high-level nuclear waste, and the thermally enhanced production of petroleum. Change of phase (evaporation/condensation and freezing/melting) of fluids within the void space as a result of temperature changes in the subsurface may serve as an additional example. Finally, in the chemical industry, most processes that take place in reactors (e.g., phase change, dissolution, chemical reactions) occur under non-isothermal conditions. Because of its importance, we have added an appendix (Appendix A) that presents and discusses modeling of phenomena of transport that take place in chemical reactors.

J. Bear, *Modeling Phenomena of Flow and Transport in Porous Media*,
Theory and Applications of Transport in Porous Media 31,
https://doi.org/10.1007/978-3-319-72826-1_8

Within a porous medium domain, thermal energy may be transported by four mechanisms:

- Advection by a fluid (or fluids) moving in the void space.
- Conduction in all solid and fluid phases (overlooking advection in a deformable solid). We shall limit the discussion to the case in which the solid matrix is composed of a single substance.
- Mass diffusion in the fluid phases.
- Thermal dispersion in the fluid phase(s).

All these modes of energy transport will be discussed in this chapter. We shall assume that radiation plays no role in energy transport in porous media.

Fluid mass moving through the void space (= advection) carries with it thermal energy, and, therefore, any complete heat transport model must also treat, simultaneously, the transport of fluid mass. As we shall see below, the coupling between the transport of mass and heat is also due to the fact that both the fluid's density and its viscosity are temperature dependent. Most partitioning and equilibrium coefficients, discussed in Chaps. 6 and 7, are strongly temperature dependent. The chemical potential discussed in Sect. 2.2.4 is also temperature-dependent. Temperature may also affect the permeability (e.g., through the swelling of clay, mineral reactions, and biological matter clogging of pores) and the capillary pressure (e.g., through the effect on surface tension). Finally, especially in chemical engineering, many processes that take place in chemical reactors involve absorption and generation of heat.

Unlike the case of fluid mass transport, where the solid is assumed to be impervious to the transport of mass, heat is also transported (by conduction) through the solid matrix. In general, the average temperatures of the solid and of the fluids that occupy the void space are not the same. For example, a hot fluid flowing through the void space can, at least over a certain period of time, have a different temperature than that of the solid matrix. However, in general, after a while, the temperature of the solid will equilibrate with that of the fluid if the average (microscopic level) distance from a point inside the solid to the fluid in the void space is sufficiently small, as long as the thermal conductivity of the solid is not too small. The discussion in this chapter will be based on the assumption that the time required to reach *thermal equilibrium* between all phases present at a point in a porous medium domain, i.e., within an REV centered at the point, including the solid, is sufficiently short so that it can be ignored. This assumption, referred to as *local thermal equilibrium*, or as *approximate equilibrium*, (Sect. 2.1) between the phases, states that at every (macroscopic) point within a porous medium domain, the (average) temperature is essentially the same for all phases present at that point, including the solid, i.e., within the REV centered at the point. An important case where the assumption of local thermal equilibrium does not apply is the flow of a hot geothermal fluid through widely spaced fractures in a solid rock domain.

In spite of the statement that in most cases we assume that all phases present in an REV are in thermal equilibrium, i.e., at the same temperature, there are certain cases of counter-current flow in chemical reactors, in which, intentionally, fluids in counter-current flow are maintained at different temperatures. The temperature of the solid matrix may also be different from that of either of the two fluids. This possibility will be discussed in Appendix A.

As suggested in the preamble of Chap. 2, following the phenomenological approach, the various thermodynamic variables and relationships presented at the microscopic level are assumed to be valid also at the macroscopic one. In this chapter, we shall make use of them to develop and discuss complete well-posed models for the phenomenon of heat transport in porous medium domains.

8.1 Microscopic Energy Fluxes

As for every extensive quantity, the total flux of energy at the microscopic level is made up of advective and diffusive (= conductive) fluxes.

8.1.1 Advective and Diffusive Fluxes; Single Species Fluid

At the microscopic level, the *total energy flux*, $\mathbf{j}^{t\mathbb{E}}$, at a point within a phase containing only a single species is the sum of the *internal* ($I\mathbb{E}$), *kinetic* ($K\mathbb{E}$), and *potential energy* ($P\mathbb{E}$) fluxes:

$$\mathbf{j}^{t\mathbb{E}} \equiv \mathbf{j}^{I\mathbb{E}} + \mathbf{j}^{K\mathbb{E}} + \mathbf{j}^{P\mathbb{E}}, \tag{8.1.1}$$

$$\mathbf{j}^{I\mathbb{E}} \equiv (\rho u)\mathbf{V}^{I\mathbb{E}}, \qquad \mathbf{j}^{K\mathbb{E}} \equiv \left(\frac{1}{2}\rho V^2\right)\mathbf{V}^{K\mathbb{E}}, \qquad \mathbf{j}^{P\mathbb{E}} = (\rho\varphi_{pot})\mathbf{V}^{P\mathbb{E}}, \tag{8.1.2}$$

with u and $\rho u (\equiv u')$ denoting the specific internal energy, i.e., internal energy per unit mass, and internal energy density, i.e., per unit volume, respectively. The advective energy flux is the total energy carried by the phase at its velocity, while the *diffusive energy fluxes* contain the remaining portions of the total energy flux:

$$\begin{aligned}
\mathbf{j}^{I\mathbb{E}}_{dif} &\equiv \rho u \left(\mathbf{V}^{I\mathbb{E}} - \mathbf{V}\right), \text{ diffusive flux of internal energy,}\\
\mathbf{j}^{K\mathbb{E}}_{dif} &\equiv \tfrac{1}{2}\rho V^2 \left(\mathbf{V}^{K\mathbb{E}} - \mathbf{V}\right), \text{ diffusive flux of kinetic energy,}\\
\mathbf{j}^{P\mathbb{E}}_{dif} &\equiv \rho\varphi_{pot} \left(\mathbf{V}^{P\mathbb{E}} - \mathbf{V}\right), \text{ diffusive flux of potential energy.}
\end{aligned}$$

The total energy flux can also be written as the sum of the total *advective energy flux* and three *diffusive energy fluxes*:

$$\mathbf{j}^{t\mathbb{E}} = \rho\left(u + \frac{1}{2}V^2 + \varphi_{pot}\right)\mathbf{V} + \mathbf{j}^{IE}_{dif} + \mathbf{j}^{K\mathbb{E}}_{dif} + \mathbf{j}^{P\mathbb{E}}_{dif}. \tag{8.1.3}$$

A. Diffusive Flux of Internal Energy

The (microscopic) *diffusive flux of internal energy* within a phase, $\mathbf{j}^{I\mathbb{E}}$, is identical to the heat flux by conduction, $\mathbf{j}^{\mathbb{H}}$, expressed by *Fourier's law*:

$$\mathbf{j}_{dif}^{I\mathbb{E}}(\equiv \mathbf{j}^{H}) = -\lambda \nabla T, \tag{8.1.4}$$

with λ and T denoting the *thermal conductivity* and the temperature of the phase, respectively. Equation (8.1.4) is valid at a point within any *isotropic* solid, or fluid phase.

B. Diffusive Flux of Kinetic Energy

The diffusive flux of kinetic energy, $\mathbf{j}^{K\mathbb{E}}$, is expressed by:

$$\mathbf{j}^{K\mathbb{E}} = -\boldsymbol{\sigma}\cdot\mathbf{V}, \tag{8.1.5}$$

in which $\boldsymbol{\sigma}$ denotes the stress tensor (e.g., Bear and Bachmat 1991, p. 81).

C. Diffusive Flux of Potential Energy

The diffusive flux of potential energy, $\mathbf{j}^{P\mathbb{E}}$, vanishes in the case of a single species phase.

D. Total Energy Flux

From the discussion above, it follows that at a point within an isotropic phase, the total energy flux is expressed as:

$$\mathbf{j}^{t\mathbb{E}} = \rho\left(u + \frac{1}{2}V^2 + \varphi_{pot}\right)\mathbf{V} - \lambda\nabla T - \boldsymbol{\sigma}\cdot\mathbf{V}. \tag{8.1.6}$$

We usually assume that $u \gg \frac{1}{2}V^2 + \varphi_{pot}$, leading to the total flux:

$$\mathbf{j}^{t\mathbb{E}} = \rho u\mathbf{V} - \lambda\nabla T - \boldsymbol{\sigma}\cdot\mathbf{V}. \tag{8.1.7}$$

8.1.2 Advective and Diffusive Fluxes; Multi-species Fluid

We consider a multiple γ-species phase. As before, the total energy flux is expressed by (8.1.3), as the sum of advective and diffusive parts.

A. Diffusive Flux of Internal Energy

From the theory of irreversible processes (e.g., De Groot and Mazur 1962, p. 26), it follows that the diffusive flux of internal energy, $\mathbf{j}_{dif}^{I\mathbb{E}}$, is the sum of the heat conductive flux, $\mathbf{j}^H$, and that of internal energy transported by the diffusive mass fluxes, $\mathbf{j}^\gamma$, in the form:

$$\mathbf{j}^{tI\mathbb{E}} = \mathbf{j}^{\mathbb{H}} + \sum_{(\gamma)} h^\gamma \mathbf{j}^\gamma = -\lambda \nabla T + \sum_{(\gamma)} h^\gamma \mathbf{j}^\gamma, \tag{8.1.8}$$

in which h^γ denotes the *specific enthalpy* of the γ-species (Sect. 2.2.2).

Other types of mass diffusion laws that depend on the temperature gradient, in addition to the concentration gradient, are also possible. If such laws are used, the above expression for $\mathbf{j}^{I\mathbb{E}}$ must be modified (De Groot and Mazur 1962), reflecting the fact that the transport of energy by mass diffusion and by conduction are interrelated.

B. Diffusive Flux of Kinetic Energy

The diffusive flux of kinetic energy is given by:

$$\mathbf{j}^{K\mathbb{E}} = -\boldsymbol{\sigma} \cdot \mathbf{V}, \quad \mathbf{V} = \sum_{(\gamma)} \mathbf{V}^\gamma. \tag{8.1.9}$$

C. Diffusive Flux of Potential Energy

The diffusive flux of potential energy, e.g., the gravitational potential, is given by:

$$\mathbf{j}^{P\mathbb{E}} = \sum_{(\gamma)} \varphi_{pot}^\gamma \, \mathbf{j}^\gamma. \tag{8.1.10}$$

It represents the work done by the diffusive mass fluxes against the conservative forces. Here, the *partial specific potential energy* with respect to mass, φ_{pot}^γ, is defined by:

$$\varphi_{pot}^\gamma \equiv \left. \frac{\partial \varphi_{pot}}{\partial m^\gamma} \right|_{p,T,m^{\delta \neq \gamma}}. \tag{8.1.11}$$

The external force, $\mathbf{f}^\gamma$, acting on the γ-species per unit mass of the species, is given by $\mathbf{f}^\gamma = -\nabla \varphi^\gamma$. Thus, the force on the species per unit mass of the phase is equal to $\rho^\gamma \mathbf{f}^\gamma = -\rho^\gamma \nabla \varphi^\gamma$. When the potential energy, φ_{pot}, is a homogeneous function of the mass of each species, then,

$$\varphi_{pot} = \sum_{(\gamma)} \omega^\gamma \varphi_{pot}^\gamma. \tag{8.1.12}$$

From the above discussion, it follows that the sum of advective and diffusive energy fluxes of a multiple species phase is given by:

$$\mathbf{j}'^{\mathbb{E}} = \rho\left(u + \frac{1}{2}V^2 + \varphi_{pot}\right)\mathbf{V} - \lambda\nabla T - \boldsymbol{\sigma}\cdot\mathbf{V} + \sum_{(\gamma)}\left(h^\gamma + \varphi^\gamma_{pot}\right)\mathbf{j}^\gamma. \tag{8.1.13}$$

Or, neglecting the kinetic and potential energy, as is commonly done:

$$\mathbf{j}'^{\mathbb{E}} = \rho u\mathbf{V} - \lambda\nabla T - \boldsymbol{\sigma}\cdot\mathbf{V} + \sum_{(\gamma)} h^\gamma\mathbf{j}^\gamma. \tag{8.1.14}$$

8.2 Microscopic Energy Balance Equation

Here, the considered extensive quantity is the energy and the state variable is the temperature.

8.2.1 *Basic Equation*

Following the discussion in Sect. 3.2.1 on the fundamental balance equation for any extensive quantity, and using the total energy flux given by (8.1.13), the microscopic-level energy balance equation for a multi-species fluid phase is:

$$\begin{aligned}\frac{\partial\rho\left(u + \frac{1}{2}V^2 + \varphi_{pot}\right)}{\partial t} = &-\nabla\cdot\left[\rho\left(u + \frac{1}{2}V^2 + \varphi_{pot}\right)\mathbf{V} + \sum_{(\gamma)}\left(h^\gamma + \varphi^\gamma_{pot}\right)\mathbf{j}^\gamma - \lambda\nabla T - \boldsymbol{\sigma}\cdot\mathbf{V}\right]\\ &+\sum_{(\gamma)}\left[\rho^\gamma\frac{\partial\varphi^\gamma_{pot}}{\partial t} + \left(h^\gamma_R + \varphi^\gamma_{pot}\right)\Gamma'^\gamma\right] + \boldsymbol{\Gamma}'^{\mathbf{M}}\cdot\mathbf{V} + \Gamma'^{\mathbb{H}},\end{aligned} \tag{8.2.1}$$

where Γ'^γ indicates a mass of γ-source. The terms on the second line denote the rates of added energy per unit volume of the phase. The first part of the first term expresses the rate at which potential energy is added. The second part expresses the energy added from a mass source, Γ'^γ. The third term is the added energy associated with a momentum source, $\boldsymbol{\Gamma}'^{\mathbf{M}}$. The last term, $\Gamma'^{\mathbb{H}}$, represents a heat source. All Γ'-sources are per unit volume of the phase. When $\Gamma'^\gamma < 0$, the quantity h^γ_R is defined as equal to the value of h^γ in the fluid at the location of the Γ'^γ-source; when $\Gamma'^\gamma > 0$, it is equal to the value of h^γ in the species flux at the source. Γ'^H denotes energy sources. These can take the form of point sources of energy carrying mass, but also in the form of point heat injection by microwave techniques.

To derive the balance equation for the internal energy of a fluid phase with multiple species, we start from the microscopic momentum balance equation:

$$\frac{\partial \rho \mathbf{V}}{\partial t} = -\nabla\cdot(\rho \mathbf{V}\mathbf{V} - \boldsymbol{\sigma}) - \sum_{(\gamma)} \rho^\gamma \nabla \varphi^\gamma_{pot} + \Gamma'^{\mathbf{M}}, \tag{8.2.2}$$

where $\mathbf{V}\mathbf{V}$ ($\equiv V_i V_j$) is the *dyadic product* of vectors. By combining this equation with the mass balance equation, we obtain the balance equation for kinetic energy, in the form:

$$\frac{\partial \frac{1}{2}\rho V^2}{\partial t} = -\nabla\cdot\left(\frac{1}{2}\rho V^2 \mathbf{V} - \boldsymbol{\sigma}\cdot\mathbf{V}\right) - \boldsymbol{\sigma} : \nabla\mathbf{V} - \sum_{(\gamma)} \rho^\gamma \nabla \varphi^\gamma_{pot}\cdot\mathbf{V} + \Gamma'^{\mathbf{M}}\cdot\mathbf{V}. \tag{8.2.3}$$

Turning to the balance equation for potential energy, we find that:

$$\begin{aligned}\frac{\partial \rho \varphi_{pot}}{\partial t} &= \sum_{(\gamma)} \varphi^\gamma_{pot} \frac{\partial \rho^\gamma}{\partial t} + \sum_{(\gamma)} \rho^\gamma \frac{\partial \varphi_{pot}}{\partial t} \\ &= -\nabla\cdot\left(\sum_{(\gamma)} \rho^\gamma \varphi^\gamma_{pot} \mathbf{V}^\gamma\right) + \sum_{(\gamma)} \rho^\gamma \nabla \varphi^\gamma_{pot}\cdot\mathbf{V}^\gamma \\ &\quad + \sum_{(\gamma)} \varphi^\gamma_{pot} \Gamma'^\gamma + \sum_{(\gamma)} \rho^\gamma \frac{\partial \varphi^\gamma_{pot}}{\partial t}.\end{aligned} \tag{8.2.4}$$

By subtracting (8.2.3) and (8.2.4) from (8.2.1), we obtain the balance equation for internal energy:

$$\begin{aligned}\frac{\partial \rho u}{\partial t} &= -\nabla\cdot\left[\rho u \mathbf{V} + \sum_{(\gamma)} h^\gamma \mathbf{j}^\gamma - \lambda \nabla T\right] \\ &\quad + \boldsymbol{\sigma} : \nabla\mathbf{V} - \sum_{(\gamma)} \nabla \varphi^\gamma_{pot} \mathbf{j}^\gamma + \sum_{(\gamma)} h^\gamma_R \Gamma'^\gamma + \Gamma'^H,\end{aligned} \tag{8.2.5}$$

in which h^γ_R is defined after (8.2.1). The first term on the second line of the above equation represents the added internal energy due to the work caused by pressure and internal stress. The second term on the same line expresses the work performed by the diffusive mass fluxes. If the only potential energy is due to gravity, then $\nabla \varphi^\gamma_{pot} = \nabla z$. This second term vanishes because $\sum_{(\gamma)} \mathbf{j}^\gamma = 0$. In general, this term will vanish if each potential energy is proportional to the mass of the respective species. Henceforth, we shall assume that this statement is true, and that the above second term vanishes. The third term on the second line of the equation is the addition of internal energy due to mass sources. The last term is the energy due to heat sources.

For a fluid phase, $\boldsymbol{\sigma} = \boldsymbol{\tau} - p\boldsymbol{\delta}$, where $\boldsymbol{\tau}$ is the *deviatoric stress*, and $\boldsymbol{\delta}$ is the unit tensor. The energy balance equation then takes the form:

$$\frac{\partial \rho u}{\partial t} = -\nabla\cdot\left(\rho h\mathbf{V} + \sum_{(\gamma)} h^{\gamma}\mathbf{j}^{\gamma} - \lambda\nabla T\right) + \mathbf{V}\cdot\nabla p + \boldsymbol{\tau} : \nabla\mathbf{V} + \sum_{(\gamma)} h_R^{\gamma}\Gamma'^{\gamma} + \Gamma'^{H}. \tag{8.2.6}$$

Assuming that:

$$|\boldsymbol{\tau} : \nabla\mathbf{V}| \ll |\nabla\cdot\mathbf{j}^{tIE}|, \quad \text{and} \quad |\mathbf{V}\cdot\nabla \mathrm{p}| \ll |\nabla\cdot\mathbf{j}^{\mathrm{tIE}}|,$$

where $\mathbf{j}^{tIE}$ is the total internal energy flux defined as:

$$\mathbf{j}^{tIE} \equiv \rho h\mathbf{V} + \sum_{(\gamma)} h^{\gamma}\mathbf{j}^{\gamma} - \lambda\nabla T, \tag{8.2.7}$$

we may approximate (8.2.6) by:

$$\frac{\partial \rho u}{\partial t} = -\nabla\cdot\left(\rho h\mathbf{V} + \sum_{(\gamma)} h^{\gamma}\mathbf{j}^{\gamma} - \lambda\nabla T\right) + \sum_{(\gamma)} h_R^{\gamma}\Gamma'^{\gamma} + \Gamma'^{H}. \tag{8.2.8}$$

The same equation can be derived by subtracting the potential energy balance equation (8.2.4) from the total energy balance equation (8.2.1), provided we assume that (1) the kinetic energy term, $\frac{1}{2}\rho V^2$, is negligible, compared to that of the internal energy, ρu, (2) that the magnitude of the term due to the *deviatoric stress* ($\boldsymbol{\tau} : \mathbf{V}$), is negligible compared to that of the total internal energy flux, $\mathbf{j}^{tIE}$, and (3) the source term related to momentum is small compared to the other source terms.

8.2.2 *For a Fluid Phase Under Simplifying Assumptions*

Let $\sum_{(\gamma)} h_R^{\gamma}\Gamma'^{\gamma} = h_R\Gamma'^{m}$, and we assume that $|\boldsymbol{\tau} : \nabla\mathbf{V}| \ll |\mathbf{j}^{tIE}|$. Then, expanding (8.2.6), and using the mass balance equation (3.2.15), we obtain:

$$\rho\frac{\partial h}{\partial t} - \frac{\partial p}{\partial t} = -\rho\mathbf{V}\cdot\nabla h - \sum_{(\gamma)} \nabla\cdot h^{\gamma}\mathbf{j}^{\gamma} - \nabla\cdot\mathbf{j}^{H} + \mathbf{V}\cdot\nabla p + \Gamma'^{H} + (h_R - h)\Gamma'^{m}, \tag{8.2.9}$$

or, equivalently,

$$\rho\frac{\mathrm{D}h}{\mathrm{D}t} - \frac{\mathrm{D}p}{\mathrm{D}t} = -\sum_{(\gamma)}\nabla\cdot h^{\gamma}\mathbf{j}^{\gamma} - \nabla\cdot\mathbf{j}^{H} + \Gamma'^{H} + (h_R - h)\Gamma'^{m}, \tag{8.2.10}$$

where $\mathrm{D}(..)/\mathrm{D}t \equiv \partial(..)/\partial t + \mathbf{V}\cdot\nabla(..)$, and, as before, $\mathbf{j}^{H} \equiv -\lambda\nabla T$.

From the first and second laws of thermodynamics, it follows that:

$$\begin{aligned} d\mathbb{H} &= Td\mathbb{S} + \mathbb{V}dp + \sum_{(\gamma)} \left.\frac{\partial G}{\partial m^{\gamma}}\right|_{p,T,m^{\delta\neq\gamma}} dm^{\gamma} \\ &= T\left.\frac{\partial \mathbb{S}}{\partial T}\right|_{p,m^{\gamma}} dT + \left(\mathbb{V} + T\left.\frac{\partial \mathbb{S}}{\partial p}\right|_{T,m^{\gamma}}\right)dp + \sum_{(\gamma)} h^{\gamma}dm^{\gamma}, \end{aligned}$$

in which $\mathbb{V}$, $\mathbb{S}$, $\mathbb{H}$ and $\mathbb{G}$ denote volume, entropy, enthalpy and Gibbs free energy, (defined in Sect. 2.2.2), respectively.

One of Maxwell's relations (Denbigh 1981, p. 91) states that:

$$\left.\frac{\partial \mathbb{S}}{\partial p}\right|_{T,m^{\gamma}} = -\left.\frac{\partial \mathbb{V}}{\partial T}\right|_{p,m^{\gamma}},$$

so that:

$$d\mathbb{H} = C_p dT + \left(\mathbb{V} - T\left.\frac{\partial \mathbb{V}}{\partial T}\right|_{p,m^{\gamma}}\right)dp + \sum_{(\gamma)} h^{\gamma}dm^{\gamma}, \tag{8.2.11}$$

where we used the following identity for the specific heat at constant pressure:

$$C_p = \frac{1}{m}\left.\frac{\partial \mathbb{H}}{\partial T}\right|_{p,m^{\gamma}}.$$

Thus, we may divide (8.2.11) by the mass, m, to obtain:

$$dh = C_p dT + (1 - T\beta_T)v dp + \sum_{(\gamma)} h^{\gamma}d\omega^{\gamma},$$

where the *thermal compressibility* of the fluid phase is defined as:

$$\beta_T \equiv \frac{1}{v}\left.\frac{\partial v}{\partial T}\right|_{p,m^{\gamma}}.$$

Then, (8.2.10) becomes:

$$\rho C_p\frac{\mathrm{D}T}{\mathrm{D}t} - T\beta_T\frac{\mathrm{D}p}{\mathrm{D}t} + \sum_{(\gamma)}\rho h^{\gamma}\frac{\mathrm{D}\omega^{\gamma}}{\mathrm{D}t} = -\sum_{(\gamma)}\nabla\cdot h^{\gamma}\mathbf{j}^{\gamma} - \nabla\cdot\mathbf{j}^{H} + \Gamma'^{H} + (h_R - h)\Gamma'^{m}. \tag{8.2.12}$$

Consider the following three cases:

CASE A. Let:

$$\frac{\partial p}{\partial t} = 0, \qquad \frac{\partial \omega^\gamma}{\partial t} = 0, \qquad \nabla \omega^\gamma = 0,$$

where the last condition implies that $\mathbf{j}^\gamma = 0$. Then, the energy balance equation is:

$$\rho C_p \frac{\partial T}{\partial t} = -\rho C_p \mathbf{V}\cdot\nabla T - \nabla\cdot\mathbf{j}^H + T\beta_T \mathbf{V}\cdot\nabla p + \Gamma'^H + (h_R - h)\Gamma'^m. \quad (8.2.13)$$

CASE B. Let:

$$\beta_T = 0, \qquad \frac{\partial \omega^\gamma}{\partial t} = 0, \qquad \nabla \omega^\gamma = 0.$$

Then, the energy balance equation takes the form:

$$\rho C_p \frac{\partial T}{\partial t} = -\rho C_p \mathbf{V}\cdot\nabla T - \nabla\cdot\mathbf{j}^H + \Gamma'^H + (h_R - h)\Gamma'^m. \quad (8.2.14)$$

CASE C. Suppose that:

$$\frac{\partial p}{\partial t} = 0, \qquad \nabla p = 0, \qquad \frac{\partial \omega^\gamma}{\partial t} = 0, \qquad \nabla \omega^\gamma = 0,$$

then:

$$\rho C_p \frac{\partial T}{\partial t} = -\rho C_p \mathbf{V}\cdot\nabla T - \nabla\cdot\mathbf{j}^H + \Gamma'^H + (h_R - h)\Gamma'^m. \quad (8.2.15)$$

Furthermore, if:

$$\left(\frac{\partial C_p}{\partial p}\right)_T = \left(\frac{\partial C_p}{\partial T}\right)_p = 0, \qquad \nabla C_p = 0,$$

then, by first multiplying both sides of the mass balance equation by $C_p T$, and adding the result to (8.2.15), we obtain the energy balance equation:

$$\frac{\partial(\rho C_p T)}{\partial t} = -\nabla\cdot\left(\rho C_p T \mathbf{V}\right) - \nabla\cdot\mathbf{j}^H + \rho\Gamma^H + C_p T\Gamma'^m + (h_R - h)\Gamma'^m. \quad (8.2.16)$$

8.2.3 *For a Deformable Elastic Solid Phase*

We consider the microscopic energy balance equation for the deformable elastic solid matrix. There are two alternative approaches to derive this balance equation. One approach is to define the heat capacity under *constant stress conditions* (which is equivalent to constant pressure conditions, if the medium is isotropic, with no shear stresses), and to write the balance equation in terms of the stress field within the solid.

The other approach is to define the heat capacity under *constant strain conditions* (which is equivalent to constant volume conditions, if the medium is isotropic with no shear strains). Both approaches are valid. Since, for most geological problems, it is the stress that is known, the derivation presented here will follow the first approach.

Considering an infinitesimal solid volume undergoing heating, the *first law of thermodynamics* implies that for a reversible transformation:

$$\delta Q = d\mathbb{U} - \mathbb{V}_o\boldsymbol{\sigma} : d\varepsilon, \tag{8.2.17}$$

where $\mathbb{V}_o$ is the initial volume, $\boldsymbol{\sigma}$ and ε denote stress and strain, respectively, and $-\boldsymbol{\sigma} : d\varepsilon$ is the *strain energy per unit volume* (Landau and Lifshitz 1986, p. 8). By extending the *enthalpy* ($\mathbb{H}$) concept to a solid, we obtain:

$$\mathbb{H} = \mathbb{U} - \mathbb{V}_o\boldsymbol{\sigma} : \varepsilon. \tag{8.2.18}$$

Substituting $\mathbb{U}$, obtained from this expression, into (8.2.17), yields:

$$\begin{aligned}\delta Q &= d\mathbb{H} + \mathbb{V}_o\varepsilon : d\boldsymbol{\sigma}\\ &= \left.\frac{\partial\mathbb{H}}{\partial T}\right|_\sigma dT + \left(\left.\frac{\partial\mathbb{H}}{\partial\sigma_{ij}}\right|_T + \mathbb{V}_o\varepsilon_{ij}\right)d\sigma_{ij},\end{aligned} \tag{8.2.19}$$

in which *Einstein's summation convention*, defined after (3.1.6), is applicable.

One result from (8.2.19) is that, under constant stress conditions, we have: $\delta Q = (\partial\mathbb{H}/\partial T)|_\sigma dT$. It then follows that:

$$C^*_\sigma \equiv \left.\frac{\partial\mathbb{H}}{\partial T}\right|_\sigma$$

is the solid's *heat capacity under constant stress*.

Next, we derive an expression for the term $(\partial\mathbb{H}/\partial\sigma_{ij})|_T$ that appears in (8.2.19). From the second law of thermodynamics, and from (8.2.19), it follows that the change in *entropy*, $\mathbb{S}$, of the solid, is given by:

$$\begin{aligned}d\mathbb{S} &= \frac{\delta Q}{T} = \frac{1}{T}\left(d\mathbb{H} + \mathbb{V}_o\varepsilon_{ij}d\sigma_{ij}\right)\\ &= \frac{1}{T}\left.\frac{\partial\mathbb{H}}{\partial T}\right|_\sigma dT + \frac{1}{T}\left(\left.\frac{\partial\mathbb{H}}{\partial\sigma_{ij}}\right|_T + \mathbb{V}_o\varepsilon_{ij}\right)d\sigma_{ij}.\end{aligned}$$

Because $d\mathbb{S}$ is an *exact differential*, we must have:

$$\left.\frac{\partial}{\partial\sigma_{ij}}\left(\frac{1}{T}\left.\frac{\partial\mathbb{H}}{\partial T}\right|_\sigma\right)\right|_T = \left.\frac{\partial}{\partial T}\left[\frac{1}{T}\left(\left.\frac{\partial\mathbb{H}}{\partial\sigma_{ij}}\right|_T + \mathbb{V}_o\varepsilon_{ij}\right)\right]\right|_\sigma.$$

By expanding the above expression, we obtain:

$$\left.\frac{\partial \mathbb{H}}{\partial \sigma_{ij}}\right|_T + \mathbb{V}_o \varepsilon_{ij} = \mathbb{V}_o T \left.\frac{\partial \varepsilon_{ij}}{\partial T}\right|_\sigma.$$

Hence, (8.2.19) becomes:

$$\delta Q = C^*_\sigma dT + T \left.\frac{\partial \varepsilon_{ij}}{\partial T}\right|_\sigma d\sigma_{ij}. \tag{8.2.20}$$

Equating with the first law, (8.2.17), we obtain

$$d\mathbb{U} = C^*_\sigma dT + T\mathbb{V} \left.\frac{\partial \varepsilon_{ij}}{\partial T}\right|_\sigma d\sigma_{ij} + \mathbb{V}\sigma_{ij} d\varepsilon_{ij}, \tag{8.2.21}$$

where $\mathbb{V}$ denotes the volume of the system, and higher terms in $d\mathbb{V}d\sigma_{ij}$ and $d\mathbb{V}d\varepsilon_{ij}$ have been dropped. Dividing by the mass of the solid, we obtain (see (2.3.20)):

$$du = C_\sigma dT + Tv\beta_{Tij} d\sigma_{ij} + \sigma_{ij} d\varepsilon_{ij}. \tag{8.2.22}$$

Here:

$$C_\sigma \equiv \frac{1}{m} C^*_\sigma = \frac{1}{m} \left.\frac{\partial \mathbb{H}}{\partial T}\right|_\sigma$$

is the *specific heat capacity under constant stress*, and the *thermal compressibility tensor* can be defined, for a thermo-elastic medium, by:

$$\beta_{Tij} \equiv \left.\frac{\partial \varepsilon_{ij}}{\partial T}\right|_\sigma.$$

We assume that energy transported by mass diffusion (if there are dissolved species) is negligible, and that there is no source of mass. Then, by employing (8.2.5) and the mass balance equation, the energy balance equation can be written in the form:

$$\rho \frac{\mathrm{D}u}{\mathrm{D}t} = -\nabla \cdot \mathbf{j}^H + \boldsymbol{\sigma} : \nabla \mathbf{V} + \Gamma'^H. \tag{8.2.23}$$

From (8.2.22), we have:

$$\rho \frac{\mathrm{D}u}{\mathrm{D}t} = \rho C_\sigma \frac{\mathrm{D}T}{\mathrm{D}t} + T\beta_T \frac{\mathrm{D}\boldsymbol{\sigma}}{\mathrm{D}t} + \boldsymbol{\sigma} : \nabla \mathbf{V}. \tag{8.2.24}$$

Here, we used the identity (Bear and Bachmat 1991, p. 82):

$$\boldsymbol{\sigma} : \frac{\mathrm{D}\varepsilon}{\mathrm{D}t} = \boldsymbol{\sigma} : \nabla \mathbf{V}.$$

Substituting (8.2.24) into (8.2.23), we obtain the following form for the energy balance equation of a deformable solid (s):

$$\rho_s C_\sigma \frac{\mathrm{D}T}{\mathrm{D}t} + T\boldsymbol{\beta}_T \frac{\mathrm{D}\boldsymbol{\sigma}}{\mathrm{D}t} = -\nabla\cdot\mathbf{j}_s^H + \Gamma_s^{\prime H}. \tag{8.2.25}$$

Consider the following three cases:
CASE A. Uniform stress, i.e., $\nabla\boldsymbol{\sigma} = 0$. We obtain:

$$\rho_s C_\sigma \frac{\mathrm{D}T}{\mathrm{D}t} + T\boldsymbol{\beta}_T \frac{\partial\boldsymbol{\sigma}}{\partial t} = -\nabla\cdot\mathbf{j}_s^H + \Gamma_s^{\prime H}. \tag{8.2.26}$$

CASE B. Steady stress, i.e., $\partial\boldsymbol{\sigma}/\partial t = 0$. We obtain:

$$\rho_s C_\sigma \frac{\mathrm{D}T}{\mathrm{D}t} = -T\boldsymbol{\beta}_T \mathbf{V}_s\cdot\nabla\boldsymbol{\sigma} - \nabla\cdot\mathbf{j}_s^H + \Gamma_s^{\prime H}. \tag{8.2.27}$$

CASE C. Steady and uniform stress, i.e., $\partial\boldsymbol{\sigma}/\partial t = \nabla\boldsymbol{\sigma} = 0$. We obtain:

$$\rho_s C_\sigma \frac{\mathrm{D}T}{\mathrm{D}t} = -\nabla\cdot\mathbf{j}_s^H + \Gamma_s^{\prime H}. \tag{8.2.28}$$

Moreover, if C_σ is a constant, then, through the use of the mass balance equation, we obtain:

$$\frac{\partial(\rho_s C_\sigma T)}{\partial t} = -\nabla\cdot(\rho_s C_\sigma T\mathbf{V}_s) - \nabla\cdot\mathbf{j}_s^H + \Gamma_s^{\prime H}. \tag{8.2.29}$$

8.3 Macroscopic Heat and Mass Fluxes

Only two kinds of energy fluxes appear in the microscopic balance equations presented in the previous section: advection and diffusion (= heat carried by the diffusive mass flux and by conduction). To obtain their macroscopic counterparts, these fluxes have to be averaged over an REV. Similar to what happens in the case of mass flux, by averaging the microscopic advective E-flux of an α-phase, at E-density e', we obtain:

$$\overline{e'_\alpha \mathbf{V}_\alpha}^\alpha = \overline{e'_\alpha}^\alpha \overline{\mathbf{V}_\alpha}^\alpha + \overline{\mathring{e}'_\alpha \mathring{\mathbf{V}}_\alpha}^\alpha, \tag{8.3.1}$$

in which the first term on the right-hand side expresses the macroscopic *advective E-flux*, carried by the macroscopic mass-weighted velocity, while the second term expresses the *dispersive E flux*.

In terms of the specific value e of E, and using intrinsic mass averaging, following (3.4.36), we obtain:

$$\overline{\rho_\alpha e_\alpha \mathbf{V}_\alpha}^\alpha = \overline{\rho_\alpha}^\alpha \widetilde{e_\alpha \mathbf{V}_\alpha}^\alpha = \overline{\rho_\alpha}^\alpha \left(\widetilde{e_\alpha}^\alpha \widetilde{\mathbf{V}_\alpha}^\alpha + \widetilde{\breve{e}_\alpha \breve{\mathbf{V}}_\alpha}^\alpha \right)$$
$$= \overline{\rho_\alpha}^\alpha \widetilde{e_\alpha}^\alpha \widetilde{\mathbf{V}_\alpha}^\alpha + \overline{\rho_\alpha}^\alpha \widetilde{\breve{e}_\alpha \breve{\mathbf{V}}_\alpha}^\alpha . \quad (8.3.2)$$

These two fluxes—advection and dispersion of E, as well as the macroscopic diffusive (for heat, conductive) flux, are discussed below.

8.3.1 *Advective and Dispersive Energy Flux*

Let us first consider the macroscopic advective internal energy ($\mathbb{U}$) flux of an α-phase, $\mathbf{J}^{\mathbb{U}}_{\alpha,adv}$ ($\equiv \mathbf{J}^{\mathbb{H}}_{\alpha,adv}$). At the microscopic level, this flux is expressed by $u'\mathbf{V}$, or $\rho u \mathbf{V}$. We use (8.3.1) to express the advective part of the average of this microscopic flux:

$$\mathbf{J}^{\mathbb{U}} = \overline{u'}^\alpha \overline{\mathbf{V}}^\alpha . \quad (8.3.3)$$

In terms of the specific value u of $\mathbb{U}$, following (8.3.2), we can write:

$$\overline{\rho_\alpha u_\alpha \mathbf{V}_\alpha}^\alpha = \overline{\rho_\alpha}^\alpha \widetilde{u_\alpha \mathbf{V}_\alpha}^\alpha = \overline{\rho_\alpha}^\alpha \left(\widetilde{u_\alpha}^\alpha \widetilde{\mathbf{V}_\alpha}^\alpha + \widetilde{\breve{u}_\alpha \breve{\mathbf{V}}_\alpha}^\alpha \right)$$
$$= \overline{\rho_\alpha}^\alpha \widetilde{u_\alpha}^\alpha \widetilde{\mathbf{V}_\alpha}^\alpha + \overline{\rho_\alpha}^\alpha \widetilde{\breve{u}_\alpha \breve{\mathbf{V}}_\alpha}^\alpha . \quad (8.3.4)$$

Thus, the average of the microscopic level advective flux is expressed as the sum of two fluxes: and advective flux and a dispersive one:

$$\mathbf{J}^{\mathbb{U}}_{\alpha,adv} = \overline{\rho_\alpha}^\alpha \widetilde{u_\alpha}^\alpha \widetilde{\mathbf{V}_\alpha}^\alpha , \quad \mathbf{J}^{U}_{\alpha,dis} = \overline{\rho_\alpha}^\alpha \widetilde{\breve{u}_\alpha \breve{\mathbf{V}}_\alpha}^\alpha , \quad (8.3.5)$$

where both fluxes are per unit area of the α-phase. To obtain the fluxes per unit area of porous medium, we have to multiply these expressions by the volumetric fraction, θ, of the phase.

A similar equation can be written for the advective and dispersive fluxes of phase enthalpy:

$$\overline{\rho_\alpha h_\alpha \mathbf{V}_\alpha}^\alpha = \overline{\rho_\alpha}^\alpha \widetilde{h_\alpha}^\alpha \widetilde{\mathbf{V}_\alpha}^\alpha + \overline{\rho_\alpha}^\alpha \widetilde{\breve{h}_\alpha \breve{\mathbf{V}}_\alpha}^\alpha . \quad (8.3.6)$$

In the above expressions, we may replace u, or h, by an appropriate expression in terms of the temperature, e.g., $h = C_p T$, where C_p is the heat capacity at constant pressure for a fluid phase. The dispersive flux is further considered in Sect. 8.3.6.

Note that although we have introduced the dispersive energy flux by REV-averaging considerations, we could have introduced this flux by the phenomenological approach, as we know and understand *phenomenologically* that always, at the macroscopic level, the total flux of any extensive quantity is made up of an advective, dispersive and diffusive fluxes. The reason is explained in Sect. 3.4.3.

8.3.2 *Advective Mass Flux*

As shown in the previous subsection, in order to calculate the advective heat flux, we need an expression for the advective mass flux.

The advective mass flux under isothermal conditions is considered in Chap. 4. In principle, the same expression, say Darcy law, e.g., in the form of (4.2.44) is still valid under nonisothermal conditions, except that we have to take into account the fact that *both the density and the viscosity depend on the temperature*, and that the pressure, through the capillary pressure relationship, which, in turn, depends on surface tension, is also related to the temperature.

For the sake of simplicity, we shall consider saturated flow in a stationary, non-deformable solid matrix, $\mathbf{V}_s = 0 \Longrightarrow \mathbf{q}_r \equiv \mathbf{q}$. Under such conditions, the advective mass flux of a fluid phase at a volumetric fraction θ, $\theta\rho\mathbf{V} (\equiv \rho\mathbf{q})$, is obtained from (4.2.44). We may write this equation in terms of a *reference density*, ρ_{ref} (at selected reference values of p, ρ^γ, T), in the form:

$$\begin{aligned}\mathbf{q}^m \equiv \rho\mathbf{q} \equiv \rho\theta\mathbf{V} &= -\frac{\rho\mathbf{k}}{\mu}(\nabla p + \rho g \nabla z)\\ &= -\frac{\rho\mathbf{k}}{\mu}\left(\nabla p + \rho_{ref} g \nabla z\right) + \frac{\rho\mathbf{k}}{\mu}(\rho_{ref} - \rho) g \nabla z,\end{aligned} \quad (8.3.7)$$

in which $\mathbf{k} = \mathbf{k}(\theta)$ is the permeability tensor of the fluid phase, $\rho = \rho(p, c^\gamma, T)$ is the mass density of the phase, $\mu = \mu(p, c^\gamma, T)$ is the dynamic viscosity of the phase, and c^γ denotes the concentration of the species in the fluid phase. The last term expresses the effect of density variations; it is often used in discussing natural convection (see Sect. 8.5).

In the case of a deformable porous medium, $\mathbf{q}$ and $\mathbf{V}$ are replaced by $\mathbf{q}_r$ and $\mathbf{V}_r = \mathbf{V} - \mathbf{V}_s$.

From (8.3.7) it follows that we may interpret the advective mass flux as produced by two driving forces: one resulting from a gradient in a *reference piezometric head*, $h_{ref} = z + p/\rho_{ref} g$, of a fictitious fluid of a reference density ρ_{ref}, and the second resulting from a *buoyancy force*, directed vertically upward, acting on a fluid of density ρ, embedded in the fluid of density ρ_{ref}.

The two forces are of the orders of magnitude:

$$O\left(\frac{k\rho_{ref} g}{\mu}\frac{(\Delta h_{ref})_c}{L_c^{(h)}}\right) \quad \text{and} \quad O\left(\frac{kg(\Delta\rho)_c}{\mu}\right),$$

respectively, where $(\Delta h_{ref})_c$ and $(\Delta\rho)_c$ are characteristic piezometric head difference and characteristic density difference, respectively, and $L_c^{(h)}$ is a length over which h_{ref} varies significantly. The ratio between the two is of order $O(R)$, with $R \equiv \{(\Delta\rho)_c/\rho_{ref}\}/\{(\Delta h_{ref})_c/L_c^{(h)}\}$.

When $R \ll 1$, the flow is governed mainly by head gradients. The flow regime is then referred to as *forced convection*. When $R \gg 1$, the flow is determined mainly by the buoyancy force, and the flow regime is referred to as *free* (or *natural*) *convection*. Altogether, the dependence of the density difference appearing in the second term on the right-hand side of (8.3.7) on temperature (in fact, also the effect of temperature on viscosity) produces *coupling* between the advective mass and heat fluxes.

From the (linear) approximate relationship (see Sect. 2.3.2) A, it follows that for $\rho = \rho(p, T, c^\gamma, \gamma = 1, ..., N^\gamma)$,

$$\Delta\rho \equiv \rho - \rho_{ref} = \beta_p \Delta p + \beta_T \Delta T + \sum_{(\gamma)} \beta_{c^\gamma} \Delta c^\gamma,$$

where the β's are coefficients. For example,

$$\rho = \rho_o \exp[\beta_p(p - p_o) - \beta_T(T - T_o)]. \tag{8.3.8}$$

The buoyancy force term in (8.3.7) is given by:

$$\frac{\rho\mathbf{k}}{\mu}(\rho_{ref} - \rho)g\nabla z = -\frac{\rho\mathbf{k}}{\mu} \cdot \left(\beta_p \Delta p + \beta_T \Delta T + \sum_{(\gamma)} \beta_{c^\gamma} \Delta c^\gamma\right) g\nabla z.$$

Usually, we assume that $\beta_p|\Delta p| \ll \sum_{(\gamma)} \beta_{c^\gamma}|\Delta c^\gamma|$, so that the buoyancy force is due primarily to variations in temperature and concentration.

Some comments on natural convection are presented in Sect. 8.5.

In a multi-phase system, the advective fluxes of heat and mass are coupled due to the dependence of the momentum transfer across fluid-fluid interfaces on surface tension, γ_{wn} (say, between a wetting fluid and a nonwetting one). The latter, in turn, depends on the temperature.

Temperature also affects the effective permeability values, again, through its effect on surface tension, which, in turn, affects the spatial distribution of the fluid phases within the void space.

Bear and Bachmat (1991, p. 186) develop the averaged momentum balance equation for each phase in two-phase flow, leading to a motion equation that, in addition to coupling between the phases that is due to momentum exchange across interphase surfaces, includes also a term (in each of the two flux equations) that is due to gradients in surface tension. The latter depends on temperature and concentration of dissolved species.

8.3.3 Diffusive Mass Flux of a γ-Species

In this subsection, we are considering diffusive heat and mass transport in a multi-species fluid phase. We shall take into account coupled phenomena as presented in Sect. 2.6: the Soret and the Dufour effects. The macroscopic diffusive flux of a γ-species in a a *multi-component fluid phase* that occupies the entire void space or part of it within a porous medium domain under nonisothermal conditions, is expressed in the form:

$$\mathbf{J}_\alpha^\gamma = \frac{\hat{\rho}_\alpha^2}{\rho_\alpha RT} \sum_{(\delta)} M^\gamma M^\delta \boldsymbol{\mathcal{D}}_\alpha^{*\gamma\delta} X_\alpha^\delta \cdot \left[\nabla \mu_\alpha^\delta \Big|_{p_\alpha, T} + M^\delta \left(v_\alpha^\delta - v_\alpha \right) \nabla p_\alpha \right.$$
$$\left. + M^\delta \left(\nabla \varphi_\alpha^\delta - \sum_{(\lambda)} \omega_\alpha^\lambda \nabla \varphi_\alpha^\lambda \right) \right] - \boldsymbol{\mathcal{D}}_\alpha^{*\gamma T} \nabla T, \tag{8.3.9}$$

(Bird et al. 1960, p. 567) in which $\boldsymbol{\mathcal{D}}_\alpha^{*\gamma\delta}$ $(\equiv \mathcal{D}_\alpha^{\gamma\delta} \mathbf{T}_\alpha^*)$ is the coefficient of molecular diffusion of the γ-species in a multi-species α-phase within the void space; $\mathcal{D}^{\gamma\delta}$ is the same coefficient for the phase, but not within a porous medium; and $\mathbf{T}_\alpha^*$ is the *tortuosity* of the phase (Bear and Bachmat 1991). Here, $\hat{\rho}_\alpha$ denotes the molar density of the α-phase, $\hat{v}_\alpha^\gamma$ denotes the partial molar volume of a γ-species in an α-phase, R $(= 8.1347\ \mathrm{J/mol^\circ K})$ is the *universal gas constant*, T is the absolute (°K) temperature, M^γ is the molar mass of a γ-species, and $X_\alpha^\gamma = n_\alpha^\gamma / n_\alpha$ is the *molar fraction* of the γ-species in the α-phase, with n_α^γ denoting the number of moles of the γ-species in the α-phase. The variables ω_α^γ and φ_α^γ are the mass fraction and the potential energy of the γ-species, respectively.

The chemical potential for a species of a phase, μ_α^γ, is defined (see (2.2.52)) by:

$$\mu_\alpha^\gamma \equiv \frac{\partial \mathbb{G}_\alpha}{\partial N_\alpha^\gamma} \bigg|_{p_\alpha, T, n^{\delta \neq \gamma}}, \tag{8.3.10}$$

where $\mathbb{G}_\alpha$ is the *Gibbs free energy* (defined in Sect. 2.2.3) of the phase. The first term in the square brackets on the right-hand side of (8.3.9) is an extension of Fick's law for a multi-species phase. The second and third terms express diffusion due to mechanical and potential energy, respectively. The coefficient $\boldsymbol{\mathcal{D}}_\alpha^{*\gamma T}$ $(\equiv \mathcal{D}_\alpha^{\gamma T} \mathbf{T}_\alpha^*)$ is the coefficient for the molecular diffusive flux due to temperature gradients, known as the *thermo-diffusive effect*, or *Soret effect* (see Sect. 2.6, or any text on irreversible thermodynamics, e.g., De Groot and Mazur 1962), and $\mathcal{D}_\alpha^{\gamma T}$ is the same coefficient, but not in a fluid continuum.

The diffusion coefficients $\mathcal{D}_\alpha^{\gamma\delta}$ satisfy the relationships:

$$\mathcal{D}_\alpha^{\gamma\gamma} = 0, \quad \text{and} \quad \sum_{(\gamma)} \left[M^\gamma M^\delta \mathcal{D}_\alpha^{\gamma\delta} - M^\gamma M^\lambda \mathcal{D}_\alpha^{\gamma\lambda} \right] = 0.$$

For a dilute solution, we have:

$$\nabla \hat{\mu}_\alpha^\gamma \bigg|_{p,T,n^{\delta \neq \gamma}} = \frac{RT}{n_\alpha^\gamma} \nabla n_\alpha^\gamma.$$

By expanding (8.3.9) for a *binary system*, $\gamma = A, B$, we obtain:

$$\begin{aligned}
\mathbf{J}_\alpha^A &= -\rho_\alpha \mathcal{D}_\alpha^{*AB} \nabla \omega_\alpha^A \\
&\quad - \frac{\hat{\rho}_\alpha \omega_\alpha^A \omega_\alpha^B}{RT} M^A M^B \mathcal{D}_\alpha^{*AB} \Big[\left(v_\alpha^A - v_\alpha^B \right) \nabla p_\alpha + \nabla (\varphi_\alpha^A - \varphi_\alpha^B) \Big] \\
&\quad - \mathcal{D}_\alpha^{*AT} \nabla T, \qquad (8.3.11) \\
\mathbf{J}_\alpha^B &= -\rho_\alpha \mathcal{D}_\alpha^{*AB} \nabla \omega_\alpha^B \\
&\quad - \frac{\hat{\rho}_\alpha \omega_\alpha^A \omega_\alpha^B}{RT} M^A M^B \mathcal{D}_\alpha^{*AB} \Big[\left(v_\alpha^B - v_\alpha^A \right) \nabla p_\alpha + \nabla (\varphi_\alpha^B - \varphi_\alpha^A) \Big] \\
&\quad - \mathcal{D}_\alpha^{*BT} \nabla T, \qquad (8.3.12)
\end{aligned}$$

where we made use of the relationship $\mathcal{D}_\alpha^{*AA} = \mathcal{D}_\alpha^{*BB} = 0$, and, for a binary system, $\mathcal{D}_\alpha^{*AB} = \mathcal{D}_\alpha^{*BA}$. We also used the identities $v = \omega^A v^A + \omega^B v^B$ and $\varphi = \omega^A \varphi^A + \omega^B \varphi^B$. When the only external forces acting on the species are gravitational, then $\varphi^A = \varphi^B = gz$, and the terms that contain the species' potential energies in the flux expressions vanish.

We note here the coupling between the diffusive mass flux of a species and the concentration gradient, as well as coupling to the pressure, to the species potential energies, and to the temperature. The flux due to the concentration gradient usually dominates over the fluxes due to pressure and temperature gradients, except when these gradients are very large. A brief discussion on coupled transport fluxes is presented in Sect. 2.6.

8.3.4 Diffusive Heat Flux (≡ Conduction)

At the microscopic level, the *diffusive heat flux* within an α-phase, also referred to as *heat conduction*, is expressed by *Fourier's law*:

$$\mathbf{j}_\alpha^H = -\lambda_\alpha \nabla T_\alpha, \qquad (8.3.13)$$

where λ_α denotes the *thermal conductivity* of the α-phase.

To obtain the macroscopic form of Fourier's law, we average the microscopic law over an REV, employing the definition of averages and the averaging rules presented in Sect. 1.1.4.

In order to apply these averaging rules to thermal conductivity in a fluid-solid system ($\alpha = f, s$), we assume that the λ_α, of each α-phase is constant within the

REV. We also assume that the gradient of the microscopic temperature on the outer boundary of the REV is approximately equal to the gradient of the macroscopic (= average) temperature assigned to the center of the REV. Let us consider two cases:

CASE A. A fluid (f) occupies the entire void space and the solid (s) is a thermal insulator. This corresponds to **CASE A** in Sect. 1.4.2 A4. From (1.4.17), we obtain:

$$\mathbf{J}_f^H = -\lambda_f \mathbf{T}_f^* \cdot \nabla \overline{T}^f = -\boldsymbol{\lambda}_f^* \cdot \nabla \overline{T}^f, \tag{8.3.14}$$

in which $\boldsymbol{\lambda}_f^* (= \lambda_f \mathbf{T}_f^*)$ is the coefficient of thermal conductivity of the f-phase that occupies the void space, and $\mathbf{T}_f^*$ is the second rank tensor of *tortuosity* of that phase (components T_{fij}^*). Note that the above coefficient is a second rank symmetric tensor.

CASE B. We assume that both phases, f and s, are thermally conductive and $\overline{T_f}^f = \overline{T_s}^s$. Then, from **CASE B** in in Sect. 1.4.2 A4, we obtain for the heat flux through the porous medium as a whole, $\mathbf{q}_{pm}^H$:

$$\begin{aligned} \mathbf{q}_{pm}^H &\equiv \phi \overline{\mathbf{j}_f^H}^f + (1-\phi)\overline{\mathbf{j}_s^H}^s = -\phi\lambda_f \overline{\nabla T}^f - (1-\phi)\lambda_s \overline{\nabla T}^s \\ &= -\left[\phi \boldsymbol{\lambda}_f^* + (1-\phi)\boldsymbol{\lambda}_s^*\right] \cdot \nabla \overline{T}^f = -\boldsymbol{\Lambda}_{pm}^H \cdot \nabla \overline{T}^f, \end{aligned} \tag{8.3.15}$$

where:

$$\boldsymbol{\Lambda}_{pm}^H = \phi \boldsymbol{\lambda}_f^* + (1-\phi)\boldsymbol{\lambda}_s^* = \phi \mathbf{T}_f^* \lambda_f + (1-\phi)\mathbf{T}_s^* \lambda_s \tag{8.3.16}$$

is the thermal conductivity of a saturated porous medium as a whole.

Omitting the averaging symbols, we can express the law of heat conduction in a saturated porous medium as a whole in the form:

$$\mathbf{q}_{pm}^H = \phi \mathbf{J}_f^H + (1-\phi)\mathbf{J}_s^H = -\boldsymbol{\Lambda}_{pm}^{\mathbb{H}} \cdot \nabla T, \quad \overline{T}^f = \overline{T}^s = T. \tag{8.3.17}$$

We recall that all $\mathbf{J}_\alpha$-fluxes are per unit area of the considered phase, while $\mathbf{q}_{pm}^H$ is per unit area of porous medium.

One of the assumptions made in the above derivation, that the approximate equality of the gradient of the microscopic temperature at the boundary of the REV with the gradient of the macroscopic temperature at its center, does not always hold. For example, while, as a consequence of the equilibrium assumption $\overline{T_f}^f \simeq \overline{T_s}^s$, the gradients of the macroscopic temperature in the two phases are the same, the gradients of the microscopic temperatures in the solid and fluid phases may be quite different from each other when their conductivities are highly contrasting. It is important to realize that when the assumption under discussion does not hold, the thermal conductive flux in a porous medium is not the sum of the individual fluxes in the two phases, as given in (8.3.15). Despite this fact, the general Fourier form of the heat conductive flux, with all phases at the same averaged temperature, T:

$$\mathbf{q}_{pm}^H = -\boldsymbol{\Lambda}_{pm}^H \cdot \nabla T, \tag{8.3.18}$$

has been found, empirically, to apply to a large variety of situations in porous medium domains.

When we consider *coupled phenomena* (Sect. 2.6), say in single phase multi-species flow, the above equation takes the form:

$$\mathbf{q}^H_{pm} = -\mathbf{\Lambda}^H_{pm} \cdot \nabla T - \sum_{\gamma=1}^{N} \boldsymbol{D}^\gamma \cdot \nabla \omega^\gamma, \tag{8.3.19}$$

The coefficient $\mathbf{\Lambda}^H_{pm}$ in (8.3.19) is referred to as the *effective thermal conductivity* (of the porous medium domain). It depends on (1) the thermal conductivities of the individual phases, and (2) the microscopic geometry of the phases distributed within the REV. In practice, since this distribution is not known (unless the medium is well-ordered), it is impossible to determine the conductivity from first principles. Instead, it is determined through experiments in which a known heat flux is forced to pass through a given porous medium domain, and the resulting temperature gradient is measured. Care must be taken that the temperature gradient is predominantly due to heat transport by conduction, and not due to advection or mass diffusion.

The lowest value of effective thermal conductivity can be obtained by visualizing the solid and fluid as alternating parallel layers, with heat being conducted normal to them. We then obtain for saturated flow:

$$\frac{1}{\Lambda^H_{pm,min}} = \frac{\phi}{\lambda_f} + \frac{1-\phi}{\lambda_s}.$$

The maximal value is obtained when heat is conducted parallel to the layers:

$$\Lambda^H_{pm,max} = \phi\lambda_f + (1-\phi)\lambda_s.$$

Many combined, parallel-series, models are given in the literature. Following are a few examples.

• For the case in which the solid is made of spheres of uniform size, without mutual influence between them, i.e., for large porosity, Maxwell (1892) suggested the relationship:

$$\frac{\Lambda^H_{pm}}{\lambda_f} = \frac{(2\lambda_f/\lambda_s + 1) - 2(1-\phi)(\lambda_f/\lambda_s - 1)}{(2\lambda_f/\lambda_s + 1) + (1-\phi)(\lambda_f/\lambda_s - 1)}. \tag{8.3.20}$$

• Kampf and Karsten (1970) suggested the relationship:

$$\frac{\Lambda^H}{\lambda_f} = 1 - \frac{(1-\phi)(\lambda_f/\lambda_s - 1)}{1 + (1-\phi)^{\frac{1}{2}}(\lambda_f/\lambda_s - 1)}. \tag{8.3.21}$$

- Schulz (1981) suggested the relationship:

$$\frac{\Lambda^H_{pm}}{\lambda_f} = \left[\frac{\lambda_s - \Lambda^H}{\phi(\lambda_s - \lambda_f)}\right]^3. \quad (8.3.22)$$

Expressions for saturated flow were also proposed by Chan and Tien (1973), Cook and Peckover (1983), Hadley (1986), and Duncan (1989).

An expression for the coefficient of thermal conductivity in a multiphase system (e.g., unsaturated flow) is discussed in Sect. 8.5.1.

8.3.5 *Diffusive Vapour Flux*

The diffusive mass flux of a species in a phase under isothermal conditions is discussed in Sect. 7.1.2. In general, unless we wish to consider fluxes due to gradients in temperature (Sect. 8.3.3), the only modification required when nonisothermal conditions prevail, is in the value of the coefficient of molecular diffusion, which is temperature dependent. However, when a liquid and a gas occupy the void space under nonisothermal conditions, such that a change of phase may take place from liquid to vapour, by *evaporation*, or from vapour to its liquid, by *condensation*, the diffusive mass flux of the vapour requires special attention. We shall focus our attention on the special case of water as the liquid and water vapour as a species of a gaseous phase, which we will take to be air.

In principle, this is a flux of the mass of a species of a phase, and as such it could be expressed by Fick's law at the macroscopic level, say, in the form of (7.2.17) Here, this equation takes the form:

$$\mathbf{J}^w_g = -\rho_g \boldsymbol{\mathcal{D}}^{*w}_g \cdot \nabla \omega^w_g, \quad (8.3.23)$$

where $\boldsymbol{\mathcal{D}}^{w*}_g (= \mathcal{D}^w_g \mathbf{T}^*_g)$ is the coefficient of molecular diffusion of the vapour in the gaseous phase in a porous medium domain, and $\omega^w_g\ (= \rho^w_g/\rho_g)$ is the mass fraction of the vapour in the gaseous phase.

De Vries and Kruger (1966) suggested an expression for $\mathcal{D}^w_g$ for a gas at atmospheric pressure and temperature T, in °C, of the form:

$$\mathcal{D}^w_g = \mathcal{D}^w_g(T) = 0.217 \left(\frac{T + T_o}{T_o}\right)^{1.88} \text{ cm}^2/\text{s}, \quad (8.3.24)$$

where T_o is a reference temperature (= 273.15 °C).

We usually assume that as long as liquid water is present in the void space, the vapour concentration in the gaseous phase, ω^w_g, is at *saturation*, $\omega^w_g|_{sat} = \omega^w_g|_{sat}(\Psi_m, \omega^\gamma_w, p_g, T)$, with ω^γ_w denoting the mass fractions of the various species (solutes) in the liquid water, and Ψ_m denoting the *matric potential* defined in

Sect. 2.5.1 B. In most cases of very low concentrations, the effect of solute concentrations is negligible. It may be significant in cases of water with a high salinity. For the sake of simplicity, we neglect this effect here. Note that $\omega_g^w|_{sat}$ defines the mass fraction of saturated water vapour that occupies the void space. This is distinct from the mass fraction of water outside a porous medium, due to the influence of the matric potential.

We shall assume that the gas pressure, p_g, is approximately constant and uniform. Then, the expression for the vapour flux can be written in form:

$$\mathbf{J}_g^w = -\rho_g \boldsymbol{\mathcal{D}}_g^{*w} \cdot \nabla \omega_g^w|_{sat} = -\rho_g \boldsymbol{\mathcal{D}}_g^{*w} \cdot \left(\frac{\partial \omega_g^w|_{sat}}{\partial \Psi_m} \nabla \Psi_m + \frac{\partial \omega_g^w|_{sat}}{\partial T} \nabla T \right). \quad (8.3.25)$$

Let us introduce the concept of *relative humidity*, defined by:

$$h_r \equiv \frac{\omega_g^w}{\omega_{go}^w\big|_{sat}(p_g, T)}$$

(see also (2.3.24)), where $\omega_{go}^w\big|_{sat}(p_g, T)$ denotes the vapour mass fraction at saturation in a gas, at the pressure p_g and temperature T, which is in contact with a *flat* water surface (outside a porous medium).

Following Edelfsen and Anderson (1943), under thermodynamic equilibrium, the relative humidity in a porous medium, is given by Kelvin's equation, (2.3.35), rewritten in the form:

$$h_r = \frac{\omega_g^w|_{sat}}{(\omega_{go}^w)\big|_{sat}} = \exp\left(\frac{\Psi_m^w M^w}{RT}\right), \quad (8.3.26)$$

where Ψ_m^w is the water matric potential for the liquid, defined in (2.3.38), for which we need the constitutive relationship:

$$\Psi_m^w = \Psi_m^w(p_g, S_w, \omega_w^\gamma, T). \quad (8.3.27)$$

The function $\omega_{go}^w\big|_{sat}(p_g, T)$ obeys the identity:

$$\omega_{go}^w\big|_{sat} = \frac{\rho_{go}^w|_{sat}}{\rho_{go}^w|_{sat} + \rho_g^a},$$

in which we can use the relationship (Kimball et al. 1976):

$$\rho_{go}^w\big|_{sat}(T) = 10^{-6} \exp\left[19.819 - \frac{4975.9}{T + T_0}\right] \quad \text{g/cm}^3, \quad (8.3.28)$$

valid for p_g = atmospheric pressure.

Another possible expression that can be used is:

$$\omega_{go}^{w}|_{sat} = \frac{M^{w} n_{go}^{w}|_{sat}}{M^{w} n_{go}^{w}|_{sat} + M^{a} n_{g}^{a}}, \tag{8.3.29}$$

with:

$$n_{go}^{w}|_{sat} = \frac{p_{sat}(T)}{p_{g}}, \qquad n_{g}^{a} = 1 - n_{go}^{w}|_{sat}, \tag{8.3.30}$$

in which $p_{sat}(T)$ is the saturated water vapour pressure given in *Steam Tables* (Meyer et al. 1968), with M^{w} and M^{a} denoting the molecular mass of water and of air, respectively.

Neglecting gradients in p_g, we can, therefore, rewrite (8.3.25) as:

$$\begin{aligned}\mathbf{J}_{g}^{w} &= -\rho_{g}\boldsymbol{\mathcal{D}}_{g}^{w*}\cdot\left[\left(h_{r}\frac{\mathrm{d}\omega_{go}^{w}|_{sat}}{\mathrm{d}T} + \frac{\partial h_{r}}{\partial T}\right)\nabla T + \omega_{go}^{w}|_{sat}\frac{\partial h_{r}}{\partial \Psi_{m}}\nabla\Psi_{m}\right]\\ &= \mathbf{J}_{g}^{wT\prime} + \mathbf{J}_{g}^{w\Psi\prime},\end{aligned} \tag{8.3.31}$$

where:

$$\mathbf{J}_{g}^{wT\prime} = -\rho_{g}\boldsymbol{\mathcal{D}}_{g}^{w*}\left(h_{r}\frac{\mathrm{d}\omega_{go}^{w}|_{sat}}{\mathrm{d}T} + \frac{\partial h_{r}}{\partial T}\right)\cdot\nabla T \tag{8.3.32}$$

is the temperature-driven part of the diffusive vapour flux, and:

$$\mathbf{J}_{g}^{w\Psi\prime} = -\rho_{g}\boldsymbol{\mathcal{D}}_{g}^{w*}\omega_{go}^{w}|_{sat}\frac{\partial h_{r}}{\partial \Psi_{m}}\cdot\nabla\Psi_{m}, \tag{8.3.33}$$

is the matric-potential-driven part.

However, it has been observed in experiments (e.g., Rollins et al. 1954) that in the presence of a temperature gradient, the actual temperature-driven part of the diffusive flux of vapour mass is *larger* than that predicted by Fick's law of mass diffusion in a gaseous phase for a non-condensible vapour. De Vries (1958) and Philip and de Vries (1957) developed a conceptual model for vapour flux that involves an additional transport of vapour through the liquid phase (= 'pure water') from low to high temperatures, due to the processes of condensation and evaporation.

To understand why the temperature-driven part of the diffusive vapour flux is underestimated by (8.3.32), let us consider the movement of vapour in the gaseous phase, say, in the x direction. As the vapour diffuses, its movement is obstructed by a pocket of water. This pocket, say, a *pendular ring* of size Δx, is bounded by gas-water interfaces at x and at $x + \Delta x$. As we have assumed that (a) the gas is saturated by the vapour, and (b) a local temperature gradient exists between the two gas-water curved surfaces (menisci) that bound the water pocket in the x-direction, vapour must condense at the lower temperature boundary, and water must evaporate at the higher temperature one. The two rates, of evaporation and condensation, must be identical if the moisture content of the liquid phase is to remain unchanged. The simultaneous condensation-evaporation processes, is then averaged to form an additional macroscopic mass flux of the vapour (through the liquid) in the unsaturated zone. Alternatively, this amounts to an increase in the cross-sectional area available to vapour diffusion, beyond the value dictated by the cross-section of the gaseous domain. For example, θ_g is increased to $\theta_g + f(\theta_g)\theta_\ell$, with $f(\theta_g) = 1$ for $\theta_g \geq \theta_{r\ell}$, and $f(\theta_g) = \theta_g/\theta_{r\ell}$ for $\theta_g \leq \theta_{r\ell}$.

Philip and de Vries introduced an additional enhancement factor that is due to the fact that in the presence of vapour, the thermal conductivity in the gaseous phase is larger than that corresponding to the average temperature. It is equal to the ratio between the gradient of the temperature in the gas and that of the averaged temperature in the soil as a whole. However, observations indicate that beyond an average temperature of 62 °C, the vapour flux is overestimated by (8.3.32).

Jury and Latey (1979) supplemented this theory by taking into account the thermal properties of the phases. Cass et al. (1984) conducted experiments at temperatures up to 35 °C, and compared results with the various theories. They concluded that

- The enhancement factor rises exponentially with moisture content up to the moisture content for which the water becomes a continuous phase. Beyond that point, the contribution of the liquid phase to vapour flux decreases.
- The enhancement factor decreases with temperature.

Bensabat (1986), modifying the theory developed by Philip and de Vries (1957), proposed a model for an isotropic porous medium that replaces (8.3.32) by the form:

$$\theta_g \mathbf{J}_g^{wT} = -\kappa(T)\zeta(\theta_g)\rho_g \boldsymbol{\mathcal{D}}_g^{w*}\left(h_r \frac{\mathrm{d}\omega_{go}^w\big|_{sat}}{\mathrm{d}T} + \frac{\partial h_r}{\partial T}\right)\cdot\nabla T, \tag{8.3.34}$$

where:

$$\zeta(\theta_g) = \theta_g \mathbf{T}_g^* + \theta_\ell \mathbf{T}_\ell^* f^T(T) f^\theta(\theta_\ell)$$

is the *enhancement coefficient* for which Childs and Malstaf (1982) suggested (for the isotropic case) the expression

$$\zeta(\theta_g) = 2.0 - 10(\phi - \theta_g),$$

and $\kappa(T)$ is a *mass flow factor* (Childs and Malstaf 1982) introduced as a correction in order to take into account the fact that the pressure in the gas is not zero, and $f^T(T)$ and $f^\theta(\theta_\ell)$ are *enhancement coefficients*. The value of f^θ vanishes both at low and at high water contents. We note the additional flux due to transport through the water in the expression for ζ. Bensabat (1986) suggested expressions for $f^\theta(\theta_\ell)$.

The correction κ should also be applied to (8.3.33), so that the pressure-driven part of the diffusive vapour flux should be written as:

$$J_g^{w\Psi} = -\kappa(T)\rho_g \boldsymbol{\mathcal{D}}_g^{w*}\omega_{go}^w\big|_{sat}\frac{\partial h_r}{\partial \Psi_m}\cdot\nabla\Psi_m. \tag{8.3.35}$$

This flux, arising from the dependence of the relative humidity of the air on the matric potential, has a significant magnitude at very low values of water content, when the tension is high (on the order of tens of bars).

Childs and Malstaf (1982) suggested an expression for the mass flow factor of the form:

$$\kappa = \kappa(T) = \frac{1}{\{[(A_\kappa T + B_\kappa)T - C_\kappa]T - D_\kappa\}T + E_\kappa}, \tag{8.3.36}$$

where $A_\kappa = -9.575 \times 10^{-9}$, $B_\kappa = 3.42 \times 10^{-7}$, $C_\kappa = 3.63 \times 10^{-5}$, $D_\kappa = 1.463 \times 10^{-4}$, and $E_\kappa = 0.99321$.

The developments mentioned above relate to 'pure water' in the unsaturated zone, i.e., water without any dissolved matter. Obviously, the presence of the latter affects vapour pressure, and (8.3.34) and (8.3.35) have to be modified to include the effects of gradients in concentration, since with this effect we have $\omega^w_{gsat} = \omega^w_{gsat}(\Psi_m, \omega^\gamma_w, p_g, T)$. Here, γ denotes all species in the gaseous phase, except water.

Bear et al. (1991) applied these developments to calculations associated with experiments on heat storage in the unsaturated zone in the soil (Sect. 8.5.1). The subject of enhanced vapour diffusion is also discussed by Clifford (2006, p. 38).

8.3.6 Dispersive Heat Flux

The phenomenon of dispersion in heat transport is analogous to that of species dispersion in mass transport. Both stem from the fact that at the microscopic level, fluid velocities vary from point to point within the void space. Hence, the average of the microscopic advective heat flux (carried by the mass averaged velocity of the phase) over an REV, yields the sum of two fluxes: a macroscopic advective flux, discussed in Sect. 8.3.1, and a *dispersive heat flux*.

The expression for the dispersive heat flux follows from (8.3.6):

$$\mathbf{J}^H_{\alpha,dis} = \overline{\rho_\alpha}^\alpha \widetilde{\check{h}_\alpha \check{\mathbf{V}}_\alpha}^\alpha. \tag{8.3.37}$$

Because of the analogy to the dispersive mass flux of a species, most of the discussion in Sect. 8.3.1, with the obvious modifications resulting from the difference in the transported extensive quantity, is applicable also to thermal dispersion. Thus, in analogy to (7.2.32), the expression for the dispersive heat flux in a fluid α-phase, with $\rho_\alpha h_\alpha\ (\equiv e^H_\alpha) = \rho_\alpha C_{\alpha,p} T$, takes the form:

$$\mathbf{J}^H_{\alpha,dis} = -\mathbf{D}'^H_\alpha \cdot \nabla(\rho_\alpha C_{\alpha,p} T), \tag{8.3.38}$$

where $\mathbf{D}'^H_\alpha$ denotes the *coefficient of thermal dispersion*, which is a second rank symmetric tensor. It is common to rewrite (8.3.38) in the form:

$$\mathbf{J}^H_{\alpha,dis} = -\mathbf{D}^H_\alpha \cdot \nabla T, \tag{8.3.39}$$

where $\mathbf{D}^H_\alpha$ denotes another *coefficient of thermal dispersion*.

To obtain an expression for $\mathbf{D}^H_\alpha$, we have to insert in (7.2.38), a *thermal Peclet number*, Pe^H_α, defined by:

$$\text{Pe}^H_\alpha = \frac{V_\alpha \Delta_\alpha}{\lambda_\alpha \rho_\alpha C_{\alpha,p}}, \tag{8.3.40}$$

in which V_α is the magnitude of the macroscopic velocity of the α-fluid, Δ_α is the hydraulic radius of the α-fluid-filled portion of the void space, λ_α is the thermal conductivity of the α-fluid phase, and $\lambda_\alpha / \rho_\alpha C_{\alpha,p}$ is its *thermal diffusivity*. This definition of the Peclet number can be understood by comparing it to the one defined by (7.2.39).

Thus, in indicial notation, omitting the subscript α, the *coefficient of thermal dispersion* of a fluid phase is given by:

$$D^H_{im} = \rho C_p a^H_{ik\ell m} \frac{V_k V_\ell}{V} f(\text{Pe}^H, \ell^H/\Delta), \tag{8.3.41}$$

where the characteristic length, ℓ^H, which indicates the distance of correlation between the velocities of energy particles, and the hydraulic radius, Δ, depend on the spatial distribution of the phase within the void space. As such, they are functions of the saturation of the phase. In this expression, $a^H_{ik\ell m}$ denotes the *thermal dispersivity*.

We usually assume that $\ell^\gamma \approx \ell^H$, and, therefore, the thermal dispersivity tensor, $\mathbf{a}^H$, and the mass of species' dispersivity tensor, $\mathbf{a}^\gamma$, are approximately the same. This means that for $\text{Pe}^H \gg 1$, $f(\text{Pe}^H, r) = \text{Pe}^H/(1 + \text{Pe}^H + \ell^H/\Delta) = O(1)$, and the coefficients of thermal and mass dispersion are the same. However, when $\text{Pe}^H \ll 1$, $f(\text{Pe}^H, r) \approx \text{Pe}^H/(1 + \ell^H/\Delta)$. Under these conditions, Bear and Bachmat (1991) examine the relative magnitudes of the advective, dispersive, and conductive heat fluxes. They found that thermal advection dominates over thermal dispersion as long as:

$$\frac{L^{(T)}_c}{a_c} \gg \frac{(\Delta T)_c}{T_c}, \tag{8.3.42}$$

where subscript c denotes characteristic values. If $\text{Pe}^H \ll 1$, advection dominates when:

$$\frac{L^{(T)}_c}{a_c} \gg \frac{(\Delta T)_c}{T_c} \text{Pe}^H. \tag{8.3.43}$$

Usually a_c and $L^{(T)}_c$ are represented by the hydraulic radius, Δ, and the characteristic length of an REV, respectively. The ratio $L^{(T)}_c/a_c$ is then often taken as 100.

This means that under most practical circumstances, thermal dispersion can be neglected with respect to thermal advection. Cheng et al. (1991) discuss thermal dispersion in porous media.

8.3.7 Coupled Transport Fluxes

The subject of coupled phenomena (in the Onsager sense) was introduced in Sect. 2.6. Here we shall repeat the case of coupling between heat transport and mass transport of chemical γ-species, in a slightly different form. Fourier's law describes the diffusive flux of heat, and Fick's law describes the diffusive flux of mass of a γ-species in a fluid phase by molecular diffusion. According to Onsager's theory, they are particular cases of a general linear law that describes the coupled (macroscopic) diffusive fluxes of heat and γ-mass:

$$\mathbf{J}^H = -\boldsymbol{\mathcal{L}}^{HH} \cdot \nabla T - \sum_{(i)} \boldsymbol{\mathcal{L}}^{H\gamma_i} \cdot \nabla \omega^{\gamma_i}, \tag{8.3.44}$$

$$\mathbf{J}^{\gamma_i} = -\sum_{(i)} \boldsymbol{\mathcal{L}}^{\gamma_i\gamma_i} \cdot \nabla \omega^{\gamma_i} - \boldsymbol{\mathcal{L}}^{\gamma_i H} \cdot \nabla T, \tag{8.3.45}$$

where the coefficients $\boldsymbol{\mathcal{L}}^{H\gamma_i}$ represent the *Dufour effect*, and the coefficients $\boldsymbol{\mathcal{L}}^{\gamma_i H}$ represent the *Soret effect*. The coefficients $\boldsymbol{\mathcal{L}}^{\gamma_i\gamma_i} \equiv \boldsymbol{\mathcal{D}}^{*\gamma_i}$ represent the coefficients of molecular diffusion, while $\boldsymbol{\mathcal{L}}^{HH} \equiv \boldsymbol{\Lambda}^*$ represents the coefficient of thermal conductivity, both in a phase within a porous medium domain. All fluxes are per unit area of the considered phase, and all coefficients are second rank symmetric tensors.

The phenomena of coupling between fluxes of heat and mass of species as described here are neglected in this.

8.4 Macroscopic Heat and Mass Transport Models

Whenever we assume *thermal equilibrium* among the solid and fluid phases at points within a porous medium domain (which is the underlying assumption here, as in most cases of heat and mass transport in the subsurface and in geological formations), the only variable, $T(\mathbf{x}, t)$, is the average temperature of the porous medium as a whole,

$$T(\mathbf{x}, t) = \frac{1}{\mathbb{V}_o} \int_{\mathbb{V}_o} \sum_{\alpha} T_\alpha(\boldsymbol{\xi}, t; \mathbf{x}) \gamma_\alpha(\boldsymbol{\xi}) d\mathbb{V}, \quad \alpha = w, n, s.$$

Thus, we need to state and solve only a *single* energy balance equation–for the porous medium as a whole. We obtain this equation by summing the equations for all the individual phases present in the system.

8.4.1 Energy Balance without Chemical Reactions

A. In Terms of Temperature

In the absence of chemical chemical reactions, it is convenient to write this equation in terms of the temperature. We start from the microscopic equation (8.2.16), in which we assumed that (a) $|\boldsymbol{\tau}_f : \nabla \mathbf{V}_f| \ll |\nabla \cdot \mathbf{j}_f^{IE}|$, (b) $C_f \equiv C_p = \text{const.}$, (c) $|\mathrm{D}p/\mathrm{D}t| \ll |\rho_f C_f \mathrm{D}T/\mathrm{D}t|$, and (d) the effect of concentration changes on the energy balance is negligible. Adding the assumption that there exist no energy and mass sources or sinks, i.e., $\Gamma^H = \Gamma^{'m} = 0$, we obtain the microscopic energy balance equation for a fluid that fills the void space, or part of it,

$$\frac{\partial}{\partial t}(\rho_f C_f T_f) = -\nabla \cdot (\rho_f C_f T_f \mathbf{V}_f + \mathbf{j}_f^H). \tag{8.4.1}$$

We have written the fluid's energy balance equation in the above form to emphasize the interpretation of $\rho_f C_f T_f$ as the *enthalpy density*. Energy sources can always be added if such sources exist.

By averaging this equation, with

$$|\overline{\mathring{\rho}_f \mathring{T}_f}^f| \ll |\overline{\rho_f}^f \overline{T_f}^f|, \quad \text{so that} \quad \overline{\rho_f T_f}^f \simeq \overline{\rho_f}^f \overline{T_f}^f,$$

and

$$\overline{\rho_f T_f \nabla \cdot \mathbf{V}_f}^f \simeq \overline{\rho_f}^f \overline{T_f}^f \overline{\nabla \cdot \mathbf{V}_f}^f,$$

which is based on the assumption that the absolute value of the average of a product of two or three deviations is much smaller than that of the products of the corresponding averages (Bear and Bachmat 1991, p. 142). Or directly by employing the phenomenological approach, we obtain:

$$\frac{\partial}{\partial t}(\phi \overline{\rho_f}^f C_f \overline{T_f}^f) = \\ -\nabla \cdot \phi \left[\overline{\rho_f}^f C_f \overline{T_f}^f \overline{\mathbf{V}_f}^f + \overline{(\rho_f \mathring{C}_f T_f) \mathring{\mathbf{V}}_f}^f + \overline{\mathbf{j}_f^H}^f \right] - f_{f\to s}^{\mathbb{H}}. \tag{8.4.2}$$

The term $\overline{(\rho_f \mathring{C}_f T_f) \mathring{\mathbf{V}}_f}^f$ $(\equiv \mathbf{J}_{dis}^H)$ expresses the *dispersive flux of heat*. We note that the total heat flux is made up of an advective, a dispersive, and a diffusive ($\equiv$ conductive) heat fluxes. The term:

$$f_{f\to s}^H \equiv \frac{1}{\mathbb{V}_o} \int_{S_{fs}} [\rho_f C_f T_f (\mathbf{V}_f - \mathbf{u}) + \mathbf{j}_f^H] \cdot \boldsymbol{\nu}_f dS_{fs},$$

represents the rate of heat transferred from the fluid to the solid, per unit volume of porous medium. We note that, in general, this transfer takes place in two modes:

by advection of mass across the (microscopic) interphase boundary, and by thermal diffusion (= conduction). In the case of saturated flow considered here, the fluid-solid interface is a material surface with respect to the fluid's mass, so that $(\mathbf{V}_f - \mathbf{u})\cdot\boldsymbol{\nu}_f \equiv 0$, and only the second mode of transport remains. However, if we consider adsorption, the energy of the adsorbate is carried across the fluid-solid interface.

Next, we average (8.2.29) for an isotropic *thermo-elastic* solid phase, assuming that (a) $C_s \equiv C_\sigma =$ constant, i.e., that the heat capacity at constant stress is a constant, (b) $|\beta_{Tij}T(\mathrm{D}\sigma_{ij}/\mathrm{D}t)| \ll |\mathrm{D}(\rho_s C_s T)/\mathrm{D}t|$, i.e., the stress is a constant or almost so, (c) $\Gamma^{'m} = 0$, i.e., that there are no sources of solid phase mass, and (d) that:

$$|\overline{\rho_s}^s\, \overline{C_s T_s}^s| \gg |\overline{\mathring{\rho}_s (\mathring{C_s T_s})}^s|, \quad \text{hence:} \quad \overline{\rho_s C_s T_s}^s \simeq \overline{\rho_s}^s\, C_s \overline{T_s}^s,$$

$$|\overline{(\rho_s C_s \mathring{T}_s)\mathring{\mathbf{V}}_s}^s| \ll |\overline{\rho_s}^s\, C_s \overline{T_s}^s\, \overline{\mathbf{V}_s}^s|.$$

Noting also that $\mathcal{S}_{sf}$ is a material surface with respect to the solid's mass, i.e., $(\mathbf{V}_s - \mathbf{u})\cdot\boldsymbol{\nu}_s \equiv 0$, the averaged, or macroscopic heat balance equation for the solid is:

$$\frac{\partial}{\partial t}(1-\phi)\overline{\rho_s}^s\, C_s \overline{T_s}^s = -\nabla\cdot(1-\phi)\left[\overline{\rho_s}^s\, C_s \overline{T_s}^s\, \overline{\mathbf{V}_s}^s + \overline{\mathbf{j}_s^H}^s\right] - \frac{1}{\mathbb{V}_o}\int_{\mathcal{S}_{fs}} \mathbf{j}_s^H\cdot\boldsymbol{\nu}_s \, d\mathcal{S}_{sf}, \tag{8.4.3}$$

in which we have made use of $\boldsymbol{\nu}_f = -\boldsymbol{\nu}_s$ and $\mathbf{j}_s^H\cdot\boldsymbol{\nu}_s = -\mathbf{j}_f^H\cdot\boldsymbol{\nu}_f$ at every point on $\mathcal{S}_{sf}$. We recall that the last terms on r.h.s. of the above equation expresses the rate of heat transferred from the fluid to the solid, per unit volume of porous medium.

Equations (8.4.2) and (8.4.3) are the two *macroscopic heat balance equations* for the fluid and for the solid phases, respectively. In order to express these equations in terms of average temperatures as the only state variables, we have to introduce appropriate expressions for the macroscopic conductive fluxes, $\overline{\mathbf{j}_f^H}^f$ and $\overline{\mathbf{j}_s^H}^s$, as well as for the surface integrals that express the (average) rate of exchange of heat between the two phases.

Usually, at the microscopic level, we *assume*, on the basis of thermodynamic considerations, a condition of *no-jump* in the temperatures of the fluid and the solid *at their common boundary*, i.e., $[\![\ T\ ]\!]_{f,s} = 0$. Even when we do invoke this condition of no-jump in temperature at the microscopic interfaces, $\mathcal{S}_{sf}$, this does not necessarily imply the equality of the *macroscopic* temperatures of the two phases *at a point*. Thus, when $\overline{T_s}^s \neq \overline{T_f}^f$, the two balance equations have to be solved simultaneously because they are linked by the terms that express the exchange of heat between the solid and fluid continua. This is, for example, the case when large solid blocks are surrounded by relatively narrow fluid filled fractures.

Under the condition of $\overline{T_f}^f \neq \overline{T_s}^s$, heat is transferred from the phase having a higher temperature to the other phase. This exchange is expressed by the surface integrals in (8.4.2) and (8.4.3). Very often, this rate of transfer is expressed as:

$$\frac{1}{\mathbb{V}_o}\int_{S_{sf}} \mathbf{j}_f^H \cdot \boldsymbol{\nu}_f \, dS = \alpha_T^\star (\overline{T_f}^f - \overline{T_s}^s), \tag{8.4.4}$$

where $\alpha_T^\star$ is referred to as a *heat transfer coefficient*

In geological formations, because solid grains, or blocks, are relatively small and the velocity of the fluid in the void space is small, the two phases are assumed to maintain *thermal equilibrium*, or *approximately so*, i.e., $\overline{T_s}^s = \overline{T_f}^f = T$. Then, by adding (8.4.2) and (8.4.3), we obtain the energy balance equation for the porous medium domain:

$$\frac{\partial}{\partial t}(\rho C)_{pm} T = -\nabla \cdot \left[\phi \overline{\rho_f}^f C_f T \overline{\mathbf{V}_f}^f + \phi \overline{(\rho_f \mathring{C}_f T) \mathring{\mathbf{V}}_f}^f + (1-\phi)\overline{\rho_s}^s C_s T \overline{\mathbf{V}_s}^s \right], \tag{8.4.5}$$

where

$$(\rho C)_{pm} = \phi \overline{\rho_f}^f C_f + (1-\phi)\overline{\rho_s}^s C_s \tag{8.4.6}$$

denotes the heat capacity of the porous medium as a whole, i.e., including the solid matrix, and we have used the equality of heat flux at points on the fluid-solid boundary, S_{fs}.

We can always add heat sources on the r.h.s. of the above equation.

Next, we assume that the solid is stationary or approximately so. More precisely, we assume that:

$$|\nabla \cdot (\rho_s T \mathbf{V}_s)| \ll |\mathrm{D}(\rho_s T)/\mathrm{D}t|.$$

Then, omitting the symbols that denote averages, (8.4.5) reduces to:

$$\frac{\partial}{\partial t}(\rho C)_{pm} T = -\nabla \cdot \left[\phi \left(\rho_f C_f T \mathbf{V}_f + \mathbf{J}_{f,dis}^H + \mathbf{J}_f^H \right) + (1-\phi)\mathbf{J}_s^H \right], \tag{8.4.7}$$

in which $\mathbf{J}_f^H$ and $\mathbf{J}_{f,dis}^{\mathbb{H}}$ are the conductive and dispersive fluxes in the fluid, respectively, and $\mathbf{J}_s^H$ denotes the conductive heat flux in the solid matrix.

Finally, by combining (8.4.7) with the mass balance equations for the fluid and the solid, and making the same assumptions as those leading to (8.4.7), we obtain:

$$\frac{\partial (\rho C)_{pm} T}{\partial t} = -\nabla \cdot (\rho_f C_f T \mathbf{q}) + \nabla \cdot (\mathbf{\Lambda}_{pm}^{*H} \cdot \nabla T), \tag{8.4.8}$$

where the coefficient $\mathbf{\Lambda}_{pm}^{*H}$ is defined by

$$\phi \mathbf{J}_f^H + (1-\phi)\mathbf{J}_s^H + \phi \mathbf{J}_{f,dis}^H = -\mathbf{\Lambda}_{pm}^H \cdot \nabla T + \phi \mathbf{J}_{f,dis}^H = -\mathbf{\Lambda}_{pm}^{*H} \cdot \nabla T, \tag{8.4.9}$$

in which $\mathbf{\Lambda}_{pm}^{*H}$ is the combined coefficient of thermal conductivity and dispersion of the saturated porous medium. Equation (8.4.8) combines both the conductive heat fluxes in the fluid and in the solid, as well as the dispersive heat flux in the fluid (unless the latter flux is neglected). If necessary, we can add an energy source term to the above equation.

B. In Terms of Enthalpy

We consider a void space occupied by two fluid phases, α, β (e.g., two liquids (w, n), or a liquid (ℓ) and a gas (g)). When we are interested in the chemical species that comprise the phases, possibly with chemical reactions, it is convenient to express the energy balance equations in terms of the relevant enthalpy, h (see Sect. 2.2.2).

By mass averaging the microscopic energy balance equation (8.2.8) for an α-phase over an REV, or directly by the phenomenological approach, we obtain the macroscopic energy ($\mathbb{E}$) balance for a porous medium domain, with two fluid phases occupying the void space. Making use of the operator defined by (7.3.80), this equation takes the form:

$$\mathcal{B}^{\mathbb{E}}_{\alpha}(\omega^{\gamma}_{\alpha}, p_{\alpha}, T_{\alpha}) = f^{\mathbb{E}}_{s\rightarrow\alpha} + f^{\mathbb{E}}_{\beta\rightarrow\alpha} + \Gamma^{\mathbb{E}}_{\alpha}, \tag{8.4.10}$$

where we have defined the *energy balance operator*,

$$\mathcal{B}^{\mathbb{E}}_{\alpha}(\omega^{\gamma}_{\alpha}, p_{\alpha}, T_{\alpha}) \equiv \frac{\partial \theta_{\alpha}\rho_{\alpha}u_{\alpha}}{\partial t} + \nabla\cdot\theta_{\alpha}\left(\rho_{\alpha}h_{\alpha}\mathbf{V}_{\alpha} + \sum_{(\gamma)} h^{\gamma}_{\alpha}\mathbf{J}^{\gamma}_{h,\alpha} + \mathbf{J}^{*H}_{\alpha}\right) - \nabla\cdot(\Lambda^{*}_{\alpha}\nabla T_{\alpha}), \tag{8.4.11}$$

in which $\mathbf{J}^{\gamma}_{h,\alpha}$ is the hydrodynamic dispersive flux of the mass of the γ-species, $\mathbf{J}^{*H}_{\alpha}$ denotes the dispersive heat flux within a fluid α-phase, and the enthalpy of a γ-species, h^{γ}_{α}, is defined by

$$h^{\gamma} \equiv u^{\gamma}_{\alpha} + p_{\alpha}v^{\gamma}_{\alpha}. \tag{8.4.12}$$

For the solid phase, the macroscopic energy balance equation is

$$\mathcal{B}^{\mathbb{E}}_{s}(T_s) = -f^{\mathbb{E}}_{s\rightarrow\alpha} - f^{\mathbb{E}}_{s\rightarrow\beta} + \Gamma^{\mathbb{E}}_{s}, \tag{8.4.13}$$

The heat exchange between the adjacent phases is represented by the term

$$f^{\mathbb{E}}_{\alpha\rightarrow\beta} = -\frac{1}{\mathrm{V}_o}\int_{S_{\alpha\beta}}\left[\rho_{\alpha}h_{\alpha}(\mathbf{V}_{\alpha} - \mathbf{u}_{\alpha\beta}) + \sum_{(\gamma)} h^{\gamma}_{\alpha}\mathbf{j}^{\gamma}_{\alpha} + p_{\alpha}\mathbf{u}_{\alpha\beta} - \lambda_{\alpha}\nabla T_{\alpha}\right]\cdot\boldsymbol{\nu}_{\alpha}\,dS, \tag{8.4.14}$$

where it is understood that the summation in this term is over all phases $\beta \neq \alpha$. The first term in the integrand is the advective flux of enthalpy across the interface, at a velocity relative to that of the interface, denoted here as $\mathbf{u}_{\alpha\beta}$. The second term represents the transport of energy by the diffusive mass fluxes. The third and fourth terms express, respectively, the work done on the interface by the fluid pressure, and the transport of heat by conduction.

By adding the balance equations (8.4.10) and (8.4.13), the terms that express the exchange of energy between the various phases cancel, and we obtain

$$\mathcal{B}_\ell^{\mathbb{E}} + \mathcal{B}_n^{\mathbb{E}} + \mathcal{B}_g^{\mathbb{E}} + \mathcal{B}_s^{\mathbb{E}} = \Gamma_{pm}^{\mathcal{E}},$$

where $\Gamma_{pm}^{\mathcal{E}} (\equiv \sum_{(\alpha)} \Gamma_\alpha^{\mathcal{E}})$ denotes the rate at which energy is added to the porous medium per unit volume of the latter from both internal (e.g., heat of wetting, or heat released in chemical reactions) and external sources (e.g., energy injected for heating the soil, and energy extracted (= sink) with extracted gas). We then obtain the *macroscopic energy balance equation for the porous medium as a whole* in the form:

$$\begin{aligned}\frac{\partial}{\partial t}&\left[\sum_{(\alpha)} (\theta_\alpha \rho_\alpha u_\alpha) + (1-\phi)\rho_s C_\sigma T\right]\\ &= -\nabla\cdot\sum_{(\alpha)} \theta_\alpha \left[\rho_\alpha h_\alpha \mathbf{V}_\alpha + \sum_{(\gamma)} h_\alpha^\gamma \mathbf{J}_{h,\alpha}^\gamma + \mathbf{J}_\alpha^{*H}\right]\\ &\quad +\nabla\cdot(\Lambda_{pm}^{*H}\nabla T) + \Gamma_{pm}^{\mathcal{E}},\end{aligned} \tag{8.4.15}$$

in which $\alpha = \ell, n, g$, $C_\sigma (\equiv C_s)$ is the specific heat capacity of the solid, and the second sum on the second line is taken over all γ-species present in the phase. The $\mathbf{J}$'s denote macroscopic fluxes (per unit phase area). Here, $\mathbf{J}_{h,\alpha}^\gamma$ is the hydrodynamic dispersive flux of the mass of the γ-species, and $\mathbf{J}_\alpha^{*H}$ denotes the dispersive heat flux within a fluid α-phase. The source term $\Gamma_{pm}^{\mathcal{E}}$ expresses the energy (= heat) added per unit volume of porous medium.

To obtain the specific enthalpies and internal energies of the three multi-species fluid phases, we need appropriate constitutive relations. These relations, which have to be determined experimentally for all fluid phases, have the functional form:

$$u_\alpha = \sum_{(\gamma)} \omega_\alpha^\gamma u_\alpha^\gamma, \qquad u_\alpha^\gamma = u_\alpha^\gamma(p_\alpha, T, \omega^\gamma, \theta_\alpha), \qquad \alpha = \ell, n, g. \tag{8.4.16}$$

$$h_\alpha = \sum_{(\gamma)} \omega_\alpha^\gamma h_\alpha^\gamma, \qquad h_\alpha^\gamma \equiv u_\alpha^\gamma + p_\alpha v_\alpha^\gamma. \tag{8.4.17}$$

We have included θ_α as a factor that affects the internal energy, because of possible surface effects. At lower saturations, the fact that the fluid's energy close to the solid surface is much higher than at a distance from that surface may dominate the internal energy per unit volume of the fluid. By including this factor, we enable the inclusion of *heat of wetting* in the model.

For dilute contaminant concentrations, *Steam Tables* (Meyer et al. 1968) may be used for the liquid phase, while for the gaseous phase we may obtain the internal energy by assuming ideal gas behavior.

* * *

In Chap. 6, we have discussed cases in which more than one phase occupies the void space. In Sect. 2.3.1 we have discussed how phase changes are represented on a phase diagram. With the above in mind, we are now ready to consider multiphase flow and heat transport when the considered substances may change phase pressure and temperature vary.

8.4.2 Energy Balance with Phase Change

Fluids that occupy the void space, as well as the solid matrix itself, may undergo phase changes in response to certain changes in pressure, temperature, and concentration of dissolved species. Phase changes may occur also as a consequence of exothermic or endothermic chemical reactions. The subject of phase change has already been introduced in Sect. 2.3.1. In principle, *all* fluid and solid phases present in a porous medium domain may undergo changes of phase. However, in this book, which deals with phenomena of transport porous medium domains, we assume that the solid matrix does not undergo a change of phase. Furthermore, we do not consider the behavior of fluids in the micro-pores inside the solid matrix. Obviously, all the basic flow and transport models, as well as the fundamental thermodynamic relationships discussed thus far in this chapter, apply also when phase changes occur.

When a problem of phase change involves a moving boundary surface that separates the two phases from each other, the problem is referred to as a *Stefan problem*. Examples of phase changes occur in hydrocarbon reservoirs, where the void space is occupied by one or two liquid phases and a gas, in geothermal reservoirs where hot water changes to steam, as evaporation from a shallow ground water table of a phreatic aquifer, and when water freezes in the subsurface.

Examples of phase change in a porous medium domain include *evaporation* of a liquid (i.e., a change from liquid to vapor/gas), *condensation* (= change from gas to liquid), *solidification* (when a liquid turns into solid), e.g., *freezing* (when water becomes ice), *melting*(like when ice turns into a liquid), and *Drying* when a liquid turns into vapor. At the microscopic level, all these changes are initiated at interphase boundaries. In a macroscopic model, when heat enters or leaves a domain through a latter's external boundary (or at point sources/sinks that act like boundaries), phase changes are initiated at such boundary and propagate into the domain's interior.

As emphasized through the book, especially in Chap. 6, at the macroscopic level, there is no sharp/macroscopic interface between the phases that occupy the void space (unless we simplify the problem and *assume*a (possibly moving) sharp interface that separates the phases in the considered domain. Thus, what we have in the macroscopic balance equation is a phase change that occurs at macroscopic points within a considered domain according to the changes in p, T and c^γ's.

Consider the sharp boundary is between two states of aggregation ($\equiv$ phases of the *same* substance within the void space. Across such a (possibly moving) boundary, a change of phase (e.g., freezing, thawing, evaporation, condensation) takes place. We assume that the solid matrix remains unchanged, although it is possible to consider cases in which the solid matrix melts, or dissolves. We assume that only a single state (rather than a mixture of states) of a fluid, is present in the void space on each side of the boundary. The *condition of no-jump* in the total flux of any considered extensive quantity in crossing the boundary. Obviously, case, $[\theta_s]_{1,2} = 0$, and we recall that $[T]_{1,2} = 0$. According to Denbigh (1981), only a small portion of the energy required for producing a change of phase is derived from the volume change.

From (3.3.5) we have for mass of a liquid-phase that changes from liquid (ℓ) to gas (g), per unit volume of porous medium:

$$f^{m_\ell}_{\ell\to g} = -\frac{1}{\mathbb{V}_o}\int_{S_{\ell\to g}} \left[\rho_\ell(\mathbf{V}_\ell - \mathbf{u}_{\ell g})\right]\cdot\boldsymbol{\nu}_\ell dS, \tag{8.4.18}$$

• A. Conditions for Mass

We recall that the interface between phases is a material surface with respect to the two phases. From (5.2.6), we obtain:

$$[\theta\rho(\mathbf{V} - \mathbf{u}) - \theta\mathbf{D}_h\cdot\nabla\rho]_{1,2}\cdot\boldsymbol{\nu} = 0, \tag{8.4.19}$$

recalling that a different state of aggregation (gas, or liquid) of the considered substance occupies the entire void space on each side of the boundary. Since $[\theta]_{1,2} \equiv [\phi]_{1,2} = 0$, and $[\mathbf{u}]_{1,2}\cdot\boldsymbol{\nu} = 0$, Eq. (8.4.19) takes the form

$$[\rho\mathbf{V}]_{1,2}\cdot\boldsymbol{\nu} - [\rho]_{1,2}\mathbf{u}\cdot\boldsymbol{\nu} - [\mathbf{D}_h\cdot\nabla\rho]_{1,2}\cdot\boldsymbol{\nu} = 0. \tag{8.4.20}$$

Equation (8.4.20) cannot be further reduced, because, in general, $[\rho]_{1,2} \neq 0$, and $[\mathbf{V}]_{1,2} \neq 0$.

The first inequality stems from the nature of the phase change, except at the *critical point*, where the densities of the two states of the considered substance are identical (e.g., Amyx et al. 1960, pp. 212–217). The jump in the normal component of the velocity arises from the change in the density, or specific volume, of the substance upon the change of state. For a change from solid to liquid, or vice versa, this effect may be negligible. However, changes from liquid to vapor, and vice versa, are associated with significant changes in the specific volume of the considered substances.

Since the mass of a considered substance does cross it, the boundary of phase change considered here *is not a material surface*, so that $(\mathbf{V} - \mathbf{u})\big|_\ell\cdot\boldsymbol{\nu} \neq 0, \ell = 1, 2$. However, we recognize that in crossing the boundary, the mass assumes a *different state* of the same substance.

• B. Conditions for Energy

Since $[\mathbf{u}]_{1,2}\cdot\boldsymbol{\nu} = 0$, and $[\theta]_{1,2} = 0$, the condition of no-jump in the total energy flux in the direction normal to the boundary, reduces to:

$$\theta[\rho u\mathbf{V}]_{1,2}\cdot\boldsymbol{\nu} - \theta[\rho u]_{1,2}\mathbf{u}\cdot\boldsymbol{\nu} - [\mathbf{\Lambda}^{*H}\cdot\nabla T]_{1,2}\cdot\boldsymbol{\nu} = 0. \tag{8.4.21}$$

in which the jump in internal energy, $[u]_{1,2} \equiv [\rho C_V T]_{1,2}$ ($\neq 0$), in (8.4.21), expresses the jump in the energetic state of the substance in the void space, on both sides of the boundary.

When we consider a change of phase from solid to liquid, or from liquid to vapor, the jump in energy represents the additional energy required to produce a more disordered state of the molecular structure, i.e., the energy required to further separate the molecules from each other. The jump in the energetic state of a substance is manifested by the fact that for the thermal flux, we have $[-\mathbf{\Lambda}^{*H}\cdot\nabla T]_{1,2} \neq 0$; this means that part of the sum of dispersive and diffusive heat fluxes entering or leaving the boundary is compensating for the energy consumed by the phase change.

An alternative form of (8.4.21) is obtained by expressing the internal energy in terms of the *enthalpy*, h:

$$[(\rho h - p)\mathbf{V}]_{1,2}\cdot\boldsymbol{\nu} - [\rho h - p]_{1,2}\mathbf{u}\cdot\boldsymbol{\nu} - [\mathbf{\Lambda}^{*H}\cdot\nabla T]_{1,2}\cdot\boldsymbol{\nu} = 0, \tag{8.4.22}$$

where $h = u + p/\rho$, with the quantity p/ρ expressing the energy associated with the volume per unit mass. Since it is often assumed that only a small part of the energy required for producing a phase change is derived from the change in volume, (8.4.22) reduces to:

$$[\rho h\mathbf{V}]_{1,2}\cdot\boldsymbol{\nu} - L_{1,2}\mathbf{u}\cdot\boldsymbol{\nu} - [\mathbf{\Lambda}^{*H}\cdot\nabla T]_{1,2}\cdot\boldsymbol{\nu} = 0, \tag{8.4.23}$$

where $L_{1,2}(= [\rho h]_{1,2})$ is the *latent heat of phase change*, defined per unit volume. It represents the energy required to produce a change in the state of a unit volume of substance. It may also be defined with respect to the density of one of the states of a considered fluid phase, e.g., in the form:

$$L_{1,2} = [\rho h]_{1,2} = \rho_1 L_1 - \rho_2 L_2, \tag{8.4.24}$$

where L_1 and L_2 are the latent heat per unit mass of states 1 and 2, present on sides 1 and 2, of a boundary, respectively. In (8.4.23), the quantity $[\rho h\mathbf{V}]_{1,2}\cdot\boldsymbol{\nu}$, is associated with the change in volume of the fluid in the void space, due to phase change, and the resulting advective energy flux that is induced across the boundary.

8.4.3 *Vaporization*

Vaporization is a phenomenon in which the stream of a considered chemical species that is dissolved in a liquid crosses the interface from that liquid into an adjacent gaseous body, in excess of the stream moving in the opposite direction, i.e., from the gaseous phase to the liquid. Boiling is a special kind of vaporization. We use this term when a liquid from an open container passes into the vapour state through the formation of bubbles. This occurs at the temperature called *boiling point*. This is the temperature at which the vapour pressure of the liquid is equal to the prevailing atmospheric pressure. At the boiling point, the vapour pressure of the liquid is sufficiently high such that the atmospheric air can be displaced; vapour bubbles can be formed in the liquid's interior, thus allowing vaporization. The boiling point depends also on the pressure to which the liquid is subjected.

The change of phase from liquid to vapour and vice versa was already discussed in Sect. 2.3.1.

The amount of heat required to vaporize one mole of liquid is the *latent heat of vaporization* mentioned earlier.

8.4.4 Initial and Boundary Conditions

As for any extensive quantity, a complete well-posed model requires information on initial and boundary conditions. These are based on the no-jump condition introduced in Sect. 5.2.3. Briefly, this condition states that for any extensive quantity the condition on the boundary is that of no-jump in the normal component of the total flux of that quantity (through all phases present in the porous medium domain) across the boundary. The condition takes into account the possibility that the boundary itself may be moving. In this chapter, the considered extensive quantity is energy, and conditions may be expressed in terms of specific internal energy, specific enthalpy, or temperature.

It is important to realize that in order to state boundary condition for a considered case, *we must know (or assume) what happens on the* external side *of the boundary* (fluxes and/or values of state variables) in that case.

In what follows, we shall continue to refer to the averaged temperature, $T(\mathbf{x}, t)$, in the porous medium as the main state variable, and assume that all (fluid and solid) phases are in *thermal equilibrium*. Thus, we are dealing with the boundary condition associated with an extensive quantity that is the *energy of the porous medium as a whole*. Boundaries will be assumed to be stationary or moving, and the solid matrix will be assumed to be rigid and stationary. We shall assume that the void space is occupied by a single fluid (f); the extension to multiple fluids is obvious.

The discussion is analogous to that on initial and boundary conditions for contaminant transport presented in Sect. 7.5.3.

Initial conditions for an energy transport problem include information on the initial distribution of temperature within the considered porous medium domain. In a coupled energy and mass transport problem, values of other state variables, such as pressure and concentration, must also be specified.

A. General No-Jump Conditions

Boundary conditions are based on the statement of "no-jump" across the boundary. Specifically, we shall make use of the no-jump condition (5.2.6). Thus, analogous to the discussion in Sect. 7.5.1, on the transport of a chemical species, we have to satisfy the condition that without phase change (see item F below), there is no temperature jump across any surface within a domain nor on its boundary, $\mathcal{B}$:

$$[\![\ T \ ]\!]_{1,2} = 0, \quad \text{on} \quad \mathcal{B}, \tag{8.4.25}$$

where 1, 2 represent the interior and exterior of the domain.

In addition, we have to satisfy the condition of no-jump in the total energy flux, unless there is a phase change (see item F. below).

The total energy flux carried by any α-phase, is given by

$$\rho_\alpha \left(u_\alpha + \frac{1}{2} V_\alpha^2 \right) \mathbf{V}_\alpha - \boldsymbol{\sigma}_\alpha \cdot \mathbf{V}_\alpha + \mathbf{J}_{\alpha,dis}^H + \mathbf{J}_{\alpha,dif}^H. \tag{8.4.26}$$

Assuming that the boundary is a *material surface* with respect to the solid matrix (although, under certain conditions, it may be displaced), and neglecting the dispersive heat flux in the solid, the condition of no-jump in the total energy flux through a *porous medium as a whole*, takes the form:

$$[\![\ \theta_f \rho_f \left(u_f + \frac{1}{2} V_f^2 \right) (\mathbf{V}_f - \mathbf{u}) - \theta_f \boldsymbol{\sigma}_f \cdot \mathbf{V}_f + \theta_f (\mathbf{J}_{f,dif}^H + \mathbf{J}_{f,dis}^H) \ ]\!]_{1,2} \cdot \boldsymbol{\nu}$$
$$+ [\![\ - \theta_s \boldsymbol{\sigma}_s \cdot \mathbf{V}_s + \theta_s \mathbf{J}_{s,dif}^H \ ]\!]_{1,2} \cdot \boldsymbol{\nu} = 0. \tag{8.4.27}$$

The expression $[\![\ \theta_f \boldsymbol{\sigma}_f \cdot \mathbf{V}_f + \theta_s \boldsymbol{\sigma}_s \cdot \mathbf{V}_s \ ]\!]_{1,2} \cdot \boldsymbol{\nu}$ appearing in (8.4.27) can be rewritten in the form

$$[\![\ \theta_f (\mathbf{V}_f - \mathbf{u}) \cdot \boldsymbol{\sigma}_f + \theta_s (\mathbf{V}_s - \mathbf{u}) \cdot \boldsymbol{\sigma}_s \ ]\!]_{1,2} \cdot \boldsymbol{\nu} + \mathbf{u} \cdot [\![\ \theta_f \boldsymbol{\sigma}_f + \theta_s \boldsymbol{\sigma}_s \ ]\!]_{1,2} \cdot \boldsymbol{\nu}.$$

We note that the first term reduces to $\theta_f (\mathbf{V}_f - \mathbf{u}) [\![\ \boldsymbol{\sigma}_f \ ]\!]_{1,2} \cdot \boldsymbol{\nu}$. However, since $[\![\ p \ ]\!]_{12} = 0$, $[\![\ \tau \ ]\!]_{1,2} \cdot^{\circ} = 0$, we have $[\![\ \boldsymbol{\sigma}_f \ ]\!]_{1,2} = 0$. The second term vanishes in view of the no-jump in solid displacement. Finally, the last term vanishes as there is no jump in stress in the solid. Altogether, the above equation reduces to:

$$[\![\ \theta_f \rho_f (u_f + \frac{1}{2} V_f^2)(\mathbf{V}_f - \mathbf{u}) + \theta_f (\mathbf{J}_{f,dif}^H + \mathbf{J}_{f,dis}^H) \ ]\!]_{1,2} \cdot \boldsymbol{\nu}$$
$$+ [\![\ \theta_s \mathbf{J}_s^H \ ]\!]_{1,2} \cdot \boldsymbol{\nu} = 0. \tag{8.4.28}$$

As stated earlier, the fluid-solid portion of the boundary is not a material surface with respect to (the conductive part of the) energy transport. Hence, as in the case of momentum, energy may be exchanged between the fluid phase and the solid one across their common portion of the boundary, and energy is not conserved within any of the phases alone. Equation (8.4.28) includes the possible exchange between the two phases, and it is impossible to separate this equation into two equations, one for each phase.

Altogether, (8.4.28) reduces to the no-jump condition:

$$\{\rho_f \mathbf{q}_{rf} [\![\ u_f \ ]\!]_{1,2} + [\![\sum_{(\alpha=f,s)} \theta_\alpha \mathbf{J}_\alpha^H + \theta_f \mathbf{J}_{f,dis}^H \ ]\!]_{1.2}\}\boldsymbol{\cdot}\boldsymbol{\nu} = 0. \tag{8.4.29}$$

Also, we have to satisfy the no-jump condition for energy flux. For a single fluid phase that occupies the entire void space, and a stationary boundary, in the absence of phase change and sources on the boundary, this condition takes the form:

$$[\![\ \rho_f h_f \mathbf{q}_f + \phi \mathbf{J}_f^{*H} + \phi \mathbf{J}_f^H + (1-\phi)\mathbf{J}_s^H \ ]\!]_{1,2} \boldsymbol{\cdot}\boldsymbol{\nu} = 0, \quad \text{on} \quad \mathcal{B}. \tag{8.4.30}$$

Or, equivalently,

$$[\![\ \rho_f h_f \mathbf{q}_f - \boldsymbol{\Lambda}_{pm}^{*H}\boldsymbol{\cdot}\nabla T \ ]\!]_{1,2} \boldsymbol{\cdot}\boldsymbol{\nu} = 0, \quad \text{on} \quad \mathcal{B}. \tag{8.4.31}$$

However, since we also have $[\![\ \rho_f \ ]\!]_{1,2} = 0$, $[\![\ h_f \ ]\!]_{1,2} = 0$, and $[\![\ \mathbf{q}_f \ ]\!]_{1,2} \boldsymbol{\cdot}\boldsymbol{\nu} = 0$, the above condition reduces to

$$[\![\ \boldsymbol{\Lambda}_{pm}^{*H}\boldsymbol{\cdot}\nabla T \ ]\!]_{1,2} \boldsymbol{\cdot}\boldsymbol{\nu} = 0, \quad \text{on} \quad \mathcal{B}. \tag{8.4.32}$$

To obtain the boundary condition in a multicomponent multiphase system, both the advective and conductive–dispersive fluxes in (8.4.30) have to be replaced by terms that express their sums over all fluid phases and species,

$$[\![\sum_{(\alpha=f,s)} \rho_\alpha h_\alpha \mathbf{q}_\alpha + \sum_{(\alpha=f,s)} \sum_{(\gamma)} h_\alpha^\gamma \theta_\alpha \mathbf{J}_\alpha^\gamma - \boldsymbol{\Lambda}_{pm}^{*H}\boldsymbol{\cdot}\nabla T \ ]\!]_{1,2} \boldsymbol{\cdot}\boldsymbol{\nu} = 0, \quad \text{on} \quad \mathcal{B}. \tag{8.4.33}$$

Boundary conditions with phase change are discussed in F. below.

B. Boundary of Prescribed Temperature

This kind of boundary condition occurs when phenomena that take place outside the considered domain impose a specified temperature, say, $f_1(\mathbf{x}, t)$, on the domain's boundary, $\mathcal{B}$. This boundary condition takes the form:

$$T(\mathbf{x}, t) = f_1(\mathbf{x}, t), \qquad \text{on} \quad \mathcal{B}. \tag{8.4.34}$$

This is a *Dirichlet*, or a *first kind* boundary condition

C. Boundary of Prescribed Heat Flux

Here, phenomena that take place in the environment impose a certain energy (in or out) flux through the boundary. Denoting this heat flux by $f_2(\mathbf{x}, t)$, and using (8.4.31), the boundary condition takes the form:

$$\left(\rho_f h_f \mathbf{q}_f - \mathbf{\Lambda}^{*H}_{pm}\boldsymbol{\cdot}\nabla T\right)\boldsymbol{\cdot}\boldsymbol{\nu} = f_2(\mathbf{x}, t), \quad \text{on} \quad \mathcal{B}. \tag{8.4.35}$$

When the single phase fluid that occupies the void space consists of multiple species, it follows from (8.4.33) that the condition at the boundary is

$$\left(\rho_f h_f \mathbf{q}_f + \sum_{(\gamma)} h^\gamma_f \theta_f \mathbf{J}^\gamma_f - \mathbf{\Lambda}^{*H}_{pm}\boldsymbol{\cdot}\nabla T\right)\boldsymbol{\cdot}\boldsymbol{\nu} = f_2(\mathbf{x}, t) \quad \text{on} \quad \mathcal{B}. \tag{8.4.36}$$

In the case of multiphase flow with multiple species, the condition is

$$\left(\sum_{(\alpha)} \rho_\alpha h_\alpha \mathbf{q}_\alpha + \sum_{(\alpha)}\sum_{(\gamma)} h^\gamma_\alpha \theta_\alpha \mathbf{J}^\gamma_\alpha - \mathbf{\Lambda}^{*H}_{pm}\boldsymbol{\cdot}\nabla T\right)\boldsymbol{\cdot}\boldsymbol{\nu} = f_2(\mathbf{x}, t), \quad \text{on} \quad \mathcal{B}. \tag{8.4.37}$$

D. Boundary Between Two Porous Medium Domains

Here, the conditions should express *both* the continuity of energy flux and the equality of temperature on the boundary. Both the temperature and the flux at the boundary are a-priori unknown. They are determined as part of the solution. This type of boundary condition is similar to that expressed by (7.5.6) and (7.5.7).

For a single fluid phase that occupies the entire void space, the above conditions take the form:

$$[\![\ \rho_f h_f \mathbf{q}_f - \mathbf{\Lambda}^{*H}_{pm}\boldsymbol{\cdot}\nabla T\ ]\!]_{1,2}\boldsymbol{\cdot}\boldsymbol{\nu} = 0, \quad \text{on} \quad \mathcal{B}. \tag{8.4.38}$$

$$[\![\ T\ ]\!]_{1,2} = 0, \quad \text{on} \quad \mathcal{B}. \tag{8.4.39}$$

When there are no fluid sources and no phase changes take place as fluids cross the boundary, the two conditions simplify to

$$[\![\ \mathbf{\Lambda}^{*H}_{pm}\boldsymbol{\cdot}\nabla T\ ]\!]_{1,2}\boldsymbol{\cdot}\boldsymbol{\nu} = 0, \quad \text{on} \quad \mathcal{B}, \tag{8.4.40}$$

$$[\![\ T\ ]\!]_{1,2} = 0, \quad \text{on} \quad \mathcal{B}. \tag{8.4.41}$$

This simplification follows from (8.4.38), when we make use of the equality

$$[\![\ \rho_f h_f \mathbf{q}_f\ ]\!]_{1,2}\boldsymbol{\cdot}\boldsymbol{\nu} = 0, \tag{8.4.42}$$

which results from $[\![\ h_f(p_f, T)\ ]\!]_{1,2} = 0$, and from the continuity of the fluid mass flux, $[\![\ \rho_f \mathbf{q}_f\ ]\!]_{1,2}\boldsymbol{\cdot}\boldsymbol{\nu} = 0$. The condition on h_f follows from (8.4.39) and from $[\![\ p_f\ ]\!]_{1,2} = 0$, which is part of the required problem boundary condition.

In the case of multiple species in multiple fluid phases, the boundary condition at the interface between two porous media is

$$[\![\sum_{(\alpha)} \rho_\alpha h_\alpha \mathbf{q}_\alpha + \sum_{(\alpha)} \sum_{(\gamma)} h_\alpha^\gamma \theta_\alpha \mathbf{J}_\alpha^\gamma - \mathbf{\Lambda}_{pm}^{*H} \cdot \nabla T]\!]_{1,2} \cdot \boldsymbol{\nu}, \quad \text{on} \quad \mathcal{B}, \tag{8.4.43}$$

and

$$[\![T]\!]_{1,2} = 0, \quad \text{on} \quad \mathcal{B}. \tag{8.4.44}$$

When there are no phase change as a fluid crosses the boundary, the conditions simplify to (8.4.40). This follows from the continuity of the specific enthalpies and of the mass fluxes at the boundary,

$$[\![h_\alpha]\!]_{1,2} = 0, \qquad [\![h_\alpha^\gamma]\!]_{1,2} = 0, \tag{8.4.45}$$

$$[\![\rho_\alpha \mathbf{q}_\alpha]\!]_{1,2} \cdot \boldsymbol{\nu} = 0, \qquad [\![\theta_\alpha \mathbf{J}_\alpha^\gamma]\!]_{1,2} \cdot \boldsymbol{\nu} = 0. \tag{8.4.46}$$

Thus, under the stated conditions, the complicated boundary conditions given by (8.4.38) and (8.4.43), which include the transport of heat by mass transport, reduce to the much simpler condition (8.4.40). The latter states that *the conductive heat flux is continuous across the boundary*.

E. Boundary with a 'Well-Mixed Zone'

From (8.4.31), and with the same approach as in Sect. 7.5.1, the flow condition for a single fluid phase with a single species across such boundary is

$$\rho_f'' h_f'' \mathbf{q}_f'' + \alpha_T^\star (T'' - T\big|_{pm}) = \left(\rho_f h_f \mathbf{q}_f - \mathbf{\Lambda}_{pm}^{*H} \cdot \nabla T\right)\big|_{pm} \cdot \boldsymbol{\nu}. \tag{8.4.47}$$

Here, $\alpha_T^\star$ is a heat transfer coefficient (for the entire porous medium), T'' is the temperature in the 'well-mixed domain', and we express the fluid flux by

$$q_f'' = \alpha_f^\star (p_f'' - p_f\big|_{pm}). \tag{8.4.48}$$

When no phase change occurs, it follows from the continuity of mass flux that $\rho_f'' q_f'' = (\rho_f \mathbf{q}_f)\big|_{pm} \cdot \boldsymbol{\nu} = q_f^m$, and the boundary condition takes the form:

$$(h_f'' - h_f\big|_{pm}) q_f^m + \alpha_T^\star (T'' - T\big|_{pm}) = -\mathbf{\Lambda}_{pm}^{*H} \cdot \nabla T\big|_{pm} \cdot \boldsymbol{\nu}. \tag{8.4.49}$$

When $h_f'' = h_f|_{pm}$, or when there is no advection, i.e., $q_f^m = 0$, Eq. (8.4.49) reduces to

$$\alpha_T^\star (T'' - T\big|_{pm}) = -\mathbf{\Lambda}_{pm}^{*H} \cdot \nabla T\big|_{pm} \cdot \boldsymbol{\nu}. \tag{8.4.50}$$

We note that we may have $T'' \neq T\big|_{pm}$ on the boundary, i.e., a jump in temperature may take place on the boundary between the porous medium domain and the fluid body. As in the case of concentration, this jump (that contradicts the condition expressed by (8.4.39)) is a consequence of introducing the transition zone and the 'well-mixed zone' approximation.

When $|(h''_f - h_f|_{pm})\, q^m_f| \gg |\alpha^\star_T(T'' - T)|$, i.e., advection dominates over conduction in the well-mixed zone, (8.4.49) reduces to

$$(h''_f - h_f\big|_{pm})\, q^m_f = -\mathbf{\Lambda}^{*H}_{pm}\boldsymbol{\cdot}\nabla T\big|_{pm}\boldsymbol{\cdot\nu}. \tag{8.4.51}$$

For multiple phases with multiple species, we use (8.4.33) to obtain the boundary condition for a well-mixed domain, in the form:

$$\sum_{(\alpha)} \rho''_\alpha h''_\alpha q''_\alpha + \sum_{(\alpha)}\sum_{(\gamma)} h''^\gamma_\alpha J''^\gamma_\alpha + \alpha^\star_T(T'' - T\big|_{pm}) =$$
$$\left(\sum_{(\alpha)} \rho_\alpha h_\alpha \mathbf{V}_\alpha + \sum_{(\alpha)}\sum_{(\gamma)} h^\gamma_\alpha \theta_\alpha \mathbf{J}^\gamma_\alpha - \mathbf{\Lambda}^{*H}_{pm}\boldsymbol{\cdot}\nabla T\big|_{pm}\right)\boldsymbol{\cdot\nu}, \tag{8.4.52}$$

where the advective fluxes in the well-mixed zone are expressed as:

$$q''_\alpha \equiv \alpha^\star_\alpha(p''_\alpha - p_\alpha\big|_{pm}), \tag{8.4.53}$$

and the diffusive mass fluxes of the species are given by

$$J''^\gamma_\alpha \equiv \alpha^{\star\gamma}_\alpha(\omega''^\gamma_\alpha - \omega^\gamma_\alpha\big|_{pm}). \tag{8.4.54}$$

When no phase changes occur as the boundary between the well-mixed zone and the porous medium domain is crossed, we write

$$\sum_{(\alpha)}(h''_\alpha - h_\alpha\big|_{pm})q^m_\alpha + \sum_{(\alpha)}\sum_{(\gamma)}(h''^\gamma_\alpha - h^\gamma_\alpha\big|_{pm})J^\gamma_\alpha$$
$$+\alpha^\star_T(T'' - T\big|_{pm}) = -\mathbf{\Lambda}^{*H}_{pm}\boldsymbol{\cdot}\nabla T\big|_{pm}\boldsymbol{\cdot\nu}, \tag{8.4.55}$$

where

$$q^m_\alpha \equiv \rho''_\alpha q''_\alpha = (\rho_\alpha \mathbf{V}_\alpha)\big|_{pm}\boldsymbol{\cdot\nu}, \qquad J^\gamma_\alpha \equiv \theta''_\alpha J''^\gamma_\alpha = (\theta \mathbf{J}^\gamma_\alpha)\big|_{pm}\boldsymbol{\cdot\nu}. \tag{8.4.56}$$

F. Fluid Phase Change at a Boundary

We consider an (assumed) sharp boundary between two states of aggregation of the fluid(s) within the void space. Across such a (possibly moving) boundary, a change of state (e.g., freezing, thawing, evaporation, condensation) of a fluid phase may take place (but the solid matrix remains unchanged!). Problems with such a (possibly moving) boundary are often referred to as *Stefan problems*.

We consider a single fluid phase that occupies the entire void space, and assume that (1) the flux of kinetic energy is negligible with respect to the thermal one, and

(2) the boundary between phases is a material surface with respect to the mass of the phases. Together with $[\![\ \mathbf{u}\ ]\!]_{1,2} \cdot \boldsymbol{\nu} = 0$, and solid phase properties, with $[\![\ \theta\ ]\!]_{1,2} = 0$, the condition of no-jump in the total energy flux in the direction normal to the boundary, reduces to:

$$\theta [\![\ \rho u \mathbf{V}\ ]\!]_{1,2} \cdot \boldsymbol{\nu} - \theta [\![\ \rho u\ ]\!]_{1,2}\, \mathbf{u} \cdot \boldsymbol{\nu} - [\![\ \mathbf{\Lambda}^{*H} \cdot \nabla T\ ]\!]_{1,2} \cdot \boldsymbol{\nu} = 0. \qquad (8.4.57)$$

Although the underlying assumptions lead to $[\![\ T\ ]\!]_{1,2} = 0$, the jump in internal energy, $[\![\ u\ ]\!]_{1,2} \equiv [\![\ \rho C_V T\ ]\!]_{1,2}\ (\neq 0)$, in (8.4.57), expresses the jump in the energetic state of the substance in the void space on both sides of the boundary.

When we consider a change of phase from solid to liquid, or from liquid to vapour, the jump in energy represents the additional energy required to produce a more disordered state of the molecular structure, i.e., the energy required to further separate the molecules from each other. The jump in the energetic state of a substance is manifested by the fact that for the thermal flux, we have $[-\mathbf{\Lambda}^{*H} \cdot \nabla T]_{1,2} \neq 0$; this means that part of the sum of dispersive and diffusive heat fluxes entering, or leaving, the boundary is compensating for the energy consumed by the phase change. This is an example of a sink, Γ^{SE}, on the boundary.

Alternative forms of (8.4.21) are obtained by expressing the internal energy in terms of the *enthalpy*, h, viz.

$$[(\rho h - p)\mathbf{V}]_{1,2} \cdot \boldsymbol{\nu} - [\rho h - p]_{1,2} \mathbf{u} \cdot \boldsymbol{\nu} - [\mathbf{\Lambda}^{*H} \cdot \nabla T]_{1,2} \cdot \boldsymbol{\nu} = 0, \qquad (8.4.58)$$

where $\mathbf{u}$ denoting the speed of displacement of the boundary and $h = u + p/\rho$, with the quantity p/ρ expressing the energy associated with the volume per unit mass. It is often assumed that:

[A2.16] Only a small part of the energy required to produce a change of phase is derived from the change in volume (Denbigh, 1981).

Then, (8.4.58) is reduces to (8.4.23). where $L_{1,2} (= [\rho h]_{1,2})$ is the *latent heat of phase change*, defined per unit volume.

In (8.4.23), the quantity $[\rho h \mathbf{V}]_{1,2} \cdot \boldsymbol{\nu}$, is associated with the change in volume of the fluid in the void space, due to phase change, and the resulting advective energy flux that is induced across the boundary.

(d) **Boundary Shape**

In the case of a solid–liquid (= liquified solid) or a liquid-vapour (= gas containing the liquid's vapour) boundary, the shape of the boundary, can be derived (Sect. 2.7.5) from the condition $[T]_{1,2} = 0$, viz.,

$$F(\mathbf{x}, t) = T(\mathbf{x}, t)\big|_1 - T(\mathbf{x}, t)\big|_2 = 0. \qquad (8.4.59)$$

The above relationship is valid also for changes from a solid state to a liquid one, and vice versa (of the material that occupies the void space).

8.5 Introduction to Natural Convection

Only a brief introduction to natural convection in a porous medium domain will be introduced here. For a comprehensive presentation on natural and forced convection, the reader is referred to Nield and Bejan (2013).

8.5.1 The Oberbeck–Boussinesq Model

As a first step, let us introduce here the *Oberbeck–Boussinesq*[1] *approximate heat and mass transport model*. The model deals with heat and mass transport of a single fluid (no solutes) saturating the void space. We added the possibility of thermal conduction through the solid, and neglected thermal dispersion in the fluid. The model involves 6 equations:

(1) An expression that relates liquid density to pressure and temperature. We may make use of the approximate equation (2.3.20). Then, for a homogeneous liquid, and $|\beta_p(p - p_o)| \ll |\beta_T(T - T_o)|, |\beta_c(c - c_o)| \ll |\beta_T(T - T_o)|$, we obtain the approximate expression for $\rho = \rho(T)$ in the form:

$$\rho = \rho(T) = \rho_o \left\{1 + \beta_T(T - T_o)\right\}, \quad \rho_o = \rho|_{T_o}. \tag{8.5.1}$$

If we take concentration changes into account, then, from (2.3.20), we may write:

$$\rho = \rho_f(T, c) = \rho_{fo} \left\{1 - \beta_T(T - T_o) + \beta_c(c - c_o)\right\}. \tag{8.5.2}$$

(2) For the momentum balance equation, we adopt the approximation presented as CASE A in Sect. 4.2.4, with $\mathbf{V}_s = 0$. This leads to Darcy's law:

$$\mathbf{q}(\equiv \phi\mathbf{V}) = -\frac{\mathbf{k}}{\mu} \cdot \left\{\nabla p + \rho(T)g\nabla z\right\}. \tag{8.5.3}$$

(3) We adopt the *Boussinesq approximation* of the fluid's mass balance equation, as discussed as **Example 1** in Sect. 3.10:

$$\nabla\cdot\mathbf{q} = 0. \tag{8.5.4}$$

(4) For the mass balance of a γ chemical species (if we wish to take solute transport into account), we use the species balance equation (7.3.2), rewritten here for the case of single phase flow, no sources and no interphase transfer, in the form:

[1] We follow the comment by Nield and Bejan (2013, p. 29) that Oberbeck (1879) suggested this approximation, which was later followed by Boussinesq (1903).

$$\phi \frac{\partial c^{\gamma}}{\partial t} = -\mathbf{q}_{\alpha} \cdot \nabla c^{\gamma} + \nabla \cdot \left(\mathbf{D}' \cdot \nabla c^{\gamma} \right). \tag{8.5.5}$$

(5) For the heat balance equation, we make use of (8.4.8) and (8.4.9), except that we neglect the dispersive heat flux:

$$(\rho c)_{pm} \frac{\partial T}{\partial t} = -\nabla \cdot (\rho_f c_f T \mathbf{q}) + \nabla \cdot \left(\mathbf{\Lambda}^{H}_{pm} \cdot \nabla T \right), \tag{8.5.6}$$

where the thermal conductivity, which includes conduction also through the solid phase, $\mathbf{\Lambda}^{H}_{pm}$, is expressed by

$$\mathbf{J}^{H}_{pm} \equiv \phi \mathbf{J}^{H}_{f} + (1 - \phi) \mathbf{J}^{H}_{s} = -\mathbf{\Lambda}^{H}_{pm} \cdot \nabla T. \tag{8.5.7}$$

In principle, the above five equations can be solved for ρ, $\mathbf{q}$, c, T, p.

8.5.2 Natural Convection

The term *natural convection* is used describes fluid motion produced by density variations in a gravity field. Such changes may be caused by changes in temperature and/or solute concentration. The non-uniform density produced by such changes causes motion due to *buoyancy effects*. The term *convective currents* describes the nature of the motion produced by such effect.

When buoyancy effects produce motion, the latter encounters resistance due to internal "friction" within the fluid and to friction at the solid-fluid (microscopic) interfaces. Both are proportional to the fluid's viscosity. The latter resistance is proportional to the inverse of the permeability. Under certain conditions, the resistance to motion is such that the initial motion produced by the disturbance will decay. Under other conditions, it will develop and grow, leading to *convective currents*.

As an example, consider an initial situation in which a layer of stationary cold (hence, heavier) fluid overlies a layer of stationary warmer (hence, lighter) one. Under certain conditions, to be discussed below, this may be an unstable situation, meaning that even a small disturbance may completely change this initial regime. Molecular diffusion and thermal conduction tend to reduce the produced currents by smoothing out density differences.

For the sake of simplicity, the discussion on this example, as throughout this subsection will be based on the following assumptions:

- The fluid contains no dissolved components.
- The fluid is a 'Boussinesq fluid', i.e., one in which the density is assumed to be independent of the temperature, except in the gravity term that appears in the expression for the advective flux.
- in order to write the mathematical model in a dimensionless form, we select $\mu = \mu_c$, and $(\rho C_V)_c = 1$. We then have

$$(\Lambda_{ij})^* = \delta_{ij}, \qquad (\rho C)^*_f = 1.$$

- The solid properties are also independent of temperature.
- The porous medium is homogeneous and isotropic. With $k = k_c$, we then have $k^*_{ij} = \delta_{ij}$.
- We neglect effect of thermal dispersion, so that $\mathbf{\Lambda}^{*H}$ is replaced by $\mathbf{\Lambda}^H$.

Making use of the methodology outlined in Sect. 3.10, the corresponding heat and fluid mass transport (Oberbeck–Boussinesq) model outlined above will involve the following set of dimensionless equations:

- The fluid's mass balance equation is:

$$\nabla^* \cdot \mathbf{q}^* = 0. \tag{8.5.8}$$

- The fluid's motion equation is (7.1.10):

$$\mathbf{q}^* = -\nabla^* \left(p^* + \frac{z^*}{\text{Eu}\,\text{Fr}^2} \right) + \text{Ra}'\, T^* \nabla^* z^*. \tag{8.5.9}$$

- The heat balance equation for the porous medium is:

$$\frac{\partial T^*}{\partial t^*} = -\mathbf{q}^* \cdot \nabla^* T^* + \nabla^{*2} T^*. \tag{8.5.10}$$

Our objective in the following paragraphs, Following Bear and Bachmat (1991, p. 474) is to investigate the conditions under which convective currents will develop in a porous medium domain. In doing so, we shall also demonstrate the methodology of such investigations.

We shall consider two typical cases of natural convection:

- An infinite horizontal fluid saturated porous medium domain heated from below.
- A fluid saturated porous medium domain in the form of an infinite strip bounded by vertical surfaces at different temperatures.

CASE A. An Infinite Horizontal Layer Heated From Below. The layer's thickness is chosen equal to the characteristic length, L_c. Hence the dimensionless thickness is 1.

The boundary conditions are:

$$\begin{aligned} &z^* = 0, z^* = 1, \; q^*_z = 0, \\ &z^* = 0, \qquad\quad T^* = 1, \\ &z^* = 1, \qquad\quad T^* = 0. \end{aligned} \tag{8.5.11}$$

We note that the governing equations and the above boundary conditions are satisfied by the no-flow solution, which also involves no convective currents:

$$q^{*(o)} \equiv 0, \quad T^{*(o)} \equiv 1 - z^*, \quad p^* + \frac{z^*}{\mathrm{Eu}\,\mathrm{Fr}^2} = \mathrm{Ra}'\, z^* \left(1 - \frac{z^*}{2}\right), \tag{8.5.12}$$

where:

$$\begin{aligned}
\mathrm{Eu} &= \frac{(\Delta p)_c T_c^*}{\rho_c V_c^2} \equiv \frac{\phi_c^2 (\Delta p)_c T_c^*}{\rho_c q_c^2}, \quad \text{(Euler number)},\\
\mathrm{Fr} = \mathrm{Fr}^{(V)} &= \frac{V_c}{\left(T_c^* g L_c^{(V)}\right)^{\frac{1}{2}}} \equiv \frac{q_c/\phi_c}{\left(T_c^* g L_c^{(V)}\right)^{\frac{1}{2}}}, \quad \text{(Froude number)},\\
\mathrm{Ra}' &= \frac{g T_c^* \beta_T (\Delta T)_c (k/\phi T^*)_c L_c}{\mathcal{D}^{\mathbb{H}}_{fl,c} \nu_c}, \quad \text{(Rayleigh number)}.
\end{aligned} \tag{8.5.13}$$

Note that all dimensionless number are *for a porous medium*.

Next we look for the existence of additional solutions to the same problem, this time with $\mathbf{q}^* \neq 0$. If we find that a solution with convective currents does exist, we shall investigate the conditions under which the stationary solutions will be unstable.

To find additional solution(s), we perturb the flow regime around the stationary solution:

$$\begin{aligned}
\mathbf{q}^* &= \mathbf{q}^{*(0)} + \varepsilon \mathbf{q}^{*\prime},\\
T^* &= T^{*(0)} + \varepsilon T^{*\prime},\\
p^* &= p^{*(0)} + \varepsilon p^{*\prime},
\end{aligned} \tag{8.5.14}$$

where $\varepsilon \mathbf{q}^{*\prime}$, $\varepsilon T^{*\prime}$ and $\varepsilon p^{*\prime}$ are the perturbations, and ε is a small parameter.

To eliminate the term involving the pressure from the motion equation, we apply the curl operator to (8.5.9), making use of the identity $\nabla \times (\nabla \times \mathbf{q}) \equiv -\nabla^2 \mathbf{q}$, valid when $\nabla \cdot \mathbf{q} = 0$. We obtain:

$$\nabla^{*2} \mathbf{q}^* = -\mathrm{Ra}'\, \nabla^* \times (\nabla^* \times T^* \nabla^* z^*). \tag{8.5.15}$$

This equation can also be written in the form:

$$\nabla^{*2} \mathbf{q}^* = -\mathrm{Ra}' \mathcal{L}^*(T^*), \tag{8.5.16}$$

where the operator $\mathcal{L}^*$ is defined as:

$$\mathcal{L}^* \equiv \mathbf{1}x^* \frac{\partial^2}{\partial x^* \partial z^*} + \mathbf{1}y^* \frac{\partial^2}{\partial y^* \partial z^*} + \mathbf{1}z^* \left(\frac{\partial^2}{\partial x^{*2}} + \frac{\partial^2}{\partial y^{*2}} \right).$$

By inserting (8.5.14) into (8.5.10) and (8.5.16), making use of the already known solution (8.5.12), and neglecting terms of order of magnitude $O(\varepsilon^2)$, we obtain the two linearized equations for the perturbed regime:

$$\nabla^{*2} \mathbf{q}^{*\prime} = -\mathrm{Ra}' \mathcal{L}^*(T^{*\prime}), \tag{8.5.17}$$

$$\nabla^{*2} T^{*\prime} = -\mathrm{Ra}' \mathcal{L}^{*}(\mathbf{q}^{*\prime}), \tag{8.5.18}$$

where we have noted from (8.5.12) that $\nabla^* T^{*(0)} \equiv -\nabla^* z^*$ and, therefore

$$q_{z^*}^{*\prime} \equiv \mathbf{q}^{*\prime} \cdot \nabla^* z^* \equiv \mathbf{q}^{*(0)} \cdot \nabla^* T^{*(0)}.$$

Thus, the solution for the perturbed temperature distribution is coupled only to the vertical component of $\mathbf{q}^{*\prime}$. We have to solve the pair of equations:

$$\nabla^{*2} q_{z^*}^{*\prime} = -\mathrm{Ra}' \nabla_{x^* y^*}^{*2} T^{*\prime}, \tag{8.5.19}$$

and (8.5.18).

We can decouple these equations, leading to the equations:

$$\left\{ \left(\frac{\partial}{\partial t^*} - \nabla^{*2} \right) \nabla^{*2} - \mathrm{Ra}' \nabla_{x^* y^*}^{*2} \right\} T^{*\prime} = 0, \tag{8.5.20}$$

$$\left\{ \left(\frac{\partial}{\partial t^*} - \nabla^{*2} \right) \nabla^{*2} - \mathrm{Ra}' \nabla_{x^* y^*}^{*2} \right\} q_{z^*}^{*\prime} = 0, \tag{8.5.21}$$

noting that the two equations are, actually, identical. The boundary conditions to be satisfied by the perturbation solution, are:

$$q_{z^*}^{*\prime} = T^{*\prime} = 0, \quad \text{on} \quad z^* = 0, 1. \tag{8.5.22}$$

The final step is to solve for $q_{z^*}^{*\prime}$ and $T^{*\prime}$, as a superposition of normal modes, and examining the stability with respect to each mode. When this is done, it is shown that the perturbations decay as long as the stability condition:

$$\mathrm{Ra}' < 4\pi^2 \tag{8.5.23}$$

is satisfied. The value of $4\pi^2$ is referred to as the *critical value* of Ra.

To summarize **CASE A**, we have seen that

- In the case of an infinite horizontal layer heated from below, a no-flow solution exists. This solution is stable, as long as the Rayleigh number does not exceed the critical value of $4\pi^2$. Then the heat is transferred only by conduction.
- When the Rayleigh number exceeds its critical value, the no-flow solution becomes unstable, disturbances will be amplified and natural convection will develop.

A linear analysis carried out for the case of an infinite layer heated from above, will reveal that the no-flow solution is unconditionally stable. Therefore, convective currents are traditionally associated with heating from below, while the case of heating from above is regarded as a state of no flow, with heat transfer taking place by conduction only.

CASE B. An Infinite Vertical Strip Between Two Impervious Walls at Different Temperatures. This case (in the vertical xz-plane) is introduced here in order to show that convective currents will *always* occur whenever a horizontal temperature gradient exists.

Because the domain here is infinite in the z-direction, we have:

$$\frac{\partial q^*_{z^*}}{\partial z^*} = 0, \quad \text{and} \quad \frac{\partial T^*}{\partial z^*} = 0. \tag{8.5.24}$$

Hence, the mass balance equation reduces to:

$$\frac{\partial q^*_{x^*}}{\partial x^*} = 0. \tag{8.5.25}$$

By combining this equation with the boundary condition:

$$q^*_{x^*}\big|_{x^*=0} = q^*_{x^*}\big|_{x^*=1} = 0, \tag{8.5.26}$$

we obtain the solution:

$$0 \le x^* \le 1, \quad -\infty < z^* < \infty, \quad q^*_{x^*} \equiv 0. \tag{8.5.27}$$

From this solution, it follows that:

$$\frac{\partial p^*}{\partial x^*} = 0, \quad q^*_{z^*} = -\frac{\partial}{\partial z^*}\left(p^* + \frac{z^*}{\text{Eu}\,\text{Fr}^2}\right) + \text{Ra}'\,T^*. \tag{8.5.28}$$

Then, $q^*_{x^*} = 0$, $\partial T^*/\partial x^* \equiv \partial^2 T^*/\partial x^{*2} = 0$, and the heat balance equation (8.5.10) reduces for the steady state to

$$\frac{\partial^2 T^*}{\partial x^{*2}} = 0. \tag{8.5.29}$$

We note that this equation indicates that in this case, *heat transport is governed by conduction only*.

The boundary conditions are

$$x^* = 0, \quad T^* = 0, \qquad x^* = 1, \quad T^* = 1. \tag{8.5.30}$$

The solution for the temperature, T^*, is

$$T^* = x^*. \tag{8.5.31}$$

In order to determine the specific discharge, we add the constraint

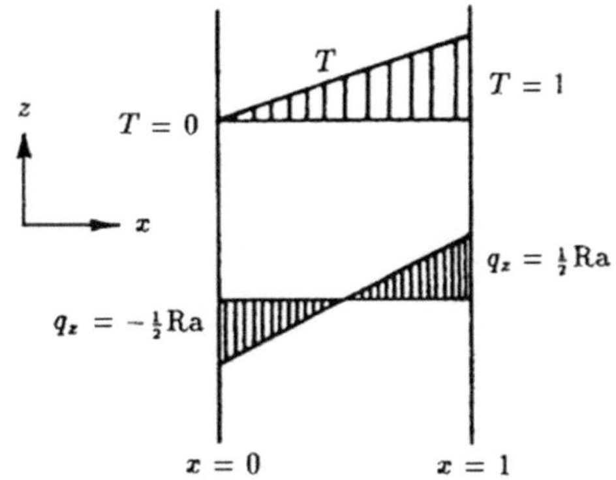

Fig. 8.1 Temperature and flux distributions in an infinite vertical strip between two walls at different temperatures

$$\int_0^1 q^*_{z^*} dx^* = 0,$$

expressing mass conservation at any horizontal cross-section. The solution for the specific discharge is then

$$q^*_{z^*} = \mathrm{Ra}' \left(x^* - \frac{1}{2} \right). \tag{8.5.32}$$

This magnitude of the convective flux, which *always* exists, is dictated by the value of the Rayleigh number. The two solutions, for the mass and heat fluxes, are shown in Fig. 8.1.

From the results of this analysis, we may conclude that

- the natural convection that develops does not affect the heat transfer which, in this case, is governed by conduction only, and
- the strength of the natural convection is determined by the value of the Rayleigh number.

We note that an analytical solution could be derived in this case, because the two equations could be decoupled.

Obviously, we have demonstrated here only a very simple case. Cheng (1985) and Boris (1987) review more complex cases. Nield and Bejabn (1998, 2013) devote a book to the subject of convection in porous media.

References

Amyx JW, Bass DM Jr, Whiting RL (1960) Petroleum reservoir engineering, physical properties. McGraw Hill, New York. 610 p

Bear J, Bensabat J, Nir A (1991) Heat and mass transfer in unsaturated porous media at the hot boundary: 1. one-dimensional analytical model. Transp Porous Media 6(3):281–298

Bear J, Bachmat Y (1991) Introduction to modeling phenomena of transport in porous media. Kluwer Publishers Company, Dordrecht, 553 p

Bensabat J (1986) Heat and mass transfer in unsaturated porous media with application to an energy storage problem. D.Sc. Thesis (in Hebrew), Technion, Israel

Bird RB, Stewart WE, Lightfoot EN (1960) Transport phenomena. Wiley, New York, 780 p

Boris SA (1987) Natural convection in porous media. In: Bear J, Corapcioglu MY (eds) Advances in transport phenomena in porous media. Martinus Nijhoff Publisher, The Netherlands, pp 77–142

Boussinesq J (1903) Théorie analytique de la chaleur, (1903), Vl. 2 Gauttier–Villars, Paris, 626 p

Cass A, Campbell GS, Jones TL (1984) Enhancement of water vapor diffusion. Soil Sci Soc Am J 48:25–32

Chan CK, Tien CL (1973) Conductance of packed spheres in vacuum. ASME J Heat Transf 95:302–308

Cheng P (1985) Natural convection in porous media: external flows. In: Kacaç S, Kiliş B, Aung W, Viskanta R (eds) Natural convection: fundamental and applications. Hemisphere, Washington D.C., pp 513–575

Cheng P, Hsiao S-W, Chen C-K (1991) Natural convection in porous media with variable porosity and thermal dispersion effects. In: Kacaç S, Kiliş B, Kulaki FA, Arinç F (eds) Convective heat and mass transfer in porous media. Kluwer Academic, The Netherlands, pp 543–562

Childs SW, Malstaf C (1982) Heat and mass transfer in unsaturated porous media, Final report, PNL-4036, Battle Pacific Northwest Labs

Clifford KH (2006) Vapor transport processes. In: Clifford KH, Webb SA (eds) Gas transport in porous media. Springer, Berlin, pp 227–446

Cook I, Peckover RS (1983) Effective thermal conductivity of debris beds. In: Proceedings of the 5th PAHR information exchange meeting on post accident debris cooling, Karlsruhe, Braun

De Vries DA, Kruger AJ (1966) On the value of the diffusion coefficient of water vapor in air. Colloques Inter. du Centre national de la Recherche Scientifique, Phénomènes de transport avec changement de phase dans les millieux poreux ou colloidaux, Paris, pp 61–69

De Vries DA (1958) Simultaneous transfer of heat and moisture in porous media. Trans Am Geophys Union 39:909–916

De Groot SR, Mazur P (1962) Non-equilibrium thermodynamics. North-Holland Publishing Company, Amsterdam, 510 p

Denbigh KG (1981) The principles of chemical equilibrium, 4th edn. Cambridge University Press, Cambridge, 494 p

Duncan AB, Peterson GP, Fletcher LS (1989) Effective thermal conductivity within packed beds of spherical particles. J Heat Transf 111:831–836

Edelfsen NE, Anderson ABC (1943) Thermodynamics of soil moisture. Hilgardia 15:31–298

Hadley GR (1986) Thermal conductivity of packed metal powders. Int J Heat Mass Transf 29:909–920

Jury WA, Latey J (1979) Water vapor movement in soil:reconciliation of theory and experiment. Soil Sci Soc Am J 43:823–827

Kampf H, Karsten G (1970) Nucl Appl Technol 9:208

Kimball BA, Jackson RD, Nakayama FS, Idso SB, Reginato RJ (1976) Soil heat flux determination: temperature gradient method with computed thermal conductivities. Soil Sci Soc Am J 40:25–28

Landau LD, Lifshitz EM (1986) Theory of elasticity, 3rd ed. Butterfield-Heinemann, London, 187 p

Maxwell JC (1892) A treatise on electricity and magnetism. Clarendon Press, Oxford

Meyer CA, McClintock RB, Silvestri GJ (1968) Thermodynamic and transport properties of steam, ASME steam tables, 2nd edn. American Society of Mechanical Engineers, New York, 328 p

Nield DA, Bejan A (1998) Convection in porous media, 2nd edn. Springer, Berlin, 546 p

Nield DA, Bejan A (2013) Convection in porous media, 4th edn. Springer, Berlin, 778 p

Oberbeck A (1879) Über die wärmeleitung der flüssigkeiten berücksichtigung der strömungen infolge von tempraturdifferenzen. Ann Phis Chem 7:271–292

Philip JR, de Vries DA (1957) Moisture movement in porous materials under temperature gradients. Trans Amer Geophys Union 38:222–232

Rollins RL, Spangler MC, Kirkham D (1954) Movement of soil moisture under thermal gradient. In: Proceedings of the highway research board, vol 33

Schulz B (1981) Thermal conductivity of porous and highly porous materials. High Temp High Press 13:649–660

Chapter 9
Poromechanics and Deformation

The objective of this chapter is to present and discuss *deformation* in porous medium domains, focussing on cases in which fluids are extracted from or injected into geological formations, possibly under non-isothermal conditions. The models presented in this chapter describe phenomena of deformation of porous medium domains in response to imposed stresses:

- Land subsidence as a consequence of pumping water from aquifers.
- Ground surface upheaval, as a consequence of injection.
- Development of fractures as a consequence of injecting fluids into a tight (e.g., rock and shale) geological formation.
- Waves in porous media.
- Induced seismicity as a result of fluid injection into confined formations.
- Soil liquefaction.

In principle, any change in the flow regime is associated with fluid pressure changes, and thus with stress changes within a considered porous medium domain. These changes produce deformation, sometime reaching a level that causes formation failure. No effort is made here to cover the subjects of Soil Mechanics, Geomechanics, or Poroelasticity. These subjects are well covered in the literature (e.g., Verruijt 2010; Cheng 2016). However, certain principles and concepts will be presented as a background for the presentation.

This is not a chapter on soil or rock mechanics. We shall focus on stress, strain and deformation associated with phenomena of flow and transport in porous medium domains. Often, the subject considered here is referred to as *poroelasticity*. Here, we are using the name *poromechanics*, as our discussion is not limited to elastic solid matrices.

Like everywhere else in this book, the fluid(s) within the void space, the solid matrix and the porous medium domain as a whole are regarded as continua. Accordingly, we start by introducing the concepts of stress and strain in a single phase continuum–first at the microscopic level and then at the macroscopic one. Then, following the *phenomenological approach*, we regard the multi-phase (solid-fluid) porous medium also as a continuum, so that we can extend the theories of (single

J. Bear, *Modeling Phenomena of Flow and Transport in Porous Media*,
Theory and Applications of Transport in Porous Media 31,
https://doi.org/10.1007/978-3-319-72826-1_9

phase) solid and fluid mechanics to (multi-phase) porous media. The core of the deformation model is the *momentum balance equation* for the (fluid-solid) system. Constitutive equations are presented for fluid-saturated (single or two-phase) domains, with *experimentally derived coefficients*. The objective is to evaluate solid matrix strain and deformation in response to changes in applied stress, primarily when such changes are a consequence of changes in fluid pressure.

Like in the case of reactive transport (Chap. 7) and nonisothermal flow (Chap. 8), we assume that the considered porous medium domain is continuously under conditions of *local (stress) equilibrium*. This means that *at every instant of time*, the processes, e.g., pressure and stress changes, as well as fluid motion and solid deformation that take place in the vicinity of a point at the macroscopic level, as a result of changes at the boundaries have reached instantaneous equilibrium. The averaged values represent what happens at the point at that instant. As everywhere else in this book, we understand that (instantaneous) equilibrium does not mean steady state. This may vary with time, while still being under equilibrium conditions at every instant.

The solid matrix itself may be composed of portions of various chemical substances, each having its own (microscopic level) stress–strain relationships. However, this aspect will not be considered here.

Actually, the subject was already introduced in Sects. 5.1.4 and 5.1.5, in connection with the definitions of *specific storativity* and *storativity* of groundwater aquifers and of the unsaturated zone. Here, we shall extend the discussion to the general case of porous medium deformation. The concept of *effective stress*, which produces deformation, was already introduced in Sect. 5.1.4, in connection with (fluid) storativity in a confined aquifer. Here, the discussion will focus on solid matrix deformation.

Consider a granular porous medium sample. For the sake of simplicity, let the void space be filled with a gas under atmospheric pressure. When this sample is loaded, it is obvious that (microscopic level) stresses are transmitted from grain to grain at the points of contact (or contact areas) between adjacent grains. There is no need to elaborate on why stresses within a granular domain cannot be calculated at this microscopic level. Instead, by averaging over an REV, or by making use of the phenomenological approach, we create a macroscopic domain, with a macroscopic (or averaged) stress at every point within that domain. This is the approach undertaken also in this chapter. Unless otherwise specified, we use here the terms stress, and strain, as well as their mathematical symbols, to indicate both their microscopic and macroscopic values.

As stated above, the objective is to determine the strain, or the deformation, of a stressed porous medium domain. This strain is produced by the *effective stress*, which is the *strain-producing stress*. The latter is affected not only by the stress imposed on the domain's boundaries, but also by the pressure of the fluid or the fluids that occupy the void space. This produces a link between the mechanical problem that involves the determination of the stress and the resulting strain within a considered domain and the hydraulic one that involves pressure variations in the (compressible) fluids that occupy and move in the void space. Solving these two problems provides the information required for determining the strain within a considered porous medium domain. The case of non-isothermal conditions will also be presented and discussed. We shall not consider the case in which the fluids in the void-space may contain chemical species which may interact with each other and with minerals comprising the solid matrix (Chap. 7). When necessary, this aspect can be added to the model considered here.

When the resulting strain exceeds certain values, failure of the solid matrix will occur. This may manifest itself, for example, as *cracks* that appear a consolidated rock and damage in the form of shear bands. Although the presentation addresses any porous medium domain, we shall focus on geological formations, with the examples of land subsidence as a result of pumping from an aquifer and CO_2 disposal in deep geological formations as typical cases of interest.

It is important to make here a comment on the various coefficients that appear in the stress–strain relationships discussed in this chapter. In the first part of the chapter, we introduce a number of Solid Mechanics concepts at the microscopic level, like stress, strain and displacement, and how they are related to each other. Then, following the phenomenological approach underlying the presentation in this book, we have assumed that the macroscopic relations, e.g., the stress–strain relationship, have the same *form* as the microscopic ones, but *with different values of the coefficients that appear in them. These have to be determined experimentally*! In principle, the stress–strain relationship of a fluid saturated porous medium should be affected also by the pressure and the compressibility of the fluid, or fluids, that occupy the void space. Thus, the constitutive relationship used for a solid are the same as those used for the solid matrix. Of course, the values of the various coefficients appearing in these relationships are different. As mentioned earlier, we have also introduced the concept of *effective stress* that does not exist at the microscopic level.

This chapter (like all other ones in this book) is written under the assumption that the reader is familiar with tensors and tensor operations. Nevertheless, some comments about tensors are presented in the following section.

9.1 Stress, Strain, and Effective Stress

In Sect. 2.3.4, for the convenience of the reader, we have reviewed some well known basic concepts of fluid and solid mechanics, like stress and strain. Here we shall make use of these concepts in connection with stresses and strains in porous media domains. As has been done throughout this book, we shall regard the porous medium domain as a *continuum*, or as multiple overlapping (solid and fluid) continua, and make use of the *phenomenological approach* to extend the single phase continuum concepts to porous medium domains.

9.1.1 Effective Stress in Two-Phase Flow

Consider a porous medium domain in which the void space is occupied by two fluid phases: a wetting fluid (w), at saturation S_w, and a non-wetting fluid (n), at S_n. From (1.1.12), it follows that the volume averaged stress (or *total stress*), $\overline{\boldsymbol{\sigma}}$, at a point in a porous medium domain, is defined by (3.3.15), repeated here for convenience, in the form:

$$\begin{aligned}\overline{\boldsymbol{\sigma}} &= \frac{1}{\mathbb{V}_o}\int_{\Omega_o} \boldsymbol{\sigma} d\mathbb{V} = \frac{1}{\mathbb{V}_o}\sum_{(\alpha=s,n,w)}\int_{\mathbb{V}_{o\alpha}} \boldsymbol{\sigma}_\alpha d\mathbb{V}_\alpha \\ &= \sum_{(\alpha=s,n,w)} \overline{\boldsymbol{\sigma}_\alpha} \equiv \overline{\boldsymbol{\sigma}_s} + \overline{\boldsymbol{\sigma}_n} + \overline{\boldsymbol{\sigma}_w},\end{aligned} \tag{9.1.1}$$

where the overbar indicates *volumetric phase average*. This means that the total stress is equal to the sum of the volumetric phase averaged stresses in the solid phase and in the two fluid phases. To facilitate the discussion on fluid stress–strain relationships, we express the averaged stress in the fluid(s) that occupy the void space, $\overline{\sigma_{f,ij}}$, in the form:

$$\overline{\sigma_{f,ij}} = \overline{\tau_{f,ij}} - \overline{p_f}\delta_{ij}, \tag{9.1.2}$$

where $\overline{\tau_{f,ij}}$ is the *viscous stress tensor*, $\overline{p_f}(= -\frac{1}{3}\sum_{(i)}\overline{\sigma_{f,ij}} - \overline{\tau_{f,ij}})$ is the pressure and δ_{ij} is the *Kronecker delta* ($\equiv$ components of the unit tensor, $\mathbf{I}$). Recall that, as is common is fluid mechanics, p_f is considered *positive for compression*, while the components of $\boldsymbol{\sigma}$ and $\boldsymbol{\tau}$ are considered *positive for tension*, as is common in solid mechanics.

Using (9.1.2), the total stress is also given by (3.3.19), repeated here for convenience:

$$\overline{\boldsymbol{\sigma}} = \overline{\boldsymbol{\sigma}_s} + \overline{\boldsymbol{\tau}_n} + \overline{\boldsymbol{\tau}_w} - \overline{p_n}\,\mathbf{I} - \overline{p_w}\,\mathbf{I}. \tag{9.1.3}$$

In the case of two fluids (w and n) that together fill up the entire void space, with negligible shear stress *within* both fluids, we define an average fluid pressure (in the two fluids) at a point in a porous medium continuum, by:

$$\overline{p_v}^v = \sum_{(\alpha=n,w)} \overline{p_\alpha} = \frac{1}{\phi}\left(\theta_w\overline{p_w}^w + \theta_n\overline{p_n}^n\right), \tag{9.1.4}$$

recalling that $\theta_\alpha = \phi S_\alpha, \quad \alpha = w, n$.

The concept of effective stress, has already been introduced in Sect. 5.1.4, and, for multiphase flow, in Sect. 6.4.1. Here, the effective stress plays a major role in analyzing porous medium deformation.

We start by expressing the total stress by (6.3.41) and (6.3.42), leading to:

$$\overline{\boldsymbol{\sigma}} = \overline{\boldsymbol{\sigma}'_s} - \overline{p_v}^v \mathbf{I}, \tag{9.1.5}$$

where:

$$\overline{\boldsymbol{\sigma}'_s} = (1-\phi)\{\overline{\boldsymbol{\sigma}_s}^s + \overline{p_v}^v \mathbf{I}\}, \quad \text{and} \quad \overline{p_v}^v = S_w \overline{p_w}^w + S_n \overline{p_n}^n, \tag{9.1.6}$$

express the *effective stress* (Sect. 5.1.4) and average fluid pressure for this case. Thus, (9.1.5) relates the (total) stress at a point in a porous medium domain to the (macroscopic) fluid pressure and to the effective stress at the point. Note that the expression for p_v in the above equation is not the only possible one (see Sect. 6.4.1). We also wish to emphasize again that *the effective stress at a point is the strain producing stress at that point.*

Henceforth, we shall remove the overbar symbols that indicate averages. We shall also make use of the dimensionless *Biot coefficient*, α_B, so that (9.1.5) will take the form:

$$\boldsymbol{\sigma} = \boldsymbol{\sigma}'_s - \alpha_B p_v \mathbf{I}, \tag{9.1.7}$$

where $\boldsymbol{\sigma}$ denotes the (total) stress at a point in the porous medium domain, $\boldsymbol{\sigma}'_s$ denotes the effective stress (tensor) at the considered point, and p_v denotes the average fluid pressure in the void space. The averaged pressure, p_v, needs not be the simple saturation-weighted pressure in the fluids. For example, Bishop and Blight (1963) and Gray and Schrefler (2007) suggested for two phase flow (w, n), a weighted average pressure:

$$p_v = a p_w + (1-a) p_n, \tag{9.1.8}$$

in which $a(S_w)$ is a weighting coefficient that depends on the saturation. In general, for two phase flow, we may write:

$$\boldsymbol{\sigma} = \boldsymbol{\sigma}'_s - \alpha_B \chi(S_w) p_w \mathbf{I}, \tag{9.1.9}$$

in which $\chi(S_w)$ expresses the relative contributions of the two fluids to the average pressure in the void-space (see Sect. 6.4.1).

9.1.2 Stress–Strain Relationship

In Sect. 2.3.5, we have presented Hooke's law, which is an example of a *constitutive equation* for a *linearly elastic solid*. Other equations of state will, obviously, describe the behavior of other kinds of materials. Originally, these EOS are at the *microscopic level*, i.e., at a point within a solid phase, regarded as a continuum. However, to investigate stress–strain relationships in a porous medium domain, with the void

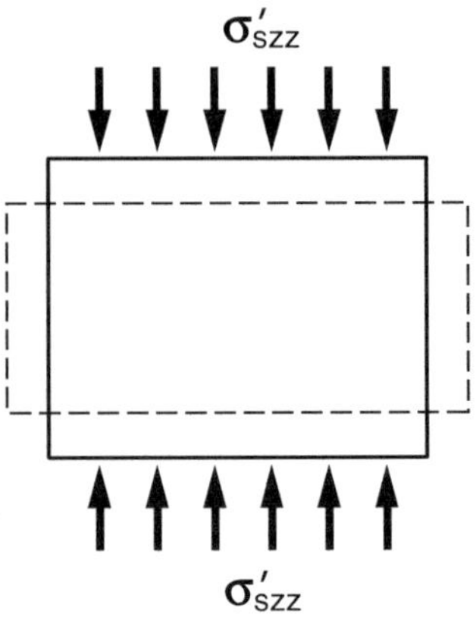

Fig. 9.1 Strain in a sample under 1-d stress in the z direction

space occupied by one or more (possibly moving) fluids, we need an EOS at the macroscopic level.

Following the *phenomenological approach* underlying the presentation in this book, we *assume* that *the macroscopic level EOS maintains the same stress–strain relationship as the macroscopic one*, except that the state variables are macroscopic and the values of the coefficients appearing in the this relationship are different. These have to be determined *experimentally* for samples of the considered porous medium. It is interesting to note that Bear and Bachmat (1991, p. 304) average the microscopic level EOS and show the approximations assumed when neglecting the terms that express solid-fluid interactions at the microscopic level.

To understand what happens when stress is applied at a point of a surface, consider the 2-d porous medium sample shown in Fig. 9.1. The only applied (here, effective) stress is σ'_{szz}. It produces two strain components: ε_{szz}, and ε_{sxx}:

$$\varepsilon_{zz} = \frac{\sigma'_{szz}}{E}, \quad \varepsilon_{xx} = \nu \frac{\sigma'_{szz}}{E}, \tag{9.1.10}$$

recalling that $\sigma'_{szz} = \sigma_{szz} - p_v$, or $= \sigma_{szz} - \alpha_B p_v$, if we use Biot's coefficient, α_B (Sect. 5.1.4). The coefficient ν is the *drained Poisson coefficient* (as water is allowed to drain during the experiment). In words, *the displacement in the stress direction is accompanied by a displacement in a direction normal to it.*

Linearly Elastic Isotropic Porous Medium

In Fig. 2.10, we have presented a number of types of solid materials, distinguishable by their elastic or non-elastic behavior. We may encounter similar stress–strain relationships when considering porous medium domains, including geological ones, i.e., at the macroscopic level. One example is the first equation in (2.3.73), rewritten at the macroscopic level for an isotropic solid matrix under isothermal conditions, in the form:

$$\sigma_{sij} = \lambda_s \varepsilon_{kk} \delta_{ij} + 2\mu_s \varepsilon_{ij}, \tag{9.1.11}$$

or:

$$\varepsilon_{sij} = \frac{1}{2\mu_s}\sigma_{ij} - \frac{\lambda_s}{2\mu_s(3\lambda_s + 2\mu_s)}\sigma_{ij}\delta_{ij}, \tag{9.1.12}$$

in which λ_s and μ_s (sometimes denoted as G_s), are two *macroscopic* coefficients known as the *Lamé constants* for the solid matrix; we are adding the subscript s to indicate that these coefficients are for the solid matrix, i.e., at the macroscopic level (see Sect. 2.3.5).

In the literature on the theory of elasticity, we find expressions for $\boldsymbol{\sigma} = \boldsymbol{\sigma}(\varepsilon)$ also in terms of other coefficients:

$$E_s = \frac{\mu_s(3\lambda_s + 2\mu_s)}{\lambda_s + \mu_s}, \quad \nu_s = \frac{\lambda_s}{2(\lambda_s + \mu_s)}, \tag{9.1.13}$$

where E_s and ν_s are *Young's modulus* and *Poisson's ratio*, and:

$$\lambda_s = \frac{\nu_s E_s}{(1+\nu_s)(1-2\nu_s)}, \quad \mu_s(=G_s) = \frac{E_s}{2(1+\nu_s)}, \tag{9.1.14}$$

in which G_s is the *shear modulus of elasticity*. Another often used coefficient is the *drained bulk modulus*, K_s:

$$K_s = \frac{E_s}{3(1-2\nu_s)}, \tag{9.1.15}$$

with:

$$\sigma_{ij} = \left(K_s - \frac{2G_s}{3}\right)\varepsilon_{kk}\delta_{ij} + 2G_s\varepsilon_{ij}. \tag{9.1.16}$$

Cheng (2016, p. 69) added on the r.h.s. of (9.1.16) a term that represents the effect of change in fluid content in the case where the drained water is replaced by air at atmospheric pressure.

Returning to the general case, let us assume that the saturated porous medium comprising an investigated geological formation behaves as a *linearly elastic porous medium*, for which Hooke's law (2.3.72), or in the form (2.3.80), serves as the constitutive equation. To obtain their porous medium equivalent, following the phenomenological approach, we assume that the shape of these relationships remains unchanged, except that the variables and the coefficients represent their macroscopic level values. Thus, for example, we apply the stress–strain relationship (2.3.80) also to a geological formation, assumed to behave as a linearly elastic porous medium. Accordingly, we obtain (in vector notation):

$$\boldsymbol{\sigma}'_s = K_s\varepsilon_v\mathbf{I} + 2G_s\left(\boldsymbol{\varepsilon} - \frac{\varepsilon_v}{3}\mathbf{I}\right) = 3K_s\varepsilon_m\mathbf{I} + 2G_s\boldsymbol{\varepsilon}_v = 2G_s\boldsymbol{\varepsilon} + \lambda\varepsilon_v\mathbf{I}, \quad (9.1.17)$$

where ε_v and ε_m, are defined in (2.3.71). Or, in the inverted form:

$$\boldsymbol{\varepsilon}_s = \frac{1+\nu_s}{E}\boldsymbol{\sigma}'_s - \frac{3\nu_s}{E_s}\sigma_m\mathbf{I} = \frac{\sigma'_m}{3K_s}\mathbf{I} + \frac{1}{2G_s}\left(\boldsymbol{\sigma}'_s - \sigma'_m\mathbf{I}\right). \quad (9.1.18)$$

In the above equations, the *bulk modulus*, K_s, is defined in (9.1.15).

Based on (2.3.80), another form of the stress–strain relationship (under isothermal conditions) for an *isotropic linearly elastic solid*, takes the form:

$$\boldsymbol{\sigma}'_s = \mu_s\left[\nabla\mathbf{w}_s + \nabla(\mathbf{w}_s)^T\right] + \lambda_s\nabla\cdot\mathbf{w}_s. \quad (9.1.19)$$

Note that the above relationship relates the displacement to the effective stress, which is the "*strain producing stress*".

9.1.3 Non-isothermal Conditions

Non-isothermal conditions occur, for example, in projects that involve injection of fluids into deep geological reservoirs, with the injected fluid being at a temperature that is different from that of the indigenous fluid in the formation. Modeling flow and transport under such conditions is discussed in Chap. 8.

When considering solid matrix deformation under non-isothermal conditions, we have to take into account the stress–strain relation of the considered solid matrix also under such conditions. A commonly encountered case is the *thermo-elastic solid.* For such solid, the stress–strain relationship *for the solid* takes any of the forms (2.3.84)–(2.3.87). Then, following the phenomenological approach, we modify the stress–strain relationship by adding the effect of thermally produced strain to (9.1.18), and expressing $\boldsymbol{\sigma}'$ by $\boldsymbol{\sigma}$, we obtain the stress–strain relationship for the saturated porous medium:

$$\boldsymbol{\varepsilon} = \frac{1+\nu_s}{E_s}\boldsymbol{\sigma} - \frac{3\nu_s}{E_s}\sigma_m\mathbf{I} - \frac{1-2\nu_s}{E_s}p_v\mathbf{I} - \alpha_T\Delta T\mathbf{I}, \quad (9.1.20)$$

in which σ_m is defined in (2.3.64), p_v denotes the average fluid pressure in the void space, α_T is the *linear thermal expansion coefficient of the thermo-elastic solid matrix*, and ΔT is the temperature change in the thermo-elastic constitutive equation. The above equation can also be written in the form:

$$\boldsymbol{\sigma}'_s = 2G_s\boldsymbol{\varepsilon} + \lambda_s\varepsilon_v\mathbf{I} - (2G_s + 3\lambda_s)\alpha_T\Delta T\mathbf{I}, \quad (9.1.21)$$

obtained from (9.1.20). In the above equation, $K_s = 2G_s + 3\lambda_s$.

9.1.4 *Anisotropic Elastic Solid Matrix*

Equation (2.3.81) expresses the stress–strain relationship for an anisotropic elastic solid matrix, assuming that (in line with the phenological approach employed in this book) the same structure of this relationship is maintained at the macroscopic level, except that the values of the various coefficients appearing in the stress–strain relationship, e.g., Young modulus, will be different.

Let us use the discussion here to introduce another common way to express the stress–strain relationship for an elastic solid. For convenience, let us repeat here the *generalized Hooke's law* (2.3.72):

$$\sigma_{ij} = C_{ijk\ell}\, \varepsilon_{k\ell}, \tag{9.1.22}$$

where $C_{ijk\ell}$ denotes the $ijk\ell$ component of the *stiffness coefficient.*

We recall that the fourth rank tensor $C_{ijk\ell}$ has 81 components. However, because σ_{ij} is symmetric in i, j, we have the relationship:

$$C_{ijk\ell} = C_{jik\ell}.$$

Since the strain, $\varepsilon_{k\ell}$ is also symmetric in k and ℓ, we also have

$$C_{ijk\ell} = C_{ij\ell k}.$$

Altogether, we reach the conclusion that the 81 components reduce to 36. Thus, each stress is linearly related to the six independent strains via six of the independent elastic constants $C_{ijk\ell}$. For example:

$$\sigma_{xx} = C_{1111}\varepsilon_{xx} + C_{1122}\varepsilon_{yy} + C_{1133}\varepsilon_{zz} + C_{1123}\varepsilon_{yz} + C_{1131}\varepsilon_{zx} + C_{1112}\varepsilon_{xy}, \tag{9.1.23}$$

usually abbreviated as:

$$\sigma_{xx} = C_{11}\varepsilon_{xx} + C_{12}\varepsilon_{yy} + C_{13}\varepsilon_{zz} + C_{14}\varepsilon_{yz} + C_{15}\varepsilon_{zx} + C_{16}\varepsilon_{xy}, \tag{9.1.24}$$

As an example of anisotropy, consider the case of an elastic solid body that has three mutually orthogonal planes of elastic symmetry. This kind of symmetry is referred to as *orthotropic symmetry*. It can be shown that for this kind of anisotropy the *elastic modulus matrix* takes the form:

$$C_{pq} = \begin{bmatrix} C_{11} & C_{12} & C_{13} & 0 & 0 & O \\ C_{21} & C_{22} & C_{23} & 0 & 0 & 0 \\ C_{31} & C_{32} & C_{33} & 0 & 0 & O \\ 0 & 0 & 0 & C_{44} & 0 & 0 \\ 0 & 0 & 0 & 0 & C_{55} & O \\ 0 & 0 & 0 & 0 & 0 & C_{66} \end{bmatrix}, \tag{9.1.25}$$

where $C_{12} = C_{21}$, and $C_{23} = C_{32}$ and $C_{13} = C_{31}$.

9.2 Modeling Non-isothermal Flow and Deformation

As stated in the preamble to this chapter, our interest here is, primarily, in models that lead to determining the strain and deformation of porous medium domains in which the void space is occupied by one or more fluids, possibly under non-isothermal conditions. In particular, we focus on geological formations. We shall not refer to the modeling of reactive transport; this part can be added when needed.

To achieve the above stated goal, we need to determine the space and time distribution of the strain in the formation, produced by the *effective stress*. The latter, defined by (9.1.5), is linked to both the total stress and to the pressure of the fluid in the void space. As outlined in Chaps. 5 and 6, the pressure can be obtained by solving the mass balance equations for the fluids in the void space. Changes in stress are produced by loading at the domains boundaries.

In the case of non-isothermal conditions, the temperature is another variable and we have to solve also the energy balance equation, written in terms of the temperature.

Altogether, the above introductory comments lead to the conclusion that in order to determine the deformation in a fluid-saturated porous medium domain under non-isothermal conditions, but (in the example considered here) without any chemistry, say, in the case of two phase flow, we have to solve a model that involves the following E-balance equations:

- The mass balance equations for the fluid, or fluids that occupy the void space. This part (presented in Chaps. 5 and 6) is usually referred to as the *hydraulic model*, or *H-model*.
- The energy balance equation for the porous medium as a whole, with porous medium temperature as a single variable. This part, discussed in Chap. 8, is referred to as the *thermal* or *T-model*.
- The momentum balance equation for the porous medium as a whole. We refer to this part as the *mechanical, or M-model*. As we shall see, because the momentum balance equation involves porosity and solid matrix velocity, we'll have to use also the solid matrix mass balance equation.
  When inertial effects may be neglected, the momentum balance equation reduces to a *force balance*. This model is considered in the current chapter.

As we have seen, these three models (and this is true also had we added the *solute transport model*) are strongly interrelated and must be solved simultaneously. Practically, the only way to solve this interrelated set of models is by numerical techniques. Efficient computer codes are available for implementing such solutions (in fact, including also the reactive transport, or C-model). However, each of the models (or sub-models), has to be stated as a *well posed* one. This means that the statement of each model (or sub-model) should include appropriate initial and boundary conditions. It is interesting to note that changes within each of the models can be triggered either by changes in boundary conditions and/or changes in any of the linked sub-models.

9.2.1 The Mechanical Model

A. The Momentum Balance Reduced to the Equilibrium Equation

We start from Sect. 3.3.2 D, where we have developed the equation that expresses the momentum balance for a porous medium as a whole, with a void space that is occupied by one fluid, or by two fluid phases: a wetting fluid and a non-wetting one. This balance is expressed by (3.3.13). By assuming continuity of traction, i.e., $[\![\ \boldsymbol{\sigma} \ ]\!]_{w,s} \cdot \boldsymbol{\nu}_w = 0$ and $[\![\ \boldsymbol{\sigma} \ ]\!]_{s,n} \cdot \boldsymbol{\nu}_s = 0$, on their common interfacial surfaces, and neglecting *surface tension* phenomena at fluid-solid interfaces, the last two surface integrals in (3.3.13) vanish. By further neglecting the force, $\mathbf{F}_c$, resulting from capillary pressure, we obtain the momentum balance equation for the porous medium as a whole:

$$\sum_{(\alpha=n,w,s)} \theta_\alpha \rho_\alpha \frac{D\mathbf{V}_\alpha}{Dt} = \nabla \cdot \overline{\boldsymbol{\sigma}} + \overline{\rho \mathbf{F}}, \tag{9.2.1}$$

where the average total stress, $\overline{\boldsymbol{\sigma}}$, and the average body force, $\overline{\rho \mathbf{F}}$, are defined by (3.3.15) and (3.3.16), respectively. The l.h.s. of the above balance equation expresses inertial forces. We shall consider them when dealing with waves in porous medium domains (Sect. 9.4). Here, however, we shall neglect them, as being much smaller that the other terms, leading to the momentum balance equation, or equilibrium equation, for the porous medium as a whole, (5.1.37), repeated here for convenience:

$$\nabla \cdot \overline{\boldsymbol{\sigma}} + \overline{\rho \mathbf{F}} = 0, \tag{9.2.2}$$

In this equation, which actually expresses a balance of forces, $\overline{\boldsymbol{\sigma}}$ is the averaged total stress, produced, for example, by loading the considered domain, and $\overline{\rho \mathbf{F}}$ represents the total *body force*. We shall use the symbol $\mathbf{b} \equiv \overline{\rho \mathbf{F}}$ to denote the body force due to gravity, per unit volume of porous medium:

$$\mathbf{b} = -g\,[(1-\phi)\rho_s + \phi S_w \rho_w + \phi S_n \rho_n]\,\nabla z, \tag{9.2.3}$$

recalling that ∇z is a unit vector directed upward.

The force $\mathbf{b}$ depends on the fluids' saturations and densities. The latter depend on fluids' pressures. This means that we have to present and solve also a two-phase flow (hydraulic) model. In the case of non-isothermal conditions, the fluid and solid densities are also affected by temperature changes.

Altogether, in the absence of inertial effects, the momentum balance equation used for deformation analysis is the *equilibrium equation*:

$$\nabla{\cdot}\boldsymbol{\sigma} + \mathbf{b} = 0, \tag{9.2.4}$$

in which we have now removed the over-bar symbol. Since, in the case of gravity as the only body force, $b_x = b_y = 0$, only $\sigma_{zz} \neq 0$, and $\sigma_{ij} = 0$ for all $i \neq j$, and for all $ij = xx$ and yy. However, we may still encounter $\sigma_{i,j} \neq 0$ for $i \neq j$ if boundary conditions impose the latter stress.

With the definition of effective stress, and (9.1.7), we can rewrite (9.2.4) in the form:

$$\nabla{\cdot}\boldsymbol{\sigma}' - \alpha_B \nabla p_v + \mathbf{b} = 0, \tag{9.2.5}$$

where p_v, denotes the *average pressure in the fluids that occupy the void space*, say $\chi(S_w)p_w$. This equation involves the tensor variable, $\boldsymbol{\sigma}'$, to be solved for. It can also be rewritten as 6 scalar equations in terms of the 6 scalar variables: $\sigma'_{xx}, \sigma'_{yy}, \sigma'_{zz}, \sigma'_{xy}, \sigma'_{xz}, \sigma'_{yz}$. However, to achieve this goal, we need one more (scalar) equation for p_v (or two equations, for p_n and p_w, in the case of two fluid phases). We also need to know the force $\mathbf{b}$, which means information on $\rho_w = \rho_w(p_w, T)$ and $\rho_n = \rho_w(p_n, T)$. We also need information on the possibly time-dependent variable porosity, $\phi = \phi(\mathbf{x}, t)$.

Next, we write (9.2.5) in terms of solid displacement $\mathbf{w}_s$. Making use of (2.3.69) and (9.1.21), we obtain:

$$G_s \nabla^2 \mathbf{w}_s + (G_s + \lambda_s)\nabla(\nabla{\cdot}\mathbf{w}_s) = \mathbf{b} + \alpha_B \nabla p_v + (2G_s + 3\lambda_s)\alpha_T \nabla(T - T_o), \tag{9.2.6}$$

in which G_s and λ_s are the Lamé constants, p_v, T, and $\mathbf{w}_s$ are state variables and $\mathbf{b}$ is expressed by (9.2.3). The above equation may be regarded as a single equation for the variable $\mathbf{w}_s$, or as three equations for the scalar variables w_{sx}, w_{sy}, w_{sz}. It is the equation that constitutes the core of the M-model.

We have assumed (for the sake of the discussion here) that changes in solute concentrations do not affect fluids' densities. Otherwise, we have to add and solve also the reactive (solute) transport ($C-$)model.

Once we solve the model for the effective stress, we can use the appropriate *stress–strain relationship* to obtain the sought distribution of strain within the domain.

B. Variable Porosity

Equation (9.2.5) contains the porosity, ϕ, as a variable: it is included in the definition of the body force **b**. In Chaps. 5 and 6 the subject of time-dependent porosity (due to porous medium compressibility) was handled by making use of the coefficient of porous medium compressibility, α_{pm}, as defined by (5.1.44), leading to the definition of specific storativity. Here, we shall take another approach.

From the solid's mass balance equation (5.1.6), repeated here for convenience in the form:

$$\frac{\partial}{\partial t}[(1-\phi)\rho_s] = -\nabla\cdot[(1-\phi)\rho_s \mathbf{V}_s], \tag{9.2.7}$$

in which $\mathbf{V}_s$ is approximated by (5.1.19). Equation (5.1.19) may be regarded as an equation for determining the time and space variations of porosity, recalling that a good approximation is $\rho_s \approx$ const., and:

$$\left|\frac{\partial \phi}{\partial t}\right| \gg |\mathbf{V}_s\cdot(\nabla\phi)|,$$

C. Possible Domain Boundaries

As in the cases of fluid flow, solute transport, and heat transport, a considered three-dimensional porous medium domain is bounded by a closed surface that separates it from its surrounding. The latter imposes certain conditions on the investigated domain. Here, we are considering only conditions that are associated with the subject of the current chapter, namely, conditions related to stresses, strains, and displacements. However, we have to keep in mind that the same boundary serves also in modeling fluid flow, solute transport, and heat transport. Accordingly, a boundary segment may be:

(a) Stationary and shape maintaining.
(b) Mobile, but shape maintaining.
(c) Mobile and free to deform. Ground surface may serve as an example. It may be regarded as a *free surface*, similar to the $F(x, y, z, t) = 0$-surface discussed in Sect. 5.2.1 D.

In each of the above cases, the external side of the boundary may be occupied by a rigid, stationary, or deformable solid body, or by a body of fluid (gas or liquid).

D. Initial and Boundary Conditions

As we have done throughout this chapter, we shall continue to consider the M-model, simplified to the *equilibrium equation* (9.2.4) and to the displacement governed by (9.2.6). To solve these equations, we have to stipulate appropriate initial and boundary conditions. Obviously, the conditions to be stipulated for the M-model are *in addition* to conditions related to the flow and energy transport models that prevail on these boundaries.

- **Initial conditions**

Initial conditions involve a statement of the stress at $t = 0$ at every point within the considered domain. Or, we may stipulate the strain, or displacement, if these are known. For example, when the considered domain is a geological formation, or part of it, we assume that initially the shear stress is zero and that the stress at every point is due to gravity only, taking into account the weight of the solid matrix and the pressure in the fluid or fluids that occupy the void space above any point within the domain. Initial pressure is often assumed hydrostatic. However, in the case of two phase flow, we have to take into account also the capillary pressure in determining initial pressure conditions. The initial displacement is taken as zero everywhere. We have to take into account also the load on ground surface due to structures (and a negative load due to excavations).

- **Boundary conditions**

As for all extensive quantities, here also, boundary conditions are based on the no-jump condition on the boundary. The condition to be specified on a boundary segment may involve any of the following cases:

(a) **A statement of the stress acting on the boundary**, i.e., force per unit area, acting at $t > 0$, if a known such stress is specified. An example is the case of flow in a confined aquifer. The ceiling of the aquifer is loaded by all soil layers up to ground surface, plus any load above ground surface, or a negative load due excavation. This (no-jump) condition can be expressed as:

$$\boldsymbol{\sigma}\cdot\boldsymbol{\nu} = \mathbf{T}(\mathbf{x}, t), \tag{9.2.8}$$

where $\mathbf{T}$ denotes a known *traction* (vector) acting on the boundary. This is a *Dirichlet*, or *traction type boundary condition*.

(b) **A statement on the velocity of the solid matrix' boundary**. This condition is based on the general statement (5.2.6) of no-jump in the flux of any considered extensive quantity. Here, we consider the momentum flux. Neglecting inertial effects, the condition reduces to that of *no-jump* in the normal momentum flux, simplified to the normal speed of displacement of the boundary. An example is a stationary boundary that confines a geological formation from above or from below. Obviously, in each case the conditions express known existing or imposed conditions. Cheng (2016, p. 216) lists a number of additional possible options.

9.2.2 The Hydraulic Model for a Deformable Matrix

As an example, consider two fluids that occupy the void space. To obtain the fluids' pressures, we have to consider and solve the two-phase flow model presented as (6.4.3)–(6.4.6) in Sect. 6.4.1. Once the pressure distributions in the two fluids has been determined at a considered time step, the densities, viscosities and other coefficients which depend on the pressure, can be updated. The details of the hydraulic model are discussed in Chap. 6, with initial and boundary conditions stated in Sect. 6.4.2. If non-isothermal conditions prevail, we have to solve the thermal model and the use the resulting pressures and temperature to update density and viscosity.

Let us add one more aspect to the flow model. Since we are considering here a deformable porous medium, this means that porosity is time-dependent, $\phi = \phi(t)$, recalling that we can always add the aspect of heterogeneity, i.e., $\phi = \phi(\mathbf{x})$. The temporal variation of porosity was presented in Sect. 5.1.2. There, we have shown that the solid's mass balance equation is expressed by (5.1.11). We also recall that $\mathbf{V}_s = d\mathbf{w}_s/dt$, where we added the subscript s as a reminder that $\mathbf{w}_s$ denotes the displacement of the solid matrix. With the material presented in that subsection, we can now write the solid's mass balance equation in the form:

$$\frac{1}{1-\phi}\frac{D_s(1-\phi)}{Dt} + \frac{1}{\rho_s}\frac{D_s\rho_s}{Dt} = -\nabla\cdot\mathbf{V}_s, \tag{9.2.9}$$

where we have made use of the approximation (5.1.19). We note here the effects of changes in both porosity and solid's density. The latter change is related to changes in pressure, $\rho_s = \rho_s(p, T)$. For example:

$$\rho_s = \rho_{so}[\beta_s(p - p_o) + \alpha_s(T - T_o)], \tag{9.2.10}$$

(e.g., Olivalla et al. 1994). Equation (9.2.9) can be used to follow porosity changes as changes in pressure, temperature and displacement occur.

We usually make the (very good) assumption that $\mathrm{D}_s\rho_s/\mathrm{Dt} = 0$, so that (9.2.9) reduces to:

$$\frac{1}{1-\phi}\frac{D_s\phi}{Dt} = \nabla\cdot\mathbf{V}_s. \tag{9.2.11}$$

Then, assuming $|\partial\phi/\partial t| \gg |\mathbf{V}_s\cdot\nabla\phi|$, the solid matrix mass balance equation takes the approximate form:

$$\frac{1}{1-\phi}\frac{\partial\phi}{\partial t} = \nabla\cdot\mathbf{V}_s. \tag{9.2.12}$$

which can be used for upgrading ϕ.

9.2.3 The Chemical Model

When the considered fluid phases carry chemical species which spread out, diffuse, and disperse within each phase, interact with the solid matrix, cross interphase boundaries, and undergo chemical reactions with each other, we use the reactive transport model described in Chap. 7. The fluid velocities appearing in this model are obtained from the hydraulic model. Exogenic and endogenic reactions produce sources and sinks of energy, which serve as input to the thermal model.

9.2.4 The Thermal Model

We assume "thermal equilibrium among all phases at a point", i.e., the temperatures of all phases within an REV are the same. The temperature, $T = T(\mathbf{x}, t)$, becomes the sole state variable to be solved for by the thermal model. To determine this temperature, we have to write and solve a *single* energy balance equation for the porous medium domain as a whole.

One option is to use (8.4.8) as the energy balance equation to be solved for $T = T(\mathbf{x}, t)$. In the case of two fluids that occupy the void space, we have to (1) modify the definition of $(\rho C)_{pm}$, to include the solid and the two fluids, (2) to replace $\rho_f C_f \mathbf{q}$ by $\sum_{\beta=w,n} \rho_\beta C_\beta \mathbf{q}_\beta$, and (3) define $\Lambda^H_{pm} = \sum_{s,w,n} \theta_\beta \lambda_\beta$, with $\theta_s = (1-\phi)$. We note that this equation includes the fluids' densities, which have to be continuously updated as temperature varies. Also, fluid viscosities (which are used to compute the fluids' specific discharges) have to be continuously updated, as they are also temperature-dependent.

Another option, especially when we consider cases with phase change, is to write the heat energy balance equation in terms of the enthalpy of the fluids and the solid (see Sect. 8.4.1 C). Repeating this energy balance equation here:

$$\begin{aligned}
&\frac{\partial}{\partial t}\left[\sum_{(\alpha=w,n)} (\theta_\alpha \rho_\alpha u_\alpha) + (1-\phi)\rho_s C_s T\right] \\
&\qquad = -\nabla\cdot \sum_{(\alpha=w,n)} \theta_\alpha \left[\rho_\alpha h_\alpha \mathbf{V}_\alpha + \sum_{(\gamma)} h^\gamma_\alpha \mathbf{J}^\gamma_{\alpha,hdis} + \mathbf{J}^{*H}_{\alpha,dis}\right] \qquad (9.2.13)\\
&\qquad +\nabla\cdot\Lambda^*_{pm}\nabla T + \Gamma^{\mathbb{E}}_{pm},
\end{aligned}$$

in which the second sum on the second line is taken over all γ-species present in the phase. The $\mathbf{J}$'s denote macroscopic fluxes (per unit phase area). Here, $\mathbf{J}^\gamma_{\alpha,hdis}$ is the hydrodynamic dispersive flux of the mass of the γ-species, and $\mathbf{J}^{*H}_{\alpha,dis}$ denotes the dispersive heat flux within a fluid α-phase.

Often, the above equation is simplified to the form:

$$\frac{\partial}{\partial t}\left\{\sum_{w,n}\phi S_\alpha \rho_\alpha h_\alpha + (1-\phi)\rho_s h_s\right\} = -\nabla\cdot\mathbf{J}^H + \sum_{w,n}\rho_\alpha S_\alpha \phi \Gamma_\alpha^{\gamma\mathbb{E}}, \qquad (9.2.14)$$

in which h_α ($\equiv u_\alpha + p_\alpha v_\alpha$) denotes the enthalpy of the α-phase, u_α denotes specific internal energy, h_s denotes the enthalpy of the solid phase, $\mathbf{J}^H$ denotes the total energy conductive fluxes (per unit area of porous medium) in *all* phases, and $\Gamma_\alpha^{\gamma\mathbb{E}}$ denotes the source of heat due to exogenic chemical reactions (in the case of a reactive transport problem).

In a numerical model, once the pressure distributions in the two fluids and the temperature have been determined at a considered time step, the densities, viscosities and other coefficients which depend on the pressure and temperature are updated. Note that we have switched here to the language of a numerical solution (like "updated"), as it is obvious that this kind of model (or set of interrelated models) can be solved only by a numerical technique.

9.2.5 The Hydro-Thermal-Mechanical (HTM) Model

One of the more complicated (but interesting!) problems of flow and transport in porous media is the Hydraulic-Thermal-Chemical-Mechanical (HTCM) one. This, for example, is the case of modeling possible leakage from a radioactive waste repository in a deep clay formation, or modeling CO_2 disposal in deep brine-containing geological formations. The latter problem involves non-isothermal two phase flow (of the indigenous saline water and the injected CO_2), with chemical reactions and with solid matrix deformation. In what follows, we shall consider only the HTM, problem, i.e., without chemical reactions. Often, one of the objectives of solving such a problem is determining whether or not a mechanical failure will occur in the geological formation into which CO_2 is injected, or at the interface between that formation and an overlying confining (and sealing) layer. The chemical aspects can easily be added, making use of the material presented in Chap. 7.

In principle, without the chemical model, to arrive at the strain, $\varepsilon(\mathbf{x}, t)$, or displacement, $\mathbf{w}$, we have to solve for the values of the following **44** space- and time-dependent *scalar* variables:

$$\phi,\ S_w,\ S_n,\ p_w,\ p_n,\ p_c,\ p_v,\ \rho_w,\ \rho_n,\ \rho_s,\ b_i,\ V_{ni},\ V_{wi},\ V_{si},\ T,\ w_{si},\ \sigma_{ij},\ \sigma'_{sij},\ \varepsilon_{ij}.$$

However, only **6** of them: e.g., S_w, p_n (or S_n, p_w), T, w_i, are *primary variables* for which we have to solve 6 PDEs (see Sect. 3.9). The other have to be obtained from definitions, equations of state, etc.

Obviously, all coefficients appearing in the model, e.g., $\mathbf{k}_w(S_w)$, $\mathbf{k}_n(S_n)$, $\mathbf{\Lambda}^*_{pm}$, must be known. Some of them may be time-dependent and have to be continuously updated. In addition, temperature and pressure-dependent coefficients, like $\mu_w(T)$ and $\mu_n(T)$, have to be continuously updated. In a heterogeneous domain, void space and solid matrix properties vary also in space.

As an example, consider the following solution routine:

(a) Assume that all variables and coefficients are known at the current (= initial) time $t = 0$ (= initial conditions). Our objective is to obtain a solution for $t + \Delta t$.
(b) **Solve** the **two mass balance PDEs** (6.4.3) to determine the *two primary variables*, e.g., p_w, S_n at the new time step, $t + \Delta t$.
(c) Use the 6 scalar equations (6.4.4)–(6.4.6), the expressions for $\rho_n, \rho_w, \rho_s, \mu_n, \mu_w$, and the energy balance equation to determine p_n, p_w, T, V_{si}, S_w, S_n, k_n, k_w. Then determine V_{wi}, V_{ni}. All updated values (also below) are functions of $(\mathbf{x}, t + \Delta t)$.
(d) Use (6.4.5) to update p_c, and (6.3.40) to update p_v.
(e) Use an appropriate EOS to determine the current values of h_w, h_n, u_w, u_n.
(f) Use (8.4.9) to update Λ^{*H},
(g) **Solve** the **energy balance PDE**, e.g., (9.2.14), or its simplified form (9.2.14), to update the primary variable $T(\mathbf{x}, t) \rightarrow T(x, t + \Delta t)$.
(h) Use (9.2.3) and the current values of $\rho_w, \rho_n, \rho_s, \phi$, to update the force components, b_i.
(i) Solve the equilibrium equation (9.2.4) for the (total) stress $\boldsymbol{\sigma}(\mathbf{x}, t)$ within the considered domain.
(j) Use (9.2.6) to update the displacement, $\mathbf{w}_s$.

In an analytical solution, all equations have to be solved simultaneously. In a numerical solution, we move one time step at a time, and adjust values of all coefficients and parameters. Often, iterations are used to improve the updated solutions until a satisfactory convergence is obtained.

9.2.6 *Failure of the Solid Matrix*

As suggested in the preamble to this chapter, often, the objective of stating and solving an M-model, which involves the strain-stress analysis outlined above, is to examine the possibility of 'failure' in the form of damage to the stressed formation, e.g., the creation of fractures in a porous rock formation, or in theca-rock bounding it.

In a fragile material, a fracture will occur when, in a certain plain, the shear stress exceeds the resistance of the rock to overcome this shear. Under such conditions, a slip will occur along the fracture plane. Accordingly, at a certain time step, we obtain the distribution of effective stress (recall that we have referred to it as the "strain producing stress"), $\boldsymbol{\sigma}'_s$ in the considered domain, we can examine whether or not this stress exceeds the value of a certain failure criterion.

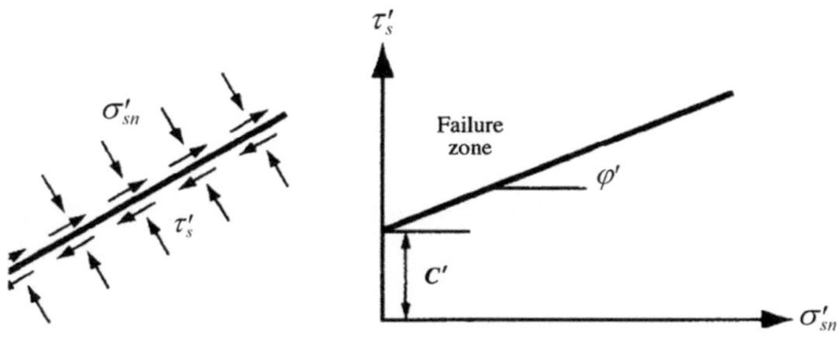

Fig. 9.2 The Mohr–Coulomb (shear) failure criterion

The Mohr–Coulomb is a commonly used criterion for failure, especially in brittle rock and granular materials (see Fig. 9.2):

$$\tau'_s = c' + \sigma'_{sn} \tan \varphi', \tag{9.2.15}$$

where τ'_s denotes the shear stress (and we recall that $\tau'_{sij} \equiv \sigma'_{sij}$, when $i \neq j$), c' denotes the internal strength of the material, or *cohesion*, σ'_{sn} denotes the normal effective stress component, acting on the considered plane, and φ' denotes the *angle of internal friction*. For sand, Verruijt (2010, p. 134) suggests that, usually, the cohesion is practically zero, i.e., $c' = 0$, and the friction angle varies in the range $30°-45°$, depending upon the angularity and the roundness of the particles. Clay soils usually have some cohesion, and a certain friction angle, but somewhat smaller than sands.

For a three-dimensional stress–strain problem, the Mohr–Coulomb failure criterion takes the form:

$$\begin{aligned}
\pm \frac{\sigma'_{s11} - \sigma'_{s22}}{2} &= \left[\frac{\sigma'_{s11} + \sigma'_{s22}}{2}\right] \sin \varphi' + c' \cos \varphi', \\
\pm \frac{\sigma'_{s22} - \sigma'_{s33}}{2} &= \left[\frac{\sigma'_{s22} + \sigma'_{s33}}{2}\right] \sin \varphi' + c' \cos \varphi', \\
\pm \frac{\sigma'_{s33} - \sigma'_{s11}}{2} &= \left[\frac{\sigma'_{s33} + \sigma'_{s11}}{2}\right] \sin \varphi' + c' \cos \varphi'.
\end{aligned} \tag{9.2.16}$$

More on the Mohr–Coulomb criterion can be found in Verruijt (2010).

9.3 Seepage Forces and Land Subsidence

9.3.1 Seepage Forces and Liquefaction

When flow takes place in a porous medium domain, a force is exerted by the moving fluid on the solid matrix. This force, referred to as *seepage force*, **F**, plays an important role in many engineering problems.

According to Terzaghi (1943), Biot (1941), *three forces act on a unit volume of the solid matrix* at a point during flow:

- The *weight* of the solid matrix (acting downward):

$$\mathbf{F}_g^* = -\rho_s g(1-\phi)\nabla z = -\rho_b g\nabla z, \quad \rho_b = (1-\phi)\rho_s, \tag{9.3.1}$$

  where ρ_s is the solid's density, and ρ_b is the bulk density of the solid matrix. Note that $\mathbf{F}^*$ (a vector) denotes *force per unit volume*.
- The *buoyancy force* (or *uplift*), which is equal to the resultant of the liquid pressure acting on the solid particles. Since the force acting on a unit volume of solid resulting from water pressure is equal to $-\nabla p$, we have:

$$\mathbf{F}_u^* = -(1-\phi)\nabla p. \tag{9.3.2}$$

- The *drag*, or shear, or *seepage force*, $\mathbf{F}_d^*$, at the solid-fluid interface, per unit volume of porous medium,

$$\mathbf{F}_d^* = -\phi(\rho_f g\nabla z + \nabla p), \tag{9.3.3}$$

  created by the fluid's motion. In other words, the porous medium exhibits a resistance to the flow through it, e.g., as expressed by (4.2.16).

For an isotropic fluid, $\mathbf{F}_d^* = \phi\mu\mathbf{q}_f/k$. In an isotropic porous medium, this force is in the direction of the normal to the equipotential surfaces. In an anisotropic porous medium, the direction of the drag force will not coincide with this normal. We shall then have:

$$\mathbf{F}_d^* = \mu\phi\mathbf{q}\cdot\mathbf{k}^T. \tag{9.3.4}$$

The resultant force acting on the solid matrix, per unit volume of porous medium, is:

$$\mathbf{F}^* = \mathbf{F}_g^* + \mathbf{F}_u^* + \mathbf{F}_d^* = -(g\rho_{pm}\nabla z + \nabla p). \tag{9.3.5}$$

in which $\rho_{pm} = \phi\rho_f + (1-\phi)\rho_s$, $\rho^* = (\rho_s - \rho_f)(1-\phi)$ is the *submerged density of the solid*, and $h = \left(z + p/g\rho_f\right)$ is the fluid's *piezometric head*. In the absence of flow, $\nabla h = 0$.

The components of the force $\mathbf{F}^*$ in the x, y, z directions are:

$$\mathbf{F}^*_x = -\frac{\partial p}{\partial x}, \quad \mathbf{F}^*_y = -\frac{\partial p}{\partial y}, \quad \mathbf{F}^*{}_z = -g\rho_{pm}\nabla z - \frac{\partial p}{\partial z}. \tag{9.3.6}$$

Note that in the above equation the $\mathbf{F}^*_i$, $i = x, y, z$, denotes a scalar.

In order to ensure a downward directed force, we must ensure that $\mathbf{F}^*_z \leq 0$, i.e., $\partial p/\partial z \geq -g\rho_{pm}$.

The force component $\mathbf{F}^*{}_z$ acting on the solid may be positive (i.e., directed upward), zero or negative (i.e., directed downward). When, in a soil profile, this force is acting downward, i.e., $\mathbf{F}^*_z \leq 0$, we encounter stable conditions in the soil. When $\mathbf{F}^*_z \geq 0$, i.e., the vertical force component is directed upward, we encounter conditions known as *quick conditions*, $\mathbf{F}^*_z \geq 0$. Since, usually, $\rho^*/\rho_f \approx 1$, the critical gradient is usually taken as unity. Sand under quick conditions is referred to as *quick-sand*. Quick conditions often occur in the vicinity of the exit of water from the soil, e.g., at the exit of flow under structures. Sand (or any assembly of unconsolidated particles) under quick conditions is also referred to as a *fluidized bed*. Under quick conditions, the strength of an unconsolidated sand becomes zero. When a domain containing an unconsolidated assembly of solid particles, with a liquid filling up the void space, is subjected to shear stresses, the assembly will collapse, such that the volume of pore space will decrease. Water is forced to evacuate the void space. When the permeability is small, this will lead to a pressure rise that leads to a reduction in the effective stress, possibly down to zero. In the case of soil, it looses its coherence. This phenomenon is referred to as *soil liquefaction*, or *quick sand* conditions (Verruijt 2010, p. 82). Bear (1972, p. 186) presents more on *quick conditions*.

The drag exerted by a moving fluid on a solid particle was already mentioned in Chap. 4.

9.3.2 Land Subsidence

Land subsidence occurs when large volumes of water have been pumped from certain types of confined aquifers, e.g., ones composed of fine-grained sediments. As water is being pumped, the stress within the formation, due to overburden, say as described by (9.2.4), is not changed. However, noting the definition (9.1.7), as water pressure is reduced, say, by pumping, the effective stress is increased, and the result is compaction of the soft layers comprising the pumped formation or part of it.

In most cases, land subsidence is small and unnoticeable. It often occurs over large areas rather than in a small location. It may take the form of lowering ground surface elevation, or in the form of local sinkholes. Land subsidence me be small per year, but if pumping continues over many years, the effect may be significant, leading to disastrous results. Famous are the cases of land subsidence in Mexico City, Venice, Bangkok, and Central Valley, California. However, the phenomenon is widespread over the world.

Under certain circumstances, water, or other fluids, including a gas, are injected into a confined formation (e.g., the injection of CO_2 into an oil reservoir). The rise in pressure will cause a reduction in the effective stress, and the result is (actually observed), or *upheaval*($\equiv$ *uplift*), of land surface.

An excellent summary of Geomechanics of subsurface water withdrawal and injection is presented by Gambolati and Teatini (2015).

Following is a presentation of the mathematical model that describes land subsidence. Actually, in the previous section, we have already presented all the ingredients of the land subsidence model. Accordingly, assuming a single fluid occupying the entire void space of a confined formation of thickness $B = B(x, y)$, the mathematical 3-d model consists of the following equations:

- Stress, pressure and effective stress relationship (9.1.7):

$$\boldsymbol{\sigma} = \boldsymbol{\sigma}'_s - p\mathbf{I}, \tag{9.3.7}$$

  where p denotes the average pressure of the fluid (or fluids) in the void space. Note that the above equation does not include the Biot coefficient, α_B, which can always be added (see (5.1.30)).
- The equilibrium equation:

$$\nabla\cdot\boldsymbol{\sigma} + \rho\mathbf{F} = 0, \quad \mathbf{F} = -g\nabla z. \tag{9.3.8}$$

  Note the difference between the forces denoted by $\mathbf{F}$ and $\mathbf{F}^*$.
- The mass balance equation for the fluid (5.1.25):

$$\phi\beta\frac{\partial p}{\partial t} + \frac{\partial \varepsilon_s}{\partial t} = -\nabla\cdot\mathbf{q}_r + \Gamma^f, \tag{9.3.9}$$

where $\mathbf{q}_r$ is the fluid's flux expressed by Darcy's law, β is the fluid's compressibility, ε_s denotes the solid matrix dilatation, and Γ^f denotes the rate of fluid injection per unit volume of porous medium.

Of course, we could continue to write and solve the model in three dimensions, but, when considering the effect of pumping from geological formations, it is more convenient, and practical, to write and solve the model as a two-dimensional one, in the horizontal (x, y)-plain. We use the procedure described in Sect. 5.3 for passing from a 3-d model to a 2-d (horizontal) one. For the sake of simplicity we'll consider the case of a confined aquifer, with Hubbert's potential Φ, defined by (4.1.7).

Our objective is to construct a model that describes land subsidence in response to pumping from a confined aquifer of thickness $B = B(x, y, t)$. As above, we start from the mass balance for a compressible fluid in a deformable porous medium in three dimensions:

$$\nabla\cdot\mathbf{q}_r + \rho g\phi\beta\frac{\partial\Phi}{\partial t} + \frac{\partial\varepsilon_{sk}}{\partial t} + P(x, y, z, t) = 0, \tag{9.3.10}$$

in which ρ denotes fluid's density, Φ is Hubbert's potential, defined by (4.1.7), and $P(x, y, z, t)$ represents distributed pumping (as volume of water extracted per unit volume of soil, per unit time).

To obtain a two-dimensional model, we integrate (9.3.10) over the aquifer thickness, $B(x, y, t)$, obtaining:

$$\int_{b_1(x,y,t)}^{b_2(x,y,t)} \left(\nabla \cdot \mathbf{q}_r + \rho g \phi \beta \frac{\partial \Phi}{\partial t} + \frac{\partial \varepsilon_{sk}}{\partial t} + P \right) dz$$
$$= \nabla' \cdot B\widetilde{\mathbf{q}_r} + \mathbf{q}_r|_{F_2} \cdot \nabla F_2 - \mathbf{q}_r|_{F_1} \cdot \nabla F_1 + B\frac{\partial \widetilde{\varepsilon_{sk}}}{\partial t} + \tilde{\rho} g \tilde{\phi} \beta B \frac{\partial \widetilde{\Phi}}{\partial t}$$
$$+ \tilde{\rho} g \tilde{\phi} \beta \left(\Phi \frac{\partial B}{\partial t} + \Phi\Big|_{F_2} \frac{\partial F_2}{\partial t} - \Phi\Big|_{F_1} \frac{\partial F_1}{\partial t} \right) + B\widetilde{P} = 0, \tag{9.3.11}$$

where $F_i = F_i(x, y, z, t) = z - b_i(x, y, t) = 0$, $i = 1, 2$, describes the bottom and top surfaces bounding the aquifer, the prime symbol over an operator indicates that the operator is in the xy-plane only, and we have made use of the approximation:

$$\int_{b_1}^{b_2} \rho g \phi \beta \frac{\partial \Phi}{\partial t}\, dz \simeq \tilde{\rho} g \tilde{\phi} \beta \int_{b_1}^{b_2} \frac{\partial \Phi}{\partial t}\, dz, \qquad \widetilde{(\,)} = \frac{1}{B} \int_{b_1}^{b_2} (\,)\, dz. \tag{9.3.12}$$

Obviously, all averaged terms are functions of x, y and possibly t.

For the impervious top and bottom bounding surfaces considered here, we use the boundary condition (5.2.12), rewritten in the form:

$$\mathbf{q}_r|_{F_1} \cdot \nabla F_1 = 0, \qquad \mathbf{q}_r|_{F_2} \cdot \nabla F_2 = 0. \tag{9.3.13}$$

Adding the assumption that equipotentials are essentially vertical, i.e., $\Phi|_{F_1} \simeq \Phi|_{F_2} = \widetilde{\Phi}$, Eq. (9.3.11) reduces to:

$$\nabla' \cdot B\widetilde{\mathbf{q}_r} + B\frac{\partial \widetilde{\varepsilon_{sk}}}{\partial t} + \tilde{\rho} g \tilde{\phi} \beta B \frac{\partial \widetilde{\Phi}}{\partial t} + B(x, y, t)\widetilde{P}(x, y, t) = 0, \tag{9.3.14}$$

where $\widetilde{P}$ represents volume of water withdrawn from the aquifer per unit horizontal area, per unit time, and $B\widetilde{\mathbf{q}_r} = -B\widetilde{\mathbf{K}} \cdot \nabla' \widetilde{h^*}$ represents integrated horizontal flux.

We note that in (9.3.14):

$$\frac{\partial \widetilde{\Phi}}{\partial t} \simeq \frac{1}{\tilde{\rho} g} \frac{\partial \tilde{p}}{\partial t} + \frac{\partial \tilde{z}}{\partial t}, \qquad \nabla' \widetilde{\Phi} \simeq + \frac{1}{\tilde{\rho} g} \nabla' \tilde{p} + \nabla' \tilde{z}, \tag{9.3.15}$$

in which $\tilde{z} = (b_1 + b_2)/2$ is the elevation of the midpoint of the aquifer.

From (5.1.18) and (5.1.19), we obtain:

$$B\widetilde{\frac{\partial \varepsilon_{sk}}{\partial t}} = B\widetilde{\nabla \cdot \mathbf{V}_s} = \nabla' \cdot B\widetilde{\mathbf{V}'_s} + \mathbf{V}_s\Big|_{F_2} \cdot \nabla F_2 - \mathbf{V}_s\Big|_{F_1} \cdot \nabla F_1. \tag{9.3.16}$$

Since the top and bottom surfaces of the aquifer are assumed to be material surfaces with respect to the solid, following (5.2.9), we have on them:

$$(\mathbf{V}_s - \mathbf{u})\big|_{F_1} \cdot \nabla F_1 = 0, \qquad (\mathbf{V}_s - \mathbf{u})\big|_{F_2} \cdot \nabla F_2 = 0. \tag{9.3.17}$$

or, in view of (5.2.1):

$$\mathbf{V}_s\big|_{F_1} \cdot \nabla F_1 = -\frac{\partial F_1}{\partial t}, \qquad \mathbf{V}_s|_{F_2} \cdot \nabla F_2 = -\frac{\partial F_2}{\partial t}. \tag{9.3.18}$$

Hence, (9.3.16) becomes:

$$B\frac{\widetilde{\partial \varepsilon_{sk}}}{\partial t} = \nabla' \cdot B\widetilde{\mathbf{V}'_s} - \frac{\partial (F_2 - F_1)}{\partial t} = \nabla' \cdot B\widetilde{\mathbf{V}'_s} + \frac{\partial B}{\partial t}. \tag{9.3.19}$$

With the solid velocity, $\mathbf{V}_s$, related to the displacement vector, $\mathbf{w}(\equiv \mathbf{w}_s)$ by (5.1.11), noting the approximation included in this equation, we obtain:

$$\begin{aligned} B\widetilde{\mathbf{V}'_s} &= \int_{(B)} \mathbf{V}'_s \, dz - \int_{(B)} \frac{\partial \mathbf{w}}{\partial t} \, dz = \frac{\partial}{\partial t}(B\widetilde{\mathbf{w}'}) + \widetilde{\mathbf{w}'}|_{F_2}\frac{\partial F_2}{\partial t} - \widetilde{\mathbf{w}'}|_{F_1}\frac{\partial F_1}{\partial t} \\ &= B\frac{\partial \widetilde{\mathbf{w}'}}{\partial t} + \left(\widetilde{\mathbf{w}'}\frac{\partial B}{\partial t} + \mathbf{w}'|_{F_2}\frac{\partial F_2}{\partial t} - \mathbf{w}'|_{F_1}\frac{\partial F_1}{\partial t}\right). \end{aligned} \tag{9.3.20}$$

At this point, we need information on $\mathbf{w}'|_{F_2}$ and $\mathbf{w}'|_{F_1}$, which are the displacement boundary conditions on F_1 and F_2, respectively. This information is not available. We circumvent this difficulty by introducing the simplifying assumption that *the horizontal displacement is constant along the vertical*, i.e.:

$$\mathbf{w}'|_{F_2} = \mathbf{w}'|_{F_1} = \widetilde{\mathbf{w}'}. \tag{9.3.21}$$

Then:

$$B\frac{\partial \widetilde{\varepsilon_{sk}}}{\partial t} = \nabla' \cdot B\frac{\partial \widetilde{\mathbf{w}'}}{\partial t} + \frac{\partial B}{\partial t}. \tag{9.3.22}$$

Following (Verruijt 1969), we now express the specific discharge and the piezometric head as a sum of initial steady state values and deviations that express excess above the latter. By averaging these expressions, we obtain:

$$\begin{aligned} \widetilde{\Phi}(x, y, t) &= \widetilde{\Phi^o}(x, y) + \widetilde{\Phi^e}(x, y, t), \\ \widetilde{\mathbf{q}'_r}(x, y, t) &= \widetilde{\mathbf{q}'^{o}_r}(x, y) + \widetilde{\mathbf{q}'^{e}_r}(x, y, t), \\ \widetilde{P}(x, y, t) &= \widetilde{P^o}(x, y) + \widetilde{P^e}(x, y, t). \end{aligned} \tag{9.3.23}$$

In terms of these variables, the averaged mass balance equation (9.3.14) is separated into two equations: a steady state mass balance equation:

$$\nabla' \cdot B\widetilde{\mathbf{q}_r'^{o}} + \widetilde{P^o} = 0, \tag{9.3.24}$$

and an unsteady one:

$$\nabla' \cdot B\widetilde{\mathbf{q}_r'^{e}} + \nabla' \cdot \left(B\frac{\partial \widetilde{\mathbf{w}}'}{\partial t} \right) + \frac{\partial B}{\partial t} + \tilde{\rho} g \tilde{\phi} \beta B \frac{\partial \widetilde{\Phi^e}}{\partial t} + B\widetilde{P^e} = 0. \tag{9.3.25}$$

The last equation can be linearized by introducing:

$$\begin{aligned} B(x, y, t) &= b_2(x, y, t) - b_1(x, y, t) \\ &= (b_2^o(x, y) + w_z|_{F_2}) - (b_1^o(x, y) + w_z|_{F_1}) = B^o(x, y) + \Delta_z, \\ \Delta_z &= w_z|_{F_2} - w_z|_{F_1} \ll B^o. \end{aligned} \tag{9.3.26}$$

We note that Δ_z, denoting compaction, is positive in the $+z$-direction. Neglecting the effect of compaction on permeability, this approximation also leads to the flux equation:

$$B\widetilde{\mathbf{q}_r'^{e}} = -B^o \widetilde{\mathbf{K}}' \cdot \nabla' \widetilde{\Phi^e}. \tag{9.3.27}$$

Altogether, by substituting (9.3.27) in (9.3.25), we obtain a single equation in the variables $\widetilde{\Phi^e}$, B, and $\widetilde{\mathbf{w}}'$. Our next step is to make use of the equilibrium equation.

9.3.3 Integrated Equilibrium Equation

The total stress tensor, $\boldsymbol{\sigma}$, at a point within an aquifer, satisfies the *equilibrium equation* (5.1.37), rewritten here for convenience in the form:

$$\nabla \cdot \boldsymbol{\sigma} + \rho \mathbf{F} = 0, \tag{9.3.28}$$

where we have omitted the average symbol; the body force acting on the porous medium, $\rho\mathbf{F}$, is assumed to remain unchanged by the compaction of the porous medium, i.e., $(\rho\mathbf{F})^e = 0$, where the superscript e denotes the increment. In (9.3.28), $\rho\,(= \rho_{pm}) = \phi\rho_f + (1-\phi)\rho_s$, represents the combined density of fluid and solid matrix.

We start from (5.1.54), which involves the incremental effective stress and effective pressure. We repeat it here for convenience as:

$$\nabla \cdot \boldsymbol{\sigma}_s'^{e} - \nabla p^e = 0. \tag{9.3.29}$$

Note that in this equation, p is positive for compression.

We then assume that the solid matrix comprising the aquifer behaves like an isotropic and (for the relatively small displacements considered here) perfectly elastic body, for which the stress–strain relationship (5.1.55) is valid. By integrating (9.3.29) (see Fig. 5.1), we obtain:

$$\int_{b_1}^{b_2} (\nabla \cdot \boldsymbol{\sigma}_s{}^e - \nabla p^e)\, dz = \nabla' \cdot B\widetilde{\boldsymbol{\sigma}_s'^e} + (\boldsymbol{\sigma}_s'{}^e - p^e\mathbf{I})\big|_{F_2} \cdot \nabla F_2$$
$$-(\boldsymbol{\sigma}_s'{}^e - p^e\mathbf{I})\big|_{F_1} \cdot \nabla F_1 - \nabla' B\widetilde{p^e} = 0. \qquad (9.3.30)$$

We assume that the porous medium on both sides of a boundary surface, say the upper one, $F_2 = 0$, is deformable. On such boundary, we have to maintain the *condition of no-jump in the total stress*, i.e.:

$$[\![\, \boldsymbol{\sigma} \,]\!]_{u,\ell} \cdot \nabla F_2 = 0, \qquad [\![\, \boldsymbol{\sigma}_s' - p\mathbf{I} \,]\!]_{u,\ell} \cdot \nabla F_2 = 0, \qquad (9.3.31)$$

where u and ℓ, respectively, denote the upper and lower sides. Following the methodology introduced earlier, this condition leads to an analogous condition related to the incremental effective stress and pressure:

$$[\![\, \boldsymbol{\sigma}_s'{}^e - p^e\mathbf{I} \,]\!]_{u,\ell} \cdot \nabla F_2 = 0. \qquad (9.3.32)$$

When the excess stress and pressure in an aquifer are due only to pumping, and not to changes in the overburden load, say, by excavation, the total stress on the upper side of the boundary, $\boldsymbol{\sigma}\big|_u$, remains unchanged. Hence, from (9.3.31), we obtain:

$$\boldsymbol{\sigma}\big|_u \cdot \nabla F_2 = \boldsymbol{\sigma}^o\big|_u \cdot \nabla F_2 = \left(\boldsymbol{\sigma}_s'{}^o\big|_\ell - p^o\big|_\ell\mathbf{I}\right) \cdot \nabla F_2, \qquad (9.3.33)$$

and:

$$\boldsymbol{\sigma}^e\big|_u \cdot \nabla F_2 = \left(\boldsymbol{\sigma}_s'{}^e\big|_\ell - p^e\big|_\ell\mathbf{I}\right) \cdot \nabla F_2 = 0. \qquad (9.3.34)$$

In view of the boundary conditions (9.3.34), Eq. (9.3.30) reduces to

$$\nabla' \cdot B\widetilde{(\boldsymbol{\sigma}_s'{}^e)} - \nabla' B\widetilde{p^e} = 0, \qquad (9.3.35)$$

in which averaged values are functions of x, y and t only.

Let us rewrite (9.3.35) in the form:

$$\frac{\partial}{\partial x} B\widetilde{(\sigma_s'{}^e)}_{xx} + \frac{\partial}{\partial y} B\widetilde{(\sigma_s'{}^e)}_{xy} - \frac{\partial}{\partial x} B\widetilde{p^e} - \frac{\partial}{\partial y} B\widetilde{p^e} = 0, \qquad (9.3.36)$$

and two analogous equations in the y and z directions (Bear and Bachmat 1991, p. 509). We then express the averaged excess effective stress tensor in terms of averaged displacements, making use of (5.1.18) and (5.1.55), and the assumption expressed by (9.3.21). We obtain:

$$\begin{aligned}\widetilde{\varepsilon_{sk}} &= \widetilde{\varepsilon}_{xx} + \widetilde{\varepsilon}_{yy} + \widetilde{\varepsilon}_{zz} \\ &= \frac{\partial \widetilde{w_x}}{\partial x} + \frac{\partial \widetilde{w_y}}{\partial y} + \frac{\partial \widetilde{w_z}}{\partial z} = \frac{\partial \widetilde{w_x}}{\partial x} + \frac{\partial \widetilde{w_y}}{\partial y} + \frac{\Delta_z}{B},\end{aligned} \tag{9.3.37}$$

$$\begin{aligned}\widetilde{(\sigma'^e_s)_{xx}} &= \lambda''_s\, \widetilde{\varepsilon_{sk}} + 2\mu'_s\, \widetilde{(\varepsilon_{sk})_{xx}} \\ &= (\lambda''_s + 2\mu'_s)\frac{\partial \widetilde{w_x}}{\partial x} + \lambda''_s\left(\frac{\partial \widetilde{w_y}}{\partial y} + \frac{\Delta_z}{B}\right),\end{aligned} \tag{9.3.38}$$

and additional analogous equations for $\widetilde{(\sigma'^e_s)_{yy}}$, $\widetilde{(\sigma'^e_s)_{xy}} = \widetilde{(\sigma'^e_s)_{yx}}$, $\widetilde{(\sigma'^e_s)_{zx}} = \widetilde{(\sigma'^e_s)_{xz}}$, $\widetilde{(\sigma'^e_s)_{yz}} = \widetilde{(\sigma'^e_s)_{zy}}$, and $\widetilde{(\sigma'^e_s)_{zz}}$. For example (Bear and Bachmat 1991, p. 510), we obtain the linearized equation:

$$\widetilde{(\sigma'^e_s)_{zz}} = \overline{\lambda''_s}\, \widetilde{\varepsilon_{sk}} + 2\overline{\mu'_s}\, \frac{\partial \widetilde{w_z}}{\partial z} = \overline{\lambda''_s}\left(\frac{\partial \widetilde{w_x}}{\partial x} + \frac{\partial \widetilde{w_y}}{\partial y}\right) + (\overline{\lambda''_s} + 2\overline{\mu'_s})\frac{\Delta_z}{B}. \tag{9.3.39}$$

By inserting these expressions into (9.3.36) and the additional, not shown, equations, and making use of (9.3.21), we obtain three equations in the four averaged variables $\widetilde{p^e}$, $\widetilde{w_x}$, $\widetilde{w_y}$, and $\widetilde{w_z}$, all functions of x, y and t only:

$$\begin{aligned}&\frac{\partial}{\partial x}\left\{B\left[(\lambda''_s + 2\mu'_s)\frac{\partial \widetilde{w_x}}{\partial x} + \lambda''_s\left(\frac{\partial \widetilde{w_y}}{\partial y} + \frac{\Delta_z}{B}\right)\right]\right\} \\ &\quad + \frac{\partial}{\partial y}\left[B\mu'_s\left(\frac{\partial \widetilde{w_x}}{\partial y} + \frac{\partial \widetilde{w_y}}{\partial x}\right)\right] - \frac{\partial}{\partial x} B\widetilde{p^e} = 0,\end{aligned} \tag{9.3.40}$$

$$\begin{aligned}&\frac{\partial}{\partial x}\left[B\mu'_s\left(\frac{\partial \widetilde{w_y}}{\partial x} + \frac{\partial \widetilde{w_x}}{\partial y}\right)\right] + \frac{\partial}{\partial y}\left\{B\left[\lambda''_s\frac{\partial \widetilde{w_x}}{\partial x}\right.\right. \\ &\quad \left.\left. + (\lambda''_s + 2\mu'_s)\frac{\partial \widetilde{w_y}}{\partial y} + \lambda''_s\frac{\Delta_z}{B}\right]\right\} - \frac{\partial}{\partial y} B\widetilde{p^e} = 0,\end{aligned} \tag{9.3.41}$$

$$\begin{aligned}&\frac{\partial}{\partial x}\left[B\mu'_s\frac{\partial \widetilde{w_z}}{\partial x} + \mu'_s\left(\widetilde{w_z} + w_z\big|_{F_2}\frac{\partial F_2}{\partial x} - w_z\big|_{F_1}\frac{\partial F_1}{\partial x}\right)\right] + \frac{\partial}{\partial y}\left[B\mu'_s\frac{\partial \widetilde{w_z}}{\partial y}\right. \\ &\quad \left. + \mu'_s\left(\widetilde{w_z} + w_z\Big|_{F_2}\frac{\partial F_2}{\partial y} - w_z\Big|_{F_1}\frac{\partial F_1}{\partial y}\right)\right] = 0.\end{aligned} \tag{9.3.42}$$

For constant λ''_s and μ'_s, and with

$$B(x, y, t) = B^o(x, y) + \Delta_z(x, y, t), \qquad \Delta_z \ll B^o, \tag{9.3.43}$$

we obtain the linearized forms of (9.3.40) and (9.3.41):

$$\mu_s' \nabla'^2 \widetilde{w_x} + (\lambda_s'' + \mu_s')\left(\frac{\partial \widetilde{w_x}}{\partial x} + \frac{\partial \widetilde{w_y}}{\partial y}\right) + \lambda_s'' \frac{\partial(\Delta_z/B^o)}{\partial x} - \frac{\partial \widetilde{p^e}}{\partial x} = 0, \quad (9.3.44)$$

$$\mu_s' \nabla'^2 \widetilde{w_y} + (\lambda_s'' + \mu_s')\left(\frac{\partial \widetilde{w_x}}{\partial x} + \frac{\partial \widetilde{w_y}}{\partial y}\right) + \lambda_s'' \frac{\partial(\Delta_z/B^o)}{\partial y} - \frac{\partial \widetilde{p^e}}{\partial y} = 0. \quad (9.3.45)$$

With the same linearization, and assuming $|\widetilde{\mathbf{V}_s'} \cdot \nabla' B| \ll |\partial B/\partial t|$, the second and third terms in (9.3.25) reduce to $B^o \partial \widetilde{\varepsilon_{sk}}/\partial t$. Thus, we may approximate the volume balance equation, (9.3.25), by (Bear and Bachmat 1991, p. 511):

$$\nabla' \cdot B^o \widetilde{K}\left(\frac{1}{\widetilde{\rho} g} \nabla' \widetilde{p^e} + \nabla' z\right) + B^o \frac{\partial \widetilde{\varepsilon_{sk}}}{\partial t} + \widetilde{\phi} \beta B^o \frac{\partial \widetilde{p^e}}{\partial t} + B^o \widetilde{P^e} = 0. \quad (9.3.46)$$

In principle, (9.3.37), (9.3.42) and (9.3.44)–(9.3.46) are five equations in the variables $\widetilde{p^e}$, $\widetilde{w_x}$, $\widetilde{w_y}$, Δ_z and $\widetilde{\varepsilon_{sk}}$. However, in (9.3.42) we still have the terms

$$w_z\big|_{F_1}, \quad \text{and} \quad w_z\big|_{F_2}, \quad \text{with} \quad \Delta_z \equiv w_z\big|_{F_2} - w_z\big|_{F_1}, \quad \text{and} \quad B = B^o + \Delta_z,$$

which are actually conditions on the surfaces $F_1 = 0$ and $F_2 = 0$, for which we have no information. In fact, in most subsidence problems, the land subsidence, as expressed by $w_z\big|_{F_2}$ is the very state variable for which a solution is sought.

At this point we may continue by introducing certain simplifying assumptions, as a substitute for the missing information. For example, we may assume that the bottom of the aquifer is fixed, i.e., $w_z\big|_{F_1} = 0$ and that w_z varies linearly with elevation, i.e., $\widetilde{w_z} = \frac{1}{2} w_z\big|_{F_2} = \Delta_z/2$, where $-\Delta_z$ denotes land subsidence (positive downward). We then end up with equations for $\widetilde{w_x}$, $\widetilde{w_y}$, Δz and $\widetilde{p^e}$. In this way, we have achieved our goal of determining the land subsidence $\Delta_z(x, y, t)$. In fact, we have solved, simultaneously, for the horizontal displacement, $\widetilde{w_x}$, as well as for the pressure in the aquifer, $\widetilde{p^e}$.

Verruijt (1969, p. 347) suggested an approach based on the assumption that consolidation occurs under *conditions of planar incremental total stress*:

$$\sigma_{zz}^e = 0, \qquad \sigma_{xz}^e = \sigma_{zx}^e = 0, \qquad \sigma_{yz}^e = \sigma_{zy}^e = 0. \quad (9.3.47)$$

This is a consequence of the assumption that displacements occur in the vertical direction only, i.e., $w_z \neq 0$, $w_x = w_y = 0$, while the total stress remains unchanged, i.e., $\boldsymbol{\sigma} \equiv \boldsymbol{\sigma}^o$ and $\boldsymbol{\sigma}^e = 0$. This assumption is justified when the aquifer is located between two soft confining layers (e.g., clay) which cannot resist shear stress. Furthermore, this assumption also justifies (9.3.21), since in a relatively thin aquifer, as implied by the planar stress assumption, lateral deformation is, more or less, uniform throughout the relatively small thickness of the layer.

From (9.3.47), it follows that the equilibrium equation (9.3.29) reduces to:

$$\nabla' \cdot \boldsymbol{\sigma}_s'^{e} - \nabla' p^e = 0, \tag{9.3.48}$$

with the boundary condition (9.3.34), and a similar one for $F_1 = 0$, also written in the xy-coordinates only.

Following the integration procedure, which led above to (9.3.36) and to analogous equations in y and z, we now obtain only (9.3.36) and an analogous equations in y; the z-equation has been eliminated.

Accordingly, we now have to solve (9.3.25), or any equivalent form of it, (9.3.44) and (9.3.45), for $\widetilde{p^e}$, $\widetilde{w_x}$, $\widetilde{w_y}$ and Δ_z. The required fourth equation is now obtained from the first condition in (9.3.47), which leads to

$$(\boldsymbol{\sigma}_s'^{e})_{zz} = p^e. \tag{9.3.49}$$

From (9.3.39) and (9.3.49), we now obtain:

$$\begin{aligned}\widetilde{p^e} &= \overline{\lambda_s''}\left(\frac{\partial \widetilde{w_x}}{\partial x} + \frac{\partial \widetilde{w_y}}{\partial y}\right) + (\overline{\lambda_s''} + 2\overline{\mu_s'})\frac{\Delta_z}{B} \\ &= \overline{\lambda_s''}\, \widetilde{\varepsilon_{sk}} + 2\overline{\mu_s'}\, \frac{\Delta_z}{B}.\end{aligned} \tag{9.3.50}$$

This completes the formulation of the mathematical model for land subsidence. Usually we assume that $w_z\big|_{F_1} = 0$, and that $-\Delta_z = -w_z\big|_{F_2}$ expresses land subsidence.

9.3.4 Terzaghi–Jacob Versus Biot Approaches

By differentiating (9.3.44) with respect to x, Eq. (9.3.45) with respect to y, linearizing both equations and then adding them, assuming constant $\overline{\lambda_s''}$, $\overline{\mu_s'}$, and B^o, we obtain

$$\nabla'^2\left\{(\overline{\lambda_s''} + 2\overline{\mu_s'})\nabla' \cdot \widetilde{\mathbf{w}'} + \overline{\lambda_s''}\,\frac{\Delta_z}{B^o} - \widetilde{p^e}\right\} = 0. \tag{9.3.51}$$

Following Verruijt (1969), we integrate (9.3.51), obtaining

$$\begin{aligned}(\overline{\lambda_s''} + 2\overline{\mu_s'})\nabla' \cdot \widetilde{\mathbf{w}'} + \overline{\lambda_s''}\frac{\Delta_z}{B^o} &= (\overline{\lambda_s''} + 2\overline{\mu_s'})\widetilde{\varepsilon_{sk}} - 2\overline{\mu_s'}\frac{\Delta_z}{B^o} \\ &= \widetilde{p^e} + \Pi'(x, y, t),\end{aligned} \tag{9.3.52}$$

where Π' satisfies

$$\nabla^2\Pi' = 0, \qquad \text{for every} \quad t.$$

The case $\Pi' = 0$, is presented after (5.1.64).

By comparing (9.3.52) with (9.3.50), obtained by introducing the *planar stress assumption*, we find that:

$$\Pi' = 2\overline{\mu'_s}\left(\nabla' \cdot \widetilde{\mathbf{w}}' - \frac{\Delta_z}{B^o}\right) \simeq 2\overline{\mu'_s}\left(\widetilde{\varepsilon_{sk}} - 2\frac{\Delta_z}{B^o}\right), \tag{9.3.53}$$

where $\widetilde{\varepsilon_{sk}}$ is defined by (9.3.37).

If we assume $\widetilde{P^e} = 0$, and *no horizontal displacement*, i.e., $\widetilde{\mathbf{w}}' \equiv 0$, Eq. (9.3.25) reduces to:

$$\nabla' \cdot B\widetilde{\mathbf{q}_r'^e} + \frac{\partial B}{\partial t} + \widetilde{\rho} g \widetilde{\phi} \beta B \frac{\partial \widetilde{h^*}}{\partial t} = 0, \tag{9.3.54}$$

where $B = B^o + \Delta_z$. Under the same conditions, (9.3.50) reduces to:

$$\widetilde{p^e} = (\overline{\lambda''_s} + 2\overline{\mu'_s})\frac{\Delta_z}{B}. \tag{9.3.55}$$

Together, (9.3.54) and (9.3.55) can now be solved for $\widetilde{p^e}$ and Δ_z.

By combining the two equations, and assuming $\Delta_z \ll B$, we obtain:

$$\nabla' \cdot B\widetilde{\mathbf{q}_r'^e} + B\left(\frac{1}{\overline{\lambda''_s} + 2\overline{\mu'_s}} + \widetilde{\phi}\beta\right)\frac{\partial \widetilde{p^e}}{\partial t} = 0. \tag{9.3.56}$$

By comparing (9.3.56) with (5.1.66), we may conclude that we could have obtained the last equation by *assuming, from the onset, that only vertical compressibility prevails*, with a coefficient of vertical compressibility:

$$\alpha = \frac{1}{\overline{\lambda''_s} + 2\overline{\mu'_s}} \tag{9.3.57}$$

as in (5.1.66). Furthermore, by comparing (9.3.53), with $\widetilde{\mathbf{w}'} = 0$, with (9.3.50), obtained by assuming (i) planar stress, and (ii) no horizontal displacement, we obtain:

$$\Pi' = -2\overline{\mu'_s}\frac{\Delta_z}{B}. \tag{9.3.58}$$

It is of interest to return at this point to the end of Sect. 5.1.6 where a comparison is made between the Terzaghi–Jacob and the Biot approaches.

9.3.5 Land Subsidence Produced by Pumping

As an example for the use of the land subsidence model developed above, consider the case of land subsidence presented by Bear and Corapcioglu (1981b). In this example, both vertical and horizontal displacements, produced by pumping from a single well in a homogeneous confined or phreatic aquifer, are considered.

With $\widetilde{P^e}$ denoting the constant pumping rate from a well of radius r_w in a confined aquifer, a land subsidence model is constructed in terms of the four variables: $\delta(r,t)(\equiv \Delta_z)$, $\widetilde{w_r}(r,t)$, $-\widetilde{p^e}$, and $\widetilde{\varepsilon^e}$, denoting (vertical) subsidence, horizontal displacement, pressure drop, and strain, or *volume dilatation*, respectively.

Based on certain simplifying assumptions, e.g.,

$$\frac{\partial B}{\partial r} \ll \frac{B}{r}, \quad \text{and} \quad \frac{1}{\Delta_z}\frac{\partial \Delta_z}{\partial t} \ll \frac{1}{B}\frac{\partial B}{\partial t},$$

the Bear and Corapcioglu (1981b) land subsidence model is composed of the following four equations:

- **(Linearized) mass balance equation**:

$$-\frac{1}{r}\frac{\partial}{\partial r}\left(r\frac{\widetilde{k^o}}{\widetilde{\mu^o}}\frac{\partial \widetilde{p^e}}{\partial r}\right) + \frac{\partial \widetilde{\varepsilon}}{\partial t} + \widetilde{\phi^o}\beta\frac{\partial \widetilde{p^e}}{\partial t} = 0. \tag{9.3.59}$$

- **Definition of dilatation**, averaged over the vertical:

$$\widetilde{\varepsilon^e} = \frac{\partial \widetilde{w}_r}{\partial t} + \frac{\widetilde{w_r}}{r} + \frac{\Delta z}{B^o}. \tag{9.3.60}$$

- **A combination of averaged (= integrated over the vertical) equilibrium equation, combined with the constitutive equations**:

$$\left(2\overline{\mu'_s} + \overline{\lambda''_s}\right)\widetilde{\varepsilon^e} - 2\overline{\mu'_s}\frac{\Delta_z}{B^o} = \widetilde{p^e} + 2g(t), \tag{9.3.61}$$

  where $g(t)$ is an arbitrary function of t.

- **Averaged constitutive relation, combined with the assumption of plane incremental total stress, suggested by** Verruijt (1969):

$$\widetilde{p^e} = 2\overline{\mu'_s}\frac{\Delta_z}{B^o} + \overline{\lambda''_s}\widetilde{\varepsilon^e}. \tag{9.3.62}$$

These four equations, in the variables: $\widetilde{p^e}$, $\widetilde{\varepsilon^e}$, $\widetilde{w_r}$, and Δ_z, are solved for the boundary and initial conditions:

$$\begin{aligned}
t \le 0,\ r \ge r_e, &\quad \widetilde{p^e}, \widetilde{\varepsilon^e}, \widetilde{w_r}, \Delta_z = 0\\
t > 0,\ r = r_w, &\quad \frac{\partial \widetilde{p^e}}{\partial r} = \frac{Q_w\widetilde{\mu^o}}{2\pi r_w B^o\widetilde{k^o}}\\
r = r_w, &\quad \widetilde{w_r} = 0\\
r \to \infty, &\quad \widetilde{p^e},\ \widetilde{w_r},\ \widetilde{\varepsilon^e},\ \Delta_z = 0.
\end{aligned} \tag{9.3.63}$$

Subject to certain simplifying assumptions, the solutions derived by Bear and Corapcioglu (1981b) for the excess pressure, expressed in terms of change in piezometric head, is:

$$\frac{\widetilde{p^e}}{\widetilde{\rho^o} g} = \Delta h = -\frac{Q_w}{4\pi T} W(u), \qquad u = \frac{r^2}{4C_v t} = \frac{Sr^2}{4Tt}, \tag{9.3.64}$$

which is the usual equation describing drawdown in a confined aquifer, as a result of pumping from a well (e.g., (Bear, 1979), p. 321). In this equation, $C_v (= 1/(\overline{\mu'_s} + \overline{\lambda''_s}) \equiv T/S)$ is a consolidation coefficient, T and S are the aquifer transmissivity and storativity, respectively, and $W(u)$ is the *exponential integral*,

$$W(u) \equiv -\text{Ei}(-u) = \int_u^\infty \frac{e^{-x}}{x}\, dx. \tag{9.3.65}$$

The vertical displacement is:

$$\delta (\equiv -\Delta_z) = \frac{Q_w}{8\pi C_v} W(u) = -\frac{S}{2} \Delta h. \tag{9.3.66}$$

This is half the value obtained by assuming that only vertical consolidation takes place (Bear and Corapcioglu 1981a). The horizontal displacement is:

$$\widetilde{w_r} = -\frac{Q_w r}{16\pi C_v B^o} \left[W(u) + \frac{1 - e^{-u}}{u} \right], \tag{9.3.67}$$

with a maximum value at

$$r|_{w_r,\max} \approx 1.1367 (C_v t)^{1/2} = 1.1367 (Tt/S)^{1/2}. \tag{9.3.68}$$

In may be of interest to note that in the four-variables model presented here, the integrated flow equation, equilibrium equation and constitutive relationship are *coupled*. A simpler approach would be to solve for the pressure drop, assuming no soil deformation, and then to estimate soil compaction, or/and subsidence from

$$\delta(x, y, t) = \int_B \frac{\partial w_z}{\partial z}\, dz = \int_B \varepsilon_{sk}(x, y, z, t)\, dz = \int_B \frac{\widetilde{p^e}}{\widetilde{\lambda''_s} + 2\widetilde{\mu'_s}}\, dz, \tag{9.3.69}$$

where B denotes the thickness of the considered layer, and we have made use of (5.1.65) to express ε_{sk} in terms of the pressure $\widetilde{p^e}$. The solution for $\widetilde{p^e}$ can be obtained by solving (9.3.56), with $\phi \approx \widetilde{\phi}$.

Some researchers (e.g., Gambolati et al. 1973, 1974; Corapcioglu and Brutsaert 1977) have presented subsidence models that take into account the time lag between measured changes in piezometric head and the observed resulting compaction. This time lag is an indication that the purely elastic constitutive relations is not appropriate for clay and silt lenses.

Analytical solutions for land subsidence produced by pumping from a well, are presented by Bear and Corapcioglu (1981a, b) and by Verruijt (2014).

9.4 Waves in Porous Media

So far, in Chaps. 1–8, we have assumed that inertial forces are negligible, leading to Darcy's law as an approximation of the fluid's momentum balance equation. In Sect. 4.3.2 we have presented the momentum balance equation for the fluid, when local acceleration may not be neglected, especially at the onset of flow and in oscillatory flows, but the advective acceleration and the internal friction in the fluid may be neglected. In this section, we shall consider the case when these effects may not be neglected.

We start from the macroscopic momentum balance equations (9.2.1) rewritten for a single fluid (f) that occupies the entire void space:

$$\phi\rho_f \frac{D_f V_{fi}}{Dt} + (1-\phi)\rho_s \frac{D_s V_{si}}{Dt} = \frac{\partial \overline{\sigma_{ij}}}{\partial x_j} + \overline{\rho F_i}, \tag{9.4.70}$$

in which $\mathbf{V}_f$ and $\mathbf{V}_s$ denote the average velocities of the fluid and the solid, $\overline{\boldsymbol{\sigma}}$ $(\equiv \phi\overline{\boldsymbol{\sigma}_f}^f + (1-\phi)\overline{\boldsymbol{\sigma}_s}^s)$ is the total stress, and $\overline{\rho\mathbf{F}}$ $(\equiv -(\phi\overline{\rho_f}^f + (1-\phi)\overline{\rho_s}^s)g\nabla z)$ is the body force, per unit volume of the porous medium, due to gravity.

Just a reminder that for any vector $\mathbf{g}$, $D_\alpha g_{\alpha i}/Dt \equiv \partial g_{\alpha i}/\partial t + V_{\alpha j}\partial g_{\alpha i}/\partial x_j$.

Since Darcy's law is written in terms of the relative velocity, $\mathbf{V}_r \equiv (\mathbf{V}_f - \mathbf{V}_s)$, we can write (9.4.70) as:

$$\phi\rho_f \frac{D_s V_{ri}}{Dt} + \rho_b \frac{D_s V_{si}}{Dt} + \phi\rho_f V_{rj}\frac{\partial}{\partial x_j}\left(V_{ri} + V_{si}\right) = \frac{\partial \overline{\sigma_{ij}}}{\partial x_j} - \rho_b g\frac{\partial z}{\partial x_i}, \tag{9.4.71}$$

where $\rho_b = \phi\overline{\rho_f}^f + (1-\phi)\overline{\rho_s}^s$ denotes the bulk density of the porous medium.

As a momentum balance equation for the fluid alone, we may use (4.2.20). Then, recalling that we may express the permeability as $\mathbf{k} = \phi\Delta^2\mathbf{R}^T$, and neglecting the term that expresses the internal friction in the fluid, we can rewrite (4.2.2) in the form:

$$\phi\rho_f \frac{D_f V_{fi}}{Dt} = -\phi\left(\frac{\partial p}{\partial x_i} + \rho_f g\frac{\partial z}{\partial x_i}\right) - \mu\phi^2 (k_{ij})^T V_{rj}. \tag{9.4.72}$$

Let us assume that the local acceleration is much larger than the convective one, i.e.,

$$\left|\frac{\partial V_{fi}}{\partial t}\right| \gg \left|V_{fj}\frac{\partial V_{fi}}{\partial x_j}\right|, \quad \left|\frac{\partial V_{si}}{\partial t}\right| \gg \left|V_{sj}\frac{\partial V_{si}}{\partial x_j}\right|, \tag{9.4.73}$$

or, making use of the methodology presented in Sect. 3.10 for deleting non-dominant effects:

$$St^{-1} = \left| \frac{(\Delta t)_c^{(V)}}{(\Delta x)_c^{(V)}/V_c} \right| \ll 1, \tag{9.4.74}$$

i.e., the time required for a local incremental change in velocity is much shorter than the time required for observing the same velocity change in space. Then (9.4.72) reduces to:

$$\phi\rho_f \frac{\partial V_{fi}}{\partial t} = -\phi\left(\frac{\partial p}{\partial x_i} + \rho_f\, g\frac{\partial z}{\partial x_i}\right) - \mu\phi^2 (k_{ij})^T V_{rj}. \tag{9.4.75}$$

Note that a high Strouhal number indicates *creeping flow*.

In terms of $\mathbf{V}_r$ and $\mathbf{V}_s$, (9.4.75) takes the form:

$$\phi\rho_f \frac{\partial V_{ri}}{\partial t} + \phi\rho_f \frac{\partial V_{si}}{\partial t} = -\phi\left(\frac{\partial p}{\partial x_i} + \rho_f g\frac{\partial z}{\partial x_i}\right) - \mu\phi^2 (k_{i\ell})^T V_{r\ell}. \tag{9.4.76}$$

Considerations similar to those leading to (9.4.73) and (9.4.74), will lead to the deletion of all the convective acceleration terms in (9.4.71). Thus, by neglecting the convective acceleration also in the solid, and making use of (5.1.28) to express the total stress, $\overline{\boldsymbol{\sigma}}$, in terms of the effective stress, $\boldsymbol{\sigma}'_s$ and fluid pressure, p, the averaged momentum balance equation (9.4.70), reduces to:

$$\phi\rho_f \frac{\partial V_{fi}}{\partial t} + (1-\phi)\rho_s \frac{\partial V_{si}}{\partial t} = \frac{\partial \sigma'_{sji}}{\partial x_j} - \left(\frac{\partial p}{\partial x_i} + \rho_b g\frac{\partial z}{\partial x_i}\right). \tag{9.4.77}$$

Or:

$$\phi\rho_f \frac{\partial V_{ri}}{\partial t} + \rho_b \frac{\partial V_{si}}{\partial t} = \frac{\partial \sigma'_{sji}}{\partial x_j} - \left(\frac{\partial p}{\partial x_i} + \rho_b g\frac{\partial z}{\partial x_i}\right). \tag{9.4.78}$$

Thus, (9.4.76) for the fluid and (9.4.78) for the porous medium as a whole, are two momentum balance equations in terms of $\mathbf{V}_r$ and $\mathbf{V}_s$. Using (9.4.76) to express $\partial V_{ri}/\partial t$ and inserting in (9.4.78), we obtain:

$$(1-\phi)\rho_s \frac{\partial V_{si}}{\partial t} - \frac{\partial \sigma'_{sji}}{\partial x_j} - (1-\phi)\left[\frac{\partial p}{\partial x_i} + \rho_s g\frac{\partial z}{\partial x_i}\right] - \mu\phi^2 (k_{i\ell})^T V_{r\ell} = 0. \tag{9.4.79}$$

Using (5.1.19) to approximate $\mathbf{V}_s \approx \partial w_s/\partial t$, $\mathbf{V}_f \approx \partial w_f/\partial t$, the above two equations can be written as

$$(1-\phi)\rho_s \frac{\partial^2 w_{si}}{\partial t^2} - \frac{\partial \sigma'_{sji}}{\partial x_j} - (1-\phi)\left[\frac{\partial p}{\partial x_i} + \rho_s g\frac{\partial z}{\partial x_i}\right] - \mu\phi^2 (k_{i\ell})^T \frac{\partial}{\partial t}(w_{f\ell} - w_{s\ell}) = 0. \tag{9.4.80}$$

Similarly, from (9.4.75), we obtain

$$\phi \rho_f \frac{\partial^2 w_{fi}}{\partial t^2} + \phi \left(\frac{\partial p}{\partial x_i} + \rho_f \, g \frac{\partial z}{\partial x_i} \right) + \mu \phi^2 (k_{i\ell})^T \frac{\partial}{\partial t} (w_{f\ell} - w_{s\ell}) = 0. \qquad (9.4.81)$$

The above two equations can be shown to be hyperbolic *wave equations* (Sorek et al. 1992). We recall that the displacements considered here are at (i.e., at the close vicinity of) the considered point.

In these two momentum balance equations, the variables are $\mathbf{w}_s$, $\mathbf{w}_f$, p, ρ_f, and $\boldsymbol{\sigma}'_s$. To obtain a complete solution for the displacements, we need three additional equations. These are:

- The constitutive relation that expresses the stress–strain relationship for the solid skeleton, e.g., (9.1.19).
- A relationship, $\rho_f = \rho_f(p)$.
- The fluid's mass balance equation, e.g., (5.1.22), and we have to use the expression (4.2.44) for $\mathbf{q}_r$.

Because the mass balance equation involves also the skeleton's dilatation, ε_{sk}, as an additional variable, we add also:

- Equation (5.1.18) that relates ε_{sk} to $\mathbf{w}_s$.

We thus have a complete set of equations to be solved for the variables of this problem. Usually, we focus our attention on the solid's displacement $\mathbf{w}_s$.

Often, the effective stress is expressed by (9.1.7), i.e., using *Biot's coefficient* α_B. Then, p in (9.4.78) and (9.4.81) is replaced by $\alpha_B p$.

Sorek et al. (1992), using a similar approach, developed a mathematical model that involves a somewhat different momentum balance equations for a thermo-elastic porous medium and show that they are waves equations.

References

Bear J (1972) Dynamics of fluids in porous media. American Elsevier, New York, p 764 (also published by Dover Publications, 1988; translated into Chinese)
Bear J (1979) Hydraulics of groundwater. McGraw-Hill, New York, p 569 (also published by Dover Publications, 2007; translated into Chinese)
Bear J, Bachmat Y (1991) Introduction to modeling phenomena of transport in porous media. Kluwer Publications Co., Dordrecht, p 553
Bear J, Corapcioglu MY (1981a) Mathematical-model for regional land subsidence due to pumping. 1. Integrated aquifer subsidence equations for vertical displacement only. Water Resour Res 17:937–946
Bear J, Corapcioglu MY (1981b) Mathematical-model for regional land subsidence due to pumping. 2. Integrated aquifer subsidence equations for vertical and horizontal displacements. Water Resour Res 17:947–958
Biot MA (1941) General theory of three-dimensional consolidation. J Appl Phys 12:155–164

Bishop AW, Blight GE (1963) Some aspects of effective stress in saturated and partly saturated soils. Geotechnique 13:177–197
Cheng AH-D (2016) Poroelascicity. Springer, Berlin, p 877
Corapcioglu MY, Brutsaert W (1977) Viscoelastic aquifer model applied to subsidence due to pumping. Water Resour Res 13:597–604
Gambolati G, Freeze RA (1973) Mathematical simulation of the subsidence of venice: 1. Theory. Water Resour Res 9(3):721–733
Gambolati G, Gatto P, Freeze RA (1974) Mathematical simulation of the subsidence of venice: 2. Results. Water Resour Res 10(3):563–577
Gambolati G, Teatini P (2015) Geomechanics of subsurface water withdrawal and injection. Water Resour Res 51:39223955
Gray WG, Schrefler BA (2007) Analysis of the solid phase stress tensor in multiphase porous media. Int J Numer Anal Methods Geomech 31:541–581
Olivella S, Carrera J, Gens A, Alonso EE (1994) Nonisothermal multiphase flow of brine and gas through saline media. Transp Porous Media 15(3):271–293
Sorek S, Bear J, Ben-Dor G, Mazor G (1992) Shock waves in saturated thermo-elastic porous medium. Transp Porous Media 9:3–13
Terzaghi K (1943) Theoretical soil mechanics. John Wiley & Sons, Inc., New York, p 510
Verruijt A (1969) Elastic storage in aquifers. In: De Wiest RJM (ed) Flow through porous media. Academic Press, New York, pp 331–376
Verruijt A (2010) Soil mechanics. Delft University, Delft, p 330
Verruijt A (2014) Theory and problems of poroelasticity. Delft Univ of Technology, Delft, p 266

Appendix A
Selected Phenomena of Transport and Processes in Chemical Engineering by Raphael Semiat[1] and Jacob Bear

Various mathematical models of phenomena of transport of mass, energy and momentum in porous media have been presented in Chaps. 5–9. However, the presentation focussed mainly on phenomena that occur in *geological formations*, as encountered in Reservoir Engineering, Geo-hydrology, Agriculture Engineering and Soil Mechanics. Phenomena that are encountered and treated by Chemical Engineers have hardly been mentioned, although such phenomena are the bread and butter of chemical engineers. The presentation in the current appendix focuses on phenomena of transport of mass energy and momentum that occur in Chemical Engineering, making use of the same kind of mathematical models that have been presented throughout this book. We shall describe and discuss the phenomena and processes of transport that occur in reactors and demonstrate appropriate mathematical models that describe them.

The *reactor*, or the *porous bed* (abbreviated 'bed') considered here is an enclosed domain. often a column, within which physical, chemical, and thermodynamic phenomena occur under control achieved by controlling conditions imposed on the domain's boundaries. It is common to use the term reactor when *chemical reactions* occur, while the term *contactor* is used when the main objective of the setup is to enable and enhance processes that require *contact* between different phases. Our objective here is to demonstrate how the models presented and discussed throughout this book are used in the important discipline of Chemical Engineering.

With the objectives of this book in mind, we shall focus on enclosed domains, e.g., reactors, that contain *both* solids and fluids (i.e., liquids and gases), such that each of them can be envisioned and treated as a *continuum*, recalling the conditions (discussed in Sect. 1.1.2), that allow us to regard and treat a domain occupied by a solid and one or more fluids as "a porous medium continuum". Altogether, porous beds/reactors considered here are composed of *multiple overlapping (solid and fluid) continua*.

Obviously, this is not a text in Chemical Engineering, nor a comprehensive review on chemical processes in chemical reactors. Our objective is to present examples of

[1]Prof. Raphael Semiat is a Professor in the Dept. of Chemical Engineering at the Technion-Israel Institute of Technology, Haifa, Israel. His main field of expertise is separation processes in the process industry, with emphasis on desalination and water purification.

J. Bear, *Modeling Phenomena of Flow and Transport in Porous Media*,
Theory and Applications of Transport in Porous Media 31,
https://doi.org/10.1007/978-3-319-72826-1

a selected number of processes and examples of phenomena of flow and transport in a specific porous medium domains–the reactor–which is of interest to chemical engineers. The presentation is limited only to the major types of reactors used in the Chemical Engineering Industry. Only reactors which involve porous medium domains are included.

Process engineering is the common name for many industrial processes that are implemented in the chemical industry. Separation, cleaning, and generation of new materials, may serve as typical examples. The different processes take place in domains of various configurations, usually called columns, beds, *reactors*, or *contactors*. A reactor contactor is a domain surrounded by an impervious envelope, with inlets and outlets through the latter. Various instruments and devices are placed within the domain to monitor, serve and enhance processes. A special type of reactors is one in which at least part of its domain is occupied by a porous medium (as defined in Sect. 1.1.2). This means that the solid matrix and each of the fluids involved may be regarded as *continua*, and the continuum models are applicable. When applied to a considered process in a specific reactor, the models guide planning and operation decisions. We shall discuss only mass and energy balance equations. We shall not present the momentum balance equation for a fluid or a gas phase, as it is usually replaced by Darcy's law (for sufficiently small Reynolds numbers). Forchheimer equation (4.3.6), or Ergun's equation (4.3.17) are usually used for flow at high Reynolds numbers.

Although the macroscopic mass, momentum and energy balance equations have been presented throughout the book, for the convenience of the reader, let us repeat some of them here. Obviously, as for all other E-balance equations, the content of the equation to be used depends on the details of the considered case, i.e., on the "assumptions" that are made in each case, or on the *conceptual model* that underlies the description of what we envision takes place in the considered domain.

- **Mass balance equation of an α-fluid phase in two phase (w, nw) flow**:

$$\frac{\partial}{\partial t}(\phi S_\alpha \rho_\alpha) = -\nabla\cdot(\rho_\alpha \mathbf{q}_\alpha) + \rho_\alpha \Gamma'_\alpha(\mathbf{x}, t), \quad \alpha = w, nw, \tag{A.1}$$

where w, nw denote the wetting and non-wetting fluids, say liquid (ℓ) and gas (g), and $\rho_\alpha \Gamma'_\alpha(\mathbf{x}, t)$ represents mass injection of an α-fluid at points in the domain, and $\mathbf{q}_\alpha$ represents α-specific discharge, expressed by Darcy's law, e.g.:

$$\mathbf{q}_\alpha = -\frac{\mathsf{k}_\alpha(S_\alpha)}{\mu_\alpha} \cdot (\nabla p_\alpha + \rho_\alpha g \nabla z), \qquad q_\alpha \equiv \phi S_\alpha (\mathbf{V}_\alpha - \mathbf{V}_s), \tag{A.2}$$

where the solid's velocity, $\mathbf{V}_s$, is usually neglected, except in a fluidized or a mixed bed. The complete two-phase flow model is expressed, for example, by (6.4.3)–(6.4.6). At high Re, Forchheimer or Ergun's flux expressions are used.

The source term includes also the cases in which, as a result of chemical reactions, or added energy (at a considered macroscopic point), mass of an α-phase is added, per unit volume of the considered phase. As examples, we may mention a gas produced in a liquid phase because of chemical reactions in the liquid, and added heat at a point by microwave energy added at a point within a domain to the extent that phase β is changed to α-phase.

• **Mass balance equation of a γ-chemical species in an α-fluid phase in two-phase flow, without phase change**:

$$\frac{\partial \phi S_\alpha \rho_\alpha \omega_\alpha^\gamma}{\partial t} = -\nabla\cdot\phi S_\alpha \left(\rho_\alpha \omega_\alpha^\gamma \mathbf{V}_\alpha + \mathbf{J}_{\alpha,dif}^\gamma + \mathbf{J}_{\alpha,dis}^\gamma\right) + \sum_{\delta=\beta,s} f_{\delta\to\alpha}^\gamma + \phi S_\alpha \rho_\alpha \Gamma_\alpha^\gamma, \tag{A.3}$$

(see (7.3.3)) where $\rho_\alpha \omega_\alpha^\gamma (\equiv c_\alpha^\gamma)$ is the advective flux, and the two J-fluxes are due to *diffusion* and *dispersion*, respectively, $f_{\delta\to\alpha}^\gamma$ represents γ-mass transfer across interphase (microscopic) boundaries, and the last term on the r.h.s. represents sources, including ones that are due to chemical reactions. For example, for every α-phase:

$$\rho_\alpha \Gamma_\alpha^\gamma = \sum_{(j)} \left.\frac{d\rho_\alpha^\gamma}{dt}\right|_{\substack{j\text{th chem.}\\ \text{reaction}}} = M^\gamma \sum_{(j)} \nu_k^\gamma R_{r,j}|_{j,\gamma \text{ in } \alpha}, \tag{A.4}$$

where M^γ denotes the *molar mass* of the γ-component, and $R_{r,j}$ denotes the rate of the jth reaction. When considering chemical reactions, it is more convenient to write the γ-mass balance equation in terms of molar concentration. Accordingly, in this appendix, the concentration c^γ is measured in γ-moles per unit volume of solution. However, we shall more often use X^γ which denotes γ mole fraction.

• **Energy balance equation for multi-phase flow, with multiple γ-species and phase change**.

A number of energy balance equations are presented in Sect. 8.4. The energy balance equation is (8.4.15), repeated here as:

$$\begin{aligned}\frac{\partial}{\partial t}&\left[\sum_{(\alpha)} (\theta_\alpha \rho_\alpha u_\alpha) + (1-\phi)\rho_s C_s T\right]\\ &= -\nabla\cdot\sum_{(\alpha)} \theta_\alpha \left[\rho_\alpha h_\alpha \mathbf{V}_\alpha + \sum_{(\gamma)} h_\alpha^\gamma \mathbf{J}_{h,\alpha}^\gamma + \mathbf{J}_\alpha^{*H}\right]\\ &\quad + \nabla\cdot(\Lambda_{pm}^{*H} \nabla T) + \Gamma_{pm}^{\mathcal{E}},\end{aligned} \tag{A.5}$$

in which $\alpha = \ell, g$, the symbol C_s denotes the specific heat capacity of the solid, and the second sum on the second line is taken over all γ-species present in the considered phase. The source term in the above equation, $\Gamma_{pm}^{\mathcal{E}}$ expresses energy from a variety of sources, e.g., heat of reaction associated with chemical reactions that occur at the point, and latent heat, associated with phase change, added at the considered macroscopic point, per unit volume of porous medium.

In what follows, we shall present the basic mass and energy balance equations mainly in the form of PDE's. Obviously, to constitute a well posed problem, these equations have to be supplemented by appropriate initial and boundary conditions. However, in the chemical industry, another kind of model is often employed–a *multi-cell*, or *multi-compartment* model. In this kind of model, the flow domain is divided into "stages", "cells" or "compartments" of finite volume, and mass and energy balance equations are written for each stage or cell, in terms of variables of state that relate average values in a cell to properties of the streams that enter and leave it. The leaving streams are considered to be in equilibrium with respect to temperature and concentration of the involved phases. Actually, the division of a porous medium column (a reactor or contactor) into "stages" is not just for computational objectives; in many cases, the column is really divided into *stages*, divided by resistance-less porous plates. The underlying assumption is that each cell (including the solid matrix and the fluids in the void space) is continuously *well mixed* so that its state can be described by a single set of variables, like pressure, temperature, solute concentration, and fluid saturation (in case of two-phase flow). Obviously, all cells are subject to specified initial conditions and boundary cells are subject to specified boundary conditions. We assume that within each *stage* all phases and components are *well mixed*. The structure of this model is similar to a *finite difference model* of a PDE, or to the *Finite Volume Model* described in Sect. 3.8. We shall also present some models of this kind.

Following is a review of the major kinds of reactors and the processes that occur in them, as encountered in the Chemical Engineering industry. The presentation is limited to reactors which involve 'porous medium domains', i.e., their active domain contains a porous medium that may be treated as a continuum, as defined and considered throughput this book.

A.1 Types of Reactors

Essentially, a reactor, or a contactor, is made up of an impervious shell inside which we have a porous solid matrix, of one kind or another, and one or two fluids that move through the void space. The fluids are comprised of chemical species that interact with each other and with the solid matrix to achieve various goals.

A.1.1 Flow Regimes in a Reactor

The flow regime in a reactor, or a porous bed, may take one of the following forms:

- **Single phase flow**. This is the more common case: a single phase liquid or gas enters through one end of the reactor and leaves through the other end. When the reactor is a vertical column, flow may enter the lower end and flow upward, or through the upper one and flow downward.

- **Co-current flow in two-phase flow**. This term is used to describe the case where two phases (two liquids, or a liquid and a gas) are fed into a bed from the *same* end of the latter, i.e., both phases flow through the reactor in the same direction. The objective is to allow interaction (e.g., mass transfer) between them, mainly across their common *microscopic* interphase surfaces. In some cases, only a small fraction of one phase dissolves in the other, or is carried away with the other phase as small droplets.

  A reactor is designed such that it ensures a good contact (at the microscopic level) between the two fluids inside it. The fluid velocities are controlled in a way that prevents 'flooding', i.e., prevents one fluid from changing the flow direction of the other fluid. As the two fluids flow, heat and/or mass is transferred from one fluid to the other across their common microscopic interface. The *two film model* (Sect. 7.4.3) can be used to describe such transfers. The interphase transfer may be implemented with one of the fluids taking the form of drops that flow through the other (continuous) fluid, or as a wetting fluid that flows next to the solid surfaces, while the non-wetting fluid is moving in a counter-current direction next to it.
- **Counter-current flow**. Here, the two *immiscible phases*, usually two liquids, or a liquid and a gas, are fed through the opposite ends of the bed. We recall that, with respect to the solid matrix, one is the wetting fluid while the other is the non-wetting one. In this case, fresh fluid entering from one end meets the fluid that enters through the other end. Heat and/or mass transfer can take place between the two fluids across their common microscopic interfaces. The heavier fluid is fed from the top of the reactor, while the lighter one is fed from the bottom. The advantage of this configuration is that the rates of heat and/or mass exchange between the two phases is higher than in the case of co-current flow, while improved performance is obtained.

  This counter-current flow regime is used to enhance (1) adsorption-desorption activities between a gas and a liquid, (2) distillation, where volatile components are extracted from the less volatile fluid, and (3) liquid-liquid extraction (often called *solvent extraction*) processes.

Following is a description of the major two types of reactors.

A.1.2 Fixed Bed Reactors

This type of reactor is composed of a shell in the form of a pipe segment, or an elongated vessel, fully packed with solid particles. Often, only part of it is occupied by solid particles. We often refer to such domain as the 'packed bed'. The void space is fully occupied by a fluid that moves parallel to the domain's axis, i.e., in a (macroscopic) 1-d flow. The flow is governed by conditions imposed on the domain's boundaries. We usually refer to the liquid as the 'solvent'. As we shall see below, sometimes the void space is occupied by two interacting immiscible fluid phases that may flow in the same direction, or in opposite directions.

The solid matrix is usually composed of solid particles of a variety of shapes and sizes. These particles, or elements, have random shapes and sizes that can be described by some statistical size distribution. Or, they may have known well-defined shapes and sizes. The solid particles, may also have a variety of relevant properties, like solubility in the flowing solvent, adsorption, ionic exchange, or catalysis properties that may contribute to the formation of new materials from the chemical species dissolved in the moving fluid. When super-saturation conditions are reached, crystals may grow on the particles' solid surfaces, thus reducing the volume of the initial void space. In such cases, to avoid clogging, the bed is also made to move. Porosity may be reduced as a consequence of crystallization, or increased as a result of dissolution of the solid matrix.

The 'bed' has known relevant geometrical properties (usually determined experimentally), like porosity, specific surface area and permeability.

A *catalyst* may be a molecule dissolved in a phase or a solid particle that participates in a catalysis process. It has properties that are tailored for a specific reaction. The solid matrix itself may be *neutral*, i.e., not interfering in the process, except for affecting the local microscopic flow regime within the void space, or it may be *active*, e.g., as an adsorbent, ion exchanger, or catalyst. The process may occur in *semi batch mode*, where one of the streams is fed continuously, or in *continuous mode*, where the changed species are continuously removed from the system. In both cases, heat may be added through domain boundaries, or from the process itself, e.g., by exogenic chemical reactions. During the process, phase changes may occur as pressure and/or temperature vary. In some cases, a non-consolidated granular bed may become fluidized (see below) as a result of pressure changes, or by mechanical mixing, using an impeller or a pump.

A.1.3 Moving Bed Reactors

Solid particles are moving in two kinds of reactors: in a *fluidized bed* and in a *stirred bed*. In both cases, the solid particles comprising the solid matrix, are free to move with respect to each other, vibrate or rotate in the fluid environment. Occasionally, particles touch each other. The movement of the solid particles is controlled by the drag produced by the moving fluid surrounding them particles, by forces acting at contact points and by gravity.

Because in both the fluidized and the stirred bed, the bed itself is moving, the turbulence in the fluid passing through the reactor is increased, leading to an increased efficiency of mass and heat transfer activities.

The solid particles may continuously stay in the reactor, or exit the latter, as in fast catalytic reactors. In the latter case, new particles are fed into the reactor in order to compensate for the ones that leave.

A.1.4 Other Characteristics of Reactors

Following are some additional characteristics of reactors:

- The type of reactor in which they take place: fixed bed, or moving bed.
- The solid matrix participation in the process. The solid matrix may 'neutral', i.e., is not interfering with nor participating in the process, or 'active'. i.e., it may serve as an adsorbent, ion exchanger, or catalyst.
- The process that takes place. It may be 'semi-batch', where one of the phases moves continuously, while the other phase is fed only at the beginning of the process, or 'continuous mode', where fluids ate fed continuously.
- The number of fluid phases passing through the reactor: one or more.
- Co-current or countercurrent flow of the fluids.
- Heat may be added from the boundaries of the bed. It may be by exogenic reactions, or it may be generated from the process itself.
- Change of phase. This may occur during the process, and so are processes like evaporation, condensation, dissolution and crystallization.
- The bed may be fluidized as a result of pressure changes and increased fluid velocity. Fluidization may also occur during certain processes, like crystallization, or by mixing of the solid particles with the fluids.
- The kind of process. It may be a *batch process*, i.e., a process that involves fluid flow, heat and heat transfer operations, all performed on a fixed quantity of solid. The process is stopped when the operation's target has been reached. This type is used in laboratories or in small scale production units, mainly in fine chemicals and pharmaceutical industries.
- The process may be a *continuous* and *steady state*. The fluid and solid matrix properties stay unchanged in time at every point along the reactor, while concentration, temperature, pressure and other transport characteristics vary along the system. Fluid phases are fed continuously at constant rates. Again, as changes occur along the reactor, the products leaving the reactor are expected to have constant properties in time.

A.2 Processes in Fixed-Bed Reactors

In a fixed-bed reactor, operations may be performed (1) with a single fluid phase passing through the void-space, or (2) two phases that pass through the void-space in co-current, and in countercurrent (or reverse) flow, i.e., flowing in opposite directions. The solid matrix may be 'neutral', i.e., it does not interfere with the process, or it may participate in the process, e.g., as an adsorbent, as an ion-exchanger, or as a catalyst. The process may take place as *semi-batch* or in *continuous mode*. In all the above options, heat may be added through the boundaries of the bed, or internally, from the reaction itself. Change of phase may occur during the process. Altogether, the above possibilities may end up in a large number of options, especially if we add

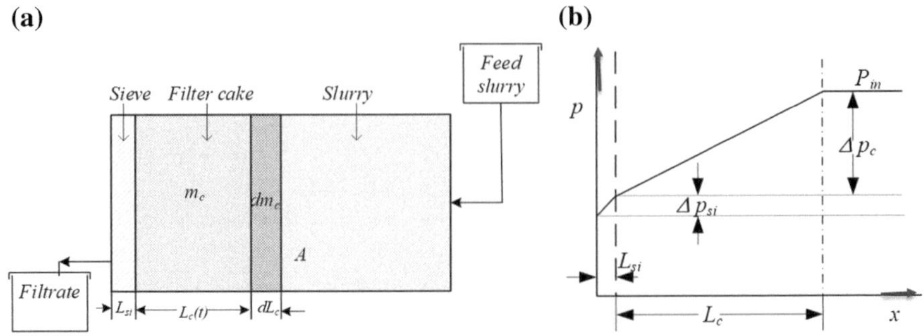

Schematic sketch of a filter. Pressure variation

Fig. A.1 Filtration

to the list the chemical species that interact inside the void-space, and sometimes with the solid matrix itself.

Following is a brief description of processes that occur in fixed-bed reactors. We shall first focus on processes that take place in single phase flow, and then present cases which involve two-phase flow.

A.2.1 *Processes in Single-Phase Flow*

A.2.2 *Filtration*

The objective of the process of *filtration* (or *cake filtration*) is to remove solid particles from a *slurry* (= a liquid with solid particles) stream. The process is performed in a horizontal or vertical column of a constant cross-sectional area A, into which the slurry is fed at one end and clear liquid (= *filtrate*) leaves at the other one.

Figure A.1a represents a typical filtration system setup. We note the slurry entering the system on the right side and leaves on the left side through a relatively thin porous medium domain-a filter, usually a kind of *cloth*, located just ahead of the outlet. It acts as a *filter domain* that prevents the passage of solid particles carried by the liquid. The stopped solid particles gradually accumulate next to the filtering domain, forming a "*filtrate cake*" of continuously growing thickness. The rate of flow of liquid and solid particles through the system is controlled by maintaining desired pressures (or piezometric heads, in the case of a vertical filter) at both ends of the column. The pressure maintained at the outlet is usually atmospheric. Sometime vacuum is applied as a boundary condition at the exit side. However, it is possible that tiny solid particles will penetrate the filter, reducing its permeability, so that the filter may have to be washed or replaced from time to time.

Altogether, the flow system within the column is composed of two porous medium layers: (1) a thin sieve (e.g., cloth, or porous metal), that prevents the solid particles carried by the slurry from passing through, and (2) a *filtrate cake* that is a saturated porous medium domain, with the solid matrix made up of the solid particles stopped by the filter, The thickness of the latter layer grows with time as more solid particles are stopped. We may regard this system as composed of two 'resistances in series' to the flow. As the thickness of the filtrate cake grows, so does the loss of (pressure or piezometric) head through it. The consequence is that the pressure at the inlet has to be increased in order to maintain an economic flow rate. It is important to note that the filtering operation can take place under constant pressure, so that the volumetric flow rate is decreasing in time, or at constant flow rate by gradually increasing the pressure at the inlet. A combination of the two modes is also possible

When the thickness of the filtrate cake reaches a certain value, or when the required pressure at the inlet becomes too high, the process is stopped, the filtrate cake is removed and the sieve is washed.

Consider the horizontal filter of constant cross-sectional area A shown in Fig. A.1, where the slurry is fed from the right side. At this end, we may either control the flux of the slurry fed into the system, or the pressure at the inlet point. In what follows, we shall assume the latter case, i.e., p_{in}, is maintained unchanged. A fixed known pressure, p_{out}, is also maintained at the downstream end of the filtration setup, where only fluid exits. Sometimes, a vacuum is maintained there.

The pressure changes along the filter (assumed horizontal) may be expressed as the sum of three pressure increments::

$$\Delta p = p_{in} - p_{out} = (\Delta p)_{si} + (\Delta p)_{ck} + (\Delta p)_{sl}, \tag{A.6}$$

where Δp represents the overall pressure drop across the filter, with $(\Delta p)_{si}$, $(\Delta p)_{ck}$ and $(\Delta p)_{sl}$ denoting the pressure drops across the sieve, the filtrate cake and the incoming slurry domains, respectively. We neglect the pressure drop in the slurry, as $(\Delta p)_{sl} \ll (\Delta p)_{ck}$. We denote the length of the filtrate cake by $L_c = L_c(t) (\gg L_{si})$. We also neglect fluid compressibility, and use x to denote length along the filtrate cake.

• **Sieve**. Along the sieve of constant thickness (ℓ_{si}) and permeability (k_{si}), the pressure drop is:

$$(\Delta p)_{si} = \ell_{si} \frac{\mu}{k_{si}} q(t), \quad q \equiv \frac{Q(t)}{A}. \tag{A.7}$$

Recall that q is taken as positive in the direction of $-x$. Usually, q is calculated by the Kozeny–Carmen law (see below).

Let us introduce the coefficient $\mathbb{R}_{si}$, associated with the flow through this sieve:

$$q = \frac{k_{si}}{\mu} \frac{(\Delta p)_{si}}{\ell_{si}} \quad \rightarrow \quad (\Delta p)_{si} = q \mathbb{R}_{si}, \quad \mathbb{R}_{si} = \frac{\mu \ell_{si}}{k_{si}}, \tag{A.8}$$

where $\mathbb{R}_{si}$ is a constant, independent of q, as long as cake porosity remains unchanged. Along the sieve, the pressure increases linearly by $p_f(t) = q(t)\mathbb{R}_f$.

• **Filtrate cake**. At any instant of time, the mass of solids, $m_c(t)$, accumulated as a *filtrate cake* of length $\ell_c(t)$, at porosity ϕ. This mass, per unit area of the reactor's cross-section, is:

$$dm_c(t) = (1-\phi)\rho_s dl_c, \quad \rightarrow \quad m_c(t) = (1-\phi)\rho_s l_c(t). \tag{A.9}$$

The cake's length is usually small, say in the range of a few millimetres or centimeters. Thus, under certain conditions, the filtrate cake is assumed to be uniform in terms of porosity and permeability.

We assume that the flow is laminar and that the fluid's flux, q, through both the filter domain and the cake is described by the *Kozeny–Carman equation* discussed in Sect. 4.1.2. With $q > 0$ in the direction of $+x$, as indicated on the figure, we have:

$$q = \frac{k_c}{\mu}\frac{\partial p}{\partial x} = \frac{C_o\phi^3}{\mu M_s^2(1-\phi)^2}\frac{\partial p}{\partial x}, \quad \text{where} \quad k_c = \frac{C_o\phi^3}{M_s^2(1-\phi)^2}, \tag{A.10}$$

in which μ denotes the fluid's viscosity, C_o is a constant coefficient, referred to as Kozeny's constant, and $M_s (= S_v/\mathbb{V}_{pm})$ is the specific surface area (= surface area of solid particles per unit volume of porous medium). The pressure gradient along the cake is:

$$\frac{\partial p_{ck}}{\partial x} = \frac{q\mu}{k_c} = q\mu\frac{M_s^2(1-\phi)^2}{C_o\phi^3}, \quad q = q(t), \tag{A.11}$$

where p_{ck} denotes the pressure along the cake.

As solid particles accumulate as a filtrate cake, the latter's length, ℓ_c, increases. With dn_c expressed by (A.9), we have

$$dm_c = \rho_s(1-\phi)d\ell_c, \quad \rightarrow \quad d\ell_c = \frac{dm_c}{\rho_s(1-\phi)} = \frac{k_c}{q\mu}dp_{ck}, \tag{A.12}$$

leading to:

$$(\Delta p)_{ck} = \frac{q\mu}{\rho_s}\frac{M_s^2(1-\phi)}{C_o\phi^3}m_c = q\mathbb{R}_c m_c, \quad \mathbb{R}_c = \frac{\mu M_s^2(1-\phi)}{\rho_s C_o\phi^3}, \tag{A.13}$$

in which $\mathbb{R}_c$ denotes a coefficient associated with the flow though the cake. This coefficient can be determined *experimentally*. Recall that here we have assumed that the filtrate cake is incompressible: its porosity, ϕ, is constant in space and time.

We can now combine the total known pressure drop at time t, across the sieve and across the filtrate cake:

$$\Delta p = p_{in} - p_{out} = (\Delta p)_{si} + (\Delta p)_{ck} = q(\mathbb{R}_{si} + \mathbb{R}_c m_c). \tag{A.14}$$

With $\mathbb{V}_{sl}$ denoting the slurry volume per unit area of reactor, its specific discharge, q_{sl}, is expressed by:

$$q_{sl} = \frac{d\mathbb{V}_{sl}}{dt}. \tag{A.15}$$

The mass of solid particles accumulating in the cake can be expressed as:

$$m_c = \mathbb{V}_{sl} c_{sl}, \tag{A.16}$$

where c_{sl} denotes the (known) concentration of solids in the incoming slurry, in terms of mass of solid per unit volume of slurry.

Expressing q_{sl} by (A.13), and m_c by (A.14), we obtain:

$$\frac{dt}{d\mathbb{V}_{sl}} \equiv \frac{1}{q_{sl}} = \frac{1}{\Delta p} \left(\mathbb{R}_{si} + \mathbb{R}_c \mathbb{V}_{sl} c_{sl} \right). \tag{A.17}$$

Integration of (A.17) yields:

$$\int_0^t dt = \frac{1}{\Delta p} \int_0^{\mathbb{V}_{sl}} \left(\mathbb{R}_{si} + \mathbb{R}_c \mathbb{V}_{sl} c_{sl} \right) dV_{sl} \tag{A.18}$$

leading to:

$$t = \frac{\mathbb{V}_{sl}}{\Delta p} \left(\mathbb{R}_f + \tfrac{1}{2} \mathbb{R}_c \mathbb{V}_{sl} c_{sl} \right) \tag{A.19}$$

This result describes the growth of the filtrate cake with time and volume of the slurry passing through the filter. By plotting $t/\mathbb{V}$ as function of $\mathbb{V}$, we get a straight line. Its intercept on the vertical $t/\mathbb{V}$-axis allows us to calculate $\mathbb{R}_{si}$, while the slope of the line allows us to calculate the value of $\mathbb{R}_c$.

Figure A.1 show the pressure variation along the filter: we note the pressure drop along the sieve and along the cake filtrate.

So far, we have assumed that the cake is incompressible, so that a constant ϕ could be employed. However, in some cases, the cake is *compressible*, so that ϕ is replaced by $\bar{\phi}_c = \phi_c(\bar{p}_{ck})$, where $\bar{p}_c$ denoting the (variable) average pressure within the cake. According to Kozeny–Carman, k_c, which depends on ϕ, also varies with x, t. The porosity may be reduced when the cake is compressible and also when newly arriving small particles enter the void space between particles already in the cake.

Designing a filtration unit means calculating the filter area, the cake's thickness and the operational pressure. When the cake is compressible, it is common to determine the dependence of $\mathbb{R}_c$ as a function of the operational pressure and take an average $\mathbb{R}_c$-value, which is already associated with an average ϕ-value.

A.2.3 Dissolution

In this process, we enhance the dissolution of a solid material (in the form of solid particles) in a liquid. The objective is to add the dissolved solid as a component in the liquid. An example is adding calcium carbonate salt to desalinated water.

Consider a liquid that flows through a cylindrical column that contains a porous medium made up of (consolidated or non-consolidated) solid particles, or solid elements of various shapes.

The liquid discharge through the porous column (also referred to as "bed") is maintained at a constant rate, Q_ℓ (= liquid volume per unit time), entering at $z = 0$ and leaving at $z = L$. This can be achieved by a pump delivering the constant discharge. Another flow regime can be achieved by maintaining fixed pressures at $x = 0$ and $x = L$. Liquid may flow upward or downward. Although there may be some effects of the space next to the side-walls of the column, we assume that the flow and transport in this bed are *one-dimensional*, upward or downward, along the bed's's axis.

The solid matrix is made up, entirely, or partly, of a chemical species γ, which can dissolve in the moving liquid. Let the concentration in the latter be denoted by $c_\ell^\gamma = c_\ell^\gamma(z, t)$. The liquid's density varies with c^γ, say, $\rho_\ell \approx \rho_{o\ell}(1 + c_\ell^\gamma)$, where c denotes the concentration of the γ-species dissolved in the liquid.

The liquid's density, $\rho_\ell(z, t)$ obeys the liquid's mass balance equation:

$$\frac{\partial \phi \rho_\ell}{\partial t} = -\frac{\partial \phi \rho_\ell V}{\partial z}, \qquad V = \frac{Q}{\phi A_{column}}, \tag{A.20}$$

$$\rho_\ell(z, t) = \rho_{o\ell}\left(1 + \frac{c_\ell^\gamma(z, t)}{c_{\ell.o}}\right). \tag{A.21}$$

The concentration of dissolved γ in the liquid obeys the γ-mass balance equation (7.3.1), which here takes the form:

$$\frac{\partial \phi c^\gamma}{\partial t} = -\frac{\partial}{\partial z}\left[\phi(c^\gamma V + J_{adv}^\gamma + J_{dif+dis}^\gamma)\right] + f_{s\rightarrow\ell}^\gamma, \tag{A.22}$$

where, the symbol $\mathbf{J}^\gamma$ denotes a flux of γ per unit *area of fluid* in the cross section.

From (7.2.17) and (7.2.33), we have:

$$J_{dif+dis}^\gamma = -(\mathcal{D}_{dif}^* + a_L V)\frac{\partial c^\gamma}{\partial z}, \tag{A.23}$$

where a_L denotes the coefficient of longitudinal hydrodynamic dispersion. The symbol $f_{s\rightarrow\ell}^\gamma$ denotes the rate at which solid mass is dissolved, i.e., γ moves from solid (s) to liquid (ℓ) per unit volume of porous medium, per unit time. Following (7.4.40), we may express this f-value in the form:

$$f^{\gamma}_{s\rightarrow\ell}(z,t) = K^{\gamma}_{s\rightarrow\ell}(c^{\gamma}_{sol} - c^{\gamma}(z,t)), \tag{A.24}$$

in which c^{γ}_{sol} denotes the solubility limit of the solid in the liquid, and the coefficient $K^{\gamma}_{s\rightarrow\ell}$ needs to be determined experimentally, or as an expression taken from a known correlation between the dimensionless mass transfer coefficient and physical and flow properties.

The initial porosity, as well as the initial column length are known.

For practical cases, the designer needs to design a column that will be operated continuously, namely, more dissolvable solid should be added, say, once a day. Solid particles are added (continuously, or in batches) to maintain (more or less) a constant column length. Then the pressure drop across the column is not significantly affected by the dissolution process. This also implies that possible changes of porosity, due to the dissolution process, may be neglected.

Assuming that the porosity is practically unchanged, the above set of equations is sufficient to determine the time-dependent concentration of the liquid leaving the bed. In fact, we have here a single degree of freedom–the concentration, $c^{\gamma}(z,t)$, for which we have to solve the single PDE (A.22).

Let us add here a comment on porous medium porosity. In practice, as the solid dissolves, both the porosity and the length of the bed are changed. We may consider two cases: one is such that when solid dissolution occurs, ϕ increases, but the total length and volume of the porous medium domain remains (practically) unchanged. This case occurs when the solid matrix is a single consolidated porous solid domain. The other is the case of unconsolidated particles (e.g., spheres). As each solid particle dissolves, its size is reduced. The solid particles continuously move and the bed shrinks and consolidates. The porosity may then vary with time. If fresh solid particles are not added, the operation is stopped at a certain length of the solid matrix.

The case of a solid matrix made up of non-consolidated granules may serve as an example. As the granules dissolve, they move, and the column consolidates, unless new grains are continuously added from above. In principle, we then have $\phi = \phi(z,t)$ and $L = L(z,t)$.

Let us assume that, initially, the solid particles are well mixed of all sizes such that we can start from a known specific surface area $s = s_o$. As dissolution occurs, and using $f_{s\rightarrow f}$ as an expression for the mass of dissolved γ-species on the solid, per unit volume of porous medium, we have:

$$f^{\gamma}_{s\rightarrow f} = \frac{\Delta m^{A}_{s}}{\mathbb{V}_{pm}\Delta t} \tag{A.25}$$

Neglecting solid and fluid compressibility, the fluid's mass balance equation reduces to:

$$\nabla\cdot\phi\rho_{\alpha}\mathbf{V}_{\alpha}(\equiv\nabla\cdot\rho_{\alpha}\mathbf{q}_{\alpha}) = 0. \tag{A.26}$$

The solute mass balance equation for $c_\alpha = c_\alpha(\mathbf{x}, t)$, reduces to:

$$\phi \frac{\partial c_\alpha^\gamma}{\partial t} = -\phi \mathbf{V}_\alpha \cdot \nabla c_\alpha^\gamma - \nabla \cdot (\mathbf{D}'_\alpha \cdot \nabla c_\alpha^\gamma) + f_{s \to \alpha}^{m^\gamma} + \phi \rho_\alpha \Gamma_\alpha^{m^\gamma}. \tag{A.27}$$

into which we can now insert appropriate expressions for the transfer of γ from the solid to the fluid, as well as source terms that express the production of γ in the fluid, say by chemical reactions.

A.2.4 Adsorption

The phenomenon of adsorption is discussed in Sect. 7.4.1. In a reactor, the objective of adsorption is to remove a dissolved species from a liquid (or a certain species from a gas) by the mechanism of adsorption of that species (= the *adsorbate*) on the solid surface (= the *adsorbent*). To achieve this goal, the chemical species in the fluid that enters the porous bed should have physical or chemical affinity to the solid comprising the bed. We assume that equilibrium is quickly reached *at the interface between the liquid and the solid's surface*. The liquid here is represented by a *thin liquid film* next to the solid (also introduced in Sect. 7.4.1). As the process progresses, more adsorbent settles on the solid until a complete or close to complete chemical saturation is achieved. At any macroscopic point within the bed, this situation develops as the concentration in the fluid changes, while satisfying the solute's mass balance equation, and that on the solid reaches equilibrium with the fresh arriving fluid. The adsorption process takes advantage of the large internal specific area of the solid matrix.

It is interesting to note that an isotherm, like Henry's law, expresses a kind of equilibrium. The adsorption driving force is proportional to the difference in concentration between the fluid and the solid, noting that one expresses solute mass per unit volume of solution, while the other is per unit area of solid surface.

The model that describes the adsorption process contains two mass balance equations: one for the considered chemical species in solution and one of that species adsorbed on the solid. We also need an equation that relates the two concentrations to each other. Usually, it is assumed that *on the fluid-solid interface* equilibrium is reached between the two concentrations (Sect. 7.4.1). This relationship is similar to that of the dissolution process, except that instead of the constant concentration on the solid surface, an equilibrium equation is used. Different *experimentally determined adsorption equilibrium relationships* may be used, according to the relationship between the adsorbate and the adsorbent, i.e., the isotherm.

In an adsorption column, liquid is pumped through the porous medium, either in a batch operation, i.e., a fixed volume of fluid moving once through the bed, or as continuous feed operation, until all (or most) of the adsorption sites on the solid's surface have been occupied by the adsorbate and the concentration in the fluid has reached the required low level.

An example is the removal of benzene from air, using a porous bed of *active carbon*. Another example is the removal of phosphate from water by using nano-particles of iron oxide/hydroxide that are preloaded on a bed of *silica gel*.

The mathematical model that describes adsorption involves the variables: c^γ_{fl}, c^γ_{fi}, F and $f^\gamma_{fl\to s}$, where we note the distinction between the concentrations c_{fl} in the fluid and c_{fi} in the film next to the solid. The model has already been presented in Sect. 7.4.1 C. Repeated here for convenience, this model includes:

• **Mass balance for the γ-adsorbate in the fluid (fl)**:

$$\phi\frac{\partial c^\gamma_{fl}}{\partial t} = -\nabla\cdot\phi\left(\mathbf{V}_{fl}c^\gamma_{fl} + \mathbf{J}^\gamma_{fl,dif} + \mathbf{J}^\gamma_{fl,dis}\right) - f^\gamma_{fl\to s}, \tag{A.28}$$

in which the dispersive+diffusive γ-flux is expressed by:

$$\phi\left(J^\gamma_{fl,dif} + J^\gamma_{fl,dis}\right) = -\phi\mathbf{D}'\nabla c^\gamma_{fl}, \tag{A.29}$$

and the fluid's flux, $\phi\mathbf{V}_{fl}$ is determined by Darcy's law.

• **Mass balance equation for the adsorbed species on the solid**:

$$\frac{\partial(\rho_b F^\gamma)}{\partial t} = f^\gamma_{fl\to s} + \rho_b\Gamma^\gamma_s, \tag{A.30}$$

where the last term on the r.h.s. denotes the rate at which mass of a γ-species is produced on the solid, per unit volume of porous medium (= volume of reactor chamber).

• **Rate of γ-mass transferred from the fluid to the solid**:

$$f^\gamma_{fl\to s} = \kappa^\gamma_{fl\to s}\left(c^\gamma_{fl} - c^\gamma_{fi}\right), \tag{A.31}$$

in which the coefficient $\kappa^\gamma_{fi\to s}$ is an experimentally determined film-to-solid mass transfer coefficient. Note that we have to be careful with the difference in concentrations *in* the fluid/film and *on* the solid, as F^γ is measured in terms of adsorbed γ mass per unit *mass* of solid matrix, whereas the concentrations c^γ_{fl}, c^γ_{fi} are measured as solute mass per unit volume of solution.

• **The isotherm that expresses the assumption of equilibrium between the concentrations in the film and on the solid**, for example:

$$c^\gamma_{fi} = K_{fi\to s}\frac{\rho_s}{\breve{e}}F^\gamma. \tag{A.32}$$

When a regeneration process of the adsorbent is conducted for the latter's reuse, a similar reverse technique may be implemented with another solvent in order to clean the adsorbent. However, in some cases, e.g., in the case of adsorption of organic chemicals on active carbon, the cleaning process is performed under high temperature and controlled oxygen in order to recover the active carbon for reuse.

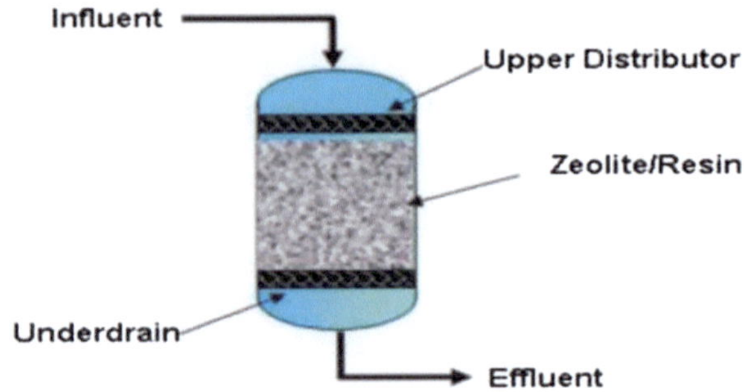

Fig. A.2 Ion exchange reactor

A.2.5 *Ion Exchange*

On exchange is discussed in Sect. 7.4.2. Figure A.2 shows, schematically, a typical ion exchange reactor.

The target is to replace an unwanted ion in an aqueous solution by another ion that is initially present (or preloaded) on the solid surface, e.g., a resin. The exchange process occurs between the solid and a liquid (e.g., an aqueous solution); the less desired ion in the solution is exchanged for a more desirable one. The process ends by washing the bed with a solution containing a high concentration of the required ion. In fact the washing/reloading process is exactly the same process with the opposite ions. This is a batch or semi continuous process; it ends when most of an unwanted ion is replaced, according to the design, by the other ion.

In a cation (*cat*) exchange process, cations are replaced by cations. It is possible to remove all kinds of cations, e.g., Ca^{+2}, or anions, e.g., F^-. An example is the exchange of cations during water treatment for hardness removal. In this process, calcium ions are exchanged by preloaded H^+ ions. This sorption-desorption reaction may be described by:

$$2HR_{cat} + Ca^{+2} \rightarrow Ca(R_{cat})_2 + 2H^+. \tag{A.33}$$

In order to remove a Cl-ion, a negatively charged ion is needed:

$$R_{an}OH + Cl^- \rightarrow R_{an}Cl + OH^-. \tag{A.34}$$

Here R^-_{cat} represents a site on the ion exchange resin that can adsorb a cation, while R^+_{an} is a site that can adsorb an anion.

Obviously, resin materials have a finite exchange capacity that is reached when all exchange sites will be occupied by the adsorbed ions. When the process ends, the resin is regenerated to its original condition.

The mathematical model of the process which involves the exchange of ion δ on the solid (s), by ion γ in the liquid (ℓ), includes the following equations (written, for simplicity, in a 1-d formulation):

• **Mass balance of γ in the liquid (ℓ)**:

We write (7.3.1) in the form:

$$\frac{\partial \phi c_\ell^\gamma}{\partial t} = -\frac{\partial}{\partial z}\left[\phi(c_\ell^\gamma V_f + J_{\ell,dif+dis}^\gamma)\right] - f_{\ell\to s}^\gamma, \tag{A.35}$$

where the symbol $J_{\ell,dif+dis}^\gamma$ denotes a flux by diffusion and dispersion in the z direction of γ per unit *area of liquid* in the cross section, and $f_{\ell\to s}^\gamma$ denotes the rate of γ-mass transferred from the liquid to the solid, per unit volume of porous medium, per unit time. The solute's sum of diffusive and dispersive flux is given by (A.35).

• **Mass balance of adsorbed γ on the solid (s)**:

We write (7.3.1) in the form:

$$(1-\phi)\rho_s\frac{\partial F^\gamma}{\partial t} = f_{\ell\to s}^\gamma, \tag{A.36}$$

where F^γ denotes the mass of γ per unit mass of solid matrix, so that each side expresses added mass of γ per unit volume of porous medium

• **Solute-to-solid mass transfer rate**:

The rate at which γ-mass moves from the liquid (ℓ) to the solid (s), per unit volume of porous medium, per unit time, is expressed by:

$$f_{\ell\to s}^\gamma(z,t) = \phi\kappa_{\ell\to s}^\gamma\left(c_\ell^\gamma - c_{fi}^\gamma\right), \tag{A.37}$$

in which c_{fi}^γ denotes film concentration, and the coefficient $\kappa_{\ell\to s}^\gamma$ is the γ-ion exchange mass transfer coefficient.

• **Equilibrium relationship**:

In addition, we assume that the concentration in the bulk liquid is in equilibrium with the concentration on the solid's surface, described, for example, by the Langmuir equilibrium isotherm (e.g., Hokanson 2004):

$$\frac{F^\delta}{F_o^\delta} = \frac{K_{Lang}^\gamma F^\gamma c_{fi}^\gamma}{1 + K_{Lang}^\gamma F^\gamma c_{fi}^\gamma}, \tag{A.38}$$

where K_{Lang}^γ is the Langmuir coefficient that depends on porosity and solid's density. Altogether, we have 4 equations for the four variables: c_ℓ^γ, c_{fi}^γ, F^γ, and $f_{s\to\ell}^\gamma$.

Obviously, the process will continue as long as there are δ-sites to be taken by adsorbing γ ions. The number of δ-sites on the solid will be part of the initial conditions.

A.2.6 Chromatography

Chromatography (or *partition chromatography*) is a separation process that is mostly used in the chemical industry for identification of different chemical species in a fluid sample, say for analytical purposes. It is also used for the separation of chemical species that are difficult to separate by common industrial techniques. The method is based on the fact that different molecules have different affinities to a solid surface (here inside a porous medium). The objective is to remove complicated chemical, or pharmaceutical compounds, from a fluid moving through the void space of a porous medium in a reactor.

The process is conducted in a reactor that has the form of a long tube containing a bed of *porous grains* that have a wide range of adsorption capabilities on the internal surfaces of the grains. A mass of fluid (= eluent) containing the kinds of molecules to be separated is fed at the entrance to the reactor tube and is directed to flow through the latter. The different components have different affinities to the *adsorbent* (= the solid surface) and, hence, under the eluent's pressure, they move along the bed at *different velocities* (see Sect. 7.4.1 C). This causes the different kinds of molecules to exit the pipe at a different times. In this way, separation is achieved.

Consider the case in which, initially we have the species γ and δ on the solid (s) surface. The objective it to replace δ ions by γ ions present in the liquid (ℓ); the replaced ions move to the liquid.

The model that describes the flow and chemical changes along the bed (i.e., in the $+x$ direction) is based on the same ideas as those underlying the ion exchange model described above. It includes the following elements:

• **Variables**: The concentrations c_ℓ^γ, c_ℓ^δ in the liquid, measured as moles per unit volume of liquid, and F_s^γ, F_s^δ on the solid, measured as moles per unit mass of the solid matrix. In the film introduced above, we have the concentrations c_f^γ, c_f^δ. In additions, we consider the rates of transfer (in moles per unit volume of liquid) $f_{\ell\to s}^\gamma$ and $f_{\ell\to S}^\delta$ also as variables that appear in the mass balance equations. Altogether, we have 6 concentration variables and 2 rates of transfer variables.

• **Mass balance equations**:
For γ in the liquid:

$$\frac{\partial \phi c_\ell^\gamma}{\partial t} = -\frac{\partial}{\partial x}\left[\phi\left(c_\ell^\gamma V_\ell + J_{\ell,dif}^\gamma + J_{\ell,dis}^\gamma\right)\right] + f_{s\to\ell}^\gamma, \tag{A.39}$$

$$\frac{\partial \phi c_\ell^\delta}{\partial t} = -\frac{\partial}{\partial x}\left[\phi\left(c_\ell^\delta V_\ell + J_{\ell,dif}^\delta + J_{\ell,dis}^\delta\right)\right] + f_{s\to\ell}^\delta, \tag{A.40}$$

where J and V denote liquid flux and liquid velocity in the $+x$-direction, and we have assumed no sources (e.g., due to chemical reactions).

For γ on the solid:

$$\frac{\partial(\rho_b F^\gamma)}{\partial t} = f_{\ell\to s}^\gamma, \qquad \frac{\partial(\rho_b F^\delta)}{\partial t} = f_{\ell\to s}^\delta. \tag{A.41}$$

• **Expressions for interphase transfer**:

$$f^{\gamma}_{\ell\to s} = f^{\gamma}_{\ell\to fi} = \phi\kappa^{\gamma}_{\ell\to s}\left(c^{\gamma}_{\ell} - c^{\gamma}_{fi}\right). \tag{A.42}$$

$$f^{\gamma}_{s\to\ell} = f^{\gamma}_{fi\to\ell} = \phi\kappa^{\delta}_{s\to\ell}\left(c^{\delta}_{fi} - c^{\delta}_{\ell}\right), \tag{A.43}$$

with $\kappa^{\gamma}_{\ell\to s} \neq \kappa^{\delta}_{s\to\ell}$.

• **Equilibrium described by a liquid-solid isotherm**.

We assume that equilibrium exists between the concentrations of γ in the film and on the solid, expressed, for example, by the *equilibrium isotherm*

$$F^{\gamma} = K^{\gamma}c^{\gamma}_{\ell}, \quad F^{\delta} = K^{\delta}c^{\delta}_{\ell}, \tag{A.44}$$

where K^{γ}, K^{δ} are experimentally determined coefficients.

Note that we have focussed here on the chemical aspect only. The liquid velocity and its dependence on pressure in not included here.

A relatively simple solution for what happens in a chromatographic column is based on dividing the latter into N cells of volume $\mathbb{V}_n, n = 1, 2, \ldots, N$, in each of which a transfer of chemical species between the solid and the eluent takes place. The γ-concentration in the liquid is denoted by c^{γ}_{ℓ}, while the γ concentration on the solid is denoted by F^{γ}.

• **The γ-mass balance in the fluid in the nth cell**. This equation takes the form:

$$\phi\mathbb{V}_n\frac{dc^{\gamma}_n}{dt} + (1-\phi)\rho_s\mathbb{V}_n\frac{dF^{\gamma}_n}{dt} = Qc^{\gamma}_{n-1} - Qc^{\gamma}_n, \tag{A.45}$$

recalling that:

$$c^{\gamma}_{\ell} = \frac{m^{\gamma}_{\ell}}{\mathbb{V}_{\ell}} = \frac{1}{\phi}\frac{m^{\gamma}}{\mathbb{V}_{pm}}, \quad F^{\gamma} = \frac{M^{\gamma}_s}{m_s} = \frac{1}{\rho_s(1-\phi)}\frac{m^{\gamma}_{\ell}}{\mathbb{V}_{pm}}.$$

where Q denotes the rate of flow through the cells, c^{γ}_n denotes the γ concentration in the liquid occupying the n'th cell, Q_n is the flow rate through cell n, while F^{γ}_n is the concentration of the same species on the solid surface, measured in γ-mass per unit cell mass.

The equilibrium of γ between the two phases is given by:

$$F^{\gamma}_n = Kc^{\gamma}_n. \tag{A.46}$$

By rearranging the two above equations, it is possible to show that

$$\frac{dc^{\gamma}_n}{d\tau} = c^{\gamma}_{n-1} - c^{\gamma}_n, \tag{A.47}$$

where

$$\tau = \frac{tQ}{[\phi + (1-\phi)K]\mathbb{V}_n \frac{dc_n^\gamma}{dt}} \tag{A.48}$$

The initial conditions are:

$$\text{For} \quad \tau < 0, \quad c_n^\gamma = 0, \quad n = 1, 2, N,$$

$$\text{For} \quad \tau = 0, \quad c_o^\gamma = c_f^\gamma, \Rightarrow \quad c_o^\gamma = c_f^\gamma \delta(\tau).$$

The solution is obtained by using the Laplace transform:

$$sY_n^\gamma = Y_{n-1}^\gamma - Y_n^\gamma. \tag{A.49}$$

Calculating from cell to cell, leads to:

$$\begin{aligned} sY_1^\gamma &= c_f^\gamma - Y_1^\gamma \Rightarrow Y_1^\gamma = \frac{c_f^\gamma}{s+1}, \\ sY_2^\gamma &= Y_1^\gamma - Y_2^\gamma \Rightarrow Y_2^s\gamma = \frac{c_f^\gamma}{(s+1)^2}, \\ \ldots \quad & \ldots, \\ Y_n^\gamma &= c_{feed}^\gamma \frac{\tau^{n-1}e^{-\tau}}{(s+1)^n}, \end{aligned} \tag{A.50}$$

or:

$$c_n^\gamma = c_{feed}^\gamma \frac{\tau^{n-1}e^{-\tau}}{(n-1)!}. \tag{A.51}$$

Since $\mathbb{V} = \sum_n \mathbb{V}^n$, we have:

$$c_n^\gamma(n,t) = \frac{c_f^\gamma}{(n-1)!}\left(\frac{tQN}{\mathbb{V}(\phi + (1-\phi)K)}\right)^{(n-1)} \exp\left(\frac{tQN}{[\mathbb{V}(\phi + (1-\phi)K)]}\right). \tag{A.52}$$

This expression provides the distribution of concentration c_n^γ at time t in every cell n along the pipe. It also represents the c_n^δ-distribution, which may be different from that of γ. On a long tube, we reach complete separation between the two components.

A.2.7 Drying

A tray of wetted solid particles, or carrying a wetted consolidated solid matrix, is dried under heating conditions. Figure A.3 shows a typical drying arrangements In this figure, we note the trays carrying wet solid particles. The trays are imbedded in a stream of hot air.

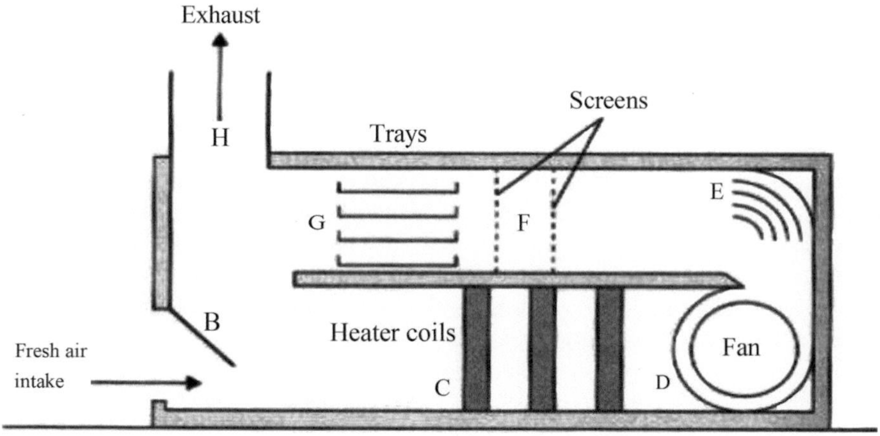

Fig. A.3 Drying tray

Figure A.3 shows a cabinet type tray drier. Initially, the layer of wet solid particles on the trays, at irreducible wetting fluid saturation, is what remains in a chemical reactor at the end of a filtration, or a sedimentation process. It still contains the liquid solution that remained (at irreducible liquid saturation) at the end of the last stage. The liquid that remains in the pore space may also be a solution that was used to wash the solids.

Within the bed, a liquid may occur as a free liquid as well as in the form of a liquid that adheres to the solid as a free unbound liquid , i.e., not chemically attached to the solid. The latter will evaporate at a certain temperature. In addition, some liquid in the cake may also be chemically attached to the solid, like in hydrates, or in clathrates.

The heat supplied to the bed causes evaporation of the liquid remaining in the void space, including the layer of liquid that adheres to the solid surfaces. The generated vapour is removed by the hot gas stream that travels around and through the bed's void space. Heat may be added by (1) conduction from the bed's external surface, (2) convection of hot gases (air) through the void space, or (3) radiation, either from above the bed's outer boundary, or directly into the bed by using micro-waves. The supplied heat will first heat the wetted cake to above the evaporation temperature at the operational pressure (sometime under vacuum). Then, vapour starts to leave the surface, and is pumped out by a vacuum pump, or by the hot gases flowing next to the tray surface.

The bulk liquid in the void space will evaporate first. Then, the liquid that is bound to the solid will start to evaporate. This takes place under a temperature that increases with time and distance from the solid surface, as it requires higher temperature and higher energy. At the end of the process, there is always some wetness that remains adhered to the solid under equilibrium with the humid air in the void space. We assume that thermal equilibrium exists between the solid matrix, the water adhered to it and the humid air in the void space. At every (macroscopic) point, thermal equilibrium exists between all phases at that point, while at that point there is still a temperature gradient along the reactor.

As the cake is heated, usually by a stream of hot gas (often air), liquid present in the cake's void space evaporates and the vapour is removed by a stream of flowing hot gas. In a batch process, the hot gas flows above a stationary trays filled with a wet cake. In a continuous operation, trays with wet cakes are made to move along a furnace, while the vapour is directed to flow *through* the cakes. In both cases the hot gases flow along the trays or against the direction of the moving trays when the latter move. When the liquid occupies only part of cakes' void space, the hot gases penetrate them and move through the void space (Fig. A.3).

Consider the case of a hot gas flowing above a stationary tray filled with a wet cake with a void space that contains vapour and water at *irreducible water saturation*.

The evaporation process depends on the temperature, on gas humidity and on vapour mass transfer from the liquid adjacent to the solid to the vapour/gas in the void-space. Assuming that the gas is air and the liquid is water, the rate of drying, N_c, of the cake on a tray per unit tray area, say, in kg/m^2/s, is expressed by:

$$N_c = k_{tray}^{\ell\to g}\left(Y_s^{\ell} - Y_g^{v}\right) = \frac{\kappa_g^H}{L_v}\left(T_g - T_s\right). \tag{A.53}$$

where $k_{tray}^{\ell\to g}$ is the (*tray specific*) liquid-to-gas mass transfer coefficient for the gas, Y_s^{ℓ} is the mass of water per unit solid mass, Y_g^{v} is the mass of water vapour per unit mass of gas (= humidity), both on a unit tray area, T_g is the gas temperatures, T_s is the solid's *wet bulb temperature*, κ_g^H is the heat transfer coefficient in the gas, and L_v is the latent heat of water evaporation.

Under constant air p, T-conditions, N_c for the tray can also be expressed as:

$$N_c = -\frac{1}{A_s}\frac{dm_s^{\ell}}{dt} = -\frac{m_s}{A_s}\frac{dX_s^{\ell}}{dt} = -\rho_s\Delta_s\frac{dX_s^{\ell}}{dt}, \tag{A.54}$$

where A_s and m_s are the area of the tray and mass of solid on it, m_s^{ℓ} is the mass of liquid in the void space of the wetted solid matrix on the tray, X_s^{ℓ} denotes the gradually diminishing liquid content on the tray, per unit mass of solid on the tray and Δ_s (a few millimeters to a few centimeters) denotes the thickness of porous medium to be dried (on the tray), assumed to remain practically unchanged during the drying process, ρ_s is the dry density of the solid matrix on the tray, and t is time.

By integration, we obtain:

$$t = \frac{\rho_s\Delta_s}{N_c}\int_{X_{s,1}^{\ell}}^{X_{s,2}^{\ell}} dX_s^{\ell} = \frac{\rho_s\Delta_s}{N_c}(X_{s,1}^{\ell} - X_{s,2}^{\ell}). \tag{A.55}$$

where subscripts 1 and 2 indicate the initial and a later points in time.

When the rate of evaporation of the bounded liquid starts to decline, the rate of drying may be described by:

$$N_c = aX + b, \quad \rightarrow \quad = N_c \frac{X_s^\ell - X_{s,3}^\ell}{X_{s,c}^\ell - X_{s,3}^\ell} = -\rho_s d_s \frac{dX_s^\ell}{dt}. \tag{A.56}$$

By integration, we obtain:

$$\ln \frac{X_{s,2}^\ell - X_{s,3}^\ell}{X_{s,1}^\ell - X_{s,3}^\ell} = -\frac{N_c t}{\rho_s d_s}(X_{s,c}^\ell - X_{s,3}^\ell), \tag{A.57}$$

where N_c is calculated by (A.54), $X_{s,c}^\ell$ is the critical point where, in drying, for example, of soil, sand, or crystals, the evaporation rate varies from constant in time to a linear reduction. Eventually, the rate reaches the value of $X_{s,3}^\ell$, which represents the water content at the end of the process, recalling that $X_{s,2}^\ell$ denotes the water content at the beginning of the beginning of the second stage. A further decline of the drying rate requires other kinds of models for the rate calculation.

A.2.8 Chemical Reactions

Here, chemical reaction occur between the material comprising the solid matrix and chemical species dissolved in the fluid that passes through the reactor. Chemical reactions are discussed in Sect. 7.3.3. The mass balance equations for processes in which the mass of a considered species is produced or consumed by chemical reactions, are discussed in detail in Sect. 7.3.4. The velocities of the fluids may be given by models of flow through a porous media. The concentration equations include also the rate of reaction, namely the rate of disappearing/production of every components in the flowing system. When heat is added or generated, an energy balance equation and all heat sources are required. Obviously, initial and boundary condition are also needed.

A.2.9 Catalytic Chemical Reactor

In power stations and in large diesel engines, the process of catalysis is implemented in a catalytic reactor, also referred to as converter. The *catalytic reactor* (also referred to as *converter*) is built as a long set of parallel square perforated solid partitioning walls. Ammonia is injected to mix with the incoming gases.

As an example, consider the *catalytic reactor* used for converting NO_x to harmless gases by reactions with ammonia (NH_3). The process takes place in an elongated chamber filled with a porous medium that provides a large specific surface area on

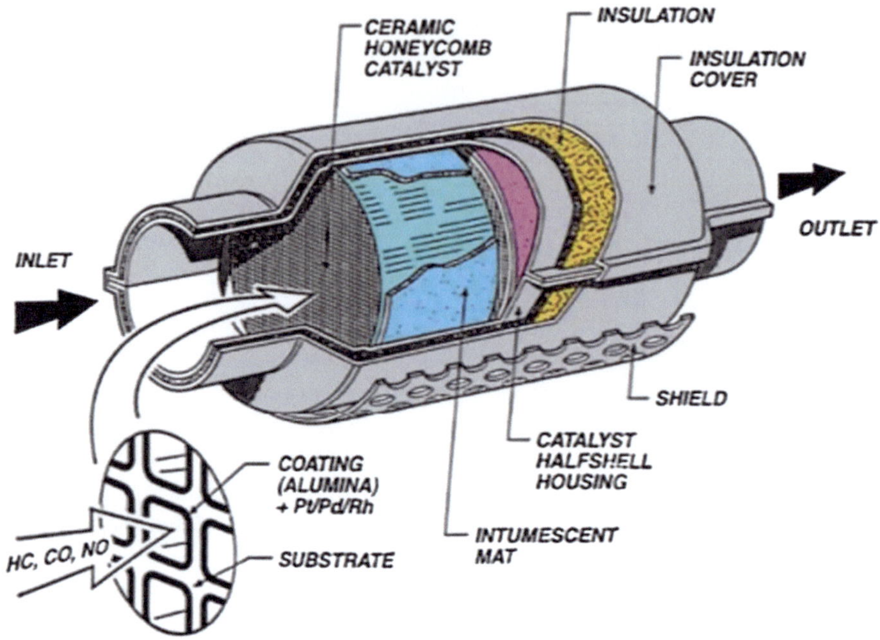

Fig. A.4 Catalytic reactor

which the chemical reactions take place. The toxic chemical species that react in order to form harmless new materials enter the reactor with the feed gas stream, while the reactions take place *on the solid surfaces* within the reactor's impervious shell. The harmless gases exit the reactor to the atmosphere.

In the converter example shown in Fig. A.4, the porous medium that fills up the chamber is made up of multiple closely spaced square-shaped tubes of 2–3 mm side, about 0.4–0.6 m long (see Fig. A.4). The objective is to achieve a large specific surface area. The tubes are made up of a special metal oxide that allows catalytic enhancement of the rate of the reactions that take place on the large internal surface area.

The reactions take place on the surface area of the solid matrix or the surface of the perforated partitioning walls inside the *catalytic chamber*. Ammonia is injected to mix with the incoming gases. The chemical reactions can be summarized in the form:

$$\begin{aligned} 4NO + 4NH_3 + O_2 &\rightarrow 4N_2 + 6H_2O, \\ 2NO_2 + 4NH_3 + O_2 &\rightarrow 3N_2 + 6H_2O, \end{aligned} \tag{A.58}$$

These reactions occur on the surface of the tubes that constitute the porous medium. This example concerns only gas flow, but, in general. we may have catalytic reactors operating with a liquid phase.

It is important to distinguish between two cases. If the process is a *homogeneous catalysis*, the reactions take place in the gas stream. If it is a *heterogenic catalysis*, the reactions take place *on the solid surfaces*. In the latter case the equations should also consider the interaction of the catalysis on the solid with the different species and the mass transfer of the different species from the fluid into the liquid film adjacent to e solid surface, where the species interact with the catalyst surface and move back to the gas stream as products.

The process includes a stage of adsorption of the components on the solid surface, followed by chemical reactions that take place on that surface. Eventually, the (harmless) products are released (desorption) from that surface and released to the atmosphere with the effluent gas.

Examples f Catalysts and Chemical Reactions

Some examples of catalysts and chemical reactions are presented below. The involved catalyst of each reaction is listed in parentheses.

- Preparation of oxygen by thermal decomposition of $KClO_3$:

$$2KClO_3 = 2KCl + 3O_2, \quad \text{(Manganese dioxide)}.$$

- Contact process for the manufacturing of sulphuric acid:

$$2SO_2 + O_2 = 2SO_3, \quad \text{(Vanadium pentoxide)}.$$

- Haber process for ammonia synthesis:

$$2N_2 + 3H_2 = 2NH_3, \quad \text{(Iron with } Al_2O_3\text{)}.$$

The Mathematical Model

The core of the mathematical model that describes what happens inside the reactor, with the reactions presented in (A.58), i.e., with no phase change, includes mass and energy balance equations. In the example below, we shall assume that only gas (g) flows through the reactor. This gas flux is governed by pressure gradient, neglecting any gravity effect. Obviously to present a complete model, we have to add initial and boundary conditions.

- **Mass balance for gas (g) flow**: Assuming steady state flow, we write:

$$\nabla \cdot \phi \rho_g \mathbf{V}_g = 0, \quad \rho_g = \sum_{(\gamma)} c_g^{\gamma}, \quad \mathbf{V}_g = -\frac{k_g}{\phi \mu_g} \nabla p_g. \tag{A.59}$$

- **Mass balance for multi-component gas (g) flow is represented by**:

$$\frac{\partial}{\partial t}\left(\sum_{(\gamma)}\phi c_g^\gamma\right) = -\nabla\cdot\sum_{(\gamma)}\phi\left(\mathbf{V}_g c_g^\gamma - D_\alpha^{'\gamma}\nabla c_g^\gamma\right) + \phi\rho_g\sum_{(\gamma)}\Gamma_g^\gamma, \tag{A.60}$$

where Γ_g^γ presents the production or disappearance of γ mass by chemical reactions. Assuming steady state operation throughout the entire cross-sectional area of the catalyst domain, we have for every γ-species:

$$-\nabla\cdot\phi\left(\mathbf{V}_g c_g^\gamma - D_\alpha^{'\gamma}\nabla c_g^\gamma\right) + \phi\rho_g\Gamma_g^\gamma = 0, \tag{A.61}$$

and,

$$-\nabla\cdot\sum_{(\gamma)}\phi\left(\mathbf{V}_g c_g^\gamma - D_\alpha^{'\gamma}\nabla c_g^\gamma\right) + \phi\rho_g\sum_{(\gamma)}\Gamma_g^\gamma = 0. \tag{A.62}$$

• **Energy balance equation for a gas saturated porous medium**:

$$\begin{aligned}\frac{\partial}{\partial t}&\left[(\phi\rho_g u_g) + (1-\phi)\rho_s C_s T\right]\\ &= -\nabla\cdot\phi\left[\rho_g h_g \mathbf{V}_g + \sum_{(\gamma)} h_g^\gamma \mathbf{J}_g^{*\gamma} + \mathbf{J}_g^{*H}\right]\\ &\quad + \nabla\cdot(\Lambda_{pm}^{*H}\nabla T) + \Gamma_{pm}^E,\end{aligned} \tag{A.63}$$

where u_g represents the internal energy of the gas, C_s is the solid's heat capacity, since the latter does not change phase, $\mathbf{V}_g$ represents the mass weighted gas velocity, h_g^γ represents the enthalpy of the γ-species in the gas, $J_{g,hdis}^\gamma$ denotes the flux due the hydrodynamic dispersion, $J_{g,hdis}^{*H}$ denotes the dispersive heat flux within the gas, Λ_{pm}^* is the coefficient of thermal conductivity of the porous medium as a whole, and Γ_{pm}^E denotes the rate of heat added by exogenic chemical reactions. Note that because of the high temperature, condensation does not occur.

• **Mass balance for a chemical γ-species in the gas flow**:

$$\frac{\partial\phi c_g^\gamma}{\partial t} = -\nabla\cdot\phi\left(c_g^\gamma\mathbf{V}_g + \mathbf{J}_{g,dif}^\gamma + \mathbf{J}_{g,dis}^\gamma\right) + f_{s\to g}^\gamma + \phi\rho_g\Gamma_g^\gamma, \tag{A.64}$$

where c_g^γ denotes the γ-concentration in the gas, with $\sum_\gamma c_g^\gamma = \rho_g$, Γ_g^γ, denotes the γ-mass source term, due to chemical reactions in the gas (= added or removed γ-mass per unit mass of gas, per unit time). Usually, we assume that gas density, ρ_g remains practically unchanged within the reactor. The gas flux, q_g is assumed to be controlled only by the gas pressure difference (Δp) between the reactor's inlet and outlet.

• **Mass balance for a γ-species on the reactor's internal surfaces of the solid matrix**: This is expressed by (A.30), repeated here for convenience:

$$\frac{\partial(\rho_b F^\gamma)}{\partial t} = f^\gamma_{g\to s} + \rho_b \Gamma^\gamma_s. \tag{A.65}$$

Assuming that the reactor operates under steady state conditions, the mass balance equations for the fluid (= gas) reduces to:

$$-\nabla\cdot\phi\left(c^\gamma_g \mathbf{V}_g + \mathbf{J}^\gamma_{g,dif} + \mathbf{J}^\gamma_{g,dis}\right) - f^\gamma_{g\to s} + \phi\rho_g\Gamma^\gamma_g = 0, \tag{A.66}$$

recalling that here and everywhere in this appendix, c^γ is molar concentration.

• **Gas mass balance**. For steady flow inside the reactor, with constant porosity, the gas mass balance reduces to:

$$\nabla\cdot(\rho_g \mathbf{V}_g) = 0, \quad \mathbf{V}_g = -\frac{k_g}{\phi\mu_g}\nabla p. \tag{A.67}$$

• **Energy balance equation**, assuming thermal equilibrium between solid matrix and gas, is given by:

$$\begin{aligned}\frac{\partial}{\partial t}&\left[\phi\rho_g u_g + (1-\phi)\rho_s C_s T\right]\\ &= -\nabla\cdot\phi\left[\rho_g h_g \mathbf{V}_g + \sum_{(\gamma)} h^\gamma_g \mathbf{J}^\gamma_{h,g} + \mathbf{J}^{*H}_g\right]\\ &\quad + \nabla\cdot(\Lambda^{*H}_{pm}\nabla T) + \Gamma^H_{pm}.\end{aligned} \tag{A.68}$$

where every term represents energy per unit volume of porous medium. In the above equation, u_g denotes internal energy of the gas, C_s is the heat capacity of the solid matrix, h^γ_g is the specific enthalpy of the gas, $\mathbf{J}^\gamma_{hyd,g}$ is the flux of hydrodynamic dispersion in the gas, Λ^{*H}_{pm} denotes the heat conduction coefficient for the porous medium as a whole, i.e., the solid matrix and the gaseous phase that occupies the void space, and $\Gamma^{\mathcal{E}}_{pm}$ denotes added energy from exogenic chemical reactions.

Usually, (1) dispersive flux term is neglected, (2) porosity is assumed constant, (3) there are no external energy sources, the energy balance equation, written for the reactor as a 1-d domain, reduces to:

$$\begin{aligned}\frac{\partial}{\partial t}[\phi\rho_g u_g + (1-\phi)\rho_s C_s]T &= -\phi\frac{\partial}{\partial x}\left[\rho_g h_g V_g + \sum_{(\gamma)} h^\gamma_g J^{*\gamma}_g + J^{*H}_g\right]\\ &\quad + \Lambda^{*H}_{pm}\frac{\partial^2 T}{\partial x^2} + \Gamma^H_{pm}.\end{aligned} \tag{A.69}$$

When, in the above one-dimensional case, heat is added at points along the reactor's boundary, we can express it in the above equation by adding a term on the r.h.s., e.g., $\dot{Q}(x,t)$, at points along the considered domain.

For the 1-d example considered here, neglecting the dispersive term, and with external heating sources along the domain, the energy balance equation simplifies to the form of:

$$\frac{\partial(\rho_{pm} C_{pm} T)}{\partial t} = -\frac{\partial(\rho_g C_{p,g} V_g T)}{\partial x} + \Lambda^{H}_{eq} \frac{\partial^2 T}{\partial x^2} + \Gamma^{*\mathbb{H}}_{pm} + \dot{Q}(x,t). \tag{A.70}$$

where C_{pm} represents the porous medium's (gas and solid) heat capacity, $\rho_g C_{p,g}$ represents the equivalent mass heat capacity, $\Lambda^{\mathbb{H}}_{eq}$ represents the equivalent thermal conductivity of the porous medium as a whole (i.e., solid matrix and gas), and $\dot{Q}(x,t)$ represent heat added at point along the reactor. These points appear as boundary conditions in a 3-d model, but are approximated in the 1-d model a point sources.

The term $\Gamma^{*\mathbb{H}}_{pm}$, that represents the net heat produced by the exogenic and endogenic chemical reactions, is expressed as:

$$\Gamma^{*\mathbb{H}}_{pm} = -R_{r1} H_{r1} - R_{r2} H_{r2}, \tag{A.71}$$

where the rates of reaction R_{r1} and R_{r2} refer to the two reactions (see (A.58)), and the H_{ri} represents the heat of reaction in the ith reaction. Recall that the rates of reaction are temperature-dependent (see (7.3.79)). Thus,

$$R_{ri} = k_{ri} P_k(c^{\gamma}), \quad k_{ri} = A_i e^{-E_{a,i} RT}, \quad i = 1, 2. \tag{A.72}$$

where R_i denotes the rate of the ith chemical reaction, k_{ri} is the Arrhenius constant for the ith reaction, and the reaction-dependent coefficient P_k is a function of the concentration.

A.2.10 Processes in Two Phase Flow

In many cases, we need to perform a chemical reaction between two chemical species under certain specified conditions. This can be achieved by using two streams, each containing a different set of chemical species. The two streams may be of the same phase or of different fluid phases. Another possibility is just to create a situation that facilitates the exchange of materials between two phases in order to achieve the separation of substances (e.g., of a pollutant) between the two phases.

A.2.11 Distillation

Distillation is a process in which a volatile chemical substance dissolved in a liquid is removed from the latter by selective evaporation and condensation. The removal of ethanol from a water-ethanol solution in the food industry, and the removal of a light

Fig. A.5 Typical porous medium particles filling a distillation column

oil component from a crude oil in the petroleum industry may serve as examples. The process is an essential feature in many industries, including oil refineries, special chemicals, pharmaceutical and food industries. It is conducted in a vertical column filled with a porous medium: bubble cups, perforated plates, or granular material. Figure A.5 shows some typical particles used as porous media in distillation columns. They are designed to allow countercurrent flow of two phases: an upward vapour flow and a downwards liquid flow, with good contact between these two streams. We note that a column filled with such particles has a very high porosity. In gas-liquid flow, they provide the opportunity for the appearance of drops of liquid streams within the vapour environment, or two phase bubble-liquid flow in continuous liquid phase flow. The passage through a long column of this kind creates also an opportunity for a number of volatile products to be separately transferred from the downward moving liquid to the rising vapour, at rates that depend on the exchange of their internal energies. The more volatile substance is removed from the rising vapour stream at the top of the column, while the less volatile one is removed at the bottom. Thus, the quality (i.e., pureness) of each product is a controllable parameter.

At its lower end, the column is connected to a reservoir (B) which contains liquid to be evaporated by a heat exchanger.

Distillation can be carried out in two modes:

(a) Batch distillation.
(b) Continuous distillation.

In both modes, the liquid phase flows downwards while the gas phase that contains similar components as the liquid, but at a higher concentration of volatiles intended for removal, flows upward.

Batch Mode

The batch-type process is initiated by placing the source liquid in the reservoir (B). This reservoir is also a boiler which evaporates the liquid. In a continuous mode of operation, the source liquid is fed at one or more points along the column. The locations of these points depend on the solutions concentration and temperature. It is essential to cause the two streams - liquid and upward vapour – to be in good contact along the column in order to allow efficient mass and heat transfer between them. The presence of a porous medium enhances this contact. A condenser (C) that includes a heat exchanger is located at the top of the column, It is fed by cooling water that liquefies the vapour exiting the column. The liquid condensate is divided into (1) a portion that constitutes the distillate product and a portion that is returned to the column. The latter liquid is returned to the top of the column so that it flows downwards through to column to the boiler. The two fluids-the downward moving liquid and the rising vapour- come into contact as they move simultaneously through the void space. this mode of operation improves the quality of the final product. The simultaneous (countercurrent) flow along the column can be described by a two-phase flow model, with variables S_ℓ, S_v, p_ℓ, p_c, Q_ℓ, q_v, etc., as discussed in Chap. 6. In what follows, we shall show a different approach, common in chemical engineering.

Two main issues that concern the gas steams along the column have to be taken care of:

- Maintaining good contact between the two fluids to allow efficient heat and mass transfer between them, and
- Preventing the possibility that the gas will force the liquid to accumulate and change flow direction, referred to as 'flooding the column'.

The two types of distillation processes, shown on Fig. A.6, will be presented and discussed below: batch distillation in Fig. A.6a and continuous operation in Fig. A.6b.

Batch Distillation

A distillation column can separate two streams. In case of binary mixture, it is possible to obtain two streams of close to pure products, with some limitations.

Following a certain initial period, the above flows and masses in the column stabilize and a quasi-steady state is established. The process is terminated when the product in the distilled product reservoir (D) reaches the quality requirements which depend on the product's concentration, density and temperature.

Initially, at $t = 0$, a liquid mass denoted as m_F moles is placed in the bottom reservoir/boiler, B. We shall use the symbols $\mathrm{m}_B(t)$, $\mathrm{m}_C(t)$, $\mathrm{m}_D(t)$ to denote the fluid masses (in moles) in the boiler (B), coming out of the condenser (C) returned from the condenser to the column (R), and accumulating in the reservoir of distilled product (D) respectively, with $\mathrm{m}_F = \mathrm{m}_B$ at $t = 0$. We shall use $X_B(t)$, $X_C(t)$, and $X_D(t)$ and $X_D^*(t)$ to denote, respectively, the mass fractions of the fluids in the boiler, coming out of the condenser, in the fluid coming out of the condenser and diverted to the top of the column, and in the reservoir of the accumulating final distilled product (of mass $\mathrm{m}_D(t)$). Actually, $X_C(t) \equiv X_D(t)$.

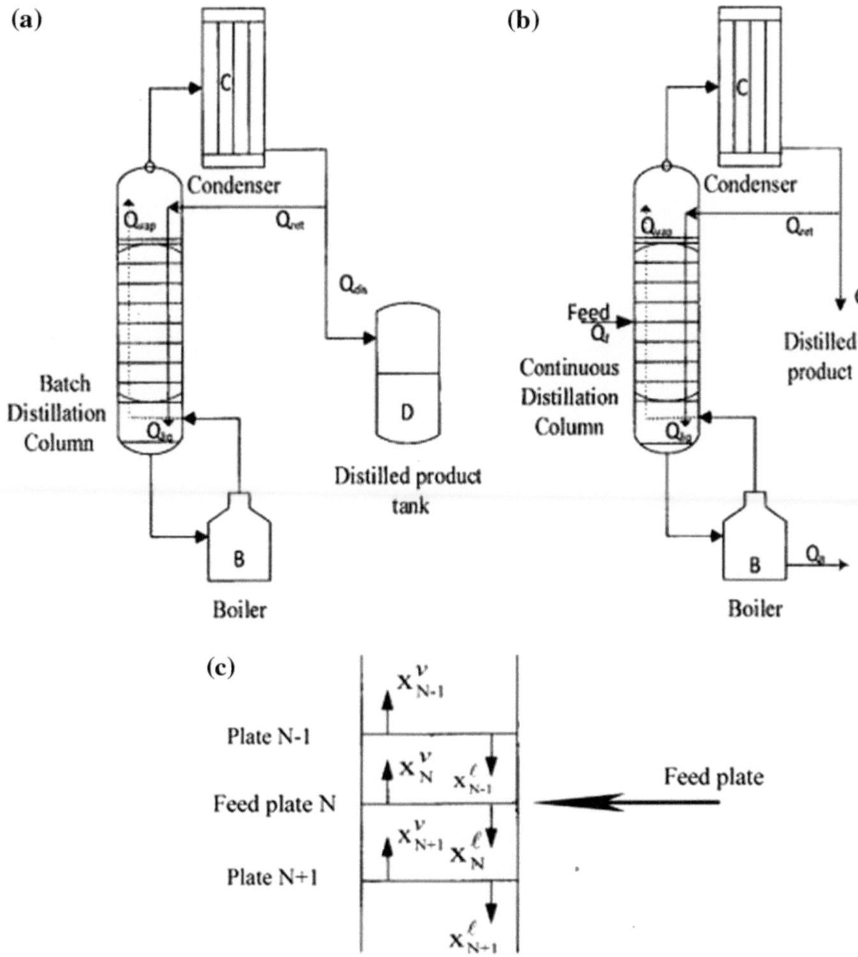

Fig. A.6 Distillation columns: **a** batch distillation column, **b** continuous distillation column, and **c** notation of the plates around the feed plate

We note that:

$$\mathrm{m}_F = \mathrm{m}_B(t) + \mathrm{m}_C(t) + \mathrm{m}_D(t) \approx const. \tag{A.73}$$

Let us start by an example that involves a very simple case. We regard the system as a closed one and neglect the volume of fluid in the column and in the pipes (= "holdup"). At any time $t > 0$, the rate of depletion of the target volatile in the boiler (B) is equal the rate of accumulation of the volatile as a distilled product in the final product reservoir (D). In fact, in practice, the volume of the C-reservoir is small and may be neglected.

Mass continuity requires that:

$$-\frac{d(X_B \mathrm{m}_B)}{dt} = \frac{(X_C d\mathrm{m}_C)}{dt} = \frac{d(X_D^* \mathrm{m}_D)}{dt}, \tag{A.74}$$

where X_B and X_D denote the molar concentration of the liquid distillate leaving the column/reboiler (B) and entering the product tank (D), respectively, m_B is the current mass left in the reboiler (B) from m_F, m_C is the integrated mass coming out of the condenser (C), and m_D is the mass accumulated in the distilled product reservoir (D). Since $d(X_B \mathrm{m}_B)/dt = d(X_C \mathrm{m}_C)/dt$, it can be shown that the time dependence can be eliminated, so that for any time t, we obtain the values of: X_B, m_B, X_C, m_C, X_D^*, m_D, with:

$$-d\mathrm{m}_B = d\mathrm{m}_C = d\mathrm{m}_D. \tag{A.75}$$

For distillation of a single volatile component from a binary solution, we have:

$$d(X_C \mathrm{m}_C) = -d(X_B \mathrm{m}_B), \quad \text{recalling that} \quad d\mathrm{m}_C = -d\mathrm{m}_B. \tag{A.76}$$

Thus,

$$X_C d\mathrm{m}_C = -\mathrm{m}_B dX_B - X_B d\mathrm{m}_B, \tag{A.77}$$

leading to:

$$\mathrm{m}_B d\mathrm{m}_B = (X_C - X_B) d\mathrm{m}_B, \tag{A.78}$$

with:

$$\begin{aligned} &\text{At } t = 0, \ \mathrm{m}_F, \ X_F \\ &\text{At } t > 0, \ \mathrm{m}_B, \ X_B, \quad d\mathrm{m}_B = -d\mathrm{m}_D. \end{aligned} \tag{A.79}$$

where X_B denotes the average concentration in B.

It can then be shown that the time dependence can be eliminated, so that we obtain:

$$-X_D d\mathrm{m}_D = d(X_B \mathrm{m}_B) = X_B d\mathrm{m}_B + \mathrm{m}_B dX_B. \tag{A.80}$$

The above equation holds also for distillation of a single component, or for a vector of components that have different equilibria, namely that for different γ-species we have different values K^γ for different pairs of X_C^γ and X_B^γ.

The solution for a single volatile component out of two is given by:

$$\int_{\mathrm{m}_F}^{\mathrm{m}_B} \frac{d\mathrm{m}_B}{\mathrm{m}_B} = \ln \frac{\mathrm{m}_F}{\mathrm{m}_B} = \int_{X_B}^{X_F} \frac{dX}{X_C - X}. \tag{A.81}$$

Assuming equilibrium between X_C and X_B of the form:

$$X_C = K X_B, \tag{A.82}$$

where K, specific for every species, is assumed constant within the considered range of concentrations, it can be shown that the solution for a single volatile component out of two may be expressed by:

$$\ln \frac{\mathrm{m}_B}{\mathrm{m}_F} = \int_{X_F}^{X_B} \frac{d\,(\ln X)}{K-1} = \frac{\ln(X_B/X_F)}{K-1}, \tag{A.83}$$

recalling that:

$$\mathrm{m}_F X_F = \mathrm{m}_D X_D^* + \mathrm{m}_B X_B, \tag{A.84}$$

$$\mathrm{m}_C = \frac{X_F - \frac{\mathrm{m}_B}{\mathrm{m}_F} X_B}{1 - \frac{\mathrm{m}_B}{\mathrm{m}_F}}, \quad X_B = f\left(\frac{\mathrm{m}_B}{\mathrm{m}_F}\right) \tag{A.85}$$

in which values of K for the equilibrium between the two phases is a function of the concentration and the temperature.

The following development leads to the time needed for completing the distillation process.

The mass balance of a volatile component is given by the relationship:

$$\mathrm{m}_B = \mathrm{m}_F \frac{X_C - X_F}{X_C - X_B}. \tag{A.86}$$

Differentiating with respect to time yields:

$$\frac{d\mathrm{m}_B}{dt} = \left[\frac{\mathrm{m}_F (X_B - X_F)}{(X_C - X_B)^2}\right] \frac{dX_F}{dt}. \tag{A.87}$$

Assuming steady state conditions throughout the system, we obtain:

$$\frac{d\mathrm{m}_D}{dt} = (Q_{m,\ell} - Q_{m,v}) = \left(1 - \frac{Q_{m,\ell}}{Q_{m,v}}\right) Q_{m,v}, \tag{A.88}$$

where $Q_{m,\ell}$ and $Q_{m,v}$ are the liquid and vapour mass discharges through the column, respectively. From the last two equations, we can get an expression for the time, t:

$$t = \int_0^t dt = -\frac{\mathrm{m}_F (X_C - X_F)}{Q_V} \int_{X_F}^{X_B} \frac{dX_B}{1 - (Q_{m,\ell}/Q_{m,v})(X_C - X_B)^2}. \tag{A.89}$$

The above equation gives the distillation time, i.e., the time needed in order to get a specified concentration of the volatile component in the final product reservoir. In order to determine the time required for producing a distillate at a specified concentration X_D^*, we need to add the mass balance equation:

$$\mathrm{m}_F X_{F,\ell} - \mathrm{m}_B X_{B.\ell} = \mathrm{m}_D X_D^*. \tag{A.90}$$

The relation between $Q_{m,\ell}$ and $Q_{m,v}$, together with the information concerning the column's cross-sectional area, are used for designing the column and the operation in it. We recall that the above development is based on the assumption that we may regard the entire column as a single stage.

An additional design parameter at the disposal of the designer is the value of the portion of the output from the condenser that is diverted back to the column.

We still have to deal with the energy/heat required for the operation described above.

The energy balance equation can be written in the form:

$$\mathrm{m}_F h_F + q_B^H - q_C^H = \mathrm{m}_D H_D + \mathrm{m}_B h_B, \tag{A.91}$$

where h_F, and h_B denote the enthalpy per unit mass in F and in B, respectively, q_B^H represents the *steam heat added to the reboiler* and q_C^H is heat removed by cooling water in the condenser (C) stream.

Continuous Distillation

In a distillation column, the flow regime involves two countercurrent streams – a downward liquid stream and an upward stream – which interact with each other. Within each stream the concentration, pressure and temperature vary along the z-axis. It is convenient to envision the column as composed of a series of segments, referred to as *stages*, in each of which the two streams are assumed to reach equilibrium.

Figure A.6c shows such stages. This stage configuration facilitates the handling of *four streams*. Each stream has its own temperature, enthalpy, mass flow-rate, and set of concentrations of the relevant dissolved species. In an ideal stage, equilibrium is reached between the two counter-current flowing streams.

A simple stage handles two phases running counter-currently as liquid and vapour streams that enter and leave it under different condition. Upon entering a stage, from the previous stage, a liquid stream will flow downward to the stage below, while a vapour will flow upward to the upper stage. Eventually, both streams leave through column's exits. The feed stream is made to enter a selected stage. This feed stream may contain liquid, vapour or both: a liquid will flow down while a gas moves up. Streams may be withdrawn from various stages along the column.

Within each stage along the column, *complete mixing* of the two streams–gas and liquid–is assumed to take place. It is further assumed that within every stage, the two fluids reach equilibrium of both concentration and temperature so that the two streams leaving the column, liquid through the stage's bottom and vapour through the latter's top, have the same temperature and their concentrations are in equilibrium with respect to all chemical components within the stage. Of course, in reality, this situation cannot always be achieved. This leads to the introduction of *stage efficiency* as a factor in the column's design. The efficiency is a measure (in percents) of how close to equilibrium is the situation reached at every stage within the column.

Furthermore, it is assumed that whatever happens in the column as a whole, as well as within any of the individual stages that comprise it, occurs under *steady state*. In the case of a system that is intended for long term operation, if not at the outset, then after a short time period, the system, here the distillation column is assumed to reach steady state conditions. Upon entering a stage, the fluid (gas or liquid) is not in equilibrium with the fluid present in the stage. It takes some time(and travel length/mixing) to reach equilibrium.

Along the column, we encounter a number of stage types:

- *The feeding stage*, where liquid is fed into the column. Sometimes, there are more than one feed stage. The feed position is chosen based on concentration and temperature, in order to minimize back-flows. It is important to note that the feed stream may be a supercooled liquid stream, saturated liquid, saturated gas, or a combination of these two streams. It may also be a superheated gas.
- A *purging stage*, where a liquid, a vapour, leaves the column.

The stream exiting a stage is assumed to be in equilibrium with the stage it has just left.

Two main fluid streams enter and leave a distillation column: a heavy product that leaves the reboiler at the bottom, and a light product that exits the condenser at the top. Although there are many types of distillation columns, here we shall consider (1) a stream that leaves the column carrying the product and (2) a stream with similar properties that is re-injected back to the column in order to supply a liquid phase to the upper stages of the column. This stream, referred to as *reflux stream*, also improves the quality of the final distilled product.

Notation

The mathematical model of a multi-stage column is based on the following system of stage numbering:

- Subscripts 1, 2, …, $n-1$ for stages from the column's upper end downward to the feeding cell (i.e., the cell which is also fed from outside).
- Subscript n used for the feeding cell.
- Subscripts $n+1, n+2, \ldots, m$ for stages from the feeding point downward to the reboiler.

The mole concentration of liquid leaving the jth cell is denoted by $X_{\ell,j}$, while the mole concentration of vapour leaving the jth cell is denoted by $X_{v,j}$.

We shall use the symbols $-\tilde{Q}_B(t)$ and $\tilde{Q}_D(t)$, to denote the time rate at which fluid mass (in moles) leaves boiler B and the condenser C at time t. $Qd(t)$ denotes the part of the condensate leaving C. The rest flows down the column as a reflux stream.

The Mathematical Model

The mathematical model of a distillation column involves the following equations:

The mass balance for the entire column states that at every instant, t, the rate $\tilde{Q}_F(t)$, at which mass (in moles) is fed into the column is equal to the rate at which mass (in moles) is leaving the column as distilled product, $\tilde{Q}_D(t)$, plus the rate at which mass of heavy product is exits to the boiler, $\tilde{Q}_B(t)$:

$$\tilde{Q}_F(t) = \tilde{Q}_B(t) + \tilde{Q}_D(t). \tag{A.92}$$

Henceforth, we shall assume that the operation has already reached a steady state, i.e., all $\partial(.)/\partial t = 0$. Under such conditions, focusing on the mass of a specific chemical species, we have at any time t:

$$\tilde{Q}_F X_F = \tilde{Q}_D X_D + \tilde{Q}_B X_B, \tag{A.93}$$

By eliminating $\tilde{Q}_B$, we obtain:

$$\frac{\tilde{Q}_D}{\tilde{Q}_F} = \frac{X_F - X_B}{X_D - X_B}. \tag{A.94}$$

By eliminating $\tilde{Q}_D$, which is the mass of the distilled product leaving the condenser at the top of the column, we obtain:

$$\frac{\tilde{Q}_B}{\tilde{Q}_F} = \frac{X_D - X_F}{X_D - X_B}. \tag{A.95}$$

The last two equations are valid for all vapour and liquid fluxes.

Under the assumed steady state, for any stages j and $j+1$ above the feeding point, the fluid mass balance can be written in the form:

$$\tilde{Q}_D = \tilde{Q}_{v,j} - \tilde{Q}_{\ell,j-1}, \quad 1 \le j < n. \tag{A.96}$$

Note that phases (e.g., g for a gas and ℓ for a liquid) are indicated by subscripts, while a component in a phase is indicated by a superscript.

Similarly, the component balance equation for any chemical species can be written for every stage in the form:

$$\tilde{Q}_{\ell,D} X_D = \tilde{Q}_{v,j} X_{v,j} - \tilde{Q}_{\ell,j-1} X_{\ell,j-1}, \quad 1 \le j < n. \tag{A.97}$$

The l.h.s. represents the most volatile component that leaves as the final product at the condensate side. It can also represent the vector of the components that leave as the final product.

In a similar way, the two equations for the part of the column below the feeding point, may be written for the jth stage:

$$\tilde{Q}_B = \tilde{Q}_{\ell,j-1} - \tilde{Q}_{v,j}, \quad n+1 \le j \le m. \tag{A.98}$$

$$\tilde{Q}_B X_B = \tilde{Q}_{\ell,j-1} X_{\ell,j-1} - \tilde{Q}_{v,j} X_{v,j}, \quad n+1 \le j \le m. \tag{A.99}$$

Equilibrium Line

The equilibrium line is the line that represents the equilibrium of a component in a two phase system as dictated by their nature. It can be found in the literature or by laboratory experiments. In the distillation system, each dissolved component has an equilibrium line.

In this case we have two operating lines, one for the domain below the feeding point and one for the domain above it. For every fluid feeding or exit point at the ends or along the column, we need to add an operating line. In the example presented here, we have one feeding point and two exit points–the condenser and the boiler.

The operating line for a stage j in the upper section $1 \geq j < n$, which is also called the *stripping section* is given by:

$$X_{v,j} = \frac{\tilde{Q}_{\ell,j+1}}{\tilde{Q}_{v,j}} X_{\ell,j+1} + \frac{\tilde{Q}_{v,j} X_{v,j} - \tilde{Q}_{\ell,j-1} X_{\ell,j-1}}{\tilde{Q}_{v,j}} = \frac{\tilde{Q}_{\ell,j+1}}{\tilde{Q}_{v,j}} X_{\ell,j+1} + \frac{\tilde{Q}_{D} X_{\ell,D}}{\tilde{Q}_{v,j}}, \tag{A.100}$$

Or, more conveniently, in the form:

$$X_{v,j} = \frac{1}{\tilde{Q}_{\ell,j} + \tilde{Q}_{\ell,D}} \left[\tilde{Q}_{\ell,j} X_{\ell,j+1} + \tilde{Q}_{\ell,D} X_{\ell,D} \right], \quad 1 \geq j < n. \tag{A.101}$$

which relates $X_{v,j+1}$ to $X_{\ell,j}$, with a slope that expresses the ratio $\tilde{Q}_\ell / \tilde{Q}_v$ in the column.

In a similar way, it is possible to show that the operating line for the lower part of the column will have the form:

$$X_{v,j} = \frac{1}{\tilde{Q}_{\ell,j} + \tilde{Q}_{\ell,B}} \left[\tilde{Q}_{\ell,j} X_{\ell,j+1} - \tilde{Q}_{\ell,B} X_{\ell,B} \right], \quad n+1 \leq j \geq m. \tag{A.102}$$

The energy balance for a stage in the upper part of the column is:

$$\tilde{Q}_{v,j} h_{v,j} - \tilde{Q}_{\ell,j-1} h_{\ell,j-1} = \tilde{Q}_{\ell,D} h_{\ell,D} + H_C, \quad 1 \leq j < n. \tag{A.103}$$

where H_C is the heat removed by the condenser.

The energy balance equation for the feed stage is:

$$\tilde{Q}_{v,n} h_{v,n} - \tilde{Q}_{\ell,n-1} h_{\ell,n-1} - \tilde{Q}_F h_F = H_C + \tilde{Q}_D h_{\ell,D}. \tag{A.104}$$

For the lower part of the column (i.e., below the feeding stage),

$$\tilde{Q}_{v,j} h_{v,j} - \tilde{Q}_{\ell,j-1} h_{\ell,j-1} = \tilde{Q}_{\ell,B} h_{\ell,B} - H_B, \tag{A.105}$$

H_B is the heat supplied to the reboiler.

The overall energy balance equation for the column is expressed by:

$$\tilde{Q}_{\ell,B} h_B - H_B + \tilde{Q}_{\ell,D} h_{\ell_D} + H_C - \tilde{Q}_F h_F, \tag{A.106}$$

where h_F denotes the molar heat per unit mass of the feed stream that may contain liquid and solid particles, as explained above.

The Reflux Ratio

As stated earlier, without the *reflux*, i.e., the stream that is returned from the condenser to become downward flow in the column, the vapour leaving the column to enter the condenser will have a concentration that is in equilibrium with that of the feed. This means a product of low quality. The term *reflux ratio* is used to express the ratio between the flux of the returned liquid to that of the distillate released as the final product. Dividing both the numerator and the denominator on the right hand-side of the upper operating line (A.101) by $\tilde{\mathrm{Q}}_{D}^{\ell}$ yields the following operating line of the column's upper section:

$$X_{v,j} = \frac{1}{1+R_d}\left[R_d X_{\ell,j+1} + X_{\ell,D}\right], \quad 1 \geq j < n. \tag{A.107}$$

in which $R_d = \tilde{\mathrm{Q}}_{\ell,j}/\tilde{\mathrm{Q}}_{\ell,D}$ is a parameter that measures the effect of separation on the product's quality. The value of R_d depends on the column's design, the number of stages, the diameter of the column, the energy supplied for the operation and the cost of the process.

The Feed Plate

As stated earlier, five feeding options are available, that depend on the nature of the feed. They cover the range from sub-cooled liquid up to supersaturated vapour. The liquid introduced as the feed joins the liquid flow that moves down the column, while the gases move upwards with the gas stream. Beside the concentration of the feed and the liquid-gaseous portions, the feed also contains enthalpy that affect the conditions in the column. With a parameter denoted by q that describes the quality of the feed (in term of how much liquid or vapor is present in the stream) according to the q-line described by the feed equation, we have:

$$X_v = -\frac{1}{\mathrm{q}-1}\left[\mathrm{q}X_\ell - X_F\right], \tag{A.108}$$

where X_F ix the feed concentration and q is the enthalpy function calculated from the two liquid streams above and below the feed plate, expressed by:

$$q = \frac{\tilde{\mathrm{Q}}_{\ell,n} - \tilde{\mathrm{Q}}_{\ell,n-1}}{\tilde{\mathrm{Q}}_F} = \frac{H_v - h_{\ell,F}}{L}, \tag{A.109}$$

where L denotes heat of phase change. Similarly, the following equation is the enthalpy function calculated from the two gas streams below and above the feed plate:

$$q - 1 = \frac{\tilde{\mathrm{Q}}_{v,n+1} - \tilde{\mathrm{Q}}_{v,n}}{\tilde{\mathrm{Q}}_F} = \frac{h_\ell - h_{\ell,F}}{L}, \tag{A.110}$$

The equilibrium condition within any jth stage is:

$$X^{\gamma}_{v,j} = K(p, T, X^{\gamma}_{\ell})X^{\gamma}_{\ell,j}, \tag{A.111}$$

So far, we have been considering the simple case of a binary system, i.e., the liquid is composed of two species only: the species of interest and another species. To simplify the presentation, we have not indicated the species of interest in the equation. In what follows, we shall assume that the liquid and the vapour contain a number of γ-species. With this in mind, following are some additional model relationships:

$$\begin{aligned}
&\sum_{(\gamma)} X^{\gamma}_{v,j} = 1.0, \quad j = 1, \ldots, mc. \\
&\sum_{(\gamma)} X^{\gamma}_{\ell,j} = 1.0, \quad j = 1, \ldots, mc. \\
&X^{\gamma}_{v,j} = K^{\gamma} X^{\gamma}_{\ell,j}, \quad K^{\gamma} = K^{\gamma}(T, p, X_{\ell}, X_{v}), \\
&H_{\ell} = H_{\ell}(T, p, X_{\ell}), \quad H_{\ell} = H_{\ell}(T, p, X_{\ell}),
\end{aligned} \tag{A.112}$$

where mc denotes the number of components in the stream.

As shown above, the full set of equations includes also the energy balance equation.

A.2.12 Stripping/Absorption

Stripping and absorption processes are two opposite counter-current, liquid-gas processes, performed in a vertical column. The process is usually used to clean a contaminant-containing gas or liquid stream. In both cases, the mass transfer is performed across the (microscopic) interfaces within the void space between the two phases. The liquid flows downward, while the gas flows upward, against the solvent's direction. The process is usually used to clean gas or liquid streams containing contaminants. Since the amounts to transferred are low, no significant energy changes occur, and an energy balance is not required as part of the process model. Note that here absorption means the passage of a specis from a gas to be dissolved in a liquid.

A typical set-up is shown in Fig. A.7.

An example of an adsorption process is the adsorption of benzene contaminant from an air stream by oil as the solvent. The oil may be cleaned and recovered by stripping the benzene with steam. The steam may be condensed to separate between the water and the liquid benzene, or fed to a high temperature direct contact boiler that is fed also with a sufficient stream of oxygen to burn the benzene while the hot steam is used for heating or power generation in the plant.

Since the process is rather similar to that of distillation, the model presented in Sect. A.2.13 on distillation is valid also here. Essentially, this is a model that describes

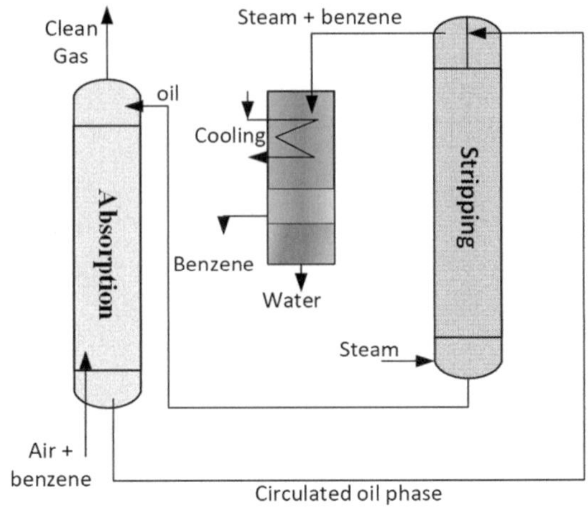

Fig. A.7 Stripping and adsorption columns

two phases-gas and liquid-that flow in opposite directions, and while doing so, a chemical species moves from one stream to the other.

The Model

As in the distillation column, the two phases flow counter-currently along the column. In both cases, stripping and absorption, the gas is introduced at the bottom of the column, as seen in the figure while the liquid is fed from the top of the column and moves downward. In most cases, it is assumed that the system operates *adiabatically*, and if heat is transferred between the phases, it is negligible. The target of this process is to move a γ-contaminant from a gas (g) stream (polluted air, say with SO_2) into a liquid ℓ absorbent like water. The objective is to achieve a low, yet permitted concentration in the exiting gas. In case of stripping, the process is based on the removal of a volatile γ-component from a liquid stream (say, oil containing natural gas) down to a low permitted concentration, with a stream of a gas (say, steam), that is appropriate for this case. The case of steam as the stripper is involved with energy transfer, yet here, for simplicity, we assume operation under constant temperature. The equilibrium assumed in the thin film layer between the two phases–gas (g) and liquid(ℓ) is considered in the form:

$$X_g^\gamma = K X_\ell^\gamma = f(X_\ell^\gamma) \tag{A.113}$$

where X_g^γ denotes the molar fraction of γ in g, and K is the equilibrium constant. This equilibrium is suited for the two cases–stripping and absorption. At low concentrations, this is the Henry's law, and the functional relationship is a straight line.

Recall that the target in absorption it to reduce a certain initial (feed) concentration from $X_{g,n+1}^\gamma$ to $X_{g,1}^\gamma$.

The mass balance equation in this case is given by:

$$X^{\gamma}_{g,i+1} = \frac{\tilde{Q}_{\ell}}{\tilde{Q}_g} X^{\gamma}_{\ell,i} + \left(X^{\gamma}_{g,1} - \frac{\tilde{Q}_{\ell}}{\tilde{Q}_g} X^{\gamma}_{\ell,0} \right), \tag{A.114}$$

where $\tilde{Q}_{\ell}$ and $\tilde{Q}_g$ are the counter-current molar flow rates of the liquid and the gas, respectively. and cell numbers vary from $i = 1$ to $i = n$.

The solution for stripping is similar to the one presented above, except that the operating line exists on the other side of the equilibrium line.

A.2.13 Solvent Extraction and Leaching

Extraction, also known as *solvent extraction for liquids* or *leaching process for solid -liquid mass transfer*, also known as *solid-liquid extraction*, liquid-liquid extraction, or just *leaching*, is the name of a process in which two phases exchange dissolved chemical species in order to separate or increase the concentration of a valuable product in one of them. This is a common process in the organic chemical industry whenever the separation is required between two nonvolatile compounds or compounds that have a close boiling temperature. It is also a well known process for the extraction of metal ions in the mining industry. In the mineral industry it is used for extracting phosphoric acid, copper sulfate, or oil from various grains, e.g., soy bean.

The target product to be separated is dissolved in a solvent or a solid. In these processes, another solvent is used to collect the desired product and allow to purify it. At least three components are involved in this process.

Figure A.8 shows a simple extraction process.

Several types of contactors are available for solvent extraction, Here, we shall focus on a vertical column that is filled with solid particles. The main function of which is to enable a close contact between the solvents involved in the process. The heavier phase is allowed to flow down while the lighter one flows counter-currently upward in a way that one of the phases is broken into drops within the other phase in order to increase the area available for mass transfer. The drops are allowed to merge and to form larger fluid particles and to brake again along the column. This causes a good mixing and re-expose the surfaces to better contact. The process may end by distillation of the solvent, if possible, or by a second stage of extraction that purifies the product.

Consider solvent A that contains a component C to be extracted. A lighter solvent B is introduced at the bottom of a column and flows upward. The feed containing liquid A + C is heavier than solvent B; it is introduced at the top of the column so that it flows downwards, as shown in Fig. A.8. Extracts C and some A leave the column with B as *Extract* at the top. The exhausted solvent A, containing some B and some C leaves at the bottom of the column as *Raffinate*. At least one *settler* that separates the drops from the two phase mixture is needed at the ends of the column. Similarly, as shown in Fig. A.8, opposite directions of the feed and solvent occur when the feed is lighter than the solvent. Leaching is another process used to achieve extraction,

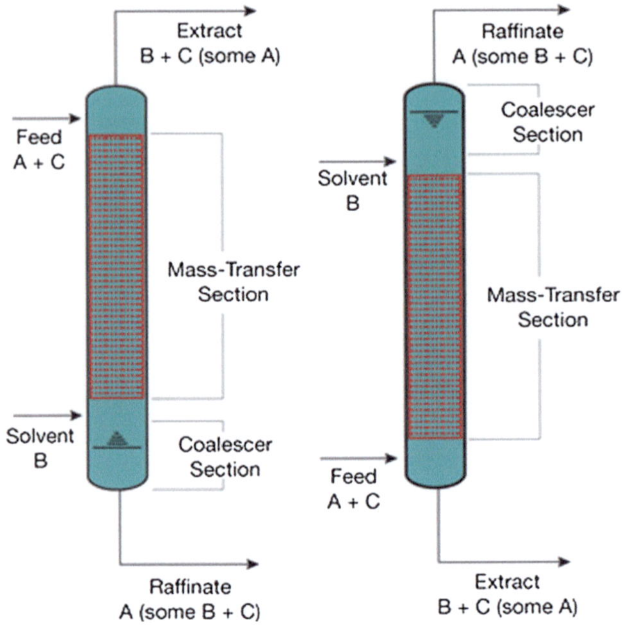

Fig. A.8 Extraction columns

e.g., of vegetable oil from grounded soybeans. In this case, the soybeans (packed in canisters) constitute a packed bed that is moving opposite to the flow of a solvent (hexane in this case) that extracts the oil. Sometime, the solvent is distilled so that the product contains very little hexane. An interesting leaching process is used with supercritical CO_2 to extract organic matter from solids, or coffee from grains with supercritical water.

The Model

As is obvious from the above description, various options are possible for implementing the leaching process. Here we shall focus on a rather simple system of *liquid-liquid extraction*, or *solvent extraction*, which is similar to the process of stripping in a vertical column (see Sect. A.2.12). The main difference is that here we introduce another solvent to extract the desired compound or range of compounds dissolved in the feed solution. Since we are dealing with two solvents, we shall refer to them as the denser and the lighter ones. Thus, the feed at the top of the column will include the heavier solvent, in order to benefit from the effect of gravity. The lighter solvent is fed at the bottom of the column. The density difference between the two solvents is usually small so that the flow inside the column is slow. In fact, we have a flow of drops in a continuous solvent. In many cases it is required to allow the separation of phases inside the column by generating stagnation regions. Usually, the top and the bottom of the columns are equipped with *settlers* to allow phase separation.

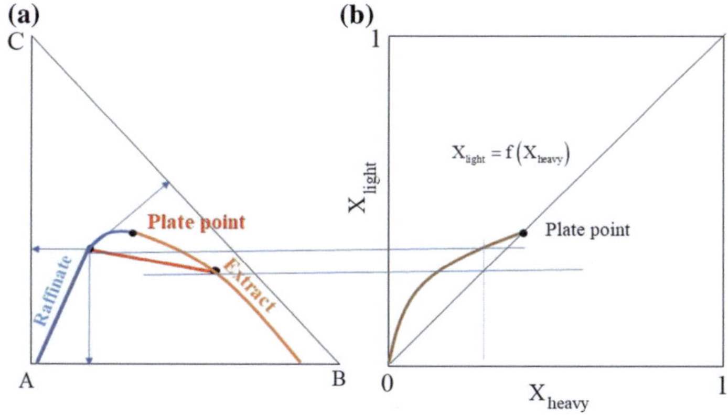

Fig. A.9 Triangular phase diagram for a mixture of two phases

In most cases, equilibrium is presented as a phase diagram, usually drawn as a *triangle*. Figure A.9a shows such triangle.

On Fig. A.9a, vertices A, B and C of the triangle represent 100% of solvent A, 100% of solvent B and 100% of solute C, respectively. The distance between the vertices indicates a linear variation from 0% to 100% of the relevant component. The three coordinates of every point on the diagram represent the mixture of the three components. The colored curve shows the relevant phase diagram. The normals from a point to the three sides indicate the concentrations. The domain E represents the extract region, while R represents the raffinate region; both end at the point P. The line between R and E represents a tie line between points in equilibrium. Figure A.9b shows the equilibrium line in terms of molar ratio concentrations. Figure A.9a demonstrates this approach through a simple case of two solvents, A, B, and a dissolved component, C, to be extracted.

We consider a simple case in which we assume that the solvents are completely immiscible. We define the final target concentrations of the two solvents and try to determine how many stages are required in order to achieve the desired separation. The size of each stage is determined by the flow rates of the two solvents and the relative velocities between them, in order to minimize entrainment and provide good contact between the phases. The velocities dictate the cross-sectional area of the stages while the thickness of each stage allows the settling of droplets carried from stage to stage.

As shown n Fig. A.9b, the equilibrium between phases may be expressed by:

$$X_{light} = K X_{heavy}, \tag{A.115}$$

where the superscripts indicate the light and heavy phases.

The mass balance at the different stages may be written as:

$$X_{light,j+1} = \frac{Q_{heavy}}{Q_{light}} X_{heavy,j} + \left(X_{light,j} - X_{heavy,F}\right), \qquad (A.116)$$

where light and heavy indicate the two phases, and j indicates the stage.

Like in the stripping model, described in Sect. A.2.12, by solving the mass balance equation for a stage (A.115), together with the equilibrium equation (A.116), we can determine the concentrations in the different stages. Calculating the stage concentration between the feed concentration and the required extract will allow us to determine the number of stages and complete the design of the entire process.

A.2.14 Chemical Reactions

Here, we discuss a simple case of reactions in counter-current flow along a porous medium column.

When two components that flow as two streams of different phases in counter-current motion through a porous bed, they may react to form a new material from the dissolved components in the two stream. The low concentration at the exit stream at both ends of the reactor, see high concentration at the surface with the other stream. This situation maintains better driving forces for the reaction than the case of co-current flow, where high concentrations take place at stream entrance and low concentration of the common exit. This allows better recovery of the product, better mass transfer and better rate of reaction. Also, the Heat needed or transferred along the reactor is better distributed along the reactor.

A.3 Processes in Moving-Bed Reactors

A *moving bed reactor* is used when we need a better contact and mixing between the solid matrix and the fluid. The solid matrix is then made up of individual grains, each completely surrounded by the flowing liquid or gas. The flow is turbulent, thus enhancing fluid-solid contact. Two main types of moving beds are used in the industry: fluidized beds and stirred beds. Batch reactions may be performed in such contactors. It is possible to add fluid/solid by mechanical means, using an impeller, or by adding and extracting fluid through the side walls, thus causing fluid circulation in the bed and continuous reactions. Examples of various cases are presented below.

A.3.1 Fluidized Bed

The concept of a fluidized porous medium was introduced and explained in Sect. 9.3.1.

Fig. A.10 A fluidized bed

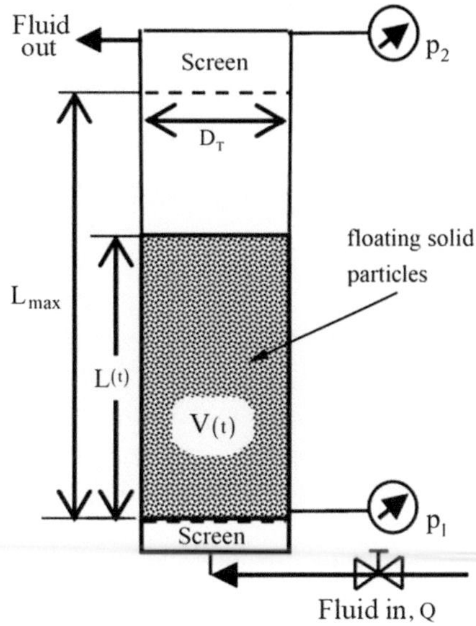

Consider a packed bed (Fig. A.10) in which the solid particles, say in the range of 0.1–0.5 cm, are unconsolidated, i.e., they just touch each other, transmitting forces through the contact points. In upward fluid (liquid or gas) flow through the bed, the solid particles remain stationary as long as the fluid's velocity remains sufficiently low. Under such conditions, the submerged weight of the solid particles overcomes the fluid's drag and the solid particles remain stationary. As the fluid's velocity is increased, the pressure loss is increased (say, according to any flux expression like Darcy's, or Ergun's law). Recalling *stokes law*, the drag acting on the solid also increases. At some fluid velocity, conditions will be such that the drag force will overcome the weight and the individual solid particles will begin to float. This is the point at which fluidization begins. We say that the bed is *"fluidized upward"* (see Sect. A.1.3) when the solid particles are heavier than the fluid (gas or liquid), or *"fluidized downward"* when they are lighter than the latter.

Various processes, e.g., chemical reaction, adsorption crystallization, and heat transfer, can take place at the (microscopic) interface between the moving fluid and the surface of the solid particles. One advantage of a fluidized bed, with respect to a stationary solid matrix, is that it significantly improves the fluid-solid heat and mass transfer due to the increased turbulence.

Consider a portion of a column of length L_o packed with unconsolidated solid particles. Let A denote the constant cross-sectional area of the column. Initially, let ϕ_o denote the porosity of the column filled with (loose, just touching) solid particles up to a (vertical) length L_o. Fluid (liquid or gas) flow is initiated through the column, and the fluid's velocity inside the void-space is gradually increased. At first, as velocity increases, the solid particles, although dragged by the fluid moving through the void space, remain at rest, as the *drag force* is smaller than the *submerged weight* of the solid particles. By the *Law of Archimedes*, the submerged weight of a solid particle fully surrounded by a fluid is given by $g(\rho_s - \rho_f)\mathbb{V}_s$ where $\mathbb{V}_s$ denotes the particles' volume. This situation will continue until a certain velocity is reached at which the solid particles begin to move, float and occupy a growing volume within the column:

$$(1-\phi_o)AL_o \rightarrow (1-\phi_1)AL_1 \rightarrow (1-\phi_2)AL_2, \ldots, \quad \text{with} \quad L_o < L_1 < L_2 < L_3 \cdots$$

Note that the subscript $m = o$ (e.g., in V_o, L_o, ϕ_o) is used for the time at which the solid particles just begin to move. Stages indicated by subscripts $1, 2, 3, \ldots$ correspond to higher fluid velocities (and corresponding higher values of V, ϕ, and L).

At every stage, the pressure drop along the fluid column containing floating solid particles can be related to the fluid's velocity or to the specific discharge. The latter can be determined by any of the fluid flux equations discussed in Chap. 4. In Chemical Engineering, it is common to use Ergun's equation (4.3.17), based on a friction factor, f_p, between a flowing fluid and the solid matrix/particles:

$$f_p = \frac{\Delta h}{L} \frac{d_p}{\rho_f q_f^2} \frac{\phi^3}{1-\phi^2}, \quad \rho g h = \rho g z + p = p^*, \tag{A.117}$$

where h is the piezometric head (for a constant density fluid), L denotes the length of the saturated packed bed, along which p^* drops by Δp^*, ϕ denotes the porosity of the bed, $q_f (= Q_f/A \equiv \phi V_f)$ denotes the fluid's specific discharge, and d_p denotes the particles' effective or mean size. In (A.117), all variable may vary with time.. In certain processes, the solid particles dissolve to the extent that they may completely disappear, or have to be filtered out. This case is not considered here.

Another form of f_p is written in terms of the Reynolds number (Re):

$$f_p = \frac{150}{Re} + 1.75, \quad Re = \frac{q_f d_p}{\nu_f}, \quad \nu_f \rho_f = \mu_f, \tag{A.118}$$

where Re for the bed is expressed by:

$$Re = \frac{\rho d_p q_f}{\mu(1-\phi_m)}. \tag{A.119}$$

At any instant, the fluid's flux, expressed by *Ergun's equation* (4.3.17), can be rewritten in the form:

$$\frac{\Delta p^*}{L} = 150 \frac{\mu}{d_p^2} \frac{(1-\phi)^2}{\phi^2} q + 1.75 \frac{\rho}{d_p} \frac{1-\phi}{\phi^3} q^2. \tag{A.120}$$

where L denotes the length of the porous medium column across which we have the p^*-drop of Δp^*, and we recall that $\Delta p^* = \delta p + \rho g \Delta z$, and z (positive upward) is taken above some datum level; $\rho \equiv \rho_f$ is assumed constant. This equation is also used for determining the fluid's specific discharge ($q = Q/A$) at which fluidization starts.

Once fluidization has started, a balance exists between the three forces that act on the solid particles, say per unit volume:

- The upward drag produced by the moving fluid, f_p expressed in (A.117), with $\phi_o \to \phi_m$, when $L_o \to L_m$.
- The upward buoyancy, or Archimedes force that depends on the fluid's density:

$$f_b = (1 - \phi_m) g \rho_f L_m. \tag{A.121}$$

- The downward weight of the solid particles:

$$f_g = (1 - \phi_m) \rho_s L_m.$$

These three forces maintain the following balance among them:

$$f_p + f_b = f_g. \tag{A.122}$$

To determine the minimum upward fluid velocity at which fluidization is initiated, we seek the point where the drag force associated with the pressure drop becomes equal to the submerged weight:

$$\Delta p_m^* = L_m (1 - \phi_m)(\rho_s - \rho_f), \tag{A.123}$$

or:

$$\frac{\Delta p_m^*}{L_m (1 - \phi_m)} = (\rho_s - \rho_f) = const., \tag{A.124}$$

where we note that the column's cross-sectional area plays no role. In the above equation, L_m is the length of the fluidized column, and ρ_s and $\rho_f (\equiv \rho)$ denote the solid and fluid densities, resp.

Altogether, for any considered q above the onset of fluidization at q_o, we can use the three equations: (A.120), (A.123) and (A.124) to determine the three variables: Δp_m^*, L_m and ϕ_m.

Various phenomena may occur during the fluidization process. For example, crystals may grow and fall down through the fluidized bed while smaller crystals are removed through the upper end of the bed where the fluid exits the column. Catalyst particles may be fouled after some time of operation and may have to be removed them the system by increasing the velocity.

A stable bed means that all the solid particles move within the same range of velocities and fluctuations. In an unstable bed, large volumes of the fluid are circulating within the bed and the spatial distribution of particles varies over a wide range of fluid volumes. Efforts are made to avoid such conditions.

Stable fluidized beds are used, for example, for crystallization and adsorption. Fast fluidization, associated with a short period of stay of solid particles in the column,

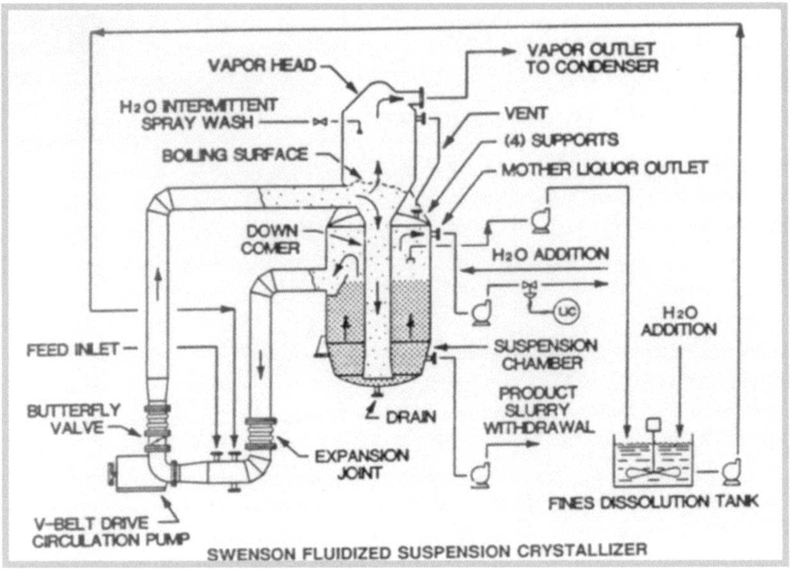

Fig. A.11 Swensen fluidized suspension crystalliser

is used when solid particles need to be removed for cleaning, e.g., in fast catalysis (FCC fast catalytic cracker in refineries).

A.3.1.1 Crystallization

Crystallization is a process in which the solutes in a fluid produce well structured solid particles in the form of crystals. These particles have a known size distribution. The crystals are created as a consequence of changes in the flow regime, e.g., change of temperature and/or concentration. In the latter case, the addition of chemicals, causes the concentration of the dissolved matter to exceeds the *saturation limit*. This limit indicates the maximum concentration of a dissolved matter before crystallization starts. Sometime, the generated solid particles are the desired products. In other cases, the target is to achieve a purer solution as the final product. In both cases, crystallization needs to take place in a moving stream to prevent internal clogging by the produced crystals.

The crystals are grown inside the reactor (see Fig. A.11). The larger crystals are moving down to the bottom of the crystallizer, while the smaller ones move upwards and continue and grow. The process may be continuous, or batch-wise.

Figure A.11 shows a double-tube crystallizer. The fluidization is achieved by circulating a solution with a pump located at the left lower part of the figure. The fluid enters the internal tube from above. The tube is maintained under vacuum that enhances evaporation of the fluid that enters from above (A on the figure). At the bottom of the crystallizer (B), the fluid changes flow direction and moves up fluidizing the solid particles (crystals). The fluidization allows the heavy particles to settle down and exit at the bottom of the crystallizer. The smaller particles are carried with the fluid that is pumped out. They dissolve in the fine dissolution tank and then are fed back to the crystallizer together with the incoming feed. The velocity of the fluid inside the vessel is dictated by the size of the particles and the flow conditions as described at the beginning of this subsection.

A.3.2 Fluidized Bed-Catalytic Process

Fast fluidized bed is used for fast chemical reactions that affect the solid particles and prevent them from regular operation. The best case to demonstrate such reaction if the case of catalytic cracking that is operated in refineries. Crude oil contains thousands of organic components. The lighter one, the most volatile are used as different fuels for our transportation system. Some heavier fuel molecules may be used in power stations for electricity production. However, heavy, large molecules, generate large amount of CO2, so there is a trend to use natural gas for power stations. The FCC (Fast Catalytic Cracker) is designed to break the large molecules and generate smaller molecules by adding hydrogen to the molecules, hence, large molecules turn into light fuels and the basis for generating polymers. The catalytic process generates a by-product, soot that covers the catalytic particles and stops their action. The fast velocity of the particles through the bed allow them to act, foul and exit at the top of the reactor. The particles are cleaned outside the reactor and returned through the bottom of the bed.

A.3.3 Stirred Moving Bed

The stirred bed reactor is basically a mixed slurry reactor where the solid matrix may be part of the process or inert. The reactor is also a type of a *continuous stirred tank reactor* (CSTR) that contains solids. The mixing allows better contact between the solids and the solution.

A.3.3.1 Dissolution

This process that is opposite to that of crystallization (Sect. A.3.1.1).

The objective here is to dissolve a certain solid material in a solution, e.g., adding needed calcium salt to desalinated water. In this case, the solids are dissolved completely or partially in the liquid present in the reactor and the loaded solvent is continuously removed. With an appropriate filtration system, this may be a continuous process.

Thus process can be used to prepare a solvent loaded with dissolved solid as a pretreatment for a chemical reaction, or addition of calcium to pure water.

A.3.3.2 Adsorption

Adsorption is a process used to remove a specific dissolved matter, e.g., a contaminant in water, from a liquid (= solvent) in a single stage, by attaching it to solid particles that have some affinity to the removed material. For example, contaminants dissolved in water can be removed by *active carbon* particles. Another example is the removal of toxic organic fumes by adsorbing them on a solid matrix.

A.3.3.3 Ion Exchange

This is a batch process that is similar to that of adsorption, except that here an ion that is preloaded on the surface of solid particles of an ion exchanger is replaced by an ion in solution (see Sect. 7.4.2) and Sect. A.2.5 above).

A.3.3.4 Crystallization

The formed crystals are the solids in the system. It is possible to allow at right mixing system to take the largest species from the bottom of the reactor while the smaller crystals stay and grow. May be either batch or continuous process.Crystallization is often used in the pharmaceutical industries and in the production of special chemicals.

A.3.3.5 Catalysis

The catalytic process was describe already in Sect. A.3.2. However, we add here the possibility of using an impeller to produce bed motion, thus mixing the reactor's content. The catalyst may be attached to the solid surface, or dissolved in the liquid,where the solid grain help to generate a better mixing. Depending on the nature of the catalytic reaction, this operation may be continuous or batch-wise.

* * *

Altogether, the above brief discussion on reactive transport processes that take place in contactors and reactors used in the chemical engineering industry demonstrate important (and very useful) applications of the theory of mass, momentum and energy transport in porous medium domains, where the latter are envisioned as continua.

Obviously, only very simple examples are presented, while the real world is much more complicated. More examples can be found in the Chemical Engineering literature.

Reference

Hokanson DR (2004) Development of ion. Exchange models for water treatment and application to the international space station water processor, Dissertation, Michigan Technological University

An Introductory Remark Concerning the Next Two Appendices

After a most interesting INTERPORE meeting in Prague, I asked myself the following question: "Since the mid-fifties of the last century, I have been watching and contributing to the development of the exciting field of *phenomena of transport in porous media*. I have just been listening for two days to interesting, some innovative, presentations about transport in porous media, but what are the real new innovations of recent years? "And almost with no hesitation my answer was" Two exciting subjects are really new:

- *The power of imaging*. The ability to actually *observe*, and often also measure phenomena that occur at the *microscopic level*—inside the void-space and on the internal surfaces.
- *The power of computing*. The exciting development and the tremendous increase in the *power of computing*. This enabled the solution of very complex problems, with multiple variables at field scale. However, it also facilitated the solution of problems at the micro- and nano-scales, describing phenomena *inside the void space*.

Indeed, during the conference, I listened to reports on actually monitoring two phase flow and adsorption on the solid. This is a real revolution; the ability to solve a huge number of PDE's in a very short time. These have revolutionized the field of modeling phenomena of transport in porous media.

I have asked colleagues who are experts on these subjects to briefly expose them within the framework of this book, which focuses on the macroscopic level.

Jacob Bear
2017

Appendix B
Recent Advances in Pore Scale Imaging by Jonathan Ajo-Franklin and Marco Voltolini, Lawrence Berkeley National Laboratory

B.1 Objectives of Pore-Scale Imaging

The primary goal of pore-scale imaging is to provide a detailed three-dimensional digital model of a porous medium sample at the smallest scale that is relevant to a considered process, e.g., diffusion or advective flow in the void space. Such digital model can be immediately used for a qualitative characterization of the sample via virtual cuts and volume rendering techniques. With a slightly increased effort, 3D-domains can also be used for a quantitative description of the sample through the use of digital morphometric analysis of the microstructure These techniques provide access to a host of structural metrics relevant to subsurface flow and transport, ranging from the simple (porosity) to complex, including 3D grain orientation anisotropy, matrix fractal properties, and pore connectivity statistics (e.g., Naverre-Sitchler et al. 2009). A secondary goal, which has only recently become feasible, is to provide a time sequence of such 3D domains to capture structural evolution in either the solid or fluid geometry which can then be utilized to validate models of the process in question, or as part of the scientific discovery process. This capacity to follow dynamic processes in real time (4D imaging) at temporal resolutions even in the sub-second range is the current frontier of pore-scale imaging.

With a static or time variable pore-scale model, an investigator can either extract microscopic geometrical information (e.g. porosity, pore-throat statistics, 2-phase saturation), or directly calculate flow properties, including permeability, from pore-scale images by solving the appropriate governing equations. The capacity to directly simulate pore-scale processes, utilizing static 3D tomographic data-sets, forms the basis of the growing field referred to as 'digital rock physics; (Kheem et al. 2001; Andrew et al. 2013). The appeal of being able to avoid performing actual measurements (such as permeability, ultrasonic wave velocities, etc.), or to simulate destructive experiments via software, thus preserving the sample (e.g. mercury intrusion porosimetry) is evident.

J. Bear, *Modeling Phenomena of Flow and Transport in Porous Media*,
Theory and Applications of Transport in Porous Media 31,
https://doi.org/10.1007/978-3-319-72826-1

B.2 Imaging Techniques

Three-dimensional pore-scale imaging targeting optically opaque materials (e.g., soils and rocks) is now conducted with a wide variety of imaging modalities including X-rays, neutrons, magnetic resonance imaging (MRI), and a variety of serial ablation/imaging techniques. Of the many relevant factors, these techniques are largely distinguished by their spatial and temporal resolution, field of view, destructiveness, and material sensitivity. We shall restrict our discussion to 3D imaging, approaches on macroscopic samples (mm to cm in size) and neglect purely 2D imaging methods as well as nano-imaging techniques targeting rock fragments below tens of microns (e.g., soft X-ray tomography, TEM tomography). For many of the tomographic imaging techniques, the 3D structure is obtained by collecting a large number of 2D projections as the sample,or source/receiver system, are rotated. The resulting set of projections (100s to 1000s of images) is then processed to yield a 3D image volume that depicts the internal components.

B.3 X-Ray Imaging

Hard X-ray micro computed tomography (XRmCT), using either synchrotron or conventional tube sources, is currently the dominant technology for both static and dynamic imaging of opaque porous samples at resolutions from roughly 0.5 microns to the centimeter range, sufficient for imaging the larger pores in sandstones and soils (Wildenschild and Sheppard 2013; Cnudde and Boone 2013). First demonstrated in the late 1980s (Flannery et al. 1987), the technique has steadily improved, taking advantage of advances in both hardware, e.g., faster and higher resolution 2D detectors and XR sources, with higher brilliance, and software, with greatly increased computational capabilities for advanced image reconstruction and processing approaches.

In conventional XRmCT, a large number of 2D X-ray projections (radiographs) are acquired and reconstructed into a 3D image volume via filtered back-projection, or more advanced algorithmic techniques, including iterative reconstruction. The volume obtained shows the structure of the sample mapping the different X-ray attenuations values in space. The attenuation values are related to the density and atomic species (Z number) present in the sample, following, for the ideal case with polychromatic XR:

$$I_{rec} = \int I_o(E) e^{[\Sigma_i (\mu_i(E) x_i)]} dE.$$

The above equation indicates that for each pixel in the projection, we obtain a recorded intensity I_{rec}, depending on the initial intensity I_o of the XR beam energy spectrum, on the linear absorption coefficient μ_i and on the path length x_i through each material i, for each energy, E, of the incident spectrum. The linear attenuation coefficient is directly related to the fraction of X-rays absorbed, or scattered, per unit volume of the material, taking into account the number and species of atoms in a unit volume of material and the probability of a photon being scattered, or absorbed, from the atoms present.

The volume obtained, after applying reconstruction procedures to the radiograph set, is populated with XR attenuation values (e.g., Kak and Slaney 1987; Stock 2008) for the constituent materials. Typically, porous materials have low values (dark) for pore space, and higher values for minerals, with dense phases, such as pyrite, being the brightest in color. This contrast allows for subsequent segmentation of pore and mineral grains (Andra et al. 2013a) as well as tracking of multiple fluid phases, assuming a sufficient difference μ_i (Porter and Wildenschild 2010); in cases where two phases have similar properties, fluid (e.g. potassium iodide) or gas (e.g., xenon) phase contrast agents can be utilized to allow effective imaging. Most recently, XR mCT techniques have been extended to higher time resolutions (>1 s), allowing dynamic monitoring of fast hydrologic processes such as *Haines jumps* (Berg et al. 2013; Armstrong et al. 2014). This class of measurement is still confined to synchrotron light sources which leverage high brilliance XR beams and superfast CCD detectors to push the boundaries of temporal resolution.

B.4 Neutron Imaging

Imaging porous systems, using neutrons (Strobl et al. 2009) has also advanced considerably, providing a secondary modality for pore/core scale imaging. Neutron radiography and tomography provide increased sensitivity to water due the higher absorption cross-section of hydrogen and are capable of penetrating large diameter samples and metal pressure vessels. Neutrons also tend to be less damaging to biological systems in comparison to high-flux X-rays, providing opportunities for imaging root/soil interactions and water uptake (e.g., Tumlinson et al. 2008; Warren et al. 2013). With these advantages also come limitations; neutron flux is several orders of magnitude below X-ray sources, resulting in considerably longer image acquisition times and, due to detector limits, lower spatial resolution. Neutron sources, both reactors and spallation sources, are also fewer in number, resulting in more limited experimental opportunities. Another challenge is the potential activation of samples, which results in radioactive material after the measurement; the resulting samples require appropriate handling and disposal.

B.5 Magnetic Resonance Imaging

Imaging techniques based on nuclear magnetic resonance (Magnetic Resonance Imaging, MRI) are especially useful in obtaining information about fluids in porous media, typically targeting the relaxation of the radio-frequency signal emitted by the nuclei of the hydrogen atoms of water and oils. Laboratory MRI machines can accommodate relatively large pressure vessels, allowing measurements of flow under non-ambient pressure conditions of samples of few centimeters in diameter; such vessels are by necessity non-magnetic and are often built from ceramics, or engineering polymers. MRI imaging can provide measurements of the invaded pore volume and a velocity of the flow (Merrill 1994; Mantle et al. 2001). Given the nature of the MR signal, this technique has been successfully employed also in the study of gas hydrates (e.g., Ersland et al. 2010). Despite the low resolution and the limited use in porous material studies, MRI imaging offers unique opportunities to study permeability and flow properties in soils and rocks.

B.6 Ablative Imaging Techniques

From a purely etymological point of view, these techniques are a true, or direct, 'tomography' approach, the word being composed by the combination of the Ancient Greek words τὸμος, meaning slice, and Υ ρ′αφ ω to write/draw. *Ablative imaging* techniques do exactly that: they rely on a serial slicing and imaging of the sample, creating a stack of images that can be converted directly into a 3D volume. As a principle, any technique capable of serial slicing with subsequent imaging can be employed. Examples where mechanical slicing by actual grinding, a small amount of sample with a diamond abrasive tool, and optical imaging (photographs) are used to obtain 3D information on rocks and similar materials can be found in different geological labs (e.g., Maloof et al. 2013). At a smaller scale, techniques based on ion milling and electron imaging are currently employed. Such techniques commonly known as Focused Ion Beam/Scanning Electron Microscopy (FIB/SEM, Holzer and Cantoni 2011, are especially useful for 3D imaging of nanoporous materials such as shales (e.g., Keller et al. 2011), where conventional XRmCT techniques cannot provide sufficient resolution. It is also worth remarking that the serial slicing techniques can, in principle, because of the direct access to the surface, be more readily coupled with different imaging techniques, e.g. EDS chemical mapping in FIB/SEM (Lemmens et al. 2011) when compared to techniques based on the acquisition of projections.

However, as is clear from the description, ablative techniques are inherently destructive, while techniques based on X-rays and neutrons are usually non -destructive in nature, and much less prone to sample preparation artifacts, which are also common with ablative techniques (e.g. curtaining in FIB/SEM). In the context of porous media studies, another significant disadvantage of ablative techniques is their inability to be employed for in-situ experiments.

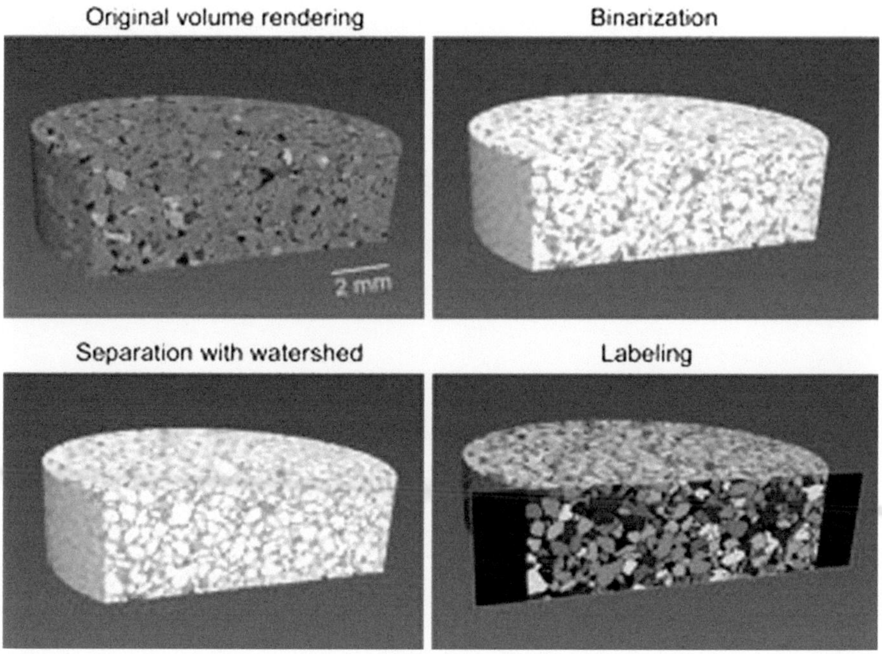

Fig. B.1 Steps in quantitative grain scale analysis

B.7 Pore-Scale Imaging for Characterization

One of the goals of tomographic techniques is to provide a 3D characterization of the microstructure of a given sample, both in a qualitative (by means of volume renderings and virtual sections) and quantitative (by means of morphometric analysis) fashion. A first qualitative observation of a tomographic dataset by the operator is usually carried out using one of the many volume rendering software available, after properly adjusting the dataset for optimal contrast and filtering volumes to minimize artifacts and noise. Virtual cutting of the sample is often employed to observe the interior of the sample. The quantitative analysis of tomographic datasets can provide a large number of morphometric characteristics of a given sample, from the computationally simple ones, such as the volume fractions of segmentable object classes (e.g., porosity), to more advanced ones such as shape-based anisotropy, grain orientation, and pore network properties.

An example of one such analysis procedure focused on the quantitative analysis of the sand grains within a sandstone sample (from Voltolini et al. 2016) is shown in Figs. B.1 and B.2.

Figure B.1 shows the initial mCT volume rendering, the binary volume (segmentation of grains and pores), the grains volume after a watershed procedure (to separate single sand grains), and the labeled grain volume (where a unique value is assigned to each sand grain, highlighted by a different gray value in figure). Once

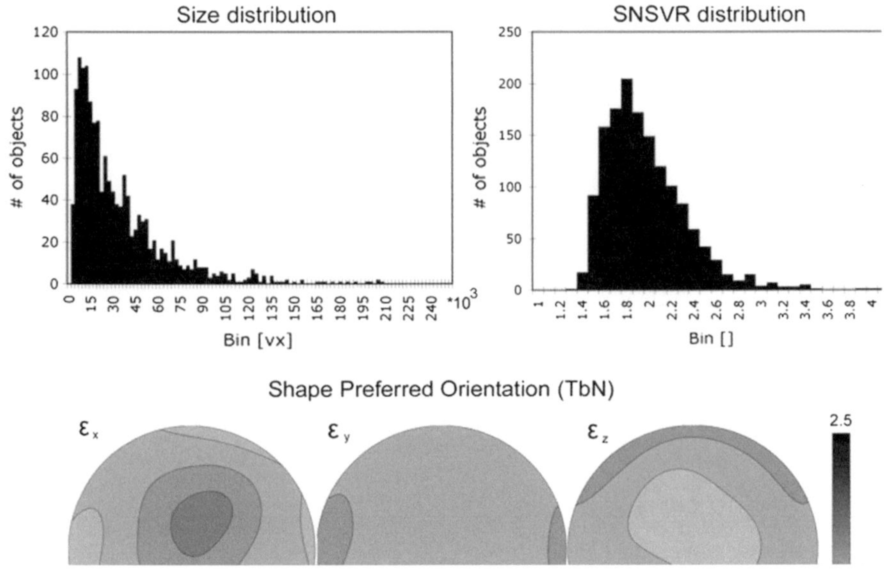

Fig. B.2 Types of statistics which can be extracted

individual grains (or pores) are separated, quantitative morphometric analysis can be employed, as shown in Fig. B.2, where the grain size histogram, the grain surface areas frequency histogram (Sphere-Normalized Surface to Volume Ratio, SNSVR, values i.e., the deviation from a sphere with the same volume of the object), and the shape preferred orientation (a measure of anisotropy) of the sand grains are displayed. These measurements are simple examples of the wide variety of quantitative spatial analysis approaches which can be applied to 3D images of porous systems; other examples include pore network attributes (e.g., Dong and Blunt 2009; Keller et al. 2011), statistical descriptions of porous materials (e.g., Jiang et al. 2012), feature identification, and, when coupled to secondary measurements, 3D mineralogical mapping (e.g., Golab et al. 2010).

B.8 Pore-Scale Imaging for Process Dynamics

With recent advances in XRmCT, particularly at synchrotron light sources, measurements on dynamic systems are now possible with temporal resolutions of a second or less. Examples of potential fields of application are extremely broad and include processes in porous media driven by flow, reaction, thermal, or mechanical evolution. Such measurements allow direct observation of pore-scale physical processes and provide opportunities to both validate numerical models parameterized at this length scale (e.g. Molins et al. 2014) as well as probe previously unobserved phenomenon at the micron length scale. New environmental cells for in situ analysis that have been used in XRmCT experiments include, but are not limited to, single- and multi- phase flow at non-ambient conditions (CO_2 geological sequestration, oil reservoir characterization), heating (drying in clays and soils, pyrolysis of oil shales, geothermal processes), freezing (permafrost, clathrates), geochemical alteration (dissolution and crystallization), and deformation (rock mechanics, fracture evolution). Multiphase flow experiments at elevated pressures, in particular, provide a rich field of investigation with recent studies using XRmCT to examine the dynamics of drainage and imbibition at deep reservoir conditions (e.g., Iglauer et al. 2011; Andrew et al. 2013, 2014; Herring et al. 2014).

The opportunities offered by 4D XRmCT comes with a cost in terms of experimental requirements and processing complexity, A first constraint for in situ measurements is the availability of an appropriate environmental cell which can recreate appropriate conditions (e.g. temperature, stress state) while maintaining sufficient x-ray transparency. A second set of constraints relate to the rate of system evolution in comparison to the time required for scanning; the process should be sufficiently slow that it can be considered close to static over the duration of a scan to avoid motion-induced reconstruction artifacts. Simultaneously, the system evolution is ideally fast enough to complete a dynamic experiment over the duration of an experimental run available at a synchrotron (2–5 days) unless an ex-situ study is selected. These requirements are quite demanding from an engineering and planning perspective, but the data acquired in such experiments can be valuable for pore-scale process observations.

One example of such a process is coupled flow and dissolution in fractured rocks; we show an example reported in Ajo-Franklin et al. (in press) and Deng et al. (2016) which utilized a high-pressure in situ cell to track fracture alteration in three dimensions due slow injection of CO_2 saturated water. Figure B.3 shows the dynamic expansion of a fracture in a sample of the Duperow dolomite, a reservoir and seal formation for natural CO_2 accumulations in north central Montana (USA). The experiment was conducted at pore pressures of 1400 psi (9.6 MPa), equivalent to depths of approximately 3181 ft (969 m) and representative of leakage from a geological carbon storage facility.

The example in Fig. B.3 shows the useful capabilities of 4D XRmCT: the ability to monitor an evolving system in situ with high spatial resolution. In this specific case it is possible to appreciate the enlargement of the fracture, a self-enhancing behavior in this case, as a function of time. The CO_2 dissolve the carbonates in the sample, generating fracture zones with an increased aperture while leaving load-bearing contacts unaltered over these time scales. The generation of a microporous zone on the surfaces of the fracture is also observed; this is due to the microstructure of the sample, with mainly dolomite crystals cemented by a small amount of calcite. The calcite dissolves more rapidly than dolomite, leaving this microporous weathered zone. Significantly, the 4D observations coupled to simultaneous fluid chemistry measurements allowed development of a new micro-continuum reactive transport model (Deng et al. 2016) which incorporated the observed geometry evolution. The synergy of 4D XRmCT measurements and pore-scale simulation is evident, with

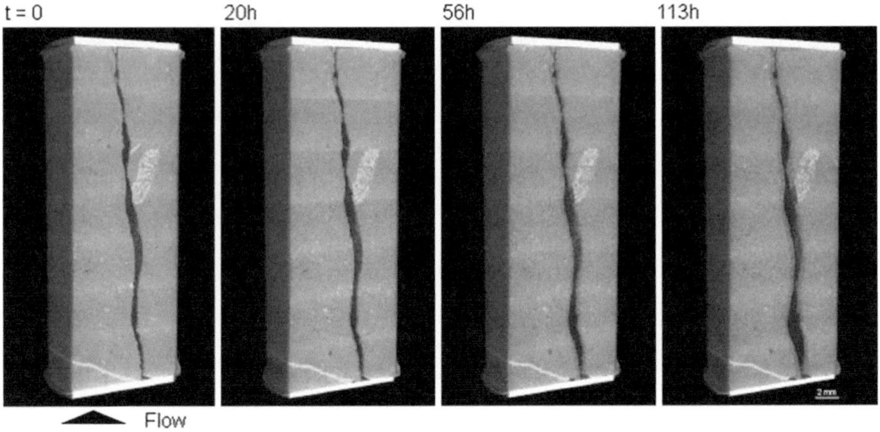

Fig. B.3 Evolution of a fracture in a Duperow dolomite sample during the flow of CO_2-saturated water, under pressure. The baseline volume and three different reaction times are shown

the experimental effort targeting the understanding of the important processes in the system studied and providing a full dataset for model validation. The model, once able to replicate the experimental data, can be used as a tool to generalize the system, and used on sample dimensions or lengths of time experimentally inaccessible.

B.9 Frontiers of Pore-Scale Imaging

Experimental developments in recent years have taken three non-exclusive directions, mainly increased spatial resolution, integration of measurement techniques, and dynamic experiments at non-ambient conditions. Advances in high resolution imagining, targeting pores well below a micron, have allowed studies on finer grained sediments including shales, chalks, and tight carbonates. The main drawback of nanometric resolution imaging in general is that the field of view is proportionately reduced, resulting in small imaging domains sometimes inappropriate for modeling an REV of material. Sub-micron resolutions are now available using high-quality laboratory cone beam CT instruments as well as most synchrotron XRmCT beamlines. Resolutions below 250 nm, using X-rays, can be obtained using a variety of XR optics, typically zone plates or parabolic mirrors (Withers 2007), or nanofocused XR beams, in a 1st generation tomographic setup approach (Suhonen et al. 2012). Increased resolution, and imaging of materials with similar attenuation values, has also accelerated research on phase imaging techniques, taking advantage of the high spatial coherence of synchrotron sources, and free electron lasers, the 4th generation light sources (Mancuso et al. 2010). Higher resolutions can also be achieved using 3D electron beam systems (e.g., TEM) tomography, FIB/SEM) with the previously mentioned limitations to very small samples (100 micron3 or less); these systems are also improving through deployment of multiple electron beams, higher ion cut rates for FIB, aberration corrected optics, and improved reconstruction algorithms for the limited angular coverage available in TEM tomography.

Another frontier in pore-scale imaging is the utilization of integrated (e.g., multi-modal) imaging techniques to capture more detailed suites of information including sub-resolution statistical datasets and/or pore-scale chemical data. Techniques that can be combined with XRmCT include X-ray scattering measurements. Simultaneous CT and small-angle scattering (SAX) has proven useful to capture sub-resolution clay properties (Suuronen et al. 2014) while CT combined with wide angle scattering (WAX) has been used for X-ray diffraction tomography to capture crystallographic properties of sample regions (Voltolini et al. 2013). Another recent multimodal imaging approach has been the combination of CT with X-ray fluorescence (e.g., Suhonen et al. 2012; Jacques et al. 2013), which, for small samples, has allowed simultaneous imaging of pore-structure and grain mineralogy.

A last frontier has been the development of dynamic (4D) imaging systems combining more sophisticated environmental cells coupled to higher temporal resolution acquisition approaches. Environmental cells have evolved in the direction of simulating extreme parameter states such as ultra high temperatures (Baker et al. 2012), pressures (Wang et al. 2005), as well as high PT combinations (Renard et al. 2016). Experiments based on environmental cells can be used in principle in lab XRmCT scanners, but the high brilliance of synchrotron sources and the high penetration of neutron sources makes those advanced facilities more suited for this kind of experiments. Software advances in 4D mCT is a second direction of advancement; iterative reconstruction algorithms are being developed and deployed on high performance computing platforms to allow faster imaging, improved reconstruction quality, and reconstruction of rapidly changing samples. A final direction of advance is the coupling of these 4D experiments with quantitative predictive models (e.g. Deng et al. 2016) with all the inherent challenges of replicating complex pore-scale systems with simultaneous evolution of flow pathways, mechanics, and geochemical properties.

B.10 Conclusions

Recent advances in pore-scale imaging have proven to be a valuable tool in investigating the properties of complex porous media. The variety of techniques and their flexibility allow a range of scientific questions to be addressed, from sample characterization, to single- and multi- phase flow, reactive transport processes, and rock mechanics. Giving their nature, 3D and 4D imaging experiments are the perfect companion to pore-scale modeling, acting both as an investigative and validation tool. The continuing effort of the scientific community to improve both hardware and software components of pore-scale imaging has resulted in significant recent advances, particularly for 4D process characterization. More complex systems at extreme conditions and requiring higher spatial and temporal resolution are within reach; during the next decade, experiments probing these limits will likely become routine and available to an even larger group of scientists.

References

Naverre-Sitchler AK, Steefel CI, Yang L, Tomutsa L, Brantley SL (2009) Evolution of porosity and diffusivity associated with chemical weathering of a basalt clast. J Geophys Res: Earth Surf 114:F2

Kheem Y, Mukerji T, Nur A (2001) Computational rock physics at the pore scale: transport properties and diagenesis in realistic pore geometries. Lead Edge 20(2):180–183

Andrew M, Bijeljic B, Blunt MJ (2013) Pore-scale imaging of geological carbon dioxide storage under in situ conditions. Geophys Res Lett 40(15):3915–3918

Wildenschild D, Sheppard AP (2013) X-ray imaging and analysis techniques for quantifying pore-scale structure and processes in subsurface porous medium systems. Adv Water Resour 51:217–246

Cnudde V, Boone MN (2013) High-resolution X-ray computed tomography in geosciences: a review of the current technology and applications. Earth Sci Rev 123:1–17

Kak C, Slaney M (1987) Principles of computerized tomographic imaging. IEEE Press, New York

Stock SR (2008) Recent advances in X-ray micro-tomography applied to materials. Int Mater Rev 53(3):129–181

Andra H, Combaret N, Dvorkin J, Glatt E, Han J, Kabel M, Keehm Y, Krzikalla F, Lee M, Madonna C, Marsh M (2013a) Digital rock physics benchmarks–part I: imaging and segmentation. Comput Geosci 50:25–32

Porter ML, Wildenschild D (2010) Image analysis algorithms for estimating porous media multiphase flow variables from computed microtomography data: a validation study. Comput Geosci 141:15–30

Berg S, Ott H, Klapp SA, Schwing A, Neiteler R, Brusse N, Makurat A, Leu L, Enzmann F, Schwarz J-O, Kersten M, Irvine S, Stanpanoni M (2013) Real-time 3D imaging of Haines jumps in porous media flow. Proc Natl Acad Sci 110(10):3755–3759

Armstrong RT, Ott H, Georgiadis A, Rucker M, Schwing A, Berg S (2014) Subsecond pore-scale displacement processes and relaxation dynamics in multiphase flow. Water Resour Res 50:9162–9176

Strobl M, Manke I, Kardjilov N, Hilger A, Dawson M, Banhart J (2009) Advances in neutron radiography and tomography. J Phys D: Appl Phys 42(24):3001–3022

Tumlinson LG, Liu H, Silk WK, Hopmans JW (2008) Thermal neutron computed tomography of soil water and plant roots. Soil Sci Soc Am J 72(5):1234–1242

Warren JM, Bilheux H, Kang M, Voisin S, Cheng CL, Horita J, Perfect E (2013) Neutron imaging reveals internal plant water dynamics. Plant Soil 3662:683–693

Merrill MR (1994) Local velocity and porosity measurements inside Casper sandstone using MRI. AIChE J 407:1262–1267

Mantle MD, Sederman AJ, Gladden LF (2001) Single-and two-phase flow in fixed-bed reactors: MRI flow visualisation and lattice-Boltzmann simulations. Chem Eng Sci 56(2):523–529

Ersland G, Husebø J, Graue A, Baldwin BA, Howard J, Stevens J (2010) Measuring gas hydrate formation and exchange with CO_2 in Bentheim sandstone using MRI tomography. Chem Eng J 158(1):25–31

Maloof AC, Samuels B, Mehra A, Spatzier A (2013) December. An automated serial grinding, imaging and reconstruction instrument (GIRI) for digital modeling of samples with weak density contrasts, AGU Fall Meeting Abstracts 12294

Holzer L, Cantoni M (2011) Review of FIB-tomography. Nano-fabrication using focused ion and electron beams: principles and applications. Oxford University Press, New York

Keller LM, Holzer L, Wepf R, Gasser P (2011) 3D geometry and topology of pore pathways in Opalinus clay: implications for mass transport. Appl Clay Sci 52(1):85–95

Lemmens HJ, Butcher AR, Botha PWSK (2011) FIB/SEM and SEM/ EDX: a New Dawn for the SEM in the core lab? Petrophysics 52(6):452–456

Voltolini M, Kwon T-H, Ajo-Franklin J (2016) Visualization and prediction of supercritical CO_2 distribution in sandstones during the invasion: contributions to geological carbon sequestration from an in situ synchrotron X-ray micro-computed tomography, under review for the Int J Greenh Gas Control

Dong H, Blunt MJ (2009) Pore-network extraction from micro-computerized tomography images. Phys Rev E 80(3):6307

Jiang Z, Van Dijke MIJ, Wu K, Couples GD, Sorbie KS, Ma J (2012) Stochastic pore network generation from 3D rock images. Transp Porous Media 94(2):571–593

Golab AN, Knackstedt MA, Averdunk H, Senden T, Butcher AR, Jaime P (2010) 3D porosity and mineralogy characterization in tight gas sandstones. Lead Edge 29(12):1476–1483

Molins S, Trebotich D, Yang L, Ajo-Franklin J, Ligocki TJ, Shen C, Steefel CI (2014) Pore-scale controls on calcite dissolution rates from flow-through laboratory and numerical experiments. Environ Sci Technol 48(13):7453–7460

Iglauer S, Paluszny A, Pentland CH, Blunt MJ (2011) Residual CO_2 imaged with X-ray microtomography. Geophys Res Lett 3821

Andrew M, Bijeljic B, Blunt M (2014) Pore-scale imaging of trapped supercritical carbon dioxide in sandstones and carbonates. Int J Greenh Gas Control 22:1–14

Herring AL, Andersson L, Newell DL, Carey JW, Wildenschild D (2014) Pore-scale observations of supercritical CO_2 drainage in Bentheimer sandstone by synchrotron X-ray imaging. Int J Greenh Gas Control 25:93–101

Deng H, Molins S, Steefel C, DePaolo D, Voltolini M, Yang L, Ajo-Franklin J (2016) A 2.5 D reactive transport model for fracture alteration simulation. Environ Sci Technol 50(14):7564–7571

Withers PJ (2007) X-ray nanotomography. Mater Today 1012:26–34

Suhonen H, Xu F, Helfen L, Ferrero C, Vladimirov P, Cloetens P (2012) X-ray phase contrast and fluorescence nanotomography for material studies. Int J Mater Res 103(2):179–183

Mancuso AP, Yefanov OM, Vartanyants IA (2010) Coherent diffractive imaging of biological samples at synchrotron and free electron laser facilities. J Biotechnol 149(4):229–237

Suuronen JP, Matusewicz M, Olin M, Serimaa R (2014) X-ray studies on the nano-and microscale anisotropy in compacted clays: comparison of bentonite and purified calcium montmorillonite. Appl Clay Sci 101:401–408

Voltolini M, Dalconi MC, Artioli G, Parisatto M, Valentini L, Russo V, Bonnin A, Tucoulou R (2013) Understanding cement hydration at the microscale: new opportunities from pencil-beam' synchrotron X-ray diffraction tomography. J Appl Crystallogr 46(1):142–152

Jacques SD, Egan CK, Wilson MD, Veale MC, Seller P, Cernik RJ (2013) A laboratory system for element specific hyperspectral X-ray imaging. Analyst 138(3):755–759

Baker DR, Brun F, O'Shaughnessy C, Mancini L, Fife JL, Rivers M (2012) A four-dimensional X-ray tomographic microscopy study of bubble growth in basaltic foam. Nat Commun 3:1135

Wang Y, Uchida T, Westferro F, Rivers ML, Nishiyama N, Gebhardt J, Lesher CE, Sutton SR (2005) High-pressure X-ray tomography microscope: synchrotron computed microtomography at high pressure and temperature. Rev Sci Instrum 76(7):073709 (1–7)

Renard F, Cordonnier B, Dysthe DK, Boller E, Tafforeau P, Rack A (2016) A deformation rig for synchrotron microtomography studies of geomaterials under conditions down to 10 km depth in the Earth. J Synchrotron Radiat 234

Appendix C
Recent Advances in High Performance Computing by David Trebotich, Lawrence Berkeley National Laboratory

High performance (*peta-scale*) computing has ushered in the ability to perform direct numerical simulation of flow and transport inside the pore space obtained from real rock samples by making use of imaging techniques described in Appendix A. We are now able to model, with high confidence, fundamental processes that determine emergent time-dependent behavior at larger scales. However, understanding phenomena that occur at pore-scale require modeling at that scale. One example of such problems is that of flow of a Newtonian fluid in the void space, described by (1.1.1)–(1.1.3). Such problems, as well as diffusive boundary layers around individual grains of rock, worm-holing originating from reactive transport in a fracture aperture, dissolution of asperities and subsequent collapse of fracture–to name a few–are now computationally accessible phenomena with pore scale modeling. Furthermore, better parameterizations of bulk properties at the larger scale can be obtained from these high resolution simulations, improving the fidelity of larger scale models.

Innovations due to high performance computing have occurred in three main areas, bringing about this new found modeling ability: (1) algorithms, (2) software, and (3) hardware.

C.1 Algorithms

Advanced algorithms have been developed that are consistent with geometry generation techniques and numerical discretizations that treat the very complex geometries obtained from arbitrary, heterogeneous pore space obtained from image data of real rock samples as presented in Appendix A (e.g., Molins et al. 2014) Traditionally, finite element methods have been used to model flow and transport problems involving complex geometry. This was changed with the introduction of adaptive, *finite volume methods* based on cut cell and embedded boundaries. Adaptive, finite volume methods are now employed to achieve high fidelity computations of reactive

J. Bear, *Modeling Phenomena of Flow and Transport in Porous Media*,
Theory and Applications of Transport in Porous Media 31,
https://doi.org/10.1007/978-3-319-72826-1

transport processes, where the reactive surface area of the mineral is resolved, making no assumptions in regard to transport fluxes. In fact, the reactive surface area obtained from geometry generation of raw image data of *real* rock is more accurate than well known and long-used BET theory (Brunauer et al. 1938) from the same experimental data, which can display up to 15% error. One particular method makes use of a sharp interface embedded boundary to resolve the surface area of the mineral (Trebotich and Graves 2015) In this cut cell approach, the pore space is intersected with a Cartesian grid resulting in 'cut' cells near the mineral boundary in the pore space and regular cells away from the boundary. A *finite volume approach* based (Sect. 3.8) can be used in the cut cells to obtain conservative discretization of operations based on diffusive or viscous and advective fluxes; the discretization in the regular cells reduce to standard *finite differences*. The *cut cell approach* also makes geometry generation from image data more automatic and, thus, more tractable than unstructured body fitted gridding approaches like finite elements. In addition, adaptive mesh refinement (AMR) is a technique that applies grid resolution only in areas of the domain where it is needed, say, near a reactive mineral or along a concentration gradient. Combining these two approaches–*embedded boundary methods* and *adaptive mesh refinement*–provides a powerful tool for multi-scale, multi-physics modeling in subsurface porous media problems.

C.2 Software

It is one thing to develop advanced algorithms for solving partial differential equations in complex geometries, as with pore scale flow and transport, and another to build these algorithms on scalable software that supports the algorithms. One high level approach is to use an object-oriented programming language like C++ to control the data and parallelism, while using an imperative language like Fortran for 'number crunching', maintaining the best of both worlds.

The software approach to incorporating cut cells into a structured grid AMR framework needs to be stencil-based, due to the irregularity of the cut cells near the boundary. The main requirements for software to support these methods are flexible and efficient data structures, operator-dependent load balancing, a mode of parallelism for both distributed computing and memory and minimization of communication. The stencil-based approach is critical to high performance; accessing the geometric information in sparse data structures for cut cells needs to be fast and not rely on accessing physical memory. Instead, pointers to arrays of data in memory and integer offsets are stored and recalled for computations in irregular cells (Trebotich et al. 2008) For *load balancing*–i.e., the procedure for distributing pieces of the computation onto multiple processors–the procedure can be as simple as evaluating an operator such as the Laplacian on a decomposed domain of boxes and measuring the time spent in each box and re-balancing as necessary. Space filling curves and sophisticated approaches to numbering, such as Morton ordering, can be used to minimize communication between boxes with load balancing. Of course, the software has to support these sophisticated features for performance scalability. And, in general, the software must be built on a formalism for parallel computing such as MPI or OpenMP, or both.

C.3 Hardware

Advanced algorithms built on scalable software are the prerequisite to high performance computing. The final link is the hardware on which to codes are run to achieve scientific results. Current supercomputers make use of semiconductor architectures that contain only a few CPU cores per computer node on the chip. Hybrid CPU-GPU systems also exist where the GPUs handle parallel, yet lightweight, algorithms. Several supercomputing facilities exist in the U.S.A., and around the world, that foster large scale scientific simulation of the type required for resolved subsurface simulation. The next generation of supercomputers is imminent. The new many-core architecture is providing creative and more efficient access to more memory, while reducing power consumption. Also, as expected with high performance computing, the data generated by a code is 'big' (e.g., one terabyte per time-step dump,) requiring extreme computing measures to accommodate the need for performance and memory, but also fast IO and storage. New off-chip hardware based on NVRAM (Non-Volatile Random Access Memory) has been developed to promote faster IO and in-transit data analysis

C.4 Pore Scale Simulation

Direct numerical simulation of flow and transport in a real pore space is an end-to-end process: from converting image data to a geometry that supports a well-posed problem, with the equations of motion, through advanced algorithms for discretizing the equations of motion and boundary and initial conditions, to high performance computing for scientific results. Figure C.1 shows some examples of complex pore structures and low permeability materials that have been modeled using massively parallel computations. The figure shows results of direct numerical simulation of flow in a fractured shale (Trebotich and Graves 2015) The computational domain has been discretized using nearly 2 billion grid points ($1920 \times 1600 \times 640$) and a grid resolution of 48 nanometers. Figure C.1 shows velocity and pressure plots. The white space in the images is the result of an optimization to reduce memory where the space that is covered by rock matrix (non-pore space) is not included in stored data. 40,000 CPU cores were used for this simulation.

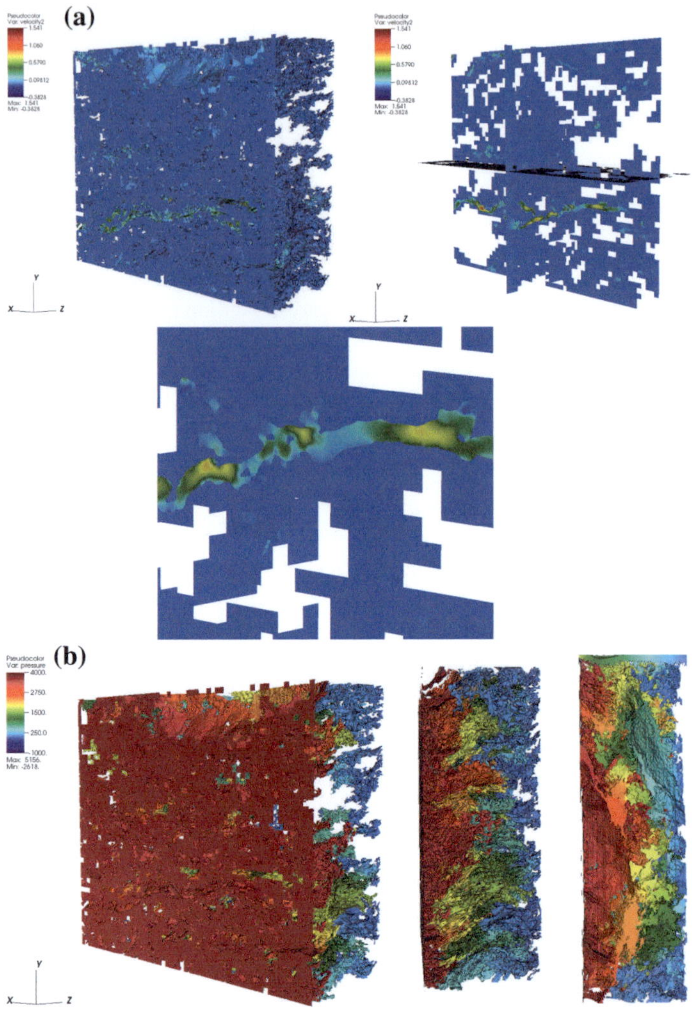

Fig. C.1 Direct numerical simulation from image data of fractured shale, using Chombo-Crunch (Trebotich and Graves 2015): **a** Velocity data mapped to surface and data slice, **b** Pressure mapped onto the mineral surface with side views. The white space in the images is due to memory optimization; it represents space occupied by rock matrix, where there is no flow. This simulation has been performed on NERSC Edison, using 40,000 CPU cores

References

Brunauer S, Emmett PH, Teller E (1938) Adsorption of gases in multimolecular layers. J Am Chem Soc 60(2):309–319

Trebotich D, Graves D (2015) An adaptive finite volume method for the incompressible Navier–Stokes equations in complex geometries. Commun Appl Math Comput Sci 10(1):43–82

Trebotich D, Van Straalen B, Graves DT, Colella P (2008) Performance of embedded boundary methods for CFD with complex geometry. J Phys: Conf Ser 125:12–83

Index

J. Bear, *Modeling Phenomena of Flow and Transport in Porous Media*,
Theory and Applications of Transport in Porous Media 31,
https://doi.org/10.1007/978-3-319-72826-1

D

S

Theory and Applications of Transport in Porous Media

Volume 31

Series editor

S. Majid Hassanizadeh, Department of Earth Sciences, Utrecht University, Utrecht, The Netherlands

Founding series editor

Jacob Bear

More information about this series at http://www.springer.com/series/6612

MIX
Papier aus verantwortungsvollen Quellen
Paper from responsible sources
FSC® C105338

If you have any concerns about our products,
you can contact us on
ProductSafety@springernature.com

In case Publisher is established outside the EU,
the EU authorized representative is:
Springer Nature Customer Service Center GmbH
Europaplatz 3, 69115 Heidelberg, Germany

Printed by Libri Plureos GmbH
in Hamburg, Germany